ADVANCED
CALCULUS

T0203629

PURE AND APPLIED MATHEMATICS

A Program of Monographs, Textbooks, and Lecture Notes

MONOGRAPHS AND TEXTBOOKS IN
PURE AND APPLIED MATHEMATICS

63. *W. L. Voxman and R. H. Goetschel, Jr.,* Advanced Calculus: An Introduction to Modern Analysis (1981)
64. *L. J. Corwin and R. H. Szczarba,* Multivariable Calculus (in press)
65. *V. I. Istrǎţescu,* Introduction to Linear Operator Theory (in press)
66. *R. D. Järvinen,* Finite and Infinite Dimensional Linear Spaces: A Comparative Study in Algebraic and Analytic Settings (in press)
67. *J. K. Beem and P. E. Ehrlich,* Global Lorentzian Geometry (in press)

Other Volumes in Preparation

ADVANCED CALCULUS

An Introduction to Modern Analysis

WILLIAM L. VOXMAN
ROY H. GOETSCHEL, JR.

Department of Mathematics
College of Letters and Science
University of Idaho
Moscow, Idaho

CRC Press
Taylor & Francis Group
Boca Raton London New York

CRC Press is an imprint of the
Taylor & Francis Group, an **informa** business

First published 1981 by Marcel Dekker, Inc.

Published 2019 by CRC Press
Taylor & Francis Group
6000 Broken Sound Parkway NW, Suite 300
Boca Raton, FL 33487-2742

First issued in paperback 2019

No claim to original U.S. Government works

ISBN 13: 978-0-367-45201-8 (pbk)
ISBN 13: 978-0-8247-6949-9 (hbk)

**Visit the Taylor & Francis Web site at
http://www.taylorandfrancis.com**

**and the CRC Press Web site at
http://www.crcpress.com**

Library of Congress Cataloging in Publication Data

Voxman, William L., [Date]
    Advanced calculus.

    (Pure and applied mathematics ; 63)
    Includes bibliographical references and index.
    1. Calculus. I. Goetschel, Jr., Roy H. [Date]
II. Title.
QA303.V68   515   80-24250
ISBN 0-8247-6949-X

To Charlie, Jane, Alex, Tanya, Mary, and our respective parents

PREFACE

This text is intended to provide students of mathematics and related disciplines, such as engineering, physics and biology, with an introduction to both the theory and applications of elementary analysis. In addition, we hope that through this book the reader will gain sufficient mathematical maturity to be able to pursue more specialized and advanced courses with greater ease and understanding. We have tried to bridge the unfortunate and unnecessary hiatus between the student of mathematics and students who use mathematics in their particular field of study. To this end emphasis is given to the genesis and resolution of a variety of applied problems—problems that serve to motivate and justify a number of the rather abstract concepts with which a student at this level must deal. It is our desire that the reader will come to view mathematics as a very viable, exciting, and integral part of the world of science rather than as an abstract entity, absorbed in its own problems.

Although the text is fairly demanding of the reader, it can be successfully used by the motivated student whose mathematical background consists of only a two- or three-semester calculus sequence. Generally, however, it is our experience that students in classes of advanced calculus have already been exposed to some work in differential equations and/or linear algebra. For those who have not, Chapter 2 of the text provides in a rather abbreviated manner the necessary background in these areas for the purposes of this book.

This chapter also serves to help ease the student into the somewhat more theoretical material that follows.

A certain degree of abstractness, not commonly found in a text at this level, is encountered in Chapters 3 and 4, as well as in certain subsequent chapters, notably Chapter 11. It is our feeling that a carefully paced introduction to metric spaces is well within the grasp of the junior-senior level student; even limited success in mastering such material will greatly facilitate the student's ability to assimilate the remainder of the text. Moreover, an early but gentle exposure to topics such as metric spaces and measure theory should prove to be of considerable use to the student who wishes to pursue more advanced courses, where comparatively little time is given to motivation and historical digressions.

The material found in the first eight chapters forms the core of the book. Chapter 9 is optional, and the succeeding chapters are relatively independent of each other and may be taken up in any order or omitted; the only exception being that Chapter 10 is a prerequisite for Chapter 13. Considerable attention is given to Fourier analysis since this material seems particularly apt for illustrating both the applied and the theoretical side of analysis and is of considerable historical import as well. It should be apparent that there is more material in the text than can be covered in a one-year course; the variety of material encompassed by the later chapters allows the instructor to select material according to his or her taste and the interests of the class.

The numbering system used in the text is self-evident and needs no elucidation here. Sections marked with an asterisk either deviate slightly from the main thrust of the book or are somewhat more difficult in nature; these sections may be omitted without unduly disturbing the continuity of the text. Problems designated with an asterisk can be quite challenging, but they should not be beyond the range of the talented student.

There are a number of persons who helped in the preparation of the manuscript. Our colleague Larry Bobisud generously read the entire manuscript and made many helpful criticisms. Ed Hewitt examined much of the text and his suggestions were of considerable value, and Arta Childears, Denise Fingerson, and Bob Read all aided (with varying degrees of enthusiasm) in different phases of the preparation. We would like to thank our students B. Weibley, G. Herbst, J. Bell, T. Heywood, P. Meier, and M. Voxman for calling our attention to a multitude of errors in a preliminary version of the text, and finally we wish to give special acknowledgment to Robin Cruz, who typed without respite many versions of the manuscript and also prepared most of the illustrations.

In spite of all the aforementioned assistance, the authors are quite aware that many errors undoubtedly remain (their entertainment value and pedagogical merit should not be underestimated by the student); we would like to encourage the reader to point out to us any such mishaps, and we welcome the reader's ideas for improving the text.

William L. Voxman
Roy H. Goetschel, Jr.

CONTENTS

PRELIMINARIES

This chapter is devoted to a rather brief review of some of the rudiments of set theory and of a few of the elementary properties of real and complex numbers and functions. No attempt is made at a complete presentation of these topics since we consider only those aspects that will be of use in the remainder of the text.

A. INDUCTION

Mathematical induction provides a useful and easily applied tool for resolving a wide range of problems. As a simple illustration, consider the following situation. Suppose we make the observation that $1 = 1 \cdot 2/2$, $1 + 2 = 2 \cdot 3/2$, and $1 + 2 + 3 = 3 \cdot 4/2$. Can we conclude that $1 + 2 + 3 + \cdots + 49 = 49 \cdot 50/2$, or more generally that $1 + 2 + \cdots + n = n(n + 1)/2$ for any positive integer n? It is precisely for solving problems of this nature that induction can be employed to good effect. The basic idea behind induction is the following. Suppose that for each positive integer n we have a statement, S_n, that we would like to verify (in our example, S_n would be the statement: $1 + 2 + \cdots + n = n(n + 1)/2$). If we can establish that

1. The statement S_1 is valid.
2. The statement S_{k+1} is valid *whenever* the statement S_k is valid.

then it can be shown that the statement S_n is valid for *each* positive integer n. This is called the *principle of induction*, and it works essentially as follows. Condition (i) asserts that the statement S_n holds for $n = 1$; since S_1 is true, it follows then from (ii) that S_2 is valid; now that S_2 is true, it follows once again from (ii) that S_3 holds. Thus, if we proceed one step at a time in this fashion, it certainly seems plausible that for any positive integer n the statement S_n will be valid [provided of course that (i) and (ii) can be verified]. Establishment of the validity of the principle of induction follows readily from the formal construction of the set of positive integers [see, for example, Pinter (1971)]. A fairly accessible proof of this principle which is based on another property of the set of positive integers, that of well ordering, is indicated in problem D.8 of this chapter.

We now apply the principle of induction to our introductory problem. It is verified directly that if $n = 1$, then $1 = 1 \cdot 2/2$; thus condition (i) is satisfied. It remains to establish condition (ii). Suppose that the statement S_k:

$$1 + 2 + \cdots + k = \frac{k(k + 1)}{2}$$

is valid. We must show (under this supposition) that S_{k+1} also holds; that is,

$$1 + 2 + \cdots + k + (k + 1) = \frac{(k + 1)(k + 2)}{2} \tag{1}$$

The validity of (1) stems easily from the observation that

$$1 + 2 + \cdots + k + (k + 1) = (1 + 2 + \cdots + k) + (k + 1)$$
$$= \frac{k(k + 1)}{2} + (k + 1)$$
$$= \frac{k(k + 1)}{2} + \frac{2(k + 1)}{2}$$
$$= \frac{(k + 1)(k + 2)}{2}$$

(1.A.1) *Observation* It should be clear that an induction process need not start with 1; for instance, if we can show that S_3 is true and that S_{k+1} is valid whenever S_k ($k \geq 3$) is valid, then S_n will hold for all $n \geq 3$.

On occasion, it will be convenient to employ a slightly different (but actually equivalent) form of the Principle of Induction:

(1.A.2) *Alternate Form of the Principle of Induction* Suppose that for each positive integer n, we have a statement S_n such that

(i) S_1 is valid.

(ii) S_{k+1} is valid whenever $S_1, S_2, \cdots, S_k$ are valid.

Then S_n is valid for each positive integer n.

(1.A.3) *Example* In the popularization of mathematics, the Fibonacci numbers have long had a special appeal. The study of these numbers was initiated by an outstanding thirteenth century Italian mathematician, Leonardo de Pisa (called Fibonacci). Although Fibonacci wrote a number of learned treatises, his present day fame rests primarily on the following problem that appeared in 1202 in his masterwork *Liber abaci*:

Assume that it is known that a newborn pair of rabbits will be able to produce a new pair of rabbits after 2 months and an additional pair every month thereafter. Suppose we start with a newborn pair and leave them (and all subsequent pairs of offspring) to their own devices for 1 year. How many pairs of rabbits will there be after 1 year?

We make the following table:

Time— beginning of month	Pairs	
1	1	
2	1	
3	2	(A new pair has been produced by the original pair.)
4	3	(A second pair has been produced by the original pair.)
5	5	(A third pair has been produced by the original pair and one pair has been produced by the first pair of offspring.)
6	8	
7	13	
8	21	
9	34	
10	55	
11	89	
12	144	
13	233	

If we denote

$$u_1 = 1$$
$$u_2 = 1$$
$$u_3 = 2$$
$$u_4 = 3$$
$$u_5 = 5$$
$$\cdots\cdots$$

then in general we find that $u_n = u_{n-1} + u_{n-2}$ (at the nth stage we have all the pairs from the previous stage together with latest offspring from the pairs present at the $(n - 2)$nd stage); thus, for $n \geq 3$, each u_n in the above sequence is the sum of the two immediately preceding numbers in this sequence. We now use (1.A.2) to find a way of calculating u_n directly. Specifically, we show that for all n

$$u_n = \frac{\alpha^n - \beta^n}{\sqrt{5}} \tag{2}$$

where $\alpha = (1 + \sqrt{5})/2$ and $\beta = (1 - \sqrt{5})/2$.

The reader may verify by direct substitution that (2) holds for $n = 1$ and $n = 2$. Suppose then that (2) is true for $n = 1, 2, \ldots, k - 1, k$; we are to show that under this supposition

$$u_{k+1} = \frac{\alpha^{k+1} - \beta^{k+1}}{\sqrt{5}}$$

By our inductive assumption, we have

$$u_{k+1} = u_k + u_{k-1} = \frac{\alpha^k - \beta^k}{\sqrt{5}} + \frac{\alpha^{k-1} - \beta^{k-1}}{\sqrt{5}}$$
$$= \frac{\alpha^{k-1}(\alpha + 1) - \beta^{k-1}(\beta + 1)}{\sqrt{5}} \tag{3}$$

Now we observe that $\alpha = (1 + \sqrt{5})/2$ and $\beta = (1 - \sqrt{5})/2$ are the roots of the equation $x^2 - x - 1 = 0$, and hence

$$\alpha^2 = \alpha + 1 \tag{4}$$
$$\beta^2 = \beta + 1 \tag{5}$$

Therefore, by (3), (4), and (5) it follows that

$$u_{k+1} = \frac{\alpha^{k-1}(\alpha^2) - \beta^{k-1}(\beta^2)}{\sqrt{5}} = \frac{\alpha^{k+1} - \beta^{k+1}}{\sqrt{5}}$$

and hence, by (1.A.2), formula (2) is valid for all n.

Additional properties of the Fibonacci numbers are considered in the problem section.

(1.A.4) *Exercise* This exercise serves to illustrate that some care must be taken when induction is applied. We use induction to "prove" that all men are 7 feet tall (in the likely event that there is at least one man in the world who is 7 feet tall). Let S_n be the statement that all men in any group of n men are of equal height. Note that S_1 is clearly valid. Suppose then that S_k is valid and let $w_1, w_2, \ldots, w_k, w_{k+1}$ be any group of $k + 1$ men. Since S_k is assumed to be true, we have that the men $w_1, w_2, \ldots, w_k$ are of the same height and also that the men $w_2, \ldots, w_k, w_{k+1}$ are of the same height. The man w_2 is common to both groups, and hence, it follows that all of the men in the group w_1, $w_2, \ldots, w_k, w_{k+1}$ are the same height; that is, S_{k+1} is valid. Therefore, by the principle of induction, all men in any group of n men are of equal height. Since there is a finite number of men in the world, and since we are assuming that there is at least one man 7 feet tall, it follows that all men are 7 feet tall. Find the error (if any) in the proof we have outlined.

B. SETS AND FUNCTIONS

From time to time it will be both convenient and necessary to employ some of the basic notions from set theory. In mathematics, a *set* is an undefined object. Intuitively, a set may be considered as a group of objects, a class of things, a conglomeration of entities, etc. But trying to define a set in these terms would only be begging the question: What is a class, group, conglomeration, etc? Examples of sets include

$$X = \{1,2,3,4\}$$
$$Y = \{y \mid y \text{ is a citizen of Moscow}\}$$
$$Z = \{z \mid z \text{ is a prime number}\}$$

(The vertical line $\mid$ is to be read "such that.") The objects found in a set are called the *elements* or *members* of the set: whenever x is a member of a set A, we write $x \in A$, and whenever x is not a member of A, we write $x \notin A$. We say that a set A is a *subset* of a set B if each member of A is a member of B, i.e., $x \in A$ implies that $x \in B$. If A is a subset of B, then we write $A \subset B$.

(1.B.1) *Example* If $B = \{-\pi, 2, e, \frac{1}{2}\}$ and $A = \{e, 2\}$, then $A \subset B$.

(1.B.2) *Definition* Two sets A and B are said to be *equal* if they contain precisely the same elements.

(1.B.3) *Observation* Note that (1.B.2) is equivalent to saying that $A = B$ if and only if $A \subset B$ and $B \subset A$. Under this definition, the sets $A = \{2,2,4,6,7\}$ and $B = \{6,4,2,4,4,7\}$ are equal.

The *cartesian product* of two sets A and B, denoted by $A \times B$, is the set consisting of all ordered pairs (a,b), where $a \in A$ and $b \in B$, i.e., $A \times B = \{(a,b) \mid a \in A, b \in B\}$. Although it is unlikely that the reader at this stage would be bothered by such technicalities, it should at least be pointed out that to define an ordered pair merely as the symbol (a,b) leaves something to be desired; to be mathematically rigorous such a symbol must be translated into set notation. In a more formal presentation, a typical element in the set $A \times B$ would be the set $\{\{a\}, \{a,b\}\}$ (which is perfectly well defined: it is a set containing two elements, the sets $\{a\}$ and $\{a,b\}$). A typical member of $B \times A$ would have the form $\{\{b\}, \{a,b\}\}$, where $a \in A$ and $b \in B$. Then the symbol (a,b) would be used in place of the well-defined object $\{\{a\}, \{a,b\}\}$, and (b,a) would be used to denote $\{\{b\}, \{a,b\}\}$. It can be shown (see problem B.10) that with this notation, $(a,b) = (c,d)$ if and only if $a = c$ and $b = d$. Note that $(a,a) = \{\{a\}, \{a,a\}\} = \{\{a\}, \{a\}\} = \{\{a\}\}$, which serves to point out that in this case order is of no consequence.

(1.B.4) *Observation* If $\mathbf{R}^1$ denotes the set of real numbers, then the *cartesian plane*, $\mathbf{R}^2$, is the set $\mathbf{R}^1 \times \mathbf{R}^1$. We define *n-dimensional space*, $\mathbf{R}^n$, to be the set of n-tuples $\{(x_1, x_2, \ldots, x_n) \mid x_i \in \mathbf{R}^1$ for each i, $1 \leq i \leq n\}$. ($\mathbf{R}^n$ may be defined somewhat more formally by inductively defining $\mathbf{R}^n$ to be $\mathbf{R}^{n-1} \times \mathbf{R}^1$.)

Cartesian products can be used to define functions. If A and B are sets, then a *function* from A into B is a subset f of $A \times B$ with the property:

If $a \in A$, there is a unique element $b \in B$ such that $(a,b) \in f$.   (1)

It is customary in this situation to write $f: A \to B$ and $f(a) = b$ whenever $(a,b) \in f$. Note that by the uniqueness condition in (1) we cannot have that $f(a) = b_1$ and $f(a) = b_2$ for distinct values b_1 and b_2.

(1.B.5) *Examples* (a) $A = \{1,2,3\}$; $B = \{\frac{1}{2},0,4,6\}$. Then $f = \{(1,0), (2,4), (3,0)\}$ is a function from A into B. This is usually denoted by $f: A \to B$, where

$$f(1) = 0 \qquad f(2) = 4 \qquad f(3) = 0$$

(b) A = real numbers; B = nonnegative real numbers. Then $f = \{(x,x^2) \mid x \in A\}$ is a function from A into B. Normally, we would write $f: A \to B$, where $f(x) = x^2$.

If $f: A \to B$, the set A is called the *domain* of f and the set $\{b \in B \mid f(a) = b$ for some $a \in A\}$ is called the *range* or *image* of f. A function $f: A \to B$ is said to be one-to-one $(1 - 1)$, (or *injective*) if whenever $a_1 \neq a_2$, then $f(a_1) \neq f(a_2)$, and f is said to be *onto* (or *surjective*) if range $f = B$. A function $f: A \to B$ which is both injective and surjective is said to be *bijective*. An injective (surjective, bijective) function is called an *injection* (*surjection*, *bijection*).

(1.B.6) *Examples* (a) $f: [0,3] \to \mathbf{R}^1, f(x) = 2x + 1$. ($\mathbf{R}^1$ denotes the set of real numbers and $[0,3]$ denotes $\{x \in \mathbf{R}^1 \mid 0 \le x \le 3\}$.) The domain of f is $[0,3]$, the range of f is $[1,7]$, and f is injective but not onto.

(b) $f: \mathbf{R}^1 \to [0,\infty)$, $f(x) = x^2$ ($[0,\infty)$ denotes $\{x \in \mathbf{R}^1 \mid x \ge 0\}$). The domain of f is $\mathbf{R}^1$, the range of f is $[0,\infty)$, and f is onto but not injective.

(c) $f: \mathbf{R}^1 \to (-1,1)$, $f(x) = x/(1 + |x|)$ (here $(-1,1)$ denotes $\{x \in \mathbf{R}^1 \mid -1 < x < 1\}$). The function f is a bijection. (Can you prove this?)

(1.B.7) *Definition* If $f: A \to B$ is a bijection, then the *inverse* of f, f^{-1}, is the function mapping B onto A, where $f^{-1}(b) = a$ whenever $f(a) = b$. Note that the domain of f^{-1} is the range of f.

(1.B.8) *Examples* (a) If $A = \{1,7,9\}$, $B = \{0, -\frac{1}{2}, 3\}$ and $f(1) = -\frac{1}{2}$, $f(7) = 0$, $f(9) = 3$, then $f^{-1}: B \to A$ is defined by $f^{-1}(0) = 7, f^{-1}(-\frac{1}{2}) = 1$, $f^{-1}(3) = 9$.

(b) If $f(x) = 2x^2 - 1$ for $x > 0$, then $f^{-1}(y) = \sqrt{(y+1)/2}$; the domain of f^{-1} is $\{y \mid y > -1\}$.

(c) The inverse of the function given in (1.B.6; c) is defined by $f^{-1}(y) = y/(1 - |y|)$ (see problem B.9).

Observe that if f is not $1 - 1$, then f^{-1} cannot be defined, since for some $a_1 \ne a_2$, we would have $f(a_1) = f(a_2) = b$, and consequently $f^{-1}(b)$ could not be uniquely determined. Furthermore, f must be onto in order that the domain of f^{-1} be B.

There is an unfortunate notational "abuse" that has become firmly rooted in the mathematical literature (and will not be uprooted here). If $f: A \to B$ (where f is neither necessarily injective nor surjective) and if $D \subset B$, then the symbol $f^{-1}(D)$ is used to denote the set $\{x \in A \mid f(x) \in D\}$. In other words, $f^{-1}(D)$ is the subset S of A that consists of all elements in A that are mapped by f into D. The notational problem stems from the fact that the inverse of f, f^{-1}, may not even exist and yet the symbol $f^{-1}(D)$ is always well defined.

(1.B.9) *Example* Suppose that $A = \{1,2,3,4,5\}$, $B = \{4,7,9,10\}$, and $f: A \to B$ is defined by $f(1) = 4$, $f(2) = 4$, $f(3) = 7$, $f(4) = 9$, $f(5) = 9$. If $D = \{4,7\}$, then $f^{-1}(D) = \{1,2,3\}$.

The reader is already well acquainted with the following three definitions.

(1.B.10) *Definition* Suppose that $f: A \to B$ and $g: B \to C$ are functions. Then the *composition* of f and g is the function $h: A \to C$ defined by $h(a) = g(f(a))$. It is customary to denote h by $g \circ f$.

(1.B.11) *Examples* (a) If $f(x) = x^2 + 1$ and $g(x) = 3x - 1$, then $(g \circ f)(x)$ $= 3(x^2 + 1) - 1$.

(b) If $f: A \to B$ has an inverse f^{-1}, then $f \circ f^{-1} = \mathrm{id}_B$ and $f^{-1} \circ f = \mathrm{id}_A$, where id_B and id_A are the identity functions [$\mathrm{id}_A(a) = a$ for each $a \in A$ and $\mathrm{id}_B(b) = b$ for each $b \in B$].

(1.B.12) *Definition* Suppose that $f: A \to X$ and $B \subset A$. Then the *restriction* of f to B, $f|_B$, is the function mapping B into X defined by $f|_B(b) = f(b)$ for each $b \in B$. The symbol $f(B)$ is used to denote the set $\{f(b) \mid b \in B\}$.

(1.B.13) *Definition* Suppose that $A \subset \mathbf{R}^1$ and $f: A \to \mathbf{R}^1$. Then f is said to be an *increasing function* on A if whenever $x, y \in A$ and $x < y$, then $f(x) \le f(y)$; f is said to be *strictly increasing* on A if whenever $x, y \in A$ and $x < y$, then $f(x) < f(y)$. The function f is a *decreasing function* on A if whenever $x, y \in A$ and $x < y$, then $f(x) \ge f(y)$, and f is *strictly decreasing* on A if whenever $x, y \in A$ and $x < y$, then $f(x) > f(y)$.

(1.B.14) *Example* The function $f: [0, \infty) \to \mathbf{R}^1$ defined by $f(x) = x^2$ is strictly increasing and the function $f: (-\infty, 0] \to \mathbf{R}^1$ defined by $f(x) = x^2$ is strictly decreasing; the function $f: \mathbf{R}^1 \to \mathbf{R}^1$ defined by $f(x) = x^2$ is neither increasing nor decreasing.

Functions are used to define sequences.

(1.B.15) *Definition* An *(infinite) sequence* in a set X is a function, $f: A \to X$, where A is the set of all integers greater than or equal to some fixed integer p (unless specifically indicated otherwise, we shall assume that $p = 1$). If A consists of the finite set of integers between two given integers, then $f: A \to X$ is said to be a *finite sequence*. In general, we shall use the word *sequence* to denote an infinite sequence.

If f is a sequence in a set X, then for each integer n in its domain, we shall denote $f(n)$ by x_n, and we denote the sequence f by either $\{x_1, x_2, \ldots\}$ or $\{x_n\}$.

(1.B.16) *Examples* (a) Let $\mathbf{Z}^+$ denote the set of positive integers. Then the function $f: \mathbf{Z}^+ \to \mathbf{R}^1$ defined by $f(n) = 2^n - 1$ is a sequence that can be written as $\{1, 3, 7, 15, \ldots\}$ or as $\{2^n - 1\}$.

(b) Let $X = \{-1, 1\}$ and $f: \mathbf{Z}^+ \to X$ be defined by $f(n) = (-1)^n$. This sequence alternates between the numbers -1 and 1 and can be written as $\{-1, 1, -1, 1, \ldots\}$ or $\{(-1)^n\}$.

Frequently, we shall wish to select certain members of a sequence to form what is called a *subsequence*. This is always done in such a way as to maintain the original order of the terms. For example, the sequence $\{2, 12, 22, 32, \ldots\}$

is a subsequence of the sequence $\{2, 4, 6, 8, \ldots\}$ while the sequence $\{12, 2, 32, 22, 52, \ldots\}$ is not. Formally, a subsequence of a sequence is determined by the composition of the original sequence and a strictly increasing function, which does the selecting of the members of the original sequence that are to form the subsequence; the fact that this function is strictly increasing ensures that order is maintained.

(1.B.17)   *Definition* Suppose that $f: \mathbf{Z}^+ \to X$ is a sequence in X and that $g: \mathbf{Z}^+ \to \mathbf{Z}^+$ is a strictly increasing function. Then the composite function $f \circ g: \mathbf{Z}^+ \to X$ is a *subsequence* of X.

(1.B.18)   *Example* Let f, defined by $f(n) = 2^n$, be a sequence in $\mathbf{R}^1$ and suppose that $g: \mathbf{Z}^+ \to \mathbf{Z}^+$ is defined by $g(n) = 4n$. Then the sequence f is $\{2, 4, 8, 16, \ldots\}$, and the subsequence of this sequence determined by g is $\{2^{4n}\} = \{16, 256, \ldots\}$.

We shall adopt the following notation for subsequences: if $\{x_n\}$ is a sequence in a set X, then $\{x_{n_k}\}$ will denote a subsequence of $\{x_n\}$. Here it is understood (but rarely stated) that $\{x_n\}$ is defined by $f: \mathbf{Z}^+ \to X$ where $f(n) = x_n$, and $\{x_{n_k}\}$ is determined by a strictly increasing function $g: \mathbf{Z}^+ \to \mathbf{Z}^+$, where $g(k) = n_k$; consequently, $f \circ g: \mathbf{Z}^+ \to X$ is defined by $f \circ g(k) = x_{n_k}$.

We conclude this section with the definitions of periodic functions and even and odd functions. These functions will assume special importance in Chapter 10.

(1.B.19)   *Definition* Suppose that $A \subset \mathbf{R}^1$ and $f: A \to \mathbf{R}^1$. Then f is *periodic* if there is a nonzero number c such that $f(x + c) = f(x)$ for each $x \in A$ (it is assumed that $x + c \in A$ whenever $x \in A$). Such a number c is called a *period* of f. The smallest positive period c is called the *fundamental period* of f.

Not all periodic functions have a fundamental period (see problem B.13).

(1.B.20)   *Examples* (a) $f(x) = \sin x$ is periodic with periods 2π, -4π, 16π, etc. The fundamental period is 2π.
(b) $f(x) = \cos 3x$ has fundamental period $2\pi/3$.
(c) $f(x) = x - n$, whenever $n \leq x < n + 1$ and n is an integer, has fundamental period 1.

(1.B.21)   *Definition* Let A be an interval of $\mathbf{R}^1$ centered at the origin. Then a function $f: A \to R^1$ is said to be an *even* function on A if $f(-x) = f(x)$ for each $x \in A$. The function f is said to be an *odd* function if $f(-x) = -f(x)$ for each $x \in A$.

(1.B.22) *Examples* (a) The functions $f(x) = \sin x$ and $f(x) = x^k$, where k is an odd integer are odd functions.

(b) The functions $f(x) = x^k$, where k is an even integer and $f(x) = \cos x$ are even functions.

(c) The functions $f(x) = (\ln x) \sin x$ and $f(x) = e^x$ are neither even nor odd.

(1.B.23) *Observation* If f is an even function, then the graph of f is symmetric with respect to the y axis, and if f is odd, then the graph of f is symmetric with respect to the origin.

(1.B.24) *Exercise* Show that the sum of two even functions is even, that the sum of two odd functions is odd, and that the product of two even functions or two odd functions is even.

C. UNIONS, INTERSECTIONS, AND COMPLEMENTS

Unions and intersections of sets are defined as follows.

(1.C.1) *Definition* If A and B are sets, then the *union* of A and B, denoted by $A \cup B$, is the set $C = \{x \mid x \in A \text{ or } x \in B\}$.

When used in a mathematical context, the word *or* is not exclusive; therefore, in the previous definition, if x belongs to both A and B, then x is in C.

(1.C.2) *Example* If $A = \{1,2,3,6\}$ and $B = \{\square,2,-\frac{1}{2},6,7\}$ then $A \cup B = \{1,2,3,6,\square,-\frac{1}{2},7\}$.

(1.C.3) *Definition* If A and B are sets, the *intersection* of A and B, denoted by $A \cap B$, is the set $D = \{x \mid x \in A \text{ and } x \in B\}$. It may be the case that $A \cap B$ has no elements at all, in which case we write $A \cap B = \varnothing$ and call $\varnothing$ the *empty set*. The sets A and B are said to be *disjoint* if $A \cap B = \varnothing$.

(1.C.4) *Example* In the previous example (1.C.2), $A \cap B = \{6,7\}$. Quite often we shall need to form unions and intersections of more than two sets. This is handled easily with the idea of an index set. A collection $\mathscr{A}$ of sets is *indexed* by a set J if there exists a bijection $f: J \to \mathscr{A}$. If $\alpha \in J$, we denote the image of f, $f(\alpha)$, by A_α, which is, of course, a set in $\mathscr{A}$; we write $\mathscr{A} = \{A_\alpha \mid \alpha \in J\}$. For instance, the collection of sets $\mathscr{A} = \{A_1, A_2, \ldots\}$ is indexed by the set $J = \mathbf{Z}^+$, where $\mathbf{Z}^+$ denotes the set of positive integers. Unions and intersections of an indexed family of sets are defined as follows:

(1.C.5) **Definition** If $\mathscr{A} = \{A_\alpha \mid \alpha \in J\}$ is a collection of sets, the *union* of the sets A_α, $\bigcup_{\alpha \in J} A_\alpha$, is $\{x \mid x \in A_\alpha$ for at least one $\alpha \in J\}$, and the *intersection* of the sets A_α, $\bigcap_{\alpha \in J} A_\alpha$, is $\{x \mid x \in A_\alpha$ for each $\alpha \in J\}$.

(1.C.6) **Examples** (a) For each $n \in \mathbf{Z}^+$, let A_n be the interval $[-1/n, n]$. Then $\bigcup_{n \in \mathbf{Z}^+} A = \bigcup_{n=1}^{\infty} A_n = [-1, \infty)$, and $\bigcap_{n \in \mathbf{Z}^+} A_n = \bigcap_{n=1}^{\infty} A_n = [0, 1]$.
     (b) If $J = (0, \infty)$ and if for each $\alpha \in J$, $A_\alpha = (1/\alpha, \alpha)$, then $\bigcap_{\alpha \in J} A_\alpha = \varnothing$ and $\bigcup_{\alpha \in J} A_\alpha = (0, \infty)$.

The complement of a set is defined next.

(1.C.7) **Definition** If A and B are sets, then the *complement of B relative to A*, denoted by $A \backslash B$, is defined to be $\{x \in A \mid x \notin B\}$.

(1.C.8) **Example** If $A = \{n \in \mathbf{Z}^+ \mid n \geq 10\}$ and $B = \{n \in \mathbf{Z}^+ \mid n \leq 15\}$, then $A \backslash B = \{n \in \mathbf{Z}^+ \mid n \geq 16\}$.

There are numerous equalities and inequalities that can be derived from the interplay of unions, intersections, and complements of sets. We present five more or less randomly selected examples. The reader may find additional examples of this type in the problem section.

(1.C.9) **Theorem**
     (i) If A, B, and C are arbitrary sets, then $A \backslash (B \cap C) = (A \backslash B) \cup (A \backslash C)$.
     (ii) If $f: X \to Y$ and $\{A_\alpha \mid \alpha \in J\}$ is a collection of subsets of Y, then $f^{-1}(\bigcup_{\alpha \in J} A_\alpha) = \bigcup_{\alpha \in J} f^{-1}(A_\alpha)$.
     (iii) If $f: X \to Y$ is $1 - 1$ and $A \subset X$, then $f(X \backslash A) \subset Y \backslash f(A)$.
     (iv) If $f: X \to Y$ and A and B are subsets of X, then $f(A \cup B) = f(A) \cup f(B)$.
     (v) If A, B, C, and D are sets, then $(A \times B) \cap (C \times D) = (A \cap C) \times (B \cap D)$.

Proof. (i) Suppose that $x \in A \backslash (B \cap C)$. Then $x \in A$, and $x \notin B$ or $x \notin C$ and, hence, $x \in A \backslash B$ or $x \in A \backslash C$. Therefore, $x \in (A \backslash B) \cup (A \backslash C)$, and we have $A \backslash (B \cap C) \subset (A \backslash B) \cup (A \backslash C)$.
     Now suppose that $x \in (A \backslash B) \cup (A \backslash C)$. Then $x \in A$, $x \notin B$ or $x \in A$, $x \notin C$ and, hence, $x \in A$, and $x \notin B \cap C$. Consequently, $x \in A \backslash (B \cap C)$, which shows that $(A \backslash B) \cup (A \backslash C) \subset A \backslash (B \cap C)$.
     (ii) If $x \in f^{-1}(\bigcup_{\alpha \in J} A_\alpha)$, then $f(x) \in \bigcup_{\alpha \in J} A_\alpha$, and hence $f(x) \in A_{\alpha^*}$ for some $\alpha^* \in J$. Consequently, $x \in f^{-1}(A_{\alpha^*}) \subset \bigcup_{\alpha \in J} f^{-1}(A_\alpha)$, and therefore, $f^{-1}(\bigcup_{\alpha \in J} A) \subset \bigcup_{\alpha \in J} f^{-1}(A_\alpha)$.
     If $x \in \bigcup_{\alpha \in J} f^{-1}(A_\alpha)$, then $x \in f^{-1}(A_{\alpha^*})$ for some $\alpha^* \in J$. Consequently,

$f(x) \in A_{\alpha^*} \subset \bigcup_{\alpha \in J} A_\alpha$, and hence, $x \in f^{-1}(\bigcup_{\alpha \in J} A_\alpha)$. Therefore, $\bigcup_{\alpha \in J} f^{-1}(A_\alpha) \subset f^{-1}(\bigcup_{\alpha \in J} A_\alpha)$.

(iii) Suppose $y \in f(X \backslash A)$. Then there is an $x \in X \backslash A$ such that $f(x) = y$. Note that $y \notin f(A)$, for otherwise we would have that $f(a) = y$ for some $a \in A$ and $f(x) = y$, where $x \in X \backslash A$, which contradicts the fact that f is $1-1$. Hence, $y \in Y \backslash f(A)$, and we have shown that $f(X \backslash A) \subset Y \backslash f(A)$. The reader should try to construct a specific example to show that in general we cannot say that $f(X \backslash A) = Y \backslash f(A)$.

(iv) Suppose that $z \in f(A \cup B)$. Then there is an $x \in A \cup B$ such that $f(x) = z$. Since $x \in A \cup B$, we have $x \in A$ or $x \in B$, and therefore, $z = f(x) \in f(A)$ or $z = f(x) \in f(B)$, which, of course, implies that $z \in f(A) \cup f(B)$.

Now suppose that $z \in f(A) \cup f(B)$. Then there is an $x \in A$ or an $x \in B$ such that $f(x) = z$. In either case, we have $x \in A \cup B$, and hence, $z = f(x) \in f(A \cup B)$.

(v) Suppose that $(x,y) \in (A \times B) \cap (C \times D)$. Then $(x,y) \in A \times B$ and $(x,y) \in (C \times D)$, which implies that $x \in A \cap C$ and $y \in B \cap D$; thus, $(x,y) \in (A \cap C) \times (B \cap D)$.

Conversely, if $(x,y) \in (A \cap C) \times (B \cap D)$, then $x \in (A \cap C)$ and $y \in (B \cap D)$, and consequently, $x \in A$, $x \in C$, $y \in B$, and $y \in D$; therefore, $(x,y) \in A \times B$ and $(x,y) \in (C \times D)$, and hence it is an element of their intersection.

To conclude this section we present the reader with a now classic dilemma, the most famous version of which was formulated by Bertrand Russell in 1901.

Suppose that we divide all of the male inhabitants of a certain town into two disjoint sets A and B, where A consists of all males who do not shave themselves and B, all those who do shave themselves (and thus $A \cap B = \varnothing$). Let x^* be the local barber (a male), who shaves *only* those males who do not shave themselves. Since $A \cup B$ is the set of all male inhabitants in the town, obviously $x^* \in A \cup B$. Does $x^* \in A$? If so, the barber does not shave himself, but hence (by dint of his work), he does shave himself and, therefore, is also in B, which is absurd since $A \cap B = \varnothing$. Does $x^* \in B$? If so, the barber shaves himself but, again, by the nature of his work he does not shave himself (he shaves *only* those who don't shave themselves), and therefore, x^* is also in A.

The "solution" to this paradox lies in introducing a broader concept than that of a set, and is dealt with in courses in set theory (see Monk, 1969). The reader should be aware that problems inherent in even a "naive" theory of sets are quite profound.

D. COUNTABILITY

In this section we briefly investigate the number of elements that sets may have. We shall say that a set A is *finite* if there is a bijection $f: \{1, 2, \dots, n\} \to$

A; in this case A has n elements. If a set A is not finite (or empty), we say that A is *infinite*. Do all infinite sets have the same "number" of elements? It might appear at first, for instance, that there are more positive integers than positive even integers. However, when we consider the lists

$$1 \quad 2 \quad 3 \quad 4 \quad 5 \quad 6 \quad 7 \cdots$$

$$2 \quad 4 \quad 6 \quad 8 \quad 10 \quad 12 \quad 14 \cdots$$

we see that there is a bijection $f: \mathbf{Z}^+ \to B$ defined by $f(n) = 2n$, where B is the set of positive even integers. Hence, in this sense $\mathbf{Z}^+$ and B have the "same number" of elements.

(1.D.1) *Definition* Two sets A and B are said to have the same *cardinality* or to be *equipotent* if there is a bijection $\phi: A \to B$. If A and B are equipotent, we write card $A =$ card B.

The cardinality of a set A is said to be less than the cardinality of a set B (card $A <$ card B) in case there is an injection from A into B but no bijection between A and B. The *cardinality* of a finite set A is the number of elements in A.

(1.D.2) *Definition* A set A is *countably infinite* if there exists a bijection $f: \mathbf{Z}^+ \to A$.

(1.D.3) *Definition* A set A is *countable* if A is finite or countably infinite. A set that is not countable is said to be *uncountable*.

The set of rational numbers is countably infinite as may be seen from the following "infinite" table that lists (with some duplication) all of the rational numbers.

(1) 0

$\downarrow$

(2) $\frac{1}{2}$ (4) $\frac{1}{2} \to$ (5) $\frac{2}{2}$ $\frac{2}{2} \to$ $\frac{3}{2}$ $\frac{3}{2}$ $\frac{4}{2}$ $\frac{4}{2}$ $\cdots$

$\downarrow \nearrow$ $\swarrow$ $\nearrow$ $\swarrow$

(3) $\frac{1}{3}$ (6) $-\frac{1}{3}$ $\frac{2}{3}$ $-\frac{2}{3}$ $\frac{3}{3}$ $-\frac{3}{3}$ $\frac{4}{3}$ $-\frac{4}{3}$ $\cdots$

$\swarrow$ $\nearrow$

(7) $\frac{1}{4}$ (9) $-\frac{1}{4}$ $\frac{2}{4}$ $-\frac{2}{4}$ $\frac{3}{4}$ $-\frac{3}{4}$ $\frac{4}{4}$ $-\frac{4}{4}$ $\cdots$

$\downarrow \nearrow$

(8) $\frac{1}{5}$ $-\frac{1}{5}$ $\frac{2}{5}$ $-\frac{2}{5}$ $\frac{3}{5}$ $-\frac{3}{5}$ $\frac{4}{5}$ $-\frac{4}{5}$ $\cdots$

$\frac{1}{6}$ $-\frac{1}{6}$ $\frac{2}{6}$ $-\frac{2}{6}$ $\frac{3}{6}$ $-\frac{3}{6}$ $\frac{4}{6}$ $-\frac{4}{6}$ $\cdots$

$\cdots$ $\cdots$ $\cdots$ $\cdots$ $\cdots$ $\cdots$ $\cdots$ $\cdots$

Actually, because of duplication in the above table we have shown that there is a bijection between the set of rational numbers and an infinite subset of $\mathbf{Z}^+$. The desired result follows from the following "obvious" but nontrivial result which says in essence that countably infinite sets are the smallest infinite sets.

(1.D.4) Theorem Suppose that A is a countably infinite set and $B \subset A$ is infinite. Then B is countably infinite.

Proof. Since A is countably infinite, there is a bijection $f: \mathbf{Z}^+ \to A$. Let $J = \{n \in \mathbf{Z}^+ \mid f(n) \in B\}$. Let n_1 be the least integer found in J. Note that since B is infinite, $J\backslash\{n_1\}$ is nonempty and therefore has a least member, n_2. In general, if we have selected $n_1, n_2, \ldots, n_k$ in this fashion, then we let n_{k+1} be the least integer found in the nonempty set $J\backslash\{n_1, n_2, \ldots, n_k\}$. Now define $g: \mathbf{Z}^+ \to A$ by setting $g(i) = f(n_i)$ for each $i \in \mathbf{Z}^+$. Clearly, by the way we selected the n_i's, g is $1-1$. We next show that the range of g is precisely B, which, of course, implies that g is a bijection between B and $\mathbf{Z}^+$ and that B is countably infinite. It follows directly from the definition of J that range $g \subset B$. To see that $B \subset$ range g, let $b \in B$ and observe that there is a $j \in J$ such that $f(j) = b$. Furthermore, the integer j must have been the least integer at some stage of the selection of $n_1, n_2, \ldots$, and, hence, there is an integer k such that $j = n_k$. Consequently, we have that $g(k) = f(n_k) = f(j) = b$, which concludes the proof.

(1.D.5) Observation Actually there is something amiss in the above proof: we have tacitly assumed that if M is any nonempty subset of $\mathbf{Z}^+$, then M must have a least element. That this is true is known as the *principle of well ordering* and is a nontrivial property of the positive integers. In the problem set the reader is asked to show that this principle is equivalent to the principle of induction.

(1.D.6) Exercises (a) Show that if $f: \mathbf{Z}^+ \to A$ is a surjection, then A is countably infinite.
　　(b) Show that $\mathbf{Z}^+ \times \mathbf{Z}^+$ is countably infinite.

(1.D.7) Theorem Suppose that $A_1, A_2, \ldots$ is a countably infinite family of sets and that for each n, A_n is countably infinite. Then $\bigcup_{n=1}^\infty A_n$ is countable, i.e., the union of a countable number of countable sets is countable.

Proof. We first suppose that each A_n is countably infinite. Let

$$A_1 = \{a_{11}, a_{12}, a_{13}, \ldots\}$$
$$A_2 = \{a_{21}, a_{22}, a_{23}, \ldots\}$$

$$A_3 = \{a_{31}, a_{32}, a_{33}, \ldots\}$$

. . . .

Then the function $\phi: \mathbf{Z}^+ \times \mathbf{Z}^+ \to \bigcup_{n=1}^{\infty} A_n$ defined by $\phi(m,n) = a_{mn}$ is a surjection, and therefore, by (1.D.6) $\bigcup_{n=1}^{\infty} A_n$ is countably infinite.

An obvious modification of this argument shows that the same result holds if some or all of the sets A_n are finite (or empty).

In view of (1.D.6) and (1.D.7) one might justifiably inquire at this stage if there are any infinite sets that fail to be countably infinite. To show that such sets do exist, we shall consider a certain subset of the interval $(0,1)$. Any real real number in this interval has a decimal expansion of the form $0.e_1e_2e_3\cdots$, where each e_i is an integer between (and including) 0 and 9. Let A be the set of all such numbers with the property that e_i is 0 or 1 for each i. We show that A is uncountable.

If A were countably infinite, then there would be a bijection $f: \mathbf{Z}^+ \to A$ that essentially would permit us to form a complete (infinite) listing of all the elements of A; an example of such a listing is indicated below.

$$1 \leftrightarrow f(1) = .10011100\cdots$$
$$2 \leftrightarrow f(2) = .00001101010\cdots$$
$$3 \leftrightarrow f(3) = .100011110\cdots$$
$$4 \leftrightarrow f(4) = .01100100\cdots$$
$$5 \leftrightarrow f(5) = .1111111011\cdots$$

. . . .

To show that A is not countably infinite, we construct a number c (in A) that fails to be in this complete list. If the first digit following the decimal point of the first number in the given list is 1, then the number c to be constructed begins with a 0; if not, c begins with a 1. If the second digit of the second number in the list is 1, then c has 0 for its second digit and 1 otherwise. In general, the nth digit of c is 0 or 1 if the nth digit of $f(n)$ is 1 or 0. Thus in the sample list above, c would begin as $.01110\cdots$. Clearly, the number c will fail to be in the original list: therefore, no such listing is possible, and consequently, A cannot be countably infinite.

As the reader well knows, the real numbers consist of rational numbers and irrational numbers. Note that it follows easily from the above discussion and (1.D.7) that there are many more irrational numbers than rational numbers; the set of irrational numbers is in fact uncountable, and hence so is the set of real numbers. Interestingly, it is not difficult to show that in spite of the uncountability of the set of irrational numbers and the countability of the set of rational numbers, there can be found between any two rational numbers an irrational number, and between any two irrational numbers a rational number (see problem D.12).

E. sup, lub, inf, AND glb

It is easy to show that the sum, difference, product, and quotient of two rational numbers is again a rational number, and even easier to demonstrate that the properties of commutativity, associativity, distributivity, etc. (which the reader has surely encountered and dutifully mastered in previous courses) hold for the rational numbers and for the real numbers as well. Can any basic distinction then be made between the rationals and the reals? The answer is yes; however, the distinguishing property that interests us is fairly subtle (but important). We begin with an example.

Let $A = \{x \mid x$ is rational and $x < \sqrt{2}\}$. Note that there is no smallest rational number that is greater than or equal to each element of A; however, there is a real (irrational) number with this property: $\sqrt{2}$. This seemingly inconsequential observation will prove to have considerable significance as we proceed.

In order to abstract this situation slightly, we make the following definitions.

(1.E.1) *Definition* A real number b is said to be an *upper bound* of a nonempty subset A of $\mathbf{R}^1$ if $a \leq b$ for each $a \in A$; b is a *lower bound* of A if $b \leq a$ for each $a \in A$.

(1.E.2) *Definition* An upper bound b of a nonempty set $A \subset \mathbf{R}^1$ is called a *least upper bound* (or *supremum*) of A if for any other upper bound c of A, $b \leq c$. The least upper bound b is denoted by lub A or sup A.

(1.E.3) *Definition* A lower bound b of a nonempty set $A \subset \mathbf{R}^1$ is called a *greatest lower bound* (or *infimum*) of A if for any other lower bound c of A, $c \leq b$. The greatest lower bound b is denoted by glb A or inf A.

Note that inf A and sup A are unique (if they exist).

(1.E.4) *Examples* (a) If $A = \{x \mid x$ is rational and $x < \sqrt{2}\}$, then A has no lower bound (and therefore, of course, no greatest lower bound); 16 is an upper bound of A and $\sqrt{2} = $ sup A.

(b) If $A = \{x \mid -2 \leq x < 3\}$, then sup $A = 3$ and inf $A = -2$.

We can now state a critical property that distinguishes the real numbers from the rational numbers. This property is often referred to as the *axiom of completeness.* The real numbers have this property; the rational numbers (and the irrational numbers) do not.

(1.E.5) *Axiom of Completeness* If a nonempty set $A \subset \mathbf{R}^1$ has an upper bound, then A has a least upper bound (in $\mathbf{R}^1$).

Note that we cannot replace $\mathbf{R}^1$ in the above axiom with the set of rational numbers $\mathbf{Q}$, since, for example, if $A = \{x \in \mathbf{Q} \mid x < \sqrt{2}\}$, then A clearly has an upper bound but has no least upper bound (in $\mathbf{Q}$).

(1.E.6) *Observation* The obvious counterpart of (1.E.5):
If a nonempty set $A \subset \mathbf{R}^1$ has a lower bound, then A has a greatest lower bound (in $\mathbf{R}^1$) can be derived from the axiom of completeness (see problem E.2).

In the next theorem we give four elementary but useful properties of sup and inf.

(1.E.7) *Theorem*

  (i) Suppose that $A \subset \mathbf{R}^1$ and $b = \sup A$. If $c < b$, then there is a number $a \in A$ such that $c < a \le b$.
  (ii) Suppose that $A \subset \mathbf{R}^1$ and $b = \inf A$. If $b < c$, then there is a number $a \in A$ such that $b \le a < c$.
  (iii) Suppose that A and B are subsets of $\mathbf{R}^1$ and that $a \le b$ for each $a \in A$ and $b \in B$. Then $\sup A \le \inf B$.
  (iv) Suppose that A and B are subsets of $\mathbf{R}^1$ and that $\sup A$ and $\sup B$ exist. Let $C = \{a + b \mid a \in A, b \in B\}$. Then $\sup C$ exists and $\sup C = \sup A + \sup B$.

Proof. Statements (i) and (ii) follow immediately from the definitions of sup A and inf A.
To prove (iii), first note that sup A and inf B exist, since any $b \in B$ serves as an upper bound for A and any $a \in A$ serves as a lower bound for B. If inf $B < \sup A$, then by (i) there is an $a \in A$ such that inf $B < a \le \sup A$. Since inf $B < a$, by (ii) there is a $b \in B$ such that inf $B \le b < a$, which contradicts the fact that $a \le b$ for all $a \in A$, $b \in B$. Hence, it must be the case that sup $A \le \inf B$.
The proof of (iv) is slightly more subtle. Let $s_1 = \sup A$ and $s_2 = \sup B$. Then for each $a \in A$ and $b \in B$, we have that $a + b \le s_1 + s_2$, and hence, sup C exists and is less than or equal to $s_1 + s_2$.
Now let $s_3 = \sup C$. To show that $s_1 + s_2 \le s_3$, it suffices to show that for *each* positive number r, $s_1 + s_2 \le s_3 + r$ (why?). Note that if r is any positive number, then by part (i) of this theorem there is a number $a \in A$ such

that $s_1 - r/2 < a \le s_1$ and a number $b \in B$ such that $s_2 - r/2 < b \le s_2$. Consequently, we have

$$s_1 + s_2 < a + \frac{r}{2} + b + \frac{r}{2} \le s_3 + r$$

which concludes the proof.

F. COMPLEX NUMBERS AND FUNCTIONS

From time to time we shall make use of complex numbers and complex-valued functions. Here we give a brief review of complex arithmetic; a fairly solid introduction to certain aspects of complex analysis is presented in Chapters 7 and 12.

The *complex number system* is defined to be the set of all ordered pairs (a,b) in $\mathbf{R}^2$ together with the operations of addition and multiplication defined by

$$(a,b) + (c,d) = (a + c, b + d)$$

and

$$(a,b)(c,d) = (ac - bd, ad + bc)$$

respectively. In this context it is customary to refer to the x axis as the *real axis*, the y axis as the *imaginary axis*, and $\mathbf{R}^2$ as the *complex plane*, $\mathbf{C}$. We shall frequently identify a real number a with its complex counterpart $(a,0)$.

Note that we have

$$(0,1)(0,1) = (-1,0) \text{ "=" } -1$$

It is customary to denote $(0,1)$ by i, and the complex number (a,b) by $a + bi$; observe that if a "=" $(a,0)$, b "=" $(b,0)$ and i "=" $(0,1)$, then $a + bi = (a,0) + (b,0)(0,1) = (a,0) + (0,b) = (a,b)$. If $z = a + bi$, then a is called the *real part* of z and is denoted by $\text{Re}(z)$ and b is called the *imaginary part* of z and is denoted by $\text{Im}(z)$.

If $a + bi$ is a complex number, then $1/(a + bi)$ is the complex number $c + di$ with the property $(a + bi)(c + di) = 1$.

(1.F.1) *Exercise* Show that if $c + di = 1/(a + bi)$, then $c = a/(a^2 + b^2)$ and $d = -b/(a^2 + b^2)$.

The quotient $(a + bi)/(c + di)$ is defined by $(a + bi)[1/(c + di)]$.

(1.F.2) *Definition* If $z = a + bi = (a,b)$ is a complex number, then the *absolute value* of z, called the *modulus* of z (and denoted $|z|$), is defined to be $\sqrt{a^2 + b^2}$.

Observe that if $z \in C$, then $|z|$ represents the usual Euclidean distance between z and the origin $(0,0)$.

If polar coordinates are used in the complex plane:

$$x = r \cos \theta \qquad y = r \sin \theta$$

where $r \geq 0$ and $\theta \in R^1$, then a complex number $z = a + bi$ may be written in the polar form $r(\cos \theta + i \sin \theta)$, where $r = |z|$ and θ is the angle between the positive x axis and the ray emanating from the origin and passing through z. If $\theta > 0$, then the angle is measured in a counterclockwise sense from the positive x axis, and if $\theta < 0$, the angle is measured in a clockwise sense. Angle measurement here is customarily in radians.

(1.F.3)   Example   If $z = -1 + \sqrt{3}i$, then in polar form, $z = 2[\cos(2\pi/3) + i \sin (2\pi/3)]$.

Finally, if A is a set and $f: A \rightarrow C$, then for each $a \in A$, $f(a)$ will have a real and an imaginary part. Thus, we can write $f(a)$ as $u(a) + iv(a)$, where u and v are real-valued functions with domain A. Occasionally, in this connection, we shall also write, $f(a) = \text{Re}(f)(a) + i\,\text{Im}(f)(a)$.

(1.F.4)   Example   Let $A = R^1$ and $f: A \rightarrow C$ be defined by $f(x) = 2x + ix^2$. Then we can write $f = u + iv$, where $u(x) = 2x$ and $v(x) = x^2$. Note that both u and v are real-valued functions.

PROBLEMS

Section A

1. Use the principle of induction to show that

   (a) $\displaystyle\sum_{i=1}^{n} \frac{1}{i(i + 1)} = \frac{n}{n + 1}$

   (b) $1^2 + 2^2 + \cdots + n^2 = \dfrac{n(n + 1)(2n + 1)}{6}$

   (c) $1 + 3 + 5 + \cdots + (2n - 1) = n^2$

2. Use the principle of induction to show that

   (a) If $x \neq 2k\pi$, where k is an integer, then

   $$\sin x + \sin 2x + \cdots + \sin nx = \frac{\sin (nx/2) \sin (\frac{1}{2}(n + 1)x)}{\sin (x/2)}$$

(b) $1^2 + 4^2 + 7^2 + \cdots + (3n + 1)^2 = \dfrac{(n + 1)(6n^2 + 9n + 2)}{2}$,

  $n = 0, 1, 2, \ldots$

3. Use the principle of induction to show that
   (a) If $x > -1$, then $(1 + x)^n \geq 1 + nx$ for each $n \in Z^+$.
   (b) If $r \neq 1$, then $1 + r + r^2 + \cdots + r^{n-1} = (1 - r^n)/(1 - r)$ for each $n \in Z^+$.
4. Show that the Fibonacci numbers satisfy
   (a) $\sum_{i=1}^n u_i = u_{n+2} - 1$
   (b) $\sum_{i=1}^n u_i^2 = u_n u_{n+1}$
   (c) $u_{2n} = u_{n-1} u_n + u_n u_{n+1}$
5. Use the alternate form of the principle of induction (1.A.2) to show that $\alpha^{n-2} \leq u_n \leq \alpha^{n-1}$ [notation as in (1.A.3)].
6. Given a set of n points, no three of which lie on a straight line, prove (by induction) that the number of straight line segments that one can draw connecting pairs of points is $(n^2 - n)/2$.
7. Formulate conjectures (and prove them) about the following sums involving Fibonacci numbers: $\sum_{i=1}^n u_{2i}$, $\sum_{i=1}^{2n-1} u_i u_{i+1}$, $\sum_{i=1}^n (-1)^{i-1} u_i$.

Section B

1. If $A = \{2,3,-\frac{1}{2}\}$ and $B = \{3,\frac{1}{4}\}$, find $A \times B$ and $B \times A$.
2. Under what conditions will $A \times B = B \times A$?
3. Find the domain, range, and if it exists, the inverse (when restricted to a suitable domain) of the following functions:

   (a) $f: [2,5] \to R^1; f(x) = 3x - 4$

   (b) $f: [0,\infty) \to R^1; f(x) = x^2 - 3$

   (c) $f(x) = \begin{cases} 2x & \text{if } 0 < x < 2 \\ 9 - x & \text{if } 2 \leq x < 4 \end{cases}$

   (d) $f: [4,9] \to [5,6]; f(x) = \sqrt{x} + 3$

4. If $f: R^1 \to R^1$ is defined by $f(x) = x^2 - 9$, find
   (a) $f(\{2,4,-2\})$
   (b) $f^{-1}(\{2,4\})$
   (c) $f^{-1}(\{0\})$
   (d) $f([-1,4))$
   (e) $f^{-1}((-2,6])$
5. Find a function $f: Z^+ \to Z^+$ that is $1 - 1$ but not onto and a function $g: Z^+ \to Z^+$ that is onto but not $1 - 1$.

6. Show that if $f: A \rightarrow B$ and $g: B \rightarrow C$ are bijections, then $g \circ f$ is a bijection.

7. Find a bijection between the intervals $(1,4)$ and $(2,7)$.

8. If $f(x) = x^2 + 1/x$ and $g(x) = 2x - 1$, find $(f \circ g)(2)$ and $(g \circ f)(2)$.

9. Show that $f: \mathbf{R}^1 \rightarrow (-1,1)$ defined by $f(x) = x/(1 + |x|)$ and $g: (-1,1) \rightarrow \mathbf{R}^1$ defined by $g(y) = y/(1 - |y|)$ are inverse functions of each other.

10.* Use (1.B.2) to show that if $(a,b) = (c,d)$, then $a = c$ and $b = d$.

11. Given the sequence $\{n^2 + 1\}$, find the subsequences that correspond to the following functions:
    (a) $g(k) = k + 1$
    (b) $g(k) = k^2 + 1$
    (c) $g(k) = k^3$

12. Find the fundamental periods of the following functions:
    (a) $\sin 4x$
    (b) $\sin x + \cos x$
    (c) $\sin x - 2 \cos x$

13. Let $f: \mathbf{R}^1 \rightarrow \mathbf{R}^1$ be defined by $f(x) = 1$ if x is rational, and $f(x) = -1$ if x is irrational. Show that f is periodic, but that f does not have a fundamental period.

14. Determine if the following functions are even, odd, or neither:
    (a) $\sin x + \cos x$
    (b) $\sin^2 x$
    (c) $\cos x^2$
    (d) $\tan x$
    (e) $3x + 1$

15. Sketch graphs of typical odd and even functions.

16.* Show that any function $f: (-a,a) \rightarrow \mathbf{R}^1$ can be written as the sum of an even and an odd function.

17. Suppose that $f: \mathbf{R}^1 \rightarrow \mathbf{R}^1$ is even and $g: \mathbf{R}^1 \rightarrow \mathbf{R}^1$ is odd. What can be said about $f + g$?

18. Show that every sequence of real numbers has either an increasing or a decreasing subsequence.

Section C

1. If $A = \{x \mid x$ is a prime number $\}$ and $B = \{x \mid x$ is an odd number and $0 < x < 20\}$, find $A \cup B$, $A \cap B$, and $B \backslash A$.

2. For each $n \in \mathbf{Z}^+$, let $A_n = (-1/n, 2 + 1/n)$. Find $\bigcup_{n=1}^{\infty} A_n$ and $\bigcap_{n=1}^{\infty} A_n$.

3. Show that if $f: X \rightarrow Y$ and $\{A_\alpha \mid \alpha \in J\}$ is a family of subsets of Y, then

$$f^{-1}\left(\bigcap_{\alpha \in J} A_\alpha\right) = \bigcap_{\alpha \in J} f^{-1}(A_\alpha)$$

4. Show that if $f: X \to Y$ and A and B are subsets of X, then $f(A \cap B) \subset f(A) \cap f(B)$. Find an example to show that in general $f(A \cap B)$ is not equal to $f(A) \cap f(B)$.
5. Show that for arbitrary sets A, B, and C,
    (a) $A \cap (B \backslash C) = (A \cap B) \backslash C$
    (b) $(A \cap B) \cup C = (A \cup C) \cap (B \cup C)$
    (c) $(A \backslash B) \times C = (A \times C) \backslash (B \times C)$
6. (a) Show that $A \times (B \cap C) = (A \times B) \cap (A \times C)$.
    (b) Does $(A \times B) \cup (C \times D) = (A \cup C) \times (B \cup D)$?
    (c) When does $A \backslash (A \backslash B) = B$?
7. Suppose that $\{A_\alpha \mid \alpha \in J\}$ is a family of subsets of a set X. Show that

    (a) $X \backslash \bigcup_{\alpha \in J} A_\alpha = \bigcap_{\alpha \in J} (X \backslash A_\alpha)$

    (b) $X \backslash \bigcap_{\alpha \in J} A_\alpha = \bigcup_{\alpha \in J} (X \backslash A_\alpha)$

8. If $A \subset X$, $B \subset Y$, and $f: X \to Y$, show that $f(f^{-1}(B) \cap A) = B \cap f(A)$.
9.* Prove or find a counterexample: Suppose that for each $n \in \mathbf{Z}^+$ and each $m \in \mathbf{Z}^+$, A_{nm} is a subset of a given set X. Then

$$\bigcup_{n=1}^{\infty} \left(\bigcap_{m=1}^{\infty} A_{nm}\right) = \bigcap_{m=1}^{\infty} \left(\bigcup_{n=1}^{\infty} A_{nm}\right)$$

Section D

1. Show that if the sets A and B are countable, then $A \times B$ is countable.
2. Show that any collection of disjoint open intervals in $\mathbf{R}^1$ is countable.
3. A real number is said to be *algebraic* if it is the zero of a polynomial with integer coefficients. Show that the set of algebraic numbers is countable.
4. Let $\mathscr{D}$ be the set of all circles in the plane that have rational radii and are centered at points whose coordinates are rational numbers. Show that $\mathscr{D}$ is a countable set.
5. Suppose that $A_1 A_2, \ldots$ is a sequence of countable sets and let $A = \{(a_1, a_2, \ldots) \mid \text{for each } i, a_i \in A_i\}$. Is A countable?
6. Prove *Cantor's theorem*: Let A be any nonempty set and $\mathscr{P}(A)$ the set of all subsets of A [$\mathscr{P}(A)$ is called the *power set* of A]. Then card $A <$ card $\mathscr{P}(A)$. [*Hint*: First show that there is an injection $\phi: A \to \mathscr{P}(A)$, and hence, card $A \leq$ card $\mathscr{P}(A)$. Now suppose that there is a bijection $\psi: A \to \mathscr{P}(A)$ and reach a contradiction as follows: for each $a \in A$,

let $A_a = \psi(a)$ and let $A^* = \{a \in A \mid a \notin A_a\}$. Since ψ is onto there is an $\hat{a} \in A$ such that $\psi(\hat{a}) = A^*$. Show that $\hat{a} \in A_{\hat{a}}$ if and only if $\hat{a} \notin A_{\hat{a}}$.]

7.* Show that the principle of induction implies the principle of well ordering. [Hint: Let M be a nonempty set of positive integers and let M^* be the set of all positive integers that are less than or equal to each integer in M. Show that there is an $n \in M^*$ such that $n + 1 \notin M^*$.]

8.* Show that the principle of well ordering implies the principle of induction. [Hint: Suppose that the statement S_1 is true and that S_{k+1} is true whenever S_k is true. Let $M = \{n \in Z^+ \mid S_n$ is not true$\}$ and show that M is empty.]

9.* Show that the principle of induction is equivalent to the alternate principle of induction.

10. Show that the function $\phi : Z^+ \times Z^+ \rightarrow Z^+$ defined $\phi(m,n) = 2^{n-1}(2m - 1)$ is a bijection. Conclude that $Z^+ \times Z^+$ is countable.

11.* Suppose that X is a countably infinite set. Show that X has an uncountable family of subsets with the property that the intersection of any two of these subsets is finite.

12. Show that between any two rational numbers there is an irrational number, and that between any two irrational numbers there is a rational number.

Section E

1. Find upper bounds, lower bounds, sup's, and inf's (if they exist) of the following sets:
   (a) $\{x \in R^1 \mid x < \sqrt{3}\}$
   (b) $\{x \in R^1 \mid x \leq \sqrt{3}\}$
   (c) $\{x \in R^1 \mid -2 < x \leq 3\}$
   (d) $\{x \in R^1 \mid x^2 < 6\}$
   (e) $\{x \in R^1 \mid x^2 \geq 1\}$
   (f) $\{x \in R^1 \mid 2x^2 - 7x + 3 < 0\}$

2. Assume the axiom of completeness to show that if a set $A \subset R^1$ has a lower bound, then it is a greatest lower bound. [Hint: Let $B = \{x \in R^1 \mid x \leq a$ for each $a \in A\}$ and consider sup B.]

3.* Suppose that A and B are nonempty subsets of R^1 and that sup A and sup B exist. Let $C = \{ab \mid a \in A, b \in B\}$. Does sup $C = (\sup A)(\sup B)$? If not, find sufficient conditions for this equality to hold.

4.* Show that if $\{a_\alpha \mid \alpha \in J\}$ is a family of real numbers and b is a fixed real number, then

$$\sup \{(b - a_\alpha) \mid \alpha \in J\} = b - \inf \{a_\alpha \mid \alpha \in J\}$$

(Assume there is a real number M such that $|a_\alpha| \leq M$ for all α.)

Section F

1. If $z_1 = 2 + 6i$ and $z_2 = 4 - 2i$, find $z_1 + z_2$, $z_1 z_2$, and z_1/z_2. Do the same for $z_1 = \frac{1}{2} - i$ and $z_2 = 3i$.
2. Find modulus z if $z = 3 - 6i$ and $z = 3 + 2i$.
3. Write the following in polar form:

   (a) $1 + \sqrt{3}i$

   (b) $\dfrac{1 + i}{1 - i}$

4. Show that if $z_1 = a + bi$ and $z_2 = c + di$ are complex numbers, then $|z_1 z_2| = |z_1||z_2|$.
5. If $z = a + bi$, then the *conjugate* of z, $\bar{z}$, is defined to be $a - bi$. Show that if z is a complex number, then $|z|^2 = z\bar{z}$.
6. Show that if w and z are complex numbers, then $|w + z| \le |w| + |z|$. [Hint: Consider $|w + z|^2$, and use the preceding problem.]
7. Suppose that w and z are complex numbers and show that
   (a) $\text{Re}(w + z) = \text{Re}(w) + \text{Re}(z)$
   (b) $\text{Re}(wz) = (\text{Re}(w))(\text{Re}(z)) - (\text{Im}(w))(\text{Im}(z))$
   (c) $\text{Im}(w + z) = \text{Im}(w) + \text{Im}(z)$
   (d) $\text{Im}(wz) = (\text{Re}(w))(\text{Im}(z)) + (\text{Im}(w))(\text{Re}(z))$
8. Show that if z is a complex number, then $|z| \le |\text{Re}(z)| + |\text{Im}(z)| \le \sqrt{2}|z|$.
9. Use induction to show that if w and z are complex numbers and if $n \in \mathbf{Z}^+$, then

$$(w + z)^n = \sum_{k=0}^{n} \binom{n}{k} w^{n-k} z^k$$

where

$$\binom{n}{k} = \frac{n(n - 1)(n - 2)\cdots(n - k + 1)}{k!}$$

10. If e^{a+bi} is defined to be $e^a(\cos b + i \sin b)$, show that $e^{z_1+z_2} = e^{z_1} e^{z_2}$, and if x is real, then $|e^{ix}| = 1$.
11. Use induction and the trigonometric identities

$$\sin(\alpha + \beta) = \sin \alpha \cos \beta + \cos \alpha \sin \beta$$
$$\cos(\alpha + \beta) = \cos \alpha \cos \beta - \sin \alpha \sin \beta$$

to establish DeMoivre's theorem: $(\cos \theta + i \sin \theta)^n = \cos n\theta + i \sin n\theta$. Use this theorem to show that

$$\left(\cos\frac{\pi}{30} + i\sin\frac{\pi}{30}\right)^{15} = i$$

12. If $w_1 = z^2 + 1/z$ and $w_2 = z + e^z$, find $\text{Re}(w_1 w_2)$ and $\text{Im}(w_1 w_2)$.

13. For what values of the real number b is $e^{1+bi} = ie$? (See problem F.10.)

REFERENCES

Monk, J. Donald (1969): *Introduction to Set Theory*, McGraw-Hill, New York.

Pinter, Charles (1971): *Set Theory*, Addison-Wesley, Reading, Mass.

2

INTRODUCTION TO LINEAR ALGEBRA AND ORDINARY DIFFERENTIAL EQUATIONS

Depending on one's mathematical background, this chapter may be viewed as a brief review of, or a cursory introduction to, certain aspects of linear algebra and differential equations. The topics covered in this chapter will provide the reader with the necessary background in linear algebra and differential equations for the purposes of this text. For the most part, proofs of theorems are omitted in order that emphasis may be given to the content and applications of the basic results.

Linear algebra serves as a powerful unifying force for many seemingly disparate areas of mathematics, and it has immense practical importance in such diverse fields as numerical analysis, graph theory, stochastic processes, linear programming, and differential equations. The rather cavalier treatment of linear algebra found in the present chapter provides little indication of its great versatility and applicability.

During the course of the text we shall encounter a variety of differential equations that arise naturally in applied problems. In (2.D) we give a rather abbreviated introduction to the study of differential equations based on a number of results in linear algebra. More thorough presentations of this nature can be found in Boyce and DiPrima (1977) or Coddington (1961).

A. VECTOR SPACES

The reader is undoubtedly familiar with vectors (n-tuples) in $\mathbf{R}^n$ (or $\mathbf{C}^n$) and the standard operations of addition and scalar multiplication on them:

$$(a_1, a_2, \ldots, a_n) + (b_1, b_2, \ldots, b_n) = (a_1 + b_1, a_2 + b_2, \ldots, a_n + b_n)$$

$$\alpha(a_1, a_2, \ldots, a_n) = (\alpha a_1, \alpha a_2, \ldots, \alpha a_n)$$

where $\alpha \in \mathbf{R}^1$ (or $\mathbf{C}$).

Recall too that an $m \times n$ *matrix* is an array of real or complex numbers that consists of m rows and n columns, and that if

$$A = \begin{pmatrix} a_{11} & a_{12} & \cdots & a_{1n} \\ a_{21} & a_{22} & \cdots & a_{2n} \\ \cdots\cdots\cdots\cdots\cdots \\ a_{m1} & a_{m2} & \cdots & a_{mn} \end{pmatrix} \qquad B = \begin{pmatrix} b_{11} & b_{12} & \cdots & b_{1n} \\ b_{21} & b_{22} & \cdots & b_{2n} \\ \cdots\cdots\cdots\cdots\cdots \\ b_{m1} & b_{m2} & \cdots & b_{mn} \end{pmatrix}$$

are two $m \times n$ matrices, then the sum $A + B$ of these matrices is defined to be the matrix:

$$A + B = \begin{pmatrix} a_{11} + b_{11} & a_{12} + b_{12} & \cdots & a_{1n} + b_{1n} \\ a_{21} + b_{21} & a_{22} + b_{22} & \cdots & a_{2n} + b_{2n} \\ \cdots\cdots\cdots\cdots\cdots\cdots\cdots \\ a_{m1} + b_{m1} & a_{m2} + b_{m2} & \cdots & a_{mn} + b_{mn} \end{pmatrix}$$

Furthermore, if $\alpha \in \mathbf{R}^1$ (or $\mathbf{C}$), then the matrix αA is defined by

$$\alpha A = \begin{pmatrix} \alpha a_{11} & \alpha a_{12} & \cdots & \alpha a_{1n} \\ \alpha a_{21} & \alpha a_{22} & \cdots & \alpha a_{2n} \\ \cdots\cdots\cdots\cdots\cdots \\ \alpha a_{m1} & \alpha a_{m2} & \cdots & \alpha a_{mn} \end{pmatrix}$$

and if

$$A = \begin{pmatrix} a_{11} & a_{12} & \cdots & a_{1p} \\ \cdots\cdots\cdots\cdots \\ a_{m1} & a_{m2} & \cdots & a_{mp} \end{pmatrix} \qquad B = \begin{pmatrix} b_{11} & b_{12} & \cdots & b_{1n} \\ \cdots\cdots\cdots\cdots \\ b_{p1} & b_{p2} & \cdots & b_{pn} \end{pmatrix}$$

are $m \times p$ and $p \times n$ matrices, respectively, then the product AB of A and B is defined by

$$AB = C = \begin{pmatrix} c_{11} & c_{12} & \cdots & c_{1n} \\ c_{21} & c_{22} & \cdots & c_{2n} \\ \cdots\cdots\cdots\cdots\cdots \\ c_{m1} & c_{m2} & \cdots & c_{mn} \end{pmatrix}$$

where for each i, $1 \leq i \leq m$ and j, $1 \leq j \leq n$, $c_{ij} = \sum_{k=1}^{p} a_{ik} b_{kj}$.

Thus, addition and scalar multiplication of matrices is quite straight-forward. The rule for multiplication may seem somewhat unusual (if not totally artificial); justification for it is found in (2.E).

(2.A.1) *Examples*

$$\begin{pmatrix} 1 & -i \\ 2 & \pi \end{pmatrix} + 4\begin{pmatrix} 0 & -3 \\ 2i & 0 \end{pmatrix} = \begin{pmatrix} 1 & -12-i \\ 2+8i & \pi \end{pmatrix}$$

$$\begin{pmatrix} 3 & -2 & 1 \\ 2 & 1 & 0 \end{pmatrix}\begin{pmatrix} 1 & 4 & 6 & 1 \\ 2 & 1 & 5 & 2 \\ 0 & 0 & 0 & 1 \end{pmatrix} = \begin{pmatrix} -1 & 10 & 8 & -1 \\ 4 & 9 & 17 & 4 \end{pmatrix}$$

Functions that map an arbitrary set X into $\mathbf{R}^1$ provide another example of mathematical objects that are amenable to the operations of addition and scalar multiplication. If $f: X \to \mathbf{R}^1$ and $g: X \to \mathbf{R}^1$, then the sum $(f+g): X \to \mathbf{R}^1$ is defined by $(f+g)(x) = f(x) + g(x)$. If $\alpha \in \mathbf{R}^1$, then the scalar product $(\alpha f): X \to \mathbf{R}^1$, is defined by $(\alpha f)(x) = \alpha f(x)$.

Thus, we see that vectors in $\mathbf{R}^n$, matrices, and certain families of functions are all subject to similar algebraic manipulations. These sets are particular examples of a more general concept, that of a vector space. Essentially, a vector space is a set on which are defined the operations of addition and scalar multiplication (multiplication of members of the set by real or complex numbers). Since addition and scalar multiplication can be defined in a variety of ways depending in part on the nature of the underlying set, these operations are formally defined as functions that satisfy a number of "obvious" conditions.

(2.A.2) *Definition* A *vector space over* $\mathbf{R}^1$ consists of a nonempty set V of objects $\mathbf{v}$, called *vectors*, with two operations (functions), $S: V \times V \to V$ and $M: \mathbf{R}^1 \times V \to V$. The operation S will be referred to as the operation of addition, and we shall denote $S(\mathbf{v}_1, \mathbf{v}_2)$ by $\mathbf{v}_1 + \mathbf{v}_2$; the operation M will be referred to as the operation of scalar multiplication and we shall denote $M(\alpha, \mathbf{v})$ by $\alpha\mathbf{v}$, where $\alpha \in \mathbf{R}^1$ and $\mathbf{v} \in V$. These operations satisfy the following conditions:

1. If $\mathbf{v}_1, \mathbf{v}_2 \in V$, then $\mathbf{v}_1 + \mathbf{v}_2 = \mathbf{v}_2 + \mathbf{v}_1$.
2. If $\mathbf{v}_1, \mathbf{v}_2, \mathbf{v}_3 \in V$, then $(\mathbf{v}_1 + \mathbf{v}_2) + \mathbf{v}_3 = \mathbf{v}_1 + (\mathbf{v}_2 + \mathbf{v}_3)$.
3. If $\alpha \in \mathbf{R}^1$, and $\mathbf{v}_1, \mathbf{v}_2 \in V$, then $\alpha(\mathbf{v}_1 + \mathbf{v}_2) = \alpha\mathbf{v}_1 + \alpha\mathbf{v}_2$.
4. If $\alpha, \beta \in \mathbf{R}^1$ and $\mathbf{v} \in V$, then $(\alpha + \beta)\mathbf{v} = \alpha\mathbf{v} + \beta\mathbf{v}$.
5. If $\alpha, \beta \in \mathbf{R}^1$ and $\mathbf{v} \in V$, then $(\alpha\beta)\mathbf{v} = \alpha(\beta\mathbf{v})$.
6. There is a unique element, denoted $\mathbf{0}$, in V with the property that $\mathbf{v} + \mathbf{0} = \mathbf{0} + \mathbf{v} = \mathbf{v}$ for each $\mathbf{v} \in V$.

7. If $v \in V$, then there is a unique element in V, denoted $-v$, with the property that $v + (-v) = 0$.
8. If $v \in V$, then $1v = v$.

If $\mathbf{R}^1$ is replaced by $\mathbf{C}$ in this definition, then V is said to be a *vector space over* $\mathbf{C}$. The elements of $\mathbf{R}^1$ (or $\mathbf{C}$) are called *scalars*.

Although the preceding list may seem somewhat forbidding, the conditions described therein have long been familiar to the reader: they represent many of the basic properties of the real numbers.

(2.A.3) *Exercise* Verify (at least mentally) that the examples discussed earlier satisfy conditions 1 through 8 of definition (2.A.2) and, hence, define vector spaces.

(2.A.4) *Observation* A number of further properties of vector spaces follow easily from the previous definition. For instance, if V is a vector space and if $v \in V$, then $0v = (0 + 0)v = 0v + 0v$; addition of $-0v$ to both sides of this equation shows that $0v$ is equal to the zero vector, $\mathbf{0}$, for each vector $v \in V$. More results of this nature are found in the problem section.

A potentially bad feature of vector spaces is that they may contain an inordinate number of vectors—a possible source of calculational and even theoretical difficulties. A good feature of vector spaces is that they have a built-in mechanism for alleviating this problem: the notion of a *basis*. Consider, for example, the vectors $e_1 = (1,0,0)$, $e_2 = (0,1,0)$, and $e_3 = (0,0,1)$ in the vector space $\mathbf{R}^3$. Observe that every vector $(x_1,x_2,x_3) \in \mathbf{R}^3$ can be expressed in terms of $e_1, e_2,$ and e_3. For instance, we have $(7,-6,\pi) = 7e_1 - 6e_2 + \pi e_3$. Notice further that it is impossible to express any one of the vectors e_1, e_2, e_3 as a combination of the other two (for example, there do not exist scalars c_1 and c_2 such that $e_3 = c_1e_1 + c_2e_2$). Consequently, $e_1, e_2,$ and e_3 form a minimal collection of vectors in $\mathbf{R}^3$ with the property that every vector in $\mathbf{R}^3$ can be expressed as an appropriate sum of these particular vectors. As we shall see shortly, this idea represents the essence of a basis.

(2.A.5) *Definition* Suppose that V is a vector space over $\mathbf{R}^1$ (or $\mathbf{C}$) and that $v_1, v_2, \ldots, v_k$ are vectors in V. Then a *linear combination* of $v_1, v_2, \ldots, v_k$ is any vector in V of the form

$$\alpha_1 v_1 + \alpha_2 v_2 + \cdots + \alpha_k v_k$$

where each $\alpha_i \in \mathbf{R}^1$ (or $\mathbf{C}$).

(2.A.6) *Examples* (a) The vector $(-2,2)$ in $\mathbf{R}^2$ can be written as a linear combination of the vectors $(0,1), (2,4),$ and $(-1,1)$ since $(-2,2) = 3(0,1) - (1/2)(2,4) + (-1,1)$.

(b) Let $V = \{f \mid f: [2,5] \to \mathbf{R}^1\}$. If $f_1(x) = x^2$, $f_2(x) = 2x - 1$, and $f_3(x) = \sin x$, then $f = 2f_1 - 3f_2 + 0f_3$ is a linear combination of f_1, f_2, and f_3. In this case we have $f(x) = 2x^2 - 6x + 3$ for each $x \in [2,5]$.

(2.A.7) Definition Vectors $v_1, v_2, \ldots, v_k$ in a vector space V are said to be *linearly independent* if none of these vectors can be written as a linear combination of the others. The vectors are said to be *linearly dependent* if they are not linearly independent.

(2.A.8) Examples (a) The vectors, $e_1 = (1,0,0)$, $e_2 = (0,1,0)$, and $e_3 = (0,0,1)$ are linearly independent in $\mathbf{R}^3$.

(b) The vectors $(1,0,2)$, $(3,-\tfrac{1}{2},0)$, and $(9,-1,6)$ are linearly dependent in $\mathbf{R}^3$ since $3(1,0,2) + 2(3,-\tfrac{1}{2},0) = (9,-1,6)$. In this case we say that the vector $(9,-1,6)$ is *dependent* on (is a *linear combination* of) the vectors $(1,0,2)$ and $(3,-\tfrac{1}{2},0)$.

(2.A.9) Observation Vectors $v_1, v_2, \ldots, v_k$ in a vector space V are linearly dependent if and only if there exist scalars $\alpha_1, \alpha_2, \ldots, \alpha_k$ not all equal to 0 such that $\alpha_1 v_1 + \alpha_2 v_2 + \cdots + \alpha_k v_k = 0$. This follows easily, since if $v_1, v_2, \ldots, v_k$ are dependent, then one of the vectors, say v_i, can be written as a linear combination of the other vectors, i.e.,

$$v_i = \alpha_1 v_1 + \cdots + \alpha_{i-1} v_{i-1} + \alpha_{i+1} v_{i+1} + \cdots + \alpha_k v_k$$

Hence, we have

$$\alpha_1 v_1 + \cdots + \alpha_{i-1} v_{i-1} - 1 v_i + \alpha_{i+1} v_{i+1} + \cdots + \alpha_k v_k = 0$$

Conversely, if $\alpha_1 v_1 + \cdots + \alpha_j v_j + \cdots + \alpha_k v_k = 0$ and, say, $\alpha_j \neq 0$, then by adding $-\alpha_j v_j$ to both sides of this equation, we obtain

$$-\alpha_j v_j = \alpha_1 v_1 + \cdots + \alpha_{j-1} v_{j-1} + \alpha_{j+1} v_{j+1} + \cdots + \alpha_k v_k$$

Division by $-\alpha_j$ shows that v_j is a linear combination of the remaining vectors, and therefore, the vectors $v_1, v_2, \ldots, v_j, \ldots, v_k$ are linearly dependent.

(2.A.10) Examples (a) The vectors $(1,1)$ and $(-1,2)$ in $\mathbf{R}^2$ are independent, since if $\alpha_1(1,1) + \alpha_2(-1,2) = 0$, then

$$\alpha_1 - \alpha_2 = 0 \qquad \alpha_1 + 2\alpha_2 = 0$$

which implies that $-3\alpha_2 = 0$, and hence, both α_1 and α_2 must be 0. Therefore, by (2.A.9) the vectors $(1,1)$ and $(-1,2)$ cannot be linearly dependent and, consequently, are linearly independent.

(b) Let V be the vector space of all polynomials of degree less than or equal to 2 (with the usual operations of addition and scalar multiplication). Then the polynomials $p(t) = 2t^2$, $q(t) = 3t$, and $r(t) = 4t^2 + t$ are linearly dependent, since $6(2t^2) + 3t - 3(4t^2 + t)$ is identically equal to 0. On the other hand, the polynomials, $p(t) = 1$, $q(t) = t$, and $r(t) = t^2$ are linearly independent. To see this, suppose that $\alpha_1 1 + \alpha_2 t + \alpha_3 t^2 = 0$ (for all t). Setting $t = 0$, we find that $\alpha_1 = 0$, and therefore, $\alpha_2 t + \alpha_3 t^2$ is identically equal to 0. Taking derivatives, we have that $\alpha_2 + 2\alpha_3 t = 0$, and again letting $t = 0$, it follows that $\alpha_2 = 0$, which in turn implies that α_3 must be 0, and consequently, the polynomials 1, t, and t^2 are independent.

(2.A.11) *Definition* Suppose that V is a vector space and that S is a subset of V. Then S is said to *span* V if each vector in V can be written as a linear combination of vectors in S.

(2.A.12) *Examples* (a) The vectors $(1,0)$ and $(0,1)$ in $\mathbf{R}^2$ clearly span $\mathbf{R}^2$ since any vector $(x,y) \in \mathbf{R}^2$ can be written as $x(1,0) + y(0,1)$.

(b) The vectors $(1,1)$ and $(-1,2)$ also span $\mathbf{R}^2$. To see this, let (x,y) be an arbitrary vector in $\mathbf{R}^2$. We need to find scalars α_1 and α_2 such that $(x,y) = \alpha_1(1,1) + \alpha_2(-1,2)$; this is equivalent to finding scalars α_1 and α_2 such that

$$x = \alpha_1 - \alpha_2 \qquad y = \alpha_1 + 2\alpha_2$$

Clearly, $\alpha_1 = (2x + y)/3$ and $\alpha_2 = -(x - y)/3$ are the desired scalars.

(c) The polynomials $p(t) = 1$, $q(t) = t$, and $r(t) = t^2$ span the vector space of polynomials of degree less than or equal to 2, since any polynomial in this space can be written in the form $\alpha_1 1 + \alpha_2 t + \alpha_3 t^2$.

(2.A.13) *Exercise* Show that if vectors $v_1, v_2, \ldots, v_n$ span a vector space V, then any family of vectors in V that contains these vectors also spans V.

(2.A.14) *Definition* Suppose that V is a vector space. A finite subset B of V is a *basis* for V if
　(i)　B spans V.
　(ii)　B is an independent set of vectors.

(2.A.15) *Examples* (a) The vectors $e_1 = (1, 0, 0, \ldots, 0)$, $e_2 = (0, 1, 0, 0, \ldots, 0), \ldots, e_n = (0, 0, 0, \ldots, 1)$ form a basis for $\mathbf{R}^n$.

(b) It follows from (2.A.10.a) and (2.A.12.b) that the vectors $(1,1)$ and $(1,-2)$ form a basis for $\mathbf{R}^2$. Since the vectors $(1,0)$ and $(0,1)$ also form a basis for $\mathbf{R}^2$, we see that a vector space may have a number of distinct bases.

(c) It follows from (2.A.10.b) and (2.A.12.c) that the vectors 1, t, and t^2 form a basis for the vector space of polynomials of degree less than or equal to 2.

(2.A.16) *Observation* If $B = \{v_1, v_2, \ldots, v_n\}$ is a basis for a vector space V, then each vector $v \in V$ can be written in a *unique* way as a linear combination of elements of B. This can be seen as follows. Since B spans V, each vector $v \in V$ can be written in the form

$$v = \alpha_1 v_1 + \cdots + \alpha_n v_n$$

Suppose that we can also write v in the form

$$v = \tilde{\alpha}_1 v_1 + \cdots + \tilde{\alpha}_n v_n$$

Then since $-v = (-\tilde{\alpha}_1)v_1 + \cdots + (-\tilde{\alpha}_n)v_n$ (see problem A.4), we have

$$0 = v - v = (\tilde{\alpha}_1 - \alpha)v_1 + \cdots + (\tilde{\alpha}_n - \alpha_n)v_n$$

and, since the vectors v_i are independent, it follows that

$$0 = (\tilde{\alpha}_1 - \alpha_1) = (\tilde{\alpha}_2 - \alpha_2) = \cdots = (\tilde{\alpha}_n - \alpha_n)$$

hence, for each i, $\tilde{\alpha}_i = \alpha_i$.

(2.A.17) *Observation* If a vector space V has a basis B consisting of n vectors, then the dimension of V is said to be n. This concept of dimension is predicated on the assumption that every basis for V consists of precisely n vectors. A proof of the validity of this assumption can be found in most elementary linear algebra texts (see Curtis, 1968).

(2.A.18) *Observation* It can be shown (see problem A.17*) that in an n-dimensional vector space, there cannot exist a set of more than n independent vectors. Thus, for example, in $\mathbf{R}^3$ any set of four vectors must be linearly dependent.

(2.A.19) *Example* Let $V = \{\{x_n\} \mid \{x_n\}$ is a sequence in $\mathbf{R}^1\}$. If $\{x_n\}$ and $\{y_n\}$ are two such sequences, define their sum to be the sequence $\{x_n + y_n\}$, and define scalar multiplication in the obvious manner: $\alpha\{x_n\} = \{\alpha x_n\}$. Then it is easy to verify that V is a vector space. Note that the sequences $e_1 = (1, 0, 0, \ldots)$, $e_2 = (0, 1, 0, 0, \ldots)$, $e_3 = (0, 0, 1, 0, \ldots)$, $\ldots$ are elements of V, although they do not form a basis for this vector space in the sense of (2.A.14). What would be a basis for V? It should be clear that any basis for V must consist of an infinite number of vectors, which brings up the problem of defining what is meant by an infinite set of linearly independent vectors. One way to settle this problem is to declare an arbitrary set of vectors $S = \{v_\alpha \mid \alpha \in J\}$ in a vector space V to be *linearly independent* if every finite subset of S is linearly independent. A *basis* for V is then defined to be any family B of independent vectors that spans V [in the sense that each vector $v \in V$ is a (finite) linear combination of vectors in B]. In general, the actual construction of a basis for an infinite dimensional vector space (a space that fails to have a finite basis) borders on the impossible, although it can be shown with the aid of fairly deep set theoretic considerations that all such vector spaces have bases.

A second example of an infinite-dimensional vector space whose basis is not readily apparent is the space $V = \{f: [a,b] \to \mathbf{R}^1 \mid f \text{ is continuous}\}$, where $f + g$ and αf are defined by $(f + g)(x) = f(x) + g(x)$ and $(\alpha f)(x) = \alpha f(x)$, respectively.

To conclude this section we briefly review the notion of the inner (or dot) product of vectors in $\mathbf{R}^n$.

(2.A.20) **Definition** If $\mathbf{x} = (x_1, x_2, \ldots, x_n)$ and $\mathbf{y} = (y_1, y_2, \ldots, y_n)$ are vectors in $\mathbf{R}^n$, then the *inner* (or *dot*) product of $\mathbf{x}$ and $\mathbf{y}$, $\mathbf{x} \cdot \mathbf{y}$, is defined by $\mathbf{x} \cdot \mathbf{y} = x_1 y_1 + x_2 y_2 + \cdots + x_n y_n$.

Note that if we define the *length* or *norm* of a vector $\mathbf{x} = (x_1, x_2, \ldots, x_n)$ in $\mathbf{R}^n$ to be $\|\mathbf{x}\| = \sqrt{x_1^2 + x_2^2 + \cdots + x_n^2}$, then $\|\mathbf{x}\|^2 = \mathbf{x} \cdot \mathbf{x}$.

The reader may recall from earlier studies that if $\mathbf{x}$ and $\mathbf{y}$ are vectors in $\mathbf{R}^n$ and θ is the angle between these vectors ($0 \le \theta \le \pi$), then

$$\cos \theta = \frac{\mathbf{x} \cdot \mathbf{y}}{\|\mathbf{x}\| \, \|\mathbf{y}\|}. \tag{1}$$

Since $|\cos \theta| \le 1$ for all θ, it follows from (1) that $|\mathbf{x} \cdot \mathbf{y}| \le \|\mathbf{x}\| \|\mathbf{y}\|$. This inequality is known as the *Cauchy-Schwarz inequality*, and it plays a significant role in obtaining a number of important results in analysis. We give next a nongeometric proof of this inequality.

(2.A.21) **Theorem** **(Cauchy-Schwarz Inequality)** Suppose that $\mathbf{a} = (a_1, a_2, \ldots, a_n)$ and $\mathbf{b} = (b_1, b_2, \ldots, b_n)$ are two vectors in $\mathbf{R}^n$. Then $|\mathbf{a} \cdot \mathbf{b}| \le \|\mathbf{a}\| \|\mathbf{b}\|$.

Proof. It suffices to show that

$$\left(\sum_{i=1}^n a_i b_i \right)^2 \le \left(\sum_{i=1}^n a_i^2 \right) \left(\sum_{i=1}^n b_i^2 \right)$$

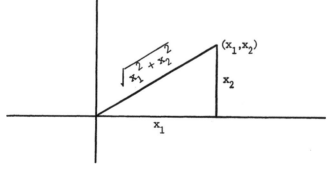

Figure 2.1

If $b_i = 0$ for each i, then clearly the above inequality holds. Suppose then that not all the b_i's are 0. If α is any real number, then

$$0 \le \sum_{i=1}^{n} (a_i + \alpha b_i)^2 = \sum_{i=1}^{n} a_i^2 + 2\alpha \sum_{i=1}^{n} a_i b_i + \alpha^2 \sum_{i=1}^{n} b_i^2$$

Let

$$\alpha = -\frac{\sum_{i=1}^{n} a_i b_i}{\sum_{i=1}^{n} b_i^2}$$

Then we have

$$0 \le \sum_{i=1}^{n} a_i^2 - 2\frac{(\sum_{i=1}^{n} a_i b_i)^2}{\sum_{i=1}^{n} b_i^2} + \frac{(\sum_{i=1}^{n} a_i b_i)^2}{\sum_{i=1}^{n} b_i^2} = \sum_{i=1}^{n} a_i^2 - \frac{(\sum_{i=1}^{n} a_i b_i)^2}{\sum_{i=1}^{n} b_i^2}$$

from which it follows that

$$\left(\sum_{i=1}^{n} a_i b_i\right)^2 \le \left(\sum_{i=1}^{n} a_i^2\right)\left(\sum_{i=1}^{n} b_i^2\right)$$

B. MATRICES, DETERMINANTS, AND SYSTEMS OF EQUATIONS

Henceforth, we shall frequently denote an $m \times n$ matrix

$$A = \begin{pmatrix} a_{11} & a_{12} & \cdots & a_{1n} \\ a_{21} & a_{22} & \cdots & a_{2n} \\ \vdots & & & \\ a_{m1} & a_{m2} & \cdots & a_{mn} \end{pmatrix}$$

by the symbol (a_{ij}), where it is understood that a_{ij} is the entry of A found in the ith row and jth column. With this notation the addition of $m \times n$ matrices $A = (a_{ij})$ and $B = (b_{ij})$ can be described by $A + B = (a_{ij}) + (b_{ij}) = (a_{ij} + b_{ij})$, and similarly, scalar multiplication can be given by $\alpha A = \alpha(a_{ij}) = (\alpha a_{ij})$.

In the sequel we shall be especially interested in *square matrices*, i.e., matrices with an equal number of rows and columns. In particular, we shall be interested in those square matrices that arise in connection with systems of equations with n equations and n unknowns of the form

$$\begin{aligned} a_{11}x_1 + a_{12}x_2 + \cdots + a_{1n}x_n &= b_1 \\ a_{21}x_1 + a_{22}x_2 + \cdots + a_{2n}x_n &= b_2 \\ \cdots\cdots\cdots\cdots\cdots\cdots\cdots\cdots\cdots & \\ a_{n1}x_1 + a_{n2}x_2 + \cdots + a_{nn}^n x_n &= b_n \end{aligned} \qquad (1)$$

The numbers a_{ij} and b_i are assumed to be known; the basic problem is to find values $x_1, \ldots, x_n$ that will simultaneously satisfy all of the equations.

(2.B.1) *Definition* The matrix

$$A = \begin{pmatrix} a_{11} & a_{12} & \cdots & a_{1n} \\ a_{21} & a_{22} & \cdots & a_{2n} \\ \cdots\cdots\cdots\cdots\cdots \\ a_{n1} & a_{n2} & \cdots & a_{nn} \end{pmatrix}$$

is called the *coefficient matrix* of the system of equations (1).

(2.B.2) *Examples* (a) The system of equations

$$\begin{aligned} 2x_1 - x_2 + x_3 &= 1 \\ -x_1 + 3x_2 + 5x_3 &= -6 \\ x_1 + x_2 + x_3 &= 0 \end{aligned}$$

has a unique solution: $x_1 = 1$, $x_2 = 0$, $x_3 = -1$.

(b) There is no solution to the system of equations

$$\begin{aligned} 2x_1 - 3x_2 + x_3 &= 1 \\ -x_1 - \tfrac{1}{2}x_2 &= 0 \\ 4x_1 - 6x_2 + 2x_3 &= 3 \end{aligned}$$

since the first and third equations are incompatible.

(c) The system of equations

$$\begin{aligned} x_1 - 2x_2 + x_3 &= -6 \\ 3x_1 - 4x_2 - x_3 &= -10 \\ -2x_1 + 4x_2 - 2x_3 &= 12 \end{aligned}$$

has an infinite number of solutions subject only to the constraint $2x_1 - 3x_2 = -8$.

The determinant of the coefficient matrix of a given system of n equations in n unknowns will enable us to determine if the system has a solution and, if so, whether the solution is unique. Furthermore, as we shall see in the next theorem, determinants may be used to calculate the unique solution (if it exists). We assume that the reader is familiar with at least one method for calculating determinants; such methods may be found in any elementary text on linear algebra (see Cullen, 1966, for example).

Note that system (1) can be expressed in the matrix form $AX = B$, where A is the coefficient matrix,

$$X = \begin{pmatrix} x_1 \\ x_1 \\ \vdots \\ x_n \end{pmatrix}, \quad \text{and} \quad B = \begin{pmatrix} b_1 \\ b_2 \\ \vdots \\ b_n \end{pmatrix}$$

Therefore, finding a solution of (1) is equivalent to finding a column matrix X such that $AX = B$.

(2.B.3) **Theorem** Suppose that A is an $n \times n$ matrix and X and B are $n \times 1$ matrices.

(i) (*Cramer's Rule*) If det $A \neq 0$, then there is a unique solution of the equation $AX = B$. Furthermore, this solution is given by

$$x_1 = \frac{\det A_1}{\det A} \qquad x_2 = \frac{\det A_2}{\det A} \qquad \cdots \qquad x_n = \frac{\det A_n}{\det A}$$

where A_i is the matrix obtained from A by the substitution of the column matrix B for the ith column of A. In particular, if $B = (0)$ and det $A \neq 0$, then the only solution of $AX = B$ is given by

$$X = \begin{pmatrix} 0 \\ 0 \\ \vdots \\ 0 \end{pmatrix} = (0)$$

(ii) If det $A = 0$, then
(a) $AX = B$ has either no solution or an infinite number of solutions.
(b) $AX = (0)$ has (an infinite number of) nonzero solutions.

(2.B.4) **Example** By Cramer's rule the solution of the system of equations

$$2x_1 + x_2 + x_3 = 3$$
$$-1x_1 - 2x_2 + 3x_3 = 1$$
$$3x_1 + 3x_2 + 3x_3 = 0$$

is given by

$$x_1 = \frac{\det \begin{pmatrix} 3 & 1 & 1 \\ 1 & -2 & 3 \\ 0 & 3 & 3 \end{pmatrix}}{\det \begin{pmatrix} 2 & 1 & 1 \\ -1 & -2 & 3 \\ 3 & 3 & 3 \end{pmatrix}} = \frac{-45}{-15} \qquad x_2 = \frac{\det \begin{pmatrix} 2 & 3 & 1 \\ -1 & 1 & 3 \\ 3 & 0 & 3 \end{pmatrix}}{\det \begin{pmatrix} 2 & 1 & 1 \\ -1 & -2 & 3 \\ 3 & 3 & 3 \end{pmatrix}} = \frac{39}{-15}$$

and

$$x_3 = \frac{\det \begin{pmatrix} 2 & 1 & 3 \\ -1 & -2 & 1 \\ 3 & 3 & 0 \end{pmatrix}}{\det \begin{pmatrix} 2 & 1 & 1 \\ -1 & -2 & 3 \\ 3 & 3 & 3 \end{pmatrix}} = \frac{6}{-15}$$

(2.B.5) *Definition* The $n \times n$ matrix

$$I_n = \begin{pmatrix} 1 & 0 & 0 & 0 & \cdots & 0 \\ 0 & 1 & 0 & 0 & \cdots & 0 \\ & & \cdots\cdots\cdots\cdots & & \\ 0 & 0 & 0 & 0 & \cdots & 1 \end{pmatrix}$$

is called the $n \times n$ *identity matrix*. When n is clear from the context, I_n will be denoted by I.

(2.B.6) *Observation* If A is any $n \times n$ matrix, then $AI = IA = A$.

(2.B.7) *Definition* If A is an $n \times n$ matrix, then a matrix B is said to be the *inverse matrix* of A in case $AB = BA = I$. The matrix B (if it exists) is denoted by A^{-1}.

(2.B.8) *Examples* (a) If $A = \begin{pmatrix} 1 & -1 \\ 2 & 1 \end{pmatrix}$, then $A^{-1} = \begin{pmatrix} \frac{1}{3} & \frac{1}{3} \\ -\frac{2}{3} & \frac{1}{3} \end{pmatrix}$.

(b) If $A = \begin{pmatrix} -1 & 1 \\ 2 & -2 \end{pmatrix}$, then A^{-1} does not exist (show that this is true).

(2.B.9) *Observation* If $AX = B$ and A^{-1} exists, then we have $A^{-1}(AX) = A^{-1}B$, and since $A^{-1}(AX) = X$, it follows that $A^{-1}B$ is the unique solution of $AX = B$.

To find the inverse of an arbitrary $n \times n$ matrix, we shall make use of the following definitions.

(2.B.10) *Definition* If $A = (a_{ij})$ is an $m \times n$ matrix, then the *transpose* of A, $A^T = (a_{ij})^T$, is the $n \times m$ matrix obtained from A by interchanging rows and columns.

(2.B.11) *Example* If

$$A = \begin{pmatrix} 2 & 1 & 0 & 0 \\ 0 & -1 & 1 & 2 \\ 1 & 1 & 1 & 1 \end{pmatrix}$$

then

$$A^T = \begin{pmatrix} 2 & 0 & 1 \\ 1 & -1 & 1 \\ 0 & 1 & 1 \\ 0 & 2 & 1 \end{pmatrix}$$

(2.B.12) *Definition*   Suppose that $A = (a_{ij})$ is an $n \times n$ matrix. For each i,j, $1 \leq i,j \leq n$, let A_{ij} be the matrix obtained from A by deleting the ith row and the jth column of A. Then for each entry a_{ij} in A, the *cofactor* of a_{ij}, cof a_{ij} is defined to be $(-1)^{i+j} \det A_{ij}$. The *comatrix* of A (often called the *adjoint* of A) is the transposed matrix of the cofactors of the entries of A.

(2.B.13) *Example*   If

$$A = \begin{pmatrix} 4 & 0 & -1 \\ 2 & 1 & 3 \\ 0 & 0 & 1 \end{pmatrix}$$

then the cofactors of the entries of A are given by

$$\text{cof } a_{11} = 1, \quad \text{cof } a_{12} = -2, \quad \text{cof } a_{13} = 0$$
$$\text{cof } a_{21} = 0, \quad \text{cof } a_{22} = 4, \quad \text{cof } a_{32} = 0$$
$$\text{cof } a_{31} = 1, \quad \text{cof } a_{32} = -14, \quad \text{cof } a_{33} = 4$$

and the comatrix of A is defined by

$$\text{comatrix } A = \begin{pmatrix} 1 & 0 & 1 \\ -2 & 4 & -14 \\ 0 & 0 & 4 \end{pmatrix}$$

(2.B.14) *Observation*   Cofactors may be used to find determinants since it can be shown (see Cullen, 1966) that if $A = (a_{ij})$ is an $n \times n$ matrix, then for each i, $\det A = \sum_{j=1}^{n} a_{ij} \text{ cof } a_{ij}$; however, for calculational purposes this method is extremely inefficient. A proof of the following theorem can be found in Curtis (1968).

(2.B.15) *Theorem*   An $n \times n$ matrix $A = (a_{ij})$ has an inverse if and only if $\det A \neq 0$. If $\det A \neq 0$, then

$$A^{-1} = \frac{1}{\det A} \text{comatrix } A$$

(2.B.16) *Example*   If A is the matrix given in Example (2.B.13), then $\det A = 4$ and, hence,

$$A^{-1} = \tfrac{1}{4} \begin{pmatrix} 1 & 0 & 1 \\ -2 & 4 & -14 \\ 0 & 0 & 4 \end{pmatrix}$$

(2.B.17) *Definition*   An $n \times n$ matrix A is said to be *nonsingular* if $\det A \neq 0$, or (equivalently) if A^{-1} exists; otherwise, A is said to be *singular*.

The calculation of determinants and inverses lies at the core of many numerical methods. Although it is a relatively simple matter to program the expressions given in (2.B.14) and (2.B.15), the time (and cost) involved in performing the calculations is astronomical. For instance, if (2.B.14) is used, then it is estimated that even with aid of a high-speed computer the calculation of the determinant of just one 20×20 matrix would take more than a century. Similar problems, of course, occur in the computation of inverses, and consequently, a great deal of mathematical research has gone into finding more efficient means for treating determinants and inverse matrices.

C.*   APPLICATIONS

In this section we consider a few examples that illustrate the utility of matrices and matrix notation. The first example involves the description of the forces involved in a mildly tangled mass-spring system.

Hooke's law states that the force exerted by a stretched spring is proportional to the displacement of the spring from its natural unstretched position, i.e.,

$$F(x) = -kx$$

where x is the deviation of the spring from its natural position and k is a positive constant, called the *spring constant*. Note that k gives the force required to stretch the spring a unit distance.

Suppose that we have a system of four masses, m_1, m_2, m_3, m_4, and seven springs whose natural equilibrium state is indicated Fig. 2.2. The springs and their respective spring constants are also given. The masses m_1, m_2, m_3, and m_4 are moved x_1, x_2, x_3, and x_4 units, respectively. We wish to find the total

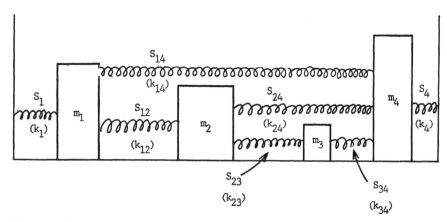

Figure 2.2

force acting on each mass. We adopt the convention that if the spring exerts a force to the right on a particular mass, then this force will be positive, and if the force acts to the left, then the force is considered to be negative. We consider the total force exerted on m_1. If $x_1 > 0$, then the spring S_1 has been stretched and, hence, will exert a force to the left on m_1. By Hooke's law (and our convention) this force is given by the negative number $-k_1 x_1$. If $x_1 < 0$, the spring S_1 has been compressed and will exert a force to the right on m_1; this force is equal to the positive number $-k_1 x_1$. Consider now the spring S_{12}. If $x_2 - x_1 > 0$, then the spring has been stretched and, hence, will exert a force to the right on m_1, given by the positive number $k_{12}(x_2 - x_1)$. If $x_2 - x_1 < 0$, then S_{12} has been compressed and will exert a force to the left on m_1 equal to the negative number $k_{12}(x_2 - x_1)$. A similar analysis shows shows that the force exerted by the spring S_{14} on m_1 is $k_{14}(x_4 - x_1)$. It follows that in all cases, the total force y_1 on m_1 is expressed by

$$y_1 = -k_1 x_1 + k_{12}(x_2 - x_1) + k_{14}(x_4 - x_1)$$
$$= (-k_1 - k_{12} - k_{14})x_1 + k_{12}x_2 + 0x_3 + k_{14}x_4$$

In a similar fashion we find that the forces y_2, y_3, and y_4 acting on the masses m_2, m_3, and m_4, respectively, are given by

$$y_2 = -k_{12}(x_2 - x_1) + k_{23}(x_3 - x_2) + k_{24}(x_4 - x_2)$$
$$= k_{12}x_1 + (-k_{12} - k_{23} - k_{24})x_2 + k_{23}x_3 + k_{24}x_4$$
$$y_3 = -k_{23}(x_3 - x_2) + k_{34}(x_4 - x_3)$$
$$= 0x_1 + k_{23}x_2 + (-k_{23} - k_{34})x_3 + k_{34}x_4$$
$$y_4 = -k_{14}(x_4 - x_1) - k_{24}(x_4 - x_2) - k_{34}(x_4 - x_3) - k_4 x_4$$
$$= k_{14}x_1 + k_{24}x_2 + k_{34}x_3 + (-k_{14} - k_{24} - k_{34} - k_4)x_4$$

If we let $X = \begin{pmatrix} x_1 \\ x_2 \\ x_3 \\ x_4 \end{pmatrix}$, $Y = \begin{pmatrix} y_1 \\ y_2 \\ y_3 \\ y_4 \end{pmatrix}$, and

$$K = \begin{pmatrix} -k_1 - k_{12} - k_{14} & k_{12} & 0 & k_{14} \\ k_{12} & -k_{12} - k_{23} - k_{24} & k_{23} & k_{24} \\ 0 & k_{23} & -k_{23} - k_{34} & k_{34} \\ k_{14} & k_{24} & k_{34} & -k_{14} - k_{24} - k_{34} - k_4 \end{pmatrix}$$

then

$$F = KX \tag{1}$$

that is,

$$
\begin{pmatrix} y_1 \\ y_2 \\ y_3 \\ y_4 \end{pmatrix} = \begin{pmatrix} -k_1 - k_{12} - k_{14} & k_{12} \\ k_{12} & -k_{12} - k_{23} - k_{24} \\ 0 & k_{23} \\ k_{14} & k_{24} \end{pmatrix}
$$

$$
\begin{pmatrix} 0 & k_{14} \\ k_{23} & k_{24} \\ -k_{23} - k_{34} & k_{34} \\ k_{34} & -k_{14} - k_{24} - k_{34} - k_4 \end{pmatrix} \begin{pmatrix} x_1 \\ x_2 \\ x_3 \\ x_4 \end{pmatrix} \quad (2)
$$

We analyze the physical significance of the matrix K. If we set $x_1 = 1$, $x_2 = 0$, $x_3 = 0$, $x_4 = 0$, then we see from (2) that

$$
y_1 = (-k_1 - k_{12} - k_{14})x_1 = (-k_1 - k_{12} - k_{14})
$$

Therefore, the entry in the first row and first column of K represents the force applied to m_1 resulting from a unit displacement of this mass. The effect of a unit displacement of, say, m_2 on m_3 may be found as follows. Let $x_1 = 0$, $x_2 = 1$, $x_3 = 0$, and $x_4 = 0$. Then

$$
y_3 = k_{23}x_2 = k_{23}
$$

and hence, the entry in the second row and third column of K, k_{23}, is precisely the force applied to the third mass due to a unit displacement of the second mass. In general, the (i,j)th entry of K represents the force applied to the jth mass arising from a unit displacement of the ith mass. Thus, is some sense, the matrix K gives an indication of the rigidity of the system; this matrix is frequently called the *stiffness matrix* of the system.

Suppose now that K is a nonsingular matrix. We examine the physical significance of

$$
K^{-1} = \begin{pmatrix} b_{11} & b_{12} & b_{13} & b_{14} \\ b_{21} & b_{22} & b_{23} & b_{24} \\ b_{31} & b_{32} & b_{33} & b_{34} \\ b_{41} & b_{42} & b_{43} & b_{44} \end{pmatrix}
$$

Since $F = KX$, we have that $X = K^{-1}F$. To interpret, for example, b_{23}, let $y_1 = 0$, $y_2 = 0$, $y_3 = 1$, $y_4 = 0$. Then

$$
\begin{pmatrix} x_1 \\ x_2 \\ x_3 \\ x_4 \end{pmatrix} = \begin{pmatrix} b_{11} & b_{12} & b_{13} & b_{14} \\ b_{21} & b_{22} & b_{23} & b_{24} \\ b_{31} & b_{32} & b_{33} & b_{34} \\ b_{41} & b_{42} & b_{43} & b_{44} \end{pmatrix} \begin{pmatrix} 0 \\ 0 \\ 1 \\ 0 \end{pmatrix}
$$

and $x_2 = b_{23}y_3 = b_{23}$. Thus b_{23} is the displacement of the second mass resulting from the application of a unit force to the third mass. In general, b_{ij}

gives the displacement of the ith mass produced by the application of a unit force to the jth mass. K^{-1} is commonly called the *elasticity matrix* of the spring system.

Next we turn to examples from probability theory. Suppose we toss a coin four times. For $i = 0, 1, 2, 3$, let A_i be the event that exactly i heads are obtained in the first three tosses, and for $j = 0, 1, 2$, let B_j be the event that exactly j heads are obtained in the last two tosses. Note, for example, that the event A_2 can occur in any of the following ways: (HHTH), (HTHT), (THHT), (HHTT), (HTHH), (THHH), and that the event B_1 occurs with any of the following combinations : (HHHT), (HTHT), (THHT), (TTHT), (HTTH), (HHTH), (THTH), (TTTH). Since there are a total of $2^4 = 16$ possible outcomes when a coin is tossed four times, we find that the event A_2 will occur with the probability $P(A_2) = \frac{6}{16} = \frac{3}{8}$; the probability that B_1 will occur is $\frac{8}{16} = \frac{1}{2}$. Direct calculation yields the following probabilities:

$$P(A_0) = \tfrac{1}{8} \qquad P(B_0) = \tfrac{1}{4}$$
$$P(A_1) = \tfrac{3}{8} \qquad P(B_2) = \tfrac{1}{4}$$
$$P(A_3) = \tfrac{1}{8}$$

Now we consider the problem of finding the probability that A_2 occurs, given that B_1 has occurred. This is called the *conditional probability* of A_2 given B_1 and is denoted $P(A_2 \mid B_1)$. Since B_1 consists of eight possibilities and three of these satisfy A_2, we have $P(A_2 \mid B_1) = \frac{3}{8}$. In general, the conditional probability that an event A occurs given that an event B occurs is defined to be $P(A \mid B) = P(A \cap B)/P(B)$. In our example, $A_2 \cap B_1$ consists of (HHTH), (HTHT), (THHT), and hence, $P_2(A_2 \mid B_1) = \frac{3}{16}/\frac{1}{2} = \frac{3}{8}$.

We make one additional observation: since B_0, B_1, B_2 give all of the possibilities of obtaining heads on the last two tosses, it follows that $P(A_2) = P(A_2 \cap B_0) + P(A_2 \cap B_1) + P(A_2 \cap B_2)$.

If we define the following matrices:

$$A = [P(A_0) \quad P(A_1) \quad P(A_2) \quad P(A_3)]$$
$$B = [P(B_0) \quad P(B_1) \quad P(B_2)]$$
$$C = \begin{pmatrix} P(A_0 \mid B_0) & P(A_1 \mid B_0) & P(A_2 \mid B_0) & P(A_3 \mid B_0) \\ P(A_0 \mid B_1) & P(A_1 \mid B_1) & P(A_2 \mid B_1) & P(A_3 \mid B_1) \\ P(A_0 \mid B_2) & P(A_1 \mid B_2) & P(A_2 \mid B_2) & P(A_3 \mid B_2) \end{pmatrix}$$

then $A = BC$. That this is true follows easily from the observation that the matrix BC has the form $(d_0 \quad d_1 \quad d_2 \quad d_3)$, where by the rule for matrix multiplication, $d_i = P(B_0)P(A_i \mid B_0) + P(B_1)P(A_i \mid B_1) + P(B_2)P(A_i \mid B_2) = P(A_i \cap B_0) + P(A_i \cap B_1) + P(A_i \cap B_2) = P(A_i)$.

To apply these ideas to a problem in genetics, we call once again on our Fibonacci rabbits. Suppose that a trait for the length of rabbit ears is carried by two genes G and g. Each rabbit is endowed with a pair of the genes (GG, Gg, or gg) and each offspring inherits one gene from each parent. For parents of the genetic-type Gg, it is assumed that the genes G and g are transmitted with equal probability. It is known that a GG rabbit has long ears, a Gg rabbit has medium ears, and a gg rabbit has short ears. Suppose that we are given a large rabbit population where the ratio of (both male and female) GG rabbits to Gg rabbits to gg rabbits is given by $p_1 : p_2 : p_3$, and p_1, p_2, and p_3 are chosen so that $p_1 + p_2 + p_3 = 1$. Under the assumption that there is random mating, we would like to predict the distributions of GG, Gg, and gg rabbits in subsequent generations. Let B_1 be the event that the male parent of an offspring is GG, B_2 the event that a rabbit's male parent is Gg, and B_3 the event that the male parent of an offspring is gg. Then $P(B_1) = p_1, P(B_2) = p_2$, and $P(B_3) = p_3$. Let A_1 be the event that a first generation offspring is GG, A_2 the event that a first generation offspring is Gg, and A_3 that a first generation offspring is gg. We wish to find $P(A_1)$, $P(A_2)$, and $P(A_3)$. First we calculate $P(A_1 \mid B_1)$. If a male GG mates with a female GG, then, of course, the offspring will be GG. This mating occurs with probability p_1. If the male GG mates with a female Gg [which occurs $(p_2 \times 100)$ percent of the time], then there is a 50: 50 chance the offspring will be GG. If the male GG successfully woos a female gg, there is no chance for a GG offspring. Hence, $P(A_1 \mid B_1) = p_1 + \tfrac{1}{2}p_2$.

Similar analyses yield

$$C = \begin{pmatrix} P(A_1 \mid B_1) & P(A_2 \mid B_1) & P(A_3 \mid B_1) \\ P(A_1 \mid B_2) & P(A_2 \mid B_2) & P(A_3 \mid B_2) \\ P(A_1 \mid B_3) & P(A_2 \mid B_3) & P(A_2 \mid B_3) \end{pmatrix}$$

$$= \begin{pmatrix} p_1 + \tfrac{1}{2}p_2 & \tfrac{1}{2}p_2 + p_3 & 0 \\ \tfrac{1}{2}p_1 + \tfrac{1}{4}p_2 & \tfrac{1}{2}p_1 + \tfrac{1}{2}p_2 + \tfrac{1}{2}p_3 & \tfrac{1}{4}p_2 + \tfrac{1}{2}p_3 \\ 0 & p_1 + \tfrac{1}{2}p_2 & \tfrac{1}{2}p_2 + p_3 \end{pmatrix}$$

If $A = (P(A_1) \quad P(A_2) \quad P(A_3))$ and $B = (P(B_1) \quad P(B_2) \quad P(B_3))$, then by our previous discussion $A = BC$.

To simplify slightly the notation, let $q_1 = p_1 + \tfrac{1}{2}p_2$ and $q_2 = \tfrac{1}{2}p_2 + p_3$. The reader may verify directly (see problem C.3) that with this notation,

$$A = BC = (q_1^2 \quad 2q_1q_2 \quad q_2^2) \tag{3}$$

and this, of course, gives the distribution of the first generation offspring. To find out what happens in the second generation, it suffices to repeat the entire process using the values $q_1^2, 2q_1q_2, q_2^2$, in place of p_1, p_2, p_3. Somewhat surprisingly, it can be shown (see problem C.4) that the distribution of the second generation as well as all subsequent generations is exactly the same as the distribution of the first generation offspring.

D. INTRODUCTION TO DIFFERENTIAL EQUATIONS

A *differential equation* is an equation that involves one or more derivatives of
an unknown function. These equations may range from the fairly simple to
the extremely complex. In the following examples it is understood that each y
is a function of t.

(2.D.1) Examples (a) $y' = t^2$ $[y'(t) = t^2]$.
    (b) $y'' + y = 0$ $[y''(t) + y(t) = 0]$.
    (c) $y' = -t/y$.
    (d) $e^{2t^2}y^{(5)} + \sin(ty)\,y'' - \cos y = t^7$.

The *order* of a differential equation is equal to the order of the highest
derivative that appears in the equation. For instance, (2.D.1.b) above is a
second-order differential equation and (2.D.1.d) is a fifth-order equation.

A function ϕ is said to be a *solution* of a differential equation on an
interval (α,β) if for each $t \in (\alpha,\beta)$, substitution of $\phi(t)$ and the appropriate
derivatives of ϕ in the given equation yields the desired equality. For example,
$\phi(t) = t^3/3 + 13$ is a solution of (1) since $(t^3/3 + 13)' = t^2$, and this solution
is valid for each $t \in \mathbf{R}^1$. Similarly, the functions $\phi_1(t) = \sin t$ and $\phi_2(t) =$
$\cos t$ are easily seen to be solutions of the equation given in (2.D.1.b); again,
these solutions are valid for the entire real line $\mathbf{R}^1$. A solution of (2.D.1.c) is
given by $\phi(t) = \sqrt{4 - t^2}$, but any open interval for which this solution is
valid cannot contain the points $|t| \geq 2$. The solution of (2.D.1.d) is left to
the reader.

Example (2.D.1.b) serves to illustrate that solutions need not be unique
(this is also true in (2.D.1.a) since $\phi(t) = t^3/3 - 6$ is clearly another solution).
However, if additional conditions are imposed, then in certain instances
uniqueness may be obtained. For example in (2.D.1.a) if, a priori, it is stip-
ulated that $y(1) = 2$, then it can be shown that the only solution satisfying
this equation (and this condition) is given by $\phi(t) = t^3/3 + 5/3$. Similarly, in
(2.D.1.b) if it is required that $y(\pi) = 2$ and $y'(\pi) = 3$, then it will follow that
$\phi(t) = 3\sin t - 2\cos t$ is the unique solution. The conditions that we have
imposed are called *initial conditions*.

In (2.D.1.d) it is not even clear that there is a solution, much less what
initial conditions would be necessary to ensure its uniqueness. Establishment
of the existence and uniqueness of solutions is frequently of considerable
importance. It is not difficult to see why it would be useful to know that a
given differential equation either has or fails to have a solution. The reasons
for dealing with uniqueness are perhaps not so transparent. Consider, though,
a differential equation that purports to describe some physical or biological

system (examples will be forthcoming shortly). A solution of such an equation can be used to give the state of the system at varying times t. If there were more than one solution to the given equation, then it would be meaningless to talk about the solution as actually describing the true physical reality of the situation. If the solution is known to be unique, this problem does not arise.

A number of differential equations can be solved by inspection. For instance, if

$$y' = ky$$

then clearly any function of the form $y(t) = Ce^{kt}$ is a solution. Note, however, that if this equation is changed just slightly,

$$y' = ky + c \tag{1}$$

then it is no longer obvious what (if any) solutions can be found.

(2.D.2) *Exercise* Show by direct substitution that $\phi(t) = -c/k + Ce^{kt}$ is a solution of (1).

Equations of the form $y' = ky$ occur frequently in problems involving growth and decay. As an example of this we discuss in some detail the problem of dating ancient artifacts, bones, etc. One of the most applicable methods for determining the age of such objects is based on the fact that when atmospheric nitrogen (with a molecular structure including 7 protons and 7 neutrons) is bombarded by radioactive cosmic neutrons, a new compound, radioactive carbon, C^{14}, (with 6 protons and 8 neutrons) is formed. The C^{14} can be oxidized and incorporated into carbon dioxide, $C^{14}O_2$, which is in turn taken up by plants during photosynthesis; animals then, by eating either the plants or other plant-eating animals, ingest and excrete the radioactive carbon. Of course, CO_2, whose carbon component C is not radioactive, is also taken up in this manner. By assuming that the ratio of C^{14} to C present in the earth's environment has long since reached a fixed (and known) equilibrium point, archaeologists are able to compute to an astonishing degree of accuracy the age of once living material. As we shall see, the necessary mathematics for these computations is surprisingly simple.

From the moment of the plant's or animal's demise, no additional C^{14} can be introduced. Furthermore, radioactive carbon is unstable, and the amount present at the time of death begins an inexorable decaying action. It has been experimentally determined that the rate of decomposition of the C^{14} is proportional to the number of molecules of C^{14} present at any particular time. This is expressed mathematically by the differential equation

$$m'(t) = -km(t)$$

where $m(t)$ represents the amount of C^{14} found at time t, and k is a positive constant of proportionality. By (1) (with $c = 0$), a solution to this equation is given by

$$m(t) = m_0 e^{-kt}$$

Clearly $m(0) = m_0$ is the quantity of C^{14} present at the time of death. The actual value of m_0 can be calculated by comparing the C^{14}/C ratio of a living organism to the C^{14}/C ratio of the object under study. To find k, use is made of the fact that the half-life of C^{14} (the time required for $\frac{1}{2}$ of a given amount of C^{14} to decompose) is known to be 5600 years. Hence, we have $m_0/2 = m_0 e^{-k5600}$, which yields $k = (-\ln \frac{1}{2})/5600$. It should be noted that this method depends on the supposition that the C^{14}/C ratio observed in living organisms today has been maintained throughout the past. One consequence of twentieth century nuclear testing is that future archaeologists will no longer be able to base their work on this assumption.

There remains, however, the question of uniqueness. If we do not know that the above solution is unique, then we cannot state categorically that the age of the object is determined by this solution and not by another. We shall deal with this problem shortly.

Equations of the type given by (1) arise often in dealing with biological and physical models. For instance, Newton's law of cooling asserts that if a cold object is introduced into a warm environment then the rate of change of the object's temperature is proportional to the difference between the temperature of the object and that of its surroundings (here we assume that the surroundings maintain a constant temperature). Thus, for example, an object at 20° placed in a room where the temperature is 70°, will heat up much more rapidly than will an object at 60° placed in the same room. If we let $y(t)$ be the temperature of the object at time t, then since a rate of change is described mathematically by the derivative, we have that the mathematical formulation of Newton's law of cooling in the case where the temperature of the surroundings is 70° can be expressed by

$$y'(t) = k(y(t) - 70) \tag{2}$$

From (2.D.2) we find that a solution of this equation is given by

$$\phi(t) = -\frac{70k}{k} + Ce^{kt} = 70 + Ce^{kt} \tag{3}$$

This solution is of limited value unless the constants k and C can be determined. If the original temperature of the object was 20°, then we have $20 = \phi(0) = 70 + C$, and hence, $C = -50$. To find k, additional information is needed; if the temperature of the object was recorded after 5 minutes and was, say 30°, then we have $30 = \phi(5) = 70 - 50e^{5k}$, and therefore, $k = 0.2 \ln 0.8$. Thus, we have obtained a solution,

$$\phi(t) = 70 - 50e^{0.2\ln 0.8t} \tag{4}$$

to (2) that permits us to calculate the temperature of the object at any time t.

Here too there is a problem of the uniqueness of the solution. To resolve this problem (as well as the general existence problem), we consider in some detail the first-order linear equation

$$y' + p(t)y = g(t) \tag{5}$$

where p and g are continuous functions on an interval (a,b). To find a solution of this equation, we first observe that if we could write the left-hand side of (5) as the derivative of a product:

$$(q(t)y)' \tag{6}$$

then integration of both sides of (5) would allow us to determine y. Although, in general, it is too much to expect that the left-hand side of (5) can be expressed by (6), it will nevertheless turn out that if both sides of (5) are multiplied by an appropriate function $\rho(t)$, then the resulting equation can be put in the desired form. Such a function ρ is called an *integrating factor*. To find the integrating factor ρ, we essentially work backward and assume its existence. Thus, we suppose that there is a function $\rho(t)$ such that

$$(\rho(t)y)' = \rho(t)g(t) \tag{7}$$

Since $(\rho(t)y)' = \rho(t)y' + \rho'(t)y$ and $\rho(t)g(t) = \rho(t)y' + \rho(t)p(t)y$, it follows from (7) that

$$\rho'(t)y + \rho(t)y' = \rho(t)y' + \rho(t)p(t)y$$

and therefore,

$$\rho'(t) = \rho(t)p(t) \tag{8}$$

Let $P(t)$ be any function such that $P'(t) = p(t)$, i.e., $P(t) = \int p(t) \, dt$. Then it is easily verified that $\rho(t) = e^{P(t)}$ satisfies equation (8) and, consequently, is an integrating factor of (5). Therefore, we have

$$(e^{P(t)}y)' = e^{P(t)}g(t) \tag{9}$$

and integration of both sides of (9) shows that

$$e^{P(t)}y = A(t) + c$$

where $A(t)$ is any function whose derivative is $e^{P(t)}g(t)$. In particular, if we set $A(t) = \int_{t_o} e^{P(x)}g(x) \, dx$, then we have that every solution of (5) has the form:

$$y(t) = e^{-P(t)} \int_{t_0}^{t} e^{P(x)}g(x)dx + ce^{-P(t)}$$

Uniqueness of the solution is obtained by assigning a value y_0 to $y(t_0)$, since this will determine the constant c. The condition $y(t_0) = y_0$ is called an *initial condition*.

(2.D.3) Example Suppose that we are given the first-order linear equation

$$y' + \left(\frac{3}{t}\right)y = t^2 + 2t \tag{10}$$

with initial condition $y(1) = 2$. An integrating factor for this equation is given by $\rho(t) = e^{\int(3/t)\,dt} = t^3$. Consequently, we have

$$(t^3 y)' = t^3(t^2 + 2t) = t^5 + 2t^4$$

and therefore,

$$y(t) = t^{-3} \int_1^t (x^5 + 2x^4)\,dx + ct^{-3}$$

To find c, note that

$$2 = y(1) = c \cdot 1^{-3} = c$$

Therefore,

$$y(t) = t^{-3}\left(\frac{t^6}{6} + \frac{2t^5}{5} - \left(\frac{1}{6} + \frac{2}{5}\right) + 2\right) = \frac{t^3}{6} + \frac{2t^2}{5} + \frac{43t^{-3}}{30}$$

is the unique solution of (10) that satisfies the initial condition $y(1) = 2$.

Are there analogous results for second and higher order differential equations? In (2.D.1.b) we saw that the functions $\sin x$ and $\cos x$ were two distinct solutions of $y'' + y = 0$. Rather amazingly, it can be shown that every solution of this equation can be expressed as a "linear combination" of these two functions. In other words, if the set $V = \{f: \mathbf{R}^1 \to \mathbf{R}^1 \mid f \text{ is a solution of } y'' + y = 0\}$ were a vector space, then the solutions $\sin x$ and $\cos x$ would span V. That V is indeed a vector space essentially follows from the next theorem.

(2.D.4) Theorem Suppose that ϕ_1 and ϕ_2 are solutions of

$$y'' + p(t)y' + q(t)y = 0 \tag{11}$$

on the interval (α, β). Then $\phi = c_1\phi_1 + c_2\phi_2$, where c_1 and c_2 are arbitrary constants, is also a solution of (11) on (α, β); hence, the set

$$V = \{\phi: (\alpha, \beta) \to \mathbf{R}^1 \mid \phi \text{ is a solution of (11)}\} \tag{12}$$

with the usual operations of addition and scalar multiplication is a vector space.

Proof. We substitute ϕ in (11) to obtain

$$\phi'' + p(t)\phi' + q(t)\phi$$
$$= (c_1\phi_1 + c_2\phi_2)'' + p(t)(c_1\phi_1 + c_2\phi_2)' + q(t)(c_1\phi_1 + c_2\phi_2)$$
$$= c_1(\phi_1'' + p(t)\phi_1' + q(t)\phi_1) + c_2(\phi_2'' + p(t)\phi_2' + q(t)\phi_2)$$
$$= c_1 \cdot 0 + c_2 \cdot 0 = 0$$

The penultimate equality follows from the fact that ϕ_1 and ϕ_2 are solutions of (11).

The reader may now readily verify that V is a vector space (either directly or via problem A.5).

(2.D.5) *Definition* An *nth-order linear differential equation* is an equation of the form

$$y^{(n)} + p_1(t)y^{(n-1)} + \cdots + p_n(t)y = g(t) \tag{13}$$

If the function g is identically 0, then the equation (13) is said to be *homogeneous*.

(2.D.6) *Examples* (a) $y^{(4)} + (\sin t)y^{(3)} + e^t t^2 y' = \cosh t$.
  (b) $y'' - (\cos t)y' + ty = 0$ (homogeneous)

A proof of the following basic result may be found in Coddington (1961).

(2.D.7) *Theorem* Suppose that

$$y'' + p(t)y' + q(t)y = g(t) \tag{14}$$

is a second-order linear differential equation and that the functions p, q, and g are continuous on an interval (α,β). Suppose further that $t_0 \in (\alpha,\beta)$ and that y_0 and y_1 are arbitrary points in $\mathbf{R}^1$. Then there is a unique solution ϕ of (14) on the interval (α,β) such that ϕ satisfies the initial conditions $\phi(t_0) = y_0$ and $\phi'(t_0) = y_1$.

With the aid of the preceding theorem we can determine the dimension of the vector space V described in (12).

(2.D.8) *Theorem* Suppose that

$$y'' + p(t)y' + q(t)y = 0 \tag{15}$$

is a second-order linear differential equation and that the functions p and q are continuous on the interval (α,β). Let V be the vector space of solutions (12). Then the dimension of V is 2.

Proof. Fix a point $t_0 \in (\alpha,\beta)$. By (2.D.7) there is a unique solution ϕ_1 of (15) such that

$$\phi_1(t_0) = 0 \qquad \phi_1'(t_0) = 1$$

and a unique solution ϕ_2 of (15) satisfying

$$\phi_2(t_0) = 1 \qquad \phi_2'(t_0) = 0$$

We show that ϕ_1 and ϕ_2 form a basis for V. To see that ϕ_1 and ϕ_2 are independent, suppose that $c_1\phi_1 + c_2\phi_2 = 0$ [that is, $c_1\phi_1(t) + c_2\phi_2(t) = 0$ for each $t \in (\alpha,\beta)$]. Then we have

$$0 = c_1\phi_1(t_0) + c_2\phi_2(t_0) = c_2$$
$$0 = c_1\phi_1'(t_0) + c_2\phi_2'(t_0) = c_1$$

and therefore, $c_1 = c_2 = 0$, which implies that ϕ_1 and ϕ_2 are independent.

Next we show that ϕ_1 and ϕ_2 span V. Suppose that ϕ is any solution of (15) on (α,β) and let $c_2 = \phi(t_0)$ and $c_1 = \phi'(t_0)$. Then by (2.D.4), $\psi = c_1\phi_1 + c_2\phi_2$ is a solution of (15); furthermore, $\psi(t_0) = c_2$ and $\psi'(t_0) = c_1$. Consequently, by the uniqueness theorem (2.D.7), we have that $\psi = \phi$, and hence ϕ_1 and ϕ_2 span V.

Note that it follows from (2.D.8) and (2.A.18) that if ϕ_1 and ϕ_2 are any two linearly independent solutions of (15), then they span the solution space. Such solutions are said to form a *fundamental set* of solutions.

(2.D.9) *Example* It is easily verified that $\phi_1(t) = e^{3t}$ and $\phi_2(t) = e^{-2t}$ are solutions of

$$y'' - y' - 6y = 0$$

To see that ϕ_1 and ϕ_2 are independent, suppose that $c_1e^{3t} + c_2e^{-2t} = 0$ for all t. Multiplication of both sides of this equation by e^{-3t} yields $c_1 + c_2e^{-5t} = 0$. If we take the derivative of both sides of the latter equation, we obtain

$$-5c_2e^{-5t} = 0$$

which implies that $c_2 = 0$, and therefore, $c_1 = 0$. Thus, ϕ_1 and ϕ_2 are linearly independent, and every solution of $y'' - y' - 6y = 0$ can be written in the form $\phi(t) = \alpha_1e^{3t} + \alpha_2e^{-2t}$.

The problem of determining whether a given pair of solutions to (15) forms a fundamental set of solutions is a nontrivial one. The following notion is frequently of use in this regard.

(2.D.10) *Definition* Suppose that $A \subset \mathbf{R}^1$ and that $\phi_1, \phi_2, \ldots, \phi_n$ is a set of functions such that for each i, $1 \le i \le n$, $\phi_i: A \to \mathbf{R}^1$ and $\phi_i^{(j)}(t)$ exists for each $t \in A$ and each integer j, $1 \le j \le n$. Then the *Wronskian* of this set of functions is the function $W: A \to \mathbf{R}^1$ defined by

$$W(t) = \det\begin{pmatrix} \phi_1(t) & \phi_2(t) & \cdots & \phi_n(t) \\ \phi_1'(t) & \phi_2'(t) & \cdots & \phi_n'(t) \\ \cdots & \cdots & \cdots & \cdots \\ \phi_1^{(n-1)}(t) & \phi_2^{(n-1)}(t) & \cdots & \phi_n^{(n-1)}(t) \end{pmatrix}$$

(2.D.11) *Example* If $\phi_1(t) = \sin t$ and $\phi_2(t) = \cos t$, then the Wronskian of ϕ_1 and ϕ_2 is given by

$$W(t) = \det \begin{pmatrix} \sin t & \cos t \\ \cos t & -\sin t \end{pmatrix} = -1$$

for each $t \in \mathbf{R}^1$.

A particularly useful theorem involving the Wronskian is the following.

(2.D.12) *Theorem* Suppose that ϕ_1 and ϕ_2 are solutions of (15) on an interval (α, β), and that for some $t_0 \in (\alpha, \beta)$, $W(t_0) \neq 0$. Then ϕ_1 and ϕ_2 form a fundamental set of solutions of (15).

Proof. It suffices to show that ϕ_1 and ϕ_2 are independent. Suppose that $c_1\phi_1 + c_2\phi_2 = 0$. Then $c_1\phi_1' + c_2\phi_2'$ is identically 0, and hence, for $t = t_0$ we have

$$c_1\phi_1(t_0) + c_2\phi_2(t_0) = 0$$
$$c_1\phi_1'(t_0) + c_2\phi_2'(t_0) = 0$$

Since $W(t_0) \neq 0$, it follows from (2.B.3) that this system has a unique solution (in terms of the unknowns c_1 and c_2), which implies that $c_1 = 0$ and $c_2 = 0$.

(2.D.13) *Exercise* Generalize (2.D.12) to nth-order homogeneous linear differential equations.

(2.D.14) *Example* Since in (2.D.11), $W(t) = -1$ (for all t), it follows that $\sin t$ and $\cos t$ span the vector space of solutions.

(2.D.15) *Exercise* Show (by direct substitution) that solutions of

$$ay'' + by' + cy = 0 \tag{16}$$

can be found as follows. Let r_1 and r_2 be the roots of the *auxiliary equation* $ar^2 + br + c = 0$.

    (a)   If r_1 and r_2 are real numbers and $r_1 \neq r_2$, the functions

$$\phi_1(t) = e^{r_1 t} \qquad \phi_2(t) = e^{r_2 t}$$

form a fundamental set of solutions of (16).

    (b)   If $r_1 = r_2$, then

$$\phi_1(t) = e^{r_1 t} \qquad \phi_2(t) = te^{r_2 t}$$

form a fundamental set of solutions of (16).

    (c)   If $r_1 = \alpha + \beta i$ and $r_2 = \alpha - \beta i$, then

$$\phi_1(t) = e^{r_1 t} = e^{\alpha t}(\cos \beta t + i \sin \beta t) \tag{17}$$

and

$$\phi_2(t) = e^{r_2 t} = e^{\alpha t}(\cos(-\beta)t + i\sin(-\beta)t$$
$$= e^{\alpha t}(\cos\beta t - i\sin\beta t) \tag{18}$$

form a fundamental set of solutions of (16). Note that if the solutions (17) and (18) are added, then by (2.D.4), the real-valued function $e^{\alpha t}\cos\beta t$ is also a solution of (16); furthermore, subtraction of (18) from (17) and division by $2i$ yields another real-valued solution $e^{\alpha t}\sin\beta t$. Show that these solutions are independent.

(2.D.16) *Examples* (a) Suppose that $4y'' + 4y' + y = 0$. The only root of the auxiliary equation $4r^2 + 4r + 1$ is $r = -\frac{1}{2}$, and hence, a fundamental set of solutions is given by $\phi_1(t) = e^{(-\frac{1}{2})t}$ and $\phi_2(t) = te^{(-\frac{1}{2})t}$.

(b) Suppose that $y'' - 4y' + 5y = 0$. The auxiliary equation is $r^2 - 4r + 5 = 0$, and roots are given by $r = [-(-4) \pm \sqrt{16 - 20}]/2$. Thus, $r_1 = 2 + i$ and $r_2 = 2 - i$, and the functions $\phi_1(t) = e^{2t}\sin t$ and $\phi_2(t) = e^{2t}\cos t$ form a fundamental set of solutions.

Although in this chapter we shall deal primarily with homogeneous equations, the following general result will be of occasional use.

(2.D.17) *Theorem* Suppose that p, q, and g are continuous functions and that ϕ^* is a solution of

$$y'' + p(t)y' + q(t)y = g(t) \tag{19}$$

Then if ϕ_1 and ϕ_2 form a fundamental set of solutions of the associated homogeneous equation, $y'' + p(t)y' + q(t)y = 0$, every solution of (19) has the form $\phi^* + c_1\phi_1 + c_2\phi_2$, where c_1 and c_2 are determined by the initial conditions

$$y(t_0) = y_0 \qquad y'(t_0) = y_1$$

Proof. Suppose that ϕ is any solution of (19). Then the reader may easily verify that $\phi - \phi^*$ is a solution of the associated homogeneous equation, and hence, by the definition of a fundamental set of solutions, $\phi - \phi^* = c_1\phi_1 + c_2\phi_2$.

(2.D.18) *Example* As may be verified directly, a particular solution of $y'' - 7y' + 6y = \sin t$ is $\phi^*(t) = \frac{5}{74}\sin t + \frac{7}{74}\cos t$. It follows from (2.D.15) that $\phi_1(t) = e^{-6t}$ and $\phi_2(t) = e^{-t}$ form a fundamental set of solutions of the associated homogeneous equation $y'' - 7y' + 6y = 0$. Hence, the general solution of the given equation is $\phi(t) = \frac{5}{74}\sin t + \frac{7}{74}\cos t + c_1 e^{-6t} + c_2 e^{-t}$.

In practice, (2.D.17) is often difficult to apply because of the problem

involved in finding one particular solution of (19). One procedure for determining such a solution, the method of the variation of parameters, is discussed in the problem section.

E. LINEAR TRANSFORMATIONS

In this section we examine functions between vector spaces. Since the essence
of a vector space lies in its algebraic structure, one might expect that special
significance would be attached to those functions that take this structure into
account. Such functions are called *linear transformations*.

(2.E.1) *Definition* Suppose that V and W are vector spaces. Then a function $f: V \to W$ is a *linear transformation* if
   (i)   For each $\mathbf{v}_1, \mathbf{v}_2 \in V, f(\mathbf{v}_1 + \mathbf{v}_2) = f(\mathbf{v}_1) + f(\mathbf{v}_2)$.
   (ii) For each $\mathbf{v} \in V$ and each $\alpha \in \mathbf{R}^1$ (or $\mathbf{C}$), $f(\alpha \mathbf{v}) = \alpha f(\mathbf{v})$.

(2.E.2.) *Observation* The preceding definition says in effect that if $f: V \to$
W is a linear transformation and if $\mathbf{v}_1$ and $\mathbf{v}_2$ are vectors in V, then we may
either sum the two vectors in V first and then map them to W via f, or we may
first send the vectors into W and then add their images under f; the end result
is the same. *A* similar observation applies to scalar multiplication. Thus, in
some sense the linear transformation f "preserves" the algebraic structure of
V in W.

(2.E.3) *Observation* From a geometrical point of view the term linear is
quite apt. For instance, suppose that $\mathbf{R}^2$ is viewed as a vector space and that
$f: \mathbf{R}^2 \to \mathbf{R}^2$ is a linear transformation. Let L be the line segment connecting
two points (x_1, y_1) and (x_2, y_2) in $\mathbf{R}^2$. Then points of L can be described parametrically by $\{t(x_1, y_1) + (1 - t)(x_2, y_2) \mid 0 \le t \le 1\}$. Since f is a linear
transformation, the image of L under f consists of the points

$$\{f(t(x_1, y_1) + (1 - t)(x_2, y_2)) \mid 0 \le t \le 1\}$$
$$= \{f(t(x_1, y_1)) + f((1 - t)(x_2, y_2)) \mid 0 \le t \le 1\}$$
$$= \{tf(x_1, y_1) + (1 - t)f(x_2, y_2) \mid 0 \le t \le 1\}$$

which is the line segment joining $f(x_1, y_1)$ and $f(x_2, y_2)$. Thus, this linear transformation maps line segments to line segments (or possibly to single points).
In a similar fashion the reader should be able to convince himself that a linear
transformation will map a triangle in $\mathbf{R}^2$ onto either a triangle or a line
segment or a point. The upshot of this observation is that from a geometrical
standpoint, a linear transformation maintains the linear nature of geometric
figures.

(2.E.4) *Examples* (a) Let $f: \mathbf{R}^2 \to \mathbf{R}^2$ be defined by $f(x,y) = (x + 2y, 3x)$. Since for any two points (x_1,y_1) and (x_2,y_2) in $\mathbf{R}^2$,

$$f((x_1,y_1) + (x_2,y_2))$$
$$= f(x_1 + x_2, y_1 + y_2) = (x_1 + x_2 + 2(y_1 + y_2), 3(x_1 + x_2))$$
$$= (x_1 + 2y_1, 3x_1) + (x_2 + 2y_2, 3x_2) = f(x_1,y_1) + f(x_2,y_2)$$

and

$$f(\alpha(x,y)) = f(\alpha x, \alpha y) = (\alpha x + \alpha y, 3\alpha x) = \alpha(x + 2y, 3x) = \alpha f(x,y)$$

it follows that f is a linear transformation.

(b) Let $V = \{f: \mathbf{R}^1 \to \mathbf{R}^1 \mid n\text{th derivative of } f \text{ exists for each } n \in \mathbf{Z}^+\}$. Define $D: V \to V$ by $D(f) = f'$. Then since the derivative of a sum is equal to the sum of the derivatives, we have that $D(f + g) = D(f) + D(g)$, and from another elementary property of derivatives it follows that $D(\alpha f) = \alpha D(f)$.

(c) Let $V = \{f: [a,b] \to \mathbf{R}^1 \mid f \text{ is continuous}\}$. Define $I: V \to \mathbf{R}^1$ by $I(f) = \int_a^b f(x)\, dx$. It follows from standard properties of the Riemann integral that I is a linear transformation.

(d) Define $T: \mathbf{R}^1 \to \mathbf{R}^1$ by $T(x) = ax$. Then T is linear; the reader might try to show that every linear transformation from $\mathbf{R}^1$ into $\mathbf{R}^1$ is of this form.

(e) Let A be the matrix

$$\begin{pmatrix} 2 & 1 \\ 3 & -1 \\ 0 & 2 \end{pmatrix}$$

and define $f: \mathbf{R}^2 \to \mathbf{R}^3$ by

$$f(x,y) = A \begin{pmatrix} x \\ y \end{pmatrix}$$

(Here $\mathbf{R}^3$ is identified with the set of 3×1 matrices over $\mathbf{R}^1$.) It is easily verified that f is linear; thus, matrices may be used to induce linear transformations.

(2.E.5) *Observation* Suppose that V and W are vector spaces and $f: V \to W$ is a linear transformation. An easy induction argument shows that if $v_1, v_2, \ldots, v_k$ are vectors in V and $\alpha_1, \alpha_2, \ldots, \alpha_k$ are scalars, then $f(\alpha_1 v_1 + \alpha_2 v_2 + \cdots + \alpha_k v_k) = \alpha_1 f(v_1) + \alpha_2 f(v_2) + \cdots + \alpha_k f(v_k)$.

Bases of vector spaces play an important role in connection with linear transformations. In fact, a linear transformation is essentially determined once its action on elements of a basis is known. To see this, suppose that $v_1, v_2, \ldots, v_n$ is a basis for a vector space V and that $f: V \to W$ is a linear transformation from V into a vector space W. If $v \in V$, then there are unique scalars $c_1, c_2, \ldots, c_n$ such that $v = c_1 v_1 + c_2 v_2 + \cdots + c_n v_n$, and since f is linear, we have

$$f(\mathbf{v}) = f(c_1\mathbf{v}_1 + c_2\mathbf{v}_2 + \cdots + c_n\mathbf{v}_n)$$
$$= c_1 f(\mathbf{v}_1) + c_2 f(\mathbf{v}_2) + \cdots + c_n f(\mathbf{v}_n)$$

Consequently, if the vectors $f(\mathbf{v}_1), f(\mathbf{v}_2), \ldots, f(\mathbf{v}_n)$ are known, the image of any vector under f may be readily calculated.

(2.E.6) *Example* Suppose that $f: \mathbf{R}^3 \to \mathbf{R}^2$ is a linear transformation and that $f(\mathbf{e}_1) = (1,1)$, $f(\mathbf{e}_2) = (0,-1)$, and $f(\mathbf{e}_3) = (2,4)$, where, as usual, $\mathbf{e}_1 = (1,0,0)$, $\mathbf{e}_2 = (0,1,0)$, and $\mathbf{e}_3 = (0,0,1)$. Then, for example, if $\mathbf{v} = (3,-4,6)$, we have $f(\mathbf{v}) = f(3,-4,6) = 3(1,1) - 4(0,-1) + 6(2,4) = (15,31)$.

With each linear transformation it is possible to associate a matrix in a very natural fashion. Suppose that V and W are vector spaces with bases $\{\mathbf{v}_1, \mathbf{v}_2, \ldots, \mathbf{v}_n\}$ and $\{\mathbf{w}_1, \mathbf{w}_2, \ldots, \mathbf{w}_m\}$, respectively. Let $\phi: V \to W$ be a linear transformation. Then we have

$$\begin{aligned}
\phi(\mathbf{v}_1) &= c_{11}\mathbf{w}_1 + c_{21}\mathbf{w}_2 + \cdots + c_{m1}\mathbf{w}_m \\
\phi(\mathbf{v}_2) &= c_{12}\mathbf{w}_1 + c_{22}\mathbf{w}_2 + \cdots + c_{m2}\mathbf{w}_m \\
&\ \cdots\cdots\cdots\cdots\cdots\cdots\cdots\cdots\cdots \\
\phi(\mathbf{v}_n) &= c_{1n}\mathbf{w}_1 + c_{2n}\mathbf{w}_2 + \cdots + c_{mn}\mathbf{w}_m
\end{aligned} \tag{1}$$

for suitable coefficients c_{ij} (every vector in W can be expressed uniquely as a linear combination of the $\mathbf{w}_i$). The matrix A_ϕ belonging to ϕ is defined to be the matrix

$$A_\phi = \begin{pmatrix} c_{11} & c_{12} & \cdots & c_{1n} \\ c_{21} & c_{22} & \cdots & c_{2n} \\ \multicolumn{4}{c}{\cdots\cdots\cdots\cdots\cdots\cdots} \\ c_{m1} & c_{m2} & \cdots & c_{mn} \end{pmatrix}$$

Note carefully the relationship between the coefficient matrix of the system of equations (1) and the entries of A_ϕ: the rows and columns have been interchanged, i.e., A_ϕ is the transpose of the coefficient matrix.

The linear transformation ϕ and its associated matrix A_ϕ are related in the following manner. Suppose that $\mathbf{v} = \alpha_1\mathbf{v}_1 + \alpha_2\mathbf{v}_2 + \cdots + \alpha_n\mathbf{v}_n$ is an arbitrary vector in V. Then

$$\begin{aligned}
\phi(\mathbf{v}) &= \phi(\alpha_1\mathbf{v}_1 + \alpha_2\mathbf{v}_2 + \cdots + \alpha_n\mathbf{v}_n) \\
&= \alpha_1\phi(\mathbf{v}_1) + \alpha_2\phi(\mathbf{v}_2) + \cdots + \alpha_n\phi(\mathbf{v}_n) \\
&= \alpha_1(c_{11}\mathbf{w}_1 + \cdots + c_{m1}\mathbf{w}_m) + \alpha_2(c_{12}\mathbf{w}_2 + \cdots \\
&\ + c_{m2}\mathbf{w}_m) + \cdots + \alpha_n(c_{1n}\mathbf{w}_1 + \cdots + c_{mn}\mathbf{w}_m) \\
&= (\alpha_1 c_{11} + \alpha_2 c_{12} + \cdots + \alpha_n c_{1n})\mathbf{w}_1 + (\alpha_1 c_{21} + \cdots \\
&\ + \alpha_n c_{2n})\mathbf{w}_2 + \cdots + (\alpha_1 c_{m1} + \cdots + \alpha_n c_{mn})\mathbf{w}_m
\end{aligned}$$

On the other hand,

$$A_\phi \begin{pmatrix} \alpha_1 \\ \alpha_2 \\ \vdots \\ \alpha_n \end{pmatrix} = \begin{pmatrix} \alpha_1 c_{11} + \cdots + \alpha_n c_{1n} \\ \alpha_1 c_{21} + \cdots + \alpha_n c_{2n} \\ \cdots\cdots\cdots\cdots\cdots\cdots \\ \alpha_1 c_{m1} + \cdots + \alpha_n c_{mn} \end{pmatrix}$$

which gives precisely the coefficients of the w_i used to express the image of $\phi(v)$ as a linear combination of the w_i. Thus, we have the following result.

(2.E.7)   *Theorem* Suppose that V and W are vector spaces and $\phi: V \to W$ is a linear transformation. If $v_1, v_2, \ldots, v_n$ and $w_1, w_2, \ldots, w_m$ are bases for V and W, respectively, and if $v = \alpha_1 v_1 + \alpha_2 v_2 + \cdots + \alpha_n v_n$, then

$$\phi(v) = \left(A_\phi \begin{pmatrix} \alpha_1 \\ \alpha_2 \\ \vdots \\ \alpha_n \end{pmatrix} \right)^T \begin{pmatrix} w_1 \\ w_2 \\ \vdots \\ w_m \end{pmatrix}$$

(2.E.8)   *Example* Let $v_1 = (1,0,0)$, $v_2 = (0,1,0)$, and $v_3 = (0,0,1)$ be a basis for $\mathbf{R}^3$ and let $w_1 = (1,1)$ and $w_2 = (0,2)$ be a basis for $\mathbf{R}^2$. Suppose that $\phi: \mathbf{R}^3 \to \mathbf{R}^2$ is a linear transformation such that

$$\phi(v_1) = (1,3) = 1(1,1) + 1(0,2)$$
$$\phi(v_2) = (0,0) = 0(1,1) + 0(0,2)$$
$$\phi(v_3) = (-1,1) = -1(1,1) + 1(0,2)$$

Then

$$A_\phi = \begin{pmatrix} 1 & 0 & -1 \\ 1 & 0 & 1 \end{pmatrix}$$

If $v = (2,3,6) = 2v_1 + 3v_2 + 6v_3$, then by (2.E.5)

$$\phi(v) = 2\phi(v_1) + 3\phi(v_2) + 6\phi(v_3) = 2(1,3) + 3(0,0) + 6(-1,1)$$
$$= -4w_1 + 8w_2$$

On the other hand,

$$A_\phi \begin{pmatrix} 2 \\ 3 \\ 6 \end{pmatrix} = \begin{pmatrix} 1 & 0 & -1 \\ 1 & 0 & 1 \end{pmatrix} \begin{pmatrix} 2 \\ 3 \\ 6 \end{pmatrix} = \begin{pmatrix} -4 \\ 8 \end{pmatrix}$$

and

$$(-4 \quad 8) \begin{pmatrix} w_1 \\ w_2 \end{pmatrix} = -4w_1 + 8w_2$$

Thus, knowledge of A_ϕ results in the easy calculation of the images under ϕ of vectors in V.

(2.E.9) Observation It is important to note that if $\phi: V \to W$ is a linear transformation, then the matrix A_ϕ associated with ϕ depends both on the bases chosen for V and W and on the *order* in which the elements of these bases are listed.

Some justification of matrix multiplication is found in the following fundamental (albeit somewhat tedious) exercise.

(2.E.10) Exercise Suppose that U, V, and W are vector spaces with bases $\{u_1, \ldots, u_m\}, \{v_1, \ldots, v_n\}$, and $\{w_1, \ldots, w_p\}$, respectively. Suppose further that $\phi_1: U \to V$ and $\phi_2: V \to W$ are linear transformations with associated matrices A_{ϕ_1} and A_{ϕ_2}, respectively. Show that

(a) The composition map $\phi_2 \circ \phi_1: U \to W$ is a linear transformation.
(b) The associated matrix of $\phi_2 \circ \phi_1$, $A_{\phi_2 \circ \phi_1}$, is equal to the matrix product $A_{\phi_2} A_{\phi_1}$.

PROBLEMS

Section A

1. Show that the set of all polynomials of degree less than or equal to n (with the obvious operations) is a vector space.
2. Show that the set consisting of only the real number 0 is a vector space over $\mathbf{R}^1$ (or $\mathbf{C}$). Assume the usual operations of addition and scalar multiplication.
3. Does the set of all matrices of the form

$$\begin{pmatrix} a & b \\ -b & c \end{pmatrix}$$

with entries from $\mathbf{R}^1$ or $\mathbf{C}$ define a vector space (with the usual operations)?
4. Show that if V is a vector space, and if $v \in V$, then $-v = (-1)v$.
5. Suppose that V is a vector space. A subset S of V is called a *vector subspace* (or a *subspace* of V) if S is a vector space (with the same operations defined for V). Show that a subset S of a vector space V over $\mathbf{R}^1$ (or $\mathbf{C}$) is a subspace of V if whenever $v_1, v_2 \in S$, then $v_1 + v_2 \in S$ and if whenever $v_1 \in S$ and $\alpha \in \mathbf{R}^1$ (or $\alpha \in \mathbf{C}$), then $\alpha v_1 \in S$.
6. Show that matrices of the form

$$\begin{pmatrix} a & 0 \\ 0 & b \end{pmatrix}$$

with $a,b \in \mathbf{R}^1$ (or $\mathbf{C}$) form a subspace of the vector space of all 2×2 matrices with real (or complex) entries (see problem A.5).

7. Determine geometrically the subspaces of $\mathbf{R}^2$.

8. Show that the vectors $(1,1)$, $(-2,1)$, $(0,1)$ are dependent vectors in $\mathbf{R}^2$ and write one of these vectors as a linear combination of the others.

9. Show that the vectors $(1,0,2)$, $(-1,1,3)$, $(2,1,4)$, and $(0,1,0)$ are dependent vectors in $\mathbf{R}^3$ and write one of these vectors as a linear combination of the others.

10. Show that the vectors $(1,-2,0)$, $(1,1,1)$ and $(0,1,0)$ form a basis for $\mathbf{R}^3$.

11. Show that the vectors 1 and i form a basis for the vector space $\mathbf{C}$ over $\mathbf{R}^1$.

12. Let P be the vector space of all polynomials. Find a basis for P.

13. Suppose that V is a vector space and that $\mathbf{v}_1, \mathbf{v}_2$, and $\mathbf{v}_3$ are independent vectors in V.
    (a) Are the vectors $\mathbf{v}_1 + \mathbf{v}_2$, $\mathbf{v}_2 + \mathbf{v}_3$, and $\mathbf{v}_3 + \mathbf{v}_1$ independent?
    (b) Are the vectors $\mathbf{v}_1$, $\mathbf{v}_1 + \mathbf{v}_2$, and $\mathbf{v}_1 + \mathbf{v}_2 + \mathbf{v}_3$ independent?

14. Let P_2 be the vector space of polynomials of degree less than or equal to 2.
    (a) Do the vectors $t^2 + 1$ and $t + 2$ span P_2?
    (b) Are the vectors $t^2 - 1$, $t + 2$, and $t - 3$ dependent or independent?

15. Let V be the vector space of all continuous functions mapping $\mathbf{R}^1$ into $\mathbf{R}^1$. Are the functions $f(t) = t^2$, $g(t) = t$, and $h(t) = e^t$ dependent or independent in V?

16. Are the matrices

$$\begin{pmatrix} 1 & 0 \\ -1 & 2 \end{pmatrix} \quad \text{and} \quad \begin{pmatrix} 2 & 1 \\ 3 & 4 \end{pmatrix}$$

dependent or independent in the vector space of all 2×2 matrices?

17.* Show that if V is an n-dimensional vector space and $S = \{\mathbf{v}_1, \mathbf{v}_2, \ldots, \mathbf{v}_n, \mathbf{v}_{n+1}\}$ is any collection of $n + 1$ vectors in V, then these vectors are linearly dependent. [*Hint*: Recall that all bases of V have precisely n vectors. Let $W = \{\mathbf{w}_1, \mathbf{w}_2, \ldots, \mathbf{w}_n\}$ be a basis for V and consider collections of vectors composed of W and vectors from S.]

Section B

1. Determine whether or not the following systems of equations have solutions; if so, find one solution and determine if this solution is unique.

(a) $x + y = 6$

  $3x - 2y + z = 1$

  $2x + y - z = 4$

(b) $x + y - z = 0$

  $2x - y + z = 1$

  $x - 2y + 3z = 2$

2. Find inverses (if they exist) of the following matrices:

(a) $\begin{pmatrix} 3 & -1 \\ 0 & 2 \end{pmatrix}$

(b) $\begin{pmatrix} 2 & 0 & 0 \\ 0 & 3 & 0 \\ 0 & 0 & 4 \end{pmatrix}$

(c) $\begin{pmatrix} 1 & 0 & -1 \\ 2 & 1 & 1 \\ 0 & 0 & 3 \end{pmatrix}$

3. Show that if A and B are nonsingular $n \times n$ matrices, then $(AB)^{-1}$ exists and is equal to $B^{-1}A^{-1}$.

4. Show that if A is an $m \times p$ matrix and B is a $p \times n$ matrix, then $(AB)^T = B^T A^T$. [Hint: Compare the (i,j)th entries of $(AB)^T$ and $B^T A^T$.]

5. Use problem B.4 to show that if A is a nonsingular matrix, then $(A^T)^{-1} = (A^{-1})^T$.

6. A matrix A is *symmetric* if $A^T = A$. Give an example of a symmetric matrix and show that a matrix A is symmetric if and only if $a_{ij} = a_{ji}$ for each i and j.

7. Show that if A is an $n \times n$ matrix, then AA^T is symmetric.

8. Show that if A is a nonsingular symmetric matrix, then A^{-1} is symmetric.

Section C*

1. Find the stiffness matrix of the following spring systems. Find the elasticity matrix of the spring system in (b):

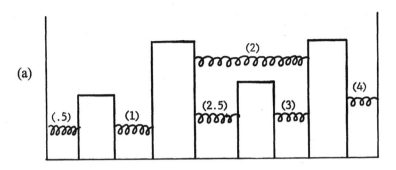

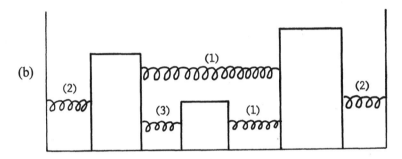

Figure 2.3

2. Given the stiffness matrix

$$\begin{pmatrix} -2.5 & 2 & 0 & 0.5 \\ 2 & -6 & 1 & 3 \\ 0 & 1 & -2.5 & 1.5 \\ 0.5 & 3 & 1.5 & 6 \end{pmatrix}$$

   for the system described in Section 2.C, find
   (a) The force acting on m_2 resulting from displacements $x_1 = 2$, $x_2 = 0$, $x_3 = -1$, $x_4 = 1$.
   (b) The force acting on m_3 resulting from the displacement of m_2 by 1 unit and of m_4 by 2 units.
3. Verify equation (3) on page 44.
4. Verify the last sentence in Section 2.C. This is known as the *Hardy-Weinberg law*.

Section D

1. Find solutions of the following equations that satisfy the given initial conditions:

  (a) $y' - 2y = t^2 + t,\ y(1) = 0$

  (b) $y' + 2ty = t,\ y(0) = 1$

  (c) $y' + \dfrac{1}{t} y = \dfrac{\sin t}{t},\ y(1) = 0$

2. Suppose that a cup of coffee has been warmed to a temperature of 175°. After 2 minutes in a room whose temperature is 65° the temperature of the coffee falls to 150°. Find how long it will take to bring the temperature of the coffee to 100°.

3. Suppose that a cup of 220° tea is placed in a room whose temperature is 60°. After 20 minutes, the temperature of the tea is 190°. At this time a heater is turned on in the room which raises the room temperature 2° per minute. What will be the temperature of the tea after 15 minutes?

4. It is frequently the case that a differential equation can be expressed in the form

$$y'f(y) = g(t)$$

(where y is a function of t). It then follows easily from the chain rule (5.A.15) that

$$\int g(t)\, dt + C = \int y'f(y)\, dt = \int f(y)\, dy$$

and this often permits calculation of y. For example, if $y' = -t/y$, then $y'y = -t$, and we have $\int y\, dy = \int -t\, dt + C$; thus $y^2/2 = (-t^2/2) + C$, and consequently, $y = \pm\sqrt{C - t^2}$. Use this procedure to solve the following differential equations:

  (a) $y' = t^3 y - t^3$

  (b) $t^2 yy' = e^y$

  (c) $y^2(t - 1)y' = -t^2(y + 1)$

  (d) $y' = \dfrac{1 + y}{1 + t}$

5. Use the procedure described in the previous problem (and discuss what goes wrong) to try to solve the following problem. Suppose that a certain colony of microorganisms increases at a rate proportional to the square of the number of microorganisms present in the colony. If at 10:00 a.m. there are 10,000 organisms and at 11:00 a.m. there are 20,000, how many will there be the following day at 10:00 a.m.?

6. A melting snowball has a radius of 3 inches at noon and a radius of 2 inches one hour later. If the rate of melting is proportional to the surface area of the snowball, find an expression for the radius as a function of time. When will the snowball be completely melted?

7. Find solutions to the following equations:
   (a) $2y'' - 3y' + y = 0 \qquad y(0) = 1 \qquad y'(0) = 0$
   (b) $4y'' - 4y' + y = 0$
   (c) $y'' + 9y = 0$
8. Verify that $\frac{1}{17}(3\cos t - 5\sin t)$ is a solution of $y'' - 3y' - 4y = 2\sin t$, and find the general solution to this equation.
9. Show that

$$\phi_1(t) = e^t \qquad \phi_2(t) = e^{-3t}$$

   are linearly independent solutions of $y'' + 2y' - 3y = 0$.
10. Determine if the following functions (with domain $\mathbf{R}^1$) are dependent or independent:
    (a) $\phi_1(x) = x \qquad \phi_2(x) = e^{ax}$
    (b) $\phi_1(x) = x \qquad \phi_2(x) = |x|$
    (c) $\phi_1(x) = e^{ix} \qquad \phi_2(x) = \cos x$
    (d) $\phi_1(x) = e^x \qquad \phi_2(x) = xe^x \qquad \phi_3(x) = x^2 e^x$
    (e) $\phi_1(t) = t^2 \qquad \phi_2(t) = t^2 - t \qquad \phi_3(t) = t^2 + t$
11. Suppose that ϕ_1 and ϕ_2 are solutions of

$$a(t)y''(t) + b(t)y'(t) + c(t)y(t) = 0$$

    (a) Show that $a(t)W'(t) + b(t)W(t) = 0$ for each t, where $W(t)$ is the Wronskian of ϕ_1 and ϕ_2.
    (b) Show that if $a(t) \neq 0$ for each t, then $W(t) = Ke^{-\int b(t)/a(t)\, dt}$ (Abel's formula). Note that this implies that $W(t)$ is either identically 0 or is never equal to 0.
    (c) Could the functions $\phi_1(t) = t^2$ and $\phi_2(t) = t|t|$ be solutions of the differential equation $y'' + p(t)y' + q(t)y = 0$, where p and q are continuous functions on $\mathbf{R}^1$?
12. Suppose that ϕ_1 and ϕ_2 are independent solutions of

$$y'' + by' + cy = 0$$

    Show that $W(t)$ is a constant for all t if and only if $b = 0$ [*Hint*: Recall that a function is constant if its derivative is 0.]
13.* Suppose that p and g are continuous functions on an interval (α, β), $t_0 \in (\alpha, \beta)$, and y_0 is an arbitrary real number. Use vector space theory to show that there is a unique solution to

$$y' + p(t)y = g(t)$$

    that satisfies the initial condition $y(t_0) = y_0$.
14. Let

$$a_n y^{(n)} + a_{n-1} y^{(n-1)} + \cdots + a_1 y + a_0 = 0 \qquad (*)$$

    be an nth-order differential equation with constant coefficients. Show that if $r_1, r_2, \ldots, r_n$ are distinct roots of the auxiliary equation

$$a_n r^n + a_{n-1} r^{n-1} + \cdots + a_1 r + a_0 = 0$$

then the functions $e^{r_1 t}$, $e^{r_2 t}$, ..., $e^{r_n t}$ form a fundamental set of solutions of (*). [*Hint:* To show that these functions are independent, suppose that $c_1 e^{r_1 t} + \cdots + c_n e^{r_n t} = 0$. Multiply this equation through by $e^{-r_1 t}$ and take the derivative of both sides of the new equation. Then multiply this result by $e^{-(r_2 - r_1)t}$ and again take derivatives. Continue in this manner to show that $c_n = 0$. Repeated applications of this procedure will show that $c_{n-1}, c_{n-2}, \ldots, c_1$ are all 0.]

15. Use problem D.14 to find a fundamental set of solutions of

$$y''' + 2y'' - y' - 2y = 0$$

16. (*Method of Variation of Parameters*). Let

$$a(t)y''(t) + b(t)y'(t) + c(t)y(t) = g(t)$$

be a second-order linear differential equation. To find a particular solution of this equation one may proceed as follows. We suppose that there is a solution of the form $\phi(t) = v_1(t)y_1(t) + v_2(t)y_2(t)$, where y_1 and y_2 form a fundamental set of solutions of the associated homogeneous equation

$$a(t)y''(t) + b(t)y'(t) + c(t)y(t) = 0 \qquad\qquad (**)$$

and the functions v_1 and v_2 are to be determined in the following manner. First assume that $v_1'(t)y_1(t) + v_2'(t)y_2(t) = 0$. Then substitute $\phi(t)$ into (**) to obtain (after suitable rearranging) $v_1'(t)y_1'(t) + v_2'(t)y_2'(t) = g(t)/a(t)$. This equation together with the assumed equation $v_1'(t)y_1(t) + v_2'(t)y_2(t) = 0$ permits calculation of v_1' and v_2' and, hence, of v_1' and v_2'. This in turn determines the solution ϕ.

17. Use the method of the variation of parameters to find particular solutions of

(a) $y'' + y = \tan t$

(b) $y'' - 3y' + 2y = \dfrac{-e^{2t}}{e^t + 1}$

(c) $t^2 y'' - ty' = e^t t^3$ (Is your solution valid for $t = 0$?)

(d) $y'' + 4y' + 4y = te^{2t}$

Section E

1. Determine whether the following mappings, $f: \mathbf{R}^2 \to \mathbf{R}^2$, are linear transformations:

(a) $f(x,y) = (x - y, x + y)$

(b) $f(x,y) = (x^2,y)$

(c) $f(x,y) = (x, y + 1)$

2. Determine whether the following functions, $f: \mathbf{R}^2 \to \mathbf{R}^3$, are linear transformations:

   (a) $f(x,y) = (x,y,2x)$

   (b) $f(x,y) = (x,0,y)$

   (c) $f(x,y) = (x,y^2,y)$

   (d) $f(x,y) = (x,0,0)$

   (e) $f(x,y) = (xy,0,0)$

3. Show that if $T: V \to W$ is a linear transformation, then $T(0) = 0$ and $T(-\mathbf{v}) = -T(\mathbf{v})$ for each $\mathbf{v} \in V$.

4. Show that $T: \mathbf{R}^2 \to \mathbf{R}^2$ defined by $T(x,y) = (-x,-y)$ is a linear transformation and give a geometrical interpretation of T.

5. Let α be any angle and define $T: \mathbf{R}^2 \to \mathbf{R}^2$ by

$$T\begin{pmatrix} x \\ y \end{pmatrix} = \begin{pmatrix} \cos\alpha & \sin\alpha \\ -\sin\alpha & \cos\alpha \end{pmatrix}\begin{pmatrix} x \\ y \end{pmatrix}$$

Show that T is a linear transformation and interpret T geometrically.

6. Let P_2 be the set of all polynomials of degree less than or equal to 2, and define $S: P_2 \to \mathbf{R}^1$ by $S(at^2 + bt + c) = \int_0^1 (at^2 + bt + c)\, dt$. Show that S is a linear transformation.

7. Show that all linear transformations from $\mathbf{R}^2$ into $\mathbf{R}^2$ have the form $T(x,y) = (ax + by, cx + dy)$. [*Hint*: Consider $T(1,0)$ and $T(0,1)$.]

8. Suppose that $T: V \to W$ is a linear transformation. The *kernel* of T, ker T, is defined to be the set $\{\mathbf{v} \in V \mid T(\mathbf{v}) = 0\}$. Show that ker T is a subspace of V.

9. Find the kernel of the linear transformations given in problems E.1(a), 2(a), 2(b), 4, and 6.

10. Suppose that $T: V \to W$ is a linear transformation. Show that the image of T, Im T, is a subspace of W.

11. It can be shown that if $T: V \to W$ is a linear transformation, then $\dim V = \dim(\ker T) + \dim(\operatorname{Im} T)$. Use this fact to show that if $T: V \to W$ is a linear transformation and if T is injective, then T is surjective, and if T is surjective, then T is injective. [*Hint*: Show that if V is a vector space of dimension n and if $S \subset V$ is a vector subspace of V also of dimension n, then $S = V$.]

12. Suppose that $\phi: P_2 \to P_2$ is a linear transformation and that $\phi(1) = t$, $\phi(t) = t^2 - 3$, $\phi(t^2) = 4$. Find $\phi(2t^2 - 2t + 1)$.

13. Suppose that the ordered bases $\{(1,0,0), (0,1,0), (0,0,1)\}$ and $\{(1,1), (0,2)\}$ are given for $\mathbf{R}^3$ and $\mathbf{R}^2$, respectively. Find the matrix associated with the linear transformation $T: \mathbf{R}^3 \to \mathbf{R}^2$ with respect to these bases if

    (a) $T(x,y,z) = (x + y, 3z)$

    (b) $T(x,y,z) = (x - y, y - z)$

14. Let $T: P_2 \to P_1$ be defined by $T(at^2 + bt + c) = 3at + c$. Find the matrix belonging to T with respect to the bases $\{1, t, 3t^2\}$ and $\{2, t\}$.

15. Compose a problem to illustrate Exercise (2.E.10)

REFERENCES

Boyce, W. E., and R. C. DiPrima (1977): *Elementary Differential Equations and Boundary Value Problems*, Wiley, New York.

Coddington, E. (1961): *An Introduction to Ordinary Differential Equations*, Prentice-Hall, Englewood Cliffs, N. J.

Cullen, C. (1966): *Matrices and Linear Transformations*, Addison-Wesley, Reading, Mass.

Curtis, C. W. (1968): *Linear Algebra: An Introductory Approach*, Allyn and Bacon, Boston.

3

LIMITS AND METRIC SPACES

A. PRELUDE

The notion of a *limit* is perhaps the most difficult concept that confronts the beginning student of mathematics. That there are initial difficulties in grasping either the idea or the import of limits is quite comprehensible when seen in a historical context. The Greek scholar Zeno of Elea (ca. 450 BC) once reported a now classic race between Achilles and a persistent but unknown tortoise which gives a hint of the problems encountered in dealing with limits and infinite processes. Since Achilles could run 10 times faster than even the fastest tortoise, he magnanimously (but perhaps mistakenly) agreed to give his slower adversary a 1-mile head start. Much to Achilles dismay, upon completing the first mile of the race he found that the tortoise had advanced an additional 0.1 mile; moreover, when Achilles reached the 1.1-mile marker, the tortoise was now 0.01 miles ahead of him, and after 1.11 miles the tortoise still maintained a slight lead of 0.001 miles. Is the defeat of Achilles inevitable? Only if one fails to take into account the nature of a limit.

In the 2000-odd years following this race the limit concept received scant clarification, and (not uncoincidentally) relatively little mathematical progress was made. Isaac Newton (1642–1727), favored with one of the keenest minds in the history of mankind, was among the first of the great mathematicians to truly grasp the significance of limits, and his contributions to calculus were (of

necessity) predicated on this idea. Still, he was never able to give an adequate
or concise explanation of the limit phenomenon, and his attempts to do so
strike the modern-day mathematician as being tortuously heroic yet pecu-
liarly imprecise. For example, we have the following excerpt from his masterwork, *Principia Mathematica* (1687):

> Ultimate ratios in which quantities vanish, are not, strictly speaking, ratios
> of ultimate quantities, but limits to which the ratio of these quantities,
> decreasing without limit, approach, and which, though they can come
> nearer than any given difference whatever, they can neither pass over nor
> attain before the quantities diminished indefinitely.

Although the seventeenth and eighteenth centuries produced an extraordinary number of mathematical geniuses, few were successful in giving
mathematical precision to profound ideas with which they were working. As
Morris Kline (1962) has pointed out:

> Some of the greatest mathematicians of the 18th century, Leonhard Euler
> and Joseph Louis Lagrange, worked on the problem of clarifying the calcu-
> lus, but without success. Both arrived at the conclusion that as it stood, the
> calculus was unsound, but that somehow errors were offsetting one another
> so that the results were correct. A more drastic opinion was offered by the
> mathematician Michel Rolle (1652–1719). He taught that the calculus was a
> collection of ingenious fallacies. Voltaire called the calculus "the art of
> numbering and measuring exactly a Thing whose existence cannot be con-
> ceived." Near the end of the 18th century the distinguished mathematician
> Jean le Rond d'Alembert (1717–1783) felt obliged to advise his students
> that they should persist in their study of the calculus; faith would even-
> tually come to them. All 18th-century attempts to supply rigorous founda-
> tions for the calculus failed.

Even as late as 1894 we have such gems of mathematical lucidity as the
Italian Vivant's observation that it is immaterial whether one calls the circle
the limit of a polygon as the sides are indefinitely decreased in length, or,
whether one looks upon the circle as a polygon with an infinite number of
infinitesimal sides. Other early mathematicians supported the view that if a
variable quantity maintains a specified property, then the limit of the variable
quantity will also have this property. The following example casts some doubt
on this interpretation of the limit (Prenowitz, 1953).

Consider an equilateral triangle whose sides are of unit length. Inside
this triangle, two more equilateral triangles are constructed (with side length
$\frac{1}{2}$); this process is repeated indefinitely as indicated in the Fig. 3.1. Note that
the polygonal paths ABC, $ADEFB$, $AGHIEJKLB$, . . . have the side AB as a
limit. However, the length of each such path is clearly 2, whereas the length of
the limiting side AB is 1.

The limit concept as it applies to derivatives and integrals was first

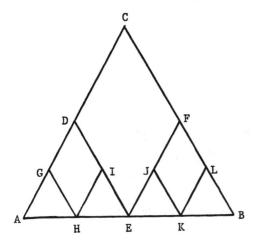

Figure 3.1

satisfactorily presented by Cauchy (1789–1857), and the use of ε and δ in defining limits (as is done today) is ascribed to Weierstrass (1815–1897). It is with the aid of the ε-δ definitions that we shall be able to make meaningful the study of limits of numbers, functions, matrices, etc., and give mathematical validity to expressions such as $\pi/4 = 1 - \frac{1}{3} + \frac{1}{5} - \frac{1}{4} + \cdots$.

The idea of a limit is inseparable from the notion of *closeness*. To say that an object x is the limit of a sequence of objects essentially means that these objects become increasingly close to x. Fine. But what is meant by close? The idea that two numbers are close is fairly clear. However, a mathematician must deal with a wide range of entities other than numbers. For instance, what do we mean if we assert that two vectors are close; are the vectors $(0,0.0001,0,0)$ and $(0,-0.0001,0,0)$ near one another even though they "point" in opposite directions? Moreover, the concept of two functions being close certainly requires a certain amount of elucidation. For example, we might say that two functions $f,g : [a,b] \to \mathbf{R}^1$ are close if $\int_a^b |f(x) - g(x)| \, dx$ is small; in this case we are measuring the area between the graphs of the functions, as indicated in Fig. 3.2. Note, however, as is seen from the next example, it is possible to construct a sequence of functions $f_1, f_2, \ldots$ such that $\int_a^b |f_n(x) - g(x)| \, dx < 1/n$ for each positive integer n, even though for each n, there is a point $x_n \in [a,b]$ such that $|f_n(x_n) - g(x_n)| > n$.

(3.A.1) *Example* For each $n \in \mathbf{Z}^+$, let $f_n : [0,1] \to \mathbf{R}^1$ be the function whose graph is indicated in Fig. 3.3. Let g be the constant function identically equal to 0. Then for each $n \in \mathbf{Z}^+$ $\int_0^1 (f_n(x) - g(x)) \, dx = 1/2n$; note, however, that if $x_n = 1/4n^2$, then $|f_n(x_n) - g(x_n)| = 2n$.

To obviate this problem we might say that a sequence of functions $f_1, f_2, \ldots$, each mapping the interval $[a,b]$ into $\mathbf{R}^1$, converges to a function g

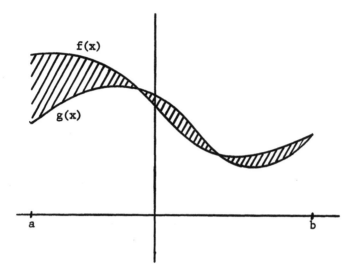

Figure 3.2

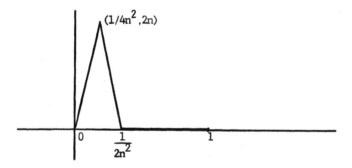

Figure 3.3

if for each $x \in [a,b]$, there is an integer N_x large enough so that for any integer n greater than N_x, $|f_n(x) - g(x)|$ is quite small. However, as is indicated in Fig. 3.4 this too leads to complications. Is the constant function $g(x) = 0$ really the "limit" of the sequence of functions $f_1, f_2, \ldots$?

It should now be apparent that in order to deal with the problem of closeness in a meaningful way, it is essential that we be able to assess distances not only between numbers, but between vectors, functions, and quite possibly other mathematical objects such as matrices as well. Thus, some theoretical considerations are in order to make precise the general notions of distance and limits. For this, we turn to metric spaces.

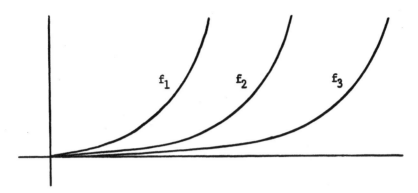

Figure 3.4

B. METRIC SPACES

We begin by considering the sequence $\{z_n\}$ in $\mathbf{R}^2$ defined by $z_1 = (5,3\frac{1}{2})$, $z_2 = (2\frac{1}{16},2\frac{7}{8})$, $z_3 = (-2,-3)$, $z_4 = (2\frac{1}{2},3)$, $z_5 = (4,1)$, and for each $n \geq 6$, $z_n = (2 + 1/\sqrt{n}, 3 - 1/2^n)$. The sequence $\{z_n\}$ clearly approaches (but in this case never reaches) the point $z = (2,3)$; z is called the *limit* of this sequence. Our first task is to translate the expression "clearly approaches z" into precise mathematical language. To say that $z = (2,3)$ is the limit of the sequence $\{z_n\}$ essentially means that "eventually" the terms of the sequence become arbitrarily close to z. This notion is given mathematical substance as follows. Given any number $\varepsilon > 0$ (think of ε as being a small positive real number), there is a suitably large positive integer, which we denote N_ε, such that for each integer $n \geq N_\varepsilon$ the (straight line) distance between the points z and z_n is less than ε. Thus, although the initial members $z_1, z_2, \ldots, z_{N_\varepsilon}$ of the sequence may be quite far from z, the remaining members of the sequence are within a distance of ε of the point z. In most cases, the smaller the number ε, the larger will be the integer N_ε. For instance, in the present example, if $\varepsilon = \frac{1}{3}$, then $N_\varepsilon = 10$ will do since for $n \geq 10$, the distance between z_n and $z = (2,3)$, given by

$$\sqrt{\left(2 + \frac{1}{\sqrt{n}} - 2\right)^2 + \left(3 - \frac{1}{2^n} - 3\right)^2} = \sqrt{\frac{1}{n} + \frac{1}{2^{2n}}}$$

is less than $\frac{1}{3}$. On the other hand, if $\varepsilon = \frac{1}{100}$, then $N_\varepsilon = 10$ will no longer suffice, but $N_\varepsilon = 100{,}000$ does have the desired property. The formal definition of the limit of a sequence in $\mathbf{R}^2$ is given next.

(3.B.1) **Definition** Suppose that $\{z_n\} = \{(x_n,y_n)\}$ is a sequence of points in $\mathbf{R}^2$. Then a point $z = (x,y)$ is said to be a *limit* of $\{z_n\}$ if for each $\varepsilon > 0$, there is a positive integer N (which may vary with ε) such that if $n \geq N$, then $\sqrt{(x_n - x)^2 + (y_n - y)^2} < \varepsilon$.

For $\mathbf{R}^1$ the definition of a limit takes on the following form.

(3.B.2) *Definition*   Suppose that $\{x_n\}$ is a sequence of points in $\mathbf{R}^1$. Then a point $x \in \mathbf{R}^1$ is a *limit* (or a *limit point*) of $\{x_n\}$ if for each $\varepsilon > 0$, there is a positive integer N (which may vary with ε) such that if $n \geq N$, then $|x_n - x| < \varepsilon$.

It should now be clear that the problem of defining limits of sequences in sets other than $\mathbf{R}^1$ and $\mathbf{R}^2$ will hinge on finding an appropriate way to determine distances. Suppose then that we are given an arbitrary set X of objects. How might we assign distances between elements of this set? Whatever distance function we choose for this task, it would be reasonable to insist that it satisfy the following rather minimal set of criteria:

1. The distance between distinct elements of X should be positive.
2. If $x \in X$, then the distance from x to itself is 0.
3. If $x,y \in X$, then the distance from x to y should be the same as that from y to x.
4. If $x,z \in X$, then the shortest route from x to z should be the direct one, i.e., for any other point $y \in X$ the distance from x to z should be less than or equal to the sum of the distances from x to y and y to z (see Fig. 3.5).

Although it is certainly possible to impose conditions other than these on a

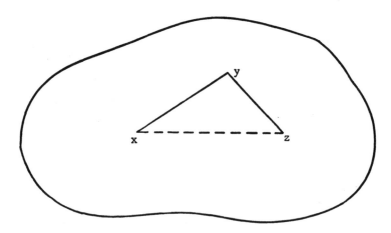

Figure 3.5

distance function, the criteria listed above are perhaps the simplest and least artificial. They lead to the following important definition.

(3.B.3) *Definition* Suppose that X is a nonempty set and $d: X \times X \to [0,\infty)$ is a function with the following properties:
  (i) $d(x,y) = 0$ if and only if $x = y$.
  (ii) $d(x,y) = d(y,x)$ for each $x,y \in X$.
  (iii) $d(x,z) \le d(x,y) + d(y,z)$ for each $x,y,z \in X$ (*triangle inequality*).
Then the pair (X,d) is called a *metric space*, and the mapping d is called the *distance function* or *metric* associated with the set X.

Note: The elements of X are commonly referred to as *points*, even though as we shall see in the following examples, points may take on a wide range of guises.

(3.B.4) *Examples* (a) Let $X = \mathbf{R}^1$ and let $d: \mathbf{R}^1 \times \mathbf{R}^1 \to [0,\infty)$ be defined by $d(x,y) = |x - y|$. The metric d is called the *usual metric* for $\mathbf{R}^1$ (some other possible metrics for $\mathbf{R}^1$ will be given later). It is trivial to verify that conditions (i) and (ii) of the definition (3.B.3) are satisfied. The triangle inequality follows from the fact that if x, y, and z are real numbers, then

$$|x - z| = |(x - y) + (y - z)| \le |x - y| + |y - z|$$

  (b) Let $X = \mathbf{R}^2$ and define $d: \mathbf{R}^2 \times \mathbf{R}^2 \to [0,\infty)$ by
$$d(\mathbf{x},\mathbf{y}) = \sqrt{(x_1 - y_1)^2 + (x_2 - y_2)^2},$$
where $\mathbf{x} = (x_1,x_2)$ and $\mathbf{y} = (y_1,y_2)$. The metric d measures the straight line distance between the points p and q. It is easy to check that conditions (i) and (ii) hold, but as is frequently the case, verification of the triangle inequality is somewhat more delicate. We shall do this in the more general example that follows.
  (c) Let $X = \mathbf{R}^n$ and define $d: \mathbf{R}^n \times \mathbf{R}^n \to [0,\infty)$ by
$$d(\mathbf{x},\mathbf{y}) = \sqrt{\textstyle\sum_{i=1}^{n} (x_i - y_i)^2},$$
where $\mathbf{x} = (x_1, x_2, \ldots, x_n)$ and $\mathbf{y} = (y_1, y_2, \ldots, y_n)$. This metric will be referred to as the *usual metric* for $\mathbf{R}^n$. That d satisfies the triangle inequality is a direct result of the following inequality.

Minkowski's Inequality

Suppose that $a_1, a_2, \ldots, a_n, b_1, b_2, \ldots, b_n$ are real numbers. Then

$$\sqrt{\sum_{i=1}^{n}(a_i + b_i)^2} \le \sqrt{\sum_{i=1}^{n} a_i^2} + \sqrt{\sum_{i=1}^{n} b_i^2}$$

Proof.

$$\sum_{i=1}^{n} (a_i + b_i)^2 = \sum_{i=1}^{n} (a_i^2 + 2a_ib_i + b_i^2)$$

$$= \sum_{i=1}^{n} a_i^2 + 2\sum_{i=1}^{n} a_ib_i + \sum_{i=1}^{n} b_i^2$$

By the Cauchy-Schwarz inequality (2.A.21) we have $\sum_{i=1}^{n} a_ib_i \leq \sqrt{\sum_{i=1}^{n} a_i^2}$ · $\sqrt{\sum_{i=1}^{n} b_i^2}$, and therefore,

$$\sum_{i=1}^{n} (a_i + b_i)^2 \leq \sum_{i=1}^{n} a_i^2 + 2\sqrt{\sum_{i=1}^{n} a_i^2}\sqrt{\sum_{i=1}^{n} b_i^2} + \sum_{i=1}^{n} b_i^2$$

$$= \left(\sqrt{\sum_{i=1}^{n} a_i^2} + \sqrt{\sum_{i=1}^{n} b_i^2}\right)^2$$

The desired inequality is now obtained by taking square roots of both ends of the preceding inequality.

To see how Minkowski's inequality implies the triangle inequality, let $\mathbf{x} = (x_1, x_2, \ldots, x_n)$, $\mathbf{y} = (y_1, y_2, \ldots, y_n)$, and $\mathbf{z} = (z_1, z_2, \ldots, z_n)$, and observe that if for each i, $1 \leq i \leq n$, we set $a_i = x_i - y_i$ and $b_i = y_i - z_i$, then it follows immediately that

$$d(x, z) = \sqrt{\sum_{i=1}^{n} (x_i - z_i)^2} = \sqrt{\sum_{i=1}^{n} (x_i - y_i + y_i - z_i)^2}$$

$$\leq \sqrt{\sum_{i=1}^{n} (x_i - y_i)^2} + \sqrt{\sum_{i=1}^{n} (y_i - z_i)^2} = d(x,y) + d(x,z)$$

(d)   Let X be an arbitrary nonempty set, and define $d(x,y) = 1$ if $x \neq y$, and $d(x,y) = 0$ if $x = y$. The reader may check that d is a metric. Note that this metric has the effect of spreading points apart. Thus, the set $\mathbf{R}^1$ with this metic is totally different from $\mathbf{R}^1$ with the usual metric. This metric is called the *discrete metric*, and the space (X,d) is called a *discrete* metric space. Although it is almost impossible to visualize this metric geometrically, it might be helpful to think of this metric as describing a telephone network, where each connection requires exactly one unit of time to complete.

(e)   Let $X = \{f \mid f: [0,1] \to \mathbf{R}^1, f \text{ is continuous}\}$. Define $d: X \times X \to [0,\infty)$ by $d(f,g) = \int_0^1 |f(x) - g(x)| \, dx$. The proof that d is a metric depends on certain basic properties of continuous functions and the Riemann integral, and will be deferred until Chapter 6, problem B.11.

(f)   Let $X = \mathbf{R}^n$ and for points $\mathbf{x} = (x_1, x_2, \ldots, x_n)$ and $\mathbf{y} = (y_1, y_2, \ldots, y_n)$ define $d(\mathbf{x},\mathbf{y}) = \max\{|x_i - y_i| \mid i = 1, 2, \ldots, n\}$. Thus, if $n = 2$, then the distance between the points indicated in Fig. 3.6 is 4. That d satisfies properties (i) and (ii) is clear. The triangle inequality is also satisfied since if

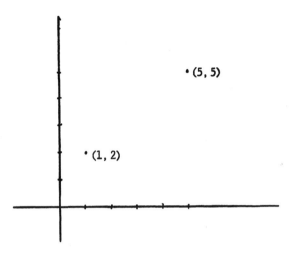

· (5, 5)

· (1, 2)

Figure 3.6

$\mathbf{x},\mathbf{y},\mathbf{z} \in \mathbf{R}^n$, then $d(\mathbf{x},\mathbf{z}) = \max\{|x_i - z_i| \,|i = 1, 2, \ldots, n\} \leq \max\{|x_i - y_i| \,|i = 1, 2, \ldots, n\} + \max\{|y_i - z_i| \,|i = 1, 2, \ldots, n\} = d(\mathbf{x},\mathbf{y}) + d(\mathbf{y},\mathbf{z})$.

(g) (*The Taxicab Metric*) Let $X = \mathbf{R}^2$ and define $d: \mathbf{R}^2 \times \mathbf{R}^2 \to [0,\infty)$ by $d(\mathbf{x},\mathbf{y}) = |x_1 - y_1| + |x_2 - y_2|$, where $\mathbf{x} = (x_1,x_2)$ and $\mathbf{y} = (y_1, y_2)$. The reader is again asked to check that d is a metric. Note that the distance between the two points indicated in Fig. 3.6 is 7.

(h) Let $X = \{f \,|\, f: Y \to \mathbf{R}^1, f \text{ is bounded}\}$. Here Y is an arbitrary fixed set; a function $f: Y \to \mathbf{R}^1$ is *bounded* if there is a positive number M such that $|f(y)| < M$ for all $y \in Y$. Define $d: X \times X \to [0,\infty)$ by $d(f,g) = \sup\{|f(y) - g(y)| \,|y \in Y\}$. The reader may verify easily that conditions (i) and (ii) of (3.B.3) are satisfied. To establish the triangle inequality observe that if $f,g,h \in X$, then

$$
\begin{aligned}
d(f,h) &= \sup\,\{|f(y) - h(y)| \,|y \in Y\} \\
&= \sup\,\{|(f(y) - g(y)) + (g(y) - h(y)|\,|\, y \in Y\} \\
&\leq \sup\,\{|f(y) - g(y)| \,|y \in Y|\} + \{\sup\,|g(y) - h(y)| \,|\, y \in Y\} \\
&= d(f, g) + d(g, h)
\end{aligned}
$$

The inequality in the above expression follows from (1.E.7.iv) (why is it not possible to replace $\leq$ with $=$?). This metric is called the *sup metric*.

(i) Let $X = \{A \,|\, A \text{ is an } m \times n \text{ matrix with real entries}\}$. Define $d: X \times X \to [0,\infty)$ by $d(A,B) = \max\{|a_{ij} - b_{ij}| \,|1 \leq i \leq m, 1 \leq j \leq n\}$. An argument completely analogous to that given in (f) can be used to show that d is a metric.

(j) If V is a vector space, then a *norm* for V is any function $\phi: V \to [0,\infty)$ satisfying the following conditions:

1. $\phi(\mathbf{v}) = 0$ if and only if $\mathbf{v} = \mathbf{0}$.
2. $\phi(\mathbf{v} + \mathbf{w}) \leq \phi(\mathbf{v}) + \phi(\mathbf{w})$ for all $\mathbf{v},\mathbf{w} \in V$.
3. $\phi(\alpha,\mathbf{v}) = |\alpha|\phi(\mathbf{v})$ for each $\alpha \in \mathbf{R}^1$ (or $\mathbf{C}$) and each $\mathbf{v} \in V$.

It is customary to denote $\phi(\mathbf{v})$ by $\|\mathbf{v}\|$. Basically, a norm measures the "distance" of a given vector to the "origin" or $\mathbf{0}$ vector. Functions such as ϕ: $\mathbf{R}^1 \to [0, \infty)$ defined by $\phi(\mathbf{x}) = |\mathbf{x}|$, ϕ: $\mathbf{R}^n \to [0, \infty)$ defined by $\phi(\mathbf{x}) = \max\{|x_i| \,| i = 1, 2, \ldots, n\}$, and ψ: $\mathbf{R}^n \to [0, \infty)$ defined by $\psi(\mathbf{x}) = (x_1^2 + x_2^2 + \cdots + x_n^2)^{\frac{1}{2}}$ where $\mathbf{x} = (x_1, x_2, \ldots, x_n)$, are norms. Other examples are given in the problem section. Every norm leads to a metric; if ϕ is a norm for a vector space V, then $d: V \times V \to [0, \infty)$, defined by $d(\mathbf{v},\mathbf{w}) = \|\mathbf{v} - \mathbf{w}\|$ is called the *metric associated with the norm* ϕ. That d is in fact a metric can be readily established by the reader [conditions (i) to (iii) of (3.B.3) follow from the above conditions 1 to 3, respectively].

   (k) Suppose that (X,d) is any metric space and that $A \subset X$. It is frequently of interest to treat A itself as a metric space. To do this we let $d_A = d|_{A \times A}$ be the restriction of d to $A \times A$. It is easily verified that d_A is a metric for A. We shall denote d_A by d, when it is clear from the context what is meant.

   (l) Suppose that (X,d) and $(Y,\hat{d})$ are metric spaces. Then the metrics d and $\hat{d}$ can be used to define various metrics for the product space $X \times Y$. Examples of such metrics include

$d_1: (X \times Y) \times (X \times Y) \to [0,\infty)$, where

$$d_1((x,y),(x',y')) = d(x,x') + \hat{d}(y,y')$$

$d_2: (X \times Y) \times (X \times Y) \to [0,\infty)$, where

$$d_2((x,y),(x',y')) = \max\{d(x,x'), \hat{d}(y,y')\}$$

$d_3: (X \times Y) \times (X \times Y) \to [0,\infty)$, where

$$d_3((x,y),(x',y')) = \sqrt{(d(x,x'))^2 + (\hat{d}(y,y'))^2}$$

(3.B.5) *Observation* If in the previous example $X = \mathbf{R}^1 = Y$, then $X \times Y$ is $\mathbf{R}^2$ and the metrics defined by d_1, d_2, and d_3, coincide with the taxicab metric, the metric given in (f), and the usual metric for $\mathbf{R}^2$, respectively. Proofs that the functions d_1, d_2, and d_3 are metrics parallel precisely the arguments used to establish that the functions given in (g), (f), and (b) are metrics.

   Although we shall encounter some additional examples of metric spaces in the chapters to come, the reader should already begin to appreciate the wide range of metric spaces that can be formed. A major advantage in passing to this level of abstraction is that any result that can be established for metric spaces in general will be valid for all specific examples of such spaces. Consequently, it will not be necessary to check out such special spaces as $\mathbf{R}^1$, subsets of $\mathbf{R}^3$, function spaces, etc., for properties that have been shown to hold for arbitrary metric spaces.

Metric spaces were first introduced in 1906 in the thesis of a French doctoral candidate in mathematics, Maurice Fréchet. Many mathematicians have continued the study of these spaces until the present day; in fact, generalizations of metric spaces, principally topological spaces, have evolved into one of the most active and significant research areas of mathematics.

Metric spaces provide us with an appropriate setting for the study of "closeness." We begin with the following definition, which generalizes the notion of an open interval in $\mathbf{R}^1$.

(3.B.6) *Definition* Suppose that (X,d) is a metric space, $x \in X$, and $\varepsilon > 0$. Then the *(open) ε-neighborhood about x* is the set $S_\varepsilon(x) = \{y \in X \mid d(x,y) < \varepsilon\}$. If we wish to emphasize the metric d, then we shall write $S_\varepsilon^d(x)$.

(3.B.7) *Examples* (a) If $X = \mathbf{R}^1$ and d is the usual metric for $\mathbf{R}^1$, then for any $x \in \mathbf{R}^1$, $S_\varepsilon(x)$ is the open interval $(x - \varepsilon, x + \varepsilon)$.

(b) If $X = \mathbf{R}^2$ and d is the usual metric for $\mathbf{R}^2$, then for each $x \in \mathbf{R}^2$, $S_\varepsilon(x)$ is an "open" disk with center at x. Note that boundary of the disk is not included. The radius of the disk is ε (Fig. 3.7).

(c) If (X,d) is any metric space with the discrete metric d, then if $x \in X$ and $\varepsilon > 1$, we have that $S_\varepsilon(x) = X$ and if $x \in X$ and $\varepsilon \le 1$, then $S_\varepsilon(x) = \{x\}$.

(d) If $X = \mathbf{R}^2$ and d is the metric given in (3.B.4.f), then for any $\mathbf{x} \in \mathbf{R}^2$, $S_\varepsilon(\mathbf{x})$ has the form shown in Fig. 3.8.

(e) If $X = \mathbf{R}^2$ and d is the metric given in (3.B.4.g), then for each $x \in X$, $S_\varepsilon(x)$ has the form shown in Fig. 3.9.

(f) *(French railway metric)* Let $X = \mathbf{R}^2$. If p and q are points of $\mathbf{R}^2$, define $d(p,q)$ to be the sum of the usual distance from p to $(0,0)$ and the distance from $(0,0)$ to q whenever p and q do not lie on the same line passing through the origin; otherwise let $d(p,q)$ be the usual distance from p to q.

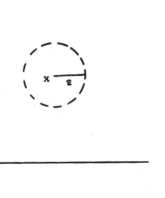

Figure 3.7

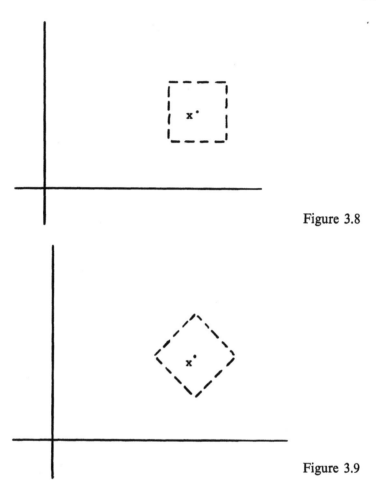

Figure 3.8

Figure 3.9

Describe, graphically, ε-neighborhoods for various points in $\mathbf{R}^2$ under this metric (Fig. 3.10).

(g) If $X = \{f: \mathbf{R}^1 \to \mathbf{R}^1 \,|\, f \text{ is bounded}\}$ is given the sup metric, then graphically an ε-neighborhood of a function f can be represented by a band of width 2ε (Fig. 3.11). Any function g whose graph lies entirely inside this band will be within a distance ε from f. The reader should compare these ε-neighborhoods with those arising from the metric described in (3.B.4.e).

We are now in a position to generalize definitions (3.B.1) and (3.B.2) with the following extremely important definition.

(3.B.8) *Definition* Suppose that (X,d) is a metric space, $\{x_n\}$ is a sequence of points in X, and that $x \in X$. Then the sequence $\{x_n\}$ *converges* to x if, given any positive number ε, there is a positive integer N (which may vary with ε)

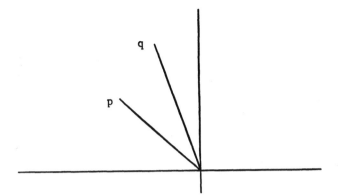

Figure 3.10

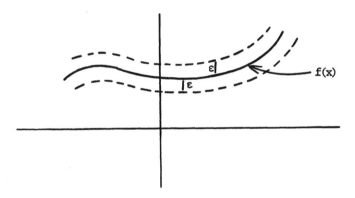

Figure 3.11

such that if $k \geq N$, then $x_k \in S_\varepsilon(x)$. The point x is called the *limit* of the sequence $\{x_n\}$. We shall frequently indicate that a sequence $\{x_n\}$ converges to a point x by writing $\{x_n\} \to x$ or $\lim_{n \to \infty} x_n = x$.

The definition says quite simply that a sequence will converge to a point x if it is eventually trapped in every ε-neighborhood of x; in most instances, the smaller the ε-neighborhood, the longer it will take before the sequence is eventually captured.

(3.B.9) *Exercise* Suppose that (X,d) is a metric space and that $\{x_n\}$ is a sequence in X. Show that $\{x_n\}$ converges to a point $x \in X$ if and only if the sequence $\{d(x_n,x)\}$ converges to 0 (in $\mathbf{R}^1$, with the usual metric).

(3.B.10) *Examples* (a) Let $X = \mathbf{R}^1$ and let d be the usual metric for $\mathbf{R}^1$. Let the sequence $\{x_n\}$ be defined by $x_n = 1/n$ for each $n \in \mathbf{Z}^+$. Then since for each given $\varepsilon > 0$, there is an integer N such that $1/N < \varepsilon$, it follows that for $n \geq N$, $d(x_n,0) = |1/n - 0| < \varepsilon$. Therefore this sequence converges to the point $x = 0$.

(b) Let $X = (0,1)$ and let d be the usual metric for X. Let $\{x_n\}$ be the sequence given in (3.B.10.a). Then this sequence fails to converge in X.

(c) Let $X = \mathbf{R}^1$ and let d be the discrete metric. Then the sequence in the two examples does not converge (consider the $\frac{1}{2}$-neighborhood of 0, $S_{\frac{1}{2}}(0)$; as indicated previously, this neighborhood consists of the single point $\{0\}$, and clearly the sequence $\{x_n\}$ is never in this neighborhood).

Examples (3.B.10.b) and (3.B.10.c) illustrate that convergence depends both on the underlying set and the metric.

(d) Let $X = \mathbf{R}^m$ and let d be the usual metric for $\mathbf{R}^m$. Let $\{\mathbf{x}_n\}$ be a sequence in X, where $\mathbf{x}_n = (x_1^{(n)}, \ldots, x_m^{(n)})$ for each n. Then $\{\mathbf{x}_n\}$ converges to a point $\mathbf{y} = (y_1, \ldots, y_m) \in \mathbf{R}^m$ if and only if for each fixed i, $1 \leq i \leq m$, the sequence $\{x_i^{(n)}\}_{n \in \mathbf{Z}^+}$ converges to y_i. In other words the sequence $\{\mathbf{x}_n\}$ converges if and only if it converges coordinatewise. Diagramatically,

$$\mathbf{x}_1 = (x_1^{(1)}, x_2^{(1)}, \ldots, x_m^{(1)})$$
$$\mathbf{x}_2 = (x_1^{(2)}, x_2^{(2)}, \ldots, x_m^{(2)})$$
$$\mathbf{x}_3 = (x_1^{(3)}, x_2^{(3)}, \ldots, x_m^{(3)})$$
$$\vdots \quad \vdots \quad \vdots \qquad \vdots$$
$$\downarrow \quad \downarrow \quad \downarrow \qquad \downarrow$$
$$\mathbf{y} = (y_1, \quad y_2, \ldots\ldots, y_m)$$

This result may be seen as follows. Suppose that $\{\mathbf{x}_n\} \to \mathbf{y}$ and that $\varepsilon > 0$ is given. We show that for each integer i, $1 \leq i \leq m$, the sequence $\{x_i^{(n)}\}_{n \in \mathbf{Z}^+}$ converges to y_i. Since $\{\mathbf{x}_n\} \to \mathbf{y}$, there is a positive integer N such that for $n \geq N$

$$d(\mathbf{x}_n,\mathbf{y}) = \sqrt{(x_1^{(n)} - y_1)^2 + \cdots + (x_m^{(n)} - y_m)^2} < \varepsilon$$

Since for each i, $1 \leq i \leq m$,

$$|x_i^{(n)} - y_i| \leq \sqrt{(x_1^{(n)} - y_1)^2 + \cdots + (x_m^{(n)} - y_m)^2}$$

it follows that the sequence $\{x_i^{(n)}\}_{n \in \mathbf{Z}^+}$ converges to y_i.

Conversely, suppose that for each i, the sequence $\{x_i^{(n)}\}_{n \in \mathbf{Z}^+}$ converges to y_i. To see that this implies that the sequence $\{\mathbf{x}_n\}$ converges to $\mathbf{y}$, first note that there is a positive integer N such that if $n \geq N$, then $|x_i^{(n)} - y_i| < \varepsilon/\sqrt{m}$ for each integer i, $1 \leq i \leq m$. Therefore, if $n \geq N$, then

$$d(\mathbf{x}_n,\mathbf{y}) = \sqrt{(x_1^{(n)} - y_1)^2 + \cdots + (x_m^{(n)} - y_m)^2}$$
$$\leq \sqrt{\varepsilon^2/m + \cdots + \varepsilon^2/m} = \sqrt{\varepsilon^2} = \varepsilon$$

and, hence, we have $\{\mathbf{x}_n\} \to \mathbf{y}$.

(e) Any bounded increasing sequence $\{x_n\}$ in $\mathbf{R}^1$ converges to sup $\{x_n \mid n \in \mathbf{Z}^+\}$. To see this, let $x^* = \sup\{x_n \mid n \in \mathbf{Z}^+\}$ and let $\varepsilon > 0$ be given. By the definition of x^*, there is a positive integer N such that $|x^* - x_N| = x^* - x_N < \varepsilon$. Since $\{x_n\}$ is an increasing sequence, it follows that if $n \geq N$, then $x_N \leq x_n \leq x^*$, and hence, $|x^* - x_n| < \varepsilon$; thus, $\{x_n\}$ converges to x^*. A similar statement obviously holds for bounded decreasing sequences.

(f) Let $X = \mathbf{R}^1$ and let d be the usual metric for $\mathbf{R}^1$. Let $\{x_n\}$ be the sequence defined by $x_n = (-1)^n(1 + 1/n)$ for each $n \in \mathbf{Z}^+$. This sequence fails to converge to either 1 or -1 (or for that matter, to any point).

(g) Let $X = \mathbf{R}^2$ and let $\{z_n\}$ be a sequence, where $z_n = (1 + 1/n, 2 + 1/n)$. Note that if we use any of the metrics given in Examples (3.B.4.b), (3.B.4.f), and (3.B.4.g), then this sequence converges to $z = (1,2)$.

(h) Suppose that Y is an arbitrary set and that $X = \{f: Y \to \mathbf{R}^1 \mid f$ is bounded$\}$. If X is given the sup metric, then whenever a sequence $\{f_n\}$ in X converges to a function $f \in X$, we have that given $\varepsilon > 0$, there is a positive integer N such that $|f_n(y) - f(y)| < \varepsilon$ for each $n \geq N$ and *each* $y \in Y$. Since the same integer N works for every $y \in Y$, this convergence is often referred to as *uniform convergence*. The reader should observe that if a sequence $\{f_n\}$ converges to a function f uniformly, then for each $y \in Y$, $\mathrm{Lim}_{n \to \infty} f_n(y) = f(y)$. To see that the converse of this observation does *not* hold, suppose that $Y = [0,1]$ and that for each n, $f_n: [0,1] \to \mathbf{R}^1$ is defined by $f_n(y) = y^n$. Suppose further that $f: [0,1] \to \mathbf{R}^1$ is defined by

$$f(y) = \begin{cases} 0 & \text{if } y \neq 1 \\ 1 & \text{if } y = 1 \end{cases}$$

Then for each $y \in [0,1]$, we have $\lim_{n \to \infty} f_n(y) = f(y)$, however, the sequence $\{f_n\}$ does not converge uniformly to f (see problem 25). (We say in this case that the sequence $\{f_n\}$ *converges pointwise* to f.)

The following theorem gives some intuitively obvious but important properties of sequences of real numbers.

(3.B.11) *Theorem* Suppose that $\{x_n\}$ and $\{y_n\}$ are sequences of real numbers that converge to x and y, respectively. Then:

(i) The sequence $\{x_n + y_n\}$ converges to $x + y$.
(ii) The sequence $\{x_n y_n\}$ converges to xy.
(iii) If $y_n \neq 0$ for each n and $y \neq 0$, then the sequence $\{x_n/y_n\}$ converges to x/y.

Proof. (i) To establish this result we have only the definition of convergence at our disposal. Let $\varepsilon > 0$ be given; we are to find a positive integer N such that if $n \geq N$, then $|(x_n + y_n) - (x + y)| < \varepsilon$. Since $\{x_n\} \to x$, there is a positive integer N_1 such that $|x_n - x| < \varepsilon/2$ whenever $n \geq N_1$, and similarly, since $\{y_n\} \to y$, there is a positive integer N_2 such that $|y_n - y| < \varepsilon/2$ whenever $n \geq N_2$. Let $N = \max\{N_1, N_2\}$. Then for $n \geq N$ we have

$$|(x_n + y_n) - (x + y)| = |(x_n - x) + (y_n - y)|$$

$$\leq |x_n - x| + |y_n - y| < \frac{\varepsilon}{2} + \frac{\varepsilon}{2} = \varepsilon$$

(ii) This part is trickier. Again let $\varepsilon > 0$ be given. As in part (i), we shall force x_n close enough to x and y_n close enough to y so that $|x_n y_n - xy|$ is less than ε. To do this we first rewrite $|x_n y_n - xy|$ as follows:

$$|x_n y_n - xy| = |x_n y_n - x_n y + x_n y - xy| = |x_n(y_n - y) + y(x_n - x)|$$

$$\leq |x_n| \, |y_n - y| + |y| \, |x_n - x|$$

We need to find a positive integer N such that if $n \geq N$, then

$$|x_n| \, |y_n - y| + |y| \, |x_n - x| < \varepsilon$$

Since $\{x_n\} \to x$ and $\{y_n\} \to y$, we can exercise control over $|x_n - x|$ and $|y_n - y|$; furthermore, multiplication of $|x_n - x|$ by the fixed number $|y|$ presents no problems. How do we deal with $|x_n|$? Since $\{x_n\} \to x$, it follows that there is an integer M such that $|x_n| < M$ for each n, i.e., the sequence $\{x_n\}$ is bounded (why?). To obtain the desired result, we now choose N so large that if $n \geq N$, then

$$|y_n - y| < \frac{\varepsilon}{2M}$$

$$|x_n - x| < \frac{\varepsilon}{2|y| + 1}$$

Thus, if $n \geq N$, we have

$$|x_n y_n - xy| \leq M|y_n - y| + |y| \, |x_n - x| < M\frac{\varepsilon}{2M} + |y|\frac{\varepsilon}{2|y| + 1} < \varepsilon$$

(iii) This part is also reasonably tricky. First we write

$$\left| \frac{x_n}{y_n} - \frac{x}{y} \right| = \left| \frac{x_n y - y_n x}{y_n y} \right| = \left| \frac{x_n y - xy + xy + y_n x}{y_n y} \right|$$

$$\leq \left| \frac{x_n - x}{y_n} \right| + \frac{|x| \, |y - y_n|}{|y_n| \, |y|}$$

The remaining portion of the proof is analogous in spirit to the proof of (ii) and is left to the reader (problem B.24); it is helpful to note that there is a positive constant r such that for large values of n, $|y_n| > r > 0$.

Let us now consider (3.B.10.f) in somewhat more detail. In this example we do not have convergence of the sequence $\{x_n\}$; nevertheless, the points 1 and -1 are of particular interest, since arbitrarily near each of these points there are an infinite number of members of the sequence. Such points are commonly (and rather appropriately) referred to as *cluster points* of the sequence. The formal definition of a cluster point is as follows.

(3.B.12) *Definition* Suppose that $\{x_n\}$ is a sequence in a metric space (X,d). Then a point $x \in X$ is called a *cluster point* of $\{x_n\}$ if for each $\varepsilon > 0$ and each integer $N \in \mathbf{Z}^+$, there is an integer $n \geq N$ such that $d(x_n,x) < \varepsilon$.

(3.B.13) *Examples* (a) If the sequence $\{x_n\}$ is defined by 1, $1 + \frac{1}{2}$, 2, $2 + \frac{1}{4}$, 3, $1 + \frac{1}{8}$, $2 + \frac{1}{8}$, $3 + \frac{1}{8}$, 4, $1 + \frac{1}{16}$, $2 + \frac{1}{16}$, $3 + \frac{1}{16}$, $4 + \frac{1}{16}$, 5, . . . , then each positive integer is a cluster point of $\{x_n\}$.

    (b) Let $\mathbf{Q}$ be the set of rational numbers and let $f: \mathbf{Z}^+ \to \mathbf{Q}$ be any 1-1, onto function. Then the sequence $\{x_n\}$ defined by $f(n) = x_n$ has an infinite number of cluster points in $\mathbf{R}^1$. In fact, every real number is a cluster point of $\{x_n\}$.

    (c) Suppose that $\{x_n\}$ is any sequence in the closed interval $[0,1]$ with the usual metric. We show that $\{x_n\}$ must have at least one cluster point. Divide $[0,1]$ into two subintervals, $[0,\frac{1}{2}]$ and $[\frac{1}{2},1]$, of equal length. Note that at least one of these subintervals has the property that x_n belongs to this interval for an infinite number of n. Let $A_1 = [a_1,b_1]$ denote this interval (if both subintervals have this property, select either one). Next subdivide $[a_1,b_1]$ into two subintervals of equal length and let A_2 be either one of these two subintervals with the property that $\{n \in \mathbf{Z}^+ \mid x_n \in A_2\}$ is infinite. Suppose now that intervals $[a_1,b_1], [a_2,b_2], \ldots, [a_k,b_k]$ have been so chosen. Define the closed interval $[a_{k+1},b_{k+1}]$ as follows. Subdivide $[a_k,b_k]$ into two subintervals (of length $1/2^{k+1}$), and let A_{k+1} be either of these two subintervals such that $\{n \in \mathbf{Z}^+ \mid x_n \in [a_{k+1},b_{k+1}]\}$ is infinite (Fig. 3.12). Thus, inductively, we have constructed a sequence of closed intervals $[a_1,b_1] \supset [a_2,b_2] \supset \ldots$ with the property that for each k, $\{n \in \mathbf{Z}^+ \mid x_n \in [a_k,b_k]\}$ is infinite. Let $w = \sup\{a_n \mid n \in \mathbf{Z}^+\}$ and $z = \inf\{b_n \mid n \in \mathbf{Z}^+\}$. We show that $w = z$, and that $w (= z)$ is a cluster point of the sequence $\{x_n\}$. First observe that if $j,k \in \mathbf{Z}^+$ and if $j < k$, then $a_j \leq a_{j+1} \leq \cdots \leq a_k \leq b_k$, and if $k < j$, then $a_j < b_j \leq$

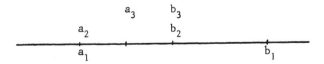

Figure 3.12

$b_{j-1} \leq b_{j-2} \leq \cdots \leq b_k$. Thus, for any pair of integers j,k, we have that $a_j < b_k$, and hence, by (1.E.7.iii), $w = \sup\{a_n \mid n \in \mathbf{Z}^+\} \leq z = \inf\{b_n \mid n \in \mathbf{Z}^+\}$. Since the sequence $\{b_n - a_n\}$ clearly converges to 0, it follows that $w = z$. That $w \; (= z)$ is a cluster point may be seen as follows. Given $\varepsilon > 0$, there is an integer N such that $1/N < \varepsilon/2$, $|a_N - w| < \varepsilon$, and $|w - b_N| < \varepsilon$; consequently $[a_N, b_N] \subset S_\varepsilon(w)$, and by our construction, there is an $n > N$ such that $x_n \in [a_N, b_N] \subset S_\varepsilon(w)$. Therefore, w is a cluster point for the sequence $\{x_n\}$.

It is perhaps surprising (at this juncture, anyway) that the property of [0,1] described in the foregoing discussion will be of considerable importance. In fact, cluster points will soon play a significant role in our development of continuous functions.

(3.B.14) *Exercise* Suppose that (X,d) is a metric space and that a sequence $\{x_n\}$ in X converges to a point $x \in X$. Show that x is the unique cluster point of $\{x_n\}$.

Example (3.B.10.g) also merits special consideration. We saw that in this example convergence of the sequence did not depend on which of the three metrics we used (of course, if we had thrown in the discrete metric the situation would have changed considerably). In other words, the shape of the ε-neighborhoods was of no particular significance in determining whether or not a sequence converged. This observation leads us to formulate a slight generalization of the ε-neighborhood, the open set. Open sets usually turn out to be more manageable than the more restrictive notion of ε-neighborhoods.

(3.B.15) *Definition* Suppose that (X,d) is a metric space. Then a subset A in X is *open* in X if for each $x \in A$, there exists an $\varepsilon > 0$, such that $S_\varepsilon(x) \subset A$.

(3.B.16) *Observations* (i) If (X,d) is a metric space, then X itself is open (in X).

(ii) Suppose that (X,d) is a metric space and that $\varepsilon > 0$. Then $S_\varepsilon(x)$ is open in X. We can see this as follows. If $y \in S_\varepsilon(x)$, we have $d(x,y) = r < \varepsilon$. Let $\hat{\varepsilon} = \varepsilon - r$. Then $S_{\hat{\varepsilon}}(y) \subset S_\varepsilon(x)$ for if $z \in S_{\hat{\varepsilon}}(y)$, we have by the triangle inequality $d(z,x) \leq d(z,y) + d(y,x) < \hat{\varepsilon} + r = \varepsilon - r + r = \varepsilon$ (Fig. 3.13).

(iii) If $\{A_\alpha \mid \alpha \in J\}$ is a family of open sets of a metric space (X,d), then $\bigcup_{\alpha \in J} A_\alpha$ is also open. This is clear since if $x \in \bigcup_{\alpha \in J} A_\alpha$, then there is an $\alpha^* \in J$ such that $x \in A_{\alpha^*}$. But since A_{α^*} is open, there exists $\varepsilon > 0$ such that $S_\varepsilon(x) \subset A_{\alpha^*} \subset \bigcup_{\alpha \in J} A_\alpha$. Note that as a consequence of this example and the previous one, it follows that a subset A of a metric space is open in X if and only if A is a union of ε-neighborhoods.

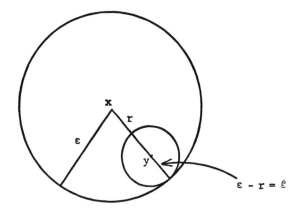

$$\varepsilon - r = \hat{\varepsilon}$$

Figure 3.13

(iv)' Observe that open sets in a metric space X depend very heavily on the set X itself. For instance, the set $[2,5)$ is not open in $X = \mathbf{R}^1$, but is open in $X = [2,10)$.

The following theorem shows how ε-neighborhoods may be replaced by open sets.

(3.B.17) Theorem Suppose that (X,d) is a metric space and that $\{x_n\}$ is a sequence in X. Then $\{x_n\}$ converges to a point $x \in X$ if and only if for each open set U that contains x, there is an integer N (which usually depends on U) such that $x_i \in U$ for each $i > N$.

Proof. Suppose that $\{x_n\}$ converges to x and that U is an open set containing x. Since U is open, there is an $\varepsilon > 0$ such that $S_\varepsilon(x) \subset U$. By Definition (3.B.15), there is a positive integer N such that if $n \geq N$, then $x_n \in S_\varepsilon(x) \subset U$.

The converse is even easier since by (3.B.16.ii), each ε-neighborhood $S_\varepsilon(x)$ is an open set.

Note that it follows from this theorem that for convergence problems, ε-neighborhoods can be replaced by open sets. Hence, the next definition is of some interest.

(3.B.18) Definition Two metrics d and $\hat{d}$ on a set X are *equivalent* provided that a set $A \subset X$ is open under the metric d if and only if A is open under the metric $\hat{d}$.

By (3.B.17) we see that if d and $\hat{d}$ are equivalent metrics for a set X, then a sequence $\{x_n\}$ converges to a point x in X under the metric d if and only if it converges to x under the metric $\hat{d}$. The next theorem gives a easily applied condition for checking whether two metrics are equivalent.

(3.B.19) Theorem Suppose that d and $\hat{d}$ are metrics for a set X. Then d and $\hat{d}$ are equivalent if for any $r > 0$, there exists $s > 0$ such that $S_s^{\hat{d}}(x) \subset S_r^d(x)$ and given $r' > 0$, there exists $s' > 0$ such that $S_{s'}^d(x) \subset S_{r'}^{\hat{d}}(x)$.

Proof. Suppose that U is open in (X,d) and that $x \in U$. Then there is an $r > 0$ such that $S_r^d(x) \subset U$. By hypothesis there is an $s > 0$ such that $S_s^{\hat{d}}(x) \subset S_r^d(x) \subset U$. Therefore U is (by definition) open in $(X,\hat{d})$. In a similar fashion it is shown that if U is open in $(X,\hat{d})$, then U is open in (X,d). Hence, the metrics d and $\hat{d}$ are equivalent.

(3.B.20) Exercise Show that the metrics given in Examples (3.B.4.b), (3.B.4.f), and (3.B.4.g) are all equivalent.

A *bounded metric* for a set X is any metric d with the property that for some positive number M, $d(x,y) < M$ for each $x,y \in X$. In the next theorem we see that all metrics are equivalent to bounded metrics.

(3.B.21) Theorem Suppose that (X,d) is a metric space. Then there is a bounded metric d^* for X that is equivalent to d.

Proof. Define $d^*(x,y) = \min\{d(x,y),1\}$. That d^* is a bounded metric for X is an easy exercise (see problem 3). To show that d and d^* are equivalent it suffices to observe that if $0 < r < 1$, then for each $x \in X$, $S_r^d(x) = S_r^{d^*}(x)$, and hence, by (3.B.19), d and d^* generate the same open sets.

PROBLEMS

1. Determine whether or not the following spaces are metric spaces:
   (a) $X = \{a,b,c\}$; $d: X \times X \to [0,\infty)$ is defined by $d(a,b) = d(b,a) = 1$, $d(b,c) = d(c,b) = 2$, $d(a,c) = d(c,a) = 3$, $d(a,a) = d(b,b) = d(c,c) = 0$.
   (b) $X = \mathbf{R}^1 \setminus \{0\}$; $d: X \times X \to [0,\infty)$ is defined by $d(x,y) = |1/x - 1/y|$.
   (c) $X = \mathbf{R}^2$; $d: X \times X \to [0,\infty)$ is defined by

$$d((x_1,y_1), (x_2,y_2)) = \begin{cases} \frac{1}{2} & \text{if } x_1 = x_2, y_1 \neq y_2, \text{ or } y_1 = y_2, x_1 \neq x_2 \\ 1 & \text{if } x_1 \neq x_2 \text{ and } y_1 \neq y_2 \\ 0 & \text{otherwise} \end{cases}$$

(d) $X = \mathbf{R}^1$; $d: X \times X \to [0, \infty)$ is defined by $d(x,y) = |x^2 - y^2|$.

2. Determine the nature of the ε-neighborhoods for those examples in problem 1 that are metric spaces.

3. Show that the function d^* given in (3.B.21) is a metric.

4. (a) Show that the following functions are norms for $\mathbf{R}^n$:

    (i) $\phi: \mathbf{R}^n \to \mathbf{R}^1$ $\phi(x_1, x_2, \ldots, x_n)$
                          $= \max\{|x_1|, |x_2|, \ldots, |x_n|\}$

    (ii) $\phi: \mathbf{R}^n \to \mathbf{R}^1$ $\phi(x_1, x_2, \ldots, x_n) = |x_1| + \cdots + |x_n|$

    (iii) $\phi: \mathbf{R}^n \to \mathbf{R}^1$ $\phi(x_1, x_2, \ldots, x_n) = \sqrt{x_1^2 + \cdots + x_n^2}$

   (b) Let X be a set and let V be the vector space $\{f: X \to \mathbf{R}^1 \mid f$ is bounded$\}$ with the usual operations. Show that $\phi: V \to [0,\infty)$ defined by $\phi(f) = \sup\{|f(x)| \mid x \in X\}$ is a norm for V.

5. Determine whether the following sequences in $\mathbf{R}^1$ (usual metric) are convergent, and if so, find their limits:

   (a) $\{x_n\}$, where for each n,

$$x_n = \frac{3n^2 - 2n + 6}{2n^2 + 4n - 2}$$

   (b) $\{y_n\}$, where

$$y_n = \begin{cases} \dfrac{1}{n} & \text{if } n \text{ is odd} \\ 2 - \dfrac{1}{n} & \text{if } n \text{ is even} \end{cases}$$

6. Let $\{x_n\}$ be a bounded decreasing sequence in $\mathbf{R}^1$. Show that $\{x_n\}$ converges to $\inf\{x_n \mid n \in \mathbf{Z}^+\}$.

7. Suppose that $0 < a < 1$. Show that the limit of the sequence $\{x_n\}$, where $x_n = a^n$ for each n converges to 0. [*Hint*: Observe that $\{x_n\}$ is a decreasing sequence and apply the preceding problem.]

8. Let a be a real number and for each n, let $u_n = 1 + a + a^2 + \cdots + a^n$. For what values of a does the sequence $\{u_n\}$ converge. [*Hint*: Recall that $u_n = (1 - a^{n+1})/(1 - a)$ if $a \neq 1$.]

9. In $\mathbf{R}^2$ let the sequence $\{u_n\}$ be defined by $u_n = (1/n, 3 - 1/n)$. For which of the metrics for $\mathbf{R}^2$ described in Sec. 3.B does this sequence converge?

10.* Use L'Hospital's rule (see Chapter 5, if necessary) to show that if for each n, $f_n: [0,1] \to \mathbf{R}^1$ is defined by $f_n(x) = n^2 x(1 - x)^n$, then the sequence $\{f_n\}$ converges pointwise to the constant function $f(x) \equiv 0$. Show that the sequence $\{\int_0^1 f_n(x)\, dx\}$ converges to 1, but that the sequence $\{f_n\}$ fails to converge in the metric defined by (3.B.4.e). (*Hint*: Make the substitution $u = 1 - x$.)

11.* Show that the sequence $\{f_n\}$ defined in the preceding problem fails to converge in the metric space (X,d), where $X = \{f: [0,1] \to \mathbf{R}^1 \mid f$ is bounded$\}$ and d is the sup metric. [*Hint*: For each n, let $x_n = 1/(1 + n)$; show that $\lim_{n \to \infty} f_n(x_n) = \infty$.]

12. Show that if (X,d) is a metric space and $\{x_n\}$ is a sequence in X that converges to points x and y, then $x = y$. Thus, the limit of a sequence (if it exists) is unique.

13. Show that if a sequence $\{x_n\}$ in a metric space (X,d) converges to a point x, then given $\varepsilon > 0$, there is a positive integer N_ε such that if $m \geq N_\varepsilon$ and $n \geq N_\varepsilon$, then $d(x_m,x_n) < \varepsilon$.

14. Show that every bounded sequence in $\mathbf{R}^1$ (usual metric) has a cluster point.

15.* Show that every bounded sequence in $\mathbf{R}^2$ (usual metric) has a cluster point.

16. What sequences in a discrete metric space have cluster points?

17. Let $\mathbf{Q}$ be the set of rational mumbers and suppose that $f: \mathbf{Z}^+ \to \mathbf{Q}$ and $g: \mathbf{Z}^+ \to \mathbf{Q}$ are arbitrary bijections. For each n, let $u_n = (f(n),g(n))$. Find all cluster points of the sequence $\{u_n\}$ in $\mathbf{R}^2$ (usual metric).

18. For each n, let $f_n: (0,1) \to \mathbf{R}^1$ be defined by $f_n(x) = x/(1 + nx)$. Show that the sequence $\{f_n\}$ converges in the set $X = \{f: (0,1) \to \mathbf{R}^1 \mid f$ is bounded$\}$, with the sup metric.

19. Suppose that (X,d) is a metric space, $x \in X$ and $r > 0$. Show that $Y = \{y \in X \mid d(x,y) > r\}$ is an open set.

20. Describe the nature of open sets in $[a,b]$ (usual metric).

21. Suppose that (X,d) is a metric space. Show that finite intersections of open sets in X are open. Does the same hold true for infinite intersections?

22. Suppose that (X,d) is a metric space. If A and B are subsets of X, then the *distance between A and B*, $\rho(A,B)$, is defined to be $\inf\{d(x,y) \mid x \in A, y \in B\}$. True or false: $\rho(A,B) = 0$ if and only if $A \cap B \neq \varnothing$.

23.* Let (X,d) be a metric space and let Y be the power set of X, $\mathscr{P}(X)$. If $C,D \in Y$, let $d_C(D) = \sup\{d(x,C) \mid x \in D\}$, where $d(x,C) = \inf\{d(x,y) \mid y \in C\}$. Define $\rho: Y \times Y \to [0,\infty)$ by $\rho(A,B) = \max\{d_A(B), d_B(A)\}$. Show that (Y,ρ) is a metric space and interpret ρ geometrically. If $x,y \in X$, compare $d(x,y)$ and $\rho(\{x\},\{y\})$.

24. Complete the proof of (3.B.11.iii).

25. Show that the sequence $\{f_n\}$ defined in (3.B.10.h) does not converge uniformly to f.

REFERENCES

Kline, Morris (1962): *Mathematics: A Cultural Approach*, Addison-Wesley, Reading, Mass.
Prenowitz, Walter (1953): *Amer. Math. Monthly*, **60**: 32.

4

CONTINUITY, COMPACTNESS, AND CONNECTEDNESS

A. CONTINUITY

In this chapter many new ideas are introduced and studied. The reader is advised to give special attention to definitions; not only should every definition be memorized, but sufficient time should be spent to ensure that a solid intuitive grasp of each new concept is attained.

In a very general way, continuous functions bear the same relationship to metric spaces that linear transformations do to vector spaces. We have seen that linear transformations are of special importance because they take the algebraic structure of vector spaces into account. In metric spaces the focus is on convergence properties of sequences rather than on algebraic considerations, and therefore, emphasis is given to those functions that take these properties into account; such functions are called *continuous functions*.

(4.A.1) *Definition* Suppose that (X,d) and $(Y,\hat{d})$ are metric spaces, $f: X \rightarrow Y$, and $x \in X$. The function f is said to be *continuous* at x if whenever a sequence of points $\{x_n\}$ in X converges to x, then the sequence of points $\{f(x_n)\}$ (in Y) converges to $f(x)$. The function f is said to be *continuous* (or *continuous on X*) if f is continuous at each point of X.

The reader should appreciate the natural parallel between vector spaces (linear transformations) and metric spaces (continuous functions): in the case of vector spaces, the algebraic structure is in some sense preserved by linear transformations [Observation (2.E.2)], while in the case of metric spaces the convergence of sequences is maintained.

The majority of functions with which the reader is familiar are continuous, e.g., $\sin x$, e^x, $\ln x$, $\sqrt{x}$, polynomial functions, etc. The continuity of many of these functions is perhaps best established in the context of infinite series and will not be attempted here. In the next theorem we do establish the continuity of three particularly simple functions.

(4.A.2) Theorem Suppose that (X,d) and $(Y,\hat{d})$ are metric spaces and that $y^* \in Y$. Then

(i)   The constant function $f\colon X \to Y$ defined by $f(x) = y^*$ for each $x \in X$ is continuous.

(ii)   The identity function $f\colon X \to X$ defined by $f(x) = x$ for each $x \in X$ is continuous.

(iii) The absolute value function $f\colon \mathbf{R}^1 \to [0,\infty)$, $f(x) = |x|$, is continuous.

Proof. (i) Suppose that a sequence $\{x_n\}$ converges to a point $x \in X$. Then since $f(x_n) = y^*$ for each x_n, clearly the constant sequence $\{f(x_n)\}$ converges to $f(x) = y^*$, and hence, f is continuous at each point $x \in X$.

(ii)   The proof is equally trivial. Suppose that $x \in X$ and that $\{x_n\}$ converges to x. Then since $f(x) = x$ and $f(x_n) = x_n$ for each n, the sequence $\{f(x_n)\}$ obviously converges to $f(x)$.

(iii) The reader may establish easily that f is continuous at each point $x \neq 0$. Suppose then that $x = 0$, and let $\{x_n\}$ be a sequence that converges to 0. We must show that the sequence $\{f(x_n)\} = \{|x_n|\}$ converges to $f(0) = 0$. Let $\varepsilon > 0$ be given. Since $\{x_n\} \to 0$, there is a positive integer N such that if $n \geq N$, then $|x_n - 0| = |x_n| < \varepsilon$. Note that for $n \geq N$, we also have

$$|f(x_n) - f(0)| = \big||x_n| - |0|\big| = \big||x_n|\big| = |x_n| < \varepsilon$$

and hence, the sequence $\{f(x_n)\}$ converges to $f(0)$.

In the next theorem we see that sums, products, and certain quotients of continuous real-valued functions are continuous.

(4.A.3) Theorem Suppose that (X,d) is a metric space and that the functions $f\colon X \to \mathbf{R}^1$ and $g\colon X \to \mathbf{R}^1$ are continuous at a point $x \in X$. Then the functions $h = f + g$, $j = f \cdot g$, and $k = f/g$ are continuous at x (assuming that k is defined in a neighborhood of x).

Proof. First we show that $f + g$ is continuous. Suppose that a sequence

$\{x_n\}$ converges to x. We are to show that the sequence $\{f(x_n) + g(x_n)\}$ converges to $f(x) + g(x)$. Since f and g are continuous at x, the sequences $\{f(x_n)\}$ and $\{g(x_n)\}$ converge to $f(x)$ and $g(x)$, respectively. It now follows immediately from (3.B.11) that the sequence $\{f(x_n) + g(x_n)\}$ converges to $f(x) + g(x)$. Similar arguments show that j and k are continuous at x.

(4.A.4) Corollary Polynomial functions $p(x) = a_n x^n + a_{n-1} x^{n-1} + \cdots + a_1 x + a_0$ are continuous.

    Proof. This follows immediately from (4.A.2) and (4.A.3).

The composition of continuous functions is continuous.

(4.A.5) Theorem Suppose that (X,d), $(Y,\hat{d})$, and $(Z,\check{d})$ are metric spaces and that $f: X \to Y$ is continuous at a point $x \in X$ and $g: Y \to Z$ is continuous at $f(x)$. Then $g \circ f$ is continuous at x.

    Proof. Suppose that $\{x_n\}$ is a sequence that converges to x. Then since f is continuous at x, the sequence $\{f(x_n)\}$ converges to $f(x)$. Since g is continuous at $f(x)$, the sequence $\{g(f(x_n))\}$ converges to $g(f(x)) = (g \circ f)(x)$ which shows that $g \circ f$ is continuous at x.

(4.A.6) Observation If $f: X \to \mathbf{R}^1$ and $g: X \to \mathbf{R}^1$ are arbitrary real-valued functions, then $h = \max\{f,g\}$ is the function mapping X into $\mathbf{R}^1$ defined by $h(x) = \max\{f(x),g(x)\}$ for each $x \in X$. It is easy to see that max $\{f,g\} = (f + g + |f - g|)/2$ and, therefore, by our preceding results if f and g are continuous, so is $\max\{f,g\}$. Since $\min\{f,g\} = (f + g - |f - g|)/2$, a similar result holds for the minimum of two functions.

Next we show that linear functions mapping $\mathbf{R}^m$ into $\mathbf{R}^k$ are continuous.

(4.A.7) Theorem Suppose that $T: \mathbf{R}^m \to \mathbf{R}^k$ is a linear transformation. Then T is continuous.

    Proof. Suppose that $\{\mathbf{x}_n\}$ is a sequence in $\mathbf{R}^m$ that converges to $\mathbf{x} \in \mathbf{R}^m$, where $\mathbf{x}_n = (x_1^{(n)}, x_2^{(n)}, \ldots, x_m^{(n)}) = x_1^{(n)}\mathbf{e}_1 + x_2^{(n)}\mathbf{e}_2 + \cdots + x_m^{(n)}\mathbf{e}_m$ and $\mathbf{x} = (x_1, x_2, \ldots, x_m) = x_1\mathbf{e}_1 + x_2\mathbf{e}_2 + \cdots + x_m\mathbf{e}_m$ [as usual, $\mathbf{e}_1 = (1,0,0,\ldots,0)$, $\mathbf{e}_2 = (0,1,0,\ldots,0), \ldots, \mathbf{e}_m = (0,0,\ldots,1)$]. We are to show that the sequence $\{T(\mathbf{x}_n)\}$ in $\mathbf{R}^k$ converges to $T(\mathbf{x})$. Since T is a linear transformation, we have

$$T(\mathbf{x}_n) = T(x_1{}^{(n)}\mathbf{e}_1 + x_2{}^{(n)}\mathbf{e}_2 + \cdots + x_m{}^{(n)}\mathbf{e}_m)$$
$$= x_1{}^{(n)}T(\mathbf{e}_1) + \cdots + x_m{}^{(n)}T(\mathbf{e}_m)$$

By (3.B.10.d), for each integer i, $1 \le i \le m$, the sequence $\{x_i^{(n)}\}_{n \in \mathbf{Z}^+}$ converges to x_i, and consequently, the sequence $\{T(\mathbf{x}_n)\}$ converges to

$$x_1 T(e_1) + \cdots + x_m T(e_m) = T(x_1 e_1 + \cdots + x_m e_m) = T(x)$$

which concludes the proof.

Recall that if V is a vector space and $\phi: V \to [0,\infty)$ is a norm for V, then ϕ determines a metric d for V, where $d(x,y) = \phi(x - y)$.

(4.A.8) Theorem Suppose that V is a vector space, ϕ is a norm for V, and V is given the metric d induced by ϕ. Then $\phi: V \to [0,\infty)$ is a continuous function.

Proof. We shall use the following inequality which is valid for any two vectors x and y in V:

$$|\phi(x) - \phi(y)| \le \phi(x - y) \tag{1}$$

To establish (1) observe that by the definition of a norm

$$\phi(x) = \phi(x - y + y) \le \phi(x - y) + \phi(y)$$

and consequently,

$$\phi(x) - \phi(y) \le \phi(x - y) \tag{2}$$

Similarly,

$$\phi(y) = \phi(y - x + x) \le \phi(y - x) + \phi(x)$$

and therefore,

$$\phi(y) - \phi(x) \le \phi(y - x) = \phi((-1)(x - y))$$
$$= |-1|\phi(x - y) = \phi(x - y) \tag{3}$$

Thus, (1) follows from (2) and (3). Suppose now that a sequence of vectors $\{x_n\}$ converges to a vector x in V. Then since

$$|\phi(x_n) - \phi(x)| \le \phi(x_n - x) = d(x_n, x) \tag{4}$$

and since the right-hand side of (4) tends to 0 with increasing n, we have that $\{\phi(x_n)\}$ converges to $\phi(x)$, which shows that ϕ is continuous.

(4.A.9) Observation Suppose that (X,d) is a metric space and that $f: X \to \mathbf{R}^n$, where $\mathbf{R}^n$ is given the usual metric. For each integer i, $1 \le i \le n$, define the function $f_i: X \to \mathbf{R}^1$ by setting $f_i(x)$ equal to the ith coordinate of $f(x)$. Then, as an immediate consequence of (3.B.10.d), we have that f is continuous if and only if f_i is continuous for each i. As a particular example, if $X = \mathbf{R}^2$, then the function $f: X \to \mathbf{R}^4$ defined by $f(x,y) = (x^3 y^2, 2x + y, x, y^4)$ is continuous.

(4.A.10) *Observation* Restrictions of continuous functions are continuous. Suppose that (X,d) and $(Y,\hat{d})$ are metric spaces, $A \subset X$, and that $f: X \rightarrow Y$ is continuous. Then $f|_A: A \rightarrow Y$ is continuous since any sequence in A that converges to a point $a \in A$ may be considered as a sequence in X that converges to a, and hence, the image of this sequence under f will converge in Y to $f(a) = f|_A(a)$.

There are a number of ways of characterizing continuity that do not directly involve sequences. One quite useful approach might be roughly described as follows. Suppose that X and Y are metric spaces, $x \in X$, and that $f: X \rightarrow Y$. Consider $f(x)$ to be the center of a circular "target" in Y of radius ε, namely $S_\varepsilon(f(x))$ (Fig. 4.1). Suppose that not only x but also all points at a distance from x of less than some positive number δ are sent by f into this target (Fig. 4.2). If this occurs for all circular targets with center $f(x)$ and of varying radii ε, then f will be continuous at x.

Conversely, if f is continuous at x, then given any circular target centered at $f(x)$ of radius ε, there is a number δ (that depends on the target size) such that f maps $S_\delta(x)$ into the target area.

This concept of *continuity* is somewhat more formally expressed in the following theorem.

(4.A.11) *Theorem* Suppose that (X,d) and $(Y,\hat{d})$ are metric spaces, $x \in X$, and $f: X \rightarrow Y$. Then f is continuous at x if and only if given $\varepsilon > 0$, there is a $\delta > 0$ such that whenever $d(x,y) < \delta$, then $\hat{d}(f(x),f(y)) < \varepsilon$ [or, equivalently, $f(S_\delta(x)) \subset S_\varepsilon(f(x))$].

Figure 4.1

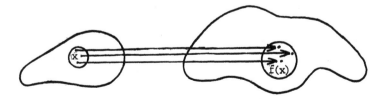

Figure 4.2

Proof. Suppose that f is continuous at x and let $\varepsilon > 0$ be given. If no $\delta > 0$ can be found that satisfies the conclusion of the theorem, then in particular for each $\delta = 1/n$, where $n \in \mathbf{Z}^+$, there is a point $x_n \in X$ such that $d(x_n,x) < 1/n$ and $\hat{d}(f(x_n),f(x)) \geq \varepsilon$. Note that the sequence $\{x_n\}$ determined in this manner converges to x, but that the corresponding sequence $\{f(x_n)\}$ fails to converge to $f(x)$, which contradicts the definition of continuity; consequently, there must exist a positive number δ such that $\hat{d}(f(x),f(y)) < \varepsilon$ whenever $d(x,y) < \delta$.

Conversely, suppose that a sequence $\{x_n\}$ converges to x. We are to show that the sequence $\{f(x_n)\}$ converges to $f(x)$. Let $\varepsilon > 0$ be given; we must find a positive integer N such that if $n \geq N$, then $f(x_n) \in S_\varepsilon(f(x))$. By hypothesis, there is a positive number δ such that if $d(x,y) < \delta$, then $\hat{d}(f(x),f(y)) < \varepsilon$. Since the sequence $\{x_n\}$ converges to x, there is an integer N such that for $n \geq N$, $x_n \in S_\delta(x)$. Then for $n \geq N$, we have $f(x_n) \in S_\varepsilon(f(x))$, and therefore, the sequence $\{f(x_n)\}$ converges to $f(x)$.

Again we emphasize that if we wish to use this theorem to check the continuity of a function f at a given point x, we must *start* with an arbitrary ε-neighborhood about the point $f(x)$, and *then* try to find a δ-neighborhood (whose size will usually depend on ε) about the point x such that any point in this latter neighborhood is mapped by f into the *given* ε-neighborhood. If for *each* ε, such a number δ can be found, then f is continuous at x. If for *some* ε, no such δ is available, then f fails to be continuous at x.

(4.A.12) *Example* Consider the continuous function $f: \mathbf{R}^1 \to [0,\infty)$ defined by $f(x) = x^2$. Let $x = 3$ and $\varepsilon = \frac{1}{10}$. We wish to find a number $\delta > 0$ such that if $d(y,x) = |y - x| < \delta$, then $|f(y) - f(x)| < \varepsilon$. No harm is done if we consider only those values y that are relatively close to x, say at a distance 1 or less. Note that for these values we have

$$|f(y) - f(x)| = |y^2 - x^2| = |y + x|\,|y - x| < 7|y - x|$$

Hence, if $\delta = \varepsilon/7$ and $|y - x| < \delta$, we are assured that $|f(y) - f(x)| < 7\delta = \varepsilon$; thus, if $\varepsilon = \frac{1}{10}$, then $\delta = \frac{1}{70}$ will do. The reader should observe, however, that if $x = 20$ and $\varepsilon = \frac{1}{10}$, then $\delta = \frac{1}{70}$ is no longer sufficiently small, since if $y = 20 + \frac{1}{80}$, then $|y - x| < \frac{1}{70}$, but $|f(20 + \frac{1}{80}) - f(20)| = 0.50 > \frac{1}{10}$. Nevertheless, it is easy to see that for $\varepsilon = \frac{1}{10}$, $\delta = 1/41(10)$ works. It therefore appears that not only does δ depend on ε, but also on the point x in question. We shall return to this point when we discuss uniformly continuous functions.

The characterization of continuity found in (4.A.11) is that which is customarily employed in introductory calculus texts, and Figs. 4.3 to 4.6 are usually given to illustrate this concept. Note that the ε-neighborhood has been

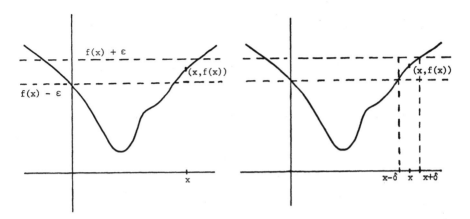

Figure 4.3

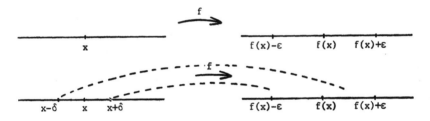

Figure 4.4

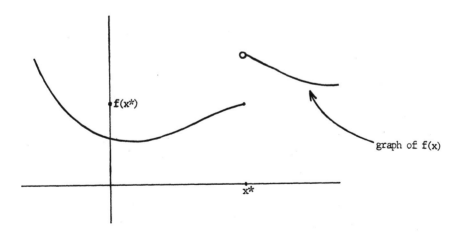

Figure 4.5

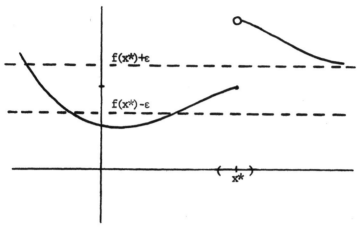

Figure 4.6

translated into an ε-strip, and the same fate has befallen the δ-neighborhood (Fig. 4.3). It would perhaps be conceptually clearer to view the same problem as indicated in Fig. 4.4, although of course the graph of f is no longer seen.

Functions mapping from $\mathbf{R}^1$ into $\mathbf{R}^1$ that have a "break" in their graph are not continuous. For instance, in Fig. 4.5 we see that f is not continuous at the point x^* since if ε is chosen to yield a band as indicated in Fig. 4.6, then no matter how small δ is chosen, there will be points in the δ-neighborhood of x^* whose images under f will not lie in the ε-neighborhood of f (in fact, any point in such a neighborhood and to the right of x^* will be mapped to a point outside of this neighborhood). Note too that if $\{x_n\}$ is any sequence of points to the right of x^* that converges to x^*, then the sequence $\{f(x_n)\}$ fails to converge to $f(x^*)$.

One additional characterization of continuity is given in the following theorem, a characterization that further reflects the usefulness of open sets.

(4.A.13) *Theorem* Suppose that (X,d) and $(Y,\hat{d})$ are metric spaces and that $f: X \to Y$. Then f is continuous if and only if $f^{-1}(V)$ is open in X for each open set V in Y.

Proof. Suppose that f is continuous, V is an open set in Y, and $x \in f^{-1}(V)$. We are to show that there is a positive number r such that $S_r(x) \subset f^{-1}(V)$. Since V is open and $f(x) \in V$, there is an $\varepsilon > 0$ such that $S_\varepsilon(f(x)) \subset V$. Since f is continuous, there is by (4.A.11) an $r > 0$ such that $f(S_r(x)) \subset S_\varepsilon(f(x)) \subset V$, and consequently, $S_r(x) \subset f^{-1}[S_\varepsilon(f(x))] \subset f^{-1}(V)$.

Suppose now that inverse images of open sets are open. Let $x \in X$ and $\varepsilon > 0$ be given. Since $f^{-1}[S_\varepsilon(f(x))]$ is open, there is a positive number δ such that $S_\delta(x) \subset f^{-1}[S_\varepsilon(f(x))]$; therefore, $f(S_\delta(x)) \subset S_\varepsilon(f(x))$, and hence, by (4.A.11), f is continuous at x.

(4.A.14) *Corollary* Suppose that d and $\hat{d}$ are equivalent metrics for a set X and that $f: X \to Y$, where Y is an arbitrary metric space. Then f is continuous with respect to the metric d if and only if f is continuous with respect to the metric $\hat{d}$.

Proof. This result follows immediately from (4.A.13) since equivalent metrics generate the same open sets.

B. CONTINUITY AND COMPACTNESS

The reader may well inquire why the concept of continuity merits such detailed examination. This question will be more adequately answered as we proceed, but a general response now would at least include the following points:

1. A very substantial number of functions that arise naturally in the study of physical phenomena, in models of biological systems, in problems in economics, psychology, etc., are continuous.
2. Continuous functions are reasonably well behaved and, hence, amenable to a variety of mathematical manipulations.
3. Continuity is the needed ingredient in the hypotheses of a number of important theorems, e.g., existence and uniqueness results involving differential equations.
4. The continuity of a function can often be used to extract considerable information about the function.

Naturally, the above statements are subject to certain qualifications, but in a general sense, each statement is sufficiently valid to justify a serious study of the properties of continuity, We begin by examining the role of continuity in the familiar problem of finding maxima and minima of functions.

Examples which illustrate the importance of determining maxima and minima are legion; they might include such complex problems as the calculation of the minimum amount of fuel necessary for a given rocket trip as the duration of the trip varies, finding maximal profits in a specialized price-cost system, and the determination of the minimal number of noxious bacteria in an experimentally controlled environment. In a later chapter we examine the problem of finding a curve connecting two fixed points in the plane along which an object will slide from one point to the other in minimum time.

Before proceeding, we define precisely what is meant by "attaining a maximum (or minimum)."

(4.B.1) *Definition* Suppose that (X,d) is a metric space and that $f: X \to \mathbf{R}^1$. Then f attains a *local maximum* at a point $x \in X$ if there is an ε-neighborhood of x such that for each $y \in S_\varepsilon(x)$, $f(y) \le f(x)$. If $f(y) \le f(x)$ for each $y \in X$, then f is said to attain an *absolute maximum* at x.

A *local minimum* and an *absolute minimum* are defined in an analogous manner (replace $\le$ with $\ge$).

We now try to determine what conditions can be imposed on f and/or (X,d) in order to guarantee the existence of maxima and minima. In the next chapter derivatives are used to locate the points where these extrema are attained.

Consider the following three functions:

1. $f: (0,2] \to \mathbf{R}^1$ defined by $f(x) = 1/x$.
2. $f: [0,2] \to \mathbf{R}^1$ defined by

$$f(x) = \begin{cases} \dfrac{1}{x} & \text{if } x \ne 0 \\ 3 & \text{if } x = 0 \end{cases}$$

3. $f: [1,2] \to \mathbf{R}^1$ defined by $f(x) = 1/x$.

Clearly, neither of the first two functions attains an absolute maximum, whereas the third function does. Furthermore, the first and third functions are continuous and the second is not. Thus, continuity by itself is not sufficient to ensure the existence of a maximum. Since the primary difference between 1 and 2 is the nature of the domain, one might be led to conjecture that if a function f is continuous on a closed and bounded interval, then f will attain a maximum (and a minimum) on this interval. This conjecture turns out to be valid, but the restriction that the domain must be a closed and bounded interval is far too stringent (it wouldn't even make sense for domains other than subsets of $\mathbf{R}^1$). Faced with this kind of situation, we abstract the properties that will yield the desired result in a setting as general as possible. To do this in the present case, we first need to generalize the notion of closed and bounded intervals.

(4.B.2) *Definition* Suppose that (X,d) is a metric space and that $A \subset X$. Then A is *closed* in X if $X \backslash A$ is open.

(4.B.3) *Examples* (a) If (X,d) is a metric space, then X and the empty set $\varnothing$ are closed (why?).

(b) If $X = \mathbf{R}^1$ and A is the set of integers $\mathbf{Z}$, then A is closed in $\mathbf{R}^1$.

(c) If $X = \mathbf{R}^n$ and $A = \{(x_1, x_2, \ldots, x_n) \mid x^2 + \cdots + x^2 \le 16\}$, then A is closed.

(d) If (X,d) has the discrete metric, then every subset of X is closed.

(e) Any interval of the form $[a, b]$, $[a, \infty)$, or $(-\infty, b]$ is closed in $\mathbf{R}^1$ (usual metric).

(f) If (X,d) is a metric space and $\{x_n\}$ is a sequence in X that converges to $x \in X$, then $A = \{x_n \mid n \in \mathbf{Z}^+\} \cup \{x\}$ is closed.

(g) The half-open interval [8,17) in $\mathbf{R}^1$ is neither open nor closed in $\mathbf{R}^1$, although it is closed in the space $X = (0,17)$.

It is clear that the interval (2,5] is not closed in $\mathbf{R}^1$, but that the interval [2,5] is. Although the number 2 is not found in (2,5], it is (loosely speaking) arbitrarily close to this set; on the other hand, outside of the interval [2,5] no such "arbitrarily close" point is to be found, Thus, we might surmise that a set A will be closed if it contains all points which essentially "adhere" to it. This idea is made rigorous in the next definition and theorem. For the remainder of this chapter the reader is encouraged to make sketches to illustrate the various concepts that are discussed.

(4.B.4) *Definition* Suppose that (X,d) is a metric space and that $A \subset X$. Then a point $x \in X$ is an *adherence point* of A is $S_\varepsilon(x) \cap A \neq \varnothing$, for each $\varepsilon > 0$.

Clearly, any point in A is an adherence point of A; however, as evidenced in the previous discussion, there may be adherence points of A lying outside of A (2 is an adherence point of (2,5]).

(4.B.5) *Theorem* Suppose that (X,d) is a metric space and that $A \subset X$. Then A is closed if and only if A contains all of its adherence points.

Proof. Suppose that A is closed and that x is an adherence point of A. If $x \in X \backslash A$, then since $X \backslash A$ is open, there is an $\varepsilon > 0$ such that $S_\varepsilon(x) \subset X \backslash A$. This, however, is impossible since $S_\varepsilon(x) \cap A \neq \varnothing$ for each $\varepsilon > 0$. Therefore, $x \in A$.

Conversely, suppose that A contains all of its adherence points. To show that A is closed, we show that $X \backslash A$ is open. Let $x \in X \backslash A$. Then x cannot be an adherence point of A, and consequently, there is an $\varepsilon > 0$ such that $S_\varepsilon(x) \subset X \backslash A$; hence, $X \backslash A$ is open and A is closed.

(4.B.6) *Exercise* Suppose that A is a subset of a metric space (X,d) and let B be the set of all adherence points of A (note that $A \subset B$). Show that B is closed. The set B is called the *closure* of A and will frequently be denoted by $\bar{A}$. It is easily seen that $\bar{A} = \overline{(\bar{A})}$, i.e., $\bar{A}$ contains all of its adherence points (problem B.3).

The notion that adherence points of a set A are either points of A or are "arbitrarily close" to A finds expression in the following result.

(4.B.7) *Theorem* Suppose that (X,d) is a metric space, $A \subset X$, and x is an adherence point of A. Then there is a sequence $\{x_n\}$ *in A* that converges to x.

Proof. If $x \in A$, let $x_n = x$ for each n; then the constant sequence $\{x_n\}$ certainly converges to x. Suppose now that $x \notin A$. Then for each $n \in \mathbf{Z}^+$, there is a point $x_n \in S_{1/n}(x) \cap A$; it is clear that the sequence $\{x_n\}$ formed by these points converges to x.

(4.B.8) *Exercises* (a) Show that if $A \subset \mathbf{R}^1$ and sup A (or inf A) exists, then sup A (or inf A) is an adherence point of A.

(b) Show that if (X,d) is a metric space, $A \subset X$, and $\{x_n\}$ is a sequence in A, then all limit and cluster points of $\{x_n\}$ are adherence points of A.

(c) Suppose that (X,d) is a metric space and that $A \subset X$. Show that a point $x \in X$ is an adherence point of A if and only if for each open set U in X that contains x, $U \cap A \neq \varnothing$.

The notion of boundedness in $\mathbf{R}^1$ is generalized as follows.

(4.B.9) *Definition* Suppose that A is a subset of a metric space (X,d). Then A is *bounded* if there exists a real number $r > 0$ such that $d(x,y) \leq r$ for each $x,y \in A$. Note that an equivalent formulation of this definition would be: A is bounded if $D = \sup\{d(x,y) \mid x,y \in A\}$ is finite; D is defined to be the *diameter* of A.

Using this new terminology, we might now conjecture: If a subset A of a metric space (X,d) is closed and bounded, and $f: A \to \mathbf{R}^1$ is continuous, then f attains an absolute maximum (and absolute minimum) in A. Reasonable enough. But suppose that X is the set of integers with the discrete metric; then X is certainly closed and bounded. Define $f: X \to \mathbf{R}^1$ by $f(n) = n$. All of the conditions of the conjecture are satisfied, but f certainly fails to have a maximum.

Further exploration is in order. What inherent properties does the interval $[1,2]$ with the usual metric have that the set of integers with the discrete metric does not? Since the properties of sequences have permeated our efforts thus far, we might continue in this vein and try to contrast the behavior of certain sequences in $[1,2]$ with those in the discrete space X given above. We have already established that any sequence in $[1,2]$ must have a cluster point (3.B.13.c); on the other hand, there are sequences $\{x_n\}$ in X that fail to have this property (for example, let $x_n = n$ for each $n \in \mathbf{Z}^+$). Consequently, we might refine out conjecture as follows: Suppose that (X,d) is a metric space and that A is a subset of X with the property that each sequence in A has a cluster point. If $f: A \to \mathbf{R}^1$ is continuous, then f attains an absolute maximum (and an absolute minimum) in A. This is what we shall be able to prove, and in fact, this result constitutes one of the most important and frequently used theorems in analysis.

As is often the case in mathematics, whenever a particular property shows signs of usefulness, it becomes the basis for a definition.

(4.B.10) Definition A subset A of a metric space (X,d) is said to be *compact* if every sequence in A has a cluster point in A.

(4.B.11) Examples (a) All closed and bounded intervals in $\mathbf{R}^1$ are compact.

(b) Any finite subset of a metric space is compact.

(c) If (X,d) is a metric space and $\{x_n\}$ is a sequence in X that converges to a point $x \in X$, then $A = \{x_n \mid n \in \mathbf{Z}^+\} \cup \{x\}$ is compact.

(d) The bounded interval $X = (0,1)$ is not compact [the sequence $\{1/n\}$ fails to have a cluster point in $(0,1)$]. Note moreover that the interval $(0,\tfrac{1}{2}]$ is closed (in X) and bounded; hence, the property of being closed and bounded is not sufficient to characterize compactness. We shall presently show, however, that closed and bounded subsets of $\mathbf{R}^n$ (with the usual metric) are compact.

The definition of compactness implies that compact sets are in some sense small, since points in any infinite subset of a compact set A must "cluster together"—there is not sufficient room in A for such sets to stretch out.

(4.B.12) Exercise Show that compact sets are bounded.

We see in the next result that compact sets are always closed.

(4.B.13) Theorem If A is a compact subset of a metric space (X,d), then A is closed.

Proof. Suppose that x is an adherence point of A. By (4.B.7) there is a sequence $\{x_n\}$ in A that converges to x, and by (3.B.14), x is the only cluster point of this sequence. Since A is compact, the sequence $\{x_n\}$ has a cluster point in A, which then must be x. Consequently, A contains all of its points of adherence, and thus is closed.

(4.B.14) Exercise Show that closed subsets of compact spaces are compact.

In order to establish our conjecture concerning maxima and minima, we need the following extremely useful result involving the interplay between continuity and compactness.

(4.B.15) *Theorem* Suppose that (X,d) and $(Y,\hat{d})$ are metric spaces and that A is a compact subset of X. If $f: A \to Y$ is continuous and onto, then Y is compact. Briefly stated, the continuous image of a compact set is compact.

Proof. Suppose that $\{y_n\}$ is a sequence in $f(A)$. We are to show that $\{y_n\}$ has a cluster point in $f(A)$. For each $n \in \mathbf{Z}^+$, select a point $x_n \in f^{-1}(y_n)$. Then the sequence $\{x_n\}$ lies in the compact set A, and hence has a cluster point, x^*, in A. We show that $f(x^*)$ is a cluster point of $\{y_n\}$. To see this, let $\varepsilon > 0$ and $N \in \mathbf{Z}^+$ be given. We find an $m \geq N$ such that $y_m \in S_\varepsilon(f(x^*))$. Since f is continuous, there is a $\delta > 0$ such that $f(S_\delta(x^*)) \subset S_\varepsilon(f(x^*))$. Furthermore, since x^* is a cluster point of $\{x_n\}$, there is an $m \geq N$ such that $x_m \in S_\delta(x^*)$. Therefore, $y_m = f(x_m) \in S_\varepsilon(f(x^*))$, and consequently, $f(x^*)$ is a cluster point of $\{y_n\}$.

Our conjecture now becomes a theorem.

(4.B.16) *Theorem* Suppose that A is a compact metric space and that $f: A \to \mathbf{R}^1$ is continuous. Then f attains an absolute maximum in A.

Proof. Since $f(A)$ is compact, it is bounded; hence, $s = \sup f(A)$ exists. By (4.B.8.a) s is an adherence point of $f(A)$, and since $f(A)$ is closed [(4.B.15) and (4.B.13)], we have that $s \in f(A)$. Therefore, there is a point $x \in A$ such that $f(x) = s$, and consequently, f attains an absolute maximum at x.

Note: The theorem obviously holds if the word maximum is replaced by minimum; the proof is exactly the same.

Another frequently used result involving compactness is the following.

(4.B.17) *Theorem* Suppose that (X,d) is a compact metric space, $(Y,\hat{d})$ is a metric space, and $f: X \to Y$ is a continuous bijection. Then f^{-1} is continuous.

Proof. By (4.A.13) it suffices to show that if U is open in X, then $(f^{-1})^{-1}(U) = f(U)$ is open in Y. Suppose then that U is an open subset of X. Since $X \setminus U$ is closed in X, it follows from (4.B.14) that $X \setminus U$ is compact. Consequently, by (4.B.15), $f(X \setminus U)$ is compact in Y, which implies that $f(X \setminus U)$ is closed in Y (4.B.13). Since f is a bijection, we have $f(X \setminus U) = Y \setminus f(U)$, and therefore, $f(U)$ is the complement of a closed set and, thus, is open.

We now establish the very important result that closed and bounded subsets of $\mathbf{R}^n$ are compact. This will follow easily from the next theorem, which is of considerable interest in itself.

(4.B.18) *Theorem* Suppose that (X,d) and $(Y,\hat{d})$ are compact metric spaces. Then $X \times Y$ with any of the metrics given in (3.B.4.d) is compact.

To prove this theorem, we make use of the following lemma.

(4.B.19) *Lemma* Suppose that $\{x_n\}$ is a sequence in a metric space (X,d). Then a point $x^* \in X$ is a cluster point of $\{x_n\}$ if and only if there is a subsequence of $\{x_n\}$ that converges to x^*.

Proof. It follows immediately from the definitions of a cluster point and a subsequence that if a subsequence of the sequence $\{x_n\}$ converges to x^*, then x^* is a cluster point of $\{x_n\}$. Conversely, suppose that x^* is a cluster point of the sequence $\{x_n\}$. We construct a subsequence of $\{x_n\}$ that converges to x^* by choosing appropriate points from the neighborhoods $S_1(x^*)$, $S_{\frac{1}{2}}(x^*)$, $S_{\frac{1}{3}}(x^*)$, ... in the following manner. Select any term of the sequence $\{x_n\}$ that lies in $S_1(x^*)$, and denote this term by x_{n_1}. In the neighborhood $S_{\frac{1}{2}}(x^*)$, select any term x_k of the sequence $\{x_n\}$ such that $k > n_1$ (why is this possible?). Denote this term by x_{n_2}. In $S_{\frac{1}{3}}(x^*)$, choose any term x_k of the sequence $\{x_n\}$ with the property that $k > n_2$, and denote this term by x_{n_3}. Continuing in this fashion, we obtain the desired convergent subsequence.

Note that it follows from the preceding lemma that a metric space (X,d) is compact *if and only if* every sequence in X has a subsequence that converges in X. We use this observation to prove (4.B.18). Let $\{(x_n,y_n)\}$ be an arbitrary sequence in $X \times Y$. Since X is compact, the sequence $\{x_n\}$ has a subsequence $\{x_{n_k}\}$ that converges in X. Since Y is compact, the subsequence $\{y_{n_k}\}$ of $\{y_n\}$ (corresponding to the subsequence $\{x_{n_k}\}$) has a convergent subsequence $\{y_{n_{k_i}}\}$. It is easily verified that the subsequence $\{(x_{n_{k_i}}, y_{n_{k_i}})\}$ of $\{(x_n,y_n)\}$ converges in $X \times Y$, and therefore, $X \times Y$ is compact.

We are now in a position to characterize compact subsets of $\mathbf{R}^n$. In (4.B.12) and (4.B.13) we have already seen that, in general, compact subsets of a metric space are closed and bounded. In $\mathbf{R}^n$, the reverse implication is also true: all closed and bounded subsets of $\mathbf{R}^n$ are compact. We prove this for the case where $n = 2$ and leave the easy generalization to the reader.

(4.B.20) *Theorem* Suppose that A is a closed and bounded subset of $\mathbf{R}^2$. Then A is compact.

Proof. Since A is bounded, there is a rectangle $S = [a,b] \times [c,d]$ in $\mathbf{R}^2$ that contains A. By the previous theorem, S is compact and since A is a closed subset of S, it follows from (4.B.14) that A is compact.

(4.B.21) *Exercise* Show that (4.B.20) is valid if $\mathbf{R}^2$ is replaced by $\mathbf{R}^1$ or more generally, by $\mathbf{R}^n$ for each $n \in \mathbf{Z}^+$.

Historically, the notion of *compactness* has undergone a long series of transformations. The most common definition of compactness (one that can be easily adapted to more general spaces than metric spaces) is given in terms of open covers. This approach is explored in Sec. 4.D.

C. CAUCHY SEQUENCES

(4.C.1) *Definition* Suppose that (X,d) is a metric space. A sequence $\{x_n\}$ in X is said to be a *Cauchy sequence* if for each $\varepsilon > 0$, there is a positive integer N such that if $m,n \geq N$, then $d(x_n,x_m) < \varepsilon$.

Clearly, convergent sequences are Cauchy since the terms of such a sequence tend toward a common limit. Conversely, it would seem likely that Cauchy sequences must converge since the distance between members of the tail end of the sequence can be made arbitrarily small. Note, however, that if $X = (0,1)$, then the sequence $\{1/n\}$ is a Cauchy sequence that fails to converge in X. This sequence does converge in $\mathbf{R}^1$, and this brings up the question of whether or not there are Cauchy sequences in $\mathbf{R}^1$ (or $\mathbf{R}^n$) that do not converge. Before resolving this problem, we show that a cluster point of a Cauchy sequence is in fact a (the) limit of the sequence.

(4.C.2) *Theorem* Suppose that (X,d) is a metric space, $\{x_n\}$ is a Cauchy sequence in X, and x is a cluster point of $\{x_n\}$. Then the sequence $\{x_n\}$ converges to x.

Proof. Let $\varepsilon > 0$ be given. We must show that there is a positive integer N such that for $n \geq N$, $d(x_n,x) < \varepsilon$. Since $\{x_n\}$ is a Cauchy sequence, there is a positive integer N such that if $k,m \geq N$, then $d(x_k,x_m) < \varepsilon/2$. Since x is a cluster point of $\{x_n\}$, there is an integer $q \geq N$ such that $d(x_q,x) < \varepsilon/2$. Note now that if $n \geq N$, then

$$d(x_n,x) < d(x_n,x_q) + d(x_q,x) < \frac{\varepsilon}{2} + \frac{\varepsilon}{2} = \varepsilon$$

and consequently, the sequence $\{x_n\}$ converges to x.

(4.C.3) *Exercise* Show that a Cauchy sequence is bounded.

Our principal theorem is the following.

(4.C.4) *Theorem* If $\{x_n\}$ is a Cauchy sequence in $\mathbf{R}^m$, then $\{x_n\}$ converges.

Proof. Let $A = \{x_n \mid n \in \mathbf{Z}^+\}$. Then by (4.B.6), the closure $\bar{A}$ of A is closed, and furthermore, by (4.B.5) and (4.B.8.b) $\bar{A}$ contains all cluster points of the sequence $\{x_n\}$. It follows easily from (4.C.3) and the definition of adherence points that $\bar{A}$ is bounded; therefore, by (4.B.21), $\bar{A}$ is compact. Hence, the sequence $\{x_n\}$ has a cluster point x in $\bar{A}$ and by (4.C.2), x is actually the limit of this sequence.

(4.C.5) *Definition* A metric space (X,d) is *complete* if each Cauchy sequence in X converges (in X).

We have shown that $\mathbf{R}^n$ is complete and that the open interval $(0,1)$ is not; moreover, it should be clear from the proof of (4.C.4) that all compact metric spaces are complete.

(4.C.6) *Exercise* Show that if (X,d) is a complete metric space and A is closed subset of X, then A is a complete metric space (with the metric inherited from X).

We conclude this section with another important example of a complete metric space.

(4.C.7) *Theorem* Suppose that X is an arbitrary set, and let $Y = \{f: X \to \mathbf{R}^1 \mid f \text{ is bounded}\}$. Then $(Y, \ )$ is a complete metric space, where d is the sup metric for Y.

Proof. Suppose that $\{f_n\}$ is an arbitrary Cauchy sequence in Y. In order to show that Y is complete, we must construct a function $f \in Y$ such that the sequence $\{f_n\}$ converges to f. To find f, we first note that for each *fixed* $x \in X$, the sequence of real numbers $\{f_n(x)\}$ is a Cauchy sequence in $\mathbf{R}^1$ (why?). Therefore, by (4.C.4), the sequence $\{f_n(x)\}$ converges to a (unique) point $r_x \in \mathbf{R}^1$. Define $f: X \to \mathbf{R}^1$ by setting $f(x) = r_x$. We show that the sequence $\{f_n\}$ converges to f in the sup metric, i.e., $\{f_n\}$ converges uniformly to f (clearly, this sequence converges pointwise to f). Let $\varepsilon > 0$ be given. Since $\{f_n\}$ is a Cauchy sequence, there is a positive integer N such that for $m,n \geq N$, $d(f_m,f_n) < \varepsilon/2$, i.e., $\sup\{|f_m(x) - f_n(x)| \mid x \in X\} < \varepsilon/2$. Let x be an arbitrary point in X; we show that if $n \geq N$, then $|f_n(x) - f(x)| < \varepsilon$, which, since x is arbitrary, implies that $d(f,f_n) \leq \varepsilon$ for each $n \geq N$. By the definition of f, the sequence $\{f_n(x)\}$ converges to $f(x)$, and hence, there is an $m \geq N$ such that $|f_m(x) - f(x)| < \varepsilon/2$ (note that m depends on the particular point x). Consequently, for each $n \geq N$, we have

$$|f_n(x) - f(x)| \leq |f_n(x) - f_m(x)| + |f_m(x) - f(x)| < \frac{\varepsilon}{2} + \frac{\varepsilon}{2} = \varepsilon$$

which concludes the proof.

D.* A CHARACTERIZATION OF COMPACTNESS

The results of this section, while of considerable importance in advanced analysis, will only be used in this text in certain proofs found in Chapter 11.

Our present goal is to obtain a characterization of compactness in terms of open covers. We need the following definitions.

(4.D.1) Definition   A *cover* of a set X is a family $\mathscr{U} = \{U_\alpha \mid \alpha \in J\}$ of subsets of X such that $X = \bigcup_{\alpha \in J} U_\alpha$. If X is a metric space, then an *open cover* of X is a cover of X by open subsets of X.

(4.D.2) Definition   Suppose that $\mathscr{U} = \{U_\alpha \mid \alpha \in J\}$ is a cover of a set X. Then a *subcover* $\mathscr{V}$ of X (with respect to $\mathscr{U}$) is any subset of $\mathscr{U}$ that is also a cover of X. When it is clear from the context what is meant, we (somewhat inaccurately) say that $\mathscr{V}$ is a subcover of $\mathscr{U}$.

(4.D.3) Example   The family $\mathscr{U} = \{(-a,a) \mid a \in \mathbf{R}^1, a > 0\}$ is an open cover of $\mathbf{R}^1$ and the family $\mathscr{V} = \{(-a,a) \mid a \in \mathbf{Z}^+\}$ forms a subcover of $\mathbf{R}^1$ (with respect to $\mathscr{U}$). Note that there is no finite subcover of $\mathbf{R}^1$ (with respect to $\mathscr{U}$).

The principal result of this section is the following.

(4.D.4) Theorem   A metric space (X,d) is compact if and only if every open cover of X has a finite subcover.

To prove this theorem we shall establish two lemmas that are of considerable significance in their own right. We begin with another definition.

(4.D.5) Definition   A metric space (X,d) is *totally bounded* if for each $\varepsilon > 0$, there are a finite number of points $\{x_1, x_2, \ldots, x_n\}$ in X such that $X = \bigcup_{i=1}^{n} S_\varepsilon(x_i)$.

(4.D.6) Examples   (a)   Any bounded subset of $\mathbf{R}^1$ is totally bounded.
   (b) $\mathbf{R}^1$ is not totally bounded.
   (c) $\mathbf{R}^1$ with the discrete metric is bounded but not totally bounded.

(4.D.7) Lemma   If (X,d) is a compact metric space, then X is totally bounded.

Proof.   Let $\varepsilon > 0$ be given. We are to find points $x_1, x_2, \ldots, x_n$ such that $X = \bigcup_{i=1}^{n} S_\varepsilon(x_i)$. Select an arbitrary point x_1. If $X = S_\varepsilon(x_1)$, we are done. If $X \neq S_\varepsilon(x_1)$, then there is a point $x_2 \notin S_\varepsilon(x_1)$; if $X = S_\varepsilon(x_1) \cup S_\varepsilon(x_2)$, again we are finished. If $X \neq S_\varepsilon(x_1) \cup S_\varepsilon(x_2)$, there is a point $x_3 \notin S_\varepsilon(x_1) \cup S_\varepsilon(x_2)$. As before, if $X = S_\varepsilon(x_1) \cup S_\varepsilon(x_2) \cup S_\varepsilon(x_3)$, then X is totally bounded. This process must terminate after a finite number of steps, for if not, we shall have constructed a sequence $\{x_n\}$ that has no cluster point [for each $i \neq j$, $d(x_i,x_j) \geq \varepsilon$]. Thus, we conclude that X is totally bounded.

(4.D.8) Lemma Suppose that $\mathcal{U}$ is an open cover of a compact metric space (X,d). Then there is a positive number γ (called a *Lebesgue* number for $\mathcal{U}$) such that for each $x \in X$, $S_\gamma(x) \subset U$ for some $U \in \mathcal{U}$ (equivalently expressed: there is a $\gamma > 0$ such that if $A \subset X$ and diameter $A < \gamma$, then A is contained in at least one member U of the open cover $\mathcal{U}$).

Proof. Suppose that no such $\gamma > 0$ exists. Then for each $n \in \mathbf{Z}^+$, there is a point x_n such that $S_{1/n}(x_n)$ is not contained in any U in $\mathcal{U}$. Since X is compact, the sequence $\{x_n\}$ has a cluster point x^*, and since $\mathcal{U}$ is a cover of X, there is a $U \in \mathcal{U}$ such that $x^* \in U$. The subset U is open; therefore, there is a positive integer N large enough so that $S_{1/N}(x^*) \subset U$. Since x^* is a cluster point of the sequence $\{x_n\}$, there is an integer $m \geq 2N$ such that $d(x_m,x^*) < 1/2N$, and hence, $S_{1/m}(x_m) \subset S_{1/N}(x^*) \subset U$ (Fig. 4.7). This contradicts the way that the members of the sequence $\{x_n\}$ were chosen, and consequently, a Lebesgue number for $\mathcal{U}$ must exist.

We are now in a position to establish Theorem (4.D.4).

Proof of Theorem (4.D.4). First suppose that X is compact, and let $\mathcal{U}$ be an open cover of X. We are to find a finite subcover of X (with respect to $\mathcal{U}$). By (4.D.8) there is a Lebesgue number γ that corresponds to $\mathcal{U}$. By (4.D.7), X is totally bounded, and hence, there exist points $x_1, x_2, \ldots, x_n$ such that $X = \bigcup_{i=1}^n S_\gamma(x_i)$. Since γ is a Lebesgue number for $\mathcal{U}$, for each i,

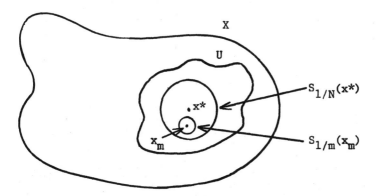

Figure 4.7

there is a set $U_i \in \mathcal{U}$ such that $S_\gamma(x_i) \subset U_i$. Clearly, the sets $U_1, U_2, \ldots, U_n$ form the desired subcover.

Now suppose that (X,d) is a metric space such that each open cover of X has a finite subcover. We are to prove that X is compact. Let $\{x_n\}$ be an arbitrary sequence in X. We must show that the sequence $\{x_n\}$ has a cluster point in X. If no point of X is a cluster point of $\{x_n\}$, then for each $x \in X$, there is an open set U_x containing x such that $\{n \mid x_n \in U_x\}$ is finite (why?). The open sets $\{U_x \mid x \in X\}$ clearly form an open cover of X, and hence, by hypothesis, there is a finite subcover $U_1, \ldots, U_k$, such that $X = U_1 \cup \cdots \cup U_k$. However, this is impossible, since $\{n \mid x_n \in U_1 \cup \cdots \cup U_k\}$ is finite; therefore, the sequence $\{x_n\}$ must have a cluster point, and consequently, X is compact.

The characterization of compactness that we have just established is often used in place of Definition (4.B.10). Although this view of compactness is perhaps less intuitive than our definition, it frequently plays a more useful role in proofs of theorems. Furthermore, the notion of compactness in terms of open covers can easily be adapted to more general spaces, in particular, to topological spaces. When this more general definition of compactness is used, then the notion of compactness as defined by cluster points (or convergent subsequences) is referred to as *sequential compactness*.

E. THE INTERMEDIATE VALUE THEOREM

In its simplest form the *intermediate value theorem* states that if f is a continuous function mapping an interval $[a,b]$ into $\mathbf{R}^1$ and if $f(a) < c < f(b)$ [or if $f(b) < c < f(a)$], then there is a point $x^* \in (a,b)$ such that $f(x^*) = c$. In other words, f takes on all values between (or intermediate to) $f(a)$ and $f(b)$. When one considers the graph of a continuous function f, the result is "obvious" (Fig. 4.8); however, as we shall see, the proof of the theorem is nontrivial.

That continuity is necessary for this result to hold is easily seen from Fig. 4.9.

Note, however, that continuity by itself is not sufficient, for if $A = [2,5] \cup [7,9]$ and if $f: A \to \mathbf{R}^1$ is defined by

$$f(x) = \begin{cases} 1 & \text{if } x \in [2,5] \\ 3 & \text{if } x \in [7,9] \end{cases}$$

then f is continuous on its domain, but corresponding to $c = 2$, there is no value x in A such that $f(x) = 2$ (Fig. 4.10).

The principal difference between A and an interval $[a,b]$ is that the latter set is "in one piece." It is this notion that we now wish to abstract. The idea of being "in one piece" depends very much on the metric involved; if $[a,b]$ is given the discrete metric, then the interval is effectively broken into an infinite

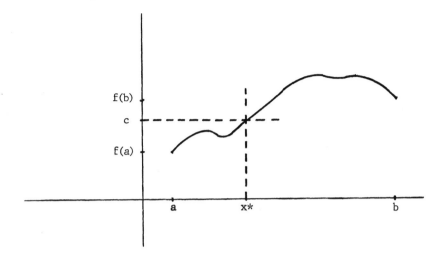

Figure 4.8

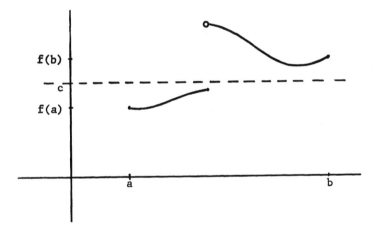

Figure 4.9

number of pieces since the distance between any two points is equal to 1. Open sets provide the key to determining whether or not a set is *connected* (the mathematical term for being "in one piece").

(4.E.1) *Definition* A metric space (X,d) is *connected* if there do not exist disjoint, nonempty open subsets of X, U, and V, such that $X = U \cup V$.

Under this definition, it is fairly clear that intervals are connected, and that a metric space consisting of just two points is not connected (each point

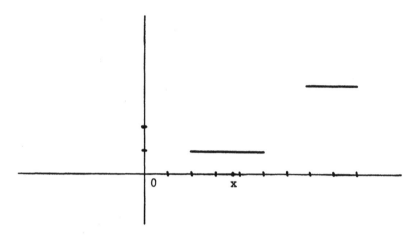

Figure 4.10

is open). On the other hand, the status of the set

$$x = \mathbf{R}^2 \backslash \{(x,y) \in \mathbf{R}^2 \mid x \text{ and } y \text{ are rational}\} \text{ (usual metric)}$$

is not so clear; is X "in one piece" in spite of all the holes (see problem E.4)? Connectivity, like compactness, is preserved under continuous maps.

(4.E.2) Theorem Suppose that (X,d) is a connected metric space, $(Y,\hat{d})$ is a metric space, and $f: X \rightarrow Y$ is continuous and onto. Then Y is connected.

Proof. Suppose that Y is not connected. Then there are nonempty open sets in Y, U and V, such that $U \cap V = \emptyset$ and $U \cup V = Y$. Since f is continuous and onto, the sets $f^{-1}(U)$ and $f^{-1}(V)$ are nonempty disjoint open subsets of X whose union is X. This, however, is impossible since X is connected, and the result follows.

(4.E.3) Theorem (Intermediate Value Theorem) Suppose that (X,d) is a connected metric space, $a,b \in X$, and $f:X \rightarrow \mathbf{R}^1$ is a continuous function. If $f(a) < c < f(b)$ [or $f(b) < c < f(a)$], then there is a point $x^* \in X$ such that $f(x^*) = c$.

Proof. If no such point x^* exists, then $U = f^{-1}(-\infty,c)$ and $V = f^{-1}(c,\infty)$ are disjoint nonempty subsets of X whose union is X, which contradicts the connectivity of X.

In problem E.5 it is seen that the converse of the intermediate value theorem is generally false.

A subset A of $\mathbf{R}^1$ is said to be an *interval* if given any two points $a,b \in A$

and a point c lying between a and b, then $c \in A$. Next we prove the obvious: intervals are connected.

(4.E.4) *Theorem* If A is an interval in $\mathbf{R}^1$ (usual metric), then A is connected.

Proof. Suppose that A is not connected. Then there are disjoint nonempty open subsets, U and V, of A whose union is A. Let $c \in U$ and $d \in V$, and suppose that $c < d$. If $s = \sup\{x \in U \mid x < d\}$, then clearly $c \leq s \leq d$. We leave it to the reader to show that $s \in U \cup V = A$, which contradicts the fact that A is an interval. [*Hint*: Use the fact that both U and V are open.]

As a corollary of (4.E.4) we have the following important special case of (4.E.3).

(4.E.5) *Corollary* If $f: [a,b] \to \mathbf{R}^1$ is continuous and if $f(a) < c < f(b)$ [or if $f(b) < c < f(a)$], then there is a point $x^* \in (a,b)$ such that $f(x^*) = c$.

(4.E.6) *Example* Suppose that $f: [3,7] \to \mathbf{R}^1$ is defined by $f(x) = x^2 + 3$. Since $f(3) = 12$ and $f(7) = 52$, there must be a number $x^* \in [3,7]$ such that $f(x^*) = 19$. In this case x^* can actually be found since $19 = x^{*2} + 3$, from which it follows that $x^* = 4$. In general, however, it may be impossible to actually calculate x^*.

In succeeding chapters we shall see that a number of proofs depend heavily on the intermediate value theorem. One immediate application of this theorem involves fixed points. A function $f: X \to X$ has a *fixed point* if there exists a point $x \in X$ such that $f(x) = x$. A metric space (X,d) is said to have the *fixed point property* if *every* continuous function $f: X \to X$ has at least one fixed point. The study of fixed points is not only aesthetically pleasing in itself but can also be put to a variety of very practical uses. For instance, many results involving the existence and uniqueness of solutions of differential equations may be obtained via fixed point theory.

In the next theorem it is shown that closed intervals in $\mathbf{R}^1$ have the fixed point property.

(4.E.7) *Theorem* Suppose that $f: [a,b] \to [a,b]$ is continuous. Then there is a point $x \in [a,b]$ such that $f(x) = x$.

Proof. If $f(a) = a$ or $f(b) = b$, then there is nothing to prove. Suppose then that $a < f(a)$ and $f(b) < b$. Define a continuous function $g: [a,b] \to \mathbf{R}^1$ by setting $g(x) = x - f(x)$. Note that $g(a) = a - f(a) < 0$ and $g(b) = b - f(b) > 0$. Since g is continuous, it follows from (4.E.5) that there is a point $x \in [a,b]$ such that $g(x) = 0$. However, this implies that $x - f(x) = 0$, i.e., $f(x) = x$.

We conclude this section with an important theorem involving fixed points and complete metric spaces.

(4.E.8) *Definition*   Suppose that (X,d) and $(Y,\hat{d})$ are metric spaces and that $f: X \to Y$. Suppose further that there is a real number α, $0 < \alpha < 1$, such that $\hat{d}(f(x),f(y)) \leq \alpha d(x,y)$ for each $x,y \in X$. Then the function f is said to be a *contraction map*.

(4.E.9) *Exercise*   Show that a contraction map is continuous.

(4.E.10) *Theorem*   Suppose that (X,d) is a complete metric space and that $f: X \to X$ is a contraction map. Then f has a unique fixed point.

Proof.   Since f is a contraction map there is a real number α, $0 < \alpha < 1$, such that $d(f(x),f(y)) \leq \alpha d(x,y)$ for each $x,y \in X$. Let x_0 be an arbitrary point in X. Set $x_1 = f(x_0)$, $x_2 = f(x_1) = f(f(x_0))$, and inductively let $x_n = f(x_{n-1})$ for each positive integer n. We shall show that the sequence $\{x_n\}$ is a Cauchy sequence and, hence, converges in X to a point x^*; the point x^* will turn out to be the unique fixed point of f.

Note that

$$d(x_1,x_2) = d(f(x_0),f(x_1)) \leq \alpha d(x_0,x_1)$$
$$d(x_2,x_3) = d(f(x_1),f(x_2)) \leq \alpha d(x_1,x_2) \leq \alpha^2 d(x_0,x_1)$$

An easy inductive argument shows that for each positive integer n,

$$d(x_n,x_{n+1}) \leq \alpha^n d(x_0,x_1)$$

Furthermore, if $m \geq n$, we have

$$d(x_n,x_m) \leq d(x_n,x_{n+1}) + d(x_{n+1},x_{n+2}) + \cdots + d(x_{m-1},x_m)$$
$$\leq (\alpha^n + \alpha^{n+1} + \cdots + \alpha^{m-1})d(x_0,x_1)$$
$$= \alpha^n(1 + \alpha + \cdots + \alpha^{m-n-1})d(x_0,x_1)$$

The sum $(1 + \alpha + \cdots + \alpha^{n-m-1})$ is easily seen to be less than $1/(1 - \alpha)$ [see (8.B.2.c)], and therefore, we have

$$d(x_n,x_m) \leq \frac{\alpha^n}{1 - \alpha} d(x_0,x_1)$$

Since $\lim_{n \to \infty} \alpha^n = 0$, it follows that $\{x_n\}$ is a Cauchy sequence and since (X,d) is a complete metric space, the sequence $\{x_n\}$ converges to a point $x^* \in X$.

A simple but rather delicate argument shows that $f(x^*) = x^*$. Since f is continuous ((4.E.9)), and since the sequence $\{x_n\}$ converges to x^*, it follows that $\lim_{n \to \infty} f(x_n) = f(x^*)$. On the other hand, $f(x_n) = x_{n+1}$ for each n, and consequently, we have

$$f(x^*) = \lim_{n \to \infty} f(x_n) = \lim_{n \to \infty} x_{n+1} = x^*$$

To see that x^* is the unique fixed point of f, suppose that x is another fixed point of f. Then we have

$$d(x,x^*) \geq \alpha d(f(x), f(x^*)) = \alpha d(x,x^*)$$

which is only possible if $x = x^*$; thus, x^* is the unique fixed point of f.

In general, it may be quite difficult to show that a given space has the fixed point property. One of the first problems involving the fixed point property was investigated by a physicist who posed the question: if a patch of oil and water are mixed on a plate, must some point of the patch remain unmoved? This is essentially equivalent to showing that the unit disk $\{(x,y) \in \mathbf{R}^2 \mid x^2 + y^2 \leq 1\}$ has the fixed point property (it does, but the proof is quite sophisticated). In problem E.14* a number of subsets of $\mathbf{R}^1$, $\mathbf{R}^2$, and $\mathbf{R}^3$ are given and the reader is invited to use his or her intuition to guess (or determine) which of these have the fixed point property.

F.* CONCLUSION

The notion of continuity was not a sudden revelation to mankind. Many early mathematicians struggled with this concept in an effort to render precise their intuitive sense of what a continuous function should be. Even the extraordinarily gifted nineteenth century mathematician Augustin-Louis Cauchy, who above all else wanted to instill mathematical analysis with a much needed rigor, could do no better than define a continuous function as follows (from Kline, 1972):

> Let $f(x)$ be a function of the variable x, and suppose that, for each value of x intermediate between two given limits [bounds], this function constantly assumes a finite and unique value. If, beginning with a value of x contained between these limits, one assigns to the variable x an infinitely small increment α, the function itself will take on as an increment the difference $f(x + \alpha) - f(x)$ which will depend at the same time on the new variable α and on the value of x. This granted, the function $f(x)$ will be, between the two limits assigned to the variable x, a *continuous* function of the variable if, for each value of x intermediate between these two limits, the numerical value of the difference $f(x + \alpha) - f(x)$ decreases indefinitely with that of α.

In other words, the function $f(x)$ will remain continuous with respect to x between the given limits, if, between these limits, an infinitely small increment of the variable always produces an infinitely small increment of the function itself.

In spite of the awkwardness of the previous passage, the reader should appreciate how close Cauchy was to the more "modern" formulations of continuity given in (4.A.1) and (4.A.11). The latter characterization of continuity was introduced by the great German mathematician Karl Weierstrass. It is important that the reader not be lulled by the simple elegance of the Weierstrass definition into believing that continuous functions are free of complications. Consider, for example, the continuous functions $\phi: [0,2\pi] \rightarrow \mathbf{R}^2$ defined by $\phi(x) = (x, \sin x)$ and $\psi: [0,1] \rightarrow \mathbf{R}^2$ defined by $\psi(x) = (x, x^2 - 1)$. The images of these functions are sketched in Fig. 4.11. Based on these two examples, one might be led to conjecture (and in fact many mathematicians once believed) that dimension is preserved under continuous functions (very roughly speaking, a one-dimensional space "looks like" a line segment, a 2-dimensional space is "similar" to $\mathbf{R}^2$ or a disk in $\mathbf{R}^2$, etc.). In 1890, Giuseppe Peano (1858–1932) startled the mathematical community by describing a continuous function $\phi: [0,1] \rightarrow \mathbf{R}^2$ that mapped the unit interval $[0,1]$ onto the unit square $S = \{(x,y) \in \mathbf{R}^2 \mid 0 \le x \le 1, 0 \le y \le 1\}$. Peano's contruction involved forming an infinite sequence of functions $f_1, f_2, \ldots$, each of which mapped $[0,1]$ into S. Hilbert modified slightly Peano's sequence; the first three functions in the Hilbert sequence are described in Fig. 4.12. The continuous surjection $f: [0,1] \rightarrow S$ is defined by setting $f(x) = \lim_{n \rightarrow \infty} f_n(x)$ for each $x \in [0,1]$. The reader might attempt to convince himself that f is indeed onto and continuous. Is f 1-1?

PROBLEMS

Section A

1. Show that $f: [0,\infty) \rightarrow \mathbf{R}^1$ defined by $f(x) = \sqrt{x}$ is continuous.
2. Let $f: \mathbf{R}^1 \rightarrow \mathbf{R}^1$ be defined by $f(x) = 4x - 7$, and let $a = 2$ and $\varepsilon = \frac{1}{3}$. Find $\delta > 0$ such that if $|x - a| < \delta$, then $|f(x) - f(a)| < \varepsilon$. Do the same for $a = 5$ and $\varepsilon = \frac{1}{3}$. Repeat the exercise for $f(x) = x^2/3$.
3. Let $f: \mathbf{R}^1 \rightarrow \mathbf{R}^1$ be defined by $f(x) = 1/x$, let $a = 1$, and $\varepsilon = \frac{1}{10}$. Find $\delta > 0$ such that if $|x - a| < \delta$, then $|f(x) - f(a)| < \varepsilon$. Do the same for $a = \frac{1}{4}$ and $\varepsilon = \frac{1}{10}$.
4. Show that if $f: X \rightarrow \mathbf{R}^1$ is continuous, $x^* \in X$, and $f(x^*) > 0$, then there exists $\varepsilon > 0$ such that $f(y) > 0$ for each $y \in S_\varepsilon(x^*)$.

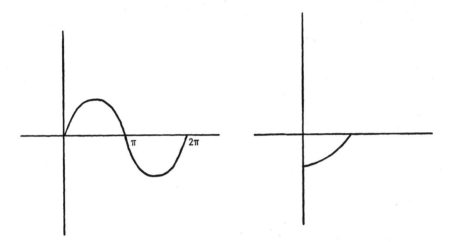

Figure 4.11

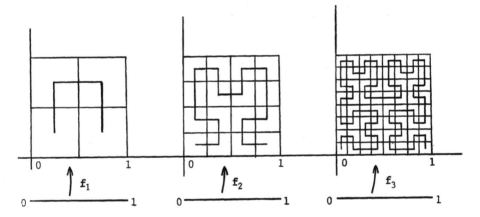

Figure 4.12

5. True or false? Suppose that X and Y are metric spaces, $f: X \to Y$, and $x^* \in X$. Then f is continuous at x^* is whenever a sequence $\{x_n\}$ of points distinct from x converges to x^*, the sequence $\{f(x_n)\}$ converges in Y. Does the condition "points distinct from x" make a difference?

6. Find an example to illustrate that if X and Y are metric spaces, $A \subset X$, and $f: X \to Y$, then $f|_A$ may be continuous, yet f is not continuous at *any* point $x \in X$.

7. Prove or give a counterexample: Suppose that X and Y are metric spaces, $f: X \to Y$, and U is an open subset of X. If f is continuous and 1-1, then $f(U)$ is open in Y.

8. Show that the function $f: \mathbf{R}^2 \to \mathbf{R}^1$ defined by

$$f(x,y) = \begin{cases} \dfrac{xy^2}{x^2 + y^4} & \text{if } (x,y) \neq (0,0) \\ 0 & \text{if } (x,y) = (0,0) \end{cases}$$

is not continuous at $(0,0)$. [*Hint*: Consider sequences of the form $\{(1/n^2, 1/n)\}$.]

9. (a) Suppose that $f: \mathbf{R}^2 \to \mathbf{R}^1$ is defined by

$$f(x,y) = \begin{cases} \dfrac{x^2 - y^2}{x^2 + y^2} & \text{if } (x,y) \neq (0,0) \\ 0 & \text{if } (x,y) = (0,0) \end{cases}$$

Is f continuous at $(0,0)$? at $(0,2)$?

   (b) Suppose that $f: \mathbf{R}^2 \to \mathbf{R}^1$ is defined by

$$f(x,y) = \begin{cases} x + y & \text{if } x \neq y \\ x^2 - 2y & \text{if } x = y \end{cases}$$

Is f continuous at $(1,2)$? at $(2,2)$? at $(4,4)$?

10. (a) Let $f: \mathbf{R}^1 \to \mathbf{R}^1$ be defined by

$$f(x) = \begin{cases} \sin \dfrac{1}{x} & \text{if } x \neq 0 \\ 0 & \text{if } x = 0 \end{cases}$$

Is f continuous at $x = 0$?

   (b) Let $f: \mathbf{R}^1 \to \mathbf{R}^1$ be defined by

$$f(x) = \begin{cases} x \sin \dfrac{1}{x} & \text{if } x \neq 0 \\ 0 & \text{if } x = 0 \end{cases}$$

Is f continuous at $x = 0$?

   (c) Let $f: \mathbf{R}^1 \to \mathbf{R}^1$ be defined by

$$f(x) = \begin{cases} \dfrac{x}{\sin (1/x)} & \text{if } x \neq 0 \\ 0 & \text{if } x = 0 \end{cases}$$

Is f continuous at $x = 0$?

11. Suppose that $f: \mathbf{R}^1 \to \mathbf{R}^1$ is continuous and that $f(x) = 0$ for each rational number x. Show that $f(x) = 0$ for all x.

12. Suppose that $f: \mathbf{R}^1 \to \mathbf{R}^1$ is defined by

$$f(x) = \begin{cases} 0 & \text{if } x \text{ is irrational} \\ \dfrac{m}{n} & \text{if } x = \dfrac{m}{n} \text{ (in reduced form)} \end{cases}$$

Show that f is continuous at all irrational points but fails to be continuous at each rational point.

13. Suppose that X is a metric space and $f: X \to \mathbf{R}^1$ is a bounded function. The *oscillation* of f on a subset A of X, $\mathrm{osc}_f\, A$, is defined to be $\sup\{|f(x) - f(y)| \mid x, y \in A\}$. The oscillation of f at a point x is defined to be $\lim_{n \to \infty} \mathrm{osc}_f\, (S_{1/n}(x))$. Show that f is continuous at x if and only if the oscillation of f at x is 0.

Section B

1. Find all adherence points of the following sets:
   (a) $A \subset \mathbf{R}^1$, where $A = [1,3) \cup (2,5) \cup (5,8]$.
   (b) $A \subset \mathbf{R}^2$, where $A = \{(x,y) \in \mathbf{R}^2 \mid x \text{ and } y \text{ are rational}\}$.
   (c) $A \subset \mathbf{R}^2$, where A is as defined in (b), but $\mathbf{R}^2$ is given the discrete metric.
   (d) $A \subset X$, where X is the set of all $n \times n$ matrices with the metric given in (3.B.4.i) and A is the set of all triangular matrices of the

   form $\begin{pmatrix} \cdots \cdots \\ \cdot \cdot \cdot \quad 0 \end{pmatrix}$

2. Suppose that X and Y are metric spaces and that $f: X \to Y$. Show that f is continuous if and only if $f^{-1}(A)$ is closed for each closed subset A of Y.

3. Suppose that X is a metric space and that $A \subset X$. Show that $\overline{(\overline{A})} = \overline{A}$.

4. Determine whether or not the following subsets of $\mathbf{R}^2$ are compact:
   (a) $\{(x,y) \in \mathbf{R}^2 \mid x^2 + y^2 = 16\}$.
   (b) For each n, let $A_n = \{(x,y) \mid 0 \le x \le 1, y = 1/n\}$ and let $A_0 = \{(x,y) \mid 0 \le x \le 1, y = 0\}$. Is $A = \bigcup_{i=0}^{\infty} A_i$ compact? Is $A = \bigcup_{i=1}^{\infty} A_i$ compact?
   (c) $A = \{(x,y) \mid x = 1/n \text{ or } x = 0, y = 1/n \text{ or } y = 0\}$.

5. Show that arbitrary intersections of closed subsets of a metric space are closed.

6. Suppose that $A_1 \supset A_2 \supset A_3 \supset \cdots$ is a sequence of nonempty compact subsets of a metric space X. Show that $\bigcap_{i=1}^{\infty} A_i \neq \varnothing$. (*Hint*: For each $n \in \mathbf{Z}^+$ select a point $x_n \in A_n$.)

7. A particularly interesting closed (and compact) subset of $\mathbf{R}^1$ can be constructed by successively removing middle thirds of subintervals of $[0,1]$ as follows. Let

$A_1 = [0,\tfrac{1}{3}] \cup [\tfrac{2}{3},1]$

$A_2 = [0,\tfrac{1}{9}] \cup [\tfrac{2}{9},\tfrac{1}{3}] \cup [\tfrac{2}{3},\tfrac{7}{9}] \cup [\tfrac{8}{9},1]$

$A_3 = [0,\tfrac{1}{27}] \cup [\tfrac{2}{27},\tfrac{1}{9}] \cup [\tfrac{2}{9},\tfrac{7}{27}] \cup [\tfrac{8}{27},\tfrac{1}{3}] \cup [\tfrac{2}{3},\tfrac{19}{27}] \cup [\tfrac{20}{27},\tfrac{2}{9}] \cup [\tfrac{8}{9},\tfrac{25}{27}] \cup [\tfrac{26}{27},1]$

Let $A = \bigcap_{n=1}^{\infty} A_n$. Then A is called the *Cantor set*.

(a) Show that each endpoint of each interval appearing in each A_n is a point in the Cantor set. Find a point in the Cantor set which is not one of these points (actually there is an uncountable number of such points). [*Hint*: Use problem B.6 and the fact that the endpoints form a countable set.]

(b) Show that the Cantor set is compact.

(c) Show that the Cantor set does not contain any interval.

(d)* Show that the Cantor set is uncountable.

(e)* Use a Cantor-like procedure to construct a closed uncountable subset of $[0,1]$ that consists solely of irrational numbers.

8. Suppose that A is a closed subset of a metric space X and that B is a compact subset of X. Show that if $A \cap B = \emptyset$, then $\rho(A,B) > 0$ (see Chapter 3, problem 22).

9. Show that if A and B are disjoint closed subsets of a metric space X, then there exist open subsets U and V of X such that $A \subset U$, $B \subset V$, and $U \cap V = \emptyset$.

10. Suppose that (X,d) and $(Y,\hat{d})$ are metric spaces. A function $f: X \to Y$ is an *isometry* if for each $w,z \in X$, $d(w,z) = \hat{d}(f(w),f(z))$.

(a) Show that isometries are continuous.

(b)* Show that if $f: X \to X$ is an isometry and X is a compact metric space, then f is onto.

11. Let $X = \{f: [0,1] \to \mathbf{R}^1 \mid f \text{ is bounded}\}$, and give X the sup metric.

(a) For each $n \in \mathbf{Z}^+$, let $f_n(x) = nx/(1 + nx)$. If $A = \{f_n \mid n \in \mathbf{Z}^+\}$ and f is the constant function $f(x) = 1$ for all $x \in [0,1]$, is f an adherence point of A?

(b) Do as in (a) if $f_n(x) = xe^{-nx}/n$ and $f(x) = 0$ for all x. [*Hint*: Determine where f_n takes on a maximum.]

(c) Let $A = \{f \in X \mid f(X) \subset [0,5]\}$. Is A compact?

12. Find a continuous function $f: \mathbf{R}^1 \to \mathbf{R}^1$ that is bounded but which attains neither a maximum nor a minimum.

13. Find all compact subsets of a discrete metric space.

14. Find an example of a continuous bijective function f that maps a metric space onto a metric space Y, but such that f^{-1} is not continuous.

Section C

1. Show that any discrete metric space is complete.

2. Suppose that X is a metric space and that $A \subset X$ is complete. Show that A is closed.

3. Suppose that X and Y are metric spaces and that $f: X \to Y$ is con-

tinuous. If $\{x_n\}$ is a Cauchy sequence in X is the sequence $\{f(x_n)\}$ necessarily a Cauchy sequence in Y?

4. Suppose that $\{x_n\}$ is a Cauchy sequence in a metric space X. Show that the sequence $\{d(x_n,x_1)\}$ is bounded.

5. Suppose that (X,d) is a complete metric space and that $A_1 \supset A_2 \supset A_3 \supset \cdots$ is a sequence of nonempty closed subsets of X such that $\lim_{i \to \infty} (\text{diam } A_i) = 0$. Show that $\bigcap_{i=1}^{\infty} A_i \neq \emptyset$. Is the hypothesis $\lim_{i \to \infty} (\text{diam } A_i) = 0$ necessary?

6. Suppose that $\{x_n\}$ is a sequence in $\mathbf{R}^1$, α is a real number such that $0 < \alpha < 1$ and $|x_{n+1} - x_n| < \alpha^n$ for each n. Show that $\{x_n\}$ is a Cauchy sequence.

7. Show that if the sequence $\{f_n\}$ in (4.C.7) is a sequence of continuous functions, then the limit function f is also continuous.

8. Let $\mathscr{A}$ be the set of all $m \times n$ matrices with real entries and define a norm $\| \ \|$ for $\mathscr{A}$ by setting $\|A\| = \sum_{i=1}^{m} \sum_{j=1}^{n} |a_{ij}|$ for each $A \in \mathscr{A}$. Let ρ be the metric associated with this norm, and show that $(\mathscr{A},\rho)$ is a complete metric space.

9. Formulate a definition of the sup metric analogous to (3.B.4.h) where $\mathbf{R}^1$ is replaced by an arbitrary metric space, and then show that Theorem (4.C.7) is valid if $\mathbf{R}^1$ is replaced by a complete metric space.

Section D*

1. Show that each bounded set in $\mathbf{R}^2$ (or $\mathbf{R}^n$) is totally bounded.

2. Show that the interval $[0,1)$ is not compact by exhibiting an open cover of $[0,1)$ that has no finite subcover.

3. Use open covers to show that if $\{x_n\}$ is a sequence in a metric space and if $\{x_n\}$ has no cluster points, then the set $A = \{x_n \mid n \in \mathbf{Z}^+\}$ fails to be compact.

4. Show that $[0,1]$ can be covered by a family of nondegenerate closed intervals that has no finite subcover.

5. Use open covers to show that $[0,1]$ is compact. [*Hint*: Let $\{U_\alpha \mid \alpha \in J\}$ be an open cover of $[0,1]$ and set $s = \sup\{b \mid [0,b]$ can be covered by a finite subcover$\}$; show that $s = 1$.]

6. Use open covers to prove (4.B.15).

7. A metric space is said to be *countably compact* if each countable open cover of X has a finite subcover. Show that countable compactness is preserved under continuous functions.

8. A family of subsets $\mathscr{A} = \{A_\alpha \mid \alpha \in J\}$ of a set X is said to have the *finite intersection property* if for each finite subset K of J, $\cap \{A_\alpha \mid \alpha \in K\} \neq \emptyset$. Show that a metric space X is compact if and only if for each family

of closed subsets $\mathscr{A} = \{A_\alpha \mid \alpha \in J\}$ of X with the finite intersection property, $\bigcap_{\alpha \in J} A_\alpha \neq \varnothing$. [*Hint*: Use the DeMorgan rules: $X \backslash \bigcup_{\alpha \in J} A_\alpha = \bigcap_{\alpha \in J} (X \backslash A_\alpha)$ and $X \backslash \bigcap_{\alpha \in J} A_\alpha = \bigcup_{\alpha \in J} (X \backslash A_\alpha)$.]

9. Show that a metric space X is compact if and only if X is complete and totally bounded.

10.* Suppose that (X,d) is a compact metric space, $(Y,\hat{d})$ is a metric space, and that $f: X \to Y$ is continuous. Show that if $\varepsilon > 0$ is given, then there is a $\delta > 0$ such that $\hat{d}(f(x_1),f(x_2)) < \varepsilon$ whenever $d(x_1,x_2) < \delta$, i.e., δ does not depend on a particular point $x \in X$.

Section E

1. Show that a connected subset A of a metric space either consists of a single point or is infinite. Could A be countably infinite?

2. Let $S = \{0,1\}$ with the discrete metric. Show that a metric space X is connected if and only if there does not exist a continuous function $f: X \to S$ that is onto.

3. Use problem E.2 to show that if $\{A_\alpha \mid \alpha \in J\}$ is a family of connected subsets of a metric space X such that $\bigcap_{\alpha \in J} A_\alpha \neq \varnothing$, then $\bigcup_{\alpha \in J} A_\alpha$ is connected.

4. Show that $\mathbf{R}^2$ and $\mathbf{R}^2 \backslash \{(x,y) \mid x \text{ and } y \text{ are rational}\}$ are connected, but that $\mathbf{R}^2 \backslash \{(x,y) \mid x \text{ or } y \text{ is irrational}\}$ is not connected. [*Hint*: Use problem E.3 and that fact that line segments are connected.]

5. Show that if $f: [0,1] \to \mathbf{R}^1$ is defined by

$$f(x) = \begin{cases} \sin \dfrac{1}{x} & \text{if } x \neq 0 \\ \tfrac{1}{2} & \text{if } x = 0 \end{cases}$$

then f is not continuous; show however, that f does satisfy the intermediate value theorem, i.e., if $a,b \in [0,1]$ and c lies between $f(a)$ and $f(b)$, then there is a point $x^* \in [0,1]$ such that $f(x^*) = c$.

6. Use the intermediate value theorem to show that $p(x) = x^3 - 3x^2 - 9x + 1$ has three distinct zeros. [*Hint*: Consider the critical points of p.]

7.* True or false: Suppose that $A_1 \supset A_2 \supset A_3 \supset \cdots$ is a sequence of closed connected subsets of $\mathbf{R}^2$. Then $\bigcap_{i=1}^{\infty} A_i$ is connected.

8. Show that if $A \subset \mathbf{R}^1$ is connected, then A is an interval.

9. Show that if $f: [0,1) \to [0,1)$ is continuous and onto, then f has a fixed point.

10. A metric space (X,d) is said to be *well chained* if for each $\varepsilon > 0$ and each pair of points x and y in X there is a finite subset $\{x_1, x_2, \ldots, x_n\}$ in X such that $x_1 = x$, $x_n = y$, and $d(x_i, x_{i+1}) < \varepsilon$ for $i = 1, 2, \ldots, n-1$.

Show that if X is connected, then X is well chained; show also that the converse of this statement is false.

11.* Suppose that $A \subset \mathbf{R}^1$ has the fixed point property. Show that A is a point or a closed and bounded interval.

12. Determine whether the following functions are contraction maps on $\mathbf{R}^1$ or on $[0,1]$:

$$\text{(a)} \quad f(x) = \frac{x}{3}$$

$$\text{(b)} \quad f(x) = x^2$$

$$\text{(c)} \quad f(x) = \sin x$$

13.* Show that if X and Y are connected metric spaces, then $X \times Y$ is connected.

14.* Try to determine (at least intuitively) which of the following sets have the fixed point property:

(a) The open interval (a,b)

(b) A circle

(c) A solid disk

(d) A cross

(e) The boundary of a sphere

(f) A solid sphere

(g) An infinite (but slightly damaged) comb

$$\left\{ (x, y) \mid x = \frac{1}{n}, 0 \leq y \leq 1 \right\} \cup \{(x, y) \mid 0 \leq x \leq 1, y = 0\}$$

(h) A snake

$$\left\{ \left(x, \sin\frac{1}{x} \right) \middle| 0 < x \leq \frac{2}{\pi} \right\} \cup \{(x, y) \mid x = 0, -1 \leq y \leq 1\}$$

REFERENCE

Kline, Morris (1972): *Mathematical Thought from Ancient to Modern Times*, Oxford University Press, New York.

5

THE DERIVATIVE: THEORY AND ELEMENTARY APPLICATIONS

A. THE DERIVATIVE

In Chapter 4 we found that if f is a continuous function mapping a compact metric space A into $\mathbf{R}^1$, then f attains a maximum and a minimum in A. Theorem (4.B.16) is a typical existence result in that it provides little or no help in pinpointing the location of the guaranteed maximum and minimum values. As the reader already knows, additional information concerning extrema can be obtained with the aid of derivatives. In the present chapter we shall examine this and other problems in which the derivative can play a significant role. Before embarking on our study of the derivative, we introduce the notion of *accumulation point*, and then we look at the limit concept from a slightly different viewpoint.

(5.A.1) *Definition* Suppose that (X,d) is a metric space and that $A \subset X$. Then a point $x \in X$ is an *accumulation point* of A if for each $\varepsilon > 0$,

$$S_\varepsilon(x) \cap (A\backslash\{x\}) \neq \varnothing$$

The reader should observe that if in the above definition, $x \notin A$, then the notions of accumulation point and adherence point coincide, since in this case $A = A\backslash\{x\}$. Note, however, that if $X = A = \mathbf{Z}^+$, then no point of A is an accumulation point of A, although by definition each point of A is an adherence point of A.

(5.A.2) *Definition*   Suppose that (X,d) and $(Y,\hat{d})$ are metric spaces, $A \subset X$, x is an accumulation point of A, and $f: A \to Y$. Then a point $y^* \in Y$ is said to be the *limit* of f as x approaches x^* (which we denote $\lim_{x \to x^*} f(x) = y^*$), if for each $\varepsilon > 0$, there is a $\delta > 0$ such that $\hat{d}(f(x),y^*) < \varepsilon$ whenever $0 < d(x,x^*) < \delta$.

It is important to note that the existence of $\lim_{x \to x^*} f(x)$ is independent of how (or whether) $f(x^*)$ is defined.

(5.A.3) *Example*   If $f: (\mathbf{R}^1 \backslash \{2\}) \to \mathbf{R}^1$ is defined by $f(x) = (x^2 - x - 2)/(x - 2)$, then $\lim_{x \to 2} f(x) = \lim_{x \to 2} [(x - 2)(x + 1)]/(x - 2) = 3$.

(5.A.4) *Definition*   Suppose that $(Y,\hat{d})$ is a metric space, (a,b) is an open interval in $\mathbf{R}^1$ and $f: (a,b) \to Y$. Let $c \in [a,b]$. Then the *right-hand limit of f at c*, $\lim_{x \to c^+} f(x)$, is equal to $y^* \in Y$ if for each $\varepsilon > 0$, there is a $\delta > 0$ such that if $|x - c| < \delta$ and if $x > c$, then $\hat{d}(f(x),y^*) < \varepsilon$. If $c \in (a,b]$, then the *left-hand limit of f at c*, $\lim_{x \to c^-} f(x)$, is equal to $y^{**} \in Y$ if for each $\varepsilon > 0$, there is a $\delta > 0$ such that if $|x - c| < \delta$ and if $x < c$, then $\hat{d}(f(x),y^{**}) < \varepsilon$.

(5.A.5) **Example**   Suppose that $f: \mathbf{R}^1 \to \mathbf{R}^1$ is defined by

$$f(x) = \begin{cases} x^2 & \text{if } x \geq 2 \\ 3x & \text{if } x < 2 \end{cases}$$

Then $\lim_{x \to 2^+} f(x) = 4$ and $\lim_{x \to 2^-} f(x) = 6$.

(5.A.6) *Observation*   Suppose that (X,d) and $(Y,\hat{d})$ are metric spaces, x^* is an accumulation point of X, and $f: X \to Y$. Then it follows immediately from (4.A.11) that f is continuous at x^* if and only if $\lim_{x \to x^*} f(x) = f(x^*)$ (see problem A.10).

(5.A.7) *Exercise*   Suppose that (X,d) and $(Y,\hat{d})$ are metric spaces, $A \subset X$, and that x^* is an accumulation point of A. Show that if $f: A \to Y$, then $\lim_{x \to x^*} f(x) = y$ if and only if for *every* sequence $\{x_n\}$ in A that converges to x^*, the sequence $\{f(x_n)\}$ converges to y. [*Hint*: Recall the proof of (4.A.11).]

The concept of the derivative is commonly motivated as follows. Consider the graph of the function f given in Fig. 5.1. With each point $(x,f(x))$ on the graph of f we wish to assign a directed line, L_x, which, as x varies from to left to right, serves to give a "sense of direction" to the graph at the point $(x,f(x))$ (Fig. 5.2).

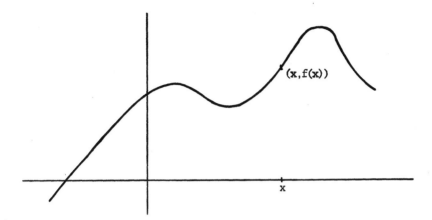

Figure 5.1

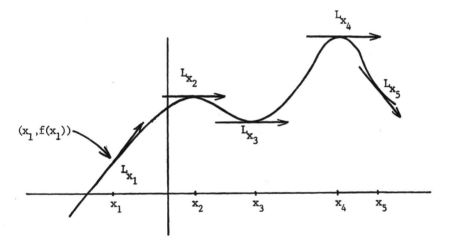

Figure 5.2

Since we assume that for each x, the point $(x, f(x))$ lies on the line L_x, in order to describe L_x it is sufficient to assign this line a slope. This we do indirectly: the desired slope is defined as a limit of slopes of lines that are presumed to approximate the (still to be determined) line L_x. Suppose then that $f: (a,b) \to \mathbf{R}^1$ and that $\hat{x} \in (a,b)$. Define $s: ([a,b]\backslash\{\hat{x}\}) \to \mathbf{R}^1$ by

$$s(x) = \frac{f(x) - f(\hat{x})}{x - \hat{x}} \tag{1}$$

Note that for each $x \in (a,b)\backslash\{\hat{x}\}$, $s(x)$ gives the slope of the line passing through the points $(x, f(x))$ and $(\hat{x}, f(\hat{x}))$. The derivative (and the slope of $L_{\hat{x}}$) is defined to be the limit of these slopes as x approaches $\hat{x}$.

(5.A.8) *Definition* Suppose that $f: (a,b) \to \mathbf{R}^1$ and $\hat{x} \in (a,b)$. Then f is *differentiable* at $\hat{x}$ if $\lim_{x \to \hat{x}} s(x)$ exists, where s is defined by (1). If f is differentiable at $\hat{x}$, we define the *derivative* of f at $\hat{x}$, $f'(\hat{x})$, by

$$f'(\hat{x}) = \lim_{x \to \hat{x}} s(x)$$

On occasion the symbols dy/dx and df/dx will also be used to denote the derivative.

If $\hat{x} \in [a,b)$, then the *right-hand derivative of f at $\hat{x}$* is defined to be $\lim_{x \to \hat{x}+} s(x)$, and similarly if $\hat{x} \in (a,b]$, then the *left-hand derivative of f at $\hat{x}$* is defined to be $\lim_{x \to \hat{x}-} s(x)$. If $f: [a,b] \to \mathbf{R}^1$, then we say that f is *differentiable on $[a,b]$* if f is differentiable on (a,b) and the right-hand and left-hand derivatives of f exist at a and b, respectively.

(5.A.9) *Definition* If $f: (a,b) \to \mathbf{R}^1$, $\hat{x} \in (a,b)$, and $f'(\hat{x})$ is defined, then the *line tangent* to the graph of f at the point $(\hat{x}, f(\hat{x}))$ is the line with slope $f'(\hat{x})$ that passes through the point $(\hat{x}, f(\hat{x}))$.

(5.A.10) *Exercise* Show that if $f: (a,b) \to \mathbf{R}^1$, $\hat{x} \in (a,b)$, and $f'(\hat{x})$ exists, then

$$f'(x) = \lim_{h \to 0} \frac{f(x+h) - f(x)}{h}$$

where $h \neq 0$ is always such that $a < \hat{x} + h < b$.

(5.A.11) *Observation* It is immediate from the definition of the derivative that if $f: (a,b) \to \mathbf{R}^1$ is a constant function, then $f'(x) = 0$ for each $x \in (a,b)$. Furthermore, if f is the identity function $f(x) = x$, then $f'(x) = 1$ for each $x \in (a,b)$.

(5.A.12) *Exercise* Show that if $f: \mathbf{R}^1 \to \mathbf{R}^1$ is defined by

$$f(x) = \begin{cases} x^2 & \text{if } x < 1 \\ (x-2)^2 & \text{if } x \geq 1 \end{cases}$$

then f is continuous at $x = 1$, but $f'(1)$ fails to exist.

Although it is clear from the preceding exercise that there are continuous functions that are not differentiable, we see next that differentiability does imply continuity.

(5.A.13) *Theorem* Suppose that $f: (a,b) \to \mathbf{R}^1$, $\hat{x} \in (a,b)$, and $f'(\hat{x})$ exists. Then f is continuous at $\hat{x}$.

Proof. Simply observe that

$$\lim_{x \to \hat{x}} f(x) = \lim_{x \to \hat{x}} \frac{f(x) - f(\hat{x})}{x - \hat{x}} (x - \hat{x}) + f(\hat{x}) = f'(\hat{x}) \cdot 0 + f(\hat{x})$$

and apply (5.A.6).

We assume that the reader is familiar with the formulas for obtaining the derivatives of sums, products, and quotients of functions (see problem A.1); furthermore, we shall suppose that the reader is quite conversant with the derivatives of such common functions as the trigonometric functions, exponential function, etc., and these will not be rederived in the text.

(5.A.14) *Exercise* Show that the derivative of an even function is odd, and that the derivative of an odd function is even.

Next we establish the chain rule, which provides a simple formula for the calculation of the derivative of a composition of two functions.

(5.A.15) *Theorem* (*The Chain Rule*) Suppose that $f: (a,b) \to \mathbf{R}^1$, $B = f(a,b)$, and $g: B \to \mathbf{R}^1$. If f is differentiable at $\hat{x}$ and g is differentiable at $\hat{y} = f(\hat{x})$ (where $\hat{y}$ lies in an interval contained in B), then $h = g \circ f$ is differentiable at $\hat{x}$ and $h'(\hat{x}) = g'(f(\hat{x})) \cdot f'(\hat{x})$.

Proof. We begin the proof in the most obvious manner, then show what goes wrong, and finally we shall add the necessary arguments to correct the "obvious" proof. If $h'(\hat{x})$ exists, then it is equal to

$$\lim_{x \to \hat{x}} \frac{h(x) - h(\hat{x})}{x - \hat{x}} = \lim_{x \to \hat{x}} \frac{g(f(x)) - g(f(\hat{x}))}{x - \hat{x}} \tag{2}$$

Since we have information concerning the derivatives of f and g at $\hat{x}$ and $f(\hat{x})$, respectively, it is natural to try to exploit this information by writing

$$\lim_{x \to \hat{x}} \frac{g(f(x)) - g(f(\hat{x}))}{x - \hat{x}} = \lim_{x \to \hat{x}} \frac{g(f(x)) - g(f(\hat{x}))}{f(x) - f(\hat{x})} \frac{f(x) - f(\hat{x})}{x - \hat{x}} \tag{3}$$

Since by (5.A.13) f is continuous at $\hat{x}$, it follows that as $x \to \hat{x}$, $f(x)$ approaches $f(\hat{x})$. Now using the fact that the limit of a product is equal to the product of the limits, we have that the right-hand side of (3) is equal to $g'(f(\hat{x})) \cdot f'(\hat{x})$, the desired result.

The flaw in this proof is that it may be the case that in each neighborhood of $\hat{x}$ there is some x such that $f(x) - f(\hat{x}) = 0$, and this renders meaningless the quotient

$$\frac{g(f(x)) - g(f(\hat{x}))}{f(x) - f(\hat{x})}$$

To get around this problem, we define a new function

$$\phi(x) = \begin{cases} \dfrac{g(f(x)) - g(f(\hat{x}))}{f(x) - f(\hat{x})} & \text{if } f(x) \ne f(\hat{x}) \\ g'(f(\hat{x})) & \text{if } f(x) = f(\hat{x}) \end{cases}$$

We observe that for *all* x, $x \ne \hat{x}$

$$\frac{h(x) - h(\hat{x})}{x - \hat{x}} = \phi(x) \cdot \frac{f(x) - f(\hat{x})}{x - \hat{x}}$$

and that $\lim_{x \to \hat{x}} \phi(x) = g'(f(\hat{x}))$ (why?).
It now follows that

$$h'(\hat{x}) = \lim_{x \to \hat{x}} \frac{h(x) - h(\hat{x})}{x - \hat{x}} = \lim_{x \to \hat{x}} \phi(x) \frac{f(x) - f(\hat{x})}{x - \hat{x}}$$

$$= \lim_{x \to \hat{x}} \phi(x) \lim_{x \to \hat{x}} \frac{f(x) - f(\hat{x})}{x - \hat{x}} = g'(f(\hat{x})) \cdot f'(\hat{x})$$

which completes the proof.

We conclude this section by establishing an important relationship between the derivative of a 1-1 function and the derivative of its inverse.

(5.A.16) Theorem Suppose that $f : [a,b] \to \mathbf{R}^1$ is continuous and 1-1 and that $f'(\hat{x})$ exists and is not equal to zero for some point $\hat{x} \in (a,b)$. Let $y = f(\hat{x})$. Then $(f^{-1})'(y)$ exists and is equal to $1/f'(\hat{x})$.

Proof. Let $B = f([a,b])$. Then by (4.B.17), $f^{-1} : B \to [a,b]$ is continuous. To find $(f^{-1})'(y)$, let $\{y_n\}$ be any sequence in B converging to y such that $y_n \ne y$ for each n. Let $x_n = f^{-1}(y_n)$ and $\hat{x} = f^{-1}(y)$. Since f^{-1} is continuous, we have by (5.A.7)

$$\hat{x} = f^{-1}(y) = \lim_{n \to \infty} f^{-1}(y_n) = \lim_{n \to \infty} x_n$$

and therefore,

$$(f^{-1})'(y) = \lim_{n \to \infty} \frac{f^{-1}(y_n) - f^{-1}(y)}{y_n - y} = \lim_{n \to \infty} \frac{x_n - \hat{x}}{f(x_n) - f(\hat{x})} = \frac{1}{f'(\hat{x})}$$

B. APPLICATIONS OF THE DERIVATIVE AND THE MEAN VALUE THEOREMS

First we see how the derivative can be employed to identify sites of local maxima and local minima.

(5.B.1) *Definition* Suppose that (X,d) is a metric space and that $A \subset X$. A point $x \in A$ is an *interior point* of A if there is an open set U in X such that $x \in U \subset A$. The set of all interior points of A is called the *interior* of A, and is denoted by int A.

Note that the interior of the subset $[a,b]$ of $\mathbf{R}^1$ is the open interval (a,b), but that the interior of $[a,b]$ considered as a subset of $\mathbf{R}^2$ is empty. Any open subset of a metric space is clearly equal to its interior.

(5.B.2) *Theorem* Suppose that $A \subset \mathbf{R}^1$ and that $f: A \to \mathbf{R}^1$ attains a local maximum (or local minimum) at a point $x^* \in$ int A. Then if $f'(x^*)$ exists, $f'(x^*) = 0$.

Proof. Since f has a local maximum at x^*, there is a positive number r such that if $x \in (x^* - r, x^* + r) \subset A$, then $f(x) \leq f(x^*)$. Note that by (5.A.7) if $\{h_n\}$ is a sequence of positive numbers in $(0,r)$ that converges to 0, then

$$f'(x^*) = \lim_{n \to \infty} \frac{f(x^* + h_n) - f(x^*)}{h_n} \leq 0$$

On the other hand, if $\{h_n\}$ is a sequence in $(-r,0)$ that converges to 0, then we have

$$f'(x^*) = \lim_{n \to \infty} \frac{f(x^* + h_n) - f(x^*)}{h_n} \geq 0$$

Consequently, $f'(x^*) = 0$.

The clause in the hypothesis of the previous theorem, "if $f'(x^*)$ exists," is clearly necessary, since even continuous functions may have local maxima or local minima at interior points where the derivative does not exist; in (5.A.12), for example, the function f has a local maximum at $x = 1$, but $f'(1)$ does not exist.

As a consequence of (5.B.2) we have the following classical result.

(5.B.3) *Theorem* (*Rolle's Theorem*) Suppose that $f: [a,b] \to \mathbf{R}^1$ is continuous and that $f'(x)$ exists for each $x \in (a,b)$. If $f(a) = f(b)$, then there is a point $c \in (a,b)$ such that $f'(c) = 0$.

Proof. By (4.B.16), f attains a maximum at some point $x_1 \in [a,b]$ and a minimum at some point $x_2 \in [a,b]$. If $x_1 = x_2$, then f is a constant function, and hence, by (5.A.11), $f'(x) = 0$ for each $x \in (a,b)$. If $x_1 \neq x_2$, then either $x_1 \in (a,b)$ or $x_2 \in (a,b)$ (or both), and therefore, by (5.B.2) either $f'(x_1) = 0$ or $f'(x_2) = 0$ (or both).

(5.B.4) *Exercise*   Suppose that $I \subset \mathbf{R}^1$ is an open interval and that $f: I \to \mathbf{R}^1$ is differentiable and has the property that $f'(x) \neq 0$ for each $x \in I$. Show that f is 1-1.

Perhaps the single most important theoretical result in elementary calculus is the *mean value theorem*. Proofs of an astonishing number of basic properties in calculus can be derived utilizing this theorem. The statement of the mean value theorem is as follows.

(5.B.5) *Theorem*   (*Mean Value Theorem*)   Suppose that $f: [a,b] \to \mathbf{R}^1$ is continuous and that f is differentiable on (a,b). Then there is a point $c \in (a,b)$ such that

$$\frac{f(b) - f(a)}{b - a} = f'(c)$$

Geometrically this result is obvious. Consider the line L passing through the points $(a,f(a))$ and $(b,f(b))$ (Fig. 5.3). If we raise or lower this line (without altering its slope), then at some point $(c,f(c))$ we obtain a line L_c tangent to the curve. Since the slope of the original line L is given by $(f(b) - f(a))/(b - a)$ and since the tangent line L_c has slope $f'(c)$, the result follows. However, this argument, while substantially correct, is too dependent on our geometrical intuition. We now prove a slightly more general result, a corollary of which is the mean value theorem.

(5.B.6) *Theorem*   (*Extended Mean Value Theorem or Cauchy Mean Value Theorem*)   Suppose that $f: [a,b] \to \mathbf{R}^1$ and $g: [a,b] \to \mathbf{R}^1$ are continuous functions such that $f'(x)$ and $g'(x)$ exist for each $x \in (a,b)$ and such that there is no point x where both $f'(x)$ and $g'(x)$ are simultaneously equal to 0. If $g(b) \neq g(a)$, then there is a point $c \in (a,b)$ such that

$$\frac{f(b) - f(a)}{g(b) - g(a)} = \frac{f'(c)}{g'(c)}$$

Proof.   The proof is based on finding a suitable function to which Rolle's theorem may be applied. Consider the image of the function $\phi: [a,b] \to \mathbf{R}^2$ defined by $\phi(x) = (g(x),f(x))$. The line L passing through the points $(g(a), f(a))$ and $(g(b),f(b))$ has slope $m^* = (f(b) - f(a))/(g(b) - g(a))$, and hence, the equation of L is given by $y = m^*z + d$, where $d = f(b) - m^*g(b)$ [since $f(b) = m^*g(b) + d$] (Fig. 5.4). We now define a function ψ that gives the vertical difference between points on L and points on the curve C (the image of ϕ); the value of this function will be 0 at the endpoints a and b. Define $\psi: [a,b] \to \mathbf{R}^1$ by $\psi(x) = [m^*g(x) + (f(b) - m^*g(b))] - f(x)$. The reader can

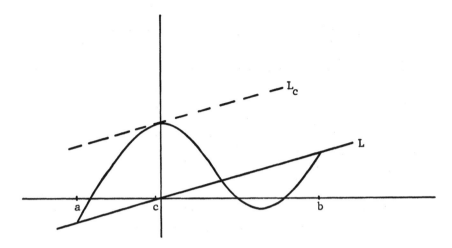

Figure 5.3

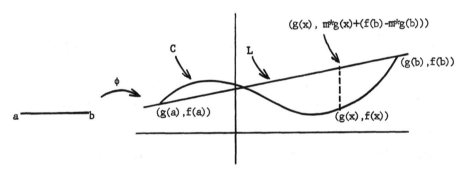

Figure 5.4

easily verify that $\psi(a) = 0$ and $\psi(b) = 0$ and that ψ satisfies the remaining hypotheses of Rolle's theorem. Consequently, there is a point $c \in (a,b)$ such that $0 = \psi'(c) = m^*g'(c) - f'(c)$. Note that $g'(c) \neq 0$, since otherwise $f'(c)$ and $g'(c)$ would be simultaneously equal to 0; thus, we have $m^* = f'(c)/g'(c)$, the desired conclusion.

(5.B.7) Corollary (Mean Value Theorem) Suppose that $f: [a,b] \to \mathbf{R}^1$ is continuous on $[a,b]$ and differentiable on (a,b). Then there is a point $c \in (a,b)$ such that

$$\frac{f(b) - f(a)}{b - a} = f'(c)$$

Proof. Let $g: [a,b] \to \mathbf{R}^1$ be defined by $g(x) = x$ and apply the preceding theorem.

C. APPLICATIONS OF THE MEAN VALUE THEOREMS

The mean value theorems constitute the principal ingredients of the proofs of the next few theorems, all of which should be familiar from the reader's earlier studies.

(5.C.1) *Theorem* Suppose that $f: (a,b) \to \mathbf{R}^1$ and $g: (a,b) \to \mathbf{R}^1$ are continuous functions such that $f'(x) = g'(x)$ for each $x \in (a,b)$. Then there is a number $C \in \mathbf{R}^1$ such that $f(x) = g(x) + C$ for each $x \in (a,b)$.

Proof. Fix a point $x^* \in (a,b)$ and let $C = f(x^*) - g(x^*)$. Suppose that $\hat{x} \in (a,b)$ is any other point in (a,b). We show that $f(\hat{x}) - g(\hat{x}) = f(x^*) - g(x^*) = C$. Note that if $h: [a,b] \to \mathbf{R}^1$ is defined by $h(x) = f(x) - g(x)$, then $h'(x) = 0$ for each $x \in (a,b)$. By the mean value theorem, we have

$$\frac{h(\hat{x}) - h(x^*)}{\hat{x} - x^*} = h'(c) = 0$$

for some c between $\hat{x}$ and x^*. Hence, $h(\hat{x}) = h(x^*)$, or equivalently, $f(\hat{x}) - g(\hat{x}) = f(x^*) - g(x^*)$, and the result follows.

We now make use of Cauchy's mean value theorem to establish L' Hospital's rules (which, incidentally, are due to Johann Bernoulli I). First, the inevitable definitions.

(5.C.2) *Definition* Suppose that (X,d) is a metric space, $A \subset X, f: A \to \mathbf{R}^1$, and x^* is an accumulation point of A. Then we say that the *limit of $f(x)$ as x approaches x^* is* ∞ (and we write $\lim_{x \to x^*} f(x) = \infty$) if corresponding to each $N \in \mathbf{Z}^+$, there is an $\varepsilon > 0$ such that $f(x) \geq N$, whenever $x \in S_\varepsilon(x^*) \cap (A\backslash\{x^*\})$.

(5.C.3) *Example* Suppose that $X = \mathbf{R}^1$, $A = (0,\infty)$, $x^* = 0$, and $f: A \to \mathbf{R}^1$ is defined by $f(x) = 1/x$. Then $\lim_{x \to x^*} f(x) = \infty$.

(5.C.4) *Definition* Suppose that (Y,d) is a metric space and $f: \mathbf{R}^1 \to Y$. Then we write $\lim_{x \to \infty} f(x) = y$ in case for each $\varepsilon > 0$, there is an $N \in \mathbf{Z}^+$ such that if $x \geq N$, then $f(x) \in S_\varepsilon(y)$. If $f: \mathbf{R}^1 \to \mathbf{R}^1$, we write $\lim_{x \to \infty} f(x) = \infty$ provided that for each $N \in \mathbf{Z}^+$, there is an $M \in \mathbf{Z}^+$ such that if $x \geq M$, then $f(x) \geq N$.

(5.C.5) *Examples* (a) If $f: \mathbf{R}^1 \to \mathbf{R}^2$ is defined by $f(x) = (2 + 3/x, -1/x)$, then $\lim_{x \to \infty} f(x) = (2,0)$.

(b) If $f: \mathbf{R}^1 \to \mathbf{R}^1$ is defined by $f(x) = x^3$, then $\lim_{x \to \infty} f(x) = \infty$.

(5.C.6) *Exercise* Suppose that $f: \mathbf{R}^1 \to \mathbf{R}^1$. Define and give examples of $\lim_{x \to -\infty} f(x) = \infty$, $\lim_{x \to \infty} f(x) = -\infty$, $\lim_{x \to -\infty} f(x) = -\infty$.

(5.C.7) *Theorem* (*L'Hospital's Rules*) Suppose that I is an open interval with an "endpoint" c, where c may either be a real number or ∞ or $-\infty$. Suppose that $f: I \to \mathbf{R}^1$ and $g: I \to \mathbf{R}^1$ are two functions with the properties:

  (i) $f'(x)$ and $g'(x)$ exist for each $x \in I$.
  (ii) $g(x) \neq 0$ and $g'(x) \neq 0$ for each $x \in I$ (and hence, by (5.B.4), g is 1-1).
  (iii) $\lim_{x \to c} f'(x)/g'(x) = q$, where q may be real or ∞ or $-\infty$.

Furthermore, suppose that either

  (a) $\lim_{x \to c} f(x) = 0 = \lim_{x \to c} g(x)$

or

  (b) $\lim_{x \to c} |g(x)| = \infty$ (where, of course, $x \in I$)

Then $\lim_{x \to c} f(x)/g(x) = q$.

Proof. We follow the proof suggested by Taylor (1952). Let $x \in I$ and define

$$m(x) = \inf \left\{ \left| \frac{f'(\alpha)}{g'(\alpha)} \right| \alpha \text{ lies between } x \text{ and } c \right\}$$

$$M(x) = \sup \left\{ \left| \frac{f'(\alpha)}{g'(\alpha)} \right| \alpha \text{ lies between } x \text{ and } c \right\}$$

Let y be any point between x and c. By Cauchy's mean value theorem we have

$$\frac{f(x) - f(y)}{g(x) - g(y)} = \frac{f'(y^*)}{g'(y^*)}$$

where y^* is a number between x and y. Hence, we have

$$m(x) \leq \frac{f(x) - f(y)}{g(x) - g(y)} \leq M(x)$$

for *each* y between x and c.
  It is easily verified that

$$m(x) \leq \frac{f(x) - f(y)}{g(x) - g(y)} = \frac{f(x)/g(x) - f(y)/g(x)}{1 - g(y)/g(x)} \leq M(x) \qquad (1)$$

and

$$m(x) \leq \frac{f(x) - f(y)}{g(x) - g(y)} = \frac{f(y)/g(y) - f(x)/g(y)}{1 - g(x)/g(y)} \leq M(x) \qquad (2)$$

Suppose now that condition (a) of the hypothesis is satisfied. Let x be fixed.

Then since both $f(y)$ and $g(y)$ tend to 0 as y approaches c, it follows from (1) that $m(x) \le f(x)/g(x) \le M(x)$. If case (b) holds and x is fixed, then letting y tend to c, we see from (2) that $m(x) \le \lim_{y \to c} f(y)/g(y) \le M(x)$. Now to complete the proof, it suffices to observe that by (iii) both $m(x)$ and $M(x)$ tend toward q as x approaches c and that y is trapped between x and c.

(5.C.8) *Examples* (a)

$$\lim_{x \to 0} \left(\frac{1}{\sin x} - \frac{1}{x} \right) = \lim_{x \to 0} \frac{x - \sin x}{x \sin x} = \lim_{x \to 0} \frac{1 - \cos x}{\sin x + (x \cos x)}$$

$$= \lim_{x \to 0} \frac{\sin x}{\cos x + (\cos x - x \sin x)} = 0$$

(b)   To find $\lim_{x \to 0} (x + e^{2x})^{1/x}$, we first observe that

$$\lim_{x \to 0} (x + e^{2x})^{1/x} = \lim_{x \to 0} e^{(1/x) \ln (x + e^{2x})}$$

Now note that

$$\lim_{x \to 0} \frac{\ln (x + e^{2x})}{x} = \lim_{x \to 0} \frac{(1/(x + e^{2x}))(1 + 2e^{2x})}{1} = 3$$

and therefore,

$$\lim_{x \to 0} (x + e^{2x})^{1/x} = e^3$$

The mean value theorem easily disposes of the next familiar result.

(5.C.9) *Theorem* Suppose that $f: I \to \mathbf{R}^1$, where I is an open interval, and that $f'(x)$ exists and is positive for each $x \in I$. Then f is a strictly increasing function on I. If $f'(x) < 0$ for each $x \in I$, then f is a strictly decreasing function on I.

Proof. Suppose that $a,b \in I$ and $a < b$. We are to show that $f(a) < f(b)$. But this is immediate since by the mean value theorem there is a point $c \in (a,b)$ such that $f'(c) = (f(b) - f(a))/(b - a)$. Since $f'(c) > 0$ and $b - a > 0$, the result follows.

The second part of the theorem is proven similarly.

(5.C.10) *Example* We use (5.C.9) to show that $1 - x \le e^{-x}$ for $0 \le x \le 1$. If $x \in [0,1)$, define $g(x) = e^{-x}/(1 - x)$. Then

$$g'(x) = \frac{(1 - x)(-e^{-x}) - (-1)(e^{-x})}{(1 - x)^2} = \frac{xe^{-x}}{(1 - x)^2}$$

consequently, g is a strictly increasing function on the interval $(0,1)$. Since $g(0) = 1$ and g is continuous at 0, we have $g(x) > 1$ for all x in $(0,1)$. Therefore, if $x \in (0,1)$, then $e^{-x}/(1 - x) > 1$, and hence $1 - x < e^{-x}$.

Theorem (5.C.9) is a prime example of how local properties of a function can yield information concerning the function's "global" behavior. That is, if we know what is happening at individual points in the domain of the function (in this case, $f'(x) > 0$), then we can often patch this data together to obtain qualitative knowledge of the entire function (in this case, f is increasing). Since, in practice, it may well happen that an investigator has only discrete information (data at certain time intervals) at his disposal, it would certainly be of interest to find a way of utilizing this local information to "generate" knowledge of what is transpiring at a less provincial level.

D. THE DIFFERENTIAL

Perhaps one of the most troublesome and least understood ideas that arises in an introductory calculus course is the notion of the *differential*. In this section we give a brief introduction to this concept; somewhat more detailed considerations of the differential and its applications are treated in subsequent chapters.

The differential of a function f at a point $\hat{x}$ is defined as follows.

(5.D.1) Definition Suppose that $f: (a,b) \to \mathbf{R}^1$, $\hat{x} \in (a,b)$, and $f'(\hat{x})$ exists. Then the *differential* of f at $\hat{x}$ is the *function* $d_{\hat{x}}f: \mathbf{R}^1 \to \mathbf{R}^1$ defined by

$$d_{\hat{x}}f(h) = f'(\hat{x})h \tag{1}$$

Geometrically, we have the situation shown in Fig. 5.5.

As is perhaps suggested by Fig. 5.5, the differential can be employed to approximate changes of f near $\hat{x}$, i.e., to approximate $f(\hat{x} + h) - f(\hat{x})$. Essentially what is happening is that the line tangent to the graph of f at the point $(\hat{x}, f(\hat{x}))$ is used to approximate this graph in the vicinity of $(\hat{x}, f(\hat{x}))$. More precisely, suppose that $f: (a,b) \to \mathbf{R}^1$ and that $f'(\hat{x})$ exists at the point $\hat{x} \in (a,b)$. Observe that if we set

$$E(h) = f'(\hat{x}) - \frac{f(\hat{x} + h) - f(\hat{x})}{h}, \qquad \hat{x} + h \in (a,b), \tag{2}$$

then $\lim_{h \to 0} E(h) = 0$. Furthermore, from (2) we have

$$d_x f(h) = f'(\hat{x})h = f(\hat{x} + h) - f(\hat{x}) + hE(h) \tag{3}$$

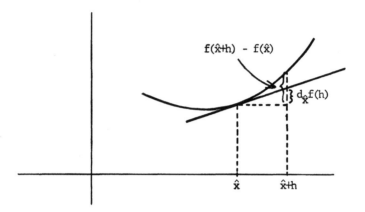

Figure 5.5

and since both factors h and $E(h)$ of $hE(h)$ tend to 0 as h approaches 0, it follows that for small values of h, $d_{\hat{x}}f(h)$ gives an excellent approximation of $f(\hat{x} + h) - f(\hat{x})$.

(5.D.2) *Observation* Frequently in the literature, the real number h in (1) is denoted by dx and dy is used to represent $d_{\hat{x}}f(h)$. With this notation we have $dy = f'(x)dx$, or $dy/dx = f'(x)$; here dy/dx is a true fraction, and not merely a symbol for the derivative. In order to emphasize that the differential is a function, we shall nevertheless continue to use the notation d_xf to denote the differential of f at the point x.

(5.D.3) *Example* To illustrate an easy application of the these ideas, we use the differential to approximate $\sqrt{4.1}$. Let $f(x) = \sqrt{x}$ and $\hat{x} = 4$. Then by (3) we have $f(4.1) - f(4) \cong f'(4) \cdot (0.1)$, and hence,

$$\sqrt{4.1} \cong \frac{1}{(2\sqrt{4})}(0.1) + 2 = 2.025$$

The actual value of $\sqrt{4.1}$ is $2.0248 \cdots$.

(5.D.4) *Observation* The differential $d_{\hat{x}}f$ is a linear transformation since $d_{\hat{x}}f(h_1 + h_2) = f'(\hat{x})(h_1 + h_2) = f'(\hat{x})h_1 + f'(\hat{x})h_2 = d_{\hat{x}}f(h_1) + d_{\hat{x}}f(h_2)$ and $d_{\hat{x}}f(\alpha h) = f'(\hat{x})(\alpha h) = \alpha f'(\hat{x})h = \alpha d_{\hat{x}}f(h)$.

(5.D.5) *Observation* Suppose that $A \subset \mathbf{R}^1$, $\hat{x} \in \text{int } A$, $f: A \to \mathbf{R}^1$ and $f'(\hat{x})$ exists. Then the graph of the function $f^{\#}: \mathbf{R}^1 \to \mathbf{R}^1$, defined by

$$f^{\#}(x) = d_{\hat{x}}f(x - \hat{x}) + f(\hat{x}) = f'(\hat{x})(x - \hat{x}) + f(\hat{x})$$

is precisely the line tangent to the graph of f at the point $(\hat{x}, f(\hat{x}))$. Furthermore, the function $f^\#$ is the best possible "linear" approximation of f in the sense that if $y(x) = a(x - \hat{x}) + f(\hat{x})$ is any other function whose graph is a line passing through the point $(\hat{x}, f(\hat{x}))$, there is an interval (c,d) containing $\hat{x}$ such that $|f^\#(x) - f(x)| < |y(x) - f(x)|$ for each $x \in (c,d)$; note that this is equivalent to saying that

$$\lim_{x \to \hat{x}} \frac{|f^\#(x) - f(x)|}{|y(x) - f(x)|} < 1$$

To see that f is indeed the best approximation in the foregoing sense, it suffices to observe that

$$\lim_{x \to \hat{x}} \frac{|f^\#(x) - f(x)|}{|y(x) - f(x)|} = \lim_{x \to \hat{x}} \frac{|[f'(\hat{x})(x - \hat{x}) + f(\hat{x}) - f(x)]/x - \hat{x}|}{|[a(x - \hat{x}) + f(\hat{x}) - f(x)]/x - \hat{x}|}$$

$$= \lim_{x \to \hat{x}} \frac{|f'(\hat{x}) - [(f(x) - f(\hat{x}))/(x - \hat{x})]|}{|a - [(f(x) - f(\hat{x}))/(x - \hat{x})]|}$$

$$= \frac{|f'(\hat{x}) - f'(\hat{x})|}{|a - f'(x)|} = 0 < 1$$

provided $a \neq f'(\hat{x})$.

This last result points out an important use of the differential; as we shall see, the "linearization" of complicated functions can be a very effective tool in achieving certain manipulative and computational advantages. The ease of dealing with functions where graphs are linear frequently offsets any problems that arise from the use of such approximations. To illustrate this we describe Newton's method for generating values close to a (usually unknown) zero of a function.

Suppose that we wish to find a solution of the equation

$$f(x) = 0$$

where the graph of f is as indicated in Fig. 5.6. Note that the line L_1, tangent to the curve at an arbitrarily chosen point $(x_1, f(x_1))$, crosses the x axis at a point x_2 that is closer to the desired value c than is x_1; moreover, the line L_2, tangent to the graph at the point $(x_2, f(x_2))$, intersects the x axis at a point x_3 still closer to c. For a wide range of functions this relationship is not merely coincidental, and in fact, under fairly reasonable conditions, it can be shown that the sequence $\{x_n\}$ constructed inductively in this manner will converge to c (see Chapter 14). Furthermore, there are a number of results that give an estimate of the error $|c - x_n|$, for each $n \in \mathbf{Z}^+$.

The sequence $\{x_n\}$ just described is defined as follows. First note that by (5.D.5) L_1 is the graph of the function $l_1(x) = d_{x_1}f(x - x_1) + f(x_1) = f'(x_1)(x - x_1) + f(x_1)$. If x_2 is such that $0 = l_1(x_2) = f'(x_1)(x_2 - x_1) + f(x_1)$, then we have $x_2 = x_1 - f(x_1)/f'(x_1)$.

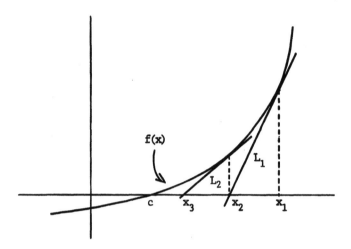

Figure 5.6

This argument may be repeated to obtain $x_3 = x_2 - f(x_2)/f'(x_2)$, and in general we have $x_n = x_{n-1} - f(x_{n-1})/f'(x_{n-1})$.

Newton's method is usually quite effective if the initial point x_1 is fairly close to c. For instance, if we wish to compute the square root of a number c using this procedure, then we define $f(x) = x^2 - c$. If $c = 38$ and $x_1 = 6$, then routine calculations show that

$$x_2 = 6.1666 \qquad x_2^2 = 38.0269$$
$$x_3 = 6.1645 \qquad x_3^2 = 33.0010$$

Newton's method is not foolproof. For instance, if we are dealing with a function such as the one whose graph is given in Fig. 5.7, then an unfortunate initial choice of x_1, could result in a sequence $\{x_n\}$ that fails to converge at all.

E.* A CONCLUDING PROBLEM

We end this review of elementary calculus with a brief consideration of a biological problem involving the cultivation of microorganisms. The presentation is based on Meyer (1977).

In microbiology, it is frequently of interest to cultivate and harvest various organisms for food, experimental investigations, medicinal purposes, etc. When one charts the growth of such a culture in a continuously replenished medium (where the original composition is kept fixed), it is frequently observed that the graph obtained is S-shaped, as illustrated in Fig. 5.8. The S shape is due to a number of factors one of which is that the volume of the medium is assumed to remain constant.

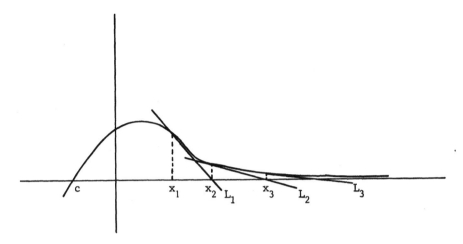

Figure 5.7

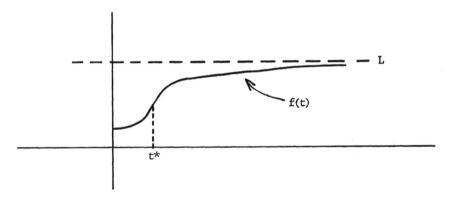

Figure 5.8

Here $f(t)$ represents the number of organisms present at time t; this function commonly satisfies the following properties:

1. f is strictly increasing [$f'(t)$ exists and is positive for each t].
2. $\lim_{t \to \infty} f(t) = L$ (the growth of the culture is eventually limited by
   the nature and amount of medium used).
3. f has an *inflection point* [a point $t^* > 0$ such that f' is strictly increasing on $(0, t^*)$, strictly decreasing on (t^*, ∞), and $f''(t^*) = 0$—note
   that f' attains a maximum at t^*].

For practical reasons we shall wish to (a) harvest the culture periodically
and (b) remove equal amounts of organisms at each harvest. Suppose that
we begin to harvest at time t_h; let r be the amount removed (where
$r < f(t_h) - f(0)$). Since $f(0) < f(t_h) - r < f(t_h)$, it follows from the intermediate

value theorem that there is a unique value $t_s < t_h$ such that $f(t_s) = f(t_h) - r$. Hence, $r = f(t_h) - f(t_s)$ and if we wait $t_h - t_s$ units of time between harvests, the conditions (a) and (b) are satisfied. The yield per unit time is clearly given by $(f(t_h) - f(t_s))/(t_h - t_s)$, and it is this value that is to be maximized. By the mean value theorem, there is a number t_h^* lying between t_s and t_h such that $(f(t_h) - f(t_s))/(t_h - t_s) = f'(t_h^*)$; consequently, our problem basically reduces to maximizing f'. From 3 above, we have that f' attains a maximum at t^*. Although the information available concerning f is too limited to enable us to define precisely a pair of values t_h and t_s that would result in a maximal harvest, it should now be clear that best results are obtained by harvesting small quantities of the organism near the inflection point, i.e., choose t_s and t_h such that $t_s < t^* < t_h$ and such that $t_h - t_s$ is small. Thus, from a practical standpoint the problem is solved.

PROBLEMS

Section A

1. Suppose that f and g are functions mapping a set A into $\mathbf{R}^1$ and that $f'(x^*)$ and $g'(x^*)$ exist at a point $x^* \in A$. Show:
    (a) $(f + g)'(x^*)$ exists and is equal to $f'(x^*) + g'(x^*)$.
    (b) $(fg)'(x^*)$ exists and is equal to $f(x^*)g'(x^*) + f'(x^*)g(x^*)$. [*Hint*: Consider $(fg)(x) - (fg)(x^*) = f(x)(g(x) - g(x^*)) + g(x^*)(f(x) - f(x^*))$.]
    (c) If $g(x^*) \neq 0$, then $(f/g)'(x^*)$ exists and is equal to

    $$\frac{g(x^*)f'(x^*) - g'(x^*)f(x^*)}{g^2(x^*)}$$

    $\left[\vphantom{\frac{1}{1}} \right.$ *Hint*: Note that

    $$\frac{(f/g)(x) - (f/g)(x^*)}{x - x^*}$$

    $$= \frac{1}{g(x)g(x^*)}\left[g(x^*)\frac{f(x) - f(x^*)}{x - x^*} - f(x^*)\frac{g(x) - g(x^*)}{x - x^*} \right].$$

2. Find the following limits:
    (a) $\lim_{(x,y)\to(2,5)} f(x,y)$, where

    $$f(x, y) = \begin{cases} x^2 + y & \text{if } y = 3x \\ 2x + y & \text{if } y \neq 3x \end{cases}$$

    (b) $\lim_{(x,y)\to(0,0)} f(x,y)$, where

$$f(x, y) = \left\{ \frac{x^2}{x^2 + y^2} \right.$$

3. Show that if $f: A \to \mathbf{R}^1$ and $f'(x^*)$ exists at some point $x^* \in A$, then there exist $\varepsilon > 0$ and $M > 0$ such that $|f(x) - f(x^*)| \le M|x - x^*|$ for each $x \in S_\varepsilon(x^*)$

4. Use problem A.3 to show that if $f'(x^*)$ exists, then f is continuous at x^*.

5. Suppose that $f: [0,1] \to \mathbf{R}^1$ is defined by

$$f(x) = \begin{cases} x \sin \dfrac{1}{x} & \text{if } x \ne 0 \\ 0 & \text{if } x = 0 \end{cases}$$

   (a) Show that f is continuous at $x = 0$.
   (b) Does $f'(0)$ exist?

6. Suppose that $f: [0,1] \to \mathbf{R}^1$ is defined by

$$f(x) = \begin{cases} x^2 \sin \dfrac{1}{x} & \text{if } x \ne 0 \\ 0 & \text{if } x = 0 \end{cases}$$

   Does $f'(0)$ exist, and if so, is f' continuous at $x = 0$?

7. Suppose that $f: \mathbf{R}^1 \to \mathbf{R}^1$ is defined by

$$f(x) = \begin{cases} x & \text{if } x \text{ is rational} \\ \sin x & \text{if } x \text{ is irrational} \end{cases}$$

   Find $f'(0)$.

8. Suppose that $f: (a,b) \to \mathbf{R}^1$ is differentiable on (a,b) and that $f'(a) < c < f'(b)$. Show that there is a point $x^* \in (a,b)$ such that $f'(x^*) = c$. [*Hint*: Consider the function $g(x) = f(x) - cx$.]

9. Find a flaw in the following proof of (5.A.16): Since $(f^{-1} \circ f)(x) = x$, we have by the chain rule $[(f^{-1})'(f(x))][f'(x)] = 1$, and therefore, $(f^{-1})'(f(x)) = 1/f'(x)$.

10. Prove Observation (5.A.6).

11. Let $u: \mathbf{R}^1 \to \mathbf{R}^1$ be defined by $u(x) = 1$ if x is a rational number and $u(x) = 0$ if x is an irrational number. Define $f: \mathbf{R}^1 \to \mathbf{R}^1$ by $f(x) = x^2 u(x)$. Show that if $x \ne 0$, then $f'(x)$ does not exist, but that $f'(0) = 0$.

12. Suppose that $f: [0,1] \to \mathbf{R}^1$ is defined by $f(x) = 0$ if x an irrational number and $f(x) = 1/q$ if $x = p/q$, where p/q is in reduced form. Show that f is continuous at each irrational point x, discontinuous at each rational point x, and that f is not differentiable at any point.

13.* True or false? Suppose that $f: \mathbf{R}^1 \to \mathbf{R}^1$ and that $g: \mathbf{R}^1 \to \mathbf{R}^1$ is defined by

$$g(x) = \lim_{h \to 0} \frac{f(x + h) - f(x - h)}{2h}$$

If f and g are continuous at a point x^*, then $f'(x^*)$ exists and $f'(x^*) = g(x^*)$.

Section B

1. Use the mean value theorem to show that if $f: (a,b) \to \mathbf{R}^1$ is such that $f'(x) = 0$ for each $x \in (a,b)$, then f is a constant function.
2. Show that $x^3 - 3x + c$ cannot have three distinct zeros in the interval $[0,1]$.[*Hint*: Use Rolle's theorem.]
3. Suppose that $f: \mathbf{R}^1 \to \mathbf{R}^1$ and that f is differentiable on $\mathbf{R}^1$. Show that if $f'(x) \neq 1$ for each $x \in \mathbf{R}^1$, then f has at most one fixed point.
4. Show that if n is even, then $x^n + bx + c$ has at most two distinct real zeros.
5. Find all values of c whose existence is guaranteed by the mean value theorem if $f(x) = 3x^2 - 5x + 7$, $1 \leq x \leq 9$.
6. Suppose that $f: [a,b] \to \mathbf{R}^1$ and that $x^* \in [a,b]$. Show that if $f'(x)$ exists for each $x \in [a,b] \setminus \{x^*\}$ and if $\lim_{x \to x^*} f'(x) = y$, then $f'(x^*)$ exists and is equal to y.
7. Suppose that $f: \mathbf{R}^1 \to \mathbf{R}^1$ and that $|f'(x)| \leq \alpha < 1$ for each $x \in \mathbf{R}^1$. Show that f is a contraction map.
8. Prove the following converse of the mean value theorem: Suppose that $f: (a,b) \to \mathbf{R}^1$, $F: (a,b) \to \mathbf{R}^1$, and that f is continuous. Suppose further that if $x,y \in (a,b)$, then there is a point c, $a < c < b$ such that

$$\frac{F(y) - F(x)}{y - x} = f(c)$$

Then F is differentiable on (a,b) and $F' = f$.

9.* Show that if $f: (-1,1) \to [0,\infty)$ is differentiable on $(-1,1)$ and if $\lim_{x \to 0^+} f(x) = \infty$, then $\lim_{x \to 0^+} f'(x) = \infty$.

Section C

1. Find the following limits:

   (a) $\lim\limits_{x \to \infty} \dfrac{x^4}{e^x}$

   (b) $\lim\limits_{x \to 0} \left(\dfrac{1}{x \sin x} - \dfrac{1}{x^2} \right)$

   (c) $\lim\limits_{x \to \infty} \dfrac{x \ln x}{e^x}$

2. Find the following limits:

   (a) $\lim\limits_{x \to 1} \dfrac{1 - \sqrt{x}}{1 - x}$

   (b) $\lim\limits_{x \to 0} \left(1 + \dfrac{c}{x}\right)^{ax}$

   (c) $\lim\limits_{x \to 0} \dfrac{x - \sin x}{2x}$

3. Suppose that $f: [a,c] \to \mathbf{R}^1$ is continuous on $[a,c]$ and differentiable on (a,c). Let $b \in [a,c]$. Show that if f' is increasing on (a,c), then $(b - a)f(c) + (c - b)f(a) \geq (c - a)f(b)$ and if f' is decreasing on (a,c), then this inequality is reversed.

4. Use problem C.3 to show that if $0 < m < n$ and $\alpha > 0$, then $(1 + \alpha/m)^m < (1 + \alpha/n)^n$. [*Hint*: Consider the function $\ln (1 + x)$.]

5. Show that if $f: (a,b) \to \mathbf{R}^1$ is continuous and differentiable at each point $x \in (a,b)$ except possibly at a point x_0, then if $\lim_{x \to x_0} f'(x)$ exists, f is differentiable at x_0 and f' is continuous at x_0.

6. Suppose that $f: [0,\infty) \to \mathbf{R}^1$ is continuous, $f(0) = 0$, $f'(x) > 0$ for each $x \in (0,\infty)$, and that f' is a strictly increasing function. Show that the function $\phi(x) = f(x)/x$ is strictly increasing.

7. Define functions f and g by $f(x) = x[\sin(1/x^4)]e^{-1/x^2}$ if $x \neq 0$, $f(0) = 0$, and $g(x) = e^{-1/x^2}$ if $x \neq 0$, $g(0) = 0$. Show that $\lim_{x \to 0} f(x)/g(x) = 0$ and that both f and g are differentiable, but that $\lim_{x \to 0} f'(x)/g'(x)$ does not exist (Rickert, 1968).

8. Suppose that $f: [a,b] \to \mathbf{R}^1$ and that f'' is continuous on $[a,b]$. Show that if $f'(c) = 0$ and $f''(c) > 0$ for some point $c \in (a,b)$, then the function f has a local minimum at c.

Section D

1. Use the appropriate differential to find approximations of

   (a) $(127)^{\frac{1}{3}}$

   (b) $\tan 0.05$

   (c) $e^{0.15}$

2. Use Newton's method to find an estimate of the zeros of the following functions:

   (a) $f(x) = x^5 - x^2 + x + 1$ \qquad (Use $x_1 = 0$, and find x_3)

   (b) $f(x) = \ln 3x - \dfrac{\sin x}{x^2}$ \qquad (Use appropriate tables.)

3. Use Newton's method to obtain an approximation of $(345)^{\frac{1}{4}}$.

4. Use Newton's method to approximate a zero of $100 - 1/x$; let $x_1 = 1$.

5.* Use Taylor's formula (see Chapter 8)

$$f(x) = f(a) + f'(a)(x - a) + (f''(x^*)/2)(x - a)^2 \qquad (*)$$

to show that the nth error term in Newton's methed, $e_n = c - x_n$, is quadratic, i.e.,

$$e_n \cong \frac{f''(c)}{2f'(c)} e_{n-1}^2$$

[*Hint*: Let $x = c$ and $a = x_{n-1}$ in (*), and recall that $f(x_{n-1}) + f'(x_{n-1})(x_n - x_{n-1}) = 0$.] Note that this result helps explain the rather amazing rapidity of convergence often observed in Newton's method.

REFERENCES

Meyer, R. W. (1977): *Amer. Math. Monthly*, **84**: 40.
Rickert, N. W. (1968): *Amer. Math. Monthly*, **75**: 166.
Taylor, A. E. (1952): *Amer. Math. Monthly*, **59**: 20.

6

A FIRST LOOK AT INTEGRATION

A. THE DEFINITE INTEGRAL

In the physical and life sciences it is frequently the case that more information is available concerning the rate of change of a process than about the process itself. Translated into mathematical terms, this means that the derivative (and perhaps an "initial value") of a function may be known; the task is to find the function itself (recall, for example, the differential equations used in connection with Newton's law of cooling and problems involving radioactive decay). Consequently, the following general problem is of considerable interest: given the derivative F' of a function F and a particular value $F(a)$ of this function, when is it mathematically possible to determine F? At present we have the following two relationships between the derivative of a function and the function itself:

1. The definition of the derivative:

$$F'(x) = \lim_{h \to 0} \frac{F(x + h) - F(x)}{h}$$

2. The mean value theorem: If $F: [a,b] \to \mathbf{R}^1$ is continuous and if F is differentiable on (a,b), then for each $x \in (a,b)$, there is a point $x^* \in (a,x)$ such that $F(x) - F(a) = F'(x^*)(x - a)$.

Note that if F' is a known function, and if the point x^* can be determined, then there is no problem in finding $F(x)$. However, except in a few very special cases, there is no direct method for computing x^*. To resolve the general problem we return once again to approximation techniques.

Recall from the previous chapter that for small values of h, the differential $d_x F(h) = F'(x)h$ is a good approximation of $F(x + h) - F(x)$. We shall base our development of the integral on this observation. Our treatment of integration in this chapter differs considerably from that commonly found in a first course in calculus. This approach is intended to give the student a somewhat different perspective of integration as well as to present (in a natural context) a number of important results such as the Weierstrass approximation theorem. A more standard (and more all-encompassing) approach to integration is given in Chapter 11. We begin with the following definition.

(6.A.1) Definition A *partition* of an interval $[a,b]$ is a finite sequence of points $P = \{x_0, x_1, \ldots, x_n\}$ in $[a,b]$ such that $a = x_0 < x_1 < \cdots < x_n = b$. The *mesh* of P is $\max\{|x_i - x_{i-1}| \mid i = 1, \ldots, n\}$. We shall denote the mesh of P by $|P|$.

Suppose that $P = \{x_0, x_1, \ldots, x_n\}$ is a partition of the interval $[a,b]$, and for each integer i, $0 \le i < n$, let $\Delta x_i = x_{i+1} - x_i$. Let $F: [a,b] \to \mathbf{R}^1$ be any function with the property that F' exists and is continuous on $[a,b]$. If the mesh of P is small, then, for each i, we would expect that the differential $d_{x_i}F(\Delta x_i) = F'(x_i)(\Delta x_i)$ would yield a good approximation of $F(x_{i+1}) - F(x_i)$, and hence, it is likely that the sum

$$d_{x_0}F(\Delta x_0) + \cdots + d_{x_{n-1}}F(\Delta x_{n-1})$$

would give a good approximation of

$$(F(x_1) - F(x_0)) + \cdots + (F(x_n) - F(x_{n-1})) = F(b) - F(a)$$

In fact, we shall be able to show that if F' is continuous on $[a,b]$, then

$$\lim_{|P|\to 0} \sum_{i=0}^{n-1} d_{x_i}F(\Delta x_i) = F(b) - F(a) \tag{1}$$

[such a limit is defined in the statement of (6.A.5)].

We shall define the integral of F' on $[a,b]$ to be

$$\int_a^b F'(x)\, dx = \lim_{|P|\to 0} \sum_{i=0}^{n-1} d_{x_i}F(\Delta x_i) = F(b) - F(a)$$

Thus, for instance, to find $\int_1^2 (x^3 - 2x)\, dx$, we observe that if $F'(x) = x^3 - 2x$, then $F(x) = x^4/4 - x^2$, and hence $\int_1^2 (x^3 - 2x)\, dx = F(2) - F(1) = \frac{3}{4}$.

(6.A.2) *Observation* Geometrically, the sum

$$d_{x_0}F(\Delta x_0) + d_{x_1}F(\Delta x_1) + \cdots + d_{x_{n-1}}F(\Delta x_{n-1})$$

may be viewed as the sum of the areas of the rectangles indicated in Fig. 6.1 (negative areas are permitted); therefore, as the mesh of the partitions tends toward 0, the limit $\int_a^b F'(x)\, dx$ (if it exists) can be interpreted as the area of the region between the graph of F' and the x axis and bounded by the lines $x = a$ and $x = b$, where area below the x axis is considered to be negative.

It is not obvious, however, that equation (1) holds. The problem is: given a suitably small mesh, $d_{x_i}F(\Delta x_i)$ will be a very good approximation of $F(x_{i+1})$ $- F(x_i)$ for each i, but when all these approximations are added together the resulting sum may differ considerably from $F(b) - F(a)$ (because of the errors arising from the large number of summands).

To establish (1) it is necessary to derive yet another important relationship between compactness and continuity. In (4.A.12), it was observed that if F is a continuous function between metric spaces X and Y and $\varepsilon > 0$ is given, then the δ needed to satisfy the continuity condition (4.A.11) may very well depend on x, i.e., for different values of x (and the same value of ε), different values of δ must be found. If X is compact, this is not the case.

(6.A.3) *Theorem* Suppose that (X,d) and $(Y,\hat{d})$ are metric spaces, X is compact, and $F: X \to Y$ is continuous. Let $\varepsilon > 0$ be given. Then there is a

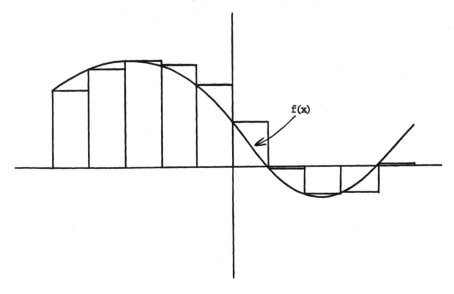

f(x)

Figure 6.1

positive number δ such that if $d(x,y) < \delta$, where x and y are *arbitrary* points in X, then $\hat{d}(F(x),F(y)) < \varepsilon$.

Proof. Suppose that for some $\varepsilon > 0$ the conclusion of the theorem is false. Then for each $n \in \mathbf{Z}^+$, there are points x_n and y_n in X such that $d(x_n,y_n) < 1/n$ and $\hat{d}(F(x_n),F(y_n)) \geq \varepsilon$. Since X is compact, the sequence $\{x_n\}$ has a cluster point $x^* \in X$. The function F is continuous at x^*; therefore, there is a $\delta > 0$ such that if $d(x^*,y) < \delta$, then $\hat{d}(F(x^*),F(y)) < \varepsilon/2$. Since x^* is a cluster point of the sequence $\{x_n\}$ and since $d(x_n,y_n) < 1/n$ for each n, there is a positive integer m sufficiently large so that $d(x_m,x^*) < \delta$ and $d(y_m,x^*) < \delta$. Hence, we have

$$\hat{d}(F(x_m), F(y_m)) \leq \hat{d}(F(x_m), F(x^*)) + \hat{d}(F(x^*), F(y_m)) < \frac{\varepsilon}{2} + \frac{\varepsilon}{2} = \varepsilon$$

which contradicts the way x_m and y_m were chosen.

(6.A.4) *Definition* If (X,d) and $(Y,\hat{d})$ are metric spaces, then a function $F: X \to Y$ is *uniformly continuous* if for each $\varepsilon > 0$ there is a $\delta > 0$ such that $\hat{d}(F(x),F(y)) < \varepsilon$ whenever $d(x,y) < \delta$.

Note that we have shown in (6.A.3) that continuous functions with compact domains are uniformly continuous.

(6.A.5) *Theorem* Suppose that $F: [a,b] \to \mathbf{R}^1$ and that F' is continuous on $[a,b]$. For each $\varepsilon > 0$ there exists a $\delta > 0$ such that if P is a partition of $[a,b]$ with mesh $< \delta$, then $|\sum_{i=0}^{n-1} d_{x_i}F(\Delta x_i) - (F(b) - F(a))| < \varepsilon$. In other words

$$\lim_{|P| \to 0} \sum_{i=0}^{n-1} d_{x_i}F(\Delta x_i) = F(b) - F(a)$$

Proof. Since F' is continuous on $[a,b]$ and $[a,b]$ is compact, it follows from (6.A.3) that F' is uniformly continuous. Therefore, corresponding to $\varepsilon/(b - a)$, there is a positive number δ such that if $w,z \in [a,b]$ and $|w - z| < \delta$, then $|F'(w) - F'(z)| < \varepsilon/(b - a)$. Let $P = \{x_0, x_1, \ldots, x_n\}$ be any partition of $[a,b]$ with mesh $< \delta$. Since $F'(w)$ exists for each w, it follows from (5.A.13) that F is continuous on $[a,b]$, and consequently the mean value theorem can be applied to F. Thus, for each i, there is a point $x_i^* \in (x_i,x_{i+1})$ such that $F(x_{i+1}) - F(x_i) = F'(x_i^*)(x_{i+1} - x_i)$. Note that the uniform continuity of F ensures that for each i, $|F'(x_i) - F'(x_i^*)| < \varepsilon/(b - a)$. We now have

$$\left| \sum_{i=0}^{n-1} d_{x_i}F(\Delta x_i) - (F(b) - F(a)) \right|$$

$$= \left| \left[\sum_{i=0}^{n-1} F'(x_i)(x_{i+1} - x_i) \right] \right.$$

$$- [(F(x_1) - F(x_0)) + (F(x_2) - F(x_1)) + \cdots + (F(x_n) - F(x_{n-1}))]\Big|$$

$$= \left| \sum_{i=0}^{n-1} F'(x_i)(x_{i+1} - x_i) - \sum_{i=0}^{n-1} F'(x_i^*)(x_{i+1} - x_i) \right|$$

$$\leq \sum_{i=0}^{n-1} |F'(x_i) - F'(x_i^*)| \, (x_{i+1} - x_i) < \frac{\varepsilon}{b-a}(b-a) = \varepsilon$$

which concludes the proof.

(6.A.6)   *Definition*   If $F: [a,b] \to \mathbf{R}^1$ and $F' = f$ is continuous on $[a,b]$, then the *definite integral* of f from a to b is defined to be the limit

$$\lim_{|P| \to 0} \sum_{i=0}^{n-1} d_{x_i} F(\Delta x_i)$$

and is denoted by $\int_a^b f(w) \, dw$. The points a and b are called the *limits* of integration, and f is called the *integrand*. [It should be noted that the letter w may be replaced by any other letter without affecting the "value" of the integral, i.e., $\int_a^b f(w) \, dw = \int_a^b f(q) \, dq = \int_a^b f(x) \, dx$, etc.]

Theorem (6.A.5) may now be restated as follows.

(6.A.7)   *Theorem*   (*The Fundamental Theorem of Calculus*)   If $f: [a,b] \to \mathbf{R}^1$ is continuous and if there is a function $F: [a,b] \to \mathbf{R}^1$ such that $F'(x) = f(x)$ for each $x \in [a,b]$, then $\int_a^b f(x) \, dx = F(b) - F(a)$.

A number of observations are in order.

(6.A.8)   *Observation*   In practice we frequently deal with a continuous function $f: [a,b] \to \mathbf{R}^1$, whose integral, $\int_a^b f(x) \, dx$, we wish to determine. By the fundamental theorem of calculus, to evaluate $\int_a^b f(x) \, dx$ it suffices to find a function $F: [a,b] \to \mathbf{R}^1$ such that $F' = f$ for then we have

$$\int_a^b f(x) \, dx = \int_a^b F'(x) \, dx = F(b) - F(a)$$

Such a function F is called a *primitive* or *antiderivative* of f. As an example, we have

$$\int_0^\pi \sin w \, dw = -\cos \pi - (-\cos 0) = 2$$

since $F(w) = -\cos w$ is a primitive of $f(w) = \sin w$.

(6.A.9) *Observation* Under the conditions of (6.A.7), it follows that for each $x \in [a,b]$,

$$F(x) = \int_a^x f(w)\, dw + F(a)$$

The expression $\int_a^x f(w)\, dw$ is often referred to as an *indefinite integral*, since one of the limits of integration is a variable (in the case of a definite integral, both limits of integration are fixed). Note that the derivative of the indefinite integral $\int_a^x f(w)\, dw$ is equal to the integrand f, i.e.,

$$\left(\int_a^x f(w)\, dw \right)' = F'(x) = f(x)$$

Thus, in this sense, integration may be considered as the "inverse" of differentiation (up to an additive constant—see the following observation). This interpretation of the integral is quite important, and although the reader is more familiar from elementary calculus with the notion that the integral "represents area" under the curve [as indicated in (6.A.2)], integration and differentiation should basically be perceived as inverse operations of each other.

(6.A.10) *Observation* If F and $\hat{F}$ are primitives of the same function f on an interval $[a,b]$, then there is a constant C such that $F(x) = \hat{F}(x) + C$ for each $x \in (a,b)$. This is merely a restatement of (5.C.1). As an example, it can be verified directly that every primitive P of the polynomial function $p(x) = a_n x^n + a_{n-1}x^{n-1} + \cdots + a_1 x + a_0$ is of the form

$$P(x) = \frac{a_n}{n+1} x^{n+1} + \frac{a_{n-1}}{n} x^n + \cdots + \frac{a_1}{2} x^2 + a_0 x + C$$

where C is a constant.

We are now faced with the problem of determining which functions have primitives. From (5.A.13) it is clear that such functions must be continuous, and we shall now show that, in fact, all continuous functions have primitives.

In (6.A.10) we observed that all polynomial functions have primitives. A truly remarkable result due to Weierstrass, the Weierstrass approximation theorem (1843), asserts that every continuous function is "almost" a polynomial in the sense that if $f: [a,b] \to \mathbf{R}^1$ is continuous and $\varepsilon > 0$ is given, then there is a polynomial p such that under the sup metric, $d(p,f) < \varepsilon$, i.e., $|p(x) - f(x)| < \varepsilon$ for each $x \in [a,b]$. From this it follows that there is a sequence of polynomials $\{p_n\}$ that converges uniformly to f. We use this fact to establish the following fundamental result. (A proof of the Weierstrass approximation theorem is presented in Sec. 6.E.)

(6.A.11)   *Theorem* If $f: [a,b] \to \mathbf{R}^1$ is continuous, then f has a primitive on $[a,b]$.

Proof.   The proof in outline form is as follows:

(a)   We let $\{p_n\}$ be a sequence of polynomials that converges to f, and we apply (6.A.10) to obtain for each n, a primitive P_n of p_n.

(b)   We show that the sequence $\{P_n\}$ of primitives is a Cauchy sequence and, therefore, by (4.C.7) converges to a function $F: [a,b] \to \mathbf{R}^1$.

(c)   We show that the function F is a primitive of f.

The details of the proof are given next.

(a)   Since f is continuous on $[a,b]$, it follows from the Weierstrass approximation theorem that there is a sequence $\{p_n\}$ of polynomials converging uniformly to f. For each n, let Q_n be a primitive of p_n, and set

$$P_n(x) = Q_n(x) - Q_n(a)$$

for each $x \in [a,b]$. Then clearly P_n is also a primitive of p_n.

(b)   Let $\varepsilon > 0$ be given. Since the sequence $\{p_n\}$ is a Cauchy sequence, there is a positive integer N such that if $m,n \geq N$, then $|p_n(w) - p_m(w)| < \varepsilon/(b - a)$ for each $w \in [a,b]$. To see that the sequence $\{P_n\}$ is a Cauchy sequence, we note that if $m,n \geq N$ and if $x \in [a,b]$, then

$$|P_n(x) - P_m(x)| = |Q_n(x) - Q_n(a) - (Q_m(x) - Q_m(a))|$$

$$= \left| \int_a^x Q_n'(w)\, dw - \int_a^x Q_m'(w)\, dw \right|$$

$$= \left| \lim_{|P| \to 0} \sum_{i=0}^{k-1} d_{x_i} Q_n(\Delta x_i) - \lim_{|P| \to 0} \sum_{i=0}^{k-1} d_{x_i} Q_m(\Delta x_i) \right|$$

$$(P = \{x_0, x_1, \ldots, x_k\} \text{ is a partition of } [a,x])$$

$$= \left| \lim_{|P| \to 0} \sum_{i=0}^{k-1} p_n(x_i)(x_{i+1} - x_i) \right.$$

$$\left. - \lim_{|P| \to 0} \sum_{i=0}^{k-1} p_m(x_i)(x_{i+1} - x_i) \right|$$

$$= \left| \lim_{|P| \to 0} \sum_{i=0}^{k-1} (p_n(x_i) - p_m(x_i))(x_{i+1} - x_i) \right|$$

$$\leq \lim_{|P| \to 0} \sum_{i=0}^{k-1} |p_n(x_i) - p_m(x_i)|(x_{i+1} - x_i)$$

$$\leq \lim_{|P| \to 0} \sum_{i=0}^{k-1} \left(\frac{\varepsilon}{b - a} \right)(x_{i+1} - x_i)$$

$$= \left(\frac{\varepsilon}{b - a} \right)(x - a) < \varepsilon$$

Consequently, the sequence $\{P_n\}$ is a Cauchy sequence (in the sup metric), and hence, by (4.C.7) there is a function $F: [a,b] \to \mathbf{R}^1$ which is the unique limit of this sequence.

(c) We now establish that $F'(x) = f(x)$ for each $x \in [a,b]$.

Suppose that $x \in [a,b]$. We consider $(F(z) - F(x))/(z - x)$, where $z \in [a,b]$. Let $\varepsilon > 0$ be given. The basic strategy is to show that if z is any fixed number sufficiently close to x, then for sufficiently large n,

$$\left| \frac{F(z) - F(x)}{z - x} - \frac{P_n(z) - P_n(x)}{z - x} \right| < \frac{\varepsilon}{3} \tag{2}$$

and

$$\left| \frac{P_n(z) - P_n(x)}{(z - x)} - f(x) \right| < \frac{2\varepsilon}{3} \tag{3}$$

and hence, as z approaches x, $(F(z) - F(x))/(z - x)$ converges to $f(x)$; since by definition

$$\lim_{z \to x} \frac{F(z) - F(x)}{z - x} = F'(x)$$

the proof of the theorem will be complete.

To establish (2) we simply note that since x and z are fixed, and since $\{P_n\}$ converges to F, $P_n(z)$ and $F(z)$ [and $P_n(x)$ and $F(x)$] are arbitrarily close for sufficiently large values of n.

Verification of (3) requires somewhat more effort.

Since f is continuous there is an interval, I, containing x in its interior so that $|f(z) - f(x)| < \varepsilon/3$ for each $z \in I$. Let $z \in I$. Having fixed z, choose n so large that inequality (2) holds and such that $|p_n(w) - f(w)| < \varepsilon/3$ for each $w \in [a,b]$. By the mean value theorem, we have that

$$\frac{P_n(z) - P_n(x)}{z - x} = P_n'(x^*) = p_n(x^*)$$

for some x^* between x and z (and therefore in I). Since $p_n(x^*) = f(x) + (f(x^*) - f(x)) + (p_n(x^*) - f(x^*))$, we have

$$\left| \frac{P_n(z) - P_n(x)}{z - x} - f(x) \right| = |p_n(x^*) - f(x)|$$

$$\leq |f(x^*) - f(x)| + |p_n(x^*) - f(x^*)| < \frac{\varepsilon}{3} + \frac{\varepsilon}{3}$$

Finally, we have

$$\left| \frac{F(z) - F(x)}{z - x} - f(x) \right| \leq \left| \frac{F(z) - F(x)}{z - x} - \frac{P_n(z) - P_n(x)}{z - x} \right|$$

$$+ \left| \frac{P_n(z) - P_n(x)}{z - x} - f(x) \right| < \frac{\varepsilon}{3} + \frac{2\varepsilon}{3}$$

and this concludes the proof.

As a consequence of (6.A.11) we obtain the following important result.

(6.A.12) *Theorem* If $f: [a,b] \to \mathbf{R}^1$ is continuous, then $\int_a^b f(x)\,dx$ exists and is equal to $\lim_{|P| \to 0} \sum_{i=0}^{n-1} f(w_i)\,\Delta x_i$, where $P = \{x_0, x_1, \ldots, x_n\}$, $\Delta x_i = x_{i+1} - x_i$, and for each i, $w_i \in [x_i, x_{i+1}]$.

Proof. By (6.A.12) f has a primitive on $[a,b]$. Hence, by (6.A.5) the integral $\int_a^b f(x)\,dx$ exists and

$$\int_a^b f(x)\,dx = \lim_{|P| \to 0} \sum_{i=1}^{n-1} f(x_i)\,\Delta x_i$$

Let $\varepsilon > 0$ be given. Since f is uniformly continuous ((6.A.3)), there is a $\delta > 0$ such that if x and $\hat{x}$ are any two points in $[a,b]$ and $|x - \hat{x}| < \delta$, then $|f(x) - f(\hat{x})| < \varepsilon/(b-a)$. Suppose that $P = \{x_0, x_1, \ldots, x_n\}$ is a partition of $[a,b]$ with mesh less than δ, and for each i, let w_i be an arbitrary point in the interval $[x_i, x_{i+1}]$. Then we have

$$\left| \sum_{i=0}^{n-1} f(x_i)\Delta x_i - \sum_{i=0}^{n-1} f(w_i)\Delta x_i \right| \leq \sum_{i=0}^{n-1} |f(x_i) - f(w_i)|\Delta x_i$$

$$\leq \frac{\varepsilon}{b-a} \sum_{i=0}^{n-1} \Delta x_i = \varepsilon$$

From this it follows easily that

$$\int_a^b f(x)\,dx = \lim_{|P| \to 0} \sum_{i=0}^{n-1} f(w_i)\Delta x_i$$

To conclude this section we remark that although the fundamental theorem of calculus can be useful in the calculation of definite integrals, the reader should realize (in spite of his previous training) that very few integration problems can actually be solved with the aid of this theorem. The difficulty lies in the impossibility of finding primitives. As a result, for many applied problems, numerical methods leading to approximations of a given definite integral must be employed. The integral itself often plays more of a theoretical role than a "practical" one. We shall give a brief introduction to some numerical methods used for approximating integrals in Chapter 14.

B. ELEMENTARY PROPERTIES OF THE DEFINITE INTEGRAL

Many of the basic properties of the definite integral follow readily from its definition (6.A.6). The next theorem illustrates one such property.

(6.B.1) *Theorem* Suppose that $f: [a,b] \to \mathbf{R}^1$ is continuous and that $c \in (a,b)$. Then

$$\int_a^b f(x)\, dx = \int_a^c f(x)\, dx + \int_c^b f(x)\, dx$$

Proof. It suffices to show that given $\varepsilon > 0$, then

$$\left| \int_a^b f(x)\, dx - \left[\int_a^c f(x)\, dx + \int_c^b f(x)\, dx \right] \right| < \varepsilon$$

Corresponding to ε, there is a $\delta > 0$ such that

(a) If $P = \{x_0, x_1, \ldots, x_n\}$ is any partition of $[a,b]$ with mesh $P < \delta$, then

$$\left| \int_a^b f(x)\, dx - \sum_{i=0}^{n-1} f(x_i)\, \Delta x_i \right| < \frac{\varepsilon}{2}$$

(b) If $P_1 = \{x_0, x_1, \ldots, x_m\}$ is any partition of $[a,c]$ with mesh $P_1 < \delta$, then

$$\left| \int_a^c f(x)\, dx - \sum_{i=0}^{m-1} f(x_i)\, \Delta x_i \right| < \frac{\varepsilon}{4}$$

(c) If $P_2 = \{x_0, x_1, \ldots, x_q\}$ is any partition of $[c,b]$ with mesh $P_2 < \delta$, then

$$\left| \int_c^b f(x)\, dx - \sum_{i=0}^{q-1} f(x_i)\, \Delta x_i \right| < \frac{\varepsilon}{4}$$

Let $P_1 = \{x_0, x_1, \ldots, x_m\}$ and $P_2 = \{x_m, x_{m+1}, \ldots, x_n\}$ be partitions of $[a,c]$ and $[c,b]$, respectively, such that mesh $P_1 < \delta$ and mesh $P_2 < \delta$. Note that $P = \{x_0, x_1, \ldots, x_m, x_{m+1}, \ldots, x_n\}$ is a partition of $[a,b]$ with mesh less than δ. Consequently, we have

$$\left| \int_a^b f(x)\, dx - \left[\int_a^c f(x)\, dx + \int_c^b f(x)\, dx \right] \right|$$

$$= \left| \int_a^b f(x)\, dx - \sum_{i=0}^{n-1} f(x_i)\, \Delta x_i + \sum_{i=0}^{m-1} f(x_i)\, \Delta x_i \right.$$

$$\left. + \sum_{i=m}^{n-1} f(x_i)\, \Delta x_i - \left[\int_a^c f(x)\, dx + \int_c^b f(x)\, dx \right] \right|$$

$$\leq \left| \int_a^b f(x)\, dx - \sum_{i=0}^{n-1} f(x_i)\, \Delta x_i \right| + \left| \sum_{i=0}^{m-1} f(x_i)\, \Delta x_i - \int_a^c f(x)\, dx \right|$$

$$+ \left| \sum_{i=m}^{n-1} f(x_i)\, \Delta x_i - \int_c^b f(x)\, dx \right| < \frac{\varepsilon}{2} + \frac{\varepsilon}{4} + \frac{\varepsilon}{4} = \varepsilon$$

and this completes the proof.

Suppose that $f: [a,b] \to \mathbf{R}^1$ is continuous. To define the integral $\int_a^b f(x)\, dx$, where $b < a$, we consider partitions of the form $P = \{x_0, \ldots, x_n\}$, where $a = x_0$, $b = x_n$, and $x_0 > x_1 > \cdots > x_n$. Then exactly as before, we define $\int_a^b f(x)\, dx$ to be $\lim_{|P| \to 0} \sum_{i=0}^{n-1} d_{x_i} F(\Delta x_i)$, where F is a primitive of f. It should be clear from the definition that $\int_a^b f(w)\, dw = -\int_b^a f(w)\, dw$. Combining this result with the previous one we have the following basic theorem.

(6.B.2)   *Theorem*   Suppose that f is a continuous function on $[a,b]$ and $c,d,e \in [a,b]$. The $\int_c^d f(w)\, dw + \int_d^e f(w)\, dw = \int_c^e f(w)\, dw$.

A number of other standard properties of the definite integral whose proofs are fairly easy consequences of the definition of the definite integral are given next.

(6.B.3)   *Theorem*   Suppose that f and g are continuous functions on $[a,b]$ and that c is a constant. Then

(i) $\int_a^b (f + g)(w)\, dw = \int_a^b f(w)\, dw + \int_a^b g(w)\, dw$.

(ii) $\int_a^b cf(w)\, dw = c \int_a^b f(w)\, dw$.

(iii) $|\int_a^b f(w)\, dw| \leq \int_a^b |f(w)|\, dw$ $(b > a)$.

(iv) If f is nonnegative, and if $f(x) > 0$ for some $x \in [a,b]$, then $\int_a^b f(w)\, dw > 0$.

(v)   If $f(x) \leq g(x)$ for each $x \in [a,b]$, then $\int_a^b f(w)\, dw \leq \int_a^b g(w)\, dw$.

Proof. See problem B.1.

The next rather specialized result is used extensively in the calculus of variations (see Sec. I, Chapter 7).

(6.B.4)   *Theorem*   Suppose that $\phi : [a,b] \to \mathbf{R}^1$ is continuous and that $\int_a^b \phi(x)g(x)\, dx = 0$ for *each* differentiable function g on $[a,b]$ for which $g(a) = g(b) = 0$. Then $\phi(x) = 0$ for each $x \in [a,b]$.

Proof. Suppose that $\phi(\hat{x}) > 0$ for some $\hat{x} \in (a,b)$. Since ϕ is continuous on $[a,b]$, there is an interval $(c,d) \subset [a,b]$ such that $\hat{x} \in (c,d)$ and $\phi(x) > 0$ for each $x \in (c,d)$. Now define $g : [a,b] \to \mathbf{R}^1$ by

$$g(x) = \begin{cases} 0 & \text{if } a \leq x \leq c \\ (x - c)^2(x - d)^2 & \text{if } c \leq x \leq d \\ 0 & \text{if } d \leq x \leq b \end{cases}$$

Then it is easily verifed that g is differentiable on $[a,b]$ and that $g(a) = g(b) = 0$. Note that $\phi(x)g(x) \geq 0$ for each $x \in [a,b]$ and that $\phi(x)g(x) > 0$ for each $x \in (c,d)$. Thus, by (6.B.3.iv) we have $\int_a^b \phi(x)g(x)\, dx = \int_c^d \phi(x)g(x)\, dx > 0$, which contradicts the hypothesis; therefore, since an analogous argument is valid if $\phi(\hat{x}) < 0$, it must be the case that ϕ is identically 0 on $[a,b]$.

Integration by parts is one of the more useful techniques employed in integration theory.

(6.B.5) Theorem (*Integration by Parts*) Suppose that f and g are continuous functions on $[a,b]$ and that f' and g' exist and are continuous on $[a,b]$. Then $\int_a^b f(t)g'(t)\,dt = f(t)g(t)|_a^b - \int_a^b g(t)f'(t)\,dt$, where $f(t)g(t)|_a^b = f(b)g(b) - f(a)g(a)$.

Proof. Integration of both sides of the equation

$$(f(t)g(t))' = f(t)g'(t) + g(t)f'(t)$$

yields

$$\int_a^b (f(t)g(t))'\,dt = \int_a^b f(t)g'(t)\,dt + \int_a^b g(t)f'(t)\,dt$$

or

$$f(t)g(t)\Big|_a^b = \int_a^b f(t)g'(t)\,dt + \int_a^b g(t)f'(t)\,dt$$

Integration by parts and the Weierstrass approximation theorem combine nicely in the proof of the following classical and rather astonishing result, the Riemann-Lebesgue lemma. This result will prove to be quite useful in succeeding chapters.

(6.B.6) Theorem (*Riemann-Lebesgue Lemma*) If $f\colon [a,b] \to \mathbf{R}^1$ is continuous, then $\lim_{c \to \infty} \int_a^b f(x) \sin cx\,dx = 0$.

Proof. We first show that the theorem is valid if f is a polynomial, P. In this case a simple integration by parts yields

$$\int_a^b P(x) \sin cx\,dx = \frac{-P(x)\cos cx}{c}\Big|_a^b + \frac{1}{c}\int_a^b P'(x)\cos cx\,dx \qquad (1)$$

and since P and P' are bounded on $[a,b]$ it follows that the right-hand side of (1) converges to 0 as $c \to \infty$. This proves the theorem in the special case where f is a polynomial.

Now suppose that $f\colon [a,b] \to \mathbf{R}^1$ is an arbitrary continuous function. By the Weierstrass approximation theorem there is a sequence of polynomials $\{P_n\}$ that converges uniformly to f on $[a,b]$. Therefore, given $\varepsilon > 0$ there is a positive integer M such that if $m \geq M$, then $|f(x) - P_m(x)| < \varepsilon/2(b - a)$ for each $x \in [a,b]$. Consequently, since $|\sin cx| \leq 1$ for all x, we have

$$\left|\int_a^b (f(x) - P_M(x)) \sin cx\,dx\right| \leq \int_a^b |f(x) - P_M(x)|\,dx$$

$$< \left(\frac{\varepsilon}{2(b - a)}\right)(b - a) = \frac{\varepsilon}{2} \qquad (2)$$

Furthermore, since P_M is a polynomial, it follows from the special case just established that for sufficiently large c,

$$\left| \int_a^b P_M(x) \sin cx \, dx \right| < \frac{\varepsilon}{2}$$

and therefore, for these values of c, it follows from (2) that

$$\left| \int_a^b f(x) \sin cx \, dx \right| \le \left| \int_a^b (f(x) - P_M(x)) \sin cx \, dx \right| + \left| \int_a^b P_M(x) \sin cx \, dx \right|$$

$$< \frac{\varepsilon}{2} + \frac{\varepsilon}{2} = \varepsilon$$

From this inequality we conclude that

$$\lim_{c \to \infty} \int_a^b f(x) \sin cx \, dx = 0$$

The mean value theorem for derivatives has an integral counterpart.

(6.B.7) *Theorem* *(Mean Value Theorem for Integrals)* Suppose that f is continuous on $[a,b]$. Then there exists $x^* \in (a,b)$ such that $\int_a^b f(t) \, dt = f(x^*)(b - a)$.

This result follows from a more general theorem, sometimes called the *weighted mean value theorem for integrals*, which we prove next. This theorem is of special use in establishing certain error bounds in numerical methods (see Chapter 14).

(6.B.8) *Theorem* Suppose that $f: [a,b] \to \mathbf{R}^1$ and $g: [a,b] \to \mathbf{R}^1$ are continuous functions and that $g(x) \ge 0$ [or $g(x) \le 0$] for each $x \in [a,b]$. Then there is a number $c \in (a,b)$ such that $\int_a^b f(x)g(x) \, dx = f(c) \int_a^b g(x) \, dx$.

Proof. Suppose that $g(x) \ge 0$ for each $x \in [a,b]$. Clearly if $g(x) = 0$ for each $x \in [a,b]$, the result is trivial. Suppose then that for some $x, g(x) > 0$, and consequently, $\int_a^b g(x) \, dx > 0$. Since f is continuous on $[a,b]$, it follows from (4.B.16) that there are points x_1 and x_2 in $[a,b]$ such that $f(x_1) = \min\{f(x) \mid x \in [a,b]\}$ and $f(x_2) = \max\{f(x) \mid x \in [a,b]\}$. Consequently, from (6.B.3) we have

$$f(x_1) \int_a^b g(x) \, dx \le \int_a^b f(x)g(x) \, dx \le f(x_2) \int_a^b g(x) \, dx$$

and therefore,

$$f(x_1) \le \frac{\int_a^b f(x)g(x) \, dx}{\int_a^b g(x) \, dx} \le f(x_2)$$

By the intermediate value theorem (4.E.5), there is a point c between x_1 and

x_2 such that

$$f(c) = \frac{\int_a^b f(x)g(x)\,dx}{\int_a^b g(x)\,dx}$$

and this concludes the proof [if $g(x) \leq 0$ for each x, then $\int_a^b f(x)g(x)\,dx = \int_a^b (-f(x))(-g(x))\,dx = -f(c)\int_a^b -g(x)\,dx = f(c)\int_a^b g(x)\,dx$].

A simple but invaluable tool used in the calculation of definite integrals is given in the next theorem.

(6.B.9) Theorem (Substitution Theorem) Suppose that f, u, and u' are continuous functions on an interval $[a,b]$. Then $\int_a^b f(u(t))u'(t)\,dt = \int_{u(a)}^{u(b)} f(t)\,dt$.

Proof. Let F be a primitive of f and let $G(x) = \int_a^x f(u(t))u'(t)\,dt = \int_a^x [F(u(t))]'\,dt$. Then by (6.A.9) we have

$$\int_a^x [F(u(t))]'\,dt = F(u(x)) - F(u(a)) = \int_{u(a)}^{u(x)} F'(t)\,dt = \int_{u(a)}^{u(x)} f(t)\,dt$$

and the result follows.

We mention once again that in Chapter 11 integration will be considered from a much broader perspective. There, for instance, we shall show that the continuity of f is not a requisite for the existence of $\int_a^b f(w)\,dw$; in fact, even with an infinite number of discontinuities, $\int_a^b f(w)\,dw$ may still exist. It should be apparent that under these circumstances, the fundamental theorem of calculus will no longer be generally valid, and obviously the present definition of the integral (6.A.6) will have to be modified (it will become a special case of a more general definition).

Integration of complex-valued functions will be taken up in some detail in Chapter 12. As a prelude to this, we conclude the present section with the consideration of functions $f: [a,b] \to \mathbf{C}$, where $\mathbf{C}$ is the complex plane. We have seen earlier (Sec. 1.F) that such functions may be written in the form $f = u + iv$, where u and v are real-valued functions with common domain $[a,b]$. For such a function f, we define the integral of f over $[a,b]$ by

$$\int_a^b f(t)\,dt = \int_a^b u(t)\,dt + i \int_a^b v(t)\,dt$$

provided that the integrals $\int_a^b u(t)\,dt$ and $\int_a^b v(t)\,dt$ exist. With this definition it is easily seen that parts (i) and (ii) of (6.B.3) hold for continuous functions $f: [a,b] \to \mathbf{C}$ and $g: [a,b] \to \mathbf{C}$. Part (iii) of this theorem is also valid, although the proof is slightly more subtle.

(6.B.10)   *Theorem*   If $f: [a,b] \to \mathbf{C}$ is a continuous function, then

$$\left| \int_a^b f(t)\, dt \right| \le \int_a^b |f(t)|\, dt.$$

Proof.   Since f is continuous on $[a,b]$, so is $|f|$, and therefore, $\int_a^b |f(t)|\, dt$ exists. For each partition $P = \{t_0, t_1, \ldots, t_n\}$ of $[a,b]$ let $R_P = \sum_{i=0}^{n-1} f(t_i^*)\, \Delta t_i$, where $t_i^* \in [t_i, t_{i+1}]$ and let $Q_P = \sum_{i=0}^{n-1} |f(t_i^*)|\, \Delta t_i$. Let $\varepsilon > 0$ be given. It follows from the definition of the integral (6.A.6) that corresponding to ε, there is a $\delta > 0$ such that if P is a partition of $[a,b]$ with mesh less than δ, then

$$\left| \int_a^b f(t)\, dt - R_P \right| < \varepsilon$$

and

$$\left| \int_a^b |f(t)|\, dt - Q_P \right| < \varepsilon$$

Thus, if mesh $P < \delta$, we have

$$\left| \int_a^b f(t)\, dt \right| \le \left| \int_a^b f(t)\, dt - R_P \right| + |R_P| \le \varepsilon + |R_P| \le \varepsilon + Q_P$$

$$\le \varepsilon + \left| Q_P - \int_a^b |f(t)|\, dt \right| + \int_a^b |f(t)|\, dt \le 2\varepsilon + \int_a^b |f(t)|\, dt$$

Since $\varepsilon > 0$ was arbitrary, the desired result follows.

C. IMPROPER INTEGRALS AND THE GAMMA AND BETA FUNCTIONS

Thus far we have confined our attention to the integration of continuous functions. In this section we see how the continuity condition can be relaxed slightly to handle cases where a function f is not continuous (or not defined) at a finite number of points in $[a,b]$. Thus, we shall investigate the existence (or nonexistence) of such integrals as

$$\int_0^1 \frac{1}{x^2}\, dx \qquad \int_2^5 \frac{3}{\sqrt{x-2}}\, dx \qquad \int_{-6}^3 g(x)\, dx$$

where

$$g(x) = \begin{cases} x^2 & \text{if } x \ne 1, 3 \\ 4 & \text{if } x = 1 \\ 7 & \text{if } x = 3 \end{cases}$$

In addition we shall define integrals over noncompact subsets of $\mathbf{R}^1$ such as $[a,\infty)$, $(-\infty,b)$, etc. All such integrals will be referred to as *improper integrals*.

(6.C.1) *Definition* Suppose that $f: [a,b) \to \mathbf{R}^1$ is continuous. Then the *improper integral* of f from a to b, $\int_a^b f(x)\, dx$, is defined to be $\lim_{c \to b^-} \int_a^c f(x)\, dx$. If this limit exists, then $\int_a^b f(x)\, dx$ is said to *converge*; otherwise, $\int_a^b f(x)\, dx$ is said to *diverge*. Similarly, if $f: (a,b] \to \mathbf{R}^1$ is continuous, then the *improper integral* $\int_a^b f(x)\, dx$ is defined by $\int_a^b f(x)\, dx = \lim_{c \to a^+} \int_c^b f(x)\, dx$.

(6.C.2) *Examples*

(a) $\displaystyle \int_0^1 \frac{1}{\sqrt{x}}\, dx = \lim_{c \to 0^+} \int_c^1 \frac{1}{\sqrt{x}}\, dx = \lim_{c \to 0^+} \left((2x^{1/2})\big|_c^1\right) = 2$

(b) $\displaystyle \int_0^1 \frac{1}{x-1}\, dx = \lim_{c \to 1^-} \int_0^c \frac{1}{x-1}\, dx = \lim_{c \to 1^-} \left(\ln|x - 1|\big|_0^c\right) = -\infty$

(6.C.3) *Observation* If $f: [a,b] \to \mathbf{R}^1$ is continuous, then $\int_a^b f(x)\, dx = \lim_{c \to b^-} \int_a^c f(x)\, dx$ (and, thus, the definitions given in (6.C.1) represent extensions of (6.A.6)). To see this let $\varepsilon > 0$ be given and let M be an upper bound for f on $[a,b]$. Then if $c \in (a,b)$ and $b - a < \varepsilon/M$, we have

$$\left| \int_a^b f(x)\, dx - \int_a^c f(x)\, dx \right| = \left| \int_c^b f(x)\, dx \right| < M(b - c) < \varepsilon$$

Note that this result implies that even if f is redefined at b, the value of the integral $\int_a^b f(x)\, dx$ will remain the same.

(6.C.4) *Exercise* Show that if $f: [a,b) \to \mathbf{R}^1$ is continuous and bounded, then $\int_a^b f(x)\, dx$ converges. [*Hint*: Let $\{c_n\}$ be an increasing sequence of real numbers that converges to b, and show that $\{\int_a^{c_n} f(x)\, dx\}$ is a Cauchy sequence.]

(6.C.5) *Definition* Suppose that $X = [a,b] \backslash \{c\}$, where $c \in (a,b)$. If $f: X \to \mathbf{R}^1$ is continuous, then the *improper integral* of f from a to b, $\int_a^b f(x)\, dx$, is defined to be $\int_a^c f(x)\, dx + \int_c^b f(x)\, dx$. If both $\int_a^c f(x)\, dx$ and $\int_c^b f(x)\, dx$ converge, then $\int_a^b f(x)\, dx$ is said to *converge*; otherwise, $\int_a^b f(x)\, dx$ is said to *diverge*.

(6.C.6) *Example* (a) Let $f(x) = (x^2 - 1)/(x + 1)$ for $x \in [-3,2]$ and $x \ne -1$. Then

$$\int_{-3}^2 f(x)\, dx = \int_{-3}^{-1} \frac{x^2 - 1}{x + 1}\, dx + \int_{-1}^2 \frac{x^2 - 1}{x + 1}\, dx$$

$$= \lim_{c \to -1^-} \int_{-3}^c (x - 1)\, dx + \lim_{c \to -1^+} \int_c^2 (x - 1)\, dx = -7\tfrac{1}{2}$$

(b) Suppose that

$$f(x) = \begin{cases} x^{-4} & \text{if } -3 \le x < 0 \\ x^{-\frac{1}{2}} & \text{if } 0 < x \le 2 \end{cases}$$

Then $\int_{-3}^{2} f(x)\,dx = \int_{-3}^{0} x^{-4}\,dx + \int_{0}^{2} x^{-\frac{1}{2}}\,dx$. It can be readily established that $\int_{-3}^{0} x^{-4}\,dx$ diverges and $\int_{0}^{2} x^{-\frac{1}{2}}\,dx = 2\sqrt{2}$. Hence, $\int_{-3}^{2} f(x)\,dx$ diverges.

(6.C.7) *Definition* Suppose that $f: [a,\infty) \to \mathbf{R}^1$ is continuous. Then the *improper integral of f from a to ∞*, $\int_{a}^{\infty} f(x)\,dx$, is defined to be $\lim_{c\to\infty} \int_{a}^{c} f(x)\,dx$. If $\lim_{c\to\infty} \int_{a}^{c} f(x)\,dx$ is finite, then $\int_{a}^{\infty} f(x)\,dx$ is said to converge; otherwise $\int_{a}^{\infty} f(x)\,dx$ diverges. The improper integral $\int_{-\infty}^{b} f(x)\,dx$ is defined in an analogous fashion.

(6.C.8) *Examples*

(a) $\displaystyle \int_{1}^{\infty} \frac{1}{x^2}\,dx = \lim_{c\to\infty} \int_{1}^{c} \frac{1}{x^2}\,dx = \lim_{c\to\infty} \left(-\frac{1}{x} \Big|_{1}^{c} \right) = 1$

(b) $\displaystyle \int_{-\infty}^{1} \frac{1}{(x-3)^2}\,dx = \lim_{c\to-\infty} \int_{c}^{1} \frac{1}{(x-3)^2}\,dx = \lim_{c\to-\infty} \left(\frac{-1}{x-3} \Big|_{c}^{1} \right) = \frac{1}{2}$

(6.C.9) *Exercise* Show that if $\int_{0}^{\infty} f(x)\,dx$ converges, then $\int_{0}^{\infty} f(x)\,dx = \lim_{n\to\infty} \int_{1/n}^{n} f(x)\,dx$.

(6.C.10) *Definition* Suppose that $f: (-\infty,\infty) \to \mathbf{R}^1$ is continuous. Then the *improper integral* $\int_{-\infty}^{\infty} f(x)\,dx$ is defined by

$$\int_{-\infty}^{\infty} f(x)\,dx = \int_{-\infty}^{0} f(x)\,dx + \int_{0}^{\infty} f(x)\,dx$$

The improper integral $\int_{-\infty}^{\infty} f(x)\,dx$ *converges* if both of the improper integrals $\int_{-\infty}^{0} f(x)\,dx$ and $\int_{0}^{\infty} f(x)\,dx$ converge; otherwise, $\int_{-\infty}^{\infty} f(x)\,dx$ diverges.

(6.C.11) *Example*

(a) $\displaystyle \int_{-\infty}^{\infty} \frac{dx}{x^2+1} = \int_{-\infty}^{0} \frac{dx}{x^2+1} + \int_{0}^{\infty} \frac{dx}{x^2+1}$

$\displaystyle \qquad = \lim_{a\to-\infty} \int_{a}^{0} \frac{dx}{x^2+1} + \lim_{b\to\infty} \int_{0}^{b} \frac{dx}{x^2+1}$

$\displaystyle \qquad = \lim_{a\to-\infty} (-\text{Arctan } a) + \lim_{b\to\infty} (\text{Arctan } b) = \frac{\pi}{2} + \frac{\pi}{2} = \pi$

(b) $\displaystyle \int_{-\infty}^{\infty} \frac{x}{\sqrt{2x^2+5}}\,dx = \int_{-\infty}^{0} \frac{x}{\sqrt{2x^2+5}}\,dx + \int_{0}^{\infty} \frac{x}{\sqrt{2x^2+5}}\,dx$

$\displaystyle \qquad = \lim_{a\to-\infty} \left(\frac{1}{2}\right)\sqrt{2x^2+5}\,\Big|_{a}^{0} + \lim_{b\to\infty} \left(\frac{1}{2}\right)\sqrt{2x^2+5}\,\Big|_{0}^{b}$

Since neither of these limits exists, $\int_{-\infty}^{\infty} (x/\sqrt{2x^2+5})\,dx$ diverges.

(6.C.12) *Observation* The number 0 in the prevîous definition can be replaced by any real number c without changing the value of $\int_{-\infty}^{\infty} f(x)\,dx$ (see problem C.2). Furthermore, it can be shown (see problem C.3) that if $\int_{-\infty}^{\infty} f(x)\,dx$ converges, then this integral is equal to $\lim_{a\to\infty} \int_{-a}^{a} f(x)\,dx$. Note, however, that if $f(x) = x$, then $\lim_{a\to\infty} \int_{-a}^{a} x\,dx = 0$, while $\int_{-\infty}^{\infty} x\,dx$ diverges. If $f: (-\infty,\infty) \to \mathbf{R}^1$, then the *Cauchy principal value* of $\int_{-\infty}^{\infty} f(x)\,dx$ is defined to be $\lim_{a\to\infty} \int_{-a}^{a} f(x)\,dx$, and will be denoted by $\mathrm{cpv}\int_{-\infty}^{\infty} f(x)\,dx$. This integral will prove to be of major importance in our study of the Fourier transform in Chapter 13.

(6.C.13) *Definition* Suppose that $c_1, c_2, \ldots, c_n$ are points lying in an interval (a,b) (where $a = -\infty$ and $b = \infty$ are allowed). If f is continuous on $(a,b)\backslash\{c_1, c_2, \ldots, c_n\}$, then the improper integral $\int_a^b f(x)\,dx$ is defined by

$$\int_a^b f(x)\,dx = \int_a^{c_1} f(x)\,dx + \int_{c_1}^{c_2} f(x)\,dx + \cdots + \int_{c_n}^b f(x)\,dx \qquad (1)$$

The integral $\int_a^b f(x)\,dx$ *converges* if each integral on the right-hand side of (1) converges; otherwise, $\int_a^b f(x)\,dx$ *diverges*.

Improper integrals are frequently used to define functions. An important example of this is the gamma function, $\Gamma: (0,\infty) \to \mathbf{R}^1$ defined by

$$\Gamma(x) = \int_0^{\infty} t^{x-1}e^{-t}\,dt \qquad x > 0$$

This function arises naturally in many areas of applied mathematics. In particular, the Γ function plays a significant role in connection with the determination and expression of solutions of many differential equations.

The convergence of $\Gamma(x)$, while easy to establish, does involve certain subtleties. It is easily verified that the derivative of the function e^x (where e^x is defined as an infinite series: $e^x = 1 + x + x^2/2! + x^3/3! + \cdots$ —see Chapter 8) is e^x. Since e^x is positively valued for all real x, it follows that e^x is a strictly increasing function, and consequently has an inverse, denoted by $\ln x$, with domain $(0,\infty)$. Recall now from elementary calculus that if a is a positive number, and x is any real number, then a^x is defined to be $e^{x\,\ln a}$. Although this latter expression may appear more formidable than a^x, it avoids logical difficulties since e^y is well defined for each real (and complex) number y.

Since $\ln 0$ is not defined, in order to establish the convergence of $\int_0^{\infty} t^{x-1}e^{-t}\,dt$, we consider the improper integrals $\int_0^1 t^{x-1}e^{-t}\,dt$ and $\int_1^{\infty} t^{x-1}e^{-t}\,dt$. Note that on the interval $(0,1]$, $g(t) = t^{x-1}e^{-t}$ is continuous, and furthermore, $0 < g(t) < t^{x-1}$ for each $t \in (0,1]$; therefore, since $\int_0^1 t^{x-1}\,dt$ converges for $x > 0$, so must $\int_0^1 g(t)\,dt$. To see that $\int_1^{\infty} t^{x-1}e^{-t}\,dt$ also converges, first observe that

$$\lim_{t \to \infty} \frac{t^{x-1}e^{-t}}{1/t^2} = \lim_{t \to \infty} \frac{t^{x+1}}{e^t} = 0$$

(prove this), and consequently, for large t, $t^{x-1}e^{-t} \leq 1/t^2$. Since $\int_1^\infty (1/t^2)\, dt$ converges, the result follows from (6.B.3.v).

Perhaps the most basic and interesting property of the Γ function is the following.

(6.C.14) *Theorem* If $x > 0$, then $\Gamma(x + 1) = x\Gamma(x)$.

Proof. By (6.C.9) and integration by parts we find

$$\Gamma(x + 1) = \lim_{n \to \infty} \int_{1/n}^n t^x e^{-t}\, dt = \lim_{n \to \infty} \left[\left(-\frac{t^x}{e^t}\Big|_{1/n}^n \right) + \int_{1/n}^n xt^{x-1}e^{-t}\, dt \right]$$

$$= \lim_{n \to \infty} x \int_{1/n}^n t^{x-1}e^{-t}\, dt = x\Gamma(x)$$

(6.C.15) *Exercise* Show that $\Gamma(1) = 1$ and hence, $\Gamma(n + 1) = n!$. Note that the formula $\Gamma(x + 1) = x\Gamma(x)$ generalizes the obvious fact that $n! = n(n - 1)!$.

The Γ function can be extended to negative real numbers as follows. If $-1 < x < 0$, we define $\Gamma(x)$ to be $(1/x)\Gamma(x + 1)$; if $-2 < x < 1$, then $\Gamma(x) = [1/x(x + 1)]\Gamma(x + 2)$; in general, if $-m < x < -m + 1$, then

$$\Gamma(x) = \frac{\Gamma(x + m)}{x(x + 1)\cdots(x + m - 1)}$$

Tables of the Γ function for values of x between 0 and 1 can be found in many elementary handbooks. Theorem (6.C.14) can be used then to find $\Gamma(x)$ for real numbers not in the interval (0,1).

The graph of the Γ function is given Fig. 6.2.

The Γ function first appeared in 1729 in correspondence between the brilliant Swiss mathematician L. Euler and the German mathematician Goldbach. Consideration of this function arose in connection with certain interpolation problems; in particular, Euler was looking for an easily expressed function f with the property that $f(n) = n!$ for each nonnegative integer n. If such a function could be found, then sense could be made of expressions such as 4.31!, namely it would be equal to $f(4.31)$.

Euler discovered that the "infinite product"

$$g(n) = \quad = \left[\left(\frac{2}{1}\right)^n \frac{1}{n + 1} \right]\left[\left(\frac{3}{2}\right)^n \frac{2}{n + 2} \right]\left[\left(\frac{4}{3}\right)^n \frac{3}{n + 3} \right]\cdots \qquad (2)$$

satisfied the condition $g(n) = n!$ for $n = 0, 1, \ldots$ (this is easily seen if one

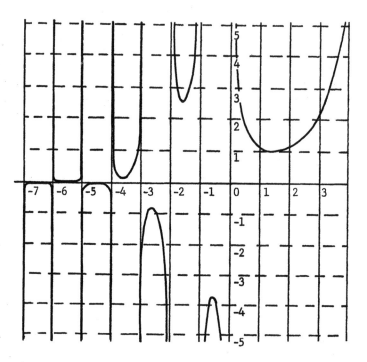

Figure 6.2

assumes that infinite products can be naturally defined as a limit of finite products and that "infinite" cancellation can be effected). Expression (2) can also be written in the form

$$g(n) = \lim_{m \to \infty} \frac{m!(m+1)^n}{(n+1)(n+2)\cdots(n+m)} \tag{3}$$

where n takes on integer values. Note that (2) and (3) are valid for all real numbers n, except for the negative integers, and hence, represent a solution to Euler's problem.

Euler went on to show that integrals could be used to resolve the same problem when he demonstrated that for $n = 0, 1, 2, \ldots,$

$$n! = \int_0^1 (-\ln x)^n \, dx \tag{4}$$

He subsequently modified (4) to obtain the function $\int_0^\infty t^{x-1} e^{-t} \, dt$ that we have just studied. All of this work was done prior to 1781; since that time a substantial amount of mathematical thought has gone into extending the notion of the Γ function to a complex setting and finding increasingly more powerful uses for it. For a quite complete discussion of the history and applications of the Γ function, the article by Davis (1959) is highly recommended.

Another important function, the *beta function* **B**, maps $(0,\infty) \times (0,\infty)$ into $\mathbf{R}^1$, and is defined by

$$\mathbf{B}(x,y) = \int_0^1 t^{x-1}(1-t)^{y-1}\,dt \tag{5}$$

This function finds many uses in statistics and other areas of applied mathematics. It is also employed in the calculation of a number of important integrals.

The **B** function is related to the Γ function by the following equation:

$$\mathbf{B}(x,y) = \frac{\Gamma(x)\Gamma(y)}{\Gamma(x+y)} \tag{6}$$

To establish (6) we first show that

$$\mathbf{B}(x,1) = \frac{1}{x} \tag{7}$$

$$\mathbf{B}(x,y) = \mathbf{B}(y,x) \tag{8}$$

and

$$\mathbf{B}(x, y+1) = \frac{y}{x+y}\,\mathbf{B}(x,y) \tag{9}$$

Equation (7) follows at once from the definition of **B**; (8) also follows easily since $\int_0^1 t^{x-1}(1-t)^{y-1}\,dt = \int_0^1 (1-w)^{x-1}w^{y-1}\,dw$, where $t = t(w) = 1 - w$. To verify (9) we first note that

$$\mathbf{B}(x,y) = \int_0^1 t^{x-1}(1-t)^{y-1}\,dt = \int_0^\infty \frac{w^{x-1}}{(1+w)^{x+y}}\,dw$$

where $t = t(w) = w/(1+w)$. Since

$$(w^x(1+w)^{-(x+y)})' = xw^{x-1}(1+w)^{-(x+y)} - (x+y)w^x(1+w)^{-(x+y)-1}$$

an application of integration by parts yields

$$0 = x\mathbf{B}(x,y) - (x+y)\mathbf{B}(x+1, y)$$

The desired result (9) now follows easily from (8).

Next we use induction to show that for each $n \in \mathbf{Z}^+$

$$\mathbf{B}(x,y) = \frac{\mathbf{B}(x, y+n)\mathbf{B}(y,n)}{\mathbf{B}(x+y,n)} \tag{10}$$

If $n = 1$, then it follows from (7) and (9) that (10) is valid. To complete the induction argument, note that from (9) we have

$$\frac{\mathbf{B}(x, y + (n + 1))\mathbf{B}(y, n + 1)}{\mathbf{B}(x + y, n + 1)}$$

$$= \frac{[(y + n)/(x + y + n)]\mathbf{B}(x, y + n)[n/(y + n)]\mathbf{B}(y,n)}{[n/(x + y + n)]\mathbf{B}(x + y, n)}$$

$$= \frac{\mathbf{B}(x, y + n)\mathbf{B}(y,n)}{\mathbf{B}(x + y, n)}$$

Finally, we determine upper and lower bounds for $B(x,z)$, $z > 1$ [the following argument is due to Nanjundiah (1969)]. Since for each $w \neq 0$, $e^w > 1 + w$, we have

$$\frac{\Gamma(x)}{(z - 1)^x} = \int_0^\infty t^{x-1}e^{-(z-1)t}\, dt > \int_0^1 t^{x-1}(1 - t)^{z-1}\, dt = B(x,z)$$

$$\frac{\Gamma(x)}{(x + z)^x} = \int_0^\infty t^{x-1}e^{-(x+z)t}\, dt < \int_0^1 \frac{t^{x-1}}{(1 + t)^{x+z}}\, dt = B(x,z)$$

Thus, as $z \to \infty$ we see that $B(x,z)$ approaches $\Gamma(x)/z^x$, and hence, as $n \to \infty$, we have from (10)

$$\mathbf{B}(x,y) = \frac{\Gamma(x)\Gamma(y)}{\Gamma(x + y)}$$

To illustrate how the beta function can arise in integration problems we consider the integral

$$\int_0^{\pi/2} \sin^{18} x\, dx \tag{11}$$

If we set $u = \sin^2 x$, then (11) becomes (via (6.B.9))

$$\int_0^1 \frac{u^9}{2u^{1/2}(1 - u)^{1/2}}\, du = \frac{1}{2}\int_0^1 u^{8\frac{1}{2}}(1 - u)^{-1/2}\, du = \frac{1}{2}\mathbf{B}\left(9\frac{1}{2},\frac{1}{2}\right)$$

D.* SOME ELEMENTARY APPLICATIONS OF THE DEFINITE INTEGRAL: BUFFON'S NEEDLE PROBLEM, ARC LENGTH, AND THE PICARD EXISTENCE THEOREM

The definite integral arises in an extraordinary variety of ways including such diverse situations as the calculation of surface areas, volumes, centers of gravity and of the work done by a variable force. In this section we shall examine briefly the role of the integral in three rather distinct contexts: a problem in probability, the calculation of arc lengths of certain curves, and the formulation of an integral equation which is used to establish the existence and uniqueness of solutions of a wide class of first-order differential equations.

We begin with a problem formulated in 1733 by the French naturalist Buffon in the Proceedings of the Paris Academy of Sciences. The statement of the problem is quite simple: A needle of length L is dropped onto a flat surface that has been marked off with a series of parallel lines, D units apart, where $D > L$ (Fig. 6.3). What is the probability that the needle will intersect one of the lines?

To resolve this problem we proceed as follows. Let c represent the center of the needle and let y denote the distance between c and the nearest line. Let α be the acute angle between the needle and the vertical line l indicated in Fig. 6.4 ($\alpha = \pi/2$ is also permitted). Note that we have

$$0 \le \alpha \le \frac{\pi}{2} \qquad 0 \le y \le \frac{D}{2}$$

It is clear from the geometry of the situation that the needle will intersect one of the lines only if

$$y \le \frac{L}{2} \cos \alpha$$

i.e., intersection occurs if and only if the point (α, y) lies in the shaded area indicated in Fig. 6.5. Since the area of the shaded region is easily seen to be

$$\int_0^{\pi/2} \frac{L}{2} \cos \alpha \, d\alpha = \frac{L}{2}$$

and since the area of the rectangle containing the possible values of α and y is

$$\left(\frac{\pi}{2}\right)\left(\frac{D}{2}\right)$$

it follows that the probability of intersection is

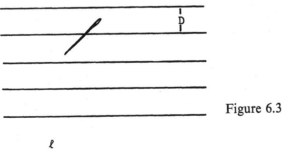

Figure 6.3

Figure 6.4

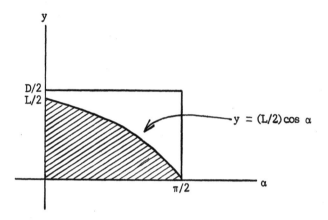

Figure 6.5

$$\frac{L/2}{\pi D/4} = \frac{2L}{\pi D}$$

(We have assumed throughout, of course, that the needle can land with equal probability anywhere on the surface.)

Next we turn to the problem of defining and calculating arc lengths. Suppose that $\phi : [a,b] \to \mathbf{R}^2$ is a continuous function. We wish to assign a length to the curve C_ϕ, which is the image of ϕ (Fig. 6.6).

To do this we form partitions $P = \{t_0, t_1, \ldots, t_n\}$ of $[a,b]$ and for each such partition, we construct a polygonal curve (a curve composed of straight-line segments) by consecutively connecting the points $\phi(t_0), \phi(t_1), \ldots, \phi(t_n)$ with line segments (Fig. 6.7).

Let L_P denote the sum of the lengths of these segments, and note that

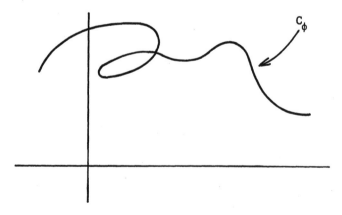

Figure 6.6

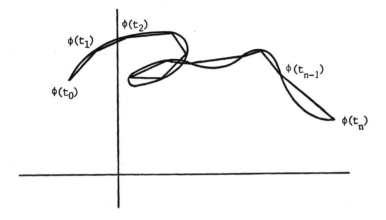

Figure 6.7

it is at least intuitively clear that L_P is less than or equal to the apparent length of C_ϕ. This leads us to define the *length* of C_ϕ to be

$$\sup \{L_P \mid P \text{ is a partition of } [a,b]\} \qquad (1)$$

For continuous functions we have the following way of determining the length of a curve.

(6.D.1) *Theorem* If $\phi \colon [a,b] \to \mathbf{R}^2$ is a continuous function, then the length of C_ϕ is equal to $\lim_{|P| \to 0} L_P$.

Proof. First suppose that the length of C_ϕ is finite, say equal to L, and let $\varepsilon > 0$ be given. Then there is a partition $Q = \{t_0, t_1, \ldots, t_M\}$ of $[a,b]$ such that $L_Q > L - \varepsilon/2$. Let $\phi = (\phi_1, \phi_2)$ and note that ϕ_1 and ϕ_2 are uniformly continuous on the compact interval $[a,b]$. Consequently, there is a $\delta > 0$ such that if t and $\hat{t}$ lie in $[a,b]$ and $|t - \hat{t}| < \delta$, then

$$d(\phi(t), \phi(\hat{t})) = \sqrt{(\phi_1(t) - \phi_1(\hat{t}))^2 + (\phi_2(t) - \phi_2(\hat{t}))^2} < \frac{\varepsilon}{4(M - 1)}$$

Suppose that P is a partition of $[a,b]$ with mesh less than δ, and let P' be the partition obtained by adding to P all the points of $Q \backslash P$ (note that there are at most $M - 1$ points in $Q \backslash P$). Since each point in $Q \backslash P$ is an endpoint for exactly two subintervals in P', the contribution to the sum $L_{P'}$ corresponding to those subintervals of $[a,b]$ that have an endpoint in $Q \backslash P$ is $2(M - 1)$ $(\varepsilon/4(M - 1)) = \varepsilon/2$. From this it follows that $L_{P'} \leq L_P + \varepsilon/2$. Furthermore, since $Q \subset P'$, it is easy to see that $L_{P'} \geq L_Q$, and therefore, we have

$$L_P + \frac{\varepsilon}{2} \geq L_{P'} \geq L_Q > L - \frac{\varepsilon}{2}$$

and hence,

$$L_P > L - \varepsilon$$

Since, $L_P \le L$, we have that if $|P| < \delta$, then $L - \varepsilon < L_P \le L$, which yields the desired result.

A similar argument may be made in the case that $\lim_{|P| \to 0} L_P = \infty$.

From a computational standpoint neither the definition of the length of a curve nor the previous result are of much use. If, however, additional conditions are imposed on the function ϕ, then the length of C_ϕ can be neatly expressed as an integral.

(6.D.2) Theorem Suppose that $\phi: [a,b] \to \mathbf{R}^2$ is defined by

$$\phi(t) = (\phi_1(t), \phi_2(t))$$

where both ϕ_1 and ϕ_2 are *continuously differentiable* on $[a,b]$, i.e., ϕ_1' and ϕ_2' are continuous on $[a,b]$. Then the length of C_ϕ is finite and is equal to

$$\int_a^b \sqrt{(\phi_1'(t))^2 + (\phi_2'(t))^2}\, dt \qquad (2)$$

Proof. Let $P = \{t_0, t_1, \ldots, t_n\}$ be an arbitrary partition of $[a,b]$. Then the length L_P of the polygonal curve associated with C_ϕ and the partition P is

$$\sum_{i=0}^{n-1} \sqrt{(\phi_1(t_{i+1}) - \phi_1(t_i))^2 + (\phi_2(t_{i+1}) - \phi_2(t_i))^2} \qquad (3)$$

Since both ϕ_1 and ϕ_2 are continuous on $[a,b]$, we can apply the mean value theorem to obtain (for each i)

$$\begin{aligned}
\phi_1(t_{i+1}) - \phi_1(t_i) &= \phi_1'(\hat{t}_i)\,\Delta t_i \\
\phi_2(t_{i+1}) - \phi_2(t_i) &= \phi_2'(\bar{t}_i)\,\Delta t_i
\end{aligned} \qquad (4)$$

for appropriate points $\hat{t}_i, \bar{t}_i \in (t_i, t_{i+1})$. It follows from (6.D.1), (3) and (4) that the length of C_ϕ is equal to

$$\lim_{|P| \to 0} \sum_{i=0}^{n-1} \sqrt{(\phi_1'(\hat{t}_i))^2 + (\phi_2'(\bar{t}_i))^2}\,\Delta t_i \qquad (5)$$

where $\hat{t}_i$ and $\bar{t}_i$ are determined from (4) (and provided, of course, that this limit exists).

Since ϕ_1' and ϕ_2' are continuous, it is immediate from (6.A.12) that the integral

$$\int_a^b \sqrt{(\phi_1'(t))^2 + (\phi_2'(t))^2}\, dt \qquad (6)$$

is finite and is equal to

$$\lim_{|P|\to 0} \sum_{i=0}^{n-1} \sqrt{(\phi_1'(t_i))^2 + (\phi_2'(t_i))^2}\, \Delta t_i \tag{7}$$

where $P = \{t_0, t_1, \ldots, t_n\}$ is a partition of $[a,b]$. From (5), (6) and (7) we see that to establish (2) we need only show that

$$\lim_{|P|\to 0} \sum_{i=0}^{n-1} \sqrt{(\phi_1'(\hat{t}_i))^2 + (\phi_2'(\hat{t}_i))^2}\, \Delta t_i = \lim_{|P|\to 0} \sum_{i=0}^{n-1} \sqrt{(\phi_1'(t_i))^2 + (\phi_2'(t_i))^2}\, \Delta t_i \tag{8}$$

This we do as follows. Let

$$L = \int_a^b \sqrt{(\phi_1'(t))^2 + (\phi_2'(t))^2}\, dt = \lim_{|P|\to 0} \sum_{i=0}^{n-1} \sqrt{(\phi_1'(t_i))^2 + (\phi_2'(t_i))^2}\, \Delta t_i$$

Clearly, to verify (8), it suffices to show that given $\varepsilon > 0$, there is a $\delta > 0$ such that if $P = \{t_0, t_1, \ldots, t_n\}$ is a partition of $[a,b]$ with mesh less than δ, then

$$\left| \left(\sum_{i=0}^{n-1} \sqrt{(\phi_1'(\hat{t}_i))^2 + (\phi_2'(\hat{t}_i))^2}\, \Delta t_i \right) - L \right| < \varepsilon \tag{9}$$

Choose δ_1 so that if P is any partition of $[a,b]$ with mesh less than δ_1, then

$$\left| \sum_{i=0}^{n-1} \sqrt{(\phi_1'(t_i))^2 + (\phi_2'(t_i))^2}\, \Delta t_i - L \right| < \frac{\varepsilon}{2} \tag{10}$$

Define $G: [a,b] \times [a,b] \to \mathbf{R}^1$ by

$$G(u,v) = \sqrt{(\phi_1'(u))^2 + (\phi_2'(v))^2}$$

Then G is a continuous function with compact domain, and therefore, by (6.A.3), G is uniformly continuous. Thus, there is a positive number δ_2 such that if (u_1,v_1) and (u_2,v_2) are any two points in $[a,b] \times [a,b]$ such that $\sqrt{(u_1 - u_2)^2 + (v_1 - v_2)^2} < \delta_2$, then $|G(u_1,v_1) - G(u_2,v_2)| < \varepsilon/2(b - a)$. Now let P be any partition of $[a,b]$ with mesh less than $\min\{\delta_1,\delta_2\}$. Then if $\hat{t}_i$ and $\bar{t}_i$ are any two points in the interval $[t_i,t_{i+1}]$, we have

$$|G(t_i, t_i) - G(\hat{t}_i, \bar{t}_i)| < \frac{\varepsilon}{2(b - a)}$$

and consequently,

$$\left| \sum_{i=0}^{n-1} \sqrt{(\phi_1'(t_i))^2 + (\phi_2'(t_i))^2}\, \Delta t_i - \sum_{i=0}^{n-1} \sqrt{(\phi_1'(\hat{t}_i))^2 + (\phi_2'(\hat{t}_i))^2}\, \Delta t_i \right|$$

$$\leq \sum_{i=0}^{n-1} |G(t_i, t_i) - G(\hat{t}_i, \bar{t}_i)|\, \Delta t_i < \frac{\varepsilon}{2(b - a)}(b - a) = \frac{\varepsilon}{2} \tag{11}$$

The desired inequality (9) now follows from (10) and (11) and the triangle inequality.

If ϕ_1' and/or ϕ_2' are not continuous (or do not exist), then some rather curious situations may arise. For instance, in the problem set the reader is asked to show that if $\phi: [0,1] \to \mathbf{R}^2$ is defined by

$$\phi(t) = \begin{cases} \left(t, t \sin \dfrac{1}{t}\right) & \text{if } t \in (0,1] \\ (0,0) & \text{if } t = 0 \end{cases}$$

then even though ϕ is continuous, the length of the curve C_ϕ is infinite. Thus, if a particle were to move at a steady pace along this curve toward the origin, at any time t its distance from the origin (along the curve) would remain infinite. Interestingly, a slight change in ϕ will result in a considerably shorter journey for the particle. In problem D.6 the reader is asked to show that if $\phi: [0,1] \to \mathbf{R}^2$ is defined by

$$\phi(t) = \begin{cases} \left(t, t^2 \sin \dfrac{1}{t}\right) & \text{if } t \in (0,1] \\ (0,0) & \text{if } t = 0 \end{cases}$$

then the distance along the curve from the point $(1, \sin 1)$ to the origin is finite.

Finally the reader should note that if $f: [a,b] \to \mathbf{R}^1$ is continuously differentiable, then it follows immediately from (6.D.2) that the length of the curve generated by the graph of f, $\{(x,f(x)) \mid x \in [a,b]\}$, is given by

$$\int_a^b \sqrt{1 + (f'(x))^2}\, dx \tag{12}$$

As our last application of the integral in this chapter, we consider the problem of establishing the existence and uniqueness of solutions of the first-order differential equation

$$y' = f(x, y) \tag{13}$$

subject to the initial condition

$$y(x_0) = y_0 \tag{14}$$

where f is a continuous function on a rectangle

$$R = [x_0 - a, x_0 + a] \times [y_0 - b, y_0 + b]$$

We shall assume, furthermore, that f satisfies the inequality

$$|f(x,y) - f(x,z)| \leq K|y - z| \tag{15}$$

for each (x,y) and (x,z) in R and for some constant K. The restriction (15) is called a *Lipschitz condition in the second argument*; it does not represent a particularly strong imposition on f, being in some sense intermediate between continuity and differentiability (cf. problems D.11 and D.13).

We shall see that under these conditions there is a (unique) solution of (13) [satisfying the initial condition (14)], which is defined on an appropriate interval $[x_0 - \delta, x_0 + \delta] \subset [x_0 - a, x_0 + a]$. In order to establish this result (known as the *Picard existence theorem*), we transform the initial value problem into an equivalent problem. Note that if u is a solution of (13) satisfying $u(x_0) = y_0$, then

$$u'(x) = f(x, u(x)) \tag{16}$$

and integration of both sides of (16) yields

$$u(x) = y_0 + \int_{x_0}^{x} f(w, u(w))\, dw \tag{17}$$

Equation (17) is called an *integral equation*.

Conversely, if u is continuous and if $u(x) = y_0 + \int_{x_0}^{x} f(w, u(w))\, dw$, then by (6.A.9), $u'(x) = f(x, u(x))$ and $u(x_0) = y_0$; consequently, u is a solution of the original initial value problem. Thus, finding a solution u of (13) such that $u(x_0) = y_0$ is equivalent to finding a function u that satisfies the integral equation (17).

The basic strategy now is as follows. We let

$$\mathscr{C} = \{u \colon [x_0 - \delta, x_0 + \delta] \to \mathbf{R}^1 \mid u \text{ is continuous}\} \tag{18}$$

for a suitable choice of δ; $\mathscr{C}$ is given the sup metric. We shall find an appropriate closed subset $\mathscr{A}$ of $\mathscr{C}$, and then define a function $T \colon \mathscr{A} \to \mathscr{A}$ such that if $u \in \mathscr{A}$, then $T(u)$ is the function in $\mathscr{A}$ defined by

$$(T(u))(x) = y_0 + \int_{x_0}^{x} f(w, u(w))\, dw \qquad x \in [x_0 - \delta, x_0 + \delta] \tag{19}$$

By (4.C.6) and (4.C.7), $\mathscr{A}$ is a complete metric space; we show that T is a contraction map and apply (4.E.10) to determine that T has a unique fixed point u^*. This means, of course, that $T(u^*) = u^*$, from which it follows that

$$u^*(x) = (T(u^*))(x) = y_0 + \int_{x_0}^{x} f(w, u^*(w))\, dw$$

and therefore, from our previous remarks, u^* is the unique solution of (13) satisfying the initial condition $u^*(x_0) = y_0$.

To find $\mathscr{A}$ we proceed as follows. Since f is continuous on the compact set R, $M = \max\{|f(x,y)| \mid (x,y) \in R\}$ exists. Let L_1 and L_2 be the lines passing through (x_0, y_0) with slopes M and $-M$, respectively. Thus, the line L_1 is defined by

$$y - y_0 = M(x - x_0) \tag{20}$$

and the line L_2 is defined by

$$y - y_0 = -M(x - x_0) \tag{21}$$

For reasons that will soon become clear, we let δ be a real number satisfying $0 < \delta < \min\{a,b/M,1/K\}$, and let D be the closed region bounded by the lines $x = x_0 - \delta$, $x = x_0 + \delta$, L_1, and L_2. Note that $D \subset R$ (Fig. 6.8).

Let $\mathscr{C}$ be defined by (18) and let $\mathscr{A}$ be the subset of $\mathscr{C}$ defined by

$$\mathscr{A} = \{u \in \mathscr{C} \mid u(x_0) = y_0 \text{ and the graph of } u \text{ lies in } D\}$$

As an easy exercise, the reader is asked in problem D.14 to show that $\mathscr{A}$ is a closed subset of $\mathscr{C}$ and, hence, is a complete metric space (with the sup metric d^*).

Let $T: \mathscr{A} \to \mathscr{A}$ be defined as in (19). We need to show that T is well defined and continuous on $[x_0 - \delta, x_0 + \delta]$ and that the graph of T lies in D. Observe that for each $w \in [x_0 - \delta, x_0 + \delta]$, $(w,u(w)) \in D$, and since $D \subset R$, it follows that $f(w,u(w))$ is well defined and continuous on $[x_0 - \delta, x_0 + \delta]$. Therefore, by (6.A.12), the function $T(u)$ is also well defined and continuous on this interval. Moreover, since for each $w \in [x_0 - \delta, x_0 + \delta]$, $|f(w,u(w))| \leq M$, we have, for each $x \in [x_0 - \delta, x_0 + \delta]$, that

$$\left| \int_{x_0}^{x} f(w, u(w)) \, dw \right| \leq M|x - x_0|$$

Consequently, if $x \in [x_0 - \delta, x_0 + \delta]$, then

$$y_0 - M|x - x_0| \leq y_0 + \int_{x_0}^{x} f(w, u(w)) \, dw \leq y_0 + M|x - x_0|$$

which shows that the graph of $T(u)$ lies in D.

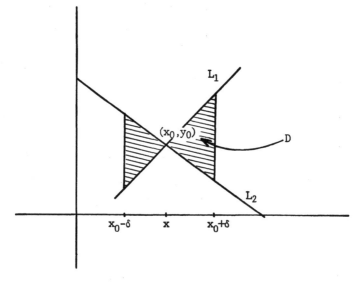

Figure 6.8

Finally, we show that T is a contraction map. Let u an v be any two functions in $\mathscr{A}$. First note that by the Lipschitz condition (15), we have for each $w \in [x_0 - \delta, x_0 + \delta]$,

$$|f(w, u(w)) - f(w, v(w))| \leq K|u(w) - v(w)|$$
$$\leq K \sup \{|u(w) - v(w)| \mid w \in [x_0 - \delta, x_0 + \delta]\}$$
$$= Kd^*(u, v)$$

Consequently,

$d^*(T(u), T(v))$

$$= \sup \left\{ \left| \int_{x_0}^x f(w, u(w))\, dw - \int_{x_0}^x f(w, v(w))\, dw \right| \, \middle| \, x \in [x_0 - \delta, x_0 + \delta] \right\}$$

$$= \sup \left\{ \left| \int_{x_0}^x [f(w, u(w)) - f(w, v(w))]\, dw \right| \, \middle| \, x \in [x_0 - \delta, x_0 + \delta] \right\}$$

$$\leq \sup \left\{ \int_{x_0}^x |f(w, u(w)) - f(w, v(w))|\, dw \, \middle| \, x \in [x_0 - \delta, x_0 + \delta] \right\}$$

$$\leq \sup \left\{ \int_{x_0}^x Kd^*(u, v)\, dw \, \middle| \, x \in [x_0 - \delta, x_0 + \delta] \right\}$$

$$= K\delta d^*(u, v)$$

Since by the definition of δ, $K\delta < 1$, it follows that T is a contraction mapping; this concludes the proof of the Picard existence theorem.

E.* THE WEIERSTRASS APPROXIMATION THEOREM

This section is devoted to establishing the remarkable Weierstrass approximation theorem that was used in Sec. 6.A. This result was first proven in 1885, and since that time a number of quite distinct proofs have been given. We shall follow the proof due to Serge Bernstein (1880–1968), which, although perhaps somewhat less elegant than many of the more "modern" proofs, has the advantage that it actually defines the polynomials that are used to approximate a given continuous function (a number of proofs only establish the existence of such polynomials). For each positive integer n, the nth Bernstein polynomial associated with a given continuous function $f: [0,1] \rightarrow \mathbf{R}^1$ is defined by

$$B_n(x) = \sum_{k=0}^n \binom{n}{k} f\left(\frac{k}{n}\right) x^k (1 - x)^{n-k} \tag{1}$$

where $\binom{n}{k} = n!/k!(n - k)!$. The following ingenious proof shows that these polynomials converge uniformly to the continuous function $f: [0,1] \rightarrow \mathbf{R}^1$.

(6.E.1) Theorem (Weierstrass Approximation Theorem) Suppose that $f: [a,b] \to \mathbf{R}^1$ is a continuous function and let $\varepsilon > 0$ be given. Then there is a polynomial $p: [a,b] \to \mathbf{R}^1$ such that $|p(x) - f(x)| < \varepsilon$ for each $x \in [a,b]$.

Proof. We first establish the theorem for the case where $[a,b] = [0,1]$. Suppose then that $f: [0,1] \to \mathbf{R}^1$ is a continuous function and let $\varepsilon > 0$ be given. We shall show that for suitably large n, $|f(x) - B_n(x)| < \varepsilon$ on $[a,b]$, where B_n is the nth Bernstein polynomial defined by (1). We shall make use of the following identities (all valid for $n \geq 2$ and $x \in [0,1]$):

(a) $\quad \displaystyle\sum_{k=0}^{n} \binom{n}{k} x^k (1 - x)^{n-k} = 1$

(b) $\quad \displaystyle\sum_{k=0}^{n} \frac{k}{n} \binom{n}{k} x^k (1 - x)^{n-k} = x$

(c) $\quad \displaystyle\sum_{k=0}^{n} \frac{k^2}{n^2} \binom{n}{k} x^k (1 - x)^{n-k} = x^2 \left(1 - \frac{1}{n}\right) + \frac{x}{n}$

(d) $\quad \displaystyle\sum_{k=0}^{n} \left(\frac{k}{n} - x\right)^2 \binom{n}{k} x^k (1 - x)^{n-k} = \frac{x(1 - x)}{n}$

Identity (a) follows easily from the binomial expansion theorem

$$(a + b)^n = \sum_{k=0}^{n} \binom{n}{k} a^k b^{n-k} \tag{2}$$

since if $a = x$ and $b = 1 - x$, then

$$1 = (x + (1 - x))^n = \sum_{k=0}^{n} \binom{n}{k} x^k (1 - x)^{n-k}$$

If we differentiate both sides of (2) with respect to a and let $a = x$ and $b = 1 - x$, then we obtain (b). The identity (c) is verified by first differentiating (2) twice with respect to a and then setting $a = x$ and $b = 1 - x$. Finally, (d) follows easily from (a), (b), and (c)—see problem E.2.

For $n \geq 2$, we now find a bound for $|f(x) - B_n(x)|$. From (1) and (a) we have

$$f(x) - B_n(x) = \sum_{k=0}^{n} \left(f(x) - f\left(\frac{k}{n}\right)\right) \binom{n}{k} x^k (1 - x)^{n-k} = \Sigma_1 + \Sigma_2$$

where Σ_1 represents the sum over all values of k such that $|x - k/n| < n^{-\frac{1}{4}}$ and Σ_2 represents the sum over the remaining k's (where $|x - k/n| \geq n^{-\frac{1}{4}}$). Let $M = \sup\{|f(x)| \mid 0 \leq x \leq 1\}$. Since $|x - k/n| < n^{-\frac{1}{4}}$ implies that $(k - nx)^2/n^{\frac{3}{2}} \geq 1$, we have

$$|\Sigma_2| \leq 2M \sum\nolimits_1 \binom{n}{k} x^k (1 - x)^{n-k} \leq 2M \sum\nolimits_2 \frac{(k - nx)^2}{n^{3/2}} \binom{n}{k} x^k (1 - x)^{n-k}$$

$$\leq \frac{2M}{n^{3/2}} \sum_{k=0}^{n} (k - nx)^2 \binom{n}{k} x^k(1 - x)^{n-k} \leq \frac{2M}{n^{3/2}} nx(1 - x) \leq \frac{2M}{n^{1/2}}$$

Let N_1 be a positive integer such that if $n \geq N_1$, then $|\sum_2| < \varepsilon/2$. By (6.A.3) f is uniformly continuous on $[0,1]$, and hence, there is a $\delta > 0$ such that if $x,x' \in [0,1]$ and $|x - x'| < \delta$, then $|f(x) - f(x')| < \varepsilon/2$. Let N_2 be a positive integer such that if $n \geq N_2$, then $n^{-\frac{1}{4}} < \delta$. Observe that if $n \geq N_2$, then $|\sum_1| \leq (\varepsilon/2) \sum_1 \binom{n}{k} x^k(1 - x)^{n-k} < \varepsilon/2$, and consequently, if $n \geq \max\{N_1,N_2\}$, then

$$|f(x) - B_n(x)| \leq |\sum_1| + |\sum_2| < \frac{\varepsilon}{2} + \frac{\varepsilon}{2} = \varepsilon$$

which proves the theorem for the case where $[a,b] = [0,1]$.

To complete the proof, let $[a,b]$ be an arbitrary closed and bounded interval and note that if $\alpha : [a,b] \to [0,1]$ is defined by

$$\alpha(x) = a + (b - a)x$$

then α has a continuous inverse $\alpha^{-1} : [0,1] \to [a,b]$ defined by

$$\alpha^{-1}(x) = \frac{x - a}{b - a}$$

Suppose now that $f : [a,b] \to \mathbf{R}^1$ is a continuous function. Then $f \circ \alpha^{-1}$ is continuous and maps the interval $[0,1]$ into $\mathbf{R}^1$; therefore, by the special case already proven there is a polynomial $q : [0,1] \to \mathbf{R}^1$ such that

$$|f \circ \alpha^{-1}(x) - q(x)| < \varepsilon$$

for each $x \in [0,1]$. Finally note that $p = q \circ \alpha$ is a polynomial with domain $[a,b]$ and that for each $x \in [a,b]$

$$|f(x) - p(x)| = |f(x) - q \circ \alpha(x)| = |(f \circ \alpha^{-1})(\alpha(x)) - q(\alpha(x))| < \varepsilon$$

which concludes the proof.

This already impressive result can be extended in two directions: the interval $[a,b]$ can be replaced by an arbitrary compact metric space and, families of functions other than polynomials can be used to approximate a given continuous function. This is a relatively recent result and is due to A. H. Stone. For a proof of the following theorem the reader may consult Smith (1971).

(6.E.2) **Theorem** (*Stone-Weierstrass Theorem*) Suppose that (X,d) is a compact metric space and that $\mathcal{F}$ is a family of continuous functions mapping X into $\mathbf{R}^1$ such that

(i) $\mathcal{F}$ contains all constant functions.

(ii) If f and g belong to $\mathcal{F}$, then so do $f + g$ and $f \cdot g$.
(iii) $\mathcal{F}$ separates points; that is, if $x, y \in X$ and $x \neq y$, then there is at least one function $f \in \mathcal{F}$ such that $f(x) \neq f(y)$.

Then every continuous function $g : X \to \mathbf{R}^1$ can be approximated uniformly by functions in $\mathcal{F}$.

Since the family of polynomials with domain $[a,b]$ satisfies the conditions of the hypothesis of (6.E.2), it follows that (6.E.1) is a corollary of (6.E.2).

As an important example of a family of functions that can be used in place of polynomials to approximate certain functions, we consider the family $\mathcal{F}$ of trigonometric functions defined on the interval $[0,2\pi]$. These are functions of the form $f(x) = \sum_{k=0}^{n} \alpha_k \cos kx + \sum_{k=0}^{n} \beta_k \sin kx$, where n can be any nonnegative integer. Note that if $c \in \mathbf{R}^1$, the function $f(x) = c \cos 0 \cdot x + 0 \sin 0 \cdot x = c$, is a constant function, and hence, all constant functions belong to $\mathcal{F}$. Clearly the sum of the two trigonometric polynomials is again a trigonometric polynomial follows from some basic trigonometric identities (see problem E.3). One minor complication arises if the domain of these functions is taken to be $[0,2\pi]$: $\mathcal{F}$ fails to separate the points 0 and 2π. To circumvent this problem, we consider the unit circle $X = \{(\cos x, \sin x) \mid 0 \leq x < 2\pi\}$, and for each function $f : [0,2\pi] \to \mathbf{R}^1$ we associate the function $\hat{f} : X \to \mathbf{R}^1$ defined by $\hat{f}(\cos x, \sin x) = f(x)$. If $\hat{\mathcal{F}} = \{\hat{f} \mid f \in \mathcal{F}\}$, then $\hat{\mathcal{F}}$ clearly satisfies the conditions of (6.E.2), and consequently, if $g : [0,2\pi] \to \mathbf{R}^1$ is a continuous function with the property that $g(0) = g(2\pi)$, then g can be approximated uniformly by functions in $\mathcal{F}$ (since $\hat{g}$ can be uniformly approximated by functions in $\hat{\mathcal{F}}$).

PROBLEMS

Section A

1. Show that if f is integrable and c is a constant, then $\int_a^b cf(x)\, dx = c \int_a^b f(x)\, dx$.
2. Determine which of the following functions are uniformly continuous:
    (a) $f(x) = x$
    (b) $f(x) = x^2$
    (c) $f(x) = \sqrt{x}$
    (d) $f(x) = \sin x$
    (e) $f(x) = e^x$
3. Show that if $f : (a,b) \to \mathbf{R}^1$ is uniformly continuous, then $\lim_{x \to a^+} f(x)$ and $\lim_{x \to b^-} f(x)$ exist.
4. Show that if $f : (a,b) \to \mathbf{R}^1$ is continuous and bounded, then f is uniformly continuous.

5. Suppose that $f: \mathbf{R}^1 \to \mathbf{R}^1$ is a continuous function and that there is a sequence of polynomials that converges uniformly to f. Show that f is a polynomial. [*Hint*: Use (and prove) the fact that a bounded polynomial must be constant.]

6. Use the fundamental theorem of calculus to evaluate

   (a) $\displaystyle\int_1^4 \left(x^3 - \frac{2}{x} \right) dx$

   (b) $\displaystyle\int_2^1 x^{-k}\, dx \qquad k \neq 1$

7. Show that if $f: [a,b] \to \mathbf{R}^1$ is continuous and $f(x) \geq 0$ for each $x \in [a,b]$, then $\int_a^b f(x)\, dx \geq 0$.

8. What is wrong with the following argument: Since $f(x) = 1/x^2$ is positive valued, $\int_{-1}^1 (1/x^2)\, dx \geq 0$ by the previous problem. However, by the fundamental theorem of calculus, $\int_{-1}^1 (1/x^2)\, dx = (-1/x)|_{-1}^1 = -2$; hence, $-2 \geq 0$.

9. Show that if $f: [0,1] \to \mathbf{R}^1$ is a continuous function and if $\int_0^x f(u)\, du \geq f(x)$ for each $x \in [0,1]$, then $f(x) = 0$ for each $x \in [0,1]$.

10. Show that if $f: (0,\infty) \to \mathbf{R}^1$ is a continuous function and if $\int_a^b f(u)\, du = \int_{ax}^{bx} f(u)\, du$ for all positive numbers a, b, and x, then there is a constant c such that $f(x) = c/x$.

Section B

1. (a) Prove (6.B.3.i).
   (b) Prove (6.B.3.ii).
   (c) Prove (6.B.3.iii).
   (d) Prove (6.B.3.iv).
   (e) Prove (6.B.3.v).

2. Calculate

   (a) $\displaystyle\int_0^\pi x \cos x\, dx$

   (b) $\displaystyle\int_1^2 \ln x\, dx$

   (c) $\displaystyle\int_0^{-3} \frac{x^2}{e^x}\, dx$

3. Give a geometrical interpretation of the mean value theorem for integrals.

4. If $f: [-a,a] \to \mathbf{R}^1$ is a continuous odd function, what can be said about $g(x) = \int_a^x f(w)\, dw$?

5. If f is a continuous function, find

$$\lim_{t \to a} \frac{1}{t-a} \int_a^t f(x)\, dx$$

6. The *average value* of a function $f: [a,b] \to \mathbf{R}^1$ is frequently defined to be $[1/(b-a)] \int_a^b f(x)\, dx$. Use the mean value for integrals to explain why the term average value is an appropriate one. If a 3-foot rod is heated to $15x^3$ degrees, where x is the distance from one end of the rod, find the average temperature of the rod.

7. Suppose that $f''(x) = c_1 f(x)$ and $g''(x) = c_2 g(x)$. Use integration by parts twice to show that

$$\int f(x)g(x)\, dx = \frac{1}{c_1 - c_2}[f'(x)g(x) - f(x)g'(x)] + C$$

where $\int f(x)g(x)\, dx$ denotes a primitive of fg. Use this result to find primitives of $e^{3x} \sin \tfrac{1}{2}x$ and $\sinh 5x \cos 2x$.

8. Use integration by parts to show that if m and n are nonnegative integers and $m > 1$, then

$$\int \sin^m \theta \cos^n \theta\, d\theta = \frac{-\sin^{m-1}\theta \cos^{n+1}\theta}{m+n} + \frac{m-1}{m+n} \int \sin^{m-2}\theta \cos^n \theta\, d\theta$$

and show that if r and s are integers, then

$$\int_{r\pi/2}^{s\pi/2} \sin^m \theta \cos^n \theta\, d\theta = \frac{m-1}{m+n} \int_{r\pi/2}^{s\pi/2} \sin^{m-2}\theta \cos^n \theta\, d\theta$$

Use this result to find $\int_\pi^{5\pi/2} \sin^6 \theta \cos^3 \theta\, d\theta$.

9. Use problem B.8 to obtain the Wallis formula

$$\frac{2}{\pi} \int_0^{\pi/2} \sin^{2n} \theta\, d\theta = \frac{1 \cdot 3 \cdots (2n-1)}{2 \cdot 4 \cdots (2n)}$$

10. Use an induction argument to show that if p is a polynomial of degree n and if f is continuous function, then a primitive of the product pf is given by $pf_1 - p'f_2 + p''f_3 - \cdots + (-1)^n p^{(n)} f_{n+1}$, where f_1 is a primitive of f, and for each $i \geq 2$, f_i is a primitive of f_{i-1}. Use this result to find a primitive of $(3x^3 - 2x^2 + 4)e^{-2x}$.

11. Let $X = \{f: [a,b] \to \mathbf{R}^1 \mid f$ is continuous$\}$ and define $f: X \times X \to \mathbf{R}^1$ by $d(f,g) = \int_a^b |f(x) - g(x)|\, dx$. Show that d is a metric for X.

12. Let $f(x) = 1/x$ and $g'(x) = 1$ and apply integration by parts to $\int (1/x)\, dx$ to show that $\int (1/x)\, dx = x(1/x) - \int x(-1/x^2)\, dx = 1 + \int (1/x)\, dx$ and, hence, that $0 = 1$.

13. Show that $\lim_{c \to \infty} \int_a^b f(x) \cos cx\, dx = 0$ for each continuous function $f: [a,b] \to \mathbf{R}^1$.

Section C

1. Calculate

   (a) $\displaystyle\int_{-1}^{1} \frac{1}{x^2}\, dx$

   (b) $\displaystyle\int_{-\infty}^{\infty} \frac{x}{x^2 + 4}\, dx$

   (c) $\displaystyle\int_{0}^{\infty} \cos x\, dx$

   (d) $\displaystyle\int_{-3}^{\infty} \frac{x}{x^2 - 9}\, dx$

   (e) $\displaystyle\int_{4}^{0} (2 - x)^{-2/3}\, dx$

2. Show that if $\int_{-\infty}^{\infty} f(x)\, dx$ converges, then $\int_{-\infty}^{\infty} f(x)\, dx = \int_{-\infty}^{c} f(x)\, dx + \int_{c}^{\infty} f(x)\, dx$ for any $c \in (-\infty,\infty)$

3. Show that if $\int_{-\infty}^{\infty} f(x)\, dx$ converges, then $\int_{-\infty}^{\infty} f(x)\, dx = \mathrm{CP}\!\int \mathrm{v}_{-\infty}^{\infty} f(x)\, dx$.

4. Find

   (a) $\displaystyle \mathrm{CP}\!\int \mathrm{v}_{-\infty}^{\infty} \sin x\, dx$

   (b) $\displaystyle \mathrm{CP}\!\int \mathrm{v}_{-\infty}^{\infty} \cos x\, dx$

   (c) $\displaystyle \mathrm{CP}\!\int \mathrm{v}_{-\infty}^{\infty} \frac{2 - x}{x^2 + 1}\, dx$

   (d) $\displaystyle \mathrm{CP}\!\int \mathrm{v}_{-\infty}^{\infty} \frac{7}{x^3 + 2}\, dx$

5. John Wallis (1616–1703) showed that

$$\left(\frac{2\cdot 2}{1\cdot 3}\right)\left(\frac{4\cdot 4}{3\cdot 5}\right)\left(\frac{6\cdot 6}{5\cdot 7}\right)\left(\frac{8\cdot 8}{7\cdot 9}\right)\cdots = \frac{\pi}{2}$$

   Use this result to show that $\Gamma(\tfrac{1}{2}) = \sqrt{\pi}$. [You may assume that $\Gamma(\tfrac{3}{2}) = g(\tfrac{1}{2})$, where g is defined by (2) in Sec. 6.C.]

6. Show that $\Gamma(x + 1) = x\Gamma(x)$ for all x except $x = 0, -1, -2, \ldots$.

7. Find $\Gamma(-\tfrac{5}{2})$.

8. Evaluate $\int_0^1 x^2 \ln^3 (1/x)\, dx$. [*Hint*: Let $u = -\ln x$, and make use of the Γ function.]

9. Show that $\Gamma(n + \tfrac{1}{2}) = (2n)!\sqrt{\pi}/2^{2n}n!$.

10. Use the B function to evaluate $\int_0^1 (x^4/\sqrt{1 - x^2})\, dx$.
11. Show that $B(n,\tfrac{1}{2}) = 2^{2n}(n!)^2/n(2n)!$.
12. Gauss defined the Γ function as follows:

$$\Gamma(\alpha) = \lim_{n \to \infty} \frac{n^{\alpha-1}n!}{\alpha(\alpha + 1)\cdots(\alpha + n - 1)}$$

Derive this formula from the fact that as $z \to \infty$, $B(x,z)$ tends towards $\Gamma(x)/z^x$.

13. Show that for each positive integer n,

$$\int_0^{\pi/2} \sin^{2n+1}(x)\, dx = \frac{\sqrt{\pi}n!}{2\Gamma(n + \tfrac{3}{2})}$$

14. Show that $B(x, 1 - x) = \int_c^\infty (w^{x-1}/(1 + w))\, dw$. It can be shown that the value of this integral is equal to $\pi/(\sin \pi x)$ for $0 < x < 1$. Use this fact to find $\Gamma(\tfrac{1}{2})$. [*Hint*: Consider $B(\tfrac{1}{2},\tfrac{1}{2})$.]

Section D

1. Suppose that a triangular-shaped piece of metal with side lengths 2, 3, and 4 is dropped on a flat surface marked off with parallel lines 5 units apart. Find the probability that some portion of the triangle will intersect one of the parallel lines.
2. In the Buffon needle problem suppose that $D < L$, and find the probability that the needle will touch or cross a line.
3. Find the lengths of the graphs of the following functions.

(a) $f(x) = \tfrac{1}{3}(x^2 + 2)^{3/2} \qquad 2 \le x \le 5$

(b) $f(x) = \dfrac{x^3}{6} + \dfrac{1}{2x} \qquad 2 \le x \le 4$

(c) $f(x) = \dfrac{x^5}{10} + \dfrac{1}{6x^3} \qquad 1 \le x \le 3$

4. Find the circumference of a circle of radius r.
5. Find the length of an arch of the cycloid described by $\phi(t) = (2t - 2 \sin t, 2 - 2 \cos t)$.
6. Find an upper bound for the length of the graph of

$$f(x) = \begin{cases} x^2 \sin \dfrac{1}{x} & \text{if } x \in (0,1] \\ 0 & \text{if } x = 0 \end{cases}$$

7. Show that the length of the graph of

$$f(x) = \begin{cases} x \sin \dfrac{1}{x} & \text{if } x \in (0,1] \\ 0 & \text{if } x = 0 \end{cases}$$

is infinite. [*Hint*: You may use that fact that if $c > 0$ and for each n, $s_n = c + c/2 + c/3 + \cdots + c/n$, then $\lim_{n \to \infty} s_n = \infty$; choose the partitions of $[0,1]$ with some care.]

8. Show that $f(x,y) = 4x^2 + 9y^2$ satisfies a Lipschitz condition in the second argument on the rectangle defined by $|x| \le 1$, $|y| \le 1$.

9. Use the fact that the function T defined in Sec. 6.D is a contraction map to show that the functions $u_0(x) = y_0$, $u_1 = T(u_0)$, $u_2 = T(u_1)$, ..., $u_n = T(u_{n-1})$, ... converge to the solution u^* of (13).

10. Use problem D.9 to obtain successive approximations to the solution of

$$y' = xy \qquad y(0) = 1 \qquad (*)$$

Compare these approximations to the true solution of (*). Do the same for

$$y' = x + y \qquad y(0) = 1$$

11. (For readers familiar with partial derivatives.) Suppose that a function f is continuous on a closed rectangle R. Suppose further that the partial derivative f_y is also continuous on R. Show that f satisfies a Lipschitz condition in the second argument on R. [*Hint*: Note that if (x,y) and (x,z) are two points in R, then $f(x,y) - f(x,z) = \int_y^t f_y(x,w)\,dw$.] Is the continuity of f_y necessary?

12. Use Problem D.11 to show that the following functions satisfy a Lipschitz condition in the second argument on any closed rectangle R.
    (a) $f(x,y) = x^2 \cos^2 y - 2x \sin y$.
    (b) $f(x,y) = a(x)y^2 + b(x)y + c(x)$, where $a(x)$, $b(x)$, and $c(x)$ are continuous functions.

13. Show that $f(x,y) = x^2|y|$ satisfies a Lipschitz condition in the second argument on $|x| \le 1$, $|y| \le 1$, but f_y is not continuous at $(x,0)$ if $x \ne 0$.

14. Show that $\mathscr{A}$ is a closed subset of $\mathscr{C}$, where $\mathscr{A}$ and $\mathscr{C}$ are as defined in the proof of the Picard existence theorem.

Section E*

1. Determine the first three Bernstein polynomials for the functions $f(x) = \sqrt{x}$, $g(x) = \sin x$, and $h(x) = x^2 - 2x + 1$ on $[0,1]$.

2. Supply the remaining details to establish identities (a), (b), (c) and (d).

3. Use the trigonometric identities

$$\sin(\alpha + \beta) = \sin \alpha \cos \beta + \sin \beta \cos \alpha$$
$$\cos(\alpha + \beta) = \cos \alpha \cos \beta - \sin \alpha \sin \beta$$

to show that the family of trigonometric polynomials described in Sec. 6.E satisfies condition (ii) of the Stone-Weierstrass theorem. [*Hint*: Consider $\sin(nx + kx) - \sin(nx - kx)$, $\cos(nx + kx) + \cos(nx - kx)$, etc.]

4. Show that every continuous function $f: [0,1] \to \mathbf{R}^1$ can be approximated uniformly by sequences of functions whose members are of the form

$$g(x) = c_0 + c_1 e^x + c_2 e^{2x} + \cdots + c_n e^{nx}$$

where $n \in \mathbf{Z}^+$ and the c_i's are arbitrary real numbers.

5. Let $X = \{1, 2, \ldots\}$ and let $\mathscr{F} = \{f: X \to \mathbf{R}^1 \mid f(n) \text{ is either 0 or 1 for each } n\}$. Show
    (a) $\mathscr{F}$ satisfies the hypothesis of (6.E.2).
    (b) There is a function $f: X \to \mathbf{R}^1$ that cannot be uniformly approximated by functions in $\mathscr{F}$. Why does this not contradict the Stone-Weierstrass theorem?

6. Suppose that $f: \mathbf{R}^1 \to \mathbf{R}^1$ is a continuous function that can be uniformly approximated by polynomials. Show that f must be a polynomial.

7. Stone has also shown that if $\{x_1, x_2, \ldots, x_n\} \subset [a,b]$ and if $P = \{p: [a,b] \to \mathbf{R}^1 \mid p \text{ is a polynomial and } p(x_i) = 0 \text{ for } 1 \le i \le n\}$, then any continuous function $f: [a,b] \to \mathbf{R}^1$ with the property that $f(x_i) = 0$, $1 \le i \le n$, can be uniformly approximated by polynomials in P. Use this result to show that if $f: [a,b] \to \mathbf{R}^1$ is continuous, and if $\{x_1, x_2, \ldots, x_n\} \subset [a,b]$, then f can be uniformly approximated by a sequence of polynomials $\{p_k\}$ with the property that $p_k(x_i) = f(x_i)$ for each i, $1 \le i \le n$ (and for each $k \in \mathbf{Z}^+$).

REFERENCES

Davis, Philip (1959): *Amer. Math. Monthly*, **66**: 849.

Nanjundiah, T. S. (1969): *Amer. Math. Monthly*, **76**: 411.

Smith, K. (1971): *Smith's Primer of Modern Analysis*, Bogden and Quigley, New York.

7

DIFFERENTIATION OF FUNCTIONS OF SEVERAL VARIABLES

Our efforts in the previous two chapters were primarily directed toward the study of functions of one variable. In physical applications, however, one often finds that it is functions of several variables that are most frequently encountered; in fact, problems involving vibrating membranes, biological systems, heat conduction, etc., all lead naturally to the consideration of multivariable functions.

We shall see shortly that many of the ideas found in Chapter 5 have natural extensions to $\mathbf{R}^n$. A good deal of the material in the present chapter should already be familiar to the reader; consequently, we give a relatively terse treatment of many of the topics introduced here.

A. PARTIAL AND DIRECTIONAL DERIVATIVES

We begin by considering functions that map subsets of the plane, $\mathbf{R}^2$, into $\mathbf{R}^1$. A typical graph of such a function, f, is illustrated in Fig. 7.1.

In the penultimate section of Chapter 5 we saw how tangent lines could be used to "linearize" a function of one variable near a given point; proceeding in this same spirit, we shall use the two-dimensional analogue of a tangent

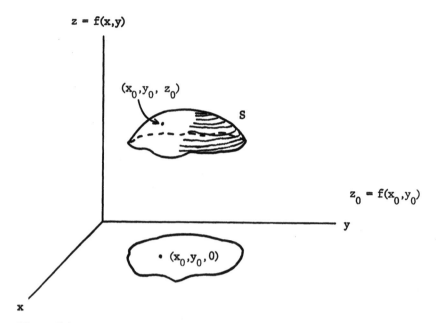

z = f(x,y)

(x_0, y_0, z_0)

S

$z_0 = f(x_0, y_0)$

y

$(x_0, y_0, 0)$

x

Figure 7.1

line, a tangent plane, as an aid in obtaining local approximations of functions of two variables (see Problem A.14).

As was the case with tangent lines, we must first decide what we mean by a tangent plane. Suppose then that $A \subset \mathbf{R}^2, f: A \to \mathbf{R}^1$, and that the graph of f is as illustrated in Fig. 7.1; this graph will be referred to as the surface S. If we slice the surface S with a plane P that is perpendicular to the xy plane and passes through the point (x_0, y_0, z_0), then we obtain a curve C_P which forms the intersection of P and S. If this curve is reasonably smooth, then we should be able to define (in P) a tangent line to C_P at the point (x_0, y_0, z_0). Thus a fairly natural criterion for determining a tangent plane T at (x_0, y_0, z_0) would be to insist that each tangent line so obtained [as we rotate the vertical plane P about the point (x_0, y_0, z_0)] lie in T.

How are we to find the equation of such a plane? Since any two distinct intersecting lines determine a plane, it suffices to consider the planes P_{x_0} and P_{y_0} determined by the equations $x = x_0$ and $y = y_0$, respectively. The slopes of the tangent lines to the curves $C_{P_{x_0}}$ and $C_{P_{y_0}}$ formed by the intersections of these planes with S are given by

$$\lim_{h \to 0} \frac{f(x_0, y_0 + h) - f(x_0, y_0)}{h}$$

and

$$\lim_{h \to 0} \frac{f(x_0 + h, y_0) - f(x_0, y_0)}{h}$$

respectively. We denote these limits (if they exist) by $f_y(x_0,y_0)$ and $f_x(x_0,y_0)$ (or $\partial f/\partial y|_{(x_0,y_0)}$ and $\partial f/\partial x|_{(x_0,y_0)}$), respectively; $f_y(x_0,y_0)$ is called the *partial derivative of f with respect to y* at the point (x_0,y_0) and $f_x(x_0,y_0)$ is called the *partial derivative of f with respect to x* at the point (x_0,y_0). Note that the tangent line at (x_0,y_0,z_0) to the curve determined by the intersection of S and P_{x_0} is given by

$$z - z_0 = f_y(x_0,y_0)(y - y_0) \qquad x = x_0 \qquad (1)$$

and the tangent line at (x_0,y_0,z_0) to the curve determined by the intersection of S and P_{y_0} is given by

$$z - z_0 = f_x(x_0,y_0)(x - x_0) \qquad y = y_0 \qquad (2)$$

Since the equation of a nonvertical plane can be written in the form:

$$a(x - x_0) + b(y - y_0) + c(z - z_0) = 0 \qquad c \neq 0$$

or, equivalently,

$$z - z_0 = -\left(\frac{a}{c}\right)(x - x_0) - \left(\frac{b}{c}\right)(y - y_0) \qquad (3)$$

we obtain by direct substitution of (1) and (2) in (3) that $f_x(x_0,y_0) = -(a/c)$ and $f_y(x_0,y_0) = -(b/c)$. Consequently, the equation of the tangent plane (if it exists) is given by

$$z = T(x,y) = f(x_0,y_0) + f_x(x_0,y_0)(x - x_0) + f_y(x_0,y_0)(y - y_0) \qquad (4)$$

(7.A.1) Example If $f(x,y) = 3x^2y - 2e^{x^2y}$, then $f_x(x,y) = 6xy - 2e^{x^2y}2xy$ and $f_y(x, y) = 3x^2 - 2e^{x^2y}x^2$. (The partial derivative $f_x(x, y)$ is calculated by treating y as a constant and taking the derivative with respect to x; $f_y(x, y)$ is calculated similarly.) At the point $(1, 2, 6 - 2e^2)$ the plane tangent to the surface generated by this function is

$$T(x,y) = 6 - 2e^2 + (12 - 8e^2)(x - 1) + (3 - 2e^2)(y - 2)$$

To establish the existence of tangent planes, we make use of the notion of the *directional derivatives*: derivatives that correspond to curves arising from intersections of the graph of a function f and arbitrary vertical planes (ones not necessarily perpendicular to either the x or y axis). Basically, a directional derivative measures the instantaneous rate of change of f when the domain of f is restricted to points lying in a specified direction.

(7.A.2) Definition Suppose that $A \subset \mathbf{R}^2$, $f: A \to \mathbf{R}^1$, and $\mathbf{w} = (x_0,y_0) \in$ int A. Let $\mathbf{u}$ be a *unit vector* in $\mathbf{R}^2$, i.e., a vector of length 1. Then the *directional derivative of f* (at $\mathbf{w}$ and in the direction of $\mathbf{u}$) is defined to be

$$\lim_{h \to 0} \frac{f(\mathbf{w} + h\mathbf{u}) - f(\mathbf{w})}{h} \qquad h \in \mathbf{R}^1$$

if this limit exists, it is denoted by $D_\mathbf{u}f(\mathbf{w})$.

(7.A.3) *Observations* (a) Note that for each given intersecting plane P, two distinct directional derivatives are obtained corresponding to the unit vectors $\mathbf{u}$ and $-\mathbf{u}$ (Fig. 7.2). In problem A.15 the reader is asked to show that $D_{\mathbf{u}}f(\mathbf{w}) = -D_{-\mathbf{u}}f(\mathbf{w})$. For example, if $f(x,y) = 2x^2y$ and $\mathbf{u} = (1/2, \sqrt{3}/2)$, then

$$
\begin{aligned}
D_{\mathbf{u}}f(-1, 1) &= \lim_{h \to 0} \frac{f((-1,1) + h(1/2, \sqrt{3}/2)) - f(-1,1)}{h} \\
&= \lim_{h \to 0} \frac{2(-1 + (1/2)h)^2(1 + h(\sqrt{3}/2)) - 2}{h} \\
&= \lim_{h \to 0} \frac{2(1 - h + (1/4)h^2)(1 + h(\sqrt{3}/2)) - 2}{h} \\
&= \lim_{h \to 0} \frac{h\sqrt{3} - 2h - h^2\sqrt{3} + h^2/2 + (h^3/4)\sqrt{3}}{h} \\
&= -2 + \sqrt{3}
\end{aligned}
$$

and

$$
\begin{aligned}
D_{-\mathbf{u}}f(-1,1) &= \lim_{h \to 0} \frac{f((-1,1) + h(-1/2, -\sqrt{3}/2)) - f(-1,1)}{h} \\
&= 2 - \sqrt{3}
\end{aligned}
$$

(b) The partial derivatives f_y and f_x defined earlier are directional derivatives with $\mathbf{u} = (0,1)$ and $\mathbf{u} = (1,0)$, respectively, since if $\mathbf{w} = (x_0, y_0)$, then

$$
\begin{aligned}
f_y(\mathbf{w}) &= \lim_{h \to 0} \frac{f(x_0, y_0 + h) - f(x_0, y_0)}{h} \\
&= \lim_{h \to 0} \frac{f(\mathbf{w} + h(0,1)) - f(\mathbf{w})}{h} = D_{(0,1)}f(\mathbf{w})
\end{aligned}
$$

and

$$
\begin{aligned}
f_x(\mathbf{w}) &= \lim_{h \to 0} \frac{f(x_0 + h, y_0) - f(x_0, y_0)}{h} \\
&= \lim_{h \to 0} \frac{f(\mathbf{w} + h(1,0)) - f(\mathbf{w})}{h} = D_{(1,0)}f(\mathbf{w})
\end{aligned}
$$

(c) In the case of functions of a single variable, the derivative

$$
\lim_{h \to 0} \frac{f(x_0 + h) - f(x_0)}{h}
$$

can be considered as a directional derivative where $\mathbf{u} = 1$ (vectors in $\mathbf{R}^1$ coincide with the real numbers).

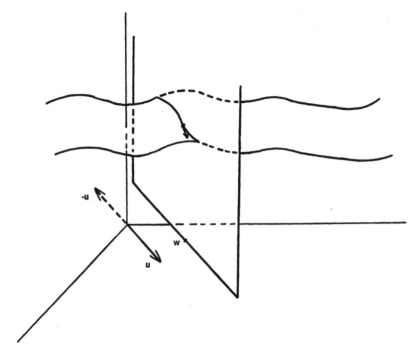

Figure 7.2

(7.A.4) *Definition* Suppose that $A \subset \mathbf{R}^2$. A function $f: A \to \mathbf{R}^1$ is *continuously differentiable* on A if f, f_x, and f_y are continuous on A.

(7.A.5) *Observation* If f is continuously differentiable in a neighborhood of a point $(x_0, y_0) \in \mathbf{R}^2$, then the directional derivatives exist and may be defined in terms of the partial derivatives. To see this, observe that if $\mathbf{u} = (u_1, u_2)$ is a given unit vector, and $\mathbf{w} = (x_0, y_0)$, then

$$\frac{f(\mathbf{w} + h\mathbf{u}) - f(\mathbf{w})}{h} = \frac{f(x_0 + hu_1, y_0 + hu_2) - f(x_0, y_0 + hu_2)}{h}$$

$$+ \frac{f(x_0, y_0 + hu_2) - f(x_0, y_0)}{h}$$

Two applications of the mean value theorem to the right-hand side of the above equation yield

$$\frac{f(\mathbf{w} + h\mathbf{u}) - f(\mathbf{w})}{h} = f_x(x_0 + h_1 u_1, y_0 + hu_2)\frac{hu_1}{h} + f_y(x_0, y_0 + h_2 u_2)\frac{hu_2}{h}$$

where h_1 and h_2 lie between 0 and h. Since f_x and f_y are continuous, we have

$$D_u f(\mathbf{w}) = \lim_{h \to 0} \frac{f(\mathbf{w} + h\mathbf{u}) - f(\mathbf{w})}{h} = f_x(x_0, y_0)u_1 + f_y(x_0, y_0)u_2$$

(7.A.6) *Exercise* Justify the use of the mean value theorem in Observation (7.A.5).

(7.A.7) *Example* In the example given in (7.A.3.a), we can now calculate $D_u f(-1,1)$ directly: since $f_x(-1,1) = -4$ and $f_y(-1,1) = 2$, we have $D_u f(-1,1) = -4(1/2) + 2(\sqrt{3}/2) = -2 + \sqrt{3}$.

(7.A.8) *Example* Suppose that

$$f(x,y) = \begin{cases} \dfrac{xy^2}{x^2 + y^4} & \text{if } x \neq 0 \\ 0 & \text{if } x = 0 \end{cases}$$

and let $\mathbf{w} = (0,0)$. Then each directional derivative of f at $\mathbf{w}$ exists, since if $\mathbf{u} = (u_1, u_2)$, where $u_1 \neq 0$, then

$$\lim_{h \to 0} \frac{f(\mathbf{w} + h\mathbf{u}) - f(\mathbf{w})}{h} = \lim_{h \to 0} \frac{f((0,0) + h(u_1, u_2)) - f(0,0)}{h}$$

$$= \lim_{h \to 0} \frac{f(hu_1, hu_2)}{h} = \lim_{h \to 0} \frac{hu_1 h^2 u_2^2}{h^3 u_1^2 + h^5 u_2^4} = \frac{u_2^2}{u_1}$$

and if $\mathbf{u} = (u_1, u_2)$, where $u_1 = 0$, then

$$\lim_{h \to 0} \frac{f(\mathbf{w} + h\mathbf{u}) - f(\mathbf{w})}{h} = 0$$

In particular, both partial derivatives f_x and f_y exist at $\mathbf{w}$. Note, however, that in spite of the existence of all the directional derivatives of f, this function fails to be continuous at $\mathbf{w}$. To see this, observe that when (x,y) tends to $(0,0)$ along the curve $x = y^2$, we have

$$\lim_{(x,y) \to (0,0)} f(x,y) = \lim_{(x,y) \to (0,0)} \frac{y^4}{2y^4} = \frac{1}{2}$$

whereas when (x,y) tends to $(0,0)$ along the curve $x = 2y^2$,

$$\lim_{(x,y) \to (0,0)} f(x,y) = \lim_{(x,y) \to (0,0)} \frac{2y^4}{5y^4} = \frac{2}{5}$$

and therefore, $\lim_{(x,y) \to (0,0)} f(x,y)$ does not exist.

(7.A.9) *Exercise* If

$$f(x,y) = \begin{cases} x + y & \text{if } x = 0 \text{ or } y = 0 \\ 1 & \text{otherwise} \end{cases}$$

show that both partial derivatives exist at $(0,0)$, but that no other directional derivatives are defined. Show also that f is not continuous at the point $(0,0)$.

As a corollary to Observation (7.A.5), we have that if the partial derivatives f_x and f_y are continuous in a neighborhood of $\mathbf{w} = (x_0, y_0)$, then the tangent plane does exist at this point. That is, all tangent lines to the surface at the point $\mathbf{w}$ lie in the plane passing through $\mathbf{w}$ and determined by the two particular tangent lines having slopes $f_x(x_0, y_0)$ and $f_y(x_0, y_0)$. To see this, suppose that a vertical "slicing" plane yields a directional derivative $s = f_x(\mathbf{w})u_1 + f_y(\mathbf{w})u_2$ (actually, of course, each such plane generates two directional derivatives). It is easily verified that the equation of the line L in this plane which is tangent to the curve determined by the intersection of the plane and the surface S is given parametrically by

$$x = x_0 + tu_1 \qquad y = y_0 + tu_2 \qquad z = z_0 + ts \qquad (5)$$

Direct substitution of a point of the form $(x_0 + tu_1, y_0 + tu_2, z_0 + ts)$ in (4) shows that L lies in the tangent plane.

As a simple illustration of the utility of tangent planes, we discuss briefly a method commonly used in numerical analysis to approximate simultaneous zeros of two functions of two variables. Recall that Newton's method was employed to find approximate solutions of equations of the form

$$f(x) = 0$$

Suppose that we are given two equations of the form

$$f_1(x,y) = 0 \qquad f_2(x,y) = 0 \qquad (6)$$

ideally, we would we like to find an ordered pair, (x^*,y^*), that satisfies (6). To obtain an approximation of (x^*,y^*), we imitate Newton's method but replace tangent lines by tangent planes. The procedure, in outline form, is as follows:

1. An arbitrary point (x_0,y_0) is chosen to approximate the true solution of (6); this point is chosen as close to the true solution as possible.
2. For $i = 1, 2$ the plane, T_i, tangent to the surface determined by the function f_i at the point $(x_0,y_0,f_i(x_0,y_0))$ is found.
3. The point of intersection, (x_1,y_1), of the tangent planes T_1 and T_2 with the plane $z = 0$ is determined. This point will generally yield a better approximation of (x^*,y^*) than the arbitrarily chosen point (x_0,y_0).
4. The entire procedure is repeated using (x_1,y_1) in place of (x_0,y_0) to obtain a point (x_2,y_2) that is the intersection point of two tangent planes and the plane $z = 0$; repeated applications of this procedure yield a sequence of points (x_0,y_0), (x_1,y_1), ... which, under fairly reasonable conditions, will converge to a true solution of (6). More details of this method are given in Chapter 14.

Suppose that $A \subset \mathbf{R}^2$, $f: A \to \mathbf{R}^1$ and that f_x exists on A. Since $f_x: A \to \mathbf{R}^1$, it makes sense to speak of the partial derivative of this function with respect to x and y, i.e., $(f_x)_x$ and $(f_x)_y$. We shall denote these functions by f_{xx} and f_{xy}, respectively. Similarly, functions such as $f_{yx}, f_{yy}, f_{xyy}, f_{yxy}$, etc., can be defined. We shall occasionally denote f_{xx} by $\partial^2 f/\partial x^2$, f_{xy} by $\partial^2 f/\partial y \partial x$, f_{yx} by $\partial^2 f/\partial x \partial y$, etc.

(7.A.10) *Example* If $f: \mathbf{R}^2 \to \mathbf{R}^1$ is defined by $f(x,y) = x^2 y + y^2 \sin x$, then $f_x(x,y) = 2xy + y^2 \cos x$, $f_y(x,y) = x^2 + 2y \sin x$, $f_{xy}(x,y) = 2x + 2y \cos x$, $f_{yx}(x,y) = 2x + 2y \cos x$, $f_{xyx}(x,y) = 2 - 2y \sin x$, etc.

Note, in the preceding example, that $f_{xy}(x,y) = f_{yx}(x,y)$ for all (x,y). That this rather remarkable result is not a coincidence follows from the next theorem.

(7.A.11) *Theorem* Suppose that f, f_x, and f_y exist in a neighborhood of a point (x_0, y_0) and that f_{xy} is continuous on this neighborhood. Then $f_{yx}(x_0, y_0)$ exists, and moreover, $f_{xy}(x_0, y_0) = f_{yx}(x_0, y_0)$.

Proof. To show that $f_{yx}(x_0, y_0)$ exists, we must show that

$$\lim_{h \to 0} \frac{f_y(x_0 + h, y_0) - f_y(x_0, y_0)}{h}$$

exists. By hypothesis, $f_{xy}(x_0, y_0)$ does exist. We shall prove that for each $\varepsilon > 0$, there is a $\delta > 0$ such that if $0 < |h| < \delta$, then

$$\left| f_{xy}(x_0, y_0) - \frac{f_y(x_0 + h, y_0) - f_y(x_0, y_0)}{h} \right| < \varepsilon \tag{7}$$

This not only establishes the existence of $f_{yx}(x_0, y_0)$ but also shows that $f_{xy}(x_0, y_0) = f_{yx}(x_0, y_0)$.

Since f_{xy} is continuous at (x_0, y_0), corresponding to $\varepsilon > 0$, there is a $\delta > 0$ such that

$$|f_{xy}(x_0, y_0) - f_{xy}(\tilde{x}, \tilde{y})| < \varepsilon \qquad \text{whenever } |x_0 - \tilde{x}| < \delta, |y_0 - \tilde{y}| < \delta \tag{8}$$

Suppose that $0 < |h| < \delta$ and $0 < |k| < \delta$, and observe that by the definition of f_y we have

$$\frac{f_y(x_0 + h, y_0) - f_y(x_0, y_0)}{h} = \lim_{k \to 0} \left(\frac{f(x_0 + h, y_0 + k) - f(x_0 + h, y_0)}{hk} \right.$$

$$\left. - \frac{f(x_0, y_0 + k) - f(x_0, y_0)}{hk} \right) \tag{9}$$

Set $q(x) = f(x, y_0 + k) - f(x, y_0)$. We apply the mean value theorem twice [to q in (a), below, and then to f_x in (b)] to obtain

(a) $q(x_0 + h) - q(x_0) = q'(\tilde{x})h = (f_x(\tilde{x}, y_0 + k) - f_x(\tilde{x}, y_0))h$, where $\tilde{x}$ lies between x_0 and $x_0 + h$ (and hence, $|x_0 - \tilde{x}| < h < \delta$).

(b) $q(x_0 + h) - q(x_0) = (f_x(\tilde{x}, y_0 + k) - f_x(\tilde{x}, y_0))h = f_{xy}(\tilde{x}, \tilde{y}))hk$, where $\tilde{y}$ lies between y_0 and $y_0 + k$ (and hence, $|y_0 - \tilde{y}| < k < \delta$).

Thus from the definition of q, (b), and (8) we have

$$\left| f_{xy}(x_0, y_0) - \left(\frac{f(x_0 + h, y_0 + k) - f(x_0 + h, y_0)}{hk} \right. \right.$$

$$\left. \left. - \frac{f(x_0, y_0 + k) - f(x_0, y_0)}{hk} \right) \right|$$

$$= \left| f_{xy}(x_0, y_0) - \frac{q(x_0 + h) - q(x_0)}{hk} \right| = |f_{xy}(x_0, y_0) - f_{xy}(\tilde{x}, \tilde{y})| < \varepsilon$$

Now let k tend to 0 and observe that the desired result then follows from (7) and (9).

Are all the hypotheses of the previous theorem necessary? In other words might it not *always* be the case that the order of taking partial derivatives is of no consequence? In the next exercise the reader is asked to show that some care must be used in this regard.

(7.A.12) *Exercise* Use the definition of partial derivatives to show that if

$$f(x,y) = \begin{cases} (xy)\dfrac{x^2 - y^2}{x^2 + y^2} & \text{if } (x,y) \neq (0,0) \\ 0 & \text{if } (x,y) = (0,0) \end{cases}$$

then $f_{xy}(0,0) = -1$ and $f_{yx}(0,0) = 1$.

We conclude this section with another exercise, a solution of which may be patterned after the proof given in (5.B.2).

(7.A.13) *Exercise* Suppose that $A \subset \mathbf{R}^2$, $\mathbf{w} \in \text{int } A$, and $f: A \to \mathbf{R}^1$ attains a maximum (or minimum) at $\mathbf{w}$. Show that if $f_x(\mathbf{w})$ and $f_y(\mathbf{w})$ exist, then they are both equal to zero.

B. THE DIFFERENTIAL AND DIFFERENTIABILITY

In this section we generalize the notions of the differential and differentiability to functions mapping subsets of $\mathbf{R}^n$ into $\mathbf{R}^1$. In order to motivate the definitions to follow, we first briefly review the ideas underlying the concept of the differentiability of a function of one variable. In Sec. 5.D we saw that if $f: (a,b) \to \mathbf{R}^1$ is differentiable at a point $\hat{x} \in (a,b)$, and if we set

$$E(h) = \frac{f(\hat{x} + h) - f(x)}{h} - f'(\hat{x})$$

then

    (a) $f(\hat{x} + h) - f(\hat{x}) = d_{\hat{x}}f(h) + hE(h)$
    (b) $\lim_{h \to 0} E(h) = 0$

From (a) and (b) it follows that if f is differentiable at $\hat{x}$, then there is a linear function L (namely, $d_{\hat{x}}f$) such that

$$\lim_{h \to 0} \frac{f(\hat{x} + h) - f(\hat{x}) - L(h)}{h} = \lim_{h \to 0} E(h) = 0$$

Conversely, suppose that L is a linear function such that

$$\lim_{h \to 0} \frac{f(\hat{x} + h) - f(\hat{x}) - L(h)}{h} = 0$$

Then this implies that f is differentiable at $\hat{x}$ and that $L = d_{\hat{x}}f$. To see this, observe that since L is a linear function, $L(h) = L(h \cdot 1) = hL(1)$, and therefore,

$$0 = \lim_{h \to 0} \frac{f(\hat{x} + h) - f(\hat{x}) - L(h)}{h} = \lim_{h \to 0} \frac{f(\hat{x} + h) - f(\hat{x})}{h} - \lim_{h \to 0} \frac{hL(1)}{h}$$

which shows that $\lim_{h \to 0} (f(\hat{x} + h) - f(\hat{x}))/h$ exists and is equal to $L(1)$, i.e., $f'(\hat{x}) = L(1)$.

    As noted in Chapter 2 a linear function is determined by its action on a basis. Since both L and $d_{\hat{x}}f$ are linear functions, since 1 is a basis for $\mathbf{R}^1$, and since $L(1) = f'(\hat{x}) = d_{\hat{x}}f(1)$, we have $L = d_{\hat{x}}f$.

    From the preceding discussion it is clear that for functions of one variable there is an alternate but equivalent definition of differentiability.

(7.B.1) *Definition* Suppose that $f: (a,b) \to \mathbf{R}^1$. Then f is *differentiable* at a point $\hat{x} \in (a,b)$ if there is a linear function $L: \mathbf{R}^1 \to \mathbf{R}^1$ such that

$$\lim_{h \to 0} \frac{f(\hat{x} + h) - f(\hat{x}) - L(h)}{h} = 0$$

    A principal advantage of this definition is that it can be readily extended to functions of more than one variable.

(7.B.2) *Definition* Suppose that $A \subset \mathbf{R}^2, f: A \to \mathbf{R}^1$, and $\mathbf{w} \in \text{int } A$. Then f is *differentiable* at $\mathbf{w}$ if there is a linear function $L: \mathbf{R}^2 \to \mathbf{R}^1$ such that

$$\lim_{\mathbf{v}\to 0} \frac{f(\mathbf{w} + \mathbf{v}) - f(\mathbf{w}) - L(\mathbf{v})}{\|\mathbf{v}\|} = 0$$

The function L is called the *differential* of f at $\mathbf{w}$ and will be denoted by $d_{\mathbf{w}}f$.

In order to speak of *the* differential of f at $\mathbf{w}$ it is necessary, of course, that L be unique. In the next theorem we show that this is the case; moreover, we see in this theorem how L can be expressed in terms of the partial derivatives of f.

(7.B.3) **Theorem** Suppose that $A \subset \mathbf{R}^2$ and that $f: A \to \mathbf{R}^1$ is differentiable at $\mathbf{w} \in \text{int } A$. Then the linear function L appearing in Definition (7.B.2) is unique, and furthermore, if $(x_1,x_2) \in \mathbf{R}^2$, then

$$L(x_1,x_2) = d_{\mathbf{w}}f(x_1,x_2) = x_1 f_x(\mathbf{w}) + x_2 f_y(\mathbf{w})$$

Proof. For appropriately small vectors, $\mathbf{v}$, let

$$E(\mathbf{v}) = \frac{f(\mathbf{w} + \mathbf{v}) - f(\mathbf{w}) - L(\mathbf{v})}{\|\mathbf{v}\|} \tag{1}$$

note that by (7.B.2) $\lim_{\mathbf{v}\to 0} E(\mathbf{v}) = 0$. If $\mathbf{u}$ is a unit vector and h is a nonzero scalar, then by (1) we have

$$L(h\mathbf{u}) = f(\mathbf{w} + h\mathbf{u}) - f(\mathbf{w}) - \|h\mathbf{u}\| E(h\mathbf{u})$$

Since L is a linear function, it follows that

$$hL(\mathbf{u}) = f(\mathbf{w} + h\mathbf{u}) - f(\mathbf{w}) - |h| \|\mathbf{u}\| E(h\mathbf{u}) \tag{2}$$

Division of both sides of (2) by h yields

$$L(\mathbf{u}) = \frac{f(\mathbf{w} + h\mathbf{u}) - f(\mathbf{w})}{h} - \frac{|h|}{h} E(h\mathbf{u}) \qquad (\|\mathbf{u}\| = 1)$$

Therefore, since $\lim_{h\to 0} E(h\mathbf{u}) = 0$, and since $L(\mathbf{u})$ is independent of h, we see that

$$L(\mathbf{u}) = \lim_{h\to 0} \frac{f(\mathbf{w} + h\mathbf{u}) - f(\mathbf{w})}{h} = D_{\mathbf{u}}f(\mathbf{w}) \tag{3}$$

Since any linear function is completely defined by its action on a basis, the mapping L is determined by its effect on the vectors $(1,0)$ and $(0,1)$ [yielding the partial derivatives, $f_x(\mathbf{w})$ and $f_y(\mathbf{w})$, respectively], and consequently, L is unique. Furthermore, by (3) and (7.A.3.b), we have $L(x_1,x_2) = L(x_1(1,0) + x_2(0,1)) = x_1 L(1,0) + x_2 L(0,1) = x_1 D_{(1,0)}f(\mathbf{w}) + x_2 D_{(0,1)}f(\mathbf{w}) = x_1 f_x(\mathbf{w}) + x_2 f_y(\mathbf{w})$, which concludes the proof.

The reader should note that it follows from (1) that for small changes in $\mathbf{w}$, the differential $L(\mathbf{v})$ yields a good approximation of the corresponding change in f, i.e., since $\|\mathbf{v}\| E(\mathbf{v}) \cong 0$,

$$L(\mathbf{v}) \cong f(\mathbf{w} + \mathbf{v}) - f(\mathbf{w})$$

This allows us to justify our remark in the previous chapter that tangent planes provide good local approximations to the graphs of functions of two variables. To see this, suppose that $f: \mathbf{R}^2 \to \mathbf{R}^1$ and that $T(x,y) = f(x_0,y_0) + f_x(x_0,y_0)(x - x_0) + f_y(x_0,y_0)(y - y_0)$ is the function defined by (1) in Sec. 7. A whose graph is the plane tangent to the graph of f at $\mathbf{w} = (x_0,y_0)$. If we set $\mathbf{v} = (x - x_0, y - y_0)$, then we have (for vectors $\mathbf{v}$ of small magnitude)

$$f(x,y) - f(x_0,y_0) = f(\mathbf{w} + \mathbf{v}) - f(\mathbf{w}) \cong L(\mathbf{v}) = d_\mathbf{w}f(x - x_0, y - y_0)$$
$$= T(x,y) - f(x_0,y_0)$$

and hence, $f(x,y) \cong T(x,y)$

(7.B.4) *Observation* As a consequence of (7.B.3), we have that differentiability implies the existence of all directional derivatives. In fact, we show in the next theorem that differentiability is strong enough to ensure continuity (which the existence of the directional derivatives does not).

(7.B.5) *Theorem* Let $A \subset \mathbf{R}^2$. If $f: A \to \mathbf{R}^1$ and f is differentiable at $\mathbf{w} \in$ int A, then f is continuous at $\mathbf{w}$.

Proof. Suppose that a sequence $\{\mathbf{w}_n\}$ converges to $\mathbf{w}$. We are to show that the sequence $\{f(\mathbf{w}_n)\}$ converges to $f(\mathbf{w})$. For each n, let $\mathbf{z}_n = \mathbf{w}_n - \mathbf{w}$; then clearly $\{\mathbf{z}_n\} \to \mathbf{0}$. Let L be the differential of f at $\mathbf{w}$ and define E as in (1). Note that

$$f(\mathbf{w}_n) - f(\mathbf{w}) = f(\mathbf{w} + \mathbf{z}_n) - f(\mathbf{w}) = L(\mathbf{z}_n) + \|\mathbf{z}_n\| E(\mathbf{z}_n)$$

Since L is continuous (4.A.7), and $\lim_{n \to \infty} E(\mathbf{z}_n) = 0$, we have

$$\lim_{n \to \infty} |f(\mathbf{w}_n) - f(\mathbf{w})| = 0$$

or, equivalently, that the sequence $\{f(\mathbf{w}_n)\}$ converges to $f(\mathbf{w})$.

(7.B.6) *Observation* The continuity of a function f at a point $\mathbf{w}$ is not sufficient to ensure the differentiability of f; however, we shall show in (7.D.13) that if the partial derivatives of f are defined in a neighborhood of $\mathbf{w}$ and are continuous at $\mathbf{w}$, then f is differentiable at $\mathbf{w}$.

To obtain the derivative of a function $f: A \to \mathbf{R}^1$ at a point $\mathbf{w} \in$ int A, where $A \subset \mathbf{R}^2$, we focus our attention on the differential of f at $\mathbf{w}$, $d_\mathbf{w}f$. By definition, $d_\mathbf{w}f$ is a linear function, and hence, we can associate a matrix $M_{d_\mathbf{w}f}$, with this function (see Sec. 2.E). The matrix $M_{d_\mathbf{w}f}$ belonging to $d_\mathbf{w}f$ [with respect to the standard bases $\mathbf{e}_1 = (1,0)$ and $\mathbf{e}_2 = (0,1)$ for $\mathbf{R}^2$, and $\hat{\mathbf{e}}_1 = 1$ for $\mathbf{R}^1$] is determined by the equations

$$d_\mathbf{w}f(\mathbf{e}_1) = f_x(\mathbf{w})1 + f_y(\mathbf{w})0 = f_x(\mathbf{w})\hat{\mathbf{e}}_1$$
$$d_\mathbf{w}f(\mathbf{e}_2) = f_x(\mathbf{w})0 + f_y(\mathbf{w})1 = f_y(\mathbf{w})\hat{\mathbf{e}}_1$$

Thus, we see that

$$M_{d_w f} = (f_x(\mathbf{w}) \quad f_y(\mathbf{w}))$$

(7.B.7) *Definition* Suppose that $A \subset \mathbf{R}^2$, $f: A \to \mathbf{R}^1$, and that the partial derivatives of f, f_x and f_y, exist at a point $\mathbf{w} \in \text{int } A$, Then the *derivative* of f at $\mathbf{w}$ is defined to be the matrix $(f_x(\mathbf{w}) \quad f_y(\mathbf{w}))$. We shall denote this matrix by $f'(\mathbf{w})$.

(7.B.8) *Observation* This definition is compatible in the following sense with the definition of the derivative of a function of one variable (5.A.8). If $f: (a,b) \to \mathbf{R}^1$ is differentiable at a point $\hat{x} \in (a,b)$, then we have $d_{\hat{x}} f(1) = f'(\hat{x}) \cdot 1$. Since 1 is a basis for $\mathbf{R}^1$, it follows that the matrix belonging to $d_{\hat{x}} f$ is the 1×1 matrix $(f'(\hat{x}))$. Thus, if the real number $f'(\hat{x})$ is regarded as a 1×1 matrix, the two definitions coincide.

(7.B.9) *Observation* In the case of a function of one variable, the derivative corresponds to the slope of the line tangent to the graph of the given function. A reasonable two-dimensional analogue would have the derivative of a function f at a point $\mathbf{w}$ correspond to the "pitch" of the tangent plane at $\mathbf{w}$. Such a "pitch" is clearly fixed once the values of the partial derivatives of f at $\mathbf{w}$ are determined. The derivative of f gives us precisely these values.

(7.B.10) *Example* If $f: \mathbf{R}^2 \to \mathbf{R}^1$ is defined by $f(x,y) = x^2 \sin xy$, then the derivative of f at $(2,\pi)$ is given by $(4 \sin 2\pi + 4\pi \cos 2\pi \quad 8 \cos 2\pi) = (4\pi \quad 8)$.

The ideas developed thus far can be readily extended to functions that map $\mathbf{R}^n$ into $\mathbf{R}^1$.

(7.B.11) *Definition* Suppose that $A \subset \mathbf{R}^n$, $\mathbf{w} = (w_1, \ldots, w_n) \in \text{int } A$, and $f: A \to \mathbf{R}^1$. Then the *ith partial derivative* of f at $\mathbf{w}$, $D_i f(\mathbf{w})$, is defined to be

$$\lim_{h \to 0} \frac{f(w_1, \ldots, w_i + h, w_{i+1}, \ldots, w_n) - f(w_1, \ldots, w_n)}{h}$$

provided this limit exists.

(7.B.12) *Definition* Suppose that $A \subset \mathbf{R}^n$, $f: A \to \mathbf{R}^1$, and that the partial derivatives of f, $D_1 f$, $D_2 f, \ldots, D_n f$ exist at a point $\mathbf{w} \in \text{int } A$. Then the *derivative* of f at $\mathbf{w}$ is defined to be the matrix $(D_1 f(\mathbf{w}) \quad D_2 f(\mathbf{w}) \quad \cdots \quad D_n f(\mathbf{w}))$. We denote this matrix by $f'(\mathbf{w})$.

(7.B.13) *Example* If $f: \mathbf{R}^4 \to \mathbf{R}^1$ is defined by

$$f(w_1, w_2, w_3, w_4) = w_1^2 w_2 + 2w_2 e^{w_4} + w_3^3 w_4$$

then $f'(w_1,w_2,w_3,w_4) = (2w_1w_2 \quad w_1^2 + 2e^{w_4} \quad 3w_3^3w_4 \quad 2w_2e^{w_4} + w_3^3)$

(7.B.14) *Definition* If $A \subset \mathbf{R}^n$, $\mathbf{w} \in$ int A, $f: A \to \mathbf{R}^1$, and $\mathbf{u}$ is a unit vector in $\mathbf{R}^n$, then the *directional derivative* of f at $\mathbf{w}$ in the direction $\mathbf{u}$ is defined by

$$D_{\mathbf{u}}f(\mathbf{w}) = \lim_{h \to 0} \frac{f(\mathbf{w} + h\mathbf{u}) - f(\mathbf{w})}{h}$$

(7.B.15) *Definition* Suppose that $A \subset \mathbf{R}^n$, $\mathbf{w} \in$ int A, and $f: A \to \mathbf{R}^1$. Then f is *differentiable* at $\mathbf{w}$ if there is a linear function $L: \mathbf{R}^n \to \mathbf{R}^1$ such that

$$\lim_{\mathbf{v} \to 0} \frac{f(\mathbf{w} + \mathbf{v}) - f(\mathbf{w}) - L(\mathbf{v})}{\|\mathbf{v}\|} = 0$$

The linear function L is called the *differential* of f at $\mathbf{w}$ and is denoted by $d_{\mathbf{w}}f$.

(7.B.16) *Definition* Suppose that $A \subset \mathbf{R}^n$, $f: A \to \mathbf{R}^1$, and $\mathbf{w} \in$ int A. Then f is *continuously differentiable* at $\mathbf{w}$ if there is an open subset U containing $\mathbf{w}$ such that the partial derivatives $D_1f, D_2f, \ldots, D_nf$ exist on U and are continuous at $\mathbf{w}$. The function f is *continuously differentiable* on A if f is continuously differentiable at each point $\mathbf{w} \in A$.

Results analogous to (7.A.5), (7.A.11), (7.A.13), (7.B.3), (7.B.4), and (7.B.5), where $\mathbf{R}^2$ is replaced by $\mathbf{R}^n$, are easily obtained by making the obvious modifications in the respective proofs.

We conclude this section with a brief discussion of the gradient.

(7.B.17) *Definition* Suppose that $D \subset \mathbf{R}^n$ and $f: D \to \mathbf{R}^1$ is continuously differentiable on D. The *gradient* of f on D is the vector-valued function $\nabla f: D \to \mathbf{R}^n$ defined by

$$\nabla f(\mathbf{z}) = (D_1f(\mathbf{z}), D_2f(\mathbf{z}), \ldots, D_nf(\mathbf{z}))$$

for each $\mathbf{z} \in D$.

Note that the difference between $\nabla f(\mathbf{z})$ and $f'(\mathbf{z})$ is that the former is a vector and the latter is a matrix.

Geometrically, the gradient of f at a point $\mathbf{z}$ can be interpreted as a vector at $\mathbf{z}$ that points in the direction in which f is increasing most rapidly. To see this, recall that the directional derivative $D_{\mathbf{u}}f$ at a point $\mathbf{z}$ measures the rate of change of the function f in the direction determined by the unit vector $\mathbf{u}$. If f is continuously differentiable and if $\mathbf{u} = (u_1, u_2, \ldots, u_n)$, then

$$D_{\mathbf{u}}f(\mathbf{z}) = u_1D_1f(\mathbf{z}) + u_2D_2f(\mathbf{z}) + \cdots + u_nD_nf(\mathbf{z}) = \nabla f(\mathbf{z}) \cdot \mathbf{u}$$

By the Cauchy-Schwarz inequality (2.A.21) it follows that

$$|\nabla f(\mathbf{z}) \cdot \mathbf{u}| \le \|\nabla f(\mathbf{z})\| \, \|\mathbf{u}\| = \|\nabla f(\mathbf{z})\| \tag{4}$$

for each unit vector $\mathbf{u}$. Equality in (4) is obtained if $\mathbf{u} = \nabla f(\mathbf{z})/\|\nabla f(\mathbf{z})\|$, since in this case,

$$|\nabla f(\mathbf{z}) \cdot \mathbf{u}| = \left| \nabla f(\mathbf{z}) \cdot \frac{\nabla f(\mathbf{z})}{\|\nabla f(\mathbf{z})\|} \right| = \frac{\|\nabla f(\mathbf{z})\|^2}{\|\nabla f(\mathbf{z})\|} = \|\nabla f(\mathbf{z})\|$$

Thus, it follows that maximum increase of f is attained by moving in the direction determined by the gradient vector.

Gradient vectors are perpendicular in the following sense to the graphs of continuously differentiable functions. Suppose that $A \subset \mathbf{R}^2$ and that $f: A \to \mathbf{R}^1$ is continuously differentiable on A. Define $g: \mathbf{R}^3 \to \mathbf{R}^1$ by

$$g(x,y,z) = z - f(x,y)$$

Then g is identically zero on the graph of f, $S_0 = \{(x,y,z) \mid g(x,y,z) = 0\}$. Furthermore, note that for each point $\mathbf{w}$ lying in the graph of f,

$$\nabla g(\mathbf{w}) = (D_1 g(\mathbf{w}), D_2 g(\mathbf{w}), D_3 g(\mathbf{w})) = (-f_x(\mathbf{w}), -f_y(\mathbf{w}), 1)$$

The parametric representation of a line passing through the point $\mathbf{w} = (x_0, y_0, z_0)$ and lying in the plane tangent to the graph of f at the point $\mathbf{w}$ is given by (5) of the previous section. If we let $\mathbf{v} = (tu_1, tu_2, tu_1 f_x(\mathbf{w}) + tu_2 f_y(\mathbf{w}))$ be parallel to this line, then direct calculation shows that

$$\nabla g(\mathbf{w}) \cdot \mathbf{v} = 0$$

which, of course, implies that these two vectors are perpendicular and, hence, that the vector $\nabla g(\mathbf{w})$ is perpendicular to the tangent plane. We say in this case that $\nabla g(\mathbf{w})$ *is perpendicular to the surface* S_0 *at the point* $\mathbf{w}$.

The above discussion can be generalized somewhat as follows. Suppose that A is an open subset of $\mathbf{R}^2$ and that $D = A \times \mathbf{R}^1$. Let $g: D \to \mathbf{R}^1$, and for each real number c, let $S_c = \{(x,y,z) \mid g(x,y,z) = c\}$. The set S_c is called a *level surface* of g in $\mathbf{R}^3$. Suppose that a level surface S_c is the graph of a continuously differentiable function $f: A \to \mathbf{R}^1$. Then if we restrict g to S_c, we have

$$c = g(x,y,z) = g(x,y,f(x,y)) = h(x,y)$$

where $h: A \to \mathbf{R}^1$. Let $\mathbf{w} = (x_0, y_0, z_0)$ be a point in S_c. We show that $\nabla g(\mathbf{w})$ is perpendicular to the surface S_c at $\mathbf{w}$. Since h_x and h_y are both identically 0 (why?), it follows easily from the chain rule (7.E.6) that $f_x(x_0, y_0) = -g_x(\mathbf{w})/g_z(\mathbf{w})$ and $f_y(x_0, y_0) = -g_y(\mathbf{w})/g_z(\mathbf{w})$. Thus, we find

$$\nabla g(\mathbf{w}) = (-f_x(x_0, y_0), -f_y(x_0, y_0), 1)g_z(\mathbf{w})$$

Since the vectors $\nabla g(\mathbf{w})$ and $(-f_x(x_0, y_0), -f_y(x_0, y_0), 1)$ lie on the same line, and since the latter vector is perpendicular to the surface S_c at $\mathbf{w}$, it follows that $\nabla g(\mathbf{w})$ is perpendicular to the surface S_c at $\mathbf{w}$, i.e., $\nabla g(\mathbf{w})$ is perpendicular to the tangent plane of S_c at $\mathbf{w}$. Consequently, this tangent plane consists of the points satisfying

$$(\nabla g(\mathbf{w})) \cdot (x - x_0, y - y_0, z - z_0) = 0$$

(7.B.18) *Example* We find the tangent plane to the surface defined by $x^2 y^3 - 2xz^2 = 6$ at the point $(1, 2, -1)$. Here $g(x,y,z) = x^2 y^3 - 2xz^2$, and hence, $\nabla g(x,y,z) = (2xy^3 - 2z^2, 3y^2 x^2, -4xz)$. Thus, $\nabla g(1, 2, -1) = (14, 12, 4)$ and the tangent plane consists of all points satisfying the equation

$$(14, 12, 4) \cdot (x - 1, y - 2, z + 1) = 0$$

which yields

$$7x + 6y + 2z = 17$$

An important application of the gradient is in the study of level curves of functions. Suppose that $D \subset \mathbf{R}^2$. We define a curve in D to be the image of any continuous function mapping a closed interval into D (for convenience however, if $\phi : [a,b] \to D$ is a continuous function, we shall often identify (incorrectly) the curve Im ϕ with ϕ itself). If $f : D \to \mathbf{R}^1$, then a *level curve* for f is a curve in D on which f is constant. If $\phi(t) = (x(t), y(t))$ is a level curve for f with the property that $x'(t)$ and $y'(t)$ exist for each t, then ∇f is perpendicular to ϕ in the sense that for each t,

$$\nabla f(x(t), y(t)) \cdot (x'(t), y'(t)) = 0 \qquad (5)$$

The vector $(x'(t), y'(t))$ is customarily referred to as the *tangent vector* to ϕ at the point $(x(t), y(t))$. To establish (5) we set $G(t) = f \circ \phi(t) = f(x(t), y(t))$ and use a result to be proven in Sec. 7.D, namely that $G'(t) = \nabla f(x(t), y(t)) \cdot (x'(t), y'(t))$; since for level curves, G is constant, we have that $G'(t) = 0$, and hence, f and ϕ are perpendicular.

Level curves are useful in the study of vector fields. If $D \subset \mathbf{R}^2$, then a *vector field* (sometimes called a *force field*) on D is a function $F : D \to \mathbf{R}^2$. Such a field is said to be *conservative* if there is a function $U : D \to \mathbf{R}^1$ such that $U_x = -F_1$ and $U_y = -F_2$, where $F(\mathbf{z}) = (F_1(\mathbf{z}), F_2(\mathbf{z}))$ for each $\mathbf{z} \in D$. The function U is called a *potential function* for the field F; note that $F = -\nabla U$. The level curves of U, usually referred to as *equipotential curves*, represent particular energy levels of the field. Observe that since $F = -\nabla U$, it follows from our earlier remarks that F is perpendicular to each level curve of U, and furthermore, F points in the direction of most rapid decrease of U (why?). As a specific example, consider the gravitational force field in $\mathbf{R}^2$ produced by a particle of unit mass located at the origin. Let $\mathbf{r} = (x,y)$; the force vector at $\mathbf{r}$ points in the direction of the unit vector $-\mathbf{r}/\|\mathbf{r}\|$. By Newton's law of gravitation, the strength of the force is $c/\|\mathbf{r}\|^2$, where c is a constant. Consequently, we have that the force field is given (in terms of $\mathbf{r}$) by $(-c/\|\mathbf{r}\|^3)\mathbf{r}$, or in terms of x and y by $F(x,y) = -c(x/(x^2 + y^2)^{3/2}, y/(x^2 + y^2)^{3/2})$. It is easily seen then that the potential function U for F is defined by $U(x,y) = -c(x^2 + y^2)^{-1/2}$ or, in terms of $\mathbf{r}$, $U(\mathbf{r}) = -c/\|\mathbf{r}\|$.

C.* DIFFERENTIABILITY IN A COMPLEX SETTING

The definition of differentiability of a complex function is completely analogous to the definition of its real counterpart (5.A.8). In the sequel the reader should keep in mind that all limits are taken with respect to the usual metric for the complex plane $\mathbf{C}$, i.e., the usual metric for $\mathbf{R}^2$.

(7.C.1) *Definition* Suppose that $A \subset \mathbf{C}$, $z_0 \in \text{int } A$, and $f: A \to \mathbf{C}$. Then f is *differentiable* at z_0 if

$$\lim_{h \to 0} \frac{f(z_0 + h) - f(z_0)}{h} \tag{1}$$

exists (here, of course, h is a complex number). If f is differentiable at z_0, then the *derivative* of f at z_0 ($f'(z_0)$ or $df(z_0)/dz$) is defined to be the value of the limit (1).

    The function $f: A \to \mathbf{C}$ is said to be *differentiable on A* if f is differentiable at each point in A.

(7.C.2) *Observation* The usual rules for differentiation: product rule, quotient rule, etc., carry over without change to the theory of complex functions—the proofs are essentially identical to those employed to establish these results for functions of a real variable and are omitted here.

    Recall from Chapter 1 that a complex function $f: A \to \mathbf{C}$ can be written in the form $f = u + iv$, where u and v map A into $\mathbf{R}^1$ (u is called the *real part* and v the *imaginary part* of f). A rather remarkable relationship between the derivative of f and the partial derivatives of u and v is brought to light in the following theorem.

(7.C.3) *Theorem* Suppose that $A \subset \mathbf{C}$, $z_0 \in \text{int } A$, and that $f = u + iv$ maps A into $\mathbf{C}$ and is differentiable at z_0. Then the partial derivatives u_x, u_y, v_x, and v_y exist at (x_0, y_0), and furthermore,

$$u_x(x_0, y_0) = v_y(x_0, y_0)$$
$$u_y(x_0, y_0) = -v_x(x_0, y_0)$$

Proof. Let h be a real number and sufficiently small so that $z_0 + h \in A$ Since f is differentiable at z_0, we have

$$f'(z_0) = \lim_{h \to 0} \frac{f(z_0 + h) - f(z_0)}{h}$$

$$= \lim_{h \to 0} \frac{u(x_0 + h, y_0) + iv(x_0 + h, y_0) - u(x_0,y_0) - iv(x_0,y_0)}{h}$$

$$= \lim_{h \to 0} \left(\frac{u(x_0 + h, y_0) - u(x_0,y_0)}{h} + i \frac{v(x_0 + h, y_0) - v(x_0,y_0)}{h} \right)$$

$$= u_x(x_0,y_0) + iv_x(x_0,y_0) \tag{2}$$

Now if we replace the real number h with the imaginary number ih, we find that

$$f'(z_0) = \lim_{h \to 0} \frac{f(z_0 + ih) - f(z_0)}{ih}$$

$$= \lim_{h \to 0} \frac{u(x_0, y_0 + h) + iv(x_0, y_0 + h) - u(x_0,y_0) - iv(x_0,y_0)}{ih}$$

$$= \frac{1}{i} \left[\lim_{h \to 0} \left(\frac{u(x_0, y_0 + h) - u(x_0,y_0)}{h} + i \frac{v(x_0, y_0 + h) - v(x_0,y_0)}{h} \right) \right]$$

$$= -iu_y(x_0,y_0) + v_y(x_0,y_0) \tag{3}$$

Thus, from (2) and (3) we obtain

$$u_x(x_0,y_0) = v_y(x_0,y_0)$$
$$u_y(x_0,y_0) = -v_x(x_0,y_0)$$

(7.C.4) Definition Suppose that $A \subset \mathbf{R}^2$. Then functions $u: A \to \mathbf{R}^1$ and $v: A \to \mathbf{R}^1$ are said to satisfy the *Cauchy-Riemann equations* on A if

$$u_x = v_y \qquad u_y = -v_x \tag{4}$$

on A.

(7.C.5) Observation In Chapter 12 it will be shown that if a function $f = u + iv$ is differentiable on an open set $U \subset \mathbf{C}$, then f has derivatives of *all* orders on U, i.e., $f^{(n)}$ exists on U for each n. As a result, in this case not only will u and v satisfy the Cauchy-Riemann equations (4), but u and v will have continuous partial derivatives on U as well. We see in the next theorem that this property characterizes differentiability.

(7.C.6) Theorem Suppose that $U \subset \mathbf{C}$ is an open set and that $f: U \to \mathbf{C}$. If $f = u + iv$ and if u_x, u_y, v_x, and v_y are continuous and satisfy the Cauchy-Riemann equations on U, then f is differentiable on U. In fact, for each $z = x + iy \in U$, $f'(z) = u_x(x,y) + iv_y(x,y)$.

Proof. Suppose that $z_0 = x_0 + iy_0 \in U$. For $h = \Delta x + i\,\Delta y$, we have

$$\frac{f(z_0 + h) - f(z_0)}{h}$$

$$= \frac{[u(x_0 + \Delta x, y_0 + \Delta y) - u(x_0,y_0)] + i[v(x_0 + \Delta x, y_0 + \Delta y) - v(x_0,y_0)]}{\Delta x + i\Delta y}$$

Judicious use of the mean value theorem (5.B.5) yields the following equalities:

$$u(x_0 + \Delta x, y_0 + \Delta y) - u(x_0,y_0)$$

$$= u(x_0 + \Delta x, y_0 + \Delta y) - u(x_0, y_0 + \Delta y) + u(x_0, y_0 + \Delta y) - u(x_0,y_0)$$

$$= u_x(x_0 + \alpha_1, y_0 + \Delta y)\,\Delta x + u_y(x_0, y_0 + \alpha_2)\,\Delta y$$

$$\text{(where } 0 < |\alpha_1| < |\Delta x|, 0 < |\alpha_2| < |\Delta y|)$$

$$= u_x(x_0,y_0)\,\Delta x + u_y(x_0,y_0)\,\Delta y + u_x(x_0 + \alpha_1, y_0 + \Delta y)\,\Delta x - u_x(x_0,y_0)\,\Delta x$$

$$+ u_y(x_0, y_0 + \alpha_2)\,\Delta y - u_y(x_0,y_0)\,\Delta y$$

Since u_x and u_y are continuous, it follows that we can write

$$u(x_0 + \Delta x, y_0 + \Delta y) - u(x_0,x_0) = u_x(x_0,y_0)\,\Delta x + u_y(x_0,y_0)\,\Delta y$$

$$+ \varepsilon_1\,\Delta x + \varepsilon_2\,\Delta y$$

where ε_1 and ε_2 tend to 0 as Δx and Δy approach 0.

Similarly, of course, we can write

$$v(x_0 + \Delta x, y_0 + \Delta y) - v(x_0,y_0) = v_x(x_0, y_0)\,\Delta x + v_y(x_0,y_0)\,\Delta y$$

$$+ \varepsilon_3\,\Delta x + \varepsilon_4\,\Delta y$$

where ε_3 and ε_4 tend to 0 as Δx and Δy approach 0.

Now note that by the Cauchy-Riemann equations we have

$$u_x + u_y + iv_x + iv_y = u_x - v_x + iv_x + iu_x = u_x + iv_x + i(u_x + iv_x)$$

and therefore, from the preceding work, we have

$$\frac{f(z_0 + h) - f(z_0)}{h} = \frac{(u_x(x_0,y_0) + iv_x(x_0,y_0))\Delta x}{\Delta x + i\,\Delta y}$$

$$+ i\,\frac{(u_x(x_0,y_0) + iv_x(x_0,y_0))\,\Delta y}{\Delta x + i\Delta y}$$

$$+ \frac{(\varepsilon_1 - i\varepsilon_3)\Delta x}{\Delta x + i\,\Delta y} + \frac{(\varepsilon_2 - i\varepsilon_4)\Delta y}{\Delta x + i\,\Delta y}$$

$$= u_x(x_0,y_0) + iv_x(x_0,y_0) + (\varepsilon_1 - i\varepsilon_3)\,\frac{\Delta x}{\Delta x + i\,\Delta y}$$

$$+ (\varepsilon_2 - i\varepsilon_4)\,\frac{\Delta y}{\Delta x + i\,\Delta y}$$

Finally, observe that $|\Delta x/(\Delta x + i\,\Delta y)| \le 1$ and $|\Delta y/(\Delta x + i\,\Delta y)| \le 1$, and hence, since $\varepsilon_1,\varepsilon_2,\varepsilon_3$, and ε_4 all tend to 0 as $h \to 0$, we have that $f'(z_0)$ exists and

$$f'(z_0) = \lim_{h\to 0} \frac{f(z_0 + h) - f(z_0)}{h} = u_x(x_0,y_0) + iv_x(x_0,y_0)$$

which concludes the proof.

Is the condition in the previous theorem that the partial derivatives be continuous necessary in order to ensure the existence of the derivative? In view of the next example it is clear that some condition must be imposed on the partial derivatives in order to establish the conclusion of Theorem (7.C.6).

(7.C.7) **Example**   Suppose that the function $f = u + iv$ is defined by

$$f(z) = \begin{cases} \dfrac{x^3 - y^3}{x^2 + y^2} + i\,\dfrac{x^3 + y^3}{x^2 + y^2} & \text{if } (x,y) \ne (0,0) \\ 0 & \text{if } (x,y) = (0,0) \end{cases}$$

Then using the definition of the partial derivative, it is easily verified that

$$u_x(0,0) = 1 = v_y(0,0)$$
$$u_y(0,0) = -1 = -v_x(0,0)$$

However, the derivative of f fails to exist at the point $z = 0$ (see problem C.6).

(7.C.8) **Example**   If $f(z) = (x^2 - y^2) + 2xyi$, then $u(x,y) = x^2 - y^2$ and $v(x,y) = 2xy$. Note that $u_x(x,y) = 2x = v_y(x,y)$ and $u_y(x,y) = -2y = -v_x(x,y)$. Since these functions are continuous, it follows that f is differentiable everywhere.

On the other hand if $f(z) = x^2 + y^2$, then $u_x(x,y) = 2x$, $u_y(x,y) = 2y$, $v_x(x,y) = 0$ and $v_y(x,y) = 0$. Hence, the derivative of f can only exist at $z = 0$.

(7.C.9) **Observation**   Of very special interest in the theory of complex variables are the analytic functions. A function is said to be *analytic* at a point z_0 if f is differentiable in an open neighborhood of z_0. In Chapter 12 we shall deal extensively with these functions. Note that thus far we have shown that if $f = u + iv$ and if u and v have continuous first partial derivatives in an open set U, then f is analytic in U.

D. GENERALIZATIONS

We now briefly discuss how the concepts studied in Secs. 7.A and 7.B may be generalized to functions mapping subsets of $\mathbf{R}^m$ into $\mathbf{R}^n$.

We begin by observing once more that if A is an arbitrary set and if a function f maps A into $\mathbf{R}^n$, then f generates coordinate functions $f_1, \ldots, f_n$, where for each i, $f_i: A \to \mathbf{R}^1$ is defined by setting $f_i(a)$ equal to the ith coordinate of $f(a)$; hence, $f(a) = (f_1(a), f_2(a), \ldots, f_n(a))$ for each $a \in A$.

(7.D.1) *Example* If $A = \mathbf{R}^2$ and $f: A \to \mathbf{R}^4$ is defined by $f(x,y) = (xy^2, -x, e^{xy}, \sin x)$, then $f_1(x,y) = xy^2$, $f_2(x,y) = -x$, $f_3(x,y) = e^{xy}$, and $f_4(x,y) = \sin x$.

The first definition presents little difficulty.

(7.D.2) *Definition* If $f: (a,b) \to \mathbf{R}^n$, then the *derivative* of f at a point $x \in (a,b)$ is defined to be the column matrix

$$f'(x) = \begin{pmatrix} f_1'(x) \\ f_2'(x) \\ \vdots \\ f_n'(x) \end{pmatrix}$$

provided that $f_i'(x)$ exists for each i, $1 \le i \le n$.

(7.D.3) *Definition* Suppose that $A \subset \mathbf{R}^m$, $f: A \to \mathbf{R}^n$, and $\mathbf{w} \in \text{int } A$. Then the ith *partial derivative* of f at $\mathbf{w}$, $D_i f(\mathbf{w})$, is defined to be the row matrix

$$D_i f(\mathbf{w}) = (D_i f_1(\mathbf{w}) \quad D_i f_2(\mathbf{w}) \quad \cdots \quad D_i f_n(\mathbf{w}))$$

provided that $D_i f_j(\mathbf{w})$ exists for each j, $1 \le j \le n$.

(7.D.4) *Example* If $f: \mathbf{R}^2 \to \mathbf{R}^3$ is defined by $f(x_1, x_2) = (x_1^3 x_2, 2x_1, x_1 \sin x_2)$, then $D_1 f(x_1, x_2) = (3x_1^2 x_2 \quad 2 \quad \sin x_2)$ and $D_2 f(x_1, x_2) = (x_1 \quad 0 \quad x_1 \cos x_2)$.

Differentiability is generalized in a natural fashion.

(7.D.5) *Definition* Suppose that $A \subset \mathbf{R}^m$, $\mathbf{w} \in \text{int } A$, and $f: A \to \mathbf{R}^n$. Then f is differentiable at $\mathbf{w}$ if there exists a linear function $L: \mathbf{R}^m \to \mathbf{R}^n$ such that

$$\lim_{\mathbf{v} \to 0} \frac{f(\mathbf{w} + \mathbf{v}) - f(\mathbf{w}) - L(\mathbf{v})}{\|\mathbf{v}\|} = 0$$

The linear function L is called the *differential* of f at w, and will generally be denoted by $d_w f$.

(7.D.6) *Exercise* If $A \subset R^m$, $w \in \text{int } A$, $f: A \to R^n$, and u is a unit vector in R^m, then the *directional derivative* of f at w in the direction u is defined by

$$D_u f(w) = \lim_{h \to 0} \frac{f(w + hu) - f(w)}{h}$$

Show that if f is differentiable at w and L is the linear function in (7.D.5), then

(a) If u is a unit vector in R^m, then $L(u) = D_u f(w)$.
(b) If $e_i = (0, 0, \ldots, 0,1,0, \ldots, 0)$, where 1 is in the ith coordinate, then $D_{e_i} f(w) = D_i f(w)$.
(c) L is unique, and if $z = (z_1, z_2, \ldots, z_m) \in R^m$, then

$$L(z) = \sum_{i=1}^{m} z_i D_i f(w) = \sum_{i=1}^{m} z_i (D_i f_1(w), \ldots, D_i f_n(w))$$

[*Hint*: Consider the proof of (7.B.3).]

(7.D.7) *Observation* If $A \subset R^m$ and f is differentiable at a point $w \in \text{int } A$, then by (7.D.6.c) the differential of f at w satisfies

$$d_w f(z) = D_1 f(w) z_1 + \cdots + D_m f(w) z_m$$

where $z = (z_1, z_2, \ldots, z_m)$. Note that $d_w f(z)$ can be expressed as the matrix product

$$d_w f(z) = (D_1 f(w) \quad \cdots \quad D_m f(w)) \begin{pmatrix} z_1 \\ \vdots \\ z_m \end{pmatrix}$$

In particular, observe that $d_w f(e_i) = D_i f(w)$, where e_i is defined as in (7.D.6).

(7.D.8) *Example* The differential of the function f given in (7.D.4) at a point $w = (w_1, w_2)$ is defined by

$$d_w f(z) = z_1(3w_1^2 w_2, 2, \sin w_2) + z_2(w_1^3, 0, w_1 \cos w_2)$$

The ideas leading to the general definition of the derivative parallel those discussed in the preceding section. Suppose that $A \subset R^m$ and that $f: A \to R^n$ is differentiable at a point $w \in \text{int } A$. Let

$$e_1 = (1, 0, \ldots, 0)$$
$$e_2 = (0, 1, \ldots, 0)$$
$$\cdots \cdots \cdots \cdots \cdots$$
$$e_m = (0, 0, \ldots, 1)$$

be the standard basis for $\mathbf{R}^m$, and

$$\hat{e}_1 = (1, 0, \ldots, 0)$$
$$\hat{e}_2 = (0, 1, \ldots, 0)$$
$$\cdots\cdots\cdots\cdots$$
$$\hat{e}_n = (0, 0, \ldots, 1)$$

be the standard basis for $\mathbf{R}^n$. Then by (7.D.6) and (7.D.7) we have

$$d_{\mathbf{w}}f(e_1) = (D_1 f_1(\mathbf{w}), \ldots, D_1 f_n(\mathbf{w})) = D_1 f_1(\mathbf{w})\hat{e}_1 + D_1 f_2(\mathbf{w})\hat{e}_2$$
$$+ \cdots + D_1 f_n(\mathbf{w})\hat{e}_n$$
$$d_{\mathbf{w}}f(e_2) = (D_2 f_1(\mathbf{w}), \ldots, D_2 f_n(\mathbf{w})) = D_2 f_1(\mathbf{w})\hat{e}_1 + D_2 f_2(\mathbf{w})\hat{e}_2$$
$$+ \cdots + D_2 f_n(\mathbf{w})\hat{e}_n$$
$$d_{\mathbf{w}}f(e_m) = (D_m f_1(\mathbf{w}), \ldots, D_m f_n(\mathbf{w})) = D_m f_1(\mathbf{w})\hat{e}_1 + D_m f_2(\mathbf{w})\hat{e}_2$$
$$+ \cdots + D_m f_n(\mathbf{w})\hat{e}_n$$

Thus, the matrix $M_{d_{\mathbf{w}}f}$ belonging to the linear transformation $d_{\mathbf{w}}f$ (with respect to the standard bases) is given by

$$\begin{pmatrix} D_1 f_1(\mathbf{w}) & D_2 f_1(\mathbf{w}) & \cdots & D_m f_1(\mathbf{w}) \\ D_1 f_2(\mathbf{w}) & D_2 f_2(\mathbf{w}) & \cdots & D_m f_2(\mathbf{w}) \\ \cdots\cdots\cdots\cdots\cdots\cdots\cdots\cdots \\ D_1 f_n(\mathbf{w}) & D_2 f_n(\mathbf{w}) & \cdots & D_m f_n(\mathbf{w}) \end{pmatrix} \qquad (1)$$

The derivative is defined in terms of this matrix.

(7.D.9) **Definition** Suppose that $A \subset \mathbf{R}^m$, $f: A \to \mathbf{R}^n$ and that all partial derivatives of f exist at a point $\mathbf{w} \in \text{int } A$. Then the *derivative* of f at $\mathbf{w}$, $f'(\mathbf{w})$, is defined to be the matrix (1), where $f = (f_1, f_2, \ldots, f_n)$.

Note that the above definition is compatible with (7.D.2).

(7.D.10) **Example** The derivative of the function f given in (7.D.4) at a point $\mathbf{w} = (w_1, w_2)$ is

$$f'(\mathbf{w}) = \begin{pmatrix} 3w_1^2 w_2 & w_1^3 \\ 2 & 0 \\ \sin w_2 & w_1 \cos w_2 \end{pmatrix}$$

The derivative matrix is frequently referred to as the *Jacobi matrix* of f at $\mathbf{w}$, and it and its determinant (called the *Jacobian* of f at $\mathbf{w}$) play an extremely interesting and significant role in advanced analysis.

Utilizing the relationship (2.E.7) between a linear transformation and its associated matrix, we obtain immediately the following substantial generalization of (1), Sec. 5.D.

(7.D.11) *Theorem* If $A \subset \mathbf{R}^m$, $f: A \to \mathbf{R}^n$, and $f'(\mathbf{w})$ exists at a point $\mathbf{w} \in$ int A, then $d_{\mathbf{w}}f(\mathbf{z}) = (q_1, q_2, \ldots, q_n)$, where

$$\begin{pmatrix} q_1 \\ q_2 \\ \vdots \\ q_n \end{pmatrix} = f'(\mathbf{w}) \begin{pmatrix} z_1 \\ z_2 \\ \vdots \\ z_m \end{pmatrix}$$

It is important to observe that contrary to the case of functions of a single variable, the definitions of differentiability and of the derivative in the context of multivariable functions allow for the derivative of a function to exist at a point $\mathbf{w}$ without implying that the function is differentiable there. In order that the derivative of a function f exist at a point $\mathbf{w}$, it is sufficient that the partial derivatives of f are defined at $\mathbf{w}$. However, if f is differentiable at $\mathbf{w}$, then by (7.D.6), all directional derivatives of f at $\mathbf{w}$ must exist. Thus, for example, the derivative of the function given in (7.A.9) exists at $\mathbf{w} = (0,0)$, even though f is not differentiable at this point. An easily applied criterion for differentiability can be given in terms of continuous differentiability.

(7.D.12) *Definition* Suppose that $A \subset \mathbf{R}^m$, $f: A \to \mathbf{R}^n$, and $\mathbf{w} \in$ int A. Then f is *continuously differentiable at* $\mathbf{w}$ if there is an open subset U containing $\mathbf{w}$ such that the partial derivatives $D_1 f, D_2 f, \ldots, D_m f$ all exist on U and are continuous at $\mathbf{w}$. The function f is *continuously differentiable on* A if f is continuously differentiable at each point $\mathbf{w} \in A$.

(7.D.13) *Theorem* Suppose that $A \subset \mathbf{R}^m$, $\mathbf{w} \in$ int A, and $f: A \to \mathbf{R}^n$ is continuously differentiable at $\mathbf{w}$. Then f is differentiable at $\mathbf{w}$.

Proof. We shall show that

$$\lim_{\mathbf{v} \to \mathbf{0}} \frac{f(\mathbf{w} + \mathbf{v}) - f(\mathbf{w}) - d_{\mathbf{w}}f(\mathbf{v})}{\|\mathbf{v}\|} = 0$$

Let U be an open subset of $\mathbf{R}^m$ such that $\mathbf{w} \in U \subset A$, and such that the partial derivatives $D_i f_j$, $1 \le i \le m$ and $1 \le j \le n$ are defined on U [as usual, $f = (f_1, f_2, \ldots, f_n)$]. First recall that if $g: U \to \mathbf{R}^n$ and $\mathbf{z}^* \in U$, then $\lim_{\mathbf{z} \to \mathbf{z}^*} g(\mathbf{z}) = \mathbf{q}$ if and only if $\lim_{\mathbf{z} \to \mathbf{z}^*} g_i(\mathbf{z}) = q_i$ for each i, where $g = (g_1, g_2, \ldots, g_n)$ and $\mathbf{q} = (q_1, q_2, \ldots, q_n)$. Consequently, we may henceforth assume that $f: U \to \mathbf{R}^1$. For any vector $\mathbf{v} \in \mathbf{R}^m$ with the property that $\mathbf{w} + \mathbf{v} = (w_1 + v_1, w_2 + v_2, \ldots, w_m + v_m) \in U$, define vectors $\mathbf{z}_0, \mathbf{z}_1, \ldots, \mathbf{z}_m$ by $\mathbf{z}_k = (w_1 + v_1, \ldots, w_k + v_k, w_{k+1}, w_{k+2}, \ldots, w_m)$ [here $\mathbf{z}_0 = (w_1, w_2, \ldots, w_m) = \mathbf{w}$ and $\mathbf{z}_m = (w_1 + v_1, w_2 + v_2, \ldots, w_m + v_m) = \mathbf{w} + \mathbf{v}$]. Then we have

$$f(\mathbf{w} + \mathbf{v}) - f(\mathbf{w}) = (f(\mathbf{z}_1) - f(\mathbf{z}_0)) + (f(\mathbf{z}_2) - f(\mathbf{z}_1))$$
$$+ \cdots + (f(\mathbf{z}_n) - f(\mathbf{z}_{n-1}))$$

Since for $1 \le k \le n$,

$$\mathbf{z}_k = (w_1 + v_1, \ldots, w_{k-1} + v_{k-1}, w_k + v_k, w_{k+1}, \ldots, w_m)$$

and

$$\mathbf{z}_{k-1} = (w_1 + v_1, \ldots, w_{k-1} + v_{k-1}, w_k, w_{k+1}, \ldots, w_m)$$

differ only in the kth coordinate, we may think of the remaining coordinates as being fixed and apply the mean value theorem (5.B.5) to obtain a number c_k between w_k and $w_k + v_k$, such that if

$$\mathbf{c}_k = (w_1 + v_1, \ldots, w_{k-1} + v_{k-1}, c_k, w_{k+1} + v_{k+1}, \ldots, w_m + v_m),$$

then $f(\mathbf{z}_k) - f(\mathbf{z}_{k-1}) = v_k D_k f(\mathbf{c}_k)$

Consequently, we have

$$f(\mathbf{w} + \mathbf{v}) - f(\mathbf{w}) = \sum_{k=1}^{m} v_k D_k f(\mathbf{c}_k)$$

Since $d_{\mathbf{w}} f(\mathbf{v}) = \sum_{k=1}^{m} v_k D_k(\mathbf{w})$, it follows that

$$|f(\mathbf{w} + \mathbf{v}) - f(\mathbf{w}) - d_{\mathbf{w}} f(\mathbf{v})| = \left| \sum_{k=1}^{m} v_k (D_k f(\mathbf{c}_k) - D_k f(\mathbf{w})) \right|$$

Clearly, $|v_k| \le \|\mathbf{v}\|$, and thus

$$|f(\mathbf{w} + \mathbf{v}) - f(\mathbf{w}) - d_{\mathbf{w}} f(\mathbf{v})| \le \|\mathbf{v}\| \sum_{k=1}^{m} |D_k f(\mathbf{c}_k) - D_k f(\mathbf{w})|$$

hence,

$$\frac{|f(\mathbf{w} + \mathbf{v}) - f(\mathbf{w}) - d_{\mathbf{w}} f(\mathbf{v})|}{\|\mathbf{v}\|} \le \sum_{k=1}^{m} |D_k f(\mathbf{c}_k) - D_k f(\mathbf{w})|$$

Note that as $\mathbf{v} \to \mathbf{0}$, the corresponding vectors $\mathbf{c}_k$ converge to $\mathbf{w}$, and therefore, by the continuity of the partial derivatives $D_k f$ at $\mathbf{w}$, $\lim_{\mathbf{v} \to \mathbf{0}} |D_k f(\mathbf{c}_k) - D_k f(\mathbf{w})| = 0$, and the theorem follows.

To conclude this section we point out that there is no general agreement with regard to the terminology used in connection with derivatives and differentiability. In particular, the function L in (7.D.5) is frequently referred to as the derivative of f at $\mathbf{w}$. The term *total derivative* is also employed in this context.

E. THE CHAIN RULE

In this section we see how the chain rule (5.A.15) can be generalized to the multidimensional case. Suppose that $f: \mathbf{R}^m \to \mathbf{R}^k$, $g: \mathbf{R}^k \to \mathbf{R}^n$, $\mathbf{x} \in \mathbf{R}^m$ and $\mathbf{w} = f(\mathbf{x})$. Let $h = g \circ f$ (Fig. 7.3).

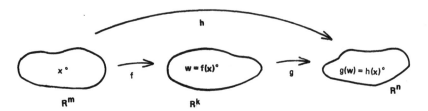

Figure 7.3

Suppose now that f is differentiable at $\mathbf{x}$ and that g is differentiable at $\mathbf{w} = f(\mathbf{x})$; then, of course, the differentials

$$d_{\mathbf{x}}f: \mathbf{R}^m \to \mathbf{R}^k \qquad d_{\mathbf{w}}g: \mathbf{R}^k \to \mathbf{R}^n$$

are defined. The chain rule asserts that $h = g \circ f$ is differentiable at $\mathbf{x}$ and, furthermore, that the differential of h at $\mathbf{x}$ is equal to the composition of the differentials of g and f, i.e.,

$$d_{\mathbf{x}}h = d_{\mathbf{w}}g \circ d_{\mathbf{x}}f$$

Note that it follows that the derivative of h at $\mathbf{x}$ also exists and by (2.E.10) is equal to the matrix product of the derivative of g and f, i.e.,

$$h'(\mathbf{x}) = M_{d_{\mathbf{x}}h} = M_{d_{\mathbf{w}}g}M_{d_{\mathbf{x}}f}$$

(7.E.1)   *Theorem* Suppose that $A \subset \mathbf{R}^m$, $f: A \to \mathbf{R}^k$, and f is differentiable at $\hat{\mathbf{x}} \in \text{int } A$. Let $B = f(A)$ and suppose that $g: B \to \mathbf{R}^n$ is differentiable at $\hat{\mathbf{w}} = f(\hat{\mathbf{x}})$. If $h = g \circ f$, then h is differentiable at $\hat{\mathbf{x}}$, and furthermore, $d_{\hat{\mathbf{x}}}h = d_{\hat{\mathbf{w}}}g \circ d_{\hat{\mathbf{x}}}f$.

Proof.   We show that

$$\lim_{\mathbf{x} \to \hat{\mathbf{x}}} \frac{h(\mathbf{x}) - h(\hat{\mathbf{x}}) - (d_{\hat{\mathbf{w}}}g \circ d_{\hat{\mathbf{x}}}f)(\mathbf{x} - \hat{\mathbf{x}})}{\|\mathbf{x} - \hat{\mathbf{x}}\|} = 0$$

Following Fadell (1973), let $\phi: B \to \mathbf{R}^n$ be defined by

$$\phi(\mathbf{w}) = \begin{cases} \dfrac{g(\mathbf{w}) - g(\hat{\mathbf{w}}) - d_{\hat{\mathbf{w}}}g(\mathbf{w} - \hat{\mathbf{w}})}{\|\mathbf{w} - \hat{\mathbf{w}}\|} & \text{if } \mathbf{w} \neq \hat{\mathbf{w}} \\ 0 & \text{if } \mathbf{w} = \hat{\mathbf{w}} \end{cases}$$

Since g is differentiable at $\hat{\mathbf{w}}$, ϕ is continuous at $\hat{\mathbf{w}}$. From the definition of ϕ we have

$$g(\mathbf{w}) - g(\hat{\mathbf{w}}) = \phi(\mathbf{w})\|\mathbf{w} - \hat{\mathbf{w}}\| + d_{\hat{\mathbf{w}}}g(\mathbf{w} - \hat{\mathbf{w}}) \tag{1}$$

Suppose that $\mathbf{x} \in A$ and $\mathbf{w} = f(\mathbf{x})$. Subtract $d_{\hat{\mathbf{w}}}g(d_{\hat{\mathbf{x}}}f(\mathbf{x} - \hat{\mathbf{x}}))$ from both sides of (1) to obtain

$$g(f(\mathbf{x})) - g(f(\hat{\mathbf{x}})) - d_{\hat{\mathbf{w}}}g(d_{\hat{\mathbf{x}}}f(\mathbf{x} - \hat{\mathbf{x}}))$$
$$= \phi(f(\mathbf{x}))\|f(\mathbf{x}) - f(\hat{\mathbf{x}})\| + d_{\hat{\mathbf{w}}}g(f(\mathbf{x}) - f(\hat{\mathbf{x}})) - d_{\hat{\mathbf{w}}}g(d_{\hat{\mathbf{x}}}f(\mathbf{x} - \hat{\mathbf{x}})) \quad (2)$$

Divide both sides of (2) by $\|\mathbf{x} - \hat{\mathbf{x}}\|$ and use the linearity of $d_{\hat{\mathbf{w}}}g$ to obtain

$$\frac{g(f(\mathbf{x})) - g(f(\hat{\mathbf{x}})) - d_{\hat{\mathbf{w}}}g(d_{\hat{\mathbf{x}}}f(\mathbf{x} - \hat{\mathbf{x}}))}{\|\hat{\mathbf{x}} - \mathbf{x}\|}$$

$$= \phi(f(\mathbf{x}))\frac{\|f(\mathbf{x}) - f(\hat{\mathbf{x}})\|}{\|\mathbf{x} - \hat{\mathbf{x}}\|} + d_{\hat{\mathbf{w}}}g\left(\frac{f(\mathbf{x}) - f(\hat{\mathbf{x}}) - d_{\hat{\mathbf{x}}}f(\mathbf{x} - \hat{\mathbf{x}})}{\|\mathbf{x} - \hat{\mathbf{x}}\|}\right) \quad (3)$$

To conclude the proof first note that as $\mathbf{x} \to \hat{\mathbf{x}}$,

(a) $\phi(f(\mathbf{x})) \to \phi(f(\hat{\mathbf{x}})) = \phi(\hat{\mathbf{w}}) = 0$

(b) $\dfrac{f(\mathbf{x}) - f(\hat{\mathbf{x}}) - d_{\hat{\mathbf{x}}}f(\mathbf{x} - \hat{\mathbf{x}})}{\|\mathbf{x} - \hat{\mathbf{x}}\|} \to 0$

Since the differential $d_{\hat{\mathbf{w}}}g$ is a linear (and continuous) function, it follows that $\lim_{\mathbf{w} \to 0} d_{\hat{\mathbf{w}}}g(\mathbf{w}) = d_{\hat{\mathbf{w}}}g(0) = 0$, and therefore, by (b), the last term on the right-hand side of (3) converges to 0 as $\mathbf{x} \to \hat{\mathbf{x}}$. Next note that $\|f(\mathbf{x}) - f(\hat{\mathbf{x}})\|/\|\mathbf{x} - \hat{\mathbf{x}}\|$ remains bounded as $\mathbf{x} \to \hat{\mathbf{x}}$, since

$$\frac{\|f(\mathbf{x}) - f(\hat{\mathbf{x}})\|}{\|\mathbf{x} - \hat{\mathbf{x}}\|} \leq \frac{\|f(\mathbf{x}) - f(\hat{\mathbf{x}}) - d_{\hat{\mathbf{x}}}f(\mathbf{x} - \hat{\mathbf{x}})\|}{\|\mathbf{x} - \hat{\mathbf{x}}\|} + \frac{\|d_{\hat{\mathbf{x}}}f(\mathbf{x} - \hat{\mathbf{x}})\|}{\|\mathbf{x} - \hat{\mathbf{x}}\|}$$

and $\|d_{\hat{\mathbf{x}}}f(\mathbf{x} - \hat{\mathbf{x}})\| \leq K\|\mathbf{x} - \hat{\mathbf{x}}\|$ for some constant K (why?). Hence, by (a) we have that the first term on the right-hand side of (3) also converges to 0 as $\mathbf{x} \to \hat{\mathbf{x}}$. Consequently,

$$\lim_{\mathbf{x} \to \hat{\mathbf{x}}} \frac{g(f(\mathbf{x})) - g(f(\hat{\mathbf{x}})) - d_{\hat{\mathbf{w}}}g(d_{\hat{\mathbf{x}}}f(\mathbf{x} - \hat{\mathbf{x}}))}{\|\mathbf{x} - \hat{\mathbf{x}}\|} = 0$$

which shows (by (7.D.5)) that the linear function $d_{\hat{\mathbf{w}}}g \circ d_{\hat{\mathbf{x}}}f$ is the differential of $h = g \circ f$ at $\hat{\mathbf{x}}$.

(7.E.2) *Example* Suppose that $f: \mathbf{R}^2 \to \mathbf{R}^4$ is defined by

$$f(x_1, x_2) = (x_1 x_2, x_2^2, \sin x_1^3 x_2, x_1)$$

and that $g: \mathbf{R}^4 \to \mathbf{R}^3$ is defined by

$$g(y_1, y_2, y_3, y_4) = (y_1^2 y_3, y_4, \cos y_2)$$

If $\mathbf{x} = (x_1, x_2)$ and $\mathbf{z} = (z_1, z_2)$, then

$$d_{\mathbf{x}}f(z_1, z_2) = z_1(x_2, 0, 3x_1^2 x_2 \cos x_1^3 x_2, 1) + z_2(x_1, 2x_2, x_1^3 \cos x_1^3 x_2, 0)$$
$$= (z_1 x_2 + z_2 x_1, 2z_2 x_2, 3z_1 x_1^2 x_2 \cos x_1^3 x_2 + z_2 x_1^3 \cos x_1^3 x_2, 1)$$

If $\mathbf{w} = (w_1, w_2, w_3, w_4)$ and $\mathbf{t} = (t_1, t_2, t_3, t_4)$, then

$$d_\mathbf{w} g(t_1, t_2, t_3, t_4) = t_1(2w_1 w_3, 0, 0) + t_2(0, 0, -\sin w_2) + t_3(w_1^2, 0, 0)$$
$$+ t_4(0, 1, 0)$$

Consequently, if $h = g \circ f$, then

$$d_x h(z_1, z_2) = (z_1 x_2 + z_2 x_1)(2w_1 w_3, 0, 0) + (2z_2 x_2)(0, 0, -\sin w_2)$$
$$+ (3z_1 x_1^2 x_2 \cos x_1^3 x_2)(w_1^2, 0, 0) + 1(0, 1, 0)$$

(7.E.3) *Exercise* In (7.E.2) find $h'(z)$.

As an important special case of (7.E.1) we have the following corollary.

(7.E.4) *Corollary* Suppose that $A \subset \mathbf{R}^1, f: A \to \mathbf{R}^2, B = f(A)$, and $g: B \to \mathbf{R}^1$. If $f = (f_1, f_2)$ and g are differentiable at $\hat{x} \in A$ and $\hat{w} = f(\hat{x}) \in B$, respectively, then the derivative of $h = g \circ f$ at $\hat{x}$ is given by

$$h'(\hat{x}) = g_x(f_1(\hat{x}), f_2(\hat{x}))f_1'(\hat{x}) + g_y(f_1(\hat{x}), f_2(\hat{x}))f_2'(\hat{x})$$

Proof. Since

$$f'(\hat{x}) = \begin{pmatrix} f_1'(\hat{x}) \\ f_2'(\hat{x}) \end{pmatrix}$$

and $g'(\hat{w}) = (g_x(\hat{w}) \quad g_y(\hat{w}))$, where $\hat{w} = f(\hat{x}) = (f_1(\hat{x}), f_2(\hat{x}))$, we have that $h'(\hat{x})$ is given by the product of these two matrices

$$h'(\hat{x}) = (g_x(f_1(\hat{x}), f_2(\hat{x})) \quad g_y(f_1(\hat{x}), f_2(\hat{x}))) \begin{pmatrix} f_1'(\hat{x}) \\ f_2'(\hat{x}) \end{pmatrix}$$
$$= g_x(f_1(\hat{x}), f_2(\hat{x}))f_1'(\hat{x}) + g_y(f_1(\hat{x}), f_2(\hat{x}))f_2'(\hat{x})$$

(7.E.5) *Example* Suppose that $f: \mathbf{R}^1 \to \mathbf{R}^2$ and $g: \mathbf{R}^2 \to \mathbf{R}^1$ are defined by $f(z) = (z^2, 2z)$ and $g(x, y) = x^3 y^2$. Then $f_1(z) = z^2$, $f_2(z) = 2z$, $g_x(x, y) = 3x^2 y^2$, and $g_y(x, y) = 2x^3 y$. Consequently, if $h = g \circ f$, then $h'(z) = (3z^4 4z^2)2z + (2z^6 2z)2 = 32z^7$.

(7.E.6) *Observation* Quite often we shall consider a function f whose domain includes the graph of another function. Suppose that $A \subset \mathbf{R}^1$, $\alpha: A \to \mathbf{R}^1$ and that C is an open subset of $\mathbf{R}^2$ containing the graph of α (Fig. 7.4).

If $g: C \to \mathbf{R}^1$, define $\psi: A \to \mathbf{R}^1$ by $\psi(x) = g(x, \alpha(x))$. Note that if $u: A \to C$ is defined by $u(x) = (x, \alpha(x))$, then $\psi = g \circ u$. Consequently, if g and α are differentiable, it follows from (7.E.4) that

$$\psi'(\hat{x}) = (g_x(\hat{x}, \alpha(\hat{x})) \quad g_y(\hat{x}, \alpha(\hat{x}))) \begin{pmatrix} 1 \\ \alpha'(\hat{x}) \end{pmatrix} = g_x(\hat{x}, \alpha(\hat{x})) + g_y(\hat{x}, \alpha(\hat{x}))\alpha'(\hat{x})$$

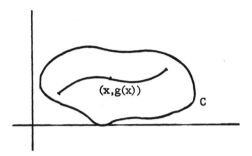

(x,g(x))

C

Figure 7.4

The chain rule as given in (7.E.4) can be applied to establish the following two-dimensional mean value theorem. A subset $A \subset \mathbf{R}^2$ is said to be *convex* if any two points in A can be connected by a straight line segment that lies entirely in A.

(7.E.7) ***Theorem*** Suppose that A is a convex subset of $\mathbf{R}^2$ and that $g: A \to \mathbf{R}^1$ is differentiable on A. If (x_0, y_0) and $(x_0 + h, y_0 + k)$ are two points in A, then there is a number c, $0 < c < 1$ such that

$$g(x_0 + h, y_0 + k) - g(x_0, y_0) = [g_x(x_0 + ch, y_0 + ck)]h$$
$$+ [g_y(x_0 + ch, y_0 + ck)]k$$

Proof. The trick is to define a function of one variable to which the mean value theorem for one variable can be applied. Let then

$$\phi(t) = g(x_0 + th, y_0 + tk)$$

Note that $\phi(1) - \phi(0) = g(x_0 + h, y_0 + k) - g(x_0, y_0)$, and furthermore, by the mean value theorem (5.B.5), $\phi(1) - \phi(0) = \phi'(c)(1 - 0)$, where $0 < c < 1$. On the other hand, by (7.E.4), we have

$$\phi'(c) = [g_x(x_0 + ch, y_0 + ck)]h + [g_y(x_0 + ch, y_0 + ck)]k$$

and this concludes the proof.

In Chapter 12 we shall need the following version of the chain rule for complex-valued functions.

(7.E.8) ***Theorem*** Suppose that D is an open subset of the complex plane $\mathbf{C}$ and that $f: D \to \mathbf{C}$. Suppose further that there is a function $F: D \to \mathbf{C}$ such that $F' = f$. If $\gamma: [a,b] \to D$ is a continuously differentiable function, then

$$\frac{d}{dt} F(\gamma(t)) = f(\gamma(t))\gamma'(t)$$

Proof. Let $F(z) = U(x,y) + iV(x,y)$ and $f(z) = u(x,y) + iv(x,y)$, where $z = x + iy$, and set $\gamma(t) = x(t) + iy(t)$. Then by (7.E.4) and the Cauchy-

Riemann equations (7.C.3), we find that

$$\frac{d}{dt} F(\gamma(t)) = \frac{d}{dt} U(x(t), y(t)) + i\frac{d}{dt} V(x(t), y(t))$$

$$= U_x(x(t), y(t))x'(t) + U_y(x(t), y(t))y'(t)$$
$$\quad + i(V_x(x(t), y(t))x'(t) + V_y(x(t), y(t))y'(t))$$
$$= U_x(x(t), y(t))x'(t) - V_x(x(t), y(t))y'(t)$$
$$\quad + i\{V_x(x(t), y(t))x'(t) + U_x(x(t), y(t))y'(t)\}$$

Since $F' = f$ on D, it follows that

$$U_x(x,y) + iV_x(x,y) = u(x,y) + iv(x,y)$$

on D, and therefore,

$$f(\gamma(t))\gamma'(t) = [u(x(t), y(t)) + iv(x(t), y(t))][x'(t) + iy'(t)]$$
$$= u(x(t), y(t))x'(t) - v(x(t), y(t))y'(t) + i[v(x(t), y(t))x'(t)$$
$$\quad + u(x(t), y(t))y'(t)]$$
$$= U_x(x(t), y(t))x'(t) - V_x(x(t), y(t))y'(t) + i[V_x(x(t), y(t))x'(t)$$
$$\quad + U_y(x(t), y(t))y'(t)] = \frac{d}{dt} F(\gamma(t))$$

F. DIFFERENTIATION UNDER THE INTEGRAL SIGN

Functions of the following type arise frequently in both theoretical and applied situations:

$$g(x) = \int_a^b f(x,y)\, dy \qquad (1)$$

where f is assumed to be a continuous function of two variables. In this section we are concerned with the problem of determining when g is differentiable and how the derivative g' can be found. Certainly one method for computing g' would be to evaluate the integral

$$\int_a^b f(x,y)\, dy$$

and take the derivative of the resulting function of x. For instance, if

$$g(x) = \int_2^4 x^2 y\, dy$$

then direct calculations show that $g(x) = 6x^2$ and $g'(x) = 12x$. Consider, however, the function

$$g(x,y) = \int_2^4 \frac{e^{xy}}{y}\, dy$$

here, the evaluation of the integral is not easily effected, and therefore, the determination of g' by the above method would be exceedingly difficult.

In the following theorem we show that if in (1) the function f_x is continuous, then g is differentiable and g' can be found without first computing the value of the integral. A useful application of this result is found in connection with the brachistochrone problem discussed in Sec. 7.I.

(7.F.1)   *Theorem*   (*Leibniz Rule for Differentiation under the Integral Sign*)
Suppose that I is a closed and bounded interval and that $f: I \times [a,b] \to \mathbf{R}^1$. Let $g: I \to \mathbf{R}^1$ be defined by $g(x) = \int_a^b f(x,y)\, dy$. If f and f_x are continuous on $I \times [a,b]$, then $g'(x) = \int_a^b f_x(x,y)\, dy$.

Proof.   Let $\hat{x} \in I$. To show that

$$g'(\hat{x}) = \lim_{h \to 0} \frac{g(\hat{x} + h) - g(\hat{x})}{h} = \int_a^b f_x(\hat{x},y)\, dy$$

it is sufficient to prove that for each $\varepsilon > 0$, there is a $\delta > 0$ such that if $0 < |h| < \delta$, then

$$\left| \frac{g(\hat{x} + h) - g(\hat{x})}{h} - \int_a^b f_x(\hat{x},y)\, dy \right| < \varepsilon$$

Let $\varepsilon > 0$ be given. Since f_x is continuous, it is uniformly continuous on the compact set $I \times [a,b]$. Therefore, corresponding to $\varepsilon/(b - a)$, there is a $\delta > 0$ such that if $|h| < \delta$ and $y \in [a,b]$, then

$$|f_x(\hat{x} + h, y) - f_x(\hat{x},y)| < \frac{\varepsilon}{b - a}$$

Consequently, if $|h| < \delta$, we have

$$\int_{\hat{x}}^{\hat{x}+h} [f_x(z,y) - f_x(\hat{x},y)]\, dz < |h|\, \frac{\varepsilon}{b - a} \tag{2}$$

For fixed y, the function $\phi(x) = f(x,y) - xf_x(\hat{x},y)$ is a primitive of $f_x(x,y) - f_x(\hat{x},y)$; thus, by the fundamental theorem of calculus

$$\left| \int_{\hat{x}}^{\hat{x}+h} (f_x(z,y) - f_x(\hat{x},y))\, dz \right| = |\phi(\hat{x} + h) - \phi(\hat{x})|$$

$$= |f(\hat{x} + h, y) - f(\hat{x},y) - hf_x(\hat{x},y)|$$

It now follows from (2) that

$$|f(\hat{x} + h, y) - f(\hat{x},y) - hf_x(\hat{x},y)| < |h| \frac{\varepsilon}{b - a}$$

and therefore if $0 < |h| < \delta$, then

$$\left|\frac{f(\hat{x} + h, y) - f(\hat{x},y)}{h} - f_x(\hat{x},y)\right| < \frac{\varepsilon}{b - a}$$

for each $y \in [a,b]$. By the definition of g and elementary properties of integration, we have

$$\left|\frac{g(\hat{x} + h) - g(\hat{x})}{h} - \int_a^b f_x(\hat{x},y)\, dy\right| = \left|\int_a^b \left(\frac{f(\hat{x} + h, y) - f(\hat{x},y)}{h} - f_x(\hat{x},y)\right) dy\right|$$

$$\leq \int_a^b \left|\frac{f(\hat{x} + h, y) - f(\hat{x},y)}{h} - f_x(\hat{x},y)\right| dy < \frac{\varepsilon}{b - a}(b - a) = \varepsilon$$

the desired inequality.

(7.F.2) *Example* Suppose that

$$g(x) = \int_1^4 \frac{e^{xy}}{y}\, dy$$

Then $f_x(x,y) = e^{xy}$, and hence, $g'(x) = (1/x)(e^{4x} - e^x)$.

G.* IMPLICIT FUNCTION THEOREMS

It can be shown (either with the aid of algebraic geometry theory or a computer) that the graph G of the equation

$$F(x,y) = 2x^4 - 3x^2y + y^2 - 2y^3 + y^4 = 0 \tag{1}$$

has the form shown in Fig. 7.5.

Although the graph G is not that of a function, it is nevertheless geometrically plausible that there are functions ϕ_1, ϕ_2, and ϕ_3 with domains (r_1,r_2), $(\hat{r}_1,\hat{r}_2)$, and (r_1,r_2), respectively, whose graphs coincide with the portions of G indicated in Fig. 7.6.

In other words, equation (1) would appear to define implicitly a number of functions: implicit in the sense that even though it may be clear that such a function ϕ is "embedded" in G, it could well be extremely difficult to obtain a specific or explicit expression for ϕ such as, say, $\phi(x) = x^3 - 2x$.

Observe that corresponding to the point (x^*,y^*) there is no function ϕ defined on any open interval containing x^* whose graph would reside in the given graph G. Hence, it becomes natural to inquire if given a point $(x_0,y_0) \in$

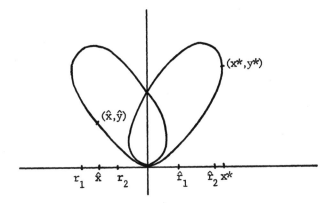

Figure 7.5

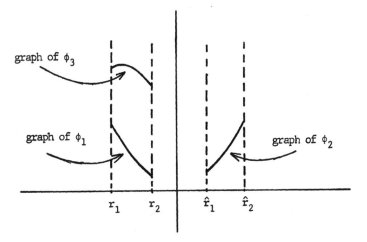

Figure 7.6

G, is there a way to determine whether or not there exists an open interval I containing x_0 and a function ϕ with domain I such that $\phi(x_0) = y_0$ and $(x,\phi(x)) \in G$ for each $x \in I$ [note that for any such function ϕ, $F(x,\phi(x)) = F(x_0,\phi(x_0))$ for each $x \in I$]. If we observe that the principal difference between the points (x^*,y^*) and (x_0,y_0) is that the tangent line to the graph at the former point has vertical slope, i.e., $F_y(x,y) = 0$, then the following result is not surprising.

(7.G.1) Theorem Suppose that $F(x,y)$ is a continuous function defined on an open set $D \subset \mathbf{R}^2$ and suppose that F_y exists on D and is continuous at a point $(x_0,y_0) \in D$. If $F_y(x_0,y_0) \neq 0$ and $F(x_0,y_0) = 0$, then there is an open interval I containing x_0 and a unique continuous function $\phi: I \to \mathbf{R}^1$ such that

(i) $\phi(x_0) = y_0$

(ii) $F(x, \phi(x)) = 0$ for each $x \in I$

Although this theorem will not be frequently used in the sequel, the reader should be aware that it and especially its natural generalization (7.G.4) constitute two of the most powerful theoretical tools at the analyst's disposal. Before going into the proof of (7.G.1) we give some indication of the versatility of this result by employing it to resolve a standard problem in differential equations.

Any first-order differential equation $y' = g(t,y)$ can be expressed in the form

$$M(t,y) + N(t,y)y' = 0 \tag{2}$$

for suitably chosen functions M and N. For instance, $y' = t^2 y - \sin ty$ may be written as $(\sin ty)/t^2 y - 1 + (1/t^2 y)y' = 0$. Suppose that corresponding to equation (2) there is a function $F(t,y)$ with the properties that

(a) $F_t = M$ and $F_y = N$ [in this case (2) is said to be an *exact differential equation*].
(b) F satisfies the hypotheses of (7.G.1) on an open set $D \subset \mathbf{R}^2$.

Then if $(t_0,y_0) \in D$, it follows that there is an open interval I containing t_0 and a function $\phi: I \to \mathbf{R}^1$ such that $\phi(t_0) = y_0$ and $F(t,\phi(t)) = 0$ for each $t \in I$. Consequently, if we set $\psi(t) = F(t,\phi(t))$, we have by (7.E.6)

$$\psi'(t) = F_t(t, \phi(t)) + F_y(t, \phi(t))\phi'(t) = M(t, \phi(t)) + N(t, \phi(t))\phi'(t)$$

On the other hand, since ψ is constant on I, it follows that $\psi'(t) = 0$ for each $t \in I$. Thus, we have $M(t,\phi(t)) + N(t,\phi(t))\phi'(t) = 0$ for each $t \in I$, and hence, ϕ is the unique solution of (2) on I such that $\phi(t_0) = y_0$. Observe that ϕ is implicitly defined by the function ψ. Necessary and sufficient conditions for the exactness of a differential equation are given in the problem set and a method for finding the function F is also discussed there.

We now turn to the proof of (7.G.1).

Proof of (7.G.1).   The proof is based to some extent on a variation of Newton's method (see problem G.10). Let $g: D \to \mathbf{R}^1$ be defined by

$$g(x,y) = y - \frac{F(x,y)}{F_y(x_0,y_0)}$$

Let $R = \{(x,y) \mid x_0 - \delta \le x \le x_0 + \delta, \ y_0 - \varepsilon \le y \le y_0 + \varepsilon\}$ be a rectangular neighborhood about (x_0,y_0) such that

$$\frac{F(x,y)}{F_y(x_0,y_0)} < \frac{\varepsilon}{2} \qquad \text{for each } x \in [x_0 - \delta, x_0 + \delta] \tag{3}$$

and

$$|g_y(x, y)| = \left|1 - \frac{F_y(x,y)}{F_y(x_0,y_0)}\right| < \frac{1}{2} \qquad \text{for each } (x,y) \in R \tag{4}$$

Note that it follows from the continuity of F and F_y that such a neighborhood exists.

Let $\mathscr{F} = \{f : [x_0 - \delta, x_0 + \delta] \to \mathbf{R}^1 \mid f \text{ is continuous}\}$ and assign $\mathscr{F}$ the sup metric $\hat{d}$ (3.B.4.h). Let $\beta : [x_0 - \delta, x_0 + \delta] \to \mathbf{R}^1$ be the constant function $\beta(x) \equiv y_0$ and let $\mathscr{M} = \{f \in \mathscr{F} \mid f(x_0) = y_0 \text{ and } \hat{d}(f,\beta) \le \varepsilon\}$. Finally, define $T : \mathscr{M} \to \mathscr{M}$ as follows: if $f \in \mathscr{M}$, then $T(f)$ is the function defined by

$$T(f)(x) = g(x, f(x))$$

[We must eventually show, of course, that $T(f) \in \mathscr{M}$.]

Our strategy is now the following. By (4.C.7), $(\mathscr{F},\hat{d})$ is a complete metric space, and it is easy to see that $\mathscr{M}$ is a closed subset of $\mathscr{F}$—see problem G.9—and thus, by (4.C.6), $\mathscr{M}$ is also complete. We show that T is a contraction map (4.E.8), and therefore by (4.E.10), T will have a fixed point ϕ. Thus, we shall have $T(\phi) = \phi$, which implies that for each $x \in [x_0 - \delta, x_0 + \delta]$

$$\phi(x) = T(\phi)(x) = g(x, \phi(x))$$

$$= \phi(x) - \frac{F(x, \phi(x))}{F_y(x_0,y_0)}$$

and consequently, $F(x,\phi(x)) = 0$ for each $x \in [x_0 - \delta, x_0 + \delta]$. Moreover, since $\phi \in \mathscr{M}$, it follows that $\phi(x_0) = y_0$, and this will complete the proof.

Now the details. First we establish that T is a contraction map. Suppose that $f_1, f_2 \in \mathscr{M}$ and let $x \in [x_0 - \delta, x_0 + \delta]$. Then, by (4) and the mean value theorem,

$$|T(f_1)(x) - T(f_2)(x)| = |g(x, f_1(x)) - g(x, f_2(x))|$$

$$= |g_y(x,y^*)(f_2(x) - f_1(x))| \le \tfrac{1}{2}|f_2(x) - f_1(x)|$$

(where y^* lies between $f_1(x)$ and $f_2(x)$); this implies that $\hat{d}(T_1(f_1),T(f_2)) \le \tfrac{1}{2}\hat{d}(f_1,f_2)$, and hence, that T is a contraction map.

To complete the proof, we show that the range of T is a subset of $\mathscr{M}$. This follows easily since if $f \in \mathscr{M}$, then $T(f)$ is clearly continuous, and furthermore,

$$\hat{d}(T(f), \beta) \le \hat{d}(T(f), T(\beta)) + \hat{d}(T(\beta), \beta)$$

$$\le \frac{1}{2}\hat{d}(f,\beta) + \hat{d}(T(\beta), \beta) < \frac{\varepsilon}{2} + \frac{\varepsilon}{2}$$

That $\hat{d}(T(\beta),\beta) < (\varepsilon/2)$ follows from (3) since

$$\tilde{d}(T(\beta), \beta) = \sup \{|T(\beta)(x) - \beta(x)| \, |x \in [x_0 - \delta, x_0 + \delta]\}$$
$$= \sup \{|g(x, \beta(x)) - \beta(x)| \, |x \in [x_0 - \delta, x_0 + \delta]\}$$
$$= \sup \left\{\left|\beta(x) - \frac{F(x, \beta(x))}{F_y(x_0, y_0)} - \beta(x)\right| x \in [x_0 - \delta, x_0 + \delta]\right\} < \varepsilon/2$$

(7.G.2) *Observation* If in the previous theorem F is continuously differentiable on D, then it can be shown that ϕ is also continuously differentiable (see Apostol, 1974). Furthermore, the roles of x and y in (7.G.1) may clearly be reversed to obtain the following result. Suppose that $F(x,y)$ is a continuously differentiable function on an open set $D \subset \mathbf{R}^2$ and suppose that $(x_0, y_0) \in D$, $F_x(x_0, y_0) \neq 0$ and $F(x_0, y_0) = 0$. Then there is an open interval I containing y_0 and a unique continuously differentiable function $\psi: I \to \mathbf{R}^1$ such that $\psi(y_0) = x_0$ and $F(\psi(y), y) = 0$ for each $y \in I$.

Theorem (7.G.1) (and its proof) can be generalized to functions mapping subsets of $\mathbf{R}^m$ into $\mathbf{R}^n$. Another proof often used in demonstrating this result is based on an equally important result, the inverse function theorem, which is stated in (7.G.3). Given a function $f: A \to \mathbf{R}^n$, where $A \subset \mathbf{R}^n$, and a point $\mathbf{w} \in \text{int } A$ it is frequently of interest to determine whether f has an inverse (at least in a small neighborhood of $f(\mathbf{w})$); furthermore, if such an inverse exists, one would like to know if it is continuous and/or differentiable. The *Jacobian of f* at $\mathbf{w}$ (the determinant of the matrix belonging to $d_{\mathbf{w}}f$) plays the critical role in this connection. The reader should pay particular attention to the rather amazing interplay that occurs in the statement of the inverse function theorem as we obtain an analytic result (involving the existence and properties of an inverse function) from a purely algebraic construct (the Jacobian).

(7.G.3) *Theorem* (*Inverse Function Theorem*) Suppose that A is an open subset of $\mathbf{R}^n$ and that $f: A \to \mathbf{R}^n$ is continuously differentiable on A. If the Jacobian of f at a point $\mathbf{w} \in A$ is nonzero, then there is an open subset U of A containing $\mathbf{w}$ such that

(i) $f(U)$ is open in $\mathbf{R}^n$.
(ii) $f|_U$ is $1 - 1$ (and hence $(f|_U)^{-1}: V \to U$ exists, where $V = f(U)$).
(iii) $(f|_U)^{-1}$ is continuously differentiable.

A proof of the preceding theorem as well as of the following theorem may be found in Apostol (1974).

(7.G.4) *Theorem* (*Implicit Function Theorem*) Suppose that $U \subset \mathbf{R}^{m+n}$ and that $f = (f_1, f_2, \ldots, f_n)$ is a continuously differentiable function mapping U in $\mathbf{R}^n$. If $f(\hat{\mathbf{x}}) = \mathbf{0}$ for some point $\hat{\mathbf{x}} = (\hat{x}_1, \hat{x}_2, \ldots, \hat{x}_m, \hat{x}_{m+1}, \ldots, \hat{x}_{m+n})$ and if

$$\det \begin{pmatrix} D_{m+1}f_1(\hat{x}) & D_{m+2}f_1(\hat{x}) & \cdots & D_{m+n}f_1(\hat{x}) \\ \cdots\cdots\cdots\cdots\cdots\cdots\cdots\cdots\cdots\cdots \\ D_{m+1}f_n(\hat{x}) & D_{m+2}f_n(\hat{x}) & \cdots & D_{m+n}f_n(\hat{x}) \end{pmatrix} \neq 0$$

then there is an open set $V \subset \mathbf{R}^m$ containing $(\hat{x}_1, \hat{x}_2, \ldots, \hat{x}_m)$ and a unique function $\phi \colon V \to \mathbf{R}^n$ such that

(i) $\phi(\hat{x}_1, \hat{x}_2, \ldots, \hat{x}_m) = (\hat{x}_{m+1}, \hat{x}_{m+2}, \ldots, \hat{x}_{m+n})$.

(ii) $f(\mathbf{t}, \phi(\mathbf{t})) = 0$ for each $\mathbf{t} = (t_1, t_2, \ldots, t_m) \in V$, where $(\mathbf{t}, \phi(\mathbf{t})) = (t_1, t_2, \ldots, t_m, \phi_1(\mathbf{t}), \phi_2(\mathbf{t}), \ldots, \phi_n(\mathbf{t}))$ and $\phi = (\phi_1, \phi_2, \ldots, \phi_n)$.

(iii) ϕ is continuously differentiable on V.

(7.G.5) *Exercise* Show that (7.G.1) is a special case of (7.G.4).

H.* AN APPLICATION OF THE IMPLICIT FUNCTION THEOREM: LAGRANGE MULTIPLIERS

Frequently, max-min–type problems include certain constraints on the function. For example, a physicist may wish to determine the maximum velocity of a particle in 3-space where the movement of the particle is restricted to a predetermined surface. As a simple illustration of this general type of problem, we see how one might determine the rectangle of largest area that can be inscribed in the ellipse $x^2/4 + y^2/9 = 1$ in such a way that the sides of the rectangle are parallel to the coordinate axes (Fig. 7.7).

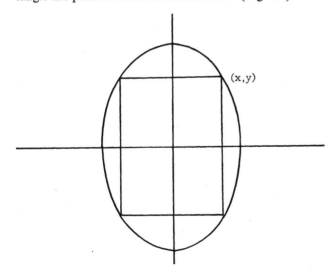

Figure 7.7

The area of the rectangle indicated in Fig. 7.7 is $2x2y$. Hence, our goal is to maximize the function $f(x,y) = 4xy$ subject to the constraint $x^2/4 + y^2/9 - 1 = 0$. This example is typical of a wide range of problems that can be resolved with the aid of Lagrange multipliers. The pertinent theorem is the following.

(7.H.1) *Theorem* Suppose that f and g are continuously differentiable functions on a set $A \subset \mathbf{R}^2$ and that f attains a local maximum (or a local minimum) at a point $(x_0,y_0) \in$ int A. If $g(x_0,y_0) = 0$ and either $g_x(x_0,y_0)$ or $g_y(x_0,y_0)$ is not equal to 0, then there exists a number λ, called a *Lagrange multiplier*, such that

$$(f_x(x_0,y_0), f_y(x_0,y_0)) = \lambda(g_x(x_0,y_0), g_y(x_0,y_0)) \tag{1}$$

Note that if gradient notation is used, then (1) can be expressed more succinctly by $\nabla f(\mathbf{w}) = \lambda \nabla g(\mathbf{w})$, where $\mathbf{w} = (x_0,y_0)$.

Before proving (7.H.1), we see how it may be applied to the problem involving the inscribed rectangle. Here $f(x,y) = 4xy$ and $g(x,y) = x^2/4 + y^2/9 - 1$. We have that $\nabla f(x,y) = (4y,4x)$ and $\nabla g(x,y) = (x/2, 2y/9)$. To find λ, we note that from the conclusion of (7.H.1) it must be the case that

$$4y = \lambda \frac{x}{2} \tag{2}$$

$$4x = \lambda \frac{2y}{9} \tag{3}$$

Multiplication of (2) by $2y$ and (3) by x yields

$$8y^2 = \lambda xy$$
$$18x^2 = \lambda xy$$

and hence, $y^2 = (9/4)x^2$.

If we substitute $y^2 = (9/4)x^2$ in the constraining condition $x^2/4 + y^2/9 - 1 = 0$, we obtain

$$\frac{x^2}{4} + \frac{x^2}{4} - 1 = 0$$

and therefore, $x = \sqrt{2}$, $y = 3/\sqrt{2}$, and the maximum area is 12.

The implicit function theorem (7.G.1) provides the essential ingredient in the proof of (7.H.1).

Proof of (7.H.1).   By hypothesis, either $g_x(x_0,y_0) \neq 0$ or $g_y(x_0,y_0) \neq 0$. Suppose that $g_y(x_0,y_0) \neq 0$ (if this is not the case, then the proof can be modified with the aid of (7.G.2)). Since $(x_0,y_0) \in$ int A, it follows from (7.G.1) that there is an open inverval I containing x_0 and a unique continuously differentiable function $\phi: I \to \mathbf{R}^1$ such that

$$\phi(x_0) = y_0 \tag{4}$$

$$g(x, \phi(x)) = g(x_0, y_0) = 0 \qquad \text{for each } x \in I. \tag{5}$$

Let $\psi: I \to \mathbf{R}^1$ be defined by $\psi(x) = g(x, \phi(x))$. Then from (5) we have $\psi(x) = 0$, and by (7.E.6) we have

$$0 = \psi'(x_0) = g_x(x_0, \phi(x_0)) + g_y(x_0, \phi(x_0))\phi'(x_0) \tag{6}$$

Let $q: I \to \mathbf{R}^1$ be defined by $q(x) = f(x, \phi(x))$. Then by hypothesis q attains a local maximum (or minimum) at x_0, and hence, by (5.B.2),

$$0 = q'(x_0) = f_x(x_0, \phi(x_0)) + f_y(x_0, \phi(x_0))\phi'(x_0) \tag{7}$$

Equations (6) and (7) may be rewritten in vector form as

$$\nabla f(x_0, y_0) \cdot (1, \phi'(x_0)) = 0$$
$$\nabla g(x_0, y_0) \cdot (1, \phi'(x_0)) = 0$$

where $\cdot$ indicates the inner product. Since two vectors $\mathbf{v}$ and $\mathbf{w}$ in $\mathbf{R}^2$ are perpendicular if and only if $\mathbf{v} \cdot \mathbf{w} = 0$, we have that $\nabla f(x_0, y_0)$ and $\nabla g(x_0, y_0)$ are perpendicular to a common vector and, hence, are collinear. Thus, since $\nabla g(x_0, y_0) \neq 0$, there is a constant λ such that $\nabla f(x_0, y_0) = \lambda \nabla g(x_0, y_0)$, which concludes the proof.

It should be observed that (7.H.1.) is only a necessary condition for the existence of extrema. Furthermore, this theorem is of little use in determining whether the extremum attained by f is a maximum or a minimum. There are some quite sophisticated results that are useful in this regard, but in most applications it is clear from the context of the problem whether the extremum is a maximum or a minimum. For instance, in the problem involving the inscribed rectangle it must be the case that at $x = \sqrt{2}$, $y = 3/\sqrt{2}$, a local maximum (not a local minimum) area is obtained.

Theorem (7.H.1) may be generalized in a number of directions. We mention (without proof) two such possibilities.

(7.H.2) Theorem Suppose that f and g are continuously differentiable real-valued functions on a set $A \subset \mathbf{R}^n$ and that f attains a maximum (or a minimum) at a point $\mathbf{w} \in \text{int } A$ such that

(i) $g(\mathbf{w}) = 0$

(ii) $\nabla g(\mathbf{w}) \neq 0$

Then there is a number λ such that $\nabla f(\mathbf{w}) = \lambda \nabla g(\mathbf{w})$.

(7.H.3) Theorem Suppose that $A \subset \mathbf{R}^n$ and $f: A \to \mathbf{R}^1$ is a continuously differentiable function that attains a local maximum (or minimum) at a point $\mathbf{w} \in \text{int } A$. Suppose that $g: A \to \mathbf{R}^m$ $(m < n)$ is continuously differentiable. If

$g(\mathbf{w}) = \mathbf{0}$ and if the vectors $\nabla g_1(\mathbf{w}), \ldots, \nabla g_m(\mathbf{w})$ [where $g = (g_1, \ldots, g_m)$] are linearly independent, then there are real numbers $\lambda_1, \ldots, \lambda_m$, called *Lagrange multipliers*, such that

$$\nabla f(\mathbf{w}) = \lambda_1 \nabla g_1(\mathbf{w}) + \cdots + \lambda_m \nabla g_m(\mathbf{w})$$

(7.H.4) *Example* In this example we use Lagrange multipliers to resolve the following problem: Given the ellipse $x^2/4 + y^2/9 = 1$ and the line $y = -2x + 10$, find points (x,y) on the ellipse and (w,z) on the line whose distance is minimal. The square of the distance between a point (x,y) on the ellipse and a point (w,z) on the line is given by $f(x,y,w,z) = (x - w)^2 + (y - z)^2$. Thus, we wish to minimize f subject to the constraints:

$$g_1(x,y,w,z) = \frac{x^2}{4} + \frac{y^2}{9} - 1 = 0$$

$$g_2(x,y,w,z) = z + 2w - 10 = 0$$

Consequently, we have that f maps a subset of $\mathbf{R}^4$ into $\mathbf{R}^1$ and $g = (g_1, g_2)$ maps $\mathbf{R}^4$ into $\mathbf{R}^2$. Routine calculations show that

$$\nabla g_1(x,y,w,z) = \left(\frac{x}{2}, \frac{2y}{9}, 0, 0\right)$$

$$\nabla g_2(x,y,w,z) = (0,0,2,1)$$

$$\nabla f(x,y,w,z) = (2(x - w), 2(y - z), -2(x - w), -2(y - z))$$

If we set $\nabla f(x,y,w,z) = \lambda_1 \nabla g_1(x,y,w,z) + \lambda_2 \nabla g_2(x,y,w,z)$, we obtain the equations

$$2(x - w) = \lambda_1 \frac{x}{2} \tag{8}$$

$$2(y - z) = \lambda_1 \frac{2y}{9} \tag{9}$$

$$-2(x - w) = 2\lambda_2 \tag{10}$$

$$-2(y - z) = \lambda_2 \tag{11}$$

These four equations together with the constraints

$$\frac{x^2}{4} + \frac{y^2}{9} = 1 \tag{12}$$

$$z + 2w = 10 \tag{13}$$

lead us to the following conclusions:

(a) Multiplication of (11) by 2 and equating with (10) shows that $(x - w) = 2(y - z)$.

(b)   From (8), (a), and (9) we see that

$$x = \frac{4(x - w)}{\lambda_1} = \frac{8(y - z)}{\lambda_1} = \frac{8\lambda_1 y/9}{\lambda_1} = \frac{8y}{9}$$

(c)   From (12) and (b) we have that $x^2/4 + 9x^2/64 = 1$, and hence, $x = \frac{8}{5}$ and $y = \frac{9}{5}$.

(d)   Since by (a), $x - 2y = w - 2z$, it follows that $w = 2z + \frac{8}{5} - \frac{18}{5} = 2z - 2$, and this equality together with (13) permits calculation of w and z: $w = \frac{18}{5}, z = \frac{14}{5}$.

I.* THE BRACHISTOCHRONE PROBLEM

This section is devoted to finding a solution to the classical brachistochrone problem. Not only is this problem of considerable interest in itself (and of historical significance as well), but its solution also provides the reader with a detailed look at how many of the ideas introduced in this chapter can be used in an applied context. The brachistochrone problem (*brachistos*, Greek for "shortest"; *chronos*, Greek for "time") may be stated informally as follows: given points P_0 and P_1 in the plane (Fig. 7.8), along which path connecting these two points will a small bead starting at P_0 arrive at P_1 in the least amount of time (Fig. 7.9)?

In more formal terms, the problem is to find a continuously differentiable

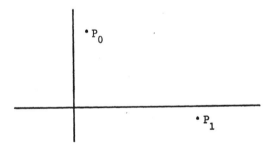

Figure 7.8

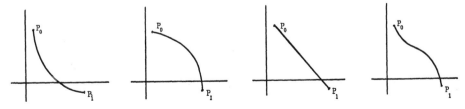

Figure 7.9

function whose graph passes through the points P_0 and P_1 and for which a freely moving (frictionless) object sliding (not rolling) along the graph from P_0 to P_1 will cover this distance in minimum time. Gravity is the only force that acts on this body.

In 1630 Galileo formulated this problem and then proceeded to give an incorrect solution to it (his solution was an arc of a circle). In 1696 J. Bernoulli challenged the "acutest mathematicians of the world" to resolve this and another somewhat less intriguing problem. Many great mathematicians worked for a number of months on these problems without success. However, Isaac Newton (1643–1727) when informed of the challenge is said to have resolved both of the problems in one evening's work. Newton's solution to the brachistochrone problem was based on the study of optics (see Simmons, 1972, for details). We shall pursue another tack by drawing on some elementary techniques from the calculus of variations, techniques which were not available to Newton (in fact, the brachistochrone problem represents the starting point for the development of this important area of mathematics).

For purposes of simplification, we shall assume that P_0 is at the origin, P_1 lies down and to the right and that the positive y axis is directed downward (Fig. 7.10). For an object sliding down the curve, the principle of the conservation of energy states that the change in the potential energy of the object is equal to the negative of the change in kinetic energy of the object. Since the potential energy of a particle of mass m at the point (x, y) is given by $-mgy$ (the negative sign is due to the fact that the positive y axis points downward), it follows that

$$mgy(t) = \tfrac{1}{2}m(v(t))^2 - \tfrac{1}{2}m(v(0))^2$$

where m is the mass of the object, g is the gravitational constant, $y(t)$ is the

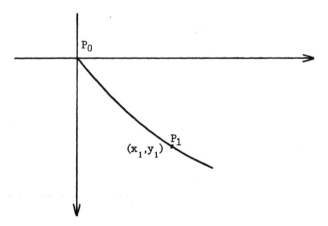

Figure 7.10

"height" of the object at time t, and $v(t)$ is the velocity of the object at time t. Thus, we have

$$v(t) = \sqrt{2g(y(t) + \alpha)} \tag{1}$$

where $\alpha = (v(0))^2/2g$.

We now introduce the following functions. Let $\phi: [0,x_1] \to \mathbf{R}^1$ $[y = \phi(x)]$ be the function on whose graph G the object is moving. Let $\psi: [0,T] \to G$ $(\psi(t) = (x(t),y(t)))$ be the function that gives the position on G of the object at time t (Fig. 7.11).

1. We assume that ϕ is a continuously differentiable function such that $(0,\phi(0)) = P_0$ and $(x_1,\phi(x_1)) = P_1$.
2. We assume that ψ is continuously differentiable, $\psi'(t) \neq 0$ on $(0,T)$ and $\psi(0) = P_0$, $\psi(T) = P_1$.

Let $\sigma: [0,T] \to [0,\infty)$ be defined by $\sigma(t) = \int_0^t \sqrt{(x'(u))^2 + (y'(u))^2}\, du$. This function gives the distance traveled on G by the object in the time t [see (6.D.2)]. In view of assumption (2), σ is a strictly increasing continuously differentiable function on $[0,T]$ such that $\sigma'(t) > 0$ on $(0,T)$. Hence, σ has a continuously differentiable inverse defined on $[0,L]$, where L is the length of the graph of G. We denote this inverse by $\tau: [0,L] \to [0,T]$; note that $\tau(s)$ gives the time required for the object to move along G a distance s from its starting position P_0.

Next we define the function $u: [0,x_1] \to [0,L]$ by

$$u(x) = \int_0^x \sqrt{1 + (\phi'(w))^2}\, dw \tag{2}$$

$u(x)$ represents the distance traveled by the object as a function of the x

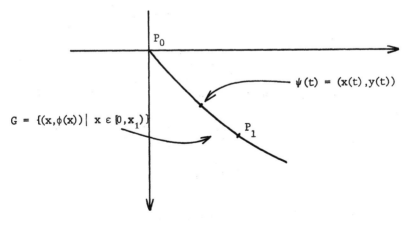

Figure 7.11

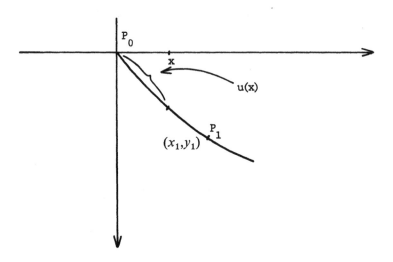

Figure 7.12

coordinate of the object's position on G (Fig. 7.12). Finally, let $t(x) = \tau \circ u(x)$. Clearly, $t(x)$ is the time required for the object to arrive at a point on the graph G whose first coordinate is x. Since both τ and u are continuously differentiable on $(0, x_1)$, so is the function t and hence, the time required for the object to move from P_0 to P_1 is $\int_0^{x_1} t'(x)\, dx$. Thus, we need to find a useful description of $t'(x)$. To this end first note that

$$t'(x) = (\tau \circ u)'(x) = \tau'(u(x))u'(x) \tag{3}$$

Since $u'(x) = \sqrt{1 + (\phi'(x))^2}$ (why?) and $\sigma(t(x)) = u(x)$, we have

$$\tau'(u(x)) = \tau'(\sigma(t(x))) = \frac{1}{\sigma'(t(x))}$$

and, consequently,

$$t'(x) = \frac{\sqrt{1 + (\phi'(x))^2}}{\sigma'(t(x))} \tag{4}$$

Since $\sigma'(t(x))$ represents the speed of the object at time $t(x)$, it follows from (1) that

$$\sigma'(t(x)) = v(t(x)) = \sqrt{2g}\sqrt{y(t(x)) + \alpha} = \sqrt{2g}\sqrt{\phi(x) + \alpha} \tag{5}$$

where

$$\alpha = \frac{(v(0))^2}{2g} \tag{6}$$

To see that the last equality in (5) is valid, note that $y(t(x))$ gives the y coordinate of the object on G at time $t(x)$, and since $t(x)$ is the time required for the

object to arrive at the location on G whose first coordinate is x, it follows that
$y(t(x)) = \phi(x)$.

From (4) and (5) we obtain

$$t'(x) = \sqrt{\frac{1 + (\phi'(x))^2}{2g(\phi(x) + \alpha)}}$$

and hence, the time T required for the object to traverse G is given by

$$\frac{1}{\sqrt{2g}} \int_0^{x_1} \sqrt{\frac{1 + (\phi'(x))^2}{\phi(x) + \alpha}} \, dx = t(x_1) - t(0) = T$$

If we set $y = \phi(x)$, then the brachistochrone problem may be formulated
as follows: Given

$$F(x,y,y') = \sqrt{\frac{1 + (y')^2}{y + \alpha}}$$

[where $\alpha = (v(0))^2/(2g)$ and $v(0)$ is the initial velocity at $P_0 = (0,0)$], does
there exist among all the continuously differentiable functions y satisfying the
boundary conditions $(0,y(0)) = (0,0)$ and $(x_1,y(x_1)) = (x_1,y_1) = P_1$ a function y_* that minimizes the value of $J(y) = \int_0^{x_1} F(x,y(x),y'(x)) \, dx$?

We proceed by assuming that such a minimum function exists and then
showing that it satisfies a differential equation (whose solution we can find).

It is helpful at this stage to reformulate the problem as follows.

Problem (*) Suppose that $F(x,y,z)$ and its first and second partial derivatives
are continuous in an open connected set $A \subset \mathbf{R}^3$ and that $y_*: [0,x_1] \to \mathbf{R}^1$
is a function which satisfies the following conditions:

(i) y_*'' is continuous.
(ii) $y_*(0) = 0$, $y_*(x_1) = y_1$.
(iii) $(x,y_*(x),y_*'(x)) \in A$ for all $x \in [0,x_1]$.
(iv) The value

$$J(y_*) = \int_0^{x_1} F(x, y_*(x), y_*'(x)) \, dx \qquad (7)$$

is a local minimum for all functions $y: [0,x_1] \to \mathbf{R}^1$ (near y_*) such that $J(y)$ is
defined and for which the conditions (i) and (ii) are satisfied. We are to find the
differential equation that y_* must satisfy.

Let $\mathscr{A}$ denote the set of all functions $y: [0,x_1] \to \mathbf{R}^1$ that satisfy (i) and (ii)
in (*) and for which $J(y)$ is defined. To say that $J(y_*)$ is a local minimal value
means that there is a δ_* neighborhood of y_* (relative to $\mathscr{A}$) such that if y
belongs to this neighborhood then $J(y) \geq J(y_*)$.

Geometrically, we can envision this δ_*-neighborhood as consisting of all those functions $y \in \mathscr{A}$ whose graphs lie within the δ_*-strip centered about the graph of y_* (Fig. 7.13).

Let $\eta: [0,x_1] \to \mathbf{R}^1$ be any function such that η' is continuous and $\eta(0) = 0 = \eta(x_1)$. For each real number ε, let $y_\varepsilon = y_* + \varepsilon\eta$, and note that y_ε satisfies conditions (i) and (ii) of (*). Since η and η' are bounded on the compact interval $[0,x_1]$, there is a constant K_η (which depends on η) such that if $|\varepsilon| \leq K_\eta$, then y_ε satisfies condition (iii) of (*), (i.e., $(x,y_\varepsilon(x),y_\varepsilon'(x)) \in A$ for each $x \in [0,x_1]$), and hence, $J(y_\varepsilon)$ is computable. We shall assume that K_η is so small that if $|\varepsilon| \leq K_\eta$ then the graph of y_ε belongs to the δ-strip about y_*. It follows that

$$J(y_\varepsilon) \geq J(y_*) \qquad\qquad (8)$$

for each ε such that $|\varepsilon| \leq K$.

By (7.F.1) we have that if $|\varepsilon| \leq K_\eta$, then

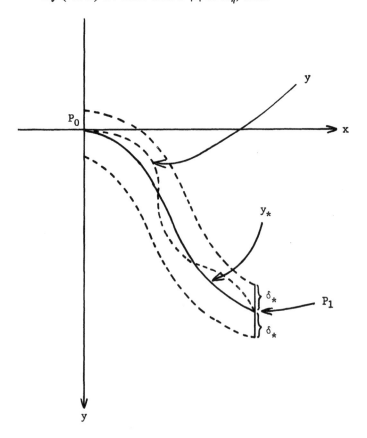

Figure 7.13

$$\frac{d}{d\varepsilon} J(y_\varepsilon) = \int_0^{x_1} \frac{d}{d\varepsilon} F(x, y_\varepsilon(x), y_\varepsilon'(x)) \, dx$$

$$= \int_0^{x_1} (F_y(x, y_\varepsilon(x), y_\varepsilon'(x))\eta(x) + F_{y'}(x, y_\varepsilon(x), y_\varepsilon'(x))\eta'(x)) \, dx$$

Integration by parts of the second term of the integrand yields (recall that $\eta(0) = 0 = \eta(x_1)$):

$$\frac{d}{d\varepsilon} J(y_\varepsilon) = F_{y'}\eta \Big|_0^{x_1} + \int_0^{x_1} \left[F_y(x, y_\varepsilon, y_\varepsilon') - \frac{d}{dx}(F_{y'}(x, y_\varepsilon, y_\varepsilon')) \right]\eta(x) \, dx$$

$$= \int_0^{x_1} \left[F_y(x, y_\varepsilon(x), y_\varepsilon'(x)) - \frac{d}{dx}(F_{y'}(x, y_\varepsilon(x), y_\varepsilon'(x))) \right]\eta(x) \, dx \quad (9)$$

Since $J(y_\varepsilon)$ is a differentiable function that maps the interval $-K_\eta < \varepsilon < K_\eta$ into $\mathbf{R}^1$ and since $J(y_0) = J(y_*)$ is a minimal value, it follows from elementary calculus that

$$0 = J'(y_0) = \int_0^{x_1} \left[F_y(x, y_*(x), y_*'(x)) - \frac{d}{dx}(F_{y'}(x, y_*(x), y_*'(x))) \right]\eta(x) \, dx \quad (10)$$

Since the expression in brackets in the integrand of (10) is continuous on $[0,x_1]$ and since η is an arbitrary function satisfying conditions (i) and (ii) of (*), it follows from (6.B.4), that on $[0,x_1]$, y_* satisfies the differential equation

$$F_y(x, y_*(x), y_*'(x)) - \frac{d}{dx}(F_{y'}(x, y_*(x), y_*'(x))) = 0 \quad (11)$$

This equation is called the *Euler-Lagrange differential equation*, and it must be satisfied by the minimal solution that we are seeking.

In the case of the brachistochrone problem, we have

$$F(x,y,y') = \sqrt{\frac{1 + (y')^2}{y + \alpha}}$$

thus, for this problem F depends on x only insofar as y and y' are functions of x, and therefore, we may regard F as a function of the variables y and y'. We exploit this fact as follows. By the chain rule and (11), we have

$$\frac{d}{dx}[F(y(x), y'(x)) - y'(x)F_{y'}(y(x), y'(x))]$$

$$= F_y(y(x), y'(x))y'(x) + F_{y'}(y(x), y'(x))y''(x) - y''(x)F_{y'}(y(x), y'(x))$$

$$- y'(x)\frac{d}{dx}(F_{y'}(y(x), y'(x)))$$

$$= y'(x)\left[F_y(y(x), y'(x)) - \frac{d}{dx}(F_{y'}(y(x), y'(x))) \right] = 0$$

and consequently, there is a constant k such that

$$F(y_*(x), y'_*(x)) - y'_*(x)F_{y'}(y_*(x), y'_*(x)) = k \qquad x \in [0, x_1] \qquad (12)$$

Since

$$F_{y'}(y,y') = \frac{y'}{\sqrt{y + \alpha}\sqrt{1 + (y')^2}}$$

it follows from (12) that

$$\frac{\sqrt{1 + (y'_*(x))^2}}{\sqrt{y_*(x) + \alpha}} - y'_*(x)\frac{y'_*(x)}{\sqrt{y_*(x) + \alpha}\sqrt{1 + (y'_*(x))^2}} = k \qquad (13)$$

and hence,

$$k = \frac{1}{\sqrt{y_*(x) + \alpha}\sqrt{1 + (y'_*(x))^2}}$$

If we set $k = 1/\sqrt{2}c$, then

$$(y_*(x) + \alpha)(1 + (y'_*(x))^2) = 2c^2$$

from which it follows that

$$y'_*(x) = \sqrt{\frac{2c^2 - (y_*(x) + \alpha)}{y_*(x) + \alpha}} \qquad (14)$$

Thus, we have reduced our original problem to that of solving the differential equation (14).

If there is a soultion y_* to this differential equation, then the derivative y'_* must be real valued and nonnegative, and this implies that y_* is an increasing function on the interval $[0, x_1]$. If we rule out the possibility that y_* is constant on some subinterval of $[0, x_1]$ (which is reasonable in the context of the brachistochrone problem), then the solution y_* must be strictly increasing on $[0, x_1]$ and, hence, has an inverse x_*. It follows from (5.A.16) and (14) that

$$x'_*(y) = \sqrt{\frac{y + \alpha}{2c^2 - (y + \alpha)}} \qquad (15)$$

To determine x_* (and to express our final answer in parametric form), we integrate (15) with the aid of the substitution $y(\theta) = c^2 \sin^2(\theta/2)$, $0 \le \theta$, and find that

$$x_*(y(\theta)) = c^2(\theta - \sin \theta)$$

If we set $\hat{x}_*(\theta) = x_*(y(\theta))$ and $\hat{y}_*(\theta) = y(\theta)$, then the solution of the brachistochrone problem is given (parametrically) by

$$\hat{x}_*(\theta) = c^2(\theta - \sin \theta)$$
$$\hat{y}_*(\theta) = -\alpha + c^2(1 - \cos \theta)$$

(see problem I.6). This solution can be shown to be unique for each pair of points $P_0 = (0,0)$ and $P_1 = (x_1, y_1)$—see Apostol (1974). If we let α tend to 0, i.e., if we let the initial velocity of the object approach zero, we see that the solution to the brachistochrone problem with initial velocity zero is given by

$$\hat{x}_*(\theta) = c^2(\theta - \sin \theta)$$
$$\hat{y}_*(\theta) = c^2(1 - \cos \theta)$$

where $c^2 = y_1/(1 - \cos \theta_1)$ $[P_1 = (x_1, y_1)$ and $y_*(\theta_1) = y_1]$.

The graph of this solution is a cycloid (Fig. 7.14); geometrically, this path is generated by the motion of a fixed point on a wheel with radius c^2 which rolls (without slipping) underneath the x axis.

In addition to representing the solution of the brachistochrone problem, the cycloid has a number of amazing properties, among them the fact that any two objects placed at distinct points on the cycloid and released simultaneously will arrive at the bottom of the trajectory at precisely the same instant.

J. THE VIBRATING STRING

We conclude this chapter with another brief departure from the realm of abstraction in order to illustrate one specific context in which partial derivatives arise naturally in a "real" (albeit somewhat idealized) setting.

Typical examples of equations involving partial derivatives that have important physical implications include

$u_{xx} = a^2 u_{tt}$ (the *wave equation*—used in describing the motion of a vibrating string)   (1)

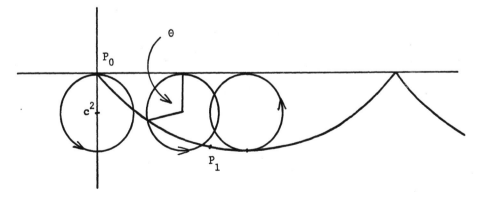

Figure 7.14

$u_{xx} = a^2 u_t$ (the *heat quation*—used in describing certain heat
flows)   (2)

$u_{xx} + u_{yy} = 0$ (the *Laplace equation*—used in describing equilibrium
displacements of a membrane fixed along its boundary)

(3)

The remainder of this section is devoted to deriving equation (1).

Suppose that a string (say a violin string) lying on a frictionless table (the *xy* plane) is stretched taut between two points (0,0) and (*L*,0) with tension T (Fig. 7.15). The string is displaced slightly in a direction parallel to the *y* axis. We wish to describe mathematically the vibratory motion of the string by finding a function that will allow us to locate a given point on the string at time *t*.

We idealize the situation by insisting that all motion occurs in a direction parallel to the *y* axis; hence, during the course of the vibratory motion, there will be changes only in the second coordinate of a given point *P* of the string. Consequently we can describe the motion of the string by a function $u(x,t)$, where $u(x,t)$ gives at time *t* the *y* coordinate of a point with first coordinate *x*. For different fixed values of *t*, typical graphs of the position of the string are indicated in Fig. 7.16.

We shall assume that the string is perfectly flexible; thus, the tensile forces acting at a point *P* will be directed along the line tangent to the curve at *P*. In essence, we are saying that there are no shearing effects (Fig. 7.17).

To begin, we consider an extremely small segment *S* of the string with endpoints P_1 and P_2. Since the string is assumed to be perfectly flexible, the tensile forces F_1 and F_2 exerted on *S* by the remaining portion of the string will act away from *S* (and, of course, along the respective tangent lines) (Fig. 7.18).

Our basic strategy is as follows. Since we have decreed that there is to be no motion other than that parallel to the *y* axis, it follows that the horizontal components of the force vectors F_1 and F_2 must cancel each other out, leaving only vertical components to consider. If we assume that the mass of the string

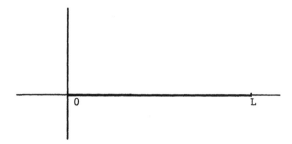

Figure 7.15

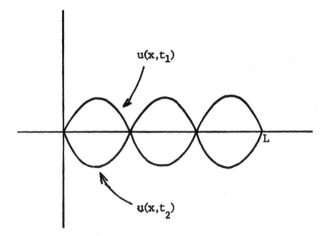

Figure 7.16

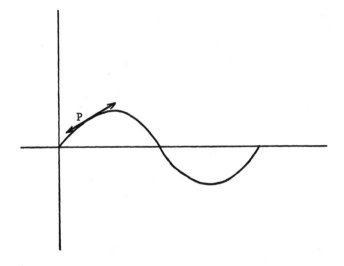

Figure 7.17

segment S is concentrated at the center of the segment i.e., that the string is homogeneous, then we can use Newton's second law of motion $F = ma$ to equate the product of this mass and its acceleration to the resultant vertical force due to F_1 and F_2. From this equation we shall eventually be able to extract a partial differential equation (an equation in which partial derivatives are present) involving $u(x,t)$.

To simplify matters we make three additional assumptions:

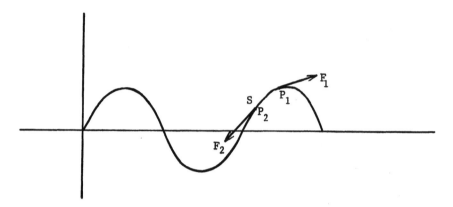

Figure 7.18

1. The density of the string is constant (and given by ρ).
2. The string is displaced very slightly so that the horizontal component of the original tension T can be assumed to remain constant throughout the period of vibration (obviously, for large displacements this assumption would be quite suspect).
3. The initial energy imparted to the string is transmitted throughout the string in such a way that only the vertical component of the tensile force is modified. Although this is not actually possible, it represents a reasonable approximation of reality provided that the string is quite flexible and the original vertical displacement is slight.

We first examine the horizontal components of the forces acting at the endpoints of the small segment S. For each point $P = (x,t)$, we let the vector $\phi(x,t)$ denote the tensile force at P caused by the portion of the string to the right of P. Thus, the forces applied to S (and acting away from the segment) are given by $\phi(x + \Delta x, t)$ at P_1 and $-\phi(x,t)$ at P_2 (Fig. 7.19). The horizontal components of these forces are easily seen to be $\|\phi(x + \Delta x, t)\| \cos \theta_1$ and $-\|\phi(x,t)\| \cos \theta_2$, where the angles θ_1 and θ_2 are measured from the indicated horizontal segment and may take on values between $-\pi/2$ and $\pi/2$. Since by assumption 2 the horizontal components are assumed to annihilate each other (only vertical motion is permitted), we have

$$\|\phi(x,t)\| \cos \theta_2 = \|\phi(x + \Delta x, t)\| \cos \theta_1 = T \qquad (4)$$

The resultant force on S is derived from summing the vertical components of the tensile force at x and $x + \Delta x$; thus we have

$$\|\phi(x + \Delta x, t)\| \sin \theta_1 - \|\phi(x,t)\| \sin \theta_2 = F \qquad (5)$$

The minus sign results from the fact that the external tensile force acting on S

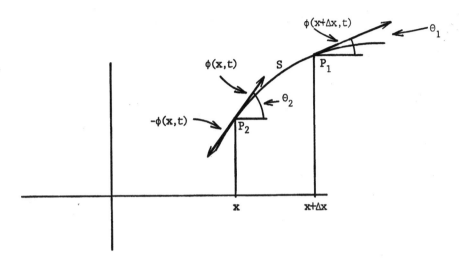

Figure 7.19

at x is directed away from S. It follows from (4) that $\|\phi(x,t)\| = T/\cos\theta_2$ and $\|\phi(x + \Delta x, t)\| = T/\cos\theta_1$, and therefore substitution in (5) yields

$$F = T\left(\frac{\sin\theta_1}{\cos\theta_1} - \frac{\sin\theta_2}{\cos\theta_2}\right) = T(\tan\theta_1 - \tan\theta_2) = T(u_x(x + \Delta x,t) - u_x(x,t))$$

The last equality holds, since for a given value of t, $\tan\theta_1$ and $\tan\theta_2$ give the slopes of the lines tangent to the curve $u(x,t)$ at the points $(x + \Delta x, t)$ and (x,t), respectively, and the partial derivative of u with respect to x at these points yields precisely these values.

Now we make use of Newton's second law of motion, $F = ma$. Since the density of the string is ρ, it follows that the mass of S is given by $\rho\,\Delta x = \int_x^{x+\Delta x} \rho\,dx$ (the mass of the string segment S does not change since there is no horizontal motion). Let $\hat{x} = x + \Delta x/2$ be the center of mass of the segment S lying between x and $x + \Delta x$. Then the acceleration at this point is given by the second derivative with respect to time, i.e., $u_{tt}(\hat{x}, t)$. Consequently, we have

$$T(u_x(x + \Delta x, t) - u_x(x,t)) = \rho\,\Delta x\, u_{tt}(\hat{x},t)$$

Division of both sides of this equation by $T\,\Delta x$ yields

$$\frac{u_x(x + \Delta x, t) - u_x(x,t)}{\Delta x} = \frac{\rho}{T} u_{tt}(\hat{x},t)$$

As Δx tends to 0, we have

$$\lim_{\Delta x \to 0} \frac{u_x(x + \Delta x, t) - u_x(x,t)}{\Delta x} = \lim_{\Delta x \to 0} \frac{\rho}{T} u_{tt}(\hat{x},t) \tag{6}$$

The limit of the left-hand side of equation (6) is $u_{xx}(x,t)$. If u_{tt} is continuous, then the limit of the right-hand side is $(\rho/T)\,u_{tt}(x,t)$, and hence, we obtain the wave equation

$$u_{xx} = a^2 u_{tt} \tag{7}$$

where $\alpha = \sqrt{\rho/T}$.

What have we accomplished? At first glance, it might appear that very little has been gained since even after a series of questionable mathematical manipulations and shaky assumptions, the best we could do was obtain a local relationship between certain partial derivatives of the desired solution $u(x,t)$. In the next chapter, however, we shall see that all this travail was not for nought as we employ infinite series of functions to find a global solution to the problem.

PROBLEMS

Section A

1. Find the directional derivatives of the following functions at the indicated points **w** and in the indicated directions **u**.
   (a) $f(x,y) = x^3 - xy$; $\mathbf{w} = (1,2)$, $\mathbf{u} = (1/\sqrt{5}, 2/\sqrt{5})$.
   (b) $f(x,y) = e^y \cos x$; $\mathbf{w} = (0,0)$, $\mathbf{u} = (-1/\sqrt{2}, -1/\sqrt{2})$.
2. If

$$f(x,y) = \begin{cases} \dfrac{xy}{x^2 + y^2} & \text{if } (x,y) \neq (0,0) \\ 0 & \text{if } (x,y) = (0,0) \end{cases}$$

   find $f_x(0,0)$ and $f_y(0,0)$.
3. Show that $\partial^3 f/\partial x \partial y \partial z = \partial^3 f/\partial y \partial x \partial z$ if $f(x,y,z) = x^2 y \sin yz - y^2 e^{zx}$.
4. Find the equations of the indicated tangent planes:
   (a) The tangent plane at the point $(0,1,1)$ to the graph of $f(x,y) = e^{x^2 y}$.
   (b) The tangent plane at the point $(-1,3,-7)$ to the graph of $f(x,y) = x^3 y^2 - 2x$.
5. Find $f_{xy}(0,1)$, $f_{yx}(0,1)$, $f_{xxy}(0,1)$, and $f_{yxy}(0,1)$ if $f(x,y) = x^2 y^3 + 2xy$.
6. Let $f(x,y) = \begin{cases} x^2 + y^2 & \text{if } x \text{ and } y \text{ are rational} \\ 0 & \text{otherwise.} \end{cases}$
   Do $f_x(0,0)$ and $f_y(0,0)$ exist? Is f continuous at $(0,0)$? What can be said about f_x, f_y and the continuity of f at a point $(a,b) \neq (0,0)$?
7. Find absolute and local maxima and minima of the following functions:
   (a) $f(x,y) = x^2 + y^2 + 14$; $x^2 + y^2 \leq 1$. (Check the boundary points independently.)
   (b) $f(x,y) = x^2 - x - 2y^2 + y + 6$; $0 \leq x \leq 1, 0 \leq y \leq 2$.
   (c) $f(x,y) = x^2 - y^2 + 2xy - 3$.

8. Suppose that a rectangular box is to be made from 25 square feet of material. If the box does not have a top, find the dimensions of the box that will have a maximum volume.

9.* Show that if f_x and f_y are differentiable at a point (x_0,y_0), then $f_{xy}(x_0,y_0) = f_{yx}(x_0,y_0)$.

10. Starting with the point $(1,1)$, use two applications of the numerical method described in this section to approximate a simultaneous solution of

$$2x^2 - 3xy + 1 = 0$$
$$x^3 - 2y^4 + 2 = 0$$

11.* Establish the following result: If all second partial derivatives of a function $f: U \to \mathbf{R}^1$ are continuous on the open set $U \subset \mathbf{R}^2$, and if
   (a) $f_x(a,b) = f_y(a,b) = 0$ at a point $(a,b) \in U$
   (b) $f_{xx}(a,b) > 0$
   (c) $f_{xx}(a,b)f_{yy}(a,b) - [f_{xy}(a,b)]^2 > 0$
   then the function f has a local minimum at the point (a,b); if in (b), $f_{xx}(a,b) < 0$, then f has a local maximum at the point (a,b). [*Hint*: Consider the function $g(t) = f(a + t \cos \theta, b + t \sin \theta)$ and use problem C.8, Chapter 5.]

12. Use problem A.11 to determine local maxima and local minima of the following functions:
   (a) $f(x,y) = 4 - x^2 - y^2$
   (b) $f(x,y) = 2y^2 - 3x^2 - 3xy + 9x$
   (c) $f(x,y) = 2xy - x^2 - 3y^2 + 8x - 2$
   (d) $f(x,y) = -2x^2 + x + y^2 - y - 5, 0 \le x \le 1, 0 \le y \le 1$

13. Show that the restriction of the function $f(x,y) = (y - x^2)(y - 2x^2)$ to any line passing through the origin has a local minimum at zero, but that the function itself does not have a local minimum at the origin.

14. Let $f(x,y) = x^2y^3$ and let $T(x,y)$ be the function defined by (4) of Sec. 7.A, where $(x_0,y_0) = (2,3)$. Find (and compare) $f(2.1,2.9)$ and $T(2.1,2.9)$. Use a calculator to compare $f(3.95,4.1)$ and $T(3.95,4.1)$ if $f(x,y) = \sqrt{x + 3y}$ and $(x_0,y_0) = (4,4)$. This exercise helps illustrate the fact that the functions defining tangent planes to graphs of functions can provide excellent local approximations.

15. Show that $D_\mathbf{u} f(\mathbf{w}) = -D_{-\mathbf{u}} f(\mathbf{w})$—see (7.A.3.a).

Section B

1. Find the derivatives of the following functions at the indicated points:
   (a) $f(x,y) = e^{x^2y}(x^2 + y)$; $\mathbf{w} = (0,2)$.
   (b) $f(x,y) = x \cos 2y - \ln xy$; $\mathbf{w} = (1,\pi)$.

2. Find $\nabla f(\mathbf{w})$ if
   (a) $f(x,y) = \sin(x\sqrt{y}) + e^{\cos x}$; $\mathbf{w} = (\pi, 4)$.
   (b) $f(\mathbf{w}) = \mathbf{w} \cdot \mathbf{w}$.
   (c) $f(\mathbf{w}) = 1/\|\mathbf{w}\|$.
   (d) $f(x,y,z) = x^2 \cos yz - e^{x^2 y}$; $\mathbf{w} = (1,0,\pi)$.

3. Suppose that $T: \mathbf{R}^2 \to \mathbf{R}^1$ is a linear function. Show that T is differentiable at each $\mathbf{w} \in \mathbf{R}^2$, and use the uniqueness of the differential to show that T is its own differential.

4. Suppose that $f: \mathbf{R}^2 \to \mathbf{R}^1$ and that C is a level curve for f on which f is identically equal to 0. Show that if f is continuously differentiable, then
   (a) ∇f lies along the line whose slope is f_y/f_x
   (b) the tangent line to the curve C at any point $\mathbf{w} \in C$ has slope $-f_x(\mathbf{w})/f_y(\mathbf{w})$.

5. Find the tangent plane to the surface defined by $x^3 \sin y - z^2 \cos 2y = 17$ at the point $(2, \pi/2, 3)$.

6. If $f(x,y) = \sin^2 xy - x^3 y^2$, find $d_\mathbf{w} f(\pi/4, 1)$, where $\mathbf{w} = (\pi/2, \pi/2)$.

7. Suppose that $f: \mathbf{R}^2 \to \mathbf{R}^1$ is defined by $f(x,y) = x^3 y^4$. Use the differential to approximate $f(1.02, 2.05)$.

8. Suppose that the temperature at a point (x,y) in a disk of radius 2 and centered at the origin is given by $T(x,y) = xy^2$. Find the direction for which the temperature is increasing most rapidly at the point $(1,1)$.

Section C*

1. Show that if $f(z) = f(x + iy) = x^2 - y^2 + 2ixy$, then the Cauchy-Riemann equations are satisfied at each point $z = x + iy$. Find $f'(z)$.

2. Use the Cauchy-Riemann equations to determine if the following functions are differentiable, and if so, find $f'(z)$:
   (a) $f(z) = f(x + iy) = e^{-x}(\cos y - i \sin y)$
   (b) $f(z) = z^2$
   (c) $f(z) = |z|^2$.
   (d) $f(z) = z^2 + 3z + i$

3. Show that if the functions f and g are differentiable at z, then
   (a) $(f + g)'(z) = f'(z) + g'(z)$
   (b) $(fg)'(z) = f'(z)g(z) + f(z)g'(z)$

4. If $z = x + iy$, then the *conjugate* of z is $x - iy$, and is denoted by $\bar{z}$. Show that if $f(z) = \bar{z}$, then $f'(z)$ does not exist for each $z \in \mathbf{C}$.

5. Show that if

$$f(z) = \begin{cases} \dfrac{z^5}{|z|^4} & \text{if } z \neq 0 \\ 0 & \text{if } z = 0 \end{cases}$$

then the Cauchy-Riemann equations are satisfied at the point $z = 0$, but $f'(0)$ does not exist. [*Hint*: Consider $f(z)/z$. You may assume that the quotient rule holds for functions of a complex variable.]

6. In Example (7.C.7) show that the derivative of f at the point $z = 0$ does not exist.

7. A function ϕ that has continuous second partial derivatives and satisfies the *Laplace equation* $\phi_{xx} + \phi_{yy} = 0$ is said to be a *harmonic function*.
   (a) Show that both the real part and the imaginary part of an analytic function (a function having derivatives of all orders) are harmonic functions.
   (b) Show that $u(x,y) = 3x - 2y + 7$ is harmonic. For what values of a, b, and c is $u(x,y) = ax^2 + bxy + cy^2$ harmonic?

8. If a function $f = u + iv$ is differentiable (and hence, analytic), then the functions u and v are said to be *conjugate harmonic functions*. Show that the function $u(x,y) = e^x \cos y$ is harmonic, and use the Cauchy-Riemann equations to find its harmonic conjugate, i.e., find $v(x,y)$ such such that $f = u + iv$ is differentiable.

Section D

1. If $f: \mathbf{R}^4 \to \mathbf{R}^2$ is defined by $f(x_1,x_2,x_3,x_4) = (x_1 x_2^2 x_3, 4x_1 x_3 + 3x_2 x_4)$, find $D_2 f(1,-1,2,1)$.

2. If $f: \mathbf{R}^4 \to \mathbf{R}^1$ is defined by $f(x_1,x_2,x_3,x_4) = x_1 x_2^3 + (\sin x_3 x_4)/x_2 - e^{x_1 x_2^3}$, find $D_2 f(-\pi,2,0,0)$ and $D_3 f(1,1,2,0)$.

3. If $f: \mathbf{R}^3 \to \mathbf{R}^1$ is defined by $(x_1^2 x_3)^3 \ln(x_2 x_3) + 6x_3^2$, find $D_2 f(1,2,1)$ and $D_3 f(2,1,1)$.

4. If f is defined as in problem D.1 and $\mathbf{w} = (1,1,2,0)$, find $d_{\mathbf{w}} f(3,0,-2,1)$.

5. If f is defined as in problem D.2 and $\mathbf{w} = (0,2,1)$, find $d_{\mathbf{w}} f(2,0,-1)$.

6. (a) If f is defined as problem D.1, find $f'(0,1,3,2)$.
   (b) If f is defined as in problem D.3, find $f'(3,1,1)$.

7. Is

$$f(x,y) = \begin{cases} \dfrac{x^3 - y^3}{x^2 + y^2} & \text{if } (x,y) \neq (0,0) \\ 0 & \text{if } (0,0) \end{cases}$$

differentiable at $(0,0)$? at $(1,0)$? Does $f'(0,0)$ exist?

Section E

1. If $f: \mathbf{R}^1 \to \mathbf{R}^2$ and $g: \mathbf{R}^2 \to \mathbf{R}^1$ are defined by $f(x) = (x^3, 2x)$ and $g(w,z) = 5wz + e^{w^2 z}$, and if $h = g \circ f$, find $h'(1)$.

2. If $f: \mathbf{R}^2 \to \mathbf{R}^1$ and $g: \mathbf{R}^1 \to \mathbf{R}^1$ are defined by $f(x,y) = 5x^2 + e^{xy}$ and $g(x) = x^4$, and if $\psi: \mathbf{R}^1 \to \mathbf{R}^1$ is defined by $\psi(x) = f(x,g(x))$, find $\psi'(2)$.

3. If $f: \mathbf{R}^3 \to \mathbf{R}^2$ and $g: \mathbf{R}^2 \to \mathbf{R}^4$ are defined by $f(x,y,z) = (x^2z,xy)$ and $g(u,v) = (u^2v,4,u^3v^2,2)$, and if $h = g \circ f$, find
    (a) $d_\mathbf{x}h(1,4,3)$, where $\mathbf{x} = (-1,1,2)$.
    (b) $h'(0,1,2)$.

4. If $f: \mathbf{R}^1 \to \mathbf{R}^3$ and $g: \mathbf{R}^3 \to \mathbf{R}^1$ are defined by $f(x) = (2x, x^2, x/(x^2 + 1))$ and $g(u,v,w) = uv^2w^3$, and if $h = g \circ f$, find $h'(2)$.

5. (a) If $f(x,y) = e^{x+y} - x^2$, show that f is differentiable at $(1,3)$.
    (b) If $g(x,y) = \ln xy + x^2y^3$, show that g is differentiable at $(1,1)$.

6. Use (7.E.7) to show that if $D \subset \mathbf{R}^2$ is an open set with the property that each pair of points p and q in D can be connected by a polygonal path

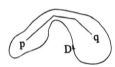

and if $f: D \to \mathbf{R}^1$ is differentiable on D and f_x and f_y are identically 0 on D, then f is a constant function.

7.* Find an open set $D \subset \mathbf{R}^2$ and a function $f: D \to \mathbf{R}^2$ such that $f_x = f_y \equiv 0$ on D but such that f is not constant.

8. A function $f: \mathbf{R}^2 \to \mathbf{R}^1$ is said to be *homogeneous of order p* if for each $(x,y) \in \mathbf{R}^2$, $f(tx,ty) = t^p f(x,y)$. Show that if f is homogeneous of order p and $(\hat{x},\hat{y}) \in \mathbf{R}^2$, then $\hat{x}f_x(\hat{x},\hat{y}) + \hat{y}f_y(\hat{x},\hat{y}) = pf(\hat{x},\hat{y})$. [*Hint*: Consider $g(t) = f(t_x,t_y)$ and compute $g'(1)$.] Verify this result if $f(x,y) = xy/(x^2 + y^2)$.

9. Use the chain rule to investigate the relationship between the differential and derivative of a function f and the differential and derivative of its inverse (assuming, of course, that f has an inverse).

10. Formulate and establish an extension of (7.E.7) to $\mathbf{R}^n$.

Section F

1. Find $g'(x)$ if

    (a) $g(x) = \displaystyle\int_0^1 \frac{\sin xy}{y}\, dy$

    (b) $g(x) = \displaystyle\int_0^1 \frac{e^{-xy^3}}{y}\, dy$

2. Show that if $g(x) = \int_{u(x)}^{v(x)} f(x,y)\, dy$, then

$$g'(x) = \int_{u(x)}^{v(x)} f_x(x,y) \, dy - f(x, u(x))u'(x) + f(x, v(x))v'(x)$$

(Assume that all the appropriate derivatives exist.) [*Hint*: Set $\phi(u,v,x) = \int_u^v f(x,y) \, dy$ and use the chain rule to show that
$g'(x) = \phi_u(u(x),v(x),x)u'(x) + \phi_v(u(x),v(x),x))v'(x) + \phi_x(u(x),v(x),x))$
and then calculate ϕ_u, ϕ_v, and ϕ_x.]

3. Use problem F.2 to find $g'(x)$ if $g(x)$ is equal to

(a) $\int_1^x \sin(x^2 y) \, dy$

(b) $\int_{x^2}^{x^4} \frac{\sin xy}{y}$

(c) $\int_{2x}^{x^3} (3y - y^3) \, dy$

4. (a) Suppose that $f(t) + g(y)y' = 0$, where y is a function of t. Interpret and justify the following equations [$\int k(t) \, dt$ represents any primitive of k].

$$\int [f(t) + g(y(t))y'(t)] \, dt = c$$

$$\int f(t) \, dt + \int g(y) \, dy = c$$

(b) Use part (a) to solve the following differential equations.

(i) $y' = \dfrac{t + 2}{y - 3}$

(ii) $y' = t^2 + \dfrac{1}{t}$

(iii) $y + ty' = 0$

(iv) $y' + ty + y = t + 1$

5. Show that the substitution $\hat{y}_*(\theta) + \alpha = 2c \sin^2(\theta/2)$ into (15) yields the solution (16).

6. Verify the solution given in (17).

7. Show that the cycloid given by (17) is generated by the motion of a fixed point on a circle of radius c^2 that rolls on the positive side of the line $y = \alpha$.

8. Given points $P_1 = (x_1, y_1)$ and $P_2 = (x_2, y_2)$ in the plane, where $x_1 < x_2$ and y_1 and y_2 are positive, find the differentiable function $f: [x_1, x_2] \to \mathbf{R}^1$ with the property that the surface area generated by rotating the

graph of f about the x axis is minimum [recall that the surface area is given by $A = 2\pi \int_{x_1}^{x_2} |f(x)| \sqrt{1 + (f'(x))^2}\, dx$].

Section G

1. Suppose that $xy^2 + ye^{x+y} = 0$. Can x be expressed in terms of y in a neighborhood of $(0,0)$? $(-1,1)$?

2. Can (7.G.4) be used to determine if the following equations can be solved (theoretically) for z?
   (a) $e^{xz} \sin xy + x^2 \cos xz = 0$ in a neighborhood of the point $(0,0,0)$.
   (b) $x^2 \cos xy + \sin xz = 0$ in a neighborhood of the point $(0,0,0)$.

3. Use the inverse function theorem to determine that $f(x) = x^3 + 4x - 2$ has an inverse in a neighborhood of the point $x = 2$.

4. Does $f(x,y) = (x^2 - y^2, 2xy)$ have an inverse in a neighborhood of $(0,0)$?

5. Show that if $F(x,y,z) = 0$ can be solved in terms of each of x and y and z, then

$$\frac{\partial z}{\partial x}\frac{\partial x}{\partial y}\frac{\partial y}{\partial z} = 0$$

[Hint: Consider $F(x,y,\varphi(x,y))$, $F(x,\psi(x,z),z)$ and $F(\eta(y,z),y,z)$ and compute appropriate partial derivatives of F.]

6. Show that if

$$M(t,y) + N(t,y)y' = 0 \qquad\qquad (*)$$

and if M, N, M_y, and N_t are continuous at $M_y = N_t$, then $(*)$ is an exact equation. [Hint: Set $\phi(t,y) = \int_0^t M(t,y)\, dt + h(y)$. Find ϕ_y and set $\phi_y = N(t,y)$ to solve for $h'(y)$. Show that $h'(y)$ is indeed a function of y only and that ϕ is the desired function for exactness.]

7. Determine whether the following equations are exact, and if exact use your proof of problem G.8 to find a solution (which may be implicit).
   (a) $(y^2 - 2ty + 6t) - (t^2 - 2ty + 2)y' = 0$
   (b) $y' + -\dfrac{3ty}{t^2 + y^2}$
   (c) $(te^{ty} + 1)y' + ye^{ty} = 0$

8. Show that the equations $x^3 - xyu + yv^4 = 0$, and $u^2 - 2v^2 + x^2y = 0$ can be solved for u and v in terms of x and y near the point $(u,v,x,y) = (1,-1,0,2)$.

9. Show the subset $\mathcal{M}$ defined in the proof of (7.G.1) is a closed subset of $\mathcal{F}$.

10.* Examine carefully the function g defined in the proof of (7.G.1), and explain how the ideas behind Newton's method are used here.

11. Let $F(x,y) = x - y^3$. Show that $F_x(0,0) = 0$ but that nevertheless corresponding to the point $(x_0,y_0) = (0,0)$ there is a function φ satisfying (i) and (ii) of (7.G.1). Thus, the obvious converse of the implicit function theorem does not hold.

Section H*

1. Use Lagrange multipliers to determine the dimensions of a box with maximal volume that is inscribed in the ellipsoid $x^2/4 + y^2/8 + z^2/16 = 1$ (assume that the edges are parallel to the axes).
2. For fixed numbers $a_1, a_2, \ldots, a_n$ maximize $\sum_{i=1}^{n} x_i^2$ subject to the constraint that $\sum_{i=1}^{n} a_i x_i = 1$.
3. Use Lagrange multipliers to determine the dimensions of a closed box with volume equal to 125 units and with minimum surface area.
4. Find the minimum distance between the origin and a point on the surface $z = x^2/2 + y^2/2 - 5$.
5. Suppose that we are to place a particle of mass m in a box with volume 1000. It can be shown that the ground-state energy of the particle is $(h^2/8m)(1/a^2 + 1/b^2 + 1/c^2)$, where h is Planck's constant and a, b, and c are the lengths of the sides of the box. Find the dimensions of the box which will minimize the ground-state energy.
6. Find the sides of a triangle with fixed perimeter and maximum area.
7. Find the distance between the parabola $y = 2 + x^2$ and the line $y = -x - 2$.
8. Use Lagrange multipliers to find the minimum of $f(x,y,z) = x^2 + y^2 + z^2$, subject to the constraint $x + 3y - 2z = 12$.

Section I

1. (a) Show that $u(x,t) = \cos 4x \sin 4t$ is a solution of $u_{xx} - u_{tt} = 0$.
   (b) Show that $u(x,t) = t^{-\frac{1}{2}} e^{-x^2/4t}$ is a solution of $u_{xx} - u_t = 0$.
2. A partial differential equation of the form
   $$a_1 u_{xx} + a_2 u_{xt} + a_3 u_{tt} + a_4 u_x + a_5 u_t + a_6 u = a_7 \qquad (*)$$
   where each a_i is a function of t and y is said to be:

   1. *Elliptic* in a region $D \subset \mathbf{R}^2$ if $a_2^2 - 4a_1 a_3 < 0$.
   2. *Hyperbolic* in a region $D \subset \mathbf{R}^2$ if $a_2^2 - 4a_1 a_3 > 0$.
   3. *Parabolic* in a region $D \subset \mathbf{R}^2$ if $a_2^2 - 4a_1 a_3 = 0$.

   Classify the following partial differential equations in these terms:
   (a) The wave equation
   (b) The heat equation

(c) Laplace equation

(d) $u_{xx} - 2u_{xt} + 3u_{tt} = 2u$

(e) $u_{xx} - u_{tt} - u_t = 0$

3. Show that the equation

$$u_{xx} - 2xu_{xt} + tu_{tt} - u = 0$$

is elliptic on one side of the parabola $t = x^2$ and hyperbolic on the other side.

4. If in problem I.2, $a_7 = 0$ and u_1 and u_2 are solutions of (*), show that $u = c_1u_1 + c_2u_2$ is also a solution (c_1 and c_2 are constants).

5. Show that if gravity is taken into account in deriving the wave equation (assume that the wire is stretched between points 0 and π), then the wave equation takes on the form $u_{tt} = a^2u_{xx} - g$, where g is the gravitational constant.

REFERENCES

Apostol, T. M. (1974): *Mathematical Analysis*, Addison-Wesley, Reading, Mass.

Bliss, G. (1944): *Calculus of Variations*, Math. Assoc. of Amer.

Fadell, A. G. (1973): *Amer. Math. Monthly*, **80**: 1134.

Simmons, G. F. (1972): *Differential Equations*, McGraw-Hill, New York.

8

SEQUENCES AND SERIES

A. THE VIBRATING STRING (CONTINUED)

In the previous chapter we found that the wave equation associated with the vibrating string problem is given by

$$u_{tt}(x,t) = a^2 u_{xx}(x,t) \qquad 0 < x < L, t > 0 \tag{1}$$

In addition, we have that the following conditions are satisfied:

$$u(0,t) = u(L,t) = 0 \qquad \text{if } t \geq 0 \tag{2}$$

$$u(x,0) = f(x) \qquad \text{if } 0 \leq x \leq L \tag{3}$$

$$u_t(x,0) = 0 \qquad \text{if } 0 \leq x \leq L \tag{4}$$

[$f(x)$ represents the initial position of the string at the instant it is released, and equation (4) follows from the fact that the initial velocity imparted to the string is 0].

With all of this information, how do we find u? Certainly a solution u to (1) that satisfies conditions (2), (3), and (4) must be a function of two variables x and t. Although at this point there is no reason to believe that the variables x and t are not thoroughly intermeshed, e.g., $u(x,t) = [(\sin xt^3)e^{tx}]/[\ln (\cos xt)]$, we might nevertheless, for purposes of simplification, first make the assumption that these variables are separated from one another, i.e., we assume that $u(x,t)$ can be written as a product $A(x)B(t)$. If we substitute $u(x,t) = A(x)B(t)$ in (1), we find that $A(x)B''(t) = a^2A''(x)B(t)$, or equivalently (provided $A(x) \neq 0$ and $B(t) \neq 0$),

$$\frac{A''(x)}{A(x)} = \frac{B''(t)}{a^2B(t)} \tag{5}$$

Equation (5) is unusual in that it is valid for *every* value of x in $(0, L)$ and *every* value of $t > 0$. Although this is not an impossibility, it does imply that both sides of the equation must be constant, say equal to λ. Hence, we obtain two homogeneous ordinary differential equations

$$A''(x) - \lambda A(x) = 0 \tag{6}$$

$$B''(t) - \lambda a^2 B(t) = 0 \tag{7}$$

each of which is readily solvable. The procedure we have just initiated is known as the *method of separation of variables*. Since we have no information regarding λ, we must now make a case by case analysis.

Case 1. $\lambda = 0$. The general solution to (6) is given by $A(x) = mx + b$. Since $0 = u(0,t) = A(0)B(t)$ for all t, it must be the case that $A(0) = 0$, and hence $b = 0$ [if $B(t) = 0$ for each t, then u would be identically 0, which is a mathematically viable possibility, but physically of no interest]. Similarly, since $u(L,t) = A(L)B(t) = 0$, it follows that $A(L) = 0$ and, thus, $m = 0$. Finally, however, these results ($m = 0$, $b = 0$) lead us to conclude that $u(x,t) = A(x)B(t)$ is identically equal to 0; thus, we reject the possibility that $\lambda = 0$.

Case 2. $\lambda > 0$. For convenience, we set $\lambda = c^2$, where $c > 0$. We are to find the general solution to $A''(x) - c^2A(x) = 0$. From (2.D.15), we have that such a solution is of the form $c_1e^{cx} + c_2e^{-cx}$. Since $A(0) = 0$, it follows that $c_1 + c_2 = 0$, or $c_2 = -c_1$. Consequently, $A(x) = c_1(e^{cx} - e^{-cx})$; as in case 1, we may assume that $A(L) = 0$ (otherwise $u \equiv 0$), and it then follows from the equation $A(L) = c_1(e^{cL} - e^{-cL})$ that $c_1 = 0$. This, however, implies that $A(x)$ (and, hence u) is identically 0. Again we reject this possibility for physical reasons.

Case 3. $\lambda < 0$. It is convenient to write $\lambda = -c^2$, where $c > 0$. We then have that $A''(x) + c^2A(x) = 0$, and by (2.D.15) the general solution can be written in the form $A(x) = c_1 \cos cx + c_2 \sin cx$. We use condition (2) again to conclude that $A(0) = 0$ and, hence, $c_1 = 0$. Therefore, $A(x) = c_2 \sin cx$, and in particular, $0 = A(L) = c_2 \sin cL$. If $c_2 = 0$, then once again we obtain the physically uninteresting trivial solution $A(x) = 0$. But this time we have another possibility: $\sin cL = 0$. This, of course, occurs only if cL is an integer multiple of π, i.e., $c = n\pi/L$ (furthermore, since $c > 0$, n must be a positive integer). Thus for each $n \in \mathbf{Z}^+$, there is a solution $A_n(x) = \beta_n \sin (n\pi x/L)$ of (6).

With this information in hand, we turn our attention to the second ordinary differential equation, $B''(t) - \lambda a^2 B(t) = 0$. Substitution of the acceptable values of c found in the above work yields

$$B''(t) + \frac{n^2 \pi^2 a^2}{L^2} B(t) = 0$$

It is an easy matter to verify that the general solution to this equation is given by

$$B(t) = b_1 \cos \frac{n\pi a t}{L} + b_2 \sin \frac{n\pi a t}{L}.$$

Since $u_t(x,0) = 0$, it follows that $A(x)B'(0) = 0$ for all x, which implies that $B'(0) = 0$. Consequently, $b_2 = 0$, and therefore, for each $n \in \mathbf{Z}^+$, we obtain a solution of (7), $B_n(t) = \gamma_n \cos(n\pi a t/L)$. Combining this result with our work in deriving A_n, we see that for each $n \in \mathbf{Z}^+$, the function $u_n(x,t) = A_n(x)B_n(t) = c_n \sin(n\pi x/L) \cos(n\pi a t/L)$ is a solution of

$$u_{tt}(x,t) = a^2 u_{xx}(x,t)$$

and satisfies the boundary conditions

$$u_n(0,t) = u_n(L,t) = 0 \qquad \text{for each } t > 0$$

and
$$(u_n)_t(x,0) = 0$$

But does $u_n(x,0)$ satisfy initial condition (3): $u_n(x,0) = f(x)$? It should be apparent that except for some very special instances, this will not be the case. Thus there is still considerable work to be done.

In 1807, a meeting of the French Academy of Science was the scene of one of the most electrifying (and controversial) moments in mathematical history: the presentation of a paper by J. Fourier in which he asserted that any function could be described as a (possibly infinite) sum of sine and cosine functions. It was soon demonstrated that Fourier's claim was too all encompassing to be totally accurate, but nevertheless he was close enough to the truth that a vast area of mathematics has evolved from his germinal observation. Giving Fourier for the moment the benefit of the doubt, we might now postulate that functions such as $f(x) = x^2$, or, in general, any "reasonable" function can be expressed in the form

$$a_0 + (a_1 \sin ax + b_1 \cos ax) + (a_2 \sin 2ax + b_2 \cos 2ax)$$
$$+ (a_3 \sin 3ax + b_3 \cos 3ax) + \cdots.$$

In our current problem we have

$$u_n(x,0) = c_n \sin \frac{n\pi x}{L} \cos \frac{n\pi a 0}{L} = c_n \sin \frac{n\pi x}{L}$$

and therefore, in view of Fourier's claim, we might conjecture that (i) $f(x)$ is equal to $c_1 \sin(\pi x/L) + c_2 \sin(2\pi x/L) + c_3 \sin(3\pi x/L) + \cdots$ for suitably chosen coefficients c_n and (ii) $u(x,t) = \sum_{n=1}^{\infty} u_n(x,t)$. These conjectures, however, are of dubious value unless

1. We can determine the coefficients c_n.
2. We can make some mathematical sense of these infinite sums.

Casting logical order aside for the moment, we first see how each c_n can be found. We begin by observing that if $v_n(x) = \sin(n\pi x/L)$, for each $n \in \mathbf{Z}^+$, then for positive integers m and n,

$$v_m'' = -\left(\frac{\pi}{L}\right)^2 m^2 v_m \tag{8}$$

and

$$v_n'' = -\left(\frac{\pi}{L}\right)^2 n^2 v_n. \tag{9}$$

If we multiply both sides of equation (8) by v_n and both sides of equation (9) by v_m and then subtract, we obtain

$$(v_n v_m'' - v_m v_n'') = \left(\frac{\pi}{L}\right)^2 (n^2 - m^2) v_m v_n$$

or, equivalently,

$$(v_n v_m' - v_m v_n')' = \left(\frac{\pi}{L}\right)^2 (n^2 - m^2) v_m v_n.$$

Therefore, integration from 0 to L yields

$$\int_0^L \left(\frac{\pi}{L}\right)^2 (n^2 - m^2) v_m(x) v_n(x)\, dx = (v_n(x) v_m'(x) - v_m(x) v_n'(x)) \Big|_0^L$$

$$= \left(\sin\left(\frac{n\pi L}{L}\right)\right)\left(\frac{m\pi}{L} \cos\left(\frac{m\pi L}{L}\right)\right)$$

$$- \left(\sin\left(\frac{m\pi L}{L}\right)\right)\left(\frac{n\pi}{L} \cos\left(\frac{n\pi L}{L}\right)\right) = 0$$

Thus, we have

$$\int_0^L \sin\left(\frac{n\pi x}{L}\right) \sin\left(\frac{m\pi x}{L}\right) dx = 0$$

whenever $m \neq n$. Now, if we throw all mathematical inhibitions to the winds, and "multiply" both sides of the "equation"

$$f(x) = c_1 \sin\frac{\pi x}{L} + c_2 \sin\frac{2\pi x}{L} + c_3 \sin\frac{3\pi x}{L} + \cdots \tag{10}$$

by $\sin(n\pi x/L)$ and then integrate both sides, we obtain

$$\int_0^L f(x) \sin\frac{n\pi x}{L}\, dx = \int_0^L c_n \sin^2\frac{n\pi x}{L}\, dx$$

Application of standard integration techniques to the right-hand side of the previous equation yields

$$\int_0^L f(x) \sin \frac{n\pi x}{L} \, dx = c_n \frac{L}{2} \tag{11}$$

and hence,

$$c_n = \frac{2}{L} f(x) \sin \frac{n\pi x}{L} \, dx.$$

(8.A.1) *Exercises* (a) Verify equation (11).

(b) Show that if it is assumed that derivatives and partial derivatives of infinite sums may be treated as if the infinite sum were a finite sum of functions, then the solution to the vibrating string problem is given by

$$u(x,t) = c_1 \sin \frac{\pi x}{L} \cos \frac{\pi a t}{L} + c_2 \sin \frac{2\pi x}{L} \cos \frac{2\pi a t}{L} + \cdots$$

where the c_n are determined as above.

But can we really get away with all this? In fact, what mathematical significance can be given to the expression $f(x) = c_1 \sin(\pi x/L) + c_2 \sin(\pi x/L) + \cdots$, and if this expression does indeed have mathematical validity, are we justified in performing the usual mathematical operations of integration and differentiation on it? These and similar questions provide the basis for the next few sections as well as for much of Chapter 10.

B. INFINITE SERIES OF NUMBERS

To simplify matters somewhat, we restrict ourselves initially to a review of infinite sums of real (and complex) numbers. Suppose that $a_1, a_2, a_3, \ldots$ is a sequence of numbers whose "sum" $a_1 + a_2 + \cdots$ we would like to find. Can this be done? Certainly in the ordinary sense of addition it is impossible; however, the addition of finite sums an infinite number of times will yield a satisfactory solution. Consider the finite sums: $s_1 = a_1$, $s_2 = a_1 + a_2$, $s_3 = a_1 + a_2 + a_3, \ldots$. The sequence $\{s_n\}$ of these well-defined sums seemingly tends toward what should correspond to the infinite sum of the a_i's. This leads to the following definition.

(8.B.1) *Definition* Suppose that $a_1, a_2, a_3, \ldots$ is a sequence of real (or complex) numbers. For each $n \in \mathbf{Z}^+$, let $s_n = a_1 + a_2 + a_3 + \cdots + a_n$. Then the *infinite sum* of the a_i's (denoted by either $a_1 + a_2 + \cdots$ or $\sum_{i=1}^{\infty} a_i$) is defined to be the limit of the sequence $\{s_n\}$, provided this limit exists. The ordered pair of sequences $(\{a_n\}, \{s_n\})$ is called an *infinite series*. If the infinite sum of the a_i's is finite, then the infinite series is said to be *convergent*. If $\lim_{n \to \infty} s_n$ does not exist (or is equal to ∞ or $-\infty$), the series is said to be *divergent*. Each sum s_n is called a *partial sum* of the series $(\{a_n\}, \{s_n\})$.

Note: In the literature there is a pervasive abuse of notation, a tradition that we shall (with some misgivings) continue to follow. The problem is that the symbol $a_1 + a_2 + \cdots = \sum_{n=1}^{\infty} a_n$ is often used in different contexts:

1. It represents the series $(\{a_n\}, \{s_n\})$.
2. It denotes the sequence $\{s_n\}$.
3. It is equal to the infinite sum, $\lim_{n\to\infty} s_n$.

As we proceed, however, the reader should be able to distinguish in which sense this symbol is being used.

(8.B.2) *Examples* (a) Let $\{a_n\}$ be defined by $a_n = 1$ for each n. Then since $s_n = n$ for each n, $\lim_{n\to\infty} s_n = \infty$, and hence the series $\sum_{n=1}^{\infty} a_n$ diverges. In this case we write $\sum_{n=1}^{\infty} a_n = \infty$.

(b) Let $\{a_n\}$ be defined by $a_n = 1$ if n is odd and $a_n = -1$ if n is even. Then $s_1 = 1$, $s_2 = 0$, $s_3 = 1, \ldots$, and the series $\sum_{n=1}^{\infty} a_n$ diverges.

(c) Let $\{a_n\}$ be defined by $a_n = cr^{n-1}$, where c is a fixed constant and r is real. To find s_n, we note that multiplication of $1 - r$ by $1 + r + r^2 + \cdots + r^n$ yields $1 - r^{n+1}$, and hence, $1 + r + r^2 + \cdots + r^n = (1 - r^{n+1})/(1 - r)$. Therefore, if $r \neq 1$,

$$s_n = c + cr + cr^2 + \cdots + cr^{n-1} = \frac{c(1 - r^n)}{1 - r}$$

and since $\lim_{n\to\infty} r^n = 0$ whenever $-1 < r < 1$, we have $\lim_{n\to\infty} s_n = c/(1 - r)$. Clearly if $r > 1$, then $\lim s_n = \infty$ (or $-\infty$), and equally clearly, if $r < -1$, the series $\sum_{n=1}^{\infty} a_n$ diverges. What happens if $r = 1$ or $r = -1$? The series $\sum_{n=1}^{\infty} cr^{n-1}$ is called a *geometric series*.

(d) Let $\{a_n\}$ be defined by $a_n = 1/n$ for each n. Although it is difficult to calculate the nth partial sum s_n for this series, we can use the fact that convergent sequences are Cauchy sequences to show that $\sum_{n=1}^{\infty} a_n = \infty$. Suppose to the contrary that $\sum_{n=1}^{\infty} a_n$ converges; then the sequence $\{s_n\}$ is Cauchy, and hence, corresponding to $\varepsilon = \frac{1}{2}$, there is an integer N such that if $m,n \geq N$, then $|s_n - s_m| < \frac{1}{2}$. Note however that $s_{2n} - s_n = 1/2n + \cdots + 1/(n + 1) > 1/2n + \cdots + 1/2n = \frac{1}{2}$. Consequently, it is clear that $\{s_n\}$ cannot be a Cauchy sequence, which proves that $\sum_{n=1}^{\infty} a_n = \infty$.

The idea of an infinite sum is an old one. Aristotle was aware that certain geometric series are convergent. In 1647, Gregory of Saint Vincent pointed out that Achilles would indeed overtake the tortoise in the race mentioned in Chapter 3 since the geometric series marking Achilles' progress in the race converges. Series were used by many seventeenth and eighteenth century mathematicians to "calculate" values such as π and e. Euler (1707–1783), for example, demonstrated that $\pi^2/6 = 1/1^2 + 1/2^2 + 1/3^2 + \cdots$ and Leibniz (1646–1716) in 1674 showed that $\pi/4 = 1 - \frac{1}{3} + \frac{1}{5} - \frac{1}{7} + \cdots$.

Questions concerning convergence of series came increasingly to the fore in the eighteenth and nineteenth centuries.

A prime use of infinite series was in the formation of trigonometric, logarithmic, and nautical tables. James Gregory in the latter part of the seventeenth century derived the expression $\tan x = x + \frac{1}{3}x^3 + \frac{2}{15}x^5 + \frac{7}{315}x^7 + \cdots$, and Mercator and Newton discovered the relationship $\log(1 + x) = x - \frac{1}{2}x^2 + \frac{1}{3}x^3 - \cdots$.

Brook Taylor published in the early eighteenth century a formula for expanding functions as infinite series; it appears however that Gregory had essentially discovered this expression some 40 years previously, but as is often the case in mathematics, the glory was to remain with another. Both Taylor polynomials and the Taylor series expansions will be dealt with presently; they represent two of the crowning early achievements of mathematical analysis.

Associated with a given infinite series, there are two central problems: does the series converge, and if so, to what value. Somewhat surprisingly, it is often the first question that is of more concern; precise answers to the second query are relatively rare and frequently of little import.

A number of tests for determining whether a given series converges or diverges have been developed. We review briefly some of the more standard ones. The reader is first asked to resolve the following easy exercises. For these exercises it is convenient to recall that convergent sequences are Cauchy sequences.

(8.B.3) *Exercises* (a) Show that if $\sum_{n=1}^{\infty} a_n$ converges, then $\lim_{n\to\infty} a_n = 0$ (consider $s_n - s_{n-1}$). In (8.B.2.d), we saw that the converse of this proposition is false.

(b) Show that a series of real numbers $\sum_{n=1}^{\infty} a_n$ converges if and only if for each $\varepsilon > 0$ there is a positive integer N such that

(i) If $m > n \geq N$, then $|s_m - s_n| = |a_{n+1} + a_{n+2} + \cdots + a_m| < \varepsilon$.
(ii) $\sum_{n=N}^{\infty} a_n < \varepsilon$.

(8.B.4) *Theorem (Comparison Test)* Suppose that $\sum_{n=1}^{\infty} a_n$ and $\sum_{n=1}^{\infty} b_n$ are infinite series of real numbers with the property that there is an integer N such that for each $n \geq N$, $0 < a_n \leq b_n$. Then

(i) If $\sum_{n=1}^{\infty} b_n$ converges, so does $\sum_{n=1}^{\infty} a_n$.
(ii) If $\sum_{n=1}^{\infty} a_n$ diverges, so does $\sum_{n=1}^{\infty} b_n$.

Proof. (i) If $\sum_{n=1}^{\infty} b_n$ converges, then its partial sums (starting with $t_N = b_1 + b_2 + \cdots + b_N$) form a bounded increasing sequence. Since this implies that the partial sums (beginning with $s_N = a_1 + a_2 + \cdots + a_N$) of the series $\sum_{n=1}^{\infty} a_n$ will also be bounded and increasing, the result follows from (3.B.10.e).

The proof of (ii) is similar.

(8.B.5) Definition A series $\sum_{n=1}^{\infty} a_n$ *converges absolutely* if the series $\sum_{n=1}^{\infty} |a_n|$ converges. The series $\sum_{n=1}^{\infty} a_n$ is said to *converge conditionally* if $\sum_{n=1}^{\infty} a_n$ converges but $\sum_{n=1}^{\infty} |a_n| = \infty$.

(8.B.6) Exercise Show that an absolutely convergent series converges. [*Hint*: Show that the sequence of partial sums is a Cauchy sequence.]

The next three tests are somewhat less transparent than (8.B.4), but frequently more useful.

(8.B.7) Theorem (Ratio Test) Given a (real or complex) series $\sum_{n=1}^{\infty} a_n$, let $R = \lim_{n \to \infty} |a_{n+1}|/|a_n|$ (we assume that for all large values of n, a_n is not 0). Then

(i) $\sum_{n=1}^{\infty} a_n$ converges absolutely if $R < 1$.
(ii) $\sum_{n=1}^{\infty} a_n$ diverges if $R > 1$.

Proof. Suppose that $R < 1$, and let q be any number satisfying $R < q < 1$. Since $R = \lim_{n \to \infty} |a_{n+1}|/|a_n|$, there is an integer N such that for $n \geq N$, $|a_{n+1}|/|a_n| < q$, or equivalently, $|a_{n+1}| < q|a_n|$. Division of both sides of this inequality by q^{n+1} yields $|a_{n+1}|/q^{n+1} < |a_n|/q^n$. Since this is true for all $n \geq N$, we have

$$\frac{|a_n|}{q^n} < \frac{|a_{n-1}|}{q^{n-1}} < \cdots < \frac{|a_N|}{q^N}$$

Consequently for $n \geq N$, $|a_n| < q^{-N}|a_N|q^n$, and since the geometric series $\sum_{n=1}^{\infty} q^{-N}|a_N|q^n$ converges, the result follows from the comparison test (note that $q^{-N}|a_N|$ is a fixed constant).

The proof of the second part of the theorem is easier. Simply observe that if $R > 1$, there is a positive integer N such that $|a_{n+1}| > |a_n|$ for all $n \geq N$. Since this implies that $\lim_{n \to \infty} a_n \neq 0$, it follows from (8.B.3.a) that $\sum_{n=1}^{\infty} a_n$ fails to converge.

The reader should note that complications may arise if $\lim_{n \to \infty} |a_{n+1}|/|a_n|$ does not exist (and is not equal to ∞). To handle cases such as this the notion of the lim sup of a sequence is quite useful. This idea and the resulting slightly stronger versions of the ratio test and the root test are presented in the problem section (see problems B.14, B.17, B.18).

(8.B.8) Theorem (Root Test) Given a (real or complex) series $\sum_{n=1}^{\infty} a_n$, let $\hat{R} = \lim_{n \to \infty} \sqrt[n]{|a_n|}$. Then

(i) If $\hat{R} < 1$, $\sum_{n=1}^{\infty} a_n$ converges absolutely.

(ii) If $\hat{R} > 1$, $\sum_{n=1}^{\infty} a_n$ diverges.

Proof. (i) Let q be any number satisfying $\hat{R} < q < 1$. There is a positive integer N such that if $n \geq N$, then $\sqrt[n]{|a_n|} < q$, or equivalently, $|a_n| < q^n$. Again the comparison test and the fact that $\sum_{n=1}^{\infty} q^n$ is a convergent geometric series can be used to conclude the proof.

(ii) If $\hat{R} > 1$, then $\sqrt[n]{|a_n|} > 1$ for sufficiently large n; thus $\lim_{n \to \infty} a_n \neq 0$, and by (8.B.3.a), $\sum_{n=1}^{\infty} a_n$ must diverge.

(8.B.9) *Examples* (a) Consider the series $\sum_{k=1}^{\infty} 3^k x^k / 2^{k+1}$, where x is an arbitrary but fixed number. Applying the ratio test, we find that

$$\lim_{k \to \infty} \left| \frac{3^{k+1} x^{k+1} / 2^{k+2}}{3^k x^k / 2^{k+1}} \right| = \left| \frac{3x}{2} \right|$$

and it follows that the series $\sum_{k=1}^{\infty} 3^k x^k / 2^{k+1}$ converges whenever $|3x/2| < 1$ ($|x| < \frac{2}{3}$) and diverges if $|x| > \frac{2}{3}$ (does this series converge if $x = \pm \frac{2}{3}$?).

(b) Consider the series $\frac{1}{4} + \frac{1}{2} + \frac{1}{8} + \frac{1}{4} + \frac{1}{16} + \frac{1}{8} + \cdots$ where $a_{2n} = 1/2^n$ and $a_{2n-1} = 1/2^{n+1}$. If we apply the root test to the even terms a_{2n}, we find that

$$\lim_{n \to \infty} a_{2n}^{1/n} = \lim_{n \to \infty} \left(\frac{1}{2^n} \right)^{1/2n} = \frac{1}{\sqrt{2}} < 1$$

Similarly, for the odd terms we have that

$$\lim_{n \to \infty} (a_{2n-1})^{1/(2n-1)} = \lim_{n \to \infty} (2^{-(n+1)})^{1/(2n-1)} = \frac{1}{\sqrt{2}}$$

and, hence, this series converges.

(8.B.10) *Observation* For both of the series $\sum_{n=1}^{\infty} 1/n$ and $\sum_{n=1}^{\infty} 1/n^2$, it easily seen that $\lim_{n \to \infty} |a_{n+1}|/|a_n| = 1$. However, the series $\sum_{n=1}^{\infty} 1/n$ diverges, and it will follow from the Cauchy condensation test below that the series $\sum_{n=1}^{\infty} 1/n^2$ converges. Hence, if the value R in the ratio test is 1, this test cannot be employed to determine convergence or divergence. Furthermore, it follows from L'Hospital's rule that $\lim_{n \to \infty} |1/n|^{1/n} = 1 = \lim_{n \to \infty} |1/n^2|^{1/n}$, and hence, if $\hat{R} = 1$, the root test also fails.

The reader may easily verify that the ratio test fails in (8.B.9.b); in general, the root test is stronger than the ratio test, but less easy to apply. The relationship between these two tests is explored further in the problem set.

The following test is slightly more esoteric.

(8.B.11)   *Theorem (Cauchy Condensation Test)*   Suppose that $\sum_{n=1}^{\infty} a_n$ is a series of real numbers with the property that $0 < a_{n+1} < a_n$ for each n. Then $\sum_{n=1}^{\infty} a_n$ converges if and only if the "condensed" series $\sum_{k=1}^{\infty} 2^k a_{2^k} = 2a_2 + 4a_4 + 8a_8 + \cdots$ converges.

Proof.   Since $0 < a_{k+1} < a_k$ for each k, it follows that for each k,

$$2^{k-1} a_{2^k} = \overbrace{a_{2^k} + \cdots + a_{2^k}}^{2^{k-1} \text{ terms}} \leq \overbrace{a_{2^{k-1}+1} + a_{2^{k-1}+2} + \cdots + a_{2^k}}^{2^{k-1} \text{ terms}}$$

Therefore, $\displaystyle\sum_{k=1}^{\infty} 2^k a_{2^k} = 2 \sum_{k=1}^{\infty} 2^{k-1} a_{2^k} = 2(a_2 + 2a_4 + 2^2 a_8 + \cdots)$

$$\leq 2[(a_2) + (a_3 + a_4) + (a_5 + a_6 + a_7 + a_8) + \cdots]$$

$$= 2 \sum_{n=2}^{\infty} a_n$$

and consequently, by the comparison test, if $\sum_{n=1}^{\infty} a_n$ converges, so does $\sum_{k=1}^{\infty} 2^k a_{2^k}$.

Since $0 < a_{k+1} < a_k$, it also follows that

$$\overbrace{a_{2^{k-1}+1} + a_{2^{k-1}+2} + \cdots + a_{2^k}}^{2^{k-1} \text{ terms}} \leq \overbrace{a_{2^{k-1}} + \cdots + a_{2^{k-1}}}^{2^{k-1} \text{ terms}} = 2^{k-1} a_{2^{k-1}}$$

Thus, we have

$$\sum_{n=2}^{\infty} a_n = a_2 + (a_3 + a_4) + (a_5 + a_6 + a_7 + a_8) + \cdots \leq \sum_{k=1}^{\infty} 2^{k-1} a_{2^{k-1}}$$

$$= a_1 + \sum_{k=1}^{\infty} 2^k a_{2^k}$$

and, hence, if $\sum_{k=1}^{\infty} 2^k a_{2^k}$ converges, so does $\sum_{n=1}^{\infty} a_n$.

(8.B.12)   *Example*   The condensed series associated with the series $\sum_{n=1}^{\infty} 1/n^k$ is

$$\frac{2}{2^k} + \frac{4}{4^k} + \frac{8}{8^k} + \cdots = \sum_{r=1}^{\infty} \frac{2^r}{(2^r)^k} = \sum_{r=1}^{\infty} \frac{1}{2^{(k-1)r}} = \sum_{r=1}^{\infty} (2^{(1-k)})^r$$

Note that the latter series is a geometric series and hence by (8.B.2.c) converges if and only if $2^{1-k} < 1$, i.e., if and only if $k > 1$. Thus, for instance, the series $\sum_{n=1}^{\infty} 1/n^2$ converges, while the series $\sum_{n=1}^{\infty} 1/\sqrt{n}$ diverges.

Before proceeding to somewhat deeper matters, we briefly consider series that have both positive and negative terms. Our principal result in this direction is the following.

(8.B.13) *Theorem (Dirichlet's Test for Series)* Suppose that $\{a_n\}$ is a decreasing sequence such that $\lim_{n \to \infty} a_n = 0$, and suppose further that the partial sums t_n of the series $\sum_{n=1}^{\infty} b_n$ are bounded. Then the series $\sum_{n=1}^{\infty} a_n b_n$ converges.

Proof. By the hypothesis there is a positive number M such that for each n, $|t_n| = |b_1 + b_2 + \cdots + b_n| \leq M$. Note that the nth partial sum of the series $\sum_{n=1}^{\infty} a_n b_n$ can be written in the form

$$s_n = \sum_{k=1}^{n} a_k b_k = a_1 t_1 + a_2(t_2 - t_1) + \cdots + a_n(t_n - t_{n-1})$$

$$= (a_1 - a_2)t_1 + (a_2 - a_3)t_2 + \cdots + (a_{n-1} - a_n)t_{n-1} + a_n t_n$$

$$= a_n t_n + \sum_{k=1}^{n-1} (a_k - a_{k+1})t_k \tag{1}$$

Therefore, if $n > m$, we find that

$$|s_n - s_m| = |a_n t_n - a_m t_m + \sum_{k=m}^{n-1} (a_k - a_{k+1})t_k|$$

$$\leq (a_n + a_m)M + \sum_{k=m}^{n-1} (a_k - a_{k+1})M$$

$$= (a_n + a_m)M + (a_m - a_n)M = 2a_m M$$

Since $\lim_{m \to \infty} 2a_m M = 0$, it follows that the sequence $\{s_n\}$ is a Cauchy sequence and, consequently, converges.

As an important corollary we have the alternating series test. A series of the form $\sum_{n=1}^{\infty} (-1)^n a_n$, where each a_n is a nonnegative real number, is called an *alternating series*.

(8.B.14) *Examples* (a) $\displaystyle\sum_{n=1}^{\infty} (-1)^n$

(b) $\displaystyle\sum_{n=1}^{\infty} (-1)^{n+1}\frac{1}{n}$

(8.B.15) *Corollary (Alternating Series Test)* Suppose that $\sum_{n=1}^{\infty} (-1)^n a_n$ is an alternating series with the property that $\{a_n\}$ is a decreasing sequence and $\lim_{n \to \infty} a_n = 0$. Then $\sum_{n=1}^{\infty} a_n$ converges.

(8.B.16) *Example* It follows from (8.B.15) and (8.B.2.d) that the series $\sum_{n=1}^{\infty} (-1)^{n+1}1/n$ is convergent (but not absolutely convergent).

The reader is asked to establish a slightly strengthened version of (8.B.13) in problem B.19.

Since the order of summation of a finite sum is immaterial $(3 + 1 = 1 + 3)$, it is natural to inquire if this property carries over to infinite sums. Dirichlet, in studying Fourier series, soon recognized that for certain series the order of summation affects drastically the convergence of the series. In fact, as we show in the next theorem a conditionally convergent series can be made to converge to any given real number (or $\pm\infty$) if the order of the terms of the series is appropriately altered.

(8.B.17) *Definition* Suppose that $\sum_{n=1}^{\infty} a_n$ is an infinite series and that ϕ: $\mathbf{Z}^+ \to \mathbf{Z}^+$ is a bijection. Then the *rearrangement* of $\sum_{n=1}^{\infty} a_n$ (defined by ϕ) is the series $\sum_{n=1}^{\infty} b_n$, where for each n, $b_n = a_{\phi(n)}$.

(8.B.18) *Example* If $\sum_{n=1}^{\infty} a_n$ is the series $\sum_{n=1}^{\infty} (-1)^{n+1} 1/n$ and ϕ: $\mathbf{Z}^+ \to \mathbf{Z}^+$ is defined by

$$\phi(n) = \begin{cases} n + 1 & \text{if } n \text{ is odd} \\ n - 1 & \text{if } n \text{ is even} \end{cases}$$

then the rearrangement of $\sum_{n=1}^{\infty} a_n$ defined by ϕ is the series

$$-\tfrac{1}{2} + 1 - \tfrac{1}{4} + \tfrac{1}{3} - \cdots$$

(8.B.19) *Theorem (Riemann)* Suppose that $\sum_{n=1}^{\infty} a_n$ is a conditionally convergent series of real numbers, and let c be either a real number or ∞ or $-\infty$. Then there is a rearrangement $\sum_{n=1}^{\infty} b_n$ of $\sum_{n=1}^{\infty} a_n$ such that $\sum_{n=1}^{\infty} b_n = c$.

Proof. We first separate the positive terms of $\sum_{n=1}^{\infty} a_n$ from the negative terms. Let p_1 be the first positive term of $\sum_{n=1}^{\infty} a_n$, p_2 the second positive term, etc. Similarly, let q_1 be the first negative term of $\sum_{n=1}^{\infty} a_n$, q_2 the second negative term, etc. Note that $\sum_{n=1}^{\infty} p_n = \infty$ and $\sum_{n=1}^{\infty} q_n = -\infty$ (if both of these series were to converge, then $\sum_{n=1}^{\infty} a_n$ would converge absolutely and if just one of these series were to converge, then $\sum_{n=1}^{\infty} a_n$ would diverge).

Suppose that c is an arbitrary nonnegative real number (the other cases where c is a negative real number, $c = \infty$, or $c = -\infty$ are handled analogously). Let n_1 be the first positive integer such that $A_1 = p_1 + p_2 + \cdots + p_{n_1} > c$. Let m_1 be the first positive integer such that if $B_1 = q_1 + q_2 + \cdots + q_{m_1}$, then $A_1 + B_1 < c$. Let n_2 be the first positive integer such that if $A_2 = p_{n_1+1} + p_{n_1+2} + \cdots + p_{n_2}$, then $A_1 + B_1 + A_2 > c$, and let m_2 be the first positive integer such that if $B_2 = q_{m_1+1} + q_{m_1+2} + \cdots + q_{m_2}$, then $A_1 + B_1 + A_2 + B_2 < c$. If we continue in this manner, it should be clear that the resulting rearrangement of $\sum_{n=1}^{\infty} a_n$ converges to c [by (8.B.3.a) $\lim_{n\to\infty} a_n = 0$ and, therefore, $\lim_{n\to\infty} p_n = 0 = \lim_{n\to\infty} q_n$].

(8.B.20) *Observation* If $\sum_{n=1}^{\infty} a_n$ is an absolutely convergent series, and $\sum_{n=1}^{\infty} a_n = A$, then any rearrangement, $\sum_{n=1}^{\infty} b_n$, of $\sum_{n=1}^{\infty} a_n$ converges to A (see problem B.10).

To conclude this section we investigate briefly the problems involved in the multiplication of infinite series. Although the addition of two series is easily effected $[\sum_{n=0}^{\infty} a_n + \sum_{n=0}^{\infty} b_n = \sum_{n=0}^{\infty} (a_n + b_n)]$, multiplication is somewhat more subtle. In the finite case we have that the product of the sums $a_0 + a_1 + \cdots + a_k$ and $b_0 + b_1 + \cdots + b_k$ contains every term of the form $a_i b_j$, where $0 \le i \le k$, and $0 \le j \le k$. Thus, in the case of infinite sums, it is reasonable to expect that the product of the series $\sum_{n=0}^{\infty} a_n$ and $\sum_{n=0}^{\infty} b_n$ should include all terms of the form $a_i b_j$ for $i = 0, 1, \ldots$ and $j = 0, 1, \ldots$; the problem is to determine in which order these terms should appear (especially in view of our work on the rearrangement of series). One fairly natural and systematic scheme for listing these terms can be extracted from the diagram below

$$
\begin{array}{lllll}
 & a_0 b_0 & a_0 b_1 & a_0 b_2 & a_0 b_3 & a_0 b_4 \cdots \\
 & \diagup & \diagup & \diagup & \diagup \\
c_0 & a_1 b_0 & a_1 b_1 & a_1 b_2 & a_1 b_3 & a_1 b_4 \cdots \\
 & \diagup & \diagup & \diagup \\
c_1 & a_2 b_0 & a_2 b_1 & a_2 b_2 & a_2 b_3 & a_2 b_4 \cdots \\
 & \diagup & \diagup \\
c_2 & a_3 b_0 & a_3 b_1 & a_3 b_2 & a_3 b_3 & a_3 b_4 \cdots \\
 & \diagup \\
c_3 & \cdots & \cdots & \cdots & \cdots & \cdots \cdots
\end{array}
$$

which helps motivate the following definition.

(8.B.21) *Definition* The *Cauchy product series* of the series $\sum_{n=0}^{\infty} a_n$ and $\sum_{n=0}^{\infty} b_n$ is the series $\sum_{n=0}^{\infty} c_n$, where for each n, $c_n = a_0 b_n + a_1 b_{n-1} + \cdots + a_n b_0$.

The obvious property that one would expect in connection with the multiplication of series is that if $A = \sum_{n=0}^{\infty} a_n$ and $B = \sum_{n=0}^{\infty} b_n$, then the Cauchy product of these series should yield $A \cdot B$. Although this property does hold in a large number of cases, there are exceptions (see problem B.11). If, however, both of the series $\sum_{n=0}^{\infty} a_n$ and $\sum_{n=0}^{\infty} b_n$ converge and at least one of these series converges absolutely, then multiplication behaves as one would anticipate.

(8.B.22) *Theorem* (*Mertens*, 1875) Suppose that the series $\sum_{n=0}^{\infty} a_n$ converges absolutely and that the series $\sum_{n=0}^{\infty} b_n$ converges. Let $A = \sum_{n=0}^{\infty} a_n$

and $B = \sum_{n=0}^{\infty} b_n$. Then the Cauchy product $\sum_{n=0}^{\infty} c_n$ of these two series converges, and furthermore, $\sum_{n=0}^{\infty} c_n = A \cdot B$.

Proof. Define the partial sums

$$A_n = \sum_{i=0}^{n} a_i \qquad B_n = \sum_{i=0}^{n} b_i \qquad C_n = \sum_{i=0}^{n} c_i$$

and for each n, let

$$r_n = B_n - B.$$

The first step in the proof is to write C_n in the form:

$$\begin{aligned}
C_n &= a_0 b_0 + (a_0 b_1 + a_1 b_0) + \cdots + (a_0 b_n + a_1 b_{n-1} + \cdots + a_n b_0) \\
&= a_0 B_n + a_1 B_{n-1} + \cdots + a_n B_0 \\
&= a_0 (B + r_n) + a_1 (B + r_{n-1}) + \cdots + a_n (B + r_0) \\
&= A_n B + a_0 r_n + a_1 r_{n-1} + \cdots + a_n r_0
\end{aligned}$$

Let $q_n = a_0 r_n + a_1 r_{n-1} + \cdots + a_n r_0$, and observe that since the sequence $\{A_n B\}$ converges to $A \cdot B$, to prove that the sequence $\{C_n\}$ converges to $A \cdot B$ (the desired result), it suffices to show that the sequence $\{q_n\}$ converges to 0. To this end observe that if N is any positive integer and if $n \geq N$, then

$$\begin{aligned}
|q_n| &= |a_0 r_n + a_1 r_{n-1} + \cdots + a_{n-N-1} r_{N+1} + a_{n-N} r_N \\
&\quad + a_{n-N+1} r_{N-1} + \cdots + a_n r_0| \\
&\leq |a_0 r_n| + |a_1 r_{n-1}| + \cdots + |a_{n-N-1} r_{N+1}| \\
&\quad + |r_N + r_{N-1} + \cdots + r_0| \left(\sum_{k=n-N}^{\infty} |a_k| \right).
\end{aligned} \qquad (4)$$

Let $\varepsilon > 0$ be given and let $A^* = \sum_{n=0}^{\infty} |a_n|$. Since the sequence $\{r_n\}$ converges to 0, there is a positive integer $\hat{N}$ such that if $n \geq \hat{N}$, then

$$|r_n| < \frac{\varepsilon}{2A^*}$$

Hence, if $n \geq \hat{N}$, we have from (4)

$$\begin{aligned}
|q_n| &< \left(\frac{\varepsilon}{2A^*} \right) (|a_0| + |a_1| + \cdots + |a_{n-\hat{N}-1}|) \\
&\quad + |r_{\hat{N}} + r_{\hat{N}-1} + \cdots + r_0| \left(\sum_{k=n-\hat{N}}^{\infty} |a_k| \right) \\
&< \varepsilon/2 + |r_{\hat{N}} + r_{\hat{N}-1} + \cdots + r_0| \left(\sum_{k=n-\hat{N}}^{\infty} |a_k| \right)
\end{aligned} \qquad (5)$$

Let $r^* = |r_{\hat{N}} + r_{\hat{N}-1} + \cdots + r_0|$. Since $\sum_{n=0}^{\infty} |a_n|$ converges, there is a positive integer $N > \hat{N}$ such that if $n \geq N$, then

$$\sum_{k=n-N}^{\infty} |a_k| < \frac{\varepsilon}{2r^*}.$$

To conclude the proof note that if $n \geq N$, then from (5) we have

$$|q_n| \leq \frac{\varepsilon}{2} + r^* \left(\frac{\varepsilon}{2r^*} \right) = \varepsilon$$

and, therefore, the sequence $\{q_n\}$ converges to 0.

We state one additional result due to Abel (1802–1829) involving the Cauchy product of infinite series. A proof of this theorem is indicated in problems E.10 and E.11.

(8.B.23) *Theorem* Suppose that the series $\sum_{n=0}^{\infty} a_n$ and $\sum_{n=0}^{\infty} b_n$ converge to A and B, respectively, and that the Cauchy product of these series converges to C. Then $A \cdot B = C$.

C. INFINITE SERIES OF FUNCTIONS

We continue our study of infinite series with the substitution of functions for numbers. If $\{f_n\}$ is a sequence of real- (or complex-) valued functions with a common domain A, then corresponding to each $x \in A$, there is a sequence of real (or complex) numbers $\{f_n(x)\}$. If for each $x \in A$, this sequence converges, a function $g: A \to \mathbf{R}^1$ can be defined by setting $g(x) = \lim_{n \to \infty} f_n(x)$; the sequence $\{f_n\}$ is then said to *converge pointwise* to g [the reader might note that this definition could be generalized to the case where each f_n maps A into a given metric space instead of into $\mathbf{R}^1$ (or $\mathbf{C}$)].

(8.C.1) *Examples* (a) Let $A = [0,1]$ and for each $n \in \mathbf{Z}^+$, define $f_n(x) = x^n$. Then the sequence $\{f_n\}$ converges pointwise to the function $g: [0,1] \to \mathbf{R}^1$ defined by

$$g(x) = \begin{cases} 0 & \text{if } x \neq 1 \\ 1 & \text{if } x = 1 \end{cases}$$

(b) Let $A = [0,1]$ and let $r_1, r_2, \ldots$ be the complete list of rationals found in A. For each n, define $f_n: [0,1] \to \mathbf{R}^1$ by

$$f_n(x) = \begin{cases} 1 & \text{if } x = r_1, r_2, \ldots, r_n \\ 0 & \text{otherwise} \end{cases}$$

Then the sequence $\{f_n\}$ converges pointwise to the function $g: [0,1] \to \mathbf{R}^1$ defined by

$$g(x) = \begin{cases} 1 & \text{if } x \text{ is rational} \\ 0 & \text{if } x \text{ is irrational} \end{cases}$$

(c)   Let $A = [-1,1]$ and define $f_n: A \to \mathbf{R}^1$ by $f_n(x) = (1 - x^2)^n$. Then the sequence $\{f_n\}$ converges pointwise to $g: [-1,1] \to \mathbf{R}^1$ defined by

$$g(x) = \begin{cases} 0 & \text{if } x \neq 0 \\ 1 & \text{if } x = 0 \end{cases}$$

These examples can be used to illustrate some of the shortcomings of pointwise convergence. For instance, in (8.C.1.a) we see that the pointwise limit of a sequence of continuous functions is not necessarily continuous. It is clear from Chapter 6 that in (8.C.1.b), for each n, $\int_0^1 f_n(x)dx$ exists and is equal to 0; it will follow from results in Chapter 11, however, that the Riemann integral $\int_0^1 g(x)dx$ fails to exist. Thus the pointwise limit of integrable functions may not be integrable. In (8.C.1.c) we have that even though $f_n(0)$ exists for each n, $g'(0)$ fails to exist (g is not even continuous at $x = 0$).

If pointwise convergence is replaced by uniform convergence, then somewhat more positive results can be obtained, as we see in the next few results. Recall that uniform convergence is equivalent to convergence in the sup metric. Note too that if $\{f_n\}$ is a sequence of complex-valued functions, where for each $n, f_n = u_n + iv_n$, then this sequence converges to a function $f = u + iv$ if and only if the sequences of real-valued functions $\{u_n\}$ and $\{v_n\}$ converge to u and v, respectively (see problem C.12).

(8.C.2) Theorem   Suppose that (Y,d) is a metric space and that $\{f_n\}$ is a sequence of continuous functions mapping Y into $\mathbf{R}^1$ (or $\mathbf{C}$). If $\{f_n\}$ converges uniformly to $g: Y \to \mathbf{R}^1$ (or $\mathbf{C}$), then g is continuous.

Proof.   Suppose that y^* is an arbitrary point in Y. We show that g is continuous at y^*. Let $\varepsilon > 0$ be given. Since $\{f_n\}$ converges uniformly to g, there is a positive integer N such that $|f_n(y) - g(y)| < \varepsilon/3$ for each $y \in Y$ and each $n \geq N$. Since f_N is continuous at y^*, corresponding to $\varepsilon/3$, there is a $\delta > 0$ such that if $y \in S_\delta(y^*)$, then $|f_N(y) - f_N(y^*)| < \varepsilon/3$. It suffices to show that $g(S_\delta(y^*)) \subset S_\varepsilon(g(y^*))$. This follows easily since if $y \in S_\delta(y^*)$, then $|g(y) - g(y^*)| \leq |g(y) - f_N(y)| + |f_N(y) - f_N(y^*)| + |f_N(y^*) - g(y^*)| < \varepsilon/3 + \varepsilon/3 + \varepsilon/3 = \varepsilon$.

(8.C.3) Exercise   Suppose that $\{f_n\}$ is a sequence of functions mapping a set X into $\mathbf{R}^1$ (or $\mathbf{C}$) and that for each $\varepsilon > 0$, there is an integer N such that $|f_n(x) - f_m(x)| < \varepsilon$ for all $m,n \geq N$ and each $x \in X$. Show that there is a unique function $f: X \to \mathbf{R}^1$ (or $\mathbf{C}$) such that the sequence $\{f_n\}$ converges uniformly to f. [*Hint*: See the proof of (4.C.7).]

Next we see that uniform convergence and integration combine well together.

(8.C.4) Theorem Suppose that $\{f_n\}$ is a sequence of continuous functions mapping an interval $[a, b]$ into $\mathbf{R}^1$. If $\{f_n\}$ converges uniformly to a function g: $[a,b] \to \mathbf{R}^1$, then g is integrable and $\lim_{n \to \infty} \int_a^b f_n(x)dx = \int_a^b g(x)dx$.

Proof. That g is integrable follows immediately from (8.C.2) and (6.A.12). The sequence $\{f_n\}$ converges uniformly to g, and hence, given $\varepsilon > 0$, there is a positive integer N such that if $n \geq N$, then

$$f_n(x) - \frac{\varepsilon}{b - a} \leq g(x) \leq f_n(x) + \frac{\varepsilon}{b - a} \tag{1}$$

for each $x \in [a,b]$. By (6.B.3.v), integration of each member of (1) yields

$$\int_a^b f_n(x)dx - \varepsilon \leq \int_a^b g(x)dx \leq \int_a^b f_n(x)dx + \varepsilon$$

and the result follows.

If $f: [a,b] \to \mathbf{C}$ and if $f = u + iv$, then the integral $\int_a^b f(x)dx$ is defined to be $\int_a^b u(x)dx + i \int_a^b v(x)dx$, provided that both of these integrals exist.

In problem C.13 the reader is asked to establish a theorem analogous to (8.C.4) for complex-valued functions.

Infinite series of functions are defined as follows.

(8.C.5) Definition Suppose that $\{f_n\}$ is a sequence of functions mapping a subset A of $\mathbf{R}^1$ into $\mathbf{R}^1$ (or $\mathbf{C}$). For each $n \in \mathbf{Z}^+$, let $s_n = f_1 + f_2 + f_3 + \cdots + f_n$. The ordered pair of sequences $(\{f_n\},\{s_n\})$ is called an *infinite series of functions* and is denoted by $\sum_{n=1}^{\infty} f_n$. The series $\sum_{n=1}^{\infty} f_n$ is said to be *pointwise convergent* if there is a function $g: A \to \mathbf{R}^1$ (or $\mathbf{C}$) such that $\lim_{n \to \infty} s_n(x) = g(x)$ for each $x \in A$. The series $\sum_{n=1}^{\infty} f_n$ is said to *converge uniformly* to g if $\{s_n\}$ converges uniformly to g. If $\sum_{n=1}^{\infty} f_n$ does not converge pointwise, then it is said to *diverge*.

The customary confusion with regard to the symbol $\sum_{n=1}^{\infty} f_n$ [see the comments following (8.B.1)] is present in this context as well.

(8.C.6) Examples (a) Let $A = \mathbf{R}^1$ and for each nonnegative integer n, define

$$f_n(x) = \frac{x^n}{n!}$$

We shall see presently that the series

$$\sum_{n=0}^{\infty} f_n(x) = \sum_{n=1}^{\infty} \frac{x^n}{n!}$$

converges uniformly on each closed interval $[a,b]$, and furthermore, for each $x \in \mathbf{R}^1$

$$\sum_{n=0}^{\infty} \frac{x^n}{n!} = e^x$$

(b) If for each $n \in \mathbf{Z}^+, f_n \colon \mathbf{R}^1 \to \mathbf{R}^1$ is defined by

$$f_n(x) = \frac{nx^2}{n^3 + x^3}$$

then it is easy to see that for each $x \in \mathbf{R}^1$, the series

$$\sum_{n=1}^{\infty} \frac{nx^2}{n^3 + x^3} \qquad (2)$$

converges (note that for purposes of convergence, this series for each $x \in \mathbf{R}^1$ is "essentially" equivalent to the series $\sum_{n=1}^{\infty} 1/n^2$).

Since infinite series of functions involve sequences of functions, it should be evident from our work thus far that the uniform convergence of a series would be of special interest. The next theorem summarizes a number of our previous results involving uniform convergence as they are applied to series. The easy proof to this theorem may be supplied by the reader (see problem C.14).

(8.C.7) **Theorem** Suppose that $\sum_{n=1}^{\infty} f_n$ is an infinite series of real- or complex-valued functions with partial sums s_n.

(i)   If $A \subset \mathbf{R}^1$ and if for each $\varepsilon > 0$, there is a positive integer N such that for all $m,n \geq N$ and for *each* $x \in A$, $|s_n(x) - s_m(x)| < \varepsilon$, then $\sum_{n=1}^{\infty} f_n$ converges uniformly on A to a function $f \colon A \to \mathbf{R}^1$ (or $\mathbf{C}$).

(ii) If $\sum_{n=1}^{\infty} f_n$ converges uniformly to a function $f \colon A \to \mathbf{R}^1$ (or $\mathbf{C}$) and if each function f_n is continuous on A, then f is continuous on A.

(iii) If $\sum_{n=1}^{\infty} f_n$ converges uniformly on $[a,b]$ to f, and if each f_n is continuous, then $\int_a^b f(x)dx = \sum_{n=1}^{\infty} \int_a^b f_n(x)dx$. In other words, $\int_a^b (\sum_{n=1}^{\infty} f_n(x))dx$ is obtained via term-by-term integration.

(8.C.8) **Example** For each $n \in \mathbf{Z}^+$, let $f_n \colon [0,1] \to \mathbf{R}^1$ be defined by $f_n(x) = x(1 - x)^n$. Note that if $0 < x < 1$, then $\sum_{n=1}^{\infty} x(1 - x)^n = x \sum_{n=1}^{\infty} (1 - x)^n = x(1/[1 - (1 - x)]) = 1$ and if $x = 0$ or $x = 1$, then $\sum_{n=1}^{\infty} x(1 - x)^n = 0$. Therefore, the series $\sum_{n=1}^{\infty} f_n(x)$ converges pointwise to the function $g \colon [0,1] \to \mathbf{R}^1$ defined by

$$g(x) = \begin{cases} 1 & \text{if } x \neq 0, 1 \\ 0 & \text{if } x = 0 \text{ or } x = 1 \end{cases}$$

Since g is not continuous, if follows from (8.C.7) that the series $\sum_{n=1}^{\infty} f_n(x)$ does not converge uniformly.

A basic test for uniform convergence of a series is given in the next theorem.

(8.C.9) Theorem (Weierstrass M-test) Suppose that $\sum_{n=1}^{\infty} f_n$ is an infinite series of real- or complex-valued functions defined on a set $A \subset \mathbf{R}^1$. Let $\{M_n\}$ be a sequence of real numbers such that for each $n \in \mathbf{Z}^+$ and each $x \in A$, $0 \le |f_n(x)| \le M_n$. Then if $\sum_{n=1}^{\infty} M_n$ converges, the series $\sum_{n=1}^{\infty} f_n$ converges uniformly.

Proof. By the comparison test $\sum_{n=1}^{\infty} |f_n(x)|$ converges for each $x \in A$, and hence, by (8.B.6) $\sum_{n=1}^{\infty} f_n(x)$ is convergent. Define $g : A \to \mathbf{R}^1$ (or C) by $g(x) = \lim_{n \to \infty} s_n(x)$. We must show that the convergence of the series $\sum_{n=1}^{\infty} f_n$ to g is uniform.

Let $\varepsilon > 0$ be given. By (8.B.3.b) there is a positive integer N such that if $n \ge N$, then $\sum_{k=n+1}^{\infty} M_k < \varepsilon$. Hence if $n \ge N$, we have

$$\left| g(x) - \sum_{k=1}^{n} f_k(x) \right| = \left| \sum_{k=1}^{\infty} f_k(x) - \sum_{k=1}^{n} f_k(x) \right| = \left| \sum_{k=n+1}^{\infty} f_k(x) \right|$$

$$\le \sum_{k=n+1}^{\infty} |f_k(x)| \le \sum_{k=n+1}^{\infty} M_k < \varepsilon,$$

and consequently, the partial sums of the series $\sum_{n=1}^{\infty} f_n$ converge uniformly to g.

(8.C.10) Examples (a) Since $|\sin nx|/n^2 \le 1/n^2$ and $\sum_{n=1}^{\infty} 1/n^2$ is convergent, the series $\sum_{n=1}^{\infty} (\sin nx)/n^2$ is uniformly convergent on $\mathbf{R}^1$.

(b) For each $n \in \mathbf{Z}^+$, define $f \colon [0,\infty) \to \mathbf{R}^1$ by $f_n(x) = e^{-nx}x^n$. Since $f_n'(x) = -ne^{-nx}x^n + e^{-nx}nx^{n-1}$, it is easy to see that f_n takes on a maximum value at $x = 1$. Hence, $f_n(x) \le e^{-n}$ for each x, and since $\sum_{n=1}^{\infty} e^{-n}$ converges (use the ratio test), the series $\sum_{n=1}^{\infty} e^{-nx}x^n$ converges uniformly.

D. TAYLOR POLYNOMIALS AND THE TAYLOR EXPANSION

Recall that the Weierstrass approximation theorem asserts that a continuous function $f \colon [a,b] \to \mathbf{R}^1$ can be uniformly approximated by a sequence of polynomial functions. We shall show presently that if, in addition, the function f satisfies certain differentiability conditions, then the polynomials used to approximate f can be chosen in a particularly elegant and useful way. It is these polynomials that we shall study in this section.

We have already seen in Chapter 5 that for any differentiable function, f, the "best linear" approximation to f near a given point x_0 is given by

$$y(x) = f'(x_0)(x - x_0) + f(x_0)$$

(by "linear" we mean here that the graph of y is a line, and by "best" we mean that given any other function of the form $\tilde{y}(x) = a(x - x_0) + f(x_0)$, there is an interval (c,d) containing x_0 such that for each $x \in (c,d)$, $|f(x) - y(x)| \le |f(x) - \tilde{y}(x)|$). A better approximation of f can be obtained from quadratic polynomials of the form

$$y(x) = f(x_0) + a(x - x_0) + b(x - x_0)^2$$

To determine the optimal value of a and b for such an approximation, we may proceed as follows. Let $y_0 = f(x_0)$ and let

$$p(x) = y_0 + a(x - x_0) + b(x - x_0)^2$$

and

$$q(x) = y_0 + c(x - x_0) + d(x - x_0)^2$$

be any two quadratic polynomials such that $a \ne c$. The polynomial p will be a better approximation of f than q if

$$\frac{f(x) - p(x)}{f(x) - q(x)} < 1$$

for all x in some interval containing x_0; this is equivalent to saying

$$\left| \lim_{x \to x_0} \frac{f(x) - p(x)}{f(x) - q(x)} \right| < 1 \tag{1}$$

From L'Hospital's rule we have

$$\left| \lim_{x \to x_0} \frac{f(x) - p(x)}{f(x) - q(x)} \right| = \left| \lim_{x \to x_0} \frac{f(x) - [y_0 + a(x - x_0) + b(x - x_0)^2]}{f(x) - [y_0 + c(x - x_0) + d(x - x_0)^2]} \right|$$

$$= \left| \lim_{x \to x_0} \frac{f'(x) - [a + 2b(x - x_0)]}{f'(x) - [c + 2d(x - x_0)]} \right| = \left| \frac{f'(x_0) - a}{f'(x_0) - c} \right|$$

Thus, since c can be chosen arbitrarily, it is clear that to ensure the desired inequality (1), we must set $a = f'(x_0)$ [which is reasonable in view of (5.D.5)].

To find the optimal value of b, we again let $y_0 = f(x_0)$ and set $y_1 = f'(x_0)$. Then two applications of L'Hospital's rule yield [assuming that $a = c = f'(x_0)$]

$$\left| \lim_{x \to x_0} \frac{f(x) - p(x)}{f(x) - q(x)} \right| = \left| \lim_{x \to x_0} \frac{f(x) - [y_0 + y_1(x - x_0) + b(x - x_0)^2]}{f(x) - [y_0 + y_1(x - x_0) + d(x - x_0)^2]} \right|$$

$$= \left| \lim_{x \to x_0} \frac{f'(x) - [y_1 + 2b(x - x_0)]}{f'(x) - [y_1 + 2d(x - x_0)]} \right| = \left| \lim_{x \to x_0} \frac{f''(x) - 2b}{f''(x) - 2d} \right|$$

$$= \frac{|f''(x_0)/2 - b|}{|f''(x_0)/2 - d|}$$

and consequently, for inequality (1) to hold we must set $b = f''(x_0)/2$ (since d can be chosen arbitrarily). Thus we have shown that the polynomial $p(x) = f(x_0) + f'(x_0)(x - x_0) + [f''(x_0)/2](x - x_0)^2$ is the best approximation of f (in the sense described above) by a quadratic polynomial, provided that f'' is continuous in a neighborhood of x_0. A similar argument may be used to establish the following more general result.

(8.D.1) Theorem Suppose that $A \subset \mathbf{R}^1$, $f: A \to \mathbf{R}^1$, $x_0 \in \text{int } A$, and $f^{(i)}$ is continuous in a neighborhood of x_0 for $1 \le i \le n$. Then the "best" nth-degree polynomial approximation of f near x_0 is given by the polynomial

$$T_n(x) = f(x_0) + \frac{f'(x_0)}{1}(x - x_0) + \cdots + \frac{f^{(n)}(x_0)}{n!}(x - x_0)^n.$$

[By "best" is meant that for each polynomial $q(x)$ of degree n satisfying $q(x_0) = y_0$, there is an interval $(c,d) \subset A$ that contains x_0 such that $|f(x) - T_n(x)| \le |f(x) - q(x)|$ for each $x \in (c,d)$.]

(8.D.2) Definition Suppose that $A \subset \mathbf{R}^1$, $f: A \to \mathbf{R}^1$, $x_0 \in \text{int } A$, and that $f^{(i)}(x_0)$ exists for each i, $1 \le i \le n$. Then the polynomial

$$T_n(x) = f(x_0) + \frac{f'(x_0)}{1}(x - x_0) + \cdots + \frac{f^{(n)}(x_0)}{n!}(x - x_0)^n \qquad (2)$$

is called the nth *Taylor polynomial* belonging to f at x_0.

(8.D.3) Examples

(a) $f(x) = \sin x \qquad x_0 = 0 \qquad T_5(x) = x - \frac{x^3}{3!} + \frac{x^5}{5!}$

(b) $f(x) = 4x^4 - 3x \qquad x_0 = 0 \qquad T_4(x) = -3x + 4x^4$

(c) $f(x) = \ln x \qquad x_0 = 1 \qquad T_4(x) = \ln 1 + (x - 1) - \frac{(x - 1)^2}{2}$

$$+ 2\frac{(x - 1)^3}{3!} - 6\frac{(x - 1)^4}{4!}$$

(8.D.4) Exercise Suppose that $A \subset \mathbf{R}^1$, $f: A \to \mathbf{R}^1$, $x_0 \in \text{int } A$, and that $f^{(i)}(x_0)$ exists for each i, $1 \le i \le n$. Show that the Taylor polynomial T_n is the unique polynomial with the property that $T_n(x_0) = f(x_0)$, $T_n'(x_0) = f'(x_0)$, $\ldots$, $T_n^{(n)}(x_0) = f^{(n)}(x_0)$.

In view of (8.D.1), one might well conclude that if $f: A \to \mathbf{R}^1$ is infinitely differentiable in a neighborhood of a point $x_0 \in \text{int } A$ [that is, $f^{(n)}(x_0)$ exists on some interval $(x_0 - \delta, x_0 + \delta)$ for each $n \in \mathbf{Z}^+$], then f is the limit of its Taylor polynomial approximations, i.e., $f(x) = \lim_{n \to \infty} T_n(x)$, or equivalently,

$$f(x) = f(x_0) + f'(x)(x - x_0) + \cdots + \frac{f^{(n)}(x_0)(x - x_0)^n}{n!} + \cdots$$

The infinite series of functions

$$f(x_0) + f'(x_0)(x - x_0) + \cdots + \frac{f^{(n)}(x_0)(x - x_0)^n}{n!} + \cdots \tag{3}$$

is called the *Taylor expansion of f about the point* x_0 and will be denoted by $T_{x_0}^f(x)$. Two immediate questions arise:

1. Does $T_{x_0}^f(x)$ converge for values other than x_0?
2. For what values of x can f be represented by its Taylor series expansion, i.e., for what values of x does $T_{x_0}^f(x) = f(x)$?

In the next example we see that question 2 is of some significance.

(8.D.5) *Example* Let $f: \mathbf{R}^1 \to \mathbf{R}^1$ be defined by

$$f(x) = \begin{cases} e^{-1/x^2} & \text{if } x \neq 0 \\ 0 & \text{if } x = 0 \end{cases}$$

Since $\lim_{x \to 0} f(x)$ is clearly 0, the function f is continuous on $\mathbf{R}^1$. Furthermore, if $x \neq 0$, then a routine calculation shows that

$$f'(x) = 2x^{-3}e^{-1/x^2}$$

In order to find $f'(0)$, we first determine

$$\lim_{x \to 0} f'(x) = \lim_{x \to 0} 2x^{-3}e^{-1/x^2}$$

Let $y = 1/x$ and observe that

$$\lim_{x \to 0} 2x^{-3}e^{-1/x^2} = \lim_{y \to \infty} 2y^3 e^{-y^2} = 0$$

(use L'Hospital's rule twice). One more application of L'Hospital's rule shows that

$$f'(0) = \lim_{x \to 0} \frac{f(x) - f(0)}{x} = \lim_{x \to 0} \frac{f'(x)}{1} = 0$$

With the aid of these techniques the reader may employ a standard induction argument to establish that

$$f^{(n)}(0) = 0$$

for each $n \in \mathbf{Z}^+$ (see problem D.9). Hence, each Taylor polynomial associated with f is identically equal to zero, and the corresponding Taylor expansion takes on the trivial form

$$T_0^f(x) = 0$$

Since $f(x) \neq 0$, whenever $x \neq 0$, we have that the sequence of partial sums $\{T_n(x)\}$ (the Taylor polynomials) fails to converge to $f(x)$ for *any* $x \neq 0$.

What has happened here? Since f is continuous, by the Weierstrass approximation theorem we know that for any closed interval containing 0, there are polynomial approximations to f which are arbitrarily and uniformly close to f. Clearly, the Taylor polynomials are not among them. On the other hand, we have shown that the Taylor polynomials (when they exist) give the best approximations to the given function f. In the next exercise the reader is invited to ponder these apparent inconsistencies.

(8.D.6) *Exercise* Examine carefully the criterion for a best approximation, and show that the function described in (8.D.5) does not lead to contradictory results.

The reader should not infer from the behavior of the rather pathological function described in (8.D.5) that Taylor expansions are of little use; this example was given only to point out that care must be exercised when trying to expand functions as infinite series. In fact, the Taylor expansion is one of the most important constructs in analysis; its virtue of yielding good approximations has long been known and the common trig tables, log tables, and many of the computational aspects of calculators and computers are based on its validity. Additional uses of this expansion, both theoretical and applied, are discussed in the next chapter.

We are still faced with the problem of determining when the Taylor expansion of a function f converges to f. In order to have convergence at a point x, it is clear that if we set $r_n(x) = f(x) - T_n(x)$, then the sequence $\{r_n(x)\}$ must converge to 0. From the next theorem we shall be able to obtain one fairly easily applied criterion for testing for this convergence.

(8.D.7) *Theorem (Taylor's Theorem with Remainder)* Suppose that f: $(a,b) \to \mathbf{R}^1$ is infinitely differentiable on (a,b), Let $x_0 \in (a,b)$ be fixed. Then for each $x \in (a,b)$,

$$f(x) = f(x_0) + f'(x_0)(x - x_0) + \frac{f''(x_0)}{2!}(x - x_0)^2 + \cdots + \frac{f^{(n)}(x_0)}{n!}(x - x_0)^n$$

$$+ \frac{f^{(n+1)}(x_{n+1}^*)}{(n + 1)!}(x - x_0)^{n+1}$$

where x_{n+1}^* is some point lying between x and x_0.

Proof. Let $x \in (a,b)$ and define $p: (a,b) \to \mathbf{R}^1$ by

$$p(w) = f(x_0) + f'(x_0)(w - x_0) + \cdots + \frac{f^{(n)}(x_0)}{n!}(w - x_0)^n + k(w - x_0)^{n+1}$$

where k is chosen so that $p(x) = f(x)$. Let $g = f - p$, and observe that $g(x_0) = 0 = g(x)$. By Rolle's theorem ((5.B.3)) there is a point x_1^* between x and x_0 such that $g'(x_1^*) = 0$. Since $g'(x_0) = 0$, another application of Rolle's theorem yields a point x_2^* between x_1^* and x_0 with the property that $g''(x_2^*) = 0$. Continuing in this manner, we eventually obtain a point x_{n+1}^* lying between x_n^* and x_0 with the property that $g^{(n+1)}(x_{n+1}^*) = 0$. Since for all $w \in (a,b)$ $g^{(n+1)}(w) = f^{(n+1)}(w) - (n + 1)!k$, we have $k = f^{(n+1)}(x_{n+1}^*)/(n + 1)!$, and the result follows.

(8.D.8) Corollary Suppose that $f: (a,b) \to \mathbf{R}^1$, $x_0 \in (a,b)$, and that f is infinitely differentiable on (a,b). Suppose further that there is a real number M such that $|f^{(n)}(x)| \le M$ for each $x \in (a,b)$ and each $n \in \mathbf{Z}^+$. Then the Taylor series expansion of f, $T_{x_0}^f$, is equal to f on (a,b).

Proof. Note that by (8.D.7) for each n,

$$|r_n(x)| = |f(x) - T_n(x)| \le \frac{|M(x - x_0)|^{n+1}}{(n + 1)!} \tag{4}$$

where T_n is the nth Taylor polynomial belonging to f. The right-hand side of (4) tends to 0 as $n \to \infty$ [apply the ratio test to $\sum_{n=1}^{\infty} M(x - x_0)^{n+1}/(n + 1)!$ and use (8.B.3.a)], and this implies that $\lim_{n \to \infty} r_n(x) = 0$ for each $x \in (a,b)$.

(8.D.9) Examples (a) If $f(x) = \sin x$ and $x_0 = 0$, then since $f^{(n)}(x) \le 1$ for each $n \in \mathbf{Z}^+$ and each $x \in \mathbf{R}^1$, we have

$$\sin x = x - \frac{x^3}{3!} + \frac{x^5}{5!} - \frac{x^7}{7!} + \cdots$$

and this expression is valid for all $x \in \mathbf{R}^1$.

(b) The Taylor series expansion of $f(x) = e^x$ about the point $x = 0$, is easily seen to be

$$T_0^f(x) = 1 + x + \frac{x^2}{2!} + \frac{x^3}{3!} + \cdots$$

To see that T_0^f converges to e^x at each point $x \in \mathbf{R}^1$, note that given a point $x^* \in \mathbf{R}^1$, the function e^x is bounded on any finite interval containing 0 and x^*; therefore, on such an interval

$$|f^{(n)}(x)| = |e^x| \le M$$

for some real number M and for each $n \in \mathbf{Z}^+$. Consequently, by the previous corollary, the Taylor series expansion of e^x converges to e^x on this entire interval and, in particular, at the point x^*.

Note too that the inequality (4) is useful in determining error estimates when Taylor polynomials are used to approximate a given function f. For instance, if we wish to find an approximation of e^x for values of x between -1 and 1, then the maximal error incurred by employing, say, the sixth Taylor polynomial associated with this function (about 0), is less than $(3/7!)1^7$ since $|f^{(7)}(x)| \leq 3$ for all x in the interval $[-1,1]$.

(8.D.10) *Observation* Taylor (and other) series expansions are possible for complex functions as well, and indeed, they play an extremely important role in complex analysis. This topic will be investigated in Chapter 12.

We now turn to the problem of determining intervals about x_0 for which the Taylor expansion of f does converge to f, and in particular, when this convergence is uniform. Resolution of this problem is best done in the context of power series.

E. POWER SERIES

(8.E.1) *Definition* An infinite series of functions of the form $\sum_{n=0}^{\infty} a_n(x - x_0)^n$ is called a *power series centered at* x_0. Here, a_n, x, and x_0 may be complex.

(8.E.2) *Examples*

(a) $\displaystyle\sum_{n=1}^{\infty} \frac{1}{n}(x - 3)^n$

(b) $\displaystyle\sum_{n=1}^{\infty} \frac{n(x + 4)^n}{2^n(n^2 - 5)}$

(c) Any Taylor series expansion.

(d) $\displaystyle\sum_{n=1}^{\infty} (2 + i)^{n^2}(x - (3 + 4i))^n$ (where x is complex)

The extent and nature of the region of convergence of a power series is handled nicely by the next theorem.

(8.E.3) *Theorem* Given the (real or complex) power series $\sum_{n=0}^{\infty} a_n(x - x_0)^n$, let

$$r = \lim_{n \to \infty} \frac{1}{|a_{n+1}/a_n|}$$

(We assume here that r exists; if $\lim_{n \to \infty} |a_{n+1}/a_n| = 0$, let r "$=$" ∞.) Then

(i) $\sum_{n=0}^{\infty} a_n(x - x_0)^n$ converges   whenever $|x - x_0| < r$,   and
$\sum_{n=0}^{\infty} a_n(x - x_0)^n$ diverges whenever $|x - x_0| > r$.

(ii) $\sum_{n=0}^{\infty} a_n(x - x_0)^n$ converges uniformly on each compact subset of
$S_r(x_0) = \{x \in \mathbf{R}^1 \mid |x - x_0| < r\}$(or $\{z \in \mathbf{C} \mid |z - x_0| < r\}$).

(iii) If $\sum_{n=0}^{\infty} |a_n| r^n$ is an absolutely convergent series of real numbers,
then $\sum_{n=0}^{\infty} a_n(x - x_0)^n$ converges uniformly on $S_r(x_0)$.

Proof.   (i)   The proof is based on an easy application of the ratio test.
By (8.B.7) we have that $\sum_{n=0}^{\infty} a_n(x - x_0)^n$ converges whenever

$$\lim_{n \to \infty} \left| \frac{a_{n+1}(x - x_0)^{n+1}}{a_n(x - x_0)^n} \right| < 1$$

and diverges if

$$\lim_{n \to \infty} \left| \frac{a_{n+1}(x - x_0)^{n+1}}{a_n(x - x_0)^n} \right| > 1$$

Clearly,

$$\lim_{n \to \infty} \left| \frac{a_{n+1}(x - x_0)^{n+1}}{a_n(x - x_0)^n} \right| = \frac{1}{r} |x - x_0|$$

and therefore, since

$$\frac{1}{r} |x - x_0| < 1 \text{ if and only if } |x - x_0| < r$$

$$\frac{1}{r} |x - x_0| > 1 \text{ if and only if } |x - x_0| > r$$

the result follows. (Note that if

$$\lim_{n \to \infty} \left| \frac{a_{n+1}(x - x_0)^{n+1}}{a_n(x - x_0)^n} \right| \neq 0$$

then $\sum_{n=1}^{\infty} a_n(x - x_0)^n$ converges for all x, and if

$$\lim_{n \to \infty} \left| \frac{a_{n+1}(x - x_0)^{n+1}}{a_n(x - x_0)^n} \right| = \infty$$

then $\sum_{n=1}^{\infty} a_n(x - x_0)^n$ converges only for $x = x_0$.)

(ii)   If A is a compact subset of $S_r(x_0)$, then there is a point q in A such
that $|q - x_0| \geq |a - x_0|$ for each $a \in A$ (see problem E.12). Hence, for each
$x \in A$, we have

$$|a_n(x - x_0)^n| \leq |a_n(q - x_0)^n|$$

If $M_n = |a_n(q - x_0)^n|$, then by (i) $\sum_{n=0}^{\infty} M_n$ converges, and therefore, by the
Weierstrass M-test, $\sum_{n=0}^{\infty} a_n(x - x_0)^n$ converges uniformly on A.

(iii)   Note that for each $x \in S_r(x_0)$, $|a_n(x - x_0)^n| \le |a_n|r^n$, and hence, if we set $M_n = |a_n|r^n$, the result is again an immediate consequence of the Weierstrass M-test.

(8.E.4)   Observation   Note that the previous theorem is invalid in the case that an infinite number of the coefficients a_n are 0 and an infinite number of these coefficients are not 0. A result involving a special case of such a situation may be found in problem E.15. Theorem (8.E.3) may be strengthened to handle *all* such cases by replacing lim with lim sup (see problem E.14).

(8.E.5)   Definition   The value r described in the previous theorem is called the *radius of convergence* for the series $\sum_{n=0}^{\infty} a_n(x - x_0)^n$. For complex power series the set $D = \{x \in C \mid |x - x_0| < r\}$ is called the *disk of convergence* and for power series of real numbers the interval $(x_0 - r, x_0 + r)$ is called the *interval of convergence*.

Note that convergence may or may not occur on the boundary of the disk of convergence or at the endpoints of the interval of convergence.

(8.E.6)   Example   The radius of convergence of the power series (x real)

$$\sum_{n=1}^{\infty} \frac{(x - 1)^n}{n2^n}$$

is given by

$$r = \lim_{n \to \infty} \left| \frac{1}{a_{n+1}/a_n} \right| = \lim_{n \to \infty} \frac{1}{n/2(n + 1)} = 2$$

Checking the endpoints, $x = 3$ and $x = -1$, we find that if $x = 3$, then

$$\sum_{n=1}^{\infty} \frac{2^n}{2^n n} = \sum_{n=1}^{\infty} \frac{1}{n} = \infty$$

and if $x = -1$, then

$$\sum_{n=1}^{\infty} \frac{(-2)^n}{2^n n} = \sum_{n=1}^{\infty} (-1)^n \frac{1}{n}$$

which converges. .Therefore, the interval of convergence of this series is $(-1,3)$ and the series converges on the interval $[-1,3)$.

(8.E.7)   Example   With the aid of the ratio test, it is easily seen that the complex power series $\sum_{n=1}^{\infty} z^n$ and $\sum_{n=1}^{\infty} z^n/n^2$ both have radius of convergence 1. It is equally easy to see that the first series fails to converge at any point on the boundary of the disk of convergence, while the second series converges at each point of the boundary.

Suppose that $f(x)$ is defined by the real power series $\sum_{n=0}^{\infty} a_n(x - x_0)^n$ with a positive radius of convergence, r, i.e., $f(x) = \sum_{n=0}^{\infty} a_n(x - x_0)^n$ for each $x \in (x_0 - r, x_0 + r)$. It is natural to inquire at this juncture if $\int_a^b f(x)dx$ and $f'(x)$ exist and if they can be found in terms of the series representation of f. As we see in the next theorem, the integral of f over any closed interval $[a,b]$ contained in the interval of convergence of $\sum_{n=0}^{\infty} a_n(x - x_0)^n$ is valid and easily effected by term-by-term integration. In the following theorem we find that the derivative of f is found by term-by-term differentiation of the power series representation of f [cf. (8.C.1.c)]. The complex counterparts of these results are discussed in Chapter 12.

(8.E.8) *Theorem*   Suppose that $\sum_{n=0}^{\infty} a_n(x - x_0)^n$ is a real power series with radius of convergence $r > 0$. Suppose that $[b,c] \subset (x_0 - r, x_0 + r)$ and let $f(x) = \sum_{n=0}^{\infty} a_n(x - x_0)^n$ for each $x \in [b,c]$. Then $\int_b^c f(x)dx$ exists and is equal to $\sum_{n=0}^{\infty} \int_b^c a_n(x - x_0)^n dx$. Thus, $\int_b^c f(x)dx$ may be found via term-by-term integration.

Proof.   This is immediate from (8.C.7.iii).
Differentiation is somewhat more difficult to deal with.

(8.E.9) *Theorem*   If the real power series $\sum_{n=0}^{\infty} a_n(x - x_0)^n$ has a positive radius of convergence r and if $f(x) = \sum_{n=0}^{\infty} a_n(x - x_0)^n$ for each $x \in (x_0 - r, x_0 + r)$, then for each $x \in (x_0 - r, x_0 + r)$, $f'(x)$ exists and is equal to $\sum_{n=0}^{\infty} na_n(x - x_0)^{n-1}$.

We prove this theorem for the case $x_0 = 0$. The reader may make the obvious modifications to establish the general case. We begin with a series of three easy lemmas that are valid for both real and complex numbers.

(8.E.10) *Lemma*   Suppose that the sequence $\{a_n q^n\}$ is bounded for some number q. Then the series $\sum_{n=0}^{\infty} a_n x^n$ converges absolutely whenever $|x| < |q|$.

Proof.   Suppose that for each n, $M \geq |a_n||q|^n$ and that $|x| < |q|$. Then

$$|a_n x^n| = |a_n||q|^n \left|\frac{x}{q}\right|^n \leq M\left|\frac{x}{q}\right|^n$$

Since $|x/q| < 1$, the geometric series $\sum_{n=0}^{\infty} M(|x/q|)^n$ converges and, hence, so does the series $\sum_{n=0}^{\infty} |a_n x^n|$.

As an immediate consequence of (8.E.10) and (8.B.3.a) we have the following lemma.

(8.E.11) *Lemma* If $\sum_{n=0}^{\infty} a_n x^n$ has radius of convergence r, then

(i) If $\rho > r$, the sequence $\{|a_n \rho^n|\}$ is not bounded.
(ii) If $\rho < r$, the sequence $\{|a_n \rho^n|\}$ is bounded.

(8.E.12) *Lemma* The series $\sum_{n=0}^{\infty} a_n x^n$ and $\sum_{n=1}^{\infty} n a_n x^{n-1}$ have the same radius of convergence.

Proof. Let r be the radius of convergence of $\sum_{n=0}^{\infty} a_n x^n$ and r' the radius of convergence of $\sum_{n=1}^{\infty} n a_n x^{n-1}$. If $\rho > r$, then by (8.E.11) the sequence $\{|a_n| \rho^n\}$ is not bounded and, hence, neither is the sequence $(n/\rho)|a_n|\rho^n = n|a_n|\rho^{n-1}$, from which it follows that $r' \leq r$.

If $\rho < r$, let ρ_1 be such that $\rho < \rho_1 < r$. Note that

$$n|a_n|\rho^{n-1} = \left(\frac{n}{\rho_1}\left(\frac{\rho}{\rho_1}\right)^{n-1}\right)(|a_n|\rho_1^n)$$

Since the first factor of the right-hand side of this equation tends to 0 with increasing n (apply the ratio test to $\sum_{n=0}^{\infty} (n/\rho_1)(\rho/\rho_1)^{n-1}$ and use (8.B.3.a)) and the second factor is bounded (why?), it follows that the sequence $\{n|a_n|\rho^{n-1}\}$ is bounded, which implies that $r \leq r'$.

Proof of (8.E.9). We wish to show that $\left(\sum_{n=0}^{\infty} a_n x^n\right)' = \sum_{n=1}^{\infty} n a_n x^{n-1}$ whenever $|x| < r$. Let $g(x) = \sum_{n=1}^{\infty} n a_n x^{n-1}$. By (8.E.12) and (8.E.8) we have that if $|x| < r$,

$$\int_0^x \left(\sum_{n=1}^{\infty} n a_n t^{n-1}\right) dt = \sum_{n=1}^{\infty} \int_0^x n a_n t^{n-1} dt$$

or, in other words,

$$\int_0^x g(t)dt = \sum_{n=0}^{\infty} a_n x^n = f(x)$$

Consequently, we have

$$f'(x) = \left(\sum_{n=0}^{\infty} a_n x^n\right)' = g(x) = \sum_{n=1}^{\infty} n a_n x^{n-1}$$

the desired result.

We conclude this section with the following interesting relationship between power series and Taylor expansions.

(8.E.13) *Theorem* Suppose that $\sum_{n=0}^{\infty} a_n(x - x_0)^n$ is a real power series with radius of convergence r and let $f(x) = \sum_{n=0}^{\infty} a_n(x - x_0)^n$ for each $x \in$

$(x_0 - r, x_0 + r)$. Then $f^{(n)}$ exists for each n, and furthermore, for each n,

$$a_n = \frac{f^{(n)}(x_0)}{n!}$$

consequently, the power series $\sum_{n=0}^{\infty} a_n(x - x_0)^n$ coincides with the Taylor expansion of f.

Proof. Take derivatives and use (8.E.9).

(8.E.14) *Corollary* Suppose that $\sum_{n=0}^{\infty} a_n(x - x_0)^n = \sum_{n=0}^{\infty} b_n(x - x_0)^n$ on some interval containing x_0. Then $a_n = b_n$ for each n.

F.   DIVERGENT SERIES

At this stage the reader may well have the impression that divergent series are of limited interest, especially since such series would appear to have minimal claim to the virtue of yielding good approximations. In this section we show that in fact divergent series can play a very significant role in obtaining approximations. We begin by considering the differential equation

$$y' - y = -\frac{1}{x} \qquad (x > 0) \tag{1}$$

Multiplying both sides of (1) by the integrating factor e^{-x}, we obtain

$$(e^{-x}y)' = -\frac{e^{-x}}{x} \tag{2}$$

and integration of both sides of (2) yields

$$y = e^x \int^x \frac{-e^{-t}}{t}\, dt \tag{3}$$

where $\int^x$ indicates that the primitive is to be considered as a function of x. It is easy to check that a particular solution y_p of (1) is given by

$$y_p(x) = e^x \int_x^\infty \frac{e^{-t}}{t}\, dt = \int_x^\infty \frac{e^{x-t}}{t}\, dt \tag{4}$$

(see problem F.1).

The function $L(x) = \int_x^\infty (e^{-t}/t)dt$ is frequently called the *incomplete gamma function*; this function has a number of important applications in advanced statistics.

The solution (4) has one drawback: it cannot be computed directly from the fundamental theorem of calculus and in fact it can only be approximated. One fairly standard method for obtaining approximations to integral expressions such as (4) is to employ repeated integration by parts. Using this procedure, we find (see problem F.2)

$$y_p(x) = \int_x^\infty \frac{e^{x-t}}{t} \, dt = \left[\frac{1}{x} - \frac{1}{x^2} + \frac{2!}{x^3} - \cdots + (-1)^{n-1} \frac{(n-1)!}{x^n} \right]$$

$$+ (-1)^n n! \int_x^\infty \frac{e^{x-t}}{t^{n+1}} \, dt \tag{5}$$

Note that the nth partial sum of the series

$$\frac{1}{x} - \frac{1}{x^2} + \frac{2!}{x^3} - \cdots + (-1)^{n-1} \frac{(n-1)!}{x^n} + \cdots \tag{6}$$

is precisely the expression that appears in the square brackets in (5). An easy application of the ratio test shows that

$$\lim_{n \to \infty} \left| \frac{n!/x^{n+1}}{(n-1)!/x^n} \right| = \left| \frac{n}{x} \right| = \infty$$

for each $x \neq 0$. Consequently, the series (6) fails to converge at any point x. Rather surprisingly, however, the partial sums s_n of this series can still be used to approximate y_p. To see this, first observe that since $|e^{x-t}| < 1$ for each $t > x$, we have that the "error," $y_p(x) - s_n(x)$, in (5) satisfies

$$|y_p(x) - s_n(x)| = \left| (-1)^n n! \int_x^\infty \frac{e^{x-t}}{t^{n+1}} \, dt \right| \leq n! \int_x^\infty \frac{dt}{t^{n+1}} = \frac{(n-1)!}{x^n} \tag{7}$$

for each $x > 0$.

Clearly, for a given point x_0,

$$\frac{(n-1)!}{x_0^n} = \frac{1}{x_0} \frac{1}{x_0} \frac{2}{x_0} \frac{3}{x_0} \cdots \frac{(n-1)}{x_0}$$

decreases with increasing n until $n - 1$ is equal to or exceeds x_0, and thereafter this error bound increases without limit as $n \to \infty$. It follows that if $[x]$ denotes the largest integer not exceeding x, then the partial sum $s_{[x_0]+1}(x_0)$ is the best estimate of $y_p(x_0)$ that can be obtained in this fashion (note, however, that this implies that the absolute error cannot be less than

$$\frac{[x_0]}{x_0^{[x_0]+1}} \,).$$

It is important to observe here that since the series (6) diverges, there is a definite limit as to how close $s_n(x_0)$ will approach $y_p(x_0)$ with increasing n, something which, of course, did not occur in the case of convergent series. Furthermore, how well s_n approximates y_p depends to a large degree on x_0. For instance, if $x_0 = 2$, then the minimum absolute error given by (7) is

$2/2^3 = 1/4$, which is rather large; however, if $x_0 = 6$, then this error is less than 0.003, and for $x_0 = 10$, the error bound is less than 3.6×10^{-5}.

Next we show that for each n, s_{n+1} provides for large values of x a more accurate estimate of y_p than does s_n. To see this first note that from (7)

$$|y_p(x) - s_{n+1}(x)| \le \frac{n!}{x^{n+1}} \tag{8}$$

Since $y_p(x) - s_n(x)$ can be written as $y_p(x) - s_{n+1}(x) + (-1)^n n!/x^{n+1}$, it follows from (8) that

$$|y_p(x) - s_n(x)| \le \frac{2(n!)}{x^{n+1}} \tag{9}$$

and, hence, for each fixed n,

$$\lim_{x \to \infty} [y_p(x) - s_n(x)]x^n = 0 \tag{10}$$

Note that from (10) and (9), we have that although

$$\lim_{x \to \infty} [y_p(x) - s_{n+1}(x)]x^{n+1} = 0$$

it may well be the case that

$$\lim_{x \to \infty} [y_p(x) - s_n(x)]x^{n+1} = \infty$$

and, therefore s_{n+1} yields a better approximation of y_p for *large* values of x than does s_n.

The sequence $\{s_n\}$ of partial sums is frequently referred to as an asymptotic expansion of y_p.

PROBLEMS

Section A

1. Use the method of separation of variables to replace the following partial differential equations with a pair of ordinary differential equations:
    (a) $u_{xx} + xu_{xt} + u_t = 0$
    (b) $tu_{xx} + xu_t = 0$
2. Find a solution to the wave equation $u_{xx} = u_{tt}$ if:
    (a) $u(x,0) = \sin(n\pi x/L)$, where L is the length of the string and $u(0,t) = u(L,t) = 0$.
    (b) The string has length 4 and $u(x,0) = \sin 4\pi x - \sin 20\pi x$ and $u(0,t) = u(4,0) = 0$.

3. Show that if the initial velocity imparted to the string of length L is given by $u_t(x,0) = v(x)$, then the Fourier coefficient A_n is given by

$$A_n = \frac{2}{L} \int_0^L v(x) \sin \frac{n\pi x}{L} \, dx$$

4. Find the solution to the wave equation if $u(x,0) = \sin(\pi x/L)$ and $u_t(x,0) = \sin(2\pi x/L)$.

5. Sketch the graphs of $u_n(x,t)$ for distinct values of t between $t = 0$ and $t = L/a$ for $n = 1, 2, 3$, where $u_n(x, t) = c_n \sin(n\pi x/L) \cos(n\pi at/L)$.

6.* Use the method of separation of variables and a case-by-case analysis to find a solution to the problem $u_t = 4u_{xx}$, $0 < x < 6$, $t > 0$ given $u(0,t) = u(6,t) = 0$ for each $t > 0$ and $u(x,0) = x^2$ for $0 \leq x \leq 6$.

Section B

1. Establish the following slight modification of the comparison test: Let $\sum_{n=1}^{\infty} a_n$ be given series of positive terms.
   (a) If there is a convergent series of positive terms $\sum_{n=1}^{\infty} b_n$ such that $\lim_{n\to\infty} a_n/b_n$ is finite, then $\sum_{n=1}^{\infty} a_n$ converges.
   (b) If there is a divergent series of positive terms $\sum_{n=1}^{\infty} d_n$ such that $\lim_{n\to\infty} d_n/a_n$ is finite, then $\sum_{n=1}^{\infty} a_n$ diverges.
   [*Hint*: Note that in (a) there is a positive real number M such that $a_n/b_n \leq M$ for each n, and that in (b) there is a positive number K such that $d_n/a_n \leq K$ for each n.]

2. Determine whether the following series converge:

   (a) $\displaystyle\sum_{n=1}^{\infty} \frac{n^2 - n}{4n^4 + 6}$

   (b) $\displaystyle\sum_{n=1}^{\infty} \frac{(1-n)^n}{n^{n+2}}$

   (c) $\displaystyle\sum_{n=1}^{\infty} \frac{3^n}{n^n}$

   (d) $\dfrac{2}{2} + \dfrac{2\cdot 5}{2\cdot 4} + \dfrac{2\cdot 5\cdot 8}{2\cdot 4\cdot 6} + \dfrac{2\cdot 5\cdot 8\cdot 11}{2\cdot 4\cdot 6\cdot 8} + \cdots$

   (e) $\dfrac{1}{2} + \dfrac{1\cdot 3}{2\cdot 4} + \dfrac{1\cdot 3\cdot 5}{2\cdot 4\cdot 6} + \cdots$

   (f) $\displaystyle\sum_{n=1}^{\infty} \frac{3^{(-1)^n} - n}{4}$

3. Use the Cauchy condensation test to show that the series $\sum_{n=2}^{\infty} 1/(n \ln^2 n)$ converges.

4. Show that if $\{a_n\}$ is a decreasing sequence of positive numbers and if

$\{a_n\}$ converges, then $\lim_{n\to\infty} na_n = 0$. [*Hint*: Show that $s_{2n} - s_n \le na_{2n}$, and that $(2n + 1)(a_{2n+1}) \le ((2n + 1)/2n)2na_{2n}$, where s_k represents the kth partial sum of the series.]

5. Show that if $\{a_n\}$ is a decreasing sequence of positive numbers and $\sum_{n=1}^{\infty} a_n$ converges, then $\lim_{n\to\infty} 2^n a_{2^n} = 0$.

6. If a ball is dropped from a height of 10 feet always rebounds 0.8 of the height it falls, find the total distance traveled by the ball. How long will the ball continue to bounce? [Recall that $s(t) = v_0 t - 16t^2$ gives the distance traveled by the object at time t, where v_0 is the initial velocity.]

7. Suppose that two people alternate in flipping a coin. The first person to obtain a head wins. What is the probability that the first person to flip the coin wins?

8. Establish the following test for convergence (the *integral test*): Suppose that $f: [1,\infty) \to (0,\infty)$ is a decreasing function.
    (a) If $\int_1^\infty f(x)dx$ converges, then the series $\sum_{n=1}^{\infty} f(n)$ converges.
    (b) If $\int_1^\infty f(x)dx$ diverges, then the series $\sum_{n=1}^{\infty} f(n)$ diverges.
    [*Hint*: Note that for each partial sum s_n,

$$s_n = f(1) + f(2) + \cdots + f(n) \le \int_1^2 f(x)dx + \int_2^3 f(x)dx + \cdots$$
$$+ \int_{n-1}^n f(x)dx$$

and also that for each n, $f(n) \ge \int_n^{n+1} f(x)dx$.]

9. Use the integral test to determine whether the following series converge:

(a) $\sum_{n=1}^{\infty} ne^{-n}$

(b) $\sum_{n=2}^{\infty} \dfrac{1}{n(\ln n)^5}$

(c) $\sum_{n=2}^{\infty} \dfrac{1}{n(\ln n)(\ln (\ln n))^3}$

10. Show that if $\sum_{n=1}^{\infty} a_n$ is absolutely convergent, then any rearrangement of this series also converges (and to the same value).

11. Suppose that $\sum_{n=1}^{\infty} a_n$ and $\sum_{n=1}^{\infty} b_n$ are series both defined by $a_n = b_n = (-1)^n(n + 1)^{-1/2}$. Show that these series converge but that the Cauchy product of the two series diverges. [*Hint*: Note that for each n,

$$c_n = \sum_{k=0}^{n} \frac{1}{(n - k + 1)(k + 1)}$$

and that $(n - k + 1)(k + 1) = (n/2 + 1)^2 - (n/2 - k)^2 \le (n/2 + 1)^2$.]

12. Suppose that $\sum_{n=1}^{\infty} a_n$ is a series such $a_n > 0$ for each n.

(a) Show that if $\sum_{n=1}^{\infty} a_n$ converges, then so do the series $\sum_{n=1}^{\infty} a_n^2$ and $\sum_{n=1}^{\infty} a_n/(1 + a_n)$.

(b) Show that if $\sum_{n=1}^{\infty} a_n$ diverges, then so do the series $\sum_{n=1}^{\infty} a_n/(1 + a_n)$ and $\sum_{n=1}^{\infty} a_n/s_n$, where s_n is the nth partial sum of the series $\sum_{n=1}^{\infty} a_n$.

13.* (a) Show that if $\lim_{n\to\infty} |a_{n+1}/a_n| = L$, then $\lim_{n\to\infty} |a_n|^{1/n} = L$, and hence, the root test is at least as strong as the ratio test.

(b) Let $\{a_n\}$ be any sequence of real numbers such that for each n $|a_n| \leq n$. Show that if $-1 < r < 1$, then the series $\sum_{n=0}^{\infty} a_n r^n$ converges. [*Hint*: Note that the ratio test is of no use here; apply the root test.]

14. Let $\{x_n\}$ be a sequence in $\mathbf{R}^1$ and for each n, let $t_n = \sup \{x_n, x_{n+1}, x_{n+2}, \ldots\}$. Then the *limit superior* of the sequence $\{x_n\}$, $\lim \sup_{n\to\infty} x_n$ is defined to be $\lim_{n\to\infty} t_n$. Find the lim sup of the following sequences:

(a) 1, 2, 3, 4, 1, 2, 3, 4, 1, 2, ...

(b) $\left\{(-1)^n\left(1 + \left(\dfrac{1}{n}\right)\right)\right\}$

(c) $\left\{\left(1 + \left(\dfrac{1}{n}\right)\right) \cos n\pi\right\}$

15. Let $\{x_n\}$ be a sequence in $\mathbf{R}^1$ and for each n, let $b_n = \inf \{x_n, x_{n+1}, x_{n+2}, \ldots\}$. Then the *limit inferior* of the sequence $\{x_n\}$, $\lim \inf_{n\to\infty} x_n$ is defined to be $\lim_{n\to\infty} b_n$. Find the lim inf of the following sequences:

(a) $\left\{\left(1 + \left(\dfrac{1}{n}\right)\right) \cos n\pi\right\}$

(b) $\{(-1)^n n\}$

(c) $\{n^3 \cos n\pi\}$

16. Show that a sequence $\{x_n\}$ of real numbers converges to $x \in \mathbf{R}^1$ if and only if $\lim \sup_{n\to\infty} x_n = x = \lim \inf_{n\to\infty} x_n$

17. Prove the following stronger version of the ratio test: Suppose that $\sum_{n=1}^{\infty} a_n$ is a series whose terms are nonzero real numbers, and let $R = \lim \sup_{n\to\infty} |a_{n+1}/a_n|$ and $r = \lim \inf_{n\to\infty} |a_{n+1}/a_n|$. If $R < 1$, then $\sum_{n=1}^{\infty} |a_n|$ converges, and if $r > 1$, then $\sum_{n=1}^{\infty} a_n$ diverges. For which values of x is convergence (and divergence) of the following series guaranteed:

$$\sum_{n=1}^{\infty} \left(3 + \sin \frac{n\pi}{2}\right) x^n?$$

18. Prove the following strengthened version of the root test: Suppose that $\sum_{n=1}^{\infty} a_n$ is an infinite series, and let $R = \lim \sup_{n\to\infty} \sqrt[n]{|a_n|}$. Then the series $\sum_{n=1}^{\infty} a_n$ converges absolutely if $R < 1$ and diverges if $R > 1$. For which values of x is convergence (and divergence) of the following

series guaranteed:

$$\sum_{n=1}^{\infty} \left(4 + \cos \frac{n\pi}{4}\right)^n x^n?$$

19. Prove the following slightly strengthened version of Dirichlet's test: Suppose that $\{a_n\}$ is a sequence such that $\lim_{n \to \infty} a_n = 0$ and such that the series $\sum_{n=1}^{\infty} |a_{n+1} - a_n|$ converges. Suppose further that the partial sums of the series $\sum_{n=1}^{\infty} b_n$ are bounded. Then the series $\sum_{n=1}^{\infty} a_n b_n$ converges.

20.* Let $\{a_n\}$ be any sequence of positive real numbers. Show that there is a constant C such that

$$\sum_{n=1}^{\infty} \frac{n}{a_1 + a_2 + \cdots + a_n} \le C \sum_{n=1}^{\infty} \frac{1}{a_n}$$

21. Show that if $b_k = (e^{i\theta})^k = e^{ik\theta}$, θ not a multiple of 2π, then $\sum_{k=1}^{\infty} e^{ik\theta}/k$ converges. [*Hint*: Note that

$$s_n = e^{i\theta}(1 + \cdots + (e^{i\theta})^{n-1}) = e^{i\theta}\left(\frac{1 - e^{in\theta}}{1 - e^{i\theta}}\right)$$

and hence, $|s_n| \le 2/|1 - e^{i\theta}|$.]

22.* Establish *Raabe's test*: Suppose that $\sum_{n=1}^{\infty} a_n$ is a positive termed series, and that $\lim_{n \to \infty} n(1 - a_{n+1}/a_n) = c$. If $c > 1$, then $\sum_{n=1}^{\infty} a_n$ converges and if $c < 1$, then $\sum_{n=1}^{\infty} a_n$ diverges. [*Hint*: Suppose that $c > 1$ and $1 < c^* < c$. Let $q = c^* - 1$ and note that for large n, $0 < a_n q \le (n-1)a_n - na_{n+1}$, which implies that the sequence $\{na_{n+1}\}$ is decreasing and converges to a positive number L. Show that for some integer N, $\sum_{n=N}^{\infty} a_n \le ((N-1)a_n - L)/q$. A similar proof shows that if $c < 1$, then $\sum_{n=1}^{\infty} a_n$ diverges.]

23. Use Raabe's test to determine if the following series converge or diverge:

(a) $\left(\frac{1}{3}\right)^2 + \left(\frac{1 \cdot 4}{3 \cdot 6}\right)^2 + \left(\frac{1 \cdot 4 \cdot 7}{3 \cdot 6 \cdot 9}\right)^2 + \cdots + \left(\frac{1 \cdot 4 \cdot 7 \cdots (3^n - 2)}{3 \cdot 6 \cdot 9 \cdots (3n)}\right)^2 + \cdots$

(b) $\sum_{n=1}^{\infty} \frac{4 \cdot 6 \cdot 8 \cdots (2n + 2)}{7 \cdot 9 \cdot 11 \cdots (2n + 3)}$.

24. Suppose that A is any uncountable set of positive numbers. Show that there is a sequence $a_1, a_2, \ldots$ in A such that $\sum_{i=1}^{\infty} a_i = \infty$.

Section C

1. Show that the series $\sum_{n=1}^{\infty} 1/(4n^2 + 6x^2)$ converges uniformly on $\mathbf{R}^1$.
2. Show that the series $\sum_{n=1}^{\infty} nx^2/(n^3 + x^3)$ converges uniformly on each closed interval $[0,b]$.

3. Determine if the following sequences converge, and if so, whether the convergence is pointwise or uniform.
   (a) $\{f_n\}$, where for each n, $f_n(x) = (\sin x)^n$, $0 \le x \le \pi$.
   (b) $\{g_n\}$, where for each n, $g_n(x) = xe^{-nx}/n$, $x \ge 0$.
   (c) $\{h_n\}$, where for each n, $h_n(x) = nx(1 - x)^n$, $0 \le x \le 1$.
4. Determine whether the following series converge uniformly on the indicated intervals:

   (a) $\displaystyle\sum_{n=1}^{\infty} \frac{x^n}{\sqrt{n}}$ $\quad -1 \le x \le 0$

   (b) $\displaystyle\sum_{n=1}^{\infty} \frac{x^n}{n(\ln n)^2}$ $\quad -1 \le x \le 1$

   (c) $\displaystyle\sum_{n=1}^{\infty} x^n(1 - x^2)$ $\quad 0 \le x \le 1$

   (d) $\displaystyle\sum_{n=1}^{\infty} \frac{\sin nx}{n^{3/2}}$ $\quad -\infty < x < \infty$.

5. Discuss possible parallels between continuity-uniform continuity and pointwise convergence-uniform convergence.
6. Prove or find a counterexample:
   (a) If the sequences $\{f_n\}$ and $\{g_n\}$ converge uniformly on a set A to the functions f and g, respectively, then the sequence $\{f_n + g_n\}$ converges uniformly to $f + g$.
   (b) If the sequences $\{f_n\}$ and $\{g_n\}$ converge uniformly on a set A to the the functions f and g, respectively, then the sequence $\{f_n g_n\}$ converges uniformly to fg.
7. Suppose that for each $n \in \mathbf{Z}^+$,

$$f_n(x) = \begin{cases} 2n^3x & \text{if } 0 \le x \le \dfrac{1}{2n} \\[2mm] n^2 - 2n^3\left(x - \dfrac{1}{2}n\right) & \text{if } \dfrac{1}{2n} \le x \le \dfrac{1}{n} \\[2mm] 0 & \text{if } \dfrac{1}{n} \le x \le 1 \end{cases}$$

Show that if $b \in (0,1)$, then $\int_0^b \lim_{n\to\infty} f_n(x)dx = 0$, but $\lim_{n\to\infty} \int_0^b f_n(x)dx = \infty$.
8. Show that if for each $n \in \mathbf{Z}^+$,

$$f_n(x) = \frac{x}{1 + n^2x^2} \qquad -1 \le x \le 1$$

then the sequence $\{f_n\}$ converges uniformly to the constant function $f(x) \equiv 0$; show, however, that the sequence $\{f_n'\}$ does not converge to $f'(x) \equiv 0$. Do the same if $f_n(x) = (\sin nx/\sqrt{n})$.

9. Show that if for each $n \in \mathbf{Z}^+$

$$f_n(x) = \begin{cases} \dfrac{1}{n} & 0 \le x \le n \\ 0 & x > n \end{cases}$$

then the sequence $\{f_n\}$ converges uniformly to the constant function $f(x) \equiv 0$, but that $\lim_{n \to \infty} \int_0^\infty f_n(x) = 1$.

10. Show that the function

$$f(x) = \frac{\cos (4n + 2)x}{(2n + 1)2n}$$

is continuous and find $\int_0^\pi f(x)\, dx$.

11. Suppose that $\{f_n\}$ is a sequence of increasing functions on $[a,b]$, $f:[a,b] \to \mathbf{R}^1$ is continuous, and that $\{f_n\}$ converges to f (pointwise convergence). Show that $\{f_n\}$ converges to f uniformly. Find an example to show that the increasing condition is necessary.

12. Suppose that $\{f_n\}$ is a sequence of complex-valued functions where for each n, $f_n = u_n + iv_n$. Show that the sequence converges (pointwise or uniformly) to a function $f = u + iv$ if and only if the sequences $\{u_n\}$ and $\{v_n\}$ converge (pointwise or uniformly) to u and v, respectively.

13. Formulate and prove a theorem analogous to (8.C.4) for complex-valued functions.

14. Prove Theorem (8.C.7).

15. The improper integral $\int_c^\infty f(x,y)\, dy$ is said to *converge uniformly* to the function $F(x)$ on the interval $[a, b]$ if given $\varepsilon > 0$, there is an $R > 0$ such that if $q \ge R$ and $a \le x \le b$, then $|\int_c^q f(x,y)\, dy - F(x)| < \varepsilon$.

   (a) Use a modification of the proof of the Weierstrass M-test to show that if $|f(x,y)| \le g(y)$, $y \ge c$, $a \le x \le b$, where g is a continuous function with the property that $\int_c^\infty g(y)\, dy$ converges, then $\int_c^\infty f(x,y)\, dy$ converges uniformly to a function $F(x)$ on $[a,b]$.

   (b) Let $A = \{(x,y) \in \mathbf{R}^2 \mid a \le x \le b, y \ge c\}$ and suppose that $f: A \to \mathbf{R}^1$ and $f_x: A \to \mathbf{R}^1$ are continuous. Show that if $\int_c^\infty f(x,y)\, dy$ and $\int_c^\infty f_x(x,y)\, dy$ converge uniformly on $[a,b]$ to functions $F(x)$ and $G(x)$, respectively, then $F'(x) = \int_c^\infty f(x,y)\, dy$ [cf. the Leibniz rule, Theorem (7.F.1)].

   (c) Show that $F(x) = \int_0^\infty e^{-xy}\sin (xy)\, dy$, $1 \le x \le 6$ converges uniformly, and find $F'(x)$.

Section D

1. Find the Taylor series expansions of the following functions about the indicated points:

(a) $f(x) = 1/x$ about $x = 1$.

(b) $f(x) = \sin x^3$ about $x = 0$. [*Hint*: First expand $\sin x$.]

(c) $f(x) = x^4 \sin \sqrt{x}$ about $x = 0$.

(d) $f(x) = \sin^2 x$ about $x = 0$.

2. Use the fourth Taylor polynomial corresponding to $f(x) = e^{x^2}$ to approximate $\int_0^{0.2} e^{x^2} \, dx$ and find an error bound for this estimate. Do the same for $\int_0^{0.1} x \cos \sqrt{x} \, dx$.

3. Show that the Taylor series expansion of $f(x) = \ln x$ about $x = 1$ converges to $\ln x$ for $0 < x < 2$. What happens at $x = 2$?

4. Suppose that $f: [a,b] \to \mathbf{R}^1$, $f^{(n)}$ exists on $[a,b]$ for each n, and that on $[a,b]$, $|f^{(n)}(x)| \le \alpha^n$ for some positive number α. Show that for each $x \in [a,b]$, $T^f(x) = f(x)$.

5. Suppose that $f: [0,1] \to \mathbf{R}^1$ and that f'' is continuous on $[0, 1]$. Show that if $f(0) = f(1) = 0$ and if $|f''(x)| \le M$ on $(0, 1)$, then $|f'(x)| \le M/2$ on $(0,1]$.

6. Suppose that $f: [a,b] \to \mathbf{R}^1$ and that $f^{(n)}$ exists for each n. Show that if $R_n(x) = [1/(n-1)!] \int_a^x (x-w)^{n-1} f^{(n)}(w) \, dw$, then for each n,

$$R_n(x) - R_{n+1}(x) = \frac{f^{(n)}(a)}{n!} (x - a)^n$$

[*Hint*: Use integration by parts on R_{n+1}.]

7. Use problem D.6 to show that if $f: [a,b] \to \mathbf{R}^1$ and $f^{(n+1)}$ is continuous on $[a,b]$, then for each $x \in [a,b]$,

$$f(x) = f(a) + f'(a)(x - a) + \frac{f''(a)}{2!} (x - a)^2 + \cdots + \frac{f^{(n)}(a)}{n!} (x - a)^n$$
$$+ R_{n+1}(x)$$

where R_n is defined as in the preceding problem.

8. Use the fact that the derivative of arctan x is $1/(1 + x^2)$ to show that

$$\pi = 4 \sum_{n=0}^{\infty} (-1)^n \frac{1}{2n + 1}$$

[*Hint*: Write $1/(1 + x^2)$ as a geometric series.]

9. Work out the remaining details in (8.D.5).

10. Show that if f is a real-valued function that is infinitely differentiable on the interval $I = [x_0 - r, x_0 + r]$ and if $R_{n+1}(x)$ (defined in D.6) converges uniformly to 0 on I as $n \to \infty$, then f may be represented by its Taylor series expansion about the point x_0.

Section E

1. Find the radius and interval of convergence of the following power

series. Check the endpoints of the interval of convergence for convergence of the given series.

(a) $1 + \dfrac{x^3}{2!} + \dfrac{x^6}{3!} + \dfrac{x^9}{4!} + \cdots$

(b) $\displaystyle\sum_{n=1}^{\infty} (-1)^n \dfrac{n^2}{3^n} (x - 2)^n$

(c) $1 + (x + 2) + \tfrac{1}{2}(x + 2)^2 + \tfrac{1}{3}(x + 2)^3 + \cdots$

2. Find the radius and interval of convergence of the following power series; find also the interval of convergence of the derivative of each power series and check the endpoints of these intervals for convergence. Note that convergence of the original series and the derivative series may differ at these endpoints.

(a) $\displaystyle\sum_{n=1}^{\infty} \dfrac{x^n}{n!}$

(b) $\displaystyle\sum_{n=1}^{\infty} \dfrac{2 \cdot 4 \cdot 6 \cdots (2n)}{1 \cdot 3 \cdot 5 \cdots (2n - 1)} x^n$

(c) $\displaystyle\sum_{n=1}^{\infty} \dfrac{(-1)^n (x - 1)^n}{n}$

3. Suppose that $\sum_{n=1}^{\infty} a_n x^n$ has a radius of convergence r. Let q be a positive integer. Find the radii of convergence of the following series:

(a) $\displaystyle\sum_{n=1}^{\infty} a_n^q x^n$

(b) $\displaystyle\sum_{n=1}^{\infty} a_n x^{qn}$

4. Find a Taylor series expansion of $f(x) = (\sin x)/(1 - x)$ about $x = 0$. What is the interval of convergence for this expansion? [*Hint*: Note that

$$\frac{\sin x}{1 - x} = (\sin x) \frac{1}{1 - x}$$

and that $1/(1 - x)$ is the sum of a geometric series.]

5. Determine if the following series converge uniformly in the indicated interval.

(a) $\displaystyle\sum_{n=0}^{\infty} (-1)^n x^n \qquad [-\tfrac{1}{2}, \tfrac{1}{2}]$

(b) $\displaystyle\sum_{n=0}^{\infty} (-1)^n x^n \qquad [0, 1)$

(c) $\displaystyle\sum_{n=0}^{\infty} (-1)^n \dfrac{n^2}{4^n} (x - 1)^n \qquad [0, 2]$

6. Show that if $\sum_{n=1}^{\infty} a_n x^n = 0$ for each $x \in [a,b]$, then $a_n = 0$ for all n.

7. Show that if $\sum_{n=1}^{\infty} a_n$ diverges and if $\{a_n\}$ is a bounded sequence, then the

radius of convergence of the power series $\sum_{n=1}^{\infty} a_n x^n$ is 1.

8. Find a Taylor series expansion of $(1 - x^2)^{-1/2}$ and use this result to expand $(1 - x^2)^{-5/2}$.

9. (a) Show that $1 + x + x^2 + \cdots = 1/(1 - x)$ $(-1 < x < 1)$.
   (b) Square both sides of the previous equation to find a Taylor series expansion of $(1 - x)^{-2}$. Is this operation valid?
   (c) Differentiate both sides of the equation given in (a) to obtain a Taylor series expansion of $(1 - x)^{-2}$. Is this operation valid?

10. Suppose that $\sum_{n=0}^{\infty} a_n$ converges to A and set $f(x) = \sum_{n=0}^{\infty} a_n x^n$ for $-1 < x < 1$. Show that $\lim_{x \to 1} f(x) = \sum_{n=0}^{\infty} a_n$. [Hint: Let $s_n = \sum_{i=0}^{n} a_i$ and show that $\sum_{n=0}^{k} a_n x^n = (1 - x)(\sum_{n=0}^{k-1} s_n x^n + s_k x^k)$ and that $f(x) = (1 - x)\sum_{n=0}^{\infty} s_n x^n$; consider $|f(x) - A|$.]

11. Use problem E.10 to prove Abel's result (8.B.23). [*Hint:* In the notation of (8.B.23), let $A(x) = \sum_{n=0}^{\infty} a_n x^n$, $B(x) = \sum_{n=0}^{\infty} b_n x^n$, and $C(x) = \sum_{n=0}^{\infty} c_n x^n$ for $0 \le x \le 1$. Show that $A(x)B(x) = C(x)$, and then apply the preceding problem.]

12. Suppose that (X,d) is a metric space, A is a compact subset of X, and that $x^* \in X$. Show that there is a point $a^* \in A$ such that for each $x \in A$, $d(x^*,x) \le d(x^*,a^*)$.

13. Suppose that the series $a_0 + a_1 x + a_2 x^2 + \cdots$ has radius of convergence r.
   (a) Show that if $a_0 = 1$, then

$$\frac{1}{\sum_{n=0}^{\infty} a_n x^n} = \sum_{n=0}^{\infty} b_n x^n$$

   where $b_0 = 1$, $b_1 = -a_1 b_0$, $b_2 = -(a_1 b_1 + a_2 b_0), \ldots, b_n = -\sum_{j=1}^{n} a_j b_{n-j}, \ldots$.
   (b) Show that if $0 < q < R$, then there is a real number $C > 1$ such that $|a_n| \le C/q^n$ for each n.
   (c) Show that for each n, $|b_n| \le 2^n C^n/q^n$ and that the radius of convergence of $\sum_{n=0}^{\infty} b_n$ is greater than or equal to $q/2C$.

14. Formulate and prove a result analogous to (8.E.3) where lim is replaced by lim sup and lim inf (see problems B.14 to B.17).

15. Suppose that the series $\sum_{n=0}^{\infty} a_n(x - x_0)^n$ has the property that for some fixed positive integers k and M

$$a_{km} \ne 0 \qquad \text{if } m = M, M + 1, \ldots$$
$$a_n = 0 \qquad \text{for all other } n > M$$

Show that if

$$\lim \left| \frac{a_{k(m+1)}}{a_{km}} \right| = \frac{1}{r}$$

then the series $\sum_{n=0}^{\infty} a_n |x - x_0|^n$ converges if $|x - x_0| < r^{1/k}$ and diverges if $|x - x_0| > r^{1/k}$.

16. (a) Show that the Taylor series expansion of $f(x) = \sqrt{1 - x}$ about 0 is given by

$$T_0^f(x) = 1 - \frac{1}{2}x - \frac{1}{2^2 \cdot 2!}x^2 - \frac{3}{2^3 \cdot 3!}x^3 - \frac{3 \cdot 5}{2^4 \cdot 4!}x^4 - \cdots$$

$$- \frac{3 \cdot 5 \cdots (2n - 3)}{2^n \cdot n!} x^n - \cdots$$

(b) Show that $T_0^f(1)$ converges and, hence, T_0^f converges uniformly on the closed interval $[-1,1]$. [*Hint:* Use Raabe's test (problem B.22*).]

(c) Show that if $\frac{3}{4} < x < 1$, then

$$|f^{(n)}(x)| > \frac{3 \cdot 5 \cdots (2n - 3)}{2}$$

and

$$\left| \frac{f^{(n)}(x)}{n!} \right| > \tfrac{1}{4}(\tfrac{2}{3})^{n-5}$$

conclude that (8.D.8) is not sufficient to show that $f(x) = T_0(x)$ on $[-1,1]$.

Section F

1. Establish equation (4).
2. Establish equation (5).
3. A formal series $\sum_{k=0}^{\infty} a_k(1/x^k)$ is said to be an *asymptotic series* of f as $x \to \infty$ if for each integer $n \geq 0$, there is a function $R_{n+1}(x)$ such that

$$f(x) \sim \sum_{k=0}^{n} \frac{a_k}{x^k} + R_{n+1}(x)$$

where $x^{n+1} R_{n+1}(x)$ remains bounded as $x \to \infty$; we write $f(x) \sim \sum_{k=0}^{\infty} a_k/x^k$ ($x \to \infty$).

(a) Show that if $f(x) \sim \sum_{k=0}^{\infty} a_k/x^k$ ($x \to \infty$), then

$$\lim_{x \to \infty} x^{n+1} \left(f(x) - \sum_{k=0}^{\infty} \frac{a_k}{x^k} \right) = a_{n+1} \qquad (*)$$

i.e., $\lim R_{n+1}(x)/(a_{n+1}/x^{n+1}) = 1$.

(b) Show that if ($*$) is true for each integer $n \geq 0$, then $f(x) \sim \sum_{k=0}^{\infty} a_k/x^k$ ($x \to \infty$).

4. Show that $e^{-x} \subset 0 + 0/x + 0/x^2 + \cdots (x \to \infty)$, and that

$$\frac{1}{1+x} \sim \sum_{k=0}^{\infty} \frac{(-1)^{n+1}}{x^n} \quad (x \to \infty)$$

5. Show that if $\sum_{k=0}^{\infty} a_k/x_k \sim f(x) \sim \sum_{k=0}^{\infty} b_k/x^k \ (x \to \infty)$, then $a_k = b_k$ for each k.

6. Show that there can be an infinite number of functions with the same asymptotic series.

7. Suppose that $F(x) = \int_0^{\infty} e^{-xt} f(t) \, dt$, where $f^{(n)}$ on $[0,\infty)$ exists for each n. Suppose further that there are constants K and a such that $|f(t)| \leq Ke^{at}$, $0 \leq t < \infty$. Show that $F(x) \sim \sum_{n=0}^{\infty} f^{(n)}(0)/x^{n+1} \ (x \to \infty)$.

8. Suppose that $f(x) \sim \sum_{n=0}^{\infty} a_n/x^n \ (x \to \infty)$ and that f is continuous on the interval $[b,\infty)$.

   (a) Show that $\displaystyle\int_x^{\infty} \left(f(t) - a_0 - \frac{a_1}{t} \right) dt$ exists if $x > b$.

   (b) Show that $\displaystyle\int_x^{\infty} \left(f(t) - a_0 - \frac{a_1}{t} \right) dt \sim \frac{a_2}{x} + \frac{a_3}{2x^2} + \frac{a_4}{3x^3} + \cdots$

   $$(x \to \infty)$$

   (c) Show that $\displaystyle\int_a^x f(t) \, dt \sim C + a_0 x + a_1 \ln x - \frac{a_2}{x} - \frac{a_3}{2x^2} + \frac{a_4}{3x^3} - \cdots$

   $(x \to \infty)$, where $C = -a_0 a - a_1 \ln a + \displaystyle\int_a^{\infty} \left(f(t) - a_0 - \frac{a_1}{t} \right) dt$.

9. Show that if $f(x) = e^{-x} \cos e^x$, then $f(x) \sim 0 + 0/x + 0/x^2 + \cdots (x \to \infty)$. Show further that $f'(x)$ does not have an asymptotic expansion as $x \to \infty$. [*Hint*: Show that $f'(x)$ oscillates as $x \to \infty$, and observe that such a function can not have an asymptotic expansion as $x \to \infty$; use problem F.3 (a).]

ELEMENTARY APPLICATIONS
OF INFINITE SERIES

Infinite series of numbers and of functions can be employed in many different contexts, some of which we have already considered. In the present chapter we continue this study by examining a number of additional but rather specialized applications of series.

A. THE IRRATIONALITY OF e

An interesting proof based on infinite series can be given to show that the number e is irrational. Suppose that e is rational, say $e = p/q$, where p and q are integers and $q > 1$. Since the Taylor expansion of e^x is given by

$$e^x = 1 + x + \frac{x^2}{2!} + \frac{x^3}{3!} + \cdots$$

we have

$$\frac{p}{q} = e^1 = 1 + 1 + \frac{1}{2!} + \frac{1}{3!} + \cdots \tag{1}$$

Multiplication of both sides of equation (1) by $q!$ yields

$$p(q-1)! = q!\left(1 + 1 + \frac{1}{2!} + \frac{1}{3!} + \cdots + \frac{1}{q!}\right) + q!\left(\frac{1}{(q+1)!}\right.$$
$$\left. + \frac{1}{(q+2)!} + \cdots\right)$$

and hence,

$$p(q-1)! - q!\left(1 + 1 + \frac{1}{2!} + \cdots + \frac{1}{q!}\right)$$

$$= q!\left(\frac{1}{(q+1)!} + \frac{1}{(q+2)!} + \cdots\right)$$

$$= \frac{1}{(q+1)} + \frac{1}{(q+1)(q+2)} + \frac{1}{(q+1)(q+2)(q+3)} + \cdots$$

$$\leq \frac{1}{(q+1)} + \frac{1}{(q+1)^2} + \frac{1}{(q+1)^3} + \cdots$$

$$= \frac{1/(q+1)}{1 - (1/(q+1))} = \frac{1}{q} < 1. \tag{2}$$

Thus, we have that

$$p(q-1)! - q!\left(1 + 1 + \frac{1}{2!} + \cdots + \frac{1}{q!}\right)$$

is a positive integer less than 1, an obvious impossibility; hence, e must be irrational.

B. CALCULATION OF LIMITS

Taylor series expansions are useful in the calculation of certain limits. As an illustration of this, we shall find

$$\lim_{x \to 0} \frac{\cos x - (1 + x^2)^{1/2}}{\sin^2 x}$$

by rewriting this limit in terms of Taylor expansions. The following Taylor expansions are easily verified:

$$\cos x = 1 - \frac{x^2}{2!} + \frac{x^4}{4!} - \frac{x^6}{6!} + \cdots$$

$$(1 + x^2)^{1/2} = 1 + \frac{x^2}{2} - \frac{x^4}{8} + \frac{x^6}{16} + \cdots$$

$$\sin^2 x = \left(x - \frac{x^3}{3!} + \frac{x^5}{5!} - \cdots\right)^2 = x^2 - \frac{x^4}{3} + \frac{2x^6}{45} - \cdots \quad \text{(Cauchy product)}$$

Thus, we have

$$\lim_{x \to 0} \frac{\cos x - (1 + x^2)^{1/2}}{\sin^2 x} = \lim_{x \to 0} \frac{\begin{array}{c}(1 - x^2/2! + x^4/4! - x^6/6! + \cdots) \\ - (1 + x^2/2 - x^4/8 + x^6/16 - \cdots)\end{array}}{x^2 - x^4/3 + 2x^6/45 - \cdots}$$

$$= \lim_{x \to 0} \frac{-x^2 + x^4/6 - 23x^6/360 + \cdots}{x^2 - x^4/3 + 2x^6/45 - \cdots}$$

$$= \lim_{x \to 0} \frac{-1 + x^2/6 - 23x^4/360 + \cdots}{1 - x^2/3 + 2x^4/45 - \cdots} = -1$$

C. THE APPROXIMATION OF DEFINITE INTEGRALS

Taylor expansions can be of decided use in approximating definite integrals in cases where it is difficult or impossible to find antiderivatives. For instance, finding a primitive of $f(x) = \cos x^3$ is virtually impossible, and hence, the fundamental theorem of calculus is of little use in evaluating $\int_a^b \cos x^3 \, dx$. The Taylor expansion of $\cos x^3$ about 0 is given by

$$T(x) = 1 - \frac{(x^3)^2}{2!} + \frac{(x^3)^4}{4!} - \frac{(x^3)^6}{6!} + \cdots$$

and it follows from (8.D.8) that $T(x) = f(x)$ for all x, and that the convergence is uniform on $[a,b]$. Therefore, by (8.C.7.iii) we have

$$\int_a^b \cos x^3 \, dx = \int_a^b T(x) \, dx = \int_a^b 1 \, dx - \int_a^b \frac{x^6}{2!} \, dx + \int_a^b \frac{x^{12}}{4!} \, dx - \cdots$$

With a suitable truncation of this series of integrals, $\int_a^b \cos x^3 \, dx$ can be calculated to an arbitrary degree of accuracy; error estimates can, of course, be determined by the remainder term [see (8.D.8) and problem D.7, Chapter 8].

D. INFINITE SERIES AND DIFFERENTIAL EQUATIONS

The use of infinite series provides the basis of an interesting and important method of finding solutions to differential equations. To illustrate this procedure, we shall deal with equations of the form

$$p(t)y''(t) + q(t)y'(t) + r(t)y(t) = 0 \tag{1}$$

where p, q, and r are polynomials. We first consider the case where $p(0) \neq 0$. Given the equation

$$y'' + ty' + 3y = 0 \tag{2}$$

we assume that there is a solution of (1) of the form

$$y(t) = \sum_{n=0}^{\infty} a_n t^n \tag{3}$$

where the coefficients a_n are to be determined. It is also assumed that y has a positive radius of convergence. From (8.E.9) we have

$$y'(t) = \sum_{n=1}^{\infty} n a_n t^{n-1}$$

$$y''(t) = \sum_{n=2}^{\infty} n(n-1) a_n t^{n-2}$$

Substitution of y, y', and y'' in (2) yields

$$0 = \sum_{n=2}^{\infty} n(n-1) a_n t^{n-2} + t \sum_{n=1}^{\infty} n a_n t^{n-1} + 3 \sum_{n=0}^{\infty} a_n t^n$$

$$= \sum_{n=2}^{\infty} n(n-1) a_n t^{n-2} + \sum_{n=1}^{\infty} n a_n t^n + \sum_{n=0}^{\infty} 3 a_n t^n \tag{4}$$

In order to combine coefficients, we reindex the series $\sum_{n=2}^{\infty} n(n-1) a_n t^{n-2}$

$$\sum_{n=2}^{\infty} n(n-1) a_n t^{n-2} = \sum_{n=0}^{\infty} (n+2)(n+1) a_{n+2} t^n$$

It follows from (4) and the preceding equation that

$$0 = (2a_2 + 3a_0) + (3 \cdot 2a_3 + 4a_1)t$$

$$+ \sum_{n=2}^{\infty} ((n+2)(n+1) a_{n+2} + n a_n + 3 a_n) t^n$$

Since the left-hand side of this equation is identically 0, we have

$$a_2 = -\frac{3}{2} a_0 \qquad a_3 = -\frac{4}{3 \cdot 2} a_1$$

and for $n \geq 2$,

$$a_{n+2} = -\frac{(n+3) a_n}{(n+2)(n+1)}$$

hence,

$$a_4 = -\frac{5a_2}{4 \cdot 3} = \frac{-5}{4 \cdot 3} \cdot \frac{-3}{2} a_0$$

$$a_5 = -\frac{6a_3}{5 \cdot 4} = \frac{-6}{5 \cdot 4} \cdot \frac{-4}{3 \cdot 2} a_1$$

$$a_6 = -\frac{7a_4}{6\cdot5} = \frac{-7}{6\cdot5}\cdot\frac{-5}{4\cdot3}\cdot\frac{-3}{2}a_0$$

$$a_7 = -\frac{8a_5}{7\cdot6} = \frac{-8}{7\cdot6}\cdot\frac{-6}{5\cdot4}\cdot\frac{-4}{3\cdot2}\cdot a_1$$

In general, we find

$$a_{2n} = (-1)^n \frac{(2n + 1)(2n - 1)\cdots3}{(2n)!}a_0 \qquad n = 1, 2, \ldots$$

$$a_{2n-1} = (-1)^{n+1} \frac{(2n)(2n - 2)\cdots4}{(2n - 1)!}a_1 \qquad n = 2, 3, \ldots$$

Since a_0 and a_1 may be chosen arbitrarily, we can set $a_0 = 1$, $a_1 = 0$ and $a_0 = 0$, $a_1 = 1$, to obtain two independent solutions of (1):

$$y_1(t) = 1 + \sum_{n=1}^{\infty} (-1)^n \frac{(2n + 1)(2n - 1)\cdots3}{(2n)!}t^{2n} \qquad (5)$$

$$y_2(t) = 1 + \sum_{n=2}^{\infty} (-1)^{n+1} \frac{(2n)(2n - 2)\cdots4}{(2n - 1)!}t^{2n-1} \qquad (6)$$

(the Wronskian of y_1 and y_2 at $t = 0$ shows that these solutions are independent). Consequently, any solution of (1) can be written in the form

$$y(t) = a_0 y_1(t) + a_1 y_2(t)$$

where y_1 and y_2 are defined by (5) and (6), and a_0 and a_1 are arbitrary constants.

It is not necessary that the proposed solution of (1) be a series centered at 0. If $p(t_0) \neq 0$, then the above procedure may be initiated using

$$y(t) = \sum_{n=0}^{\infty} a_n(t - t_0)^n$$

In the case that we wish to use a series solution about a point t_0, where $p(t_0) = 0$, then more care must be exercised. It can be shown that if in (1)

$$(t - t_0)\frac{q(t)}{p(t)}$$

and

$$(t - t_0)^2 \frac{r(t)}{p(t)}$$

have convergent Taylor series expansions about the point t_0 (such a point is called a *regular singular point*), then a slight modification of the previous procedure (called the *Frobenius method*) will work. To illustrate this method, we consider the *Bessel equation* of order p (p a nonnegative real number)

$$t^2 y'' + t y' + (t^2 - p^2)y = 0 \qquad (7)$$

(Here $t_0 = 0$.)

We first consider the special case of (7) where $p = 0$. The basic procedure for finding a solution of this Bessel equation goes as follows. It is supposed that there is a solution of (7) of the form

$$y(t) = t^r \sum_{n=0}^{\infty} a_n t^n = \sum_{n=0}^{\infty} a_n t^{r+n}$$

where the coefficients a_n and the exponent r are to be determined; it is also assumed that $a_0 \neq 0$ and that $\sum_{n=0}^{\infty} a_n t^{r+n}$ has a positive radius of convergence.

From (8.E.9) we have

$$y'(t) = \sum_{n=0}^{\infty} (r + n) a_n t^{r+n-1}$$

$$y''(t) = \sum_{n=0}^{\infty} (r + n)(r + n - 1) a_n t^{r+n-2}$$

Substitution of y, y', and y'' in (7) yields

$$0 = t^2 \sum_{n=0}^{\infty} (r + n)(r + n - 1) a_n t^{r+n-2} + t \sum_{n=0}^{\infty} (r + n) a_n t^{r+n-1}$$

$$+ t^2 \sum_{n=0}^{\infty} a_n t^{r+n}$$

$$= \sum_{n=0}^{\infty} (r + n)(r + n - 1) a_n t^{r+n} + \sum_{n=0}^{\infty} (r + n) a_n t^{r+n} + \sum_{n=0}^{\infty} a_n t^{r+n+2} \quad (8)$$

In order to combine coefficients we reindex the series $\sum_{n=0}^{\infty} a_n t^{r+n+2}$.

$$\sum_{n=0}^{\infty} a_n t^{r+n+2} = \sum_{n=2}^{\infty} a_{n-2} t^{r+n}$$

It follows from (8) that

$$0 = (r(r - 1) + r) a_0 t^r + ((r + 1)r + (r + 1)) a_1 t^{r+1}$$

$$+ \sum_{n=2}^{\infty} [((r + n)(r + n - 1) + (r + n)) a_n + a_{n-2}] t^{r+n} \quad (9)$$

Since the left-hand side of this equation is 0 and $a_0 \neq 0$, we have

$$r(r - 1) + r = 0 \quad (10)$$

Equation (10) is called the *indicial equation*; from this equation we see that $r = 0$. Since the coefficient of t must be 0 and since $r = 0$, it follows that $a_1 = 0$. For $n \geq 2$ (and $r = 0$), equation (9) yields the "recurrence formula"

$$0 = (n(n - 1) + n) a_n + a_{n-2} \quad (11)$$

which permits calculation of a_2 in terms of a_0, calculation of a_3 in terms of a_1, calculation of a_4 in terms of a_2 and, hence, in terms of a_0, etc. More precisely, from (11) we have, for $n \geq 2$,

$$a_n = \frac{-a_{n-2}}{n^2}$$

and therefore,

$$a_2 = -\frac{a_0}{2^2}$$

$$a_3 = -\frac{a_1}{3^2} = 0$$

$$a_4 = -\frac{a_2}{4^2} = \frac{a_0}{2^2 4^2}$$

$$a_5 = -\frac{a_3}{5^2} = 0$$

$$a_6 = -\frac{a_4}{6^2} = \frac{-a_0}{2^2 4^2 6^2}$$

Consequently, for odd positive integers we have $a_1 = a_3 = a_5 = \cdots = 0$, and for even positive integers

$$a_{2n} = \frac{(-1)^n a_0}{2^{2n}(n!)^2}$$

Thus, one solution of the Bessel equation of order 0 is given by

$$J_0(t) = a_0 \left(1 + \sum_{n=1}^{\infty} \frac{(-1)^n t^{2n}}{2^{2n}(n!)^2} \right) \tag{12}$$

where $a_0 \neq 0$ is an arbitrary constant. It can be shown that $K_0(t) = \sum_{k=0}^{\infty} b_k t^{k+1} + J_0(t) \ln t$ for suitably defined coefficients b_n yields another solution such that K_0 and J_0 are independent (on any interval not containing 0). [See Brauer and Nohel (1973).]

(9.D.1) Exercise Use the ratio test to show that the series given in (12) converges for all t.

In the general case, where p is a nonnegative real number, substitution of $\sum_{n=0}^{\infty} a_n t^{r+n}$ for y in (7) yields

$$0 = t^2 \sum_{n=0}^{\infty} (r+n)(r+n-1)a_n t^{r+n-2} + t \sum_{n=0}^{\infty} (r+n)a_n t^{r+n-1}$$

$$+ t^2 \sum_{n=0}^{\infty} a_n t^{r+n} - p^2 \sum_{n=0}^{\infty} a_n t^{r+n}$$

and after the usual reindexing, we obtain

$$0 = (r(r - 1) + r - p^2)a_0 + ((r + 1)r + (r + 1) - p^2)a_1 t$$

$$+ \sum_{n=2}^{\infty} ((r + n)(r + n - 1) + (r + n) - p^2)a_n + a_{n-2})t^n \quad (13)$$

The indicial equation $r^2 - p^2 = 0$ has roots $r = \pm p$; and if we set $r = p$, then equation (13) becomes

$$0 = (2p + 1)a_1 t + \sum_{n=2}^{\infty} (n(2p + n)a_n + a_{n-2})t^n$$

Consequently, we see that

$$a_1 = 0$$

and, in general, for $n \geq 2$,

$$a_n = \frac{-a_{n-2}}{n(2p + n)}.$$

Therefore,

$$a_1 = a_3 = a_5 = \cdots = 0$$

and

$$a_2 = \frac{-a_0}{2(2p + 2)}$$

$$a_4 = \frac{a_0}{2 \cdot 4(2p + 2)(2p + 4)}$$

$$\cdots\cdots\cdots\cdots\cdots\cdots\cdots\cdots$$

$$a_{2n} = (-1)^n \frac{a_0}{2 \cdot 4 \cdots (2n)(2p + 2)(2p + 4) \cdots (2p + 2n)}$$

$$= (-1)^n \frac{a_0}{2^{2n} n!(p + 1)(p + 2) \cdots (p + n)}$$

and hence, a solution to the Bessel equation of order p is given by

$$y(t) = a_0 \left(t^p + \sum_{n=1}^{\infty} \frac{(-1)^n t^{2n+p}}{2^{2n} n!(p + 1)(p + 2) \cdots (p + n)} \right)$$

Note that if Γ denotes the Γ function, then

$$\Gamma(p + n + 1) = (p + n)\Gamma(p + n - 1 + 1)$$

$$= (p + n)(p + n - 1)\Gamma(p + n - 1) = \cdots$$

$$= (p + n)(p + n - 1) \cdots (p + 1)\Gamma(p + 1)$$

and therefore,

$$(p + 1)(p + 2) \cdots (p + n) = \frac{\Gamma(p + n + 1)}{\Gamma(p + 1)}$$

Since $n! = \Gamma(n + 1)$, if we set

$$a_0 = \frac{1}{2^p \Gamma(p + 1)}$$

then a solution to (7) takes on the form

$$y(t) = \sum_{n=0}^{\infty} \frac{(-1)^n}{\Gamma(n+1)\Gamma(n+p+1)}\left(\frac{t}{2}\right)^{2n+p} \tag{14}$$

The function defined in (14) is called the *Bessel function of order p of the first kind* and is customarily denoted by $J_p(t)$. In the problem set the reader is asked to show that if p is not an integer, then a second independent solution of (7) is given by

$$J_{-p}(t) = \sum_{n=0}^{\infty} \frac{(-1)^n}{\Gamma(n+1)\Gamma(n-p+1)}\left(\frac{t}{2}\right)^{2n-p} \qquad (t > 0)$$

and hence, for p not an integer, the general solution of Bessel's equation of order p is

$$y = c_1 J_p + c_2 J_{-p} \qquad (t > 0)$$

If p is a positive integer then J_{-p} is a solution of (7), but it is not independent of J_p. Finding a second independent solution to (7) is possible but fairly difficult; the interested reader may consult Rabenstein (1972) for details.

Solutions to (7) are called *Bessel functions*. Although these functions were utilized by Daniel Bernoulli as early as 1732, it was Friedrich Bessel (1784–1846), a self-taught astronomer, who, in his investigations of planetary motion, first undertook a systematic study of such functions. Among Bessel's more notable accomplishments was the calculation of the orbit of Halley's comet; he also was the first astronomer to calculate the earth's distance from a particular star (the star now known as 61 Cygnus). Bessel functions arise naturally in studies involving fluid motion, wave propagation, elasticity, etc. The literature on Bessel functions is enormous; Watson's "A Treatise on the Theory of Bessel Functions," itself a book of 752 pages, includes a 36-page bibliography.

E. INFINITE SERIES OF MATRICES AND LINEAR SYSTEMS OF DIFFERENTIAL EQUATIONS

Infinite series of matrices prove to be of interest in a number of contexts. In particular, in this section we shall see how such series may be used to obtain solutions to linear systems of differential equations.

(9.E.1) Definition Suppose that $A_0 = (a_{ij}^{(0)})$, $A_1 = (a_{ij}^{(1)})$, $A_2 = (a_{ij}^{(2)})$, ... is a sequence of $m \times n$ matrices and that for each pair of integers i, j the series $\sum_{n=0}^{\infty} a_{ij}^{(n)}$ converges. Then the *infinite sum of the matrices* A_0, A_1, A_2, ... is defined to be the $m \times n$ matrix, $C = (c_{ij})$, where $c_{ij} = \sum_{n=0}^{\infty} a_{ij}^{(n)}$ for each i, j.

(9.E.2) *Example*

$$\begin{pmatrix} \frac{1}{2} & 1 \\ 0 & -\frac{1}{3} \end{pmatrix} + \begin{pmatrix} \frac{1}{4} & 1 \\ 0 & \frac{1}{9} \end{pmatrix} + \begin{pmatrix} \frac{1}{2^3} & \frac{1}{2!} \\ 0 & -\frac{1}{3^3} \end{pmatrix} + \begin{pmatrix} \frac{1}{2^4} & \frac{1}{3!} \\ 0 & \frac{1}{3^4} \end{pmatrix} + \cdots = \begin{pmatrix} 1 & e \\ 0 & -\frac{1}{4} \end{pmatrix}$$

(9.E.3) *Definition* If A is an $n \times n$ matrix, then e^A is defined to be the matrix

$$I + A + \frac{A^2}{2!} + \frac{A^3}{3!} + \cdots$$

(9.E.4) *Observation* In order for e^A to make sense, it must be established that for each pair i, j the (i,j)th entry c_{ij} of the matrix $I + A + A^2/2! + A^3/3! + \cdots$ is a convergent series. To see that this is the case, let q be any number greater than max $\{|a_{ij}| \mid 1 \le i, j \le n\}$. An easy inductive argument shows that if $a_{ij}^{(m)}$ is the (i,j)th element of A^m, then $a_{ij}^{(m)} \le n^m q^m$, and, therefore, since

$$c_{ij} = a_{ij}^{(0)} + a_{ij}^{(1)} + \frac{a_{ij}^{(2)}}{2!} + \frac{a_{ij}^{(3)}}{3!} + \cdots \le 1 + nq + \frac{(nq)^2}{2!} + \frac{(nq)^3}{3!} + \cdots = e^{nq}$$

$$(1)$$

we have that e^A is well defined for any matrix A.

The derivative of a matrix of functions is defined as follows.

(9.E.5) *Definition* Suppose that for $0 \le i, j \le n$, $a_{ij}(t)$ is a function mapping a subset B of $\mathbf{R}^1$ into $\mathbf{R}^1$. For each $t \in B$, let $A(t)$ be the matrix $(a_{ij}(t))$. If $a'_{ij}(t_0)$ exists for each i, j, then $A'(t_0)$ is defined to be the matrix whose (i,j)th entry is $a'_{ij}(t_0)$.

(9.E.6) *Example* If

$$A(t) = \begin{pmatrix} \sin t & t^2 \\ e^{-t} & 4 \end{pmatrix}$$

then

$$A'(t) = \begin{pmatrix} \cos t & 2t \\ -e^{-t} & 0 \end{pmatrix}$$

(9.E.7) *Exercise* (a) Show that if $A(t) = (a_{ij}(t))$ and $B(t) = (b_{ij}(t))$ are two matrices of functions (with a common domain), and if $A'(t)$ and $B'(t)$ exist, then the derivative of the matrix $C(t) = A(t) + B(t)$ is equal to $A'(t) + B'(t)$, and the derivative of the matrix $D(t) = A(t)B(t)$ is $A'(t)B(t) + A(t)B'(t)$.

(b) Show that if $g(t) = e^{tA}$, where $t \in \mathbf{R}^1$ and A is an $n \times n$ matrix with scalar entries, then $g'(t) = Ae^{At}$. [*Hint*: Note that $(tA)^m = t^m A^m$ and use (8.E.9).]

(c) Show that if $AB = BA$, then $e^A e^B = e^{A+B}$.

(d) Show that $e^{-tA}e^{tA} = I$ for all constant $n \times n$ matrices.

(9.E.8) *Definition* If $A(t) = (a_{ij}(t))$, then $\int_a^b A(t)\, dt$ is the matrix whose (i,j)th entry is $\int_a^b a_{ij}(t)\, dt$, provided that $\int_a^b a_{ij}(t)\, dt$ exists for each i, j.

We now see how these ideas can be applied to linear systems of differential equations.

The general form of a linear system of first-order ordinary differential equations for unknown functions $y_1, y_2, \ldots, y_n$ is given by

$$
\begin{aligned}
y_1' &= a_{11}(t)y_1 + \cdots + a_{1n}(t)y_n + b_1(t) \\
y_2' &= a_{21}(t)y_1 + \cdots + a_{2n}(t)y_n + b_2(t) \\
&\cdots\cdots\cdots\cdots\cdots\cdots\cdots\cdots\cdots\cdots\cdots \\
y_n' &= a_{n1}(t)y_1 + \cdots + a_{nn}(t)y_n + b_n(t)
\end{aligned}
\tag{2}
$$

A *solution* of (2) on an interval (α,β) is an ordered set of n functions $(\phi_1, \phi_2, \ldots, \phi_n)$ with common domain (α,β) and with the property that for each i, $1 \leq i \leq n$, and for each $t \in (\alpha,\beta)$

$$
\phi_i'(t) = a_{i1}(t)\phi_1(t) + \cdots + a_{in}(t)\phi_n(t) + b_i(t)
$$

The *initial value problem* associated with (2) is that of finding a solution of (2) that satisfies the given initial conditions

$$
\begin{aligned}
y_1(t_0) &= c_1 \\
y_2(t_0) &= c_2 \\
&\cdots\cdots \\
y_n(t_0) &= c_n
\end{aligned}
\tag{3}
$$

where $t_0 \in (\alpha,\beta)$ and $c_1, c_2, \ldots, c_n$ are arbitrary real numbers.

A proof of the following theorem may be found in Coddington (1961).

(9.E.9) *Theorem* If in (2) each of the functions a_{ij} and b_i is continuous on an interval (α,β), then there is a unique solution of (2) that satisfies the initial conditions (3).

(9.E.10) *Exercise* Show by direct substitution that

$$
\phi_1(t) = \tfrac{4}{5}e^{4t} + \tfrac{11}{5}e^{-t} + 3t - 2
$$
$$
\phi_2(t) = \tfrac{6}{5}e^{4t} - \tfrac{11}{5}e^{-t} - 2t + 3
$$

is a solution to the linear system

$$y_1' = y_1 + 2y_2 + t - 1$$
$$y_2' = 3y_1 + 2y_2 - 5t - 2$$

that satisfies the initial conditions

$$y_1(0) = 1 \qquad y_2(0) = 2$$

In much of the sequel we shall be primarily interested in systems of differential equations where the coefficients $a_{ij}(t)$ are constant. Consider such a system:

$$y_1'(t) = a_{11}y_1(t) + a_{12}y_2(t) + \cdots + a_{1n}y_n(t) + b_1(t)$$
$$y_2'(t) = a_{21}y_1(t) + a_{22}y_2(t) + \cdots + a_{2n}y_n(t) + b_2(t)$$
$$\cdots\cdots\cdots\cdots\cdots\cdots\cdots\cdots\cdots\cdots\cdots\cdots\cdots$$
$$y_n'(t) = a_{n1}y_1(t) + a_{n2}y_2(t) + \cdots + a_{nn}y_n(t) + b_n(t)$$

with initial conditions $y_1(0) = c_1$, $y_2(0) = c_2, \ldots, y_n(0) = c_n$. This system is expressed in matrix notation by

$$Y'(t) = AY(t) + B(t) \qquad Y(0) = C \tag{4}$$

where

$$A = \begin{pmatrix} a_{11} & a_{12} & \cdots & a_{1n} \\ \cdots\cdots\cdots\cdots\cdots \\ a_{n1} & a_{n2} & \cdots & a_{nn} \end{pmatrix} \qquad B(t) = \begin{pmatrix} b_1(t) \\ \cdots \\ b_n(t) \end{pmatrix}$$

$$C = \begin{pmatrix} c_1 \\ \cdots \\ c_n \end{pmatrix} \qquad Y(t) = \begin{pmatrix} y_1(t) \\ \cdots \\ y_n(t) \end{pmatrix}$$

The *homogeneous equation* associated with (4) is

$$Y'(t) = AY(t) \qquad Y(0) = C \tag{5}$$

It follows from (9.E.7) and (9.E.9) that $Y(t) = e^{tA}C$ is the unique solution of (5). To find a solution of (4) if B is continuous, we try a solution of the form $e^{tA}Z(t)$. Since

$$(e^{tA}Z(t))' = Ae^{tA}Z(t) + e^{tA}Z'(t)$$

substitution of $e^{tA}Z(t)$ for $Y(t)$ in (4) yields

$$Ae^{tA}Z(t) + e^{tA}Z'(t) = Ae^{tA}Z(t) + B(t)$$

and hence, by (9.E.7.d), $e^{tA}Z'(t) = B(t)$ and $Z'(t) = e^{-tA}B(t)$.
Note that if $Y(t) = e^{tA}Z(t)$, then $Y(0) = Z(0)$ and, therefore, $Z(0) = C$.

Clearly,

$$Z(t) = \int_0^t e^{-xA}B(x)dx + C$$

is a solution of $Z'(t) = e^{-tA}B(t)$ such that $Z(0) = C$. Consequently, by (9.E.7.c) we have

$$\phi(t) = e^{tA}Z(t) = e^{tA}\left(\int_0^t e^{-xA}B(x)dx + C\right) = \int_0^t e^{(t-x)A}B(x)dx + e^{tA}C \quad (6)$$

is the desired solution of (4).

(9.E.11) Example Consider the equation

$$Y'(t) = \begin{pmatrix} 1 & 0 \\ 0 & 2 \end{pmatrix}Y(t) + \begin{pmatrix} e^{-2t} \\ -2e^t \end{pmatrix} \qquad Y(0) = \begin{pmatrix} 1 \\ 1 \end{pmatrix} \quad (7)$$

The matrix e^{tA}, where

$$A = \begin{pmatrix} 1 & 0 \\ 0 & 2 \end{pmatrix}$$

is found as follows:

$$e^{tA} = I + \sum_{n=1}^{\infty} \frac{t^n}{n!}\begin{pmatrix} 1 & 0 \\ 0 & 2 \end{pmatrix}^n = I + \sum_{n=1}^{\infty} \frac{1}{n!}\begin{pmatrix} t^n & 0 \\ 0 & (2t)^n \end{pmatrix}$$

$$= I + \begin{pmatrix} \sum_{n=1}^{\infty}\frac{t^n}{n!} & 0 \\ 0 & \sum_{n=1}^{\infty}\frac{(2t)^n}{n!} \end{pmatrix} = \begin{pmatrix} e^t & 0 \\ 0 & e^{2t} \end{pmatrix}$$

To find $Z(t)$, we observe that

$$Z(t) = \int_0^t e^{-xA}B(x)dx + \begin{pmatrix} 1 \\ 1 \end{pmatrix}$$

$$= \int_0^t \begin{pmatrix} e^{-x} & 0 \\ 0 & e^{-2x} \end{pmatrix}\begin{pmatrix} e^{-2x} \\ -2e^x \end{pmatrix}dx + \begin{pmatrix} 1 \\ 1 \end{pmatrix}$$

$$= \int_0^t \begin{pmatrix} e^{-3x} \\ -2e^{-x} \end{pmatrix}dx + \begin{pmatrix} 1 \\ 1 \end{pmatrix} = \begin{pmatrix} -(\tfrac{1}{3})e^{-3x} \\ 2e^{-x} \end{pmatrix}\Big|_0^t + \begin{pmatrix} 1 \\ 1 \end{pmatrix}$$

$$= \begin{pmatrix} -(\tfrac{1}{3})e^{-3t} + \tfrac{1}{3} \\ 2e^{-t} - 2 \end{pmatrix} + \begin{pmatrix} 1 \\ 1 \end{pmatrix}$$

Consequently, the unique solution of (7) is

$$\phi(t) = \begin{pmatrix} e^t & 0 \\ 0 & e^{2t} \end{pmatrix}\begin{pmatrix} -(\tfrac{1}{3})e^{-3t} + \tfrac{1}{3} \\ 2e^{-t} - 2 \end{pmatrix} + \begin{pmatrix} e^t & 0 \\ 0 & e^{2t} \end{pmatrix}\begin{pmatrix} 1 \\ 1 \end{pmatrix}$$

$$= \begin{pmatrix} -(\tfrac{1}{3})e^{-2t} + (\tfrac{4}{3})e^t \\ 2e^t - e^{2t} \end{pmatrix}$$

In practice, the success of finding the solution $\phi(t)$ via equation (5) depends in large part on being able to calculate e^{tA}. A number of results in this direction are based on the notion of eigenvalues and eigenvectors.

(9.E.12) *Definition*   Suppose that A is an $n \times n$ matrix (real or complex). The *characteristic polynomial* ϕ of A is defined by $\phi(\lambda) = \det (A - \lambda I)$. The zeros of the characteristic polynomial are called the *eigenvalues* of A.

Note that ϕ is a polynomial of degree n and of the form

$$\phi(\lambda) = (-1)^n \lambda^n + c_{n-1} \lambda^{n-1} + \cdots + c_1 \lambda + c_0$$

(9.E.13) *Example*   If

$$A = \begin{pmatrix} 3 & 4 \\ 4 & -3 \end{pmatrix}$$

then $\phi(\lambda) = \det \begin{pmatrix} 3 - \lambda & 4 \\ 4 & -3 - \lambda \end{pmatrix} = (3 - \lambda)(-3 - \lambda) - 16 = \lambda^2 - 25$

$$= (\lambda - 5)(\lambda + 5)$$

and the eigenvalues of A are $\lambda_1 = 5$ and $\lambda_2 = 5$.

(9.E.14) *Observation*   If λ is an eigenvalue of a matrix A, then by definition $\det (A - \lambda I) = 0$, and hence it follows from (2.B.3) that the matrix equation

$$(A - \lambda I)\mathbf{x} = \mathbf{0}$$

has nonzero solutions. Any such nonzero solution $\mathbf{x}$ is called an *eigenvector* belonging to the eigenvalue λ. Since $(A - \lambda I)\mathbf{x} = \mathbf{0}$ is equivalent to the equation $A\mathbf{x} = \lambda\mathbf{x}$, it follows that eigenvectors are precisely those nonzero vectors that are transformed by A into scalar multiples of themselves.

(9.E.15) *Example*   The characteristic polynomial of the matrix

$$A = \begin{pmatrix} 0 & 1 & 0 & 0 \\ -10 & 0 & 6 & 0 \\ 0 & 0 & 0 & 1 \\ 6 & 0 & -10 & 0 \end{pmatrix}$$

can be seen by direct calculation to be

$$\phi(\lambda) = \det (A - \lambda I) = \det \begin{pmatrix} -\lambda & 1 & 0 & 0 \\ -10 & -\lambda & 6 & 0 \\ 0 & 0 & -\lambda & 1 \\ 6 & 0 & -10 & -\lambda \end{pmatrix} = \lambda^4 + 20\lambda^2 + 64$$

and it is easy to see that the eigenvalues are

$$\lambda_1 = 2i \qquad \lambda_2 = -2i \qquad \lambda_3 = 4i \qquad \lambda_4 = -4i$$

To find an eigenvector belonging to λ_1, we solve the equation

$$\begin{pmatrix} 0 & 1 & 0 & 0 \\ -10 & 0 & 6 & 0 \\ 0 & 0 & 0 & 1 \\ 6 & 0 & -10 & 0 \end{pmatrix} \begin{pmatrix} c_1 \\ c_2 \\ c_3 \\ c_4 \end{pmatrix} = 2i \begin{pmatrix} c_1 \\ c_2 \\ c_3 \\ c_4 \end{pmatrix} \tag{8}$$

Routine calculations show that

$$c_2 = 2ic_1$$
$$-10c_1 + 6c_3 = 2ic_2$$
$$c_4 = 2ic_3$$
$$6c_1 - 10c_3 = 2ic_4$$

and that

$$\mathbf{C}_1 = \begin{pmatrix} c_1 \\ c_2 \\ c_3 \\ c_4 \end{pmatrix} = \begin{pmatrix} 1 \\ 2i \\ 1 \\ 2i \end{pmatrix}$$

is a solution of (8).

In a similar fashion, eigenvectors

$$\mathbf{C}_2 = \begin{pmatrix} 1 \\ -2i \\ 1 \\ -2i \end{pmatrix} \qquad \mathbf{C}_3 = \begin{pmatrix} 1 \\ 4i \\ -1 \\ -4i \end{pmatrix} \qquad \mathbf{C}_4 = \begin{pmatrix} 1 \\ -4i \\ -1 \\ 4i \end{pmatrix}$$

that correspond to the eigenvalues $\lambda_2 = -2i$, $\lambda_3 = 4i$, and $\lambda_4 = -4i$, respectively, may be obtained (see problem E.4).

It may happen that an $n \times n$ matrix fails to have n distinct eigenvalues, in which case there may not exist n independent eigenvectors. However, if n independent eigenvectors can be found, the following theorem will prove to be of considerable use in calculating e^{tA}.

(9.E.16)   Theorem   Suppose that $A = (a_{ij})$ is an $n \times n$ matrix with n *independent* eigenvectors

$$\mathbf{q}_1 = \begin{pmatrix} q_{11} \\ \cdots \\ q_{1n} \end{pmatrix} \qquad \mathbf{q}_2 = \begin{pmatrix} q_{21} \\ \cdots \\ q_{2n} \end{pmatrix} \qquad \cdots \qquad \mathbf{q}_n = \begin{pmatrix} q_{n1} \\ \cdots \\ q_{nn} \end{pmatrix}$$

belonging to the eigenvalues, $\lambda_1, \lambda_2, \ldots, \lambda_n$, respectively. Let Q and Λ be the $n \times n$ matrices defined by

$$Q = \begin{pmatrix} q_{11} & q_{21} & \cdots & q_{n1} \\ \cdots\cdots\cdots\cdots\cdots \\ q_{1n} & q_{2n} & \cdots & q_{nn} \end{pmatrix}$$

$$\Lambda = \begin{pmatrix} \lambda_1 & 0 & 0 & \cdots & 0 \\ 0 & \lambda_2 & 0 & \cdots & 0 \\ \cdots\cdots\cdots\cdots\cdots \\ 0 & 0 & 0 & \cdots & \lambda_n \end{pmatrix}$$

Then $Q^{-1}AQ = \Lambda$. (Two $n \times n$ matrices A and B are said to be *similar* if there is an invertible matrix Q such that $Q^{-1}AQ = B$; thus, under the hypotheses of the theorem we have that A is similar to the diagonal matrix whose entries along the diagonal are precisely the eigenvalues of A.)

Proof.   Note that the columns of AQ are equal to $A\mathbf{q}_1, A\mathbf{q}_2, \ldots, A\mathbf{q}_n$. Since for each i, $\mathbf{q}_i$ is an eigenvector belonging to λ_i, we have

$$A\mathbf{q}_i = \lambda_i\mathbf{q}_i$$

Now observe that the columns of the matrix $Q\Lambda$ are defined by $\lambda_1\mathbf{q}_1$, $\lambda_2\mathbf{q}_2, \ldots, \lambda_n\mathbf{q}_n$ and, therefore, $AQ = Q\Lambda$. It can be shown (see Cullen, 1966, for example) that an $n \times n$ matrix M is nonsingular if and only if the columns (or rows) of M, considered as vectors, are independent. Thus, since the columns of Q are assumed to be independent, we have that Q^{-1} exists and, consequently,

$$Q^{-1}AQ = \Lambda$$

(9.E.17)   *Exercise*   (a)   Show that if

$$D = \begin{pmatrix} d_1 & 0 & \cdots & 0 \\ 0 & d_2 & \cdots & 0 \\ \cdots\cdots\cdots\cdots\cdots \\ 0 & 0 & \cdots & d_n \end{pmatrix}$$

is a diagonal matrix, then

$$e^D = \begin{pmatrix} e^{d_1} & 0 & \cdots & 0 \\ 0 & e^{d_2} & \cdots & 0 \\ \cdots\cdots\cdots\cdots\cdots \\ 0 & 0 & \cdots & e^{d_n} \end{pmatrix}$$

(b)   Use part (a) of this exercise and the fact that if $A = BDB^{-1}$, then for each $n \in \mathbf{Z}^+$, $A^n = BD^nB^{-1}$, to show that if D is a diagonal matrix and $A = BDB^{-1}$, then $e^{tA} = Be^{tD}B^{-1}$, $t \in \mathbf{R}^1$.

Theorem (9.E.16) and Exercise (9.E.17) can be applied to the problem of finding solutions to systems of differential equations as follows. Suppose that we are given the equations

$$Y'(t) = AY(t) + B(t) \qquad Y(0) = C \qquad (9)$$

where

$$A = \begin{pmatrix} 5 & -1 \\ 3 & 1 \end{pmatrix} \qquad B(t) = \begin{pmatrix} e^t \\ e^{-t} \end{pmatrix} \qquad C = \begin{pmatrix} 1 \\ 0 \end{pmatrix}$$

Then the usual calculations show that

$$\lambda_1 = 2 \qquad \lambda_2 = 4$$

are the eigenvalues of A and the corresponding eigenvectors are

$$q_1 = \begin{pmatrix} 1 \\ 3 \end{pmatrix} \qquad q_2 = \begin{pmatrix} 1 \\ 1 \end{pmatrix}$$

Let

$$Q = \begin{pmatrix} 1 & 1 \\ 3 & 1 \end{pmatrix}$$

Then

$$Q^{-1} = \begin{pmatrix} -\frac{1}{2} & \frac{1}{2} \\ \frac{3}{2} & -\frac{1}{2} \end{pmatrix}$$

and, by (9.E.16),

$$Q^{-1}AQ = \begin{pmatrix} 2 & 0 \\ 0 & 4 \end{pmatrix} = \Lambda$$

It follows from (9.E.17) that since $A = Q\Lambda Q^{-1}$,

$$e^{tA} = Qe^{t\Lambda}Q^{-1} = \begin{pmatrix} 1 & 1 \\ 3 & 1 \end{pmatrix}\begin{pmatrix} e^{2t} & 0 \\ 0 & e^{4t} \end{pmatrix}\begin{pmatrix} -\frac{1}{2} & \frac{1}{2} \\ \frac{3}{2} & -\frac{1}{2} \end{pmatrix}$$
$$= \begin{pmatrix} -(\frac{1}{2})e^{2t} + (\frac{3}{2})e^{4t} & (\frac{1}{2})e^{2t} - (\frac{1}{2})e^{4t} \\ -(\frac{3}{2})e^{2t} + (\frac{3}{2})e^{4t} & (\frac{3}{2})e^{2t} - (\frac{1}{2})e^{4t} \end{pmatrix}$$

The function $Z(t)$ may now be calculated as in (9.E.11), and then $Z(t)$ may be used as before to find a solution $\phi(t)$ to (9).

In the next section we present a relatively simple (albeit tedious) procedure for calculating e^{tA} when no restrictions are placed on the eigenvectors.

When are we assured of independent eigenvectors? One criterion can be formulated in terms of distinct eigenvalues.

(9.E.18) **Theorem** Suppose that $\lambda_1, \ldots, \lambda_k$ are distinct eigenvalues of an $n \times n$ matrix A and let $v_1, \ldots, v_k$ be any eigenvectors belonging to $\lambda_1, \ldots, \lambda_k$, respectively. Then the vectors $v_1, \ldots, v_k$ are linearly independent.

Proof. Suppose that the vectors $v_1, \ldots, v_k$ are dependent. Let m be the minimal number of vectors from this collection that are dependent; relabeling if necessary, we can assume that $v_1, v_2, \ldots, v_m$ are dependent (and any proper subset of these vectors are independent). By (2.A.9) there are scalars $\alpha_1, \ldots, \alpha_m$ not all zero such that

$$\alpha_1 v_1 + \alpha_2 v_2 + \cdots + \alpha_m v_m = 0 \qquad (10)$$

Note, however, that in this case none of the α_i can be 0, for otherwise we would have a still smaller number of dependent vectors. Multiplication of both sides of (10) by $(A - \lambda_m I)$ yields

$$(A - \lambda_m I)\alpha_1 v_1 + \cdots + (A - \lambda_m I)\alpha_m v_m = 0$$

Since $A v_1 = \lambda_1 v_1, A v_2 = \lambda_2 v_2, \ldots, A v_m = \lambda_m v_m$, we have

$$\alpha_1(\lambda_1 - \lambda_m)v_1 + \cdots + \alpha_{m-1}(\lambda_{m-1} - \lambda_m)v_{m-1} + \alpha_m(\lambda_m - \lambda_m)v_m = 0$$

Since

1. The λ_i are distinct.
2. $\alpha_i \neq 0$ for each i.
3. The coefficient of v_m is 0.

it follows that the vectors $v_1, \ldots, v_{m-1}$ are linearly dependent, which contradicts the minimality of m. Therefore, we conclude that the original eigenvectors were independent.

We conclude this section with two examples that illustrate how systems of equations can arise in physical problems. A number of interesting models have been developed through the application of mathematics to biology. First we consider a simplified model that might be used in connection with the diffusion of a compound across a series of cell walls (see Fig. 9.1). Suppose that a unit amount of a drug has been injected into cell C_1, and that it then begins to diffuse throughout the remaining cells. (We assume that there is no diffusion across noncontiguous walls.)

We would like to determine the concentrations of the drug in each cell at any time t. These concentrations will obviously depend on the ability of the drug to pass through each wall; suppose then that the rates of diffusion

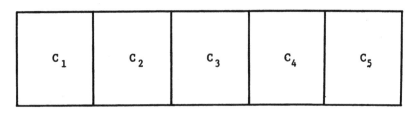

Figure 9.1

across contiguous cell walls (in both directions) are known. We let $r_{i,i+1}$ denote the rate of diffusion of the drug as it passes from cell C_i to cell C_{i+1} and $r_{i,i-1}$ will denote the rate of diffusion of the drug as it enters cell C_{i-1} from cell C_i.

For each i, let $y_i(t)$ be the concentration of the drug in cell C_i at time t. Then the rate of change of concentration of the drug in cell C_i is given by y_i'. Thus, we obtain the following equations:

$$y_1'(t) = -r_{12}y_1(t) + r_{21}y_2(t) \quad \text{(the rate out plus the rate in)}$$
$$y_2'(t) = r_{12}y_1(t) + r_{32}y_3(t) - (r_{21}y_2(t) + r_{23}y_2(t))$$
$$y_3'(t) = r_{23}y_2(t) + r_{43}y_4(t) - (r_{32}y_3(t) + r_{34}y_3(t))$$
$$y_4'(t) = r_{34}y_3(t) + r_{54}y_5(t) - (r_{43}y_4(t) + r_{45}y_4(t))$$
$$y_5'(t) = r_{45}y_4(t) - r_{54}y_5(t)$$

In matrix form this system of equations is expressed by

$$\mathbf{y}' = A\mathbf{y}$$

where

$$\mathbf{y} = \begin{pmatrix} y_1 \\ y_2 \\ y_3 \\ y_4 \\ y_5 \end{pmatrix}$$

and

$$A = \begin{pmatrix} -r_{12} & r_{21} & 0 & 0 & 0 \\ r_{12} & -r_{21} - r_{23} & r_{32} & 0 & 0 \\ 0 & r_{23} & -r_{32} - r_{34} & r_{43} & 0 \\ 0 & 0 & r_{34} & -r_{43} - r_{45} & r_{54} \\ 0 & 0 & 0 & r_{45} & -r_{54} \end{pmatrix}$$

As a second example we investigate (rather superficially) certain aspects of the kinetics of enzymatic reactions. Enzymes enter into almost all chemical reactions within cells and serve as regulators of many biochemical processes. Compounds with which enzymes react are commonly referred to as *substrates*.

In one basic model it is assumed that an enzyme E, and a substrate S, react to form a complex C, which in turn can either break down once again into E and S or can dissociate into E and a new product P; schematically, we have

$$\text{E} + \text{S} \rightleftharpoons \text{C} \quad \text{or} \quad \text{C} \longrightarrow \text{E} + \text{P}$$

It is assumed that the reaction between E and S takes place at specific parts of the enzymes, called *active centers* or *sites*. Suppose that E consists of n identical subunits each containing one active site. The sites are presumed to

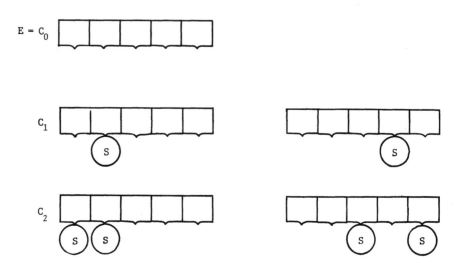

Figure 9.2

operate independently of one another. Then E can combine with j substrates S to form a complex C_j, where j takes on values from 0 to n (Fig. 9.2). Each C_j can subsequently combine with another S to form C_{j+1}, or C_j can break down into either C_{j-1} and S, or into C_{j-1} and the product P:

$$S + C_j \rightleftharpoons C_{j+1} \qquad j = 0, 1, \ldots, n-1$$
$$C_j \longrightarrow C_{j-1} + P \qquad j = 1, 2, \ldots, n$$

We assume throughout that the concentration s_0 of S remains constant (not an unreasonable assumption provided that there is an excess of S). The concentration of C_j at time t is denoted by $c_j(t)$ or just c_j when the context is clear. It is experimentally defensible to assert that the rate of association of the substrate S and the complex C_j at any time t is proportional to the amount of S and C_j present at time t. Hence, in particular, the rate of change of the concentration of $E = C_0$ due to the association of C_0 and S is given by $-ks_0c_0(t)$, where k is a positive constant and the minus sign is used to designate a rate of reduction (as opposed to a rate of increase) in the concentration of C_0. If, in general, to each active site we assign a fixed association rate constant k_{+1} (which corresponds to the reaction $C_j + S \rightarrow C_{j+1}$), then since there are n sites (which act independently) $nk_{+1}s_0$ represents the rate that a particular C_0 will change to C_1; clearly, $k = nk_{+1}$.

At the same time that C_0 is combining with S, there is also activity that involves the reaction of C_1 dissociating into S and C_0, or into P and C_0. To each active site we assign a dissociation rate constant k_{-1} (corresponding to the reaction $C_{j+1} \rightarrow C_j + S$) and a conversion rate constant k_{+2} (corresponding to the reaction $C_j \to C_{j-1} + P$). With this notation we see that the rate of change of $c_0(t)$ due to the dissociation of C_1 into C_0 and S, or the conversion of C_1 into C_0 and P is given by $(k_{-1} + k_{+2})c_1(t)$. It follows that the total rate of change of $c_0(t)$—$c_0'(t)$—that is a result of association, dissociation, and conversion can be expressed by

$$c_0'(t) = (-nk_{+1}s_0)c_0(t) + (k_{-1} + k_{+2})c_1(t)$$

More generally, if E is an enzyme with n independent binding sites, then the rate of converting C_j into C_{j+1} due to association is given by

$$((n - j)k_{+1}s_0)c_j(t)$$

since there are $n - j$ unoccupied sites in C_j. Similarly, the rate of conversion from C_{j-1} to C_j due to association is

$$((n - j + 1)k_{+1}s_0)c_{j-1}(t)$$

In a like fashion one sees that the conversion rate from C_{j+1} to C_j due to dissociation is given by

$$(k_{-1} + k_{+2})(j + 1)c_{j+1}(t)$$

Consequently, the resultant rate of change in C_j is described by the equation

$$c_j'(t) = -((n - j)k_{+1}s_0)c_j(t) + ((n - j + 1)k_{+1}s_0)c_{j-1}(t)$$
$$+ (k_{-1} + k_{+2})(j + 1)c_{j+1}(t) - (k_{-1} + k_{+2})jc_j(t)$$

The diagram in Fig. 9.3 helps summarize this discussion.

The pertinent system of equations is

$$c_0' = -(nk_{+1}s_0)c_0 + (k_{-1} + k_{+2})c_1$$
$$\cdots\cdots\cdots\cdots\cdots\cdots\cdots\cdots\cdots\cdots\cdots$$
$$c_j' = ((n - j + 1)k_{+1}s_0)c_{j-1} - j(k_{-1} + k_{+2})c_j - ((n - j)k_{+1}s_0c_j$$
$$+ ((j + 1)(k_{-1} + k_{+2}))c_{j+1}$$
$$\cdots\cdots\cdots\cdots\cdots\cdots\cdots\cdots\cdots\cdots\cdots$$
$$c_n' = (k_{+1}s_0)c_{n-1} - (n(k_{-1} + k_{+2}))c_n$$

If $n = 4$, then the preceding system of equations takes on the matrix form

$$c' = Ac$$

where

$$\mathbf{c} = \begin{pmatrix} c_0 \\ c_1 \\ c_2 \\ c_3 \\ c_4 \end{pmatrix}$$

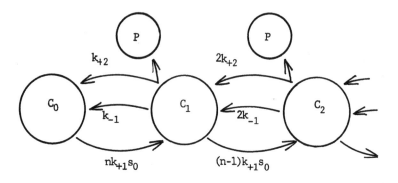

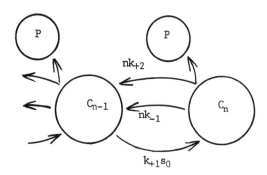

Figure 9.3

and

$$
A = \begin{pmatrix}
-4k_{+1}s_0 & k_{-1} + k_{+2} & 0 \\
4k_{+1}s_0 & -(k_{-1} + k_{+2}) - 3k_{+1}s_0 & 2(k_{-1} + k_{+2}) \\
0 & 3k_{+1}s_0 & -2(k_{-1} + k_{+2}) - 2k_{+1}s_0 \\
0 & 0 & 2k_{+1}s_0 \\
0 & 0 & 0
\end{pmatrix}
$$

$$
\begin{matrix}
0 & 0 \\
0 & 0 \\
3(k_{-1} + k_{+2}) & 0 \\
-3(k_{-1} + k_{+2}) - k_{+1}s_0 & 4(k_{-1} + k_{+2}) \\
k_{+1}s_0 & -4(k_{+1} + k_{+2})
\end{matrix}
$$

If we set $k_{+1} = 1, k_{-1} = \frac{1}{2}, k_{+2} = \frac{1}{2}$, and $s_0 = 1$, then the characteristic polynomial of A is

$$\phi(\lambda) = \det \begin{pmatrix} -4 - \lambda & 1 & 0 & 0 & 0 \\ 4 & 4 - \lambda & 2 & 0 & 0 \\ 0 & 3 & -4 - \lambda & 3 & 0 \\ 0 & 0 & 2 & -4 - \lambda & 4 \\ 0 & 0 & 0 & 1 & -4 - \lambda \end{pmatrix}$$

$$= -\lambda^5 - 20\lambda^4 - 140\lambda^3 - 400\lambda^2 - 268\lambda + 48$$

In order now to apply our previous results, we would like to calculate the eigenvalues and eigenvectors of A by finding the zeros of the characteristic polynomial given above. Although it is a relatively simple matter to show that up to five distinct zeros of this polynomial must exist, it is impossible to find the precise values of these zeros (not difficult; impossible!). In general, systems of differential equations are computationally quite difficult to handle, and in most instances numerical methods (in conjunction with a high-speed computer) are used to obtain approximations to the true solutions.

Why is there no formula available for determining the zeros of the above polynomial? Finding formulas for the zeros of polynomials has long intrigued mathematicians. In fact, rather amazingly, the quadratic formula $x = (-b \pm \sqrt{b^2 - 4ac})/2a$ for determining the zeros of second-degree polynomials of the form $ax^2 + bx + c$ was essentially known to the Babylonians as early as 300 BC. Literally hundreds of years of work were necessary before Italian mathematicians in the sixteenth century finally derived the corresponding formulas for polynomials of degree 3 and 4 (unfortunately, the formula for fourth-degree polynomials is so involved that from a computational standpoint it is practically worthless).

In spite of prodigious efforts on the part of many mathematicians, no formula was ever found for the extraction of zeros of polynomials of degree greater than 4—and rightly so. At the age of 22, the Norwegian mathematician Abel (1802–1829) demonstrated the impossibility of using algebraic operations to solve fifth-degree equations. This work was followed by that of a brilliant young French mathematician, Evariste Galois (1811–1832), who presented a paper in which he claimed to have shown that no general polynomial equation of degree greater than 4 could be solved. The paper was rejected as being unintelligible by a noted compatriot, Poisson, who however did request that Galois rewrite it. For Galois, however, this represented just one more in a series of seemingly unjust setbacks, and he soon turned to other endeavors. He became caught up in an amorous entanglement that led to a duel of honor in which, at the age of 21, he was killed.

Perhaps a premonition of his fate drove him on the eve of the duel to hastily jot down many of his mathematical discoveries, including his work on algebraic equations. These notes comprise an incredible legacy that has given inspiration to many mathematicians up to the present time, and the germinal ideas of Galois still permeate huge areas of mathematics today.

F. CALCULATION OF e^{tA}

As was indicated in the preceding section, the principal difficulty encountered in using matrices to solve systems of equations is that of evaluating e^{tA}. The customary resolution of this problem is based on a number of fairly sophis-ticated results in linear algebra. Rather than pursue this approach, we shall outline here a somewhat different method of attack [due to Putzer (1966)], whose theoretical justification relies on only one moderately deep proposi-tion, the Cayley-Hamilton theorem. (In the problem set, another approach involving what is known as the Jordan canonical form of a matrix is intro-duced.) The statement of the Cayley-Hamilton theorem is as follows [the reader is referred to Franklin (1968) for two distinct and quite readable proofs].

(9.F.1) *Theorem (Cayley-Hamilton Theorem)*   Suppose that A is an $n \times n$ matrix with characteristic polynomial $\psi(\lambda) = \lambda^n + c_1\lambda^{n-1} + c_2\lambda^{n-2} + \cdots + c_{n-1}\lambda + c_n$. Then $A^n + c_1A^{n-1} + c_2A^{n-2} + \cdots + c_nI$ is equal to the zero matrix (0). Briefly expressed,

$$\psi(A) = (0)$$

(9.F.2) *Theorem (Putzer)*   Suppose that A is an $n \times n$ matrix with (not necessarily distinct) eigenvalues $\lambda_1, \lambda_2, \ldots, \lambda_n$. Then $e^{tA} = \sum_{j=0}^{n-1} r_{j+1}(t)P_j$, where

$$P_0 = I$$
$$P_j = (A - \lambda_1I)(A - \lambda_2I)\cdots(A - \lambda_jI) \qquad (j = 1, 2, \ldots, n)$$

and the functions $r_1(t), r_2(t), \ldots, r_n(t)$ form a solution to the system of differential equations

$$r_1' = \lambda_1 r_1$$
$$r_2' = r_1 + \lambda_2 r_2$$
$$r_3' = r_2 + \lambda_3 r_3 \qquad\qquad (1)$$
$$\cdots\cdots\cdots\cdots$$
$$r_n' = r_{n-1} + \lambda_n r_n$$

that satisfies the initial condition

$$r_1(0) = 1 \qquad r_2(0) = r_3(0) = \cdots = r_n(0) = 0$$

Proof.   Let $r_0 : \mathbf{R}^1 \to \mathbf{R}^1$ be the constant function $r_0(t) \equiv 0$, and let $r_1, r_2, \ldots, r_n$ be as given in the hypothesis. Define a function ϕ from $\mathbf{R}^1$ into the set of $n \times n$ matrices by

$$\phi(t) = \sum_{j=0}^{n-1} r_{j+1}(t) P_j \tag{2}$$

Note that from (1)

$$\phi'(t) = \sum_{j=0}^{n-1} r'_{j+1}(t) P_j = \sum_{j=0}^{n-1} (r_j(t) + \lambda_{j+1} r_{j+1}(t)) P_j \tag{3}$$

and, hence,

$$\phi'(t) - \lambda_n \phi(t) = \sum_{j=0}^{n-1} [r_j(t) P_j + \lambda_{j+1} r_{j+1}(t) P_j - \lambda_n r_{j+1}(t) P_j] \tag{4}$$

Since r_0 is identically 0, $\sum_{j=0}^{n-1} r_j(t) P_j$ may be rewritten as $\sum_{j=0}^{n-2} r_{j+1}(t) P_{j+1}$. Then the right-hand side of (4) becomes

$$\sum_{j=0}^{n-2} r_{j+1}(t) P_{j+1} + \left(\sum_{j=0}^{n-2} \lambda_{j+1} r_{j+1}(t) P_j \right) + \lambda_n r_n(t) P_{n-1} - \left(\lambda_n \sum_{j=0}^{n-2} r_{j+1}(t) P_j \right)$$

$$- \lambda_n r_n(t) P_{n-1} = \sum_{j=0}^{n-2} [P_{j+1} + (\lambda_{j+1} - \lambda_n) P_j] r_{j+1}(t).$$

Since $P_{j+1} = (A - \lambda_{j+1} I) P_j$, we have

$$\phi'(t) - \lambda_n \phi(t) = \sum_{j=0}^{n-2} [(A - \lambda_{j+1} I) P_j + (\lambda_{j+1} - \lambda_n) P_j] r_{j+1}(t)$$

$$= (A - \lambda_n I) \sum_{j=0}^{n-2} P_j r_{j+1}(t) = (A - \lambda_n I)[\phi(t) - r_n(t) P_{n-1}]$$

$$= A\phi(t) - \lambda_n I \phi(t) - r_n(t)(A - \lambda_n I) P_{n-1}$$

$$= A\phi(t) - \lambda_n I \phi(t) - r_n(t) P_n$$

By the Cayley-Hamilton theorem, $P_n = (0)$, and therefore,

$$\phi'(t) = A\phi(t)$$

Since $\phi(0) = I$, the jth column of $\phi(t)$ satisfies the initial value problem

$$y'(t) = Ay(t) \qquad y(0) = \begin{pmatrix} 0 \\ 0 \\ \vdots \\ 1 \\ \vdots \\ 0 \end{pmatrix}$$

where the value 1 is in the jth row. Since the jth column of e^{tA} also satisfies this initial value problem, it follows from (9.E.9) that the jth column of $\phi(t)$ is equal to the jth column of e^{tA}, and therefore, $\phi(t) = e^{tA}$.

(9.F.3) *Example* Suppose that A is a 3×3 matrix such that all of the eigenvalues of A are equal. Then we have

$$
\begin{aligned}
r_1' &= \lambda r_1 & r_1(0) &= 1 \\
r_2' &= r_1 + \lambda r_2 & r_2(0) &= 0 \\
r_3' &= r_2 + \lambda r_3 & r_3(0) &= 0
\end{aligned}
$$

The solution of this system can easily be read off: $r_1 = e^{\lambda t}$, $r_2 = te^{\lambda t}$, $r_3 = (t^2/2)e^{\lambda t}$. Therefore, by (9.F.2)

$$
e^{tA} = \tfrac{1}{2}e^{\lambda t}[2I + 2t(A - \lambda I) + t^2(A - \lambda I)^2]
$$

(9.F.4) *Example* If

$$
A = \begin{pmatrix} 1 & 0 & -1 \\ 0 & 2 & 1 \\ 0 & 0 & 2 \end{pmatrix}
$$

then it is easily verified that the eigenvalues of A are $\lambda_1 = 1$, $\lambda_2 = 2$, $\lambda_3 = 2$. Therefore, we must solve the initial value problem

$$
\begin{aligned}
r_1' &= r_1 & r_1(0) &= 1 \\
r_2' &= r_1 + 2r_2 & r_2(0) &= 0 \\
r_3' &= r_2 + 2r_3 & r_3(0) &= 0
\end{aligned}
$$

Clearly, $r_1(t) = e^t$, and $r_2' = e^t + 2r_2$. The solution to this first-order linear differential equation satisfying $r_2(0) = 0$ is easily seen to be

$$
r_2(t) = -e^t + e^{2t}
$$

and just as easily, the solution of $r_3' = -e^t + e^{2t} + 2r_3$ satisfying the initial condition, $r_3(0) = 0$, is found to be

$$
r_3(t) = e^t + e^{2t} + te^{2t}
$$

Consequently,

$$
\begin{aligned}
e^{tA} &= e^t I + (-e^t + e^{2t})(A - I) + (e^t + e^{2t} + te^{2t})(A - I)(A - 2I) \\
&= \begin{pmatrix} e^t & 0 & e^t - e^{2t} \\ 0 & e^{2t} & te^{2t} \\ 0 & 0 & e^{2t} \end{pmatrix}
\end{aligned}
$$

G. INFINITE SERIES, INFINITE PRODUCTS, AND PROBABILITY

Many problems in probability theory lead naturally to infinite series and infinite products. As an example we consider the following situation which is discussed in Thomas (1957).

Suppose that a sequence of trials is run where the probability of success on the nth trial is p_n. We assume that the trials are independent, i.e., success or failure on any given trial does not depend on the results of the other trials. We are interested in finding the probability Q of eventually obtaining a success; this probability is given by

$$Q = \sum_{n=1}^{\infty} q_n$$

where q_n is the probability that the first success occurs on the nth trial. Note that

$$q_1 = p_1$$
$$q_2 = (1 - p_1)p_2$$
$$\dots\dots\dots\dots$$
$$q_n = (1 - p_1)(1 - p_2)\cdots(1 - p_{n-1})p_n$$

In particular, we are interested in determining conditions under which $\sum_{n=1}^{\infty} q_n$ is 1, i.e., the probability of having at least one successful trial is 1. We shall make use of the following concept.

(9.G.1) Definition Suppose that $c_1, c_2, \ldots$ is a sequence of numbers. For each n, let $t_n = c_1 \cdot c_2 \cdots c_n = \prod_{i=1}^{n} c_i$. The *infinite product of the c_i*, $\prod_{i=1}^{\infty} c_i$, is defined to be $\lim_{n \to \infty} t_n$ (provided that this limit exists). We write $\prod_{n=1}^{\infty} c_n = \infty$ if $\lim_{n \to \infty} t_n = \infty$ and $\prod_{n=1}^{\infty} c_n = -\infty$ if $\lim_{n \to \infty} t_n = -\infty$.

A very useful result that connects infinite products and infinite series is the following.

(9.G.2) Theorem Suppose that $\{p_n\}$ is a sequence of real numbers such that $0 \le p_n \le 1$ for each n. Then $\prod_{n=1}^{\infty} (1 - p_n) = 0$ if and only if $p_n = 1$ for some n, or $\sum_{n=1}^{\infty} p_n = \infty$.

Proof. Suppose that $\prod_{n=1}^{\infty} (1 - p_n) = 0$ and that $p_n \ne 1$ for each n. Observe that it follows from the fact that $p_n \ge 0$ for each n that

$$(1 - p_i)(1 - p_{i+1}) = 1 - p_i - p_{i+1} + p_i p_{i+1} \ge 1 - p_i - p_{i+1}$$

Furthermore, it is clear that

$$(1 - p_i)(1 - p_{i+1})(1 - p_{i+2}) \geq (1 - p_i - p_{i+1})(1 - p_{i+2})$$
$$\geq 1 - p_i - p_{i+1} - p_{i+2}$$

and the reader may readily establish by induction that if $k > i$, then

$$(1 - p_i)(1 - p_{i+1})\cdots(1 - p_k) \geq 1 - p_i - p_{i+1} - \cdots - p_k \qquad (1)$$

If the series $\sum_{n=1}^{\infty} p_n$ converges, then by (8.B.3), corresponding to $\varepsilon > 0$, there is an integer N such that $\sum_{n=N}^{\infty} p_n < \varepsilon$. Hence, for any $k \geq N$, it follows from (1) that

$$(1 - p_N)(1 - p_{N+1})\cdots(1 - p_k) \geq 1 - p_N - p_{N+1} - \cdots - p_k > 1 - \varepsilon$$

Let $r = (1 - p_1)(1 - p_2)\cdots(1 - p_{N-1})$. Note that, if $i \geq N$ and

$$t_i = (1 - p_1)(1 - p_2)\cdots(1 - p_{N-1})(1 - p_N)\cdots(1 - p_i)$$

then we have $t_i > r(1 - \varepsilon)$. Consequently, the sequence $\{t_i\}$ fails to converge to 0, which contradicts the fact that $\prod_{n=1}^{\infty} (1 - p_n) = 0$. Therefore, $\sum_{n=1}^{\infty} p_n = \infty$.

To prove the converse, we first observe that if $p_n = 1$ for some n, then $\prod_{n=1}^{\infty} (1 - p_n) = 0$. Recall that in (5.C.10) it was shown that $(1 - x) \leq e^{-x}$ for $0 \leq x \leq 1$. Therefore, if $t_n = (1 - p_1)(1 - p_2)\cdots(1 - p_n)$, we have

$$t_n \leq e^{-(p_1 + \cdots + p_n)}.$$

Since $\sum_{n=1}^{\infty} p_n = \infty$ implies that $\lim_{n \to \infty} e^{-(p_1 + \cdots + p_n)} = 0$, the result follows.

Resolution of the probability problem is now quite straightforward.

(9.G.3) *Theorem* Suppose that $\{p_n\}$ is a sequence of numbers such that $0 \leq p_n < 1$ for each n. Let $q_1 = p_1$, and for each $n \geq 1$, let $q_n = (1 - p_1)(1 - p_2)\cdots(1 - p_{n-1})p_n$. Then $\sum_{n=1}^{\infty} q_n \leq 1$, and moreover, $\sum_{n=1}^{\infty} q_n = 1$ if and only if $\prod_{n=1}^{\infty} (1 - p_n) = 0$. In other words, the following three conditions are equivalent:

(i) $\displaystyle\sum_{n=1}^{\infty} q_n = 1$

(ii) $\displaystyle\prod_{n=1}^{\infty} (1 - p_n) = 0$

(iii) $\displaystyle\sum_{n=1}^{\infty} p_n = \infty$

Proof. For each positive integer N, set

$$Q_N = \sum_{n=1}^{N} q_n \qquad R_N = \prod_{n=1}^{N} (1 - p_n)$$

Note that in terms of the problem discussed at the beginning of this section Q_N gives the probability of at least one success in the first N trials (why?) and R_N is the probability of no success in these trials. Therefore, $Q_N + R_N = 1$. Moreover, $\{R_N\}$ is a decreasing sequence that is bounded below and $\{Q_N\}$ is an increasing sequence with an upper bound. Consequently, both of these sequences are convergent, say with limits Q and R, respectively. Clearly, $Q = \sum_{n=1}^{\infty} q_n$, $R = \prod_{n=1}^{\infty} (1 - p_n)$, and $Q + R = 1$. Thus, if $Q = 1$, then $R = 0$, and we have that (i) implies (ii). If (ii) holds, then since $p_n \neq 1$ for each n, it follows from (9.G.2) that $\sum_{n=1}^{\infty} p_n = \infty$; therefore, (ii) implies (iii). Finally, if $\sum_{n=1}^{\infty} p_n = \infty$, then again by (9.G.2), $R = \prod_{n=1}^{\infty} (1 - p_n) = 0$, and consequently, $Q = 1$.

PROBLEMS

Section B

1. Use Taylor series expansions to find the following limits:
    (a) $\lim\limits_{x \to 0} (\sin x) \ln |x|$
    (b) $\lim\limits_{x \to 0} \dfrac{\tan x - \sin x}{\sin^2 x}$.

2. Use Taylor series expansions to find the following limits:
    (a) $\lim\limits_{x \to 0} \dfrac{e^{x^2} - 1 - \sin x^2}{x^4}$
    (b) $\lim\limits_{x \to 0} \dfrac{\ln (x + 1)}{x}$

3. Use Taylor series expansions to find the following limits:
    (a) $\lim\limits_{x \to 0+} \dfrac{x^{5/2} e^x}{\ln (1 + x)}$
    (b) $\lim\limits_{x \to 0} \dfrac{e^x + e^{-x} - 2}{x^4}$

Section C

1. Approximate $\int_0^1 \sin 2x^3 \, dx$ with an error of less than 10^{-4}.
2. Approximate $\int_0^1 [(\sin x)/x] \, dx$ with an error of less than 10^{-4}.
3. Approximate $\int_0^1 [(\ln |1 + x|)/x] \, dx$ with an error of less than 10^{-5}.

Section D

1. Use infinite series to find the general solution of:
    (a) $y'' - ty' - y = 0$
    (b) $(t^2 - 1)y'' + 6ty' + 3y = 0$
    (c) $y'' + y = \sin t$ (Use the Taylor expansion of $\sin t$.)
    (d) $y'' - ty' + y = t$
2. Find two independent series solutions of the Hermite equation

$$y'' - 2ty' + \alpha y = 0$$

3. (a) Find a series solution of $(1 + t^2)y'' + y = 0$ that satisfies the
        initial conditions $\phi(0) = 0$, $\phi'(0) = 1$.
    (b) Find two independent series solutions of $y'' - ty' - y = 0$ of the
        form $\sum_{n=0}^{\infty} a_n(t - 1)^n$.
4. Use the method of Frobenius to find solutions of:
    (a) $3ty'' + (3t + 2)y' - 4y = 0$
    (b) $t^2y'' - ty' + (1 + t^2)y = 0$
    (c) $4t^2y'' - 4ty' + (3 - 4t^2)y = 0$
    (d) $t^2y'' + t(t - \tfrac{1}{2})y' + \tfrac{1}{2}y = 0$.
5. Show that

$$J_{-p}(t) = \sum_{n=0}^{\infty} \frac{(-1)^n}{\Gamma(n + 1)\Gamma(n - p + 1)}\left(\frac{t}{2}\right)^{2n-p}$$

    is a solution of Bessel's equation of order p.
6. Show that the Bessel function of order 0 (of the first kind)

$$a_0\left(1 + \sum_{n=1}^{\infty} \frac{(-1)^n t^{2n}}{2^{2n}(n!)^2}\right)$$

    converges for all t.
7. Show that if p is not an integer, then J_p and J_{-p} are linearly independent.
8. It can be shown that if

$$t^2y'' + tp(t)y' + q(t)y = 0 \qquad\qquad (*)$$

    where p and q can be represented by their Taylor series expansions in
    a neighborhood of 0, and if r_1 is a double root (the only root) of the
    indicial equation

$$r(r - 1) + p(0)r + q(0) = 0$$

then

$$y_1(t) = |t|^{r_1} \sum_{n=0}^{\infty} a_n t^n \qquad a_0 \neq 0$$

and

$$y_2(t) = |t|^{r_1} \sum_{n=0}^{\infty} b_n t^n + y_1(t) \ln |t|$$

are linearly independent solutions of (∗); the coefficients a_n and b_n can be found by direct substitution of y_1 and y_2 into (∗). [A proof of this result may be found in Myint-U (1978).] Use this result to find two independent solutions of

(a) $t^2 y'' + t(t - 1)y' + (1 - t)y = 0$
(b) $t^2 y'' + 3ty' + (1 - 2t)y = 0$

9. It can be shown that if

$$t^2 y'' + tp(t)y' + q(t)y = 0 \qquad (\ast\ast)$$

where p and q can be represented by their Taylor series expansions in a neighborhood of 0, and if r_1 and r_2 are roots of the indicial equation

$$r(r - 1) + p(0)r + q(0) = 0$$

such that $r_1 - r_2$ is a positive integer, then

$$y_1 = |t|^{r_1} \sum_{n=0}^{\infty} a_n t^n \qquad a_0 \neq 0$$

and

$$y_2 = |t|^{r_2} \sum_{n=0}^{\infty} b_n t^n + \alpha y_1(t) \ln |t| \qquad b_0 \neq 0$$

are linearly independent solutions of (∗∗); the coefficients a_n and b_n can be found by direct substitution of y_1 and y_2 in (∗∗). Use this result to find two independent solutions of

(a) $t^2 y'' - 3ty' + (3 + 4t)y = 0$
(b) $t^2 y'' + 4ty' + (2 + t)y = 0$
(c) $t^2 y'' - (t + 2)y = 0$

10. Given the Bessel equation of order $p = \frac{1}{2}$,

$$t^2 y'' + ty' + (t^2 - \tfrac{1}{4})y = 0 \qquad t > 0, \qquad (\ast\ast\ast)$$

show that the substitution $y = t^{-1/2}u(t)$ transforms this equation into

$$u'' + u = 0$$

Find two independent solutions of (∗∗∗).

Section E

1. If A is an $n \times n$ matrix, define $\sin A$, $\cos A$, and $\ln A$.
2. Show that if $\{A_i\}$ is a sequence of $m \times n$ matrices, and if $\sum_{i=1}^{\infty} A_i$ converges, then $\lim_{i \to \infty} A_i$ is the zero matrix, (0).
3. Use the material in this section to solve the following systems of equations:

(a) $\begin{cases} y_1'(t) = 2y_2(t) + e^{-t} \\ y_2'(t) = 3y_1(t) - y_2(t) - e^{2t} \end{cases}$

with the initial conditions $y_1(0) = 1$, $y_2(0) = 2$

(b) $y'(t) = \begin{pmatrix} 4 & -1 \\ 2 & 1 \end{pmatrix} y(t) + \begin{pmatrix} 3e^{2t} \\ 2t \end{pmatrix}$ $y(0) = \begin{pmatrix} -\frac{5}{18} \\ \frac{47}{9} \end{pmatrix}$

(c) $y'(t) = \begin{pmatrix} 1 & 1 \\ 1 & 1 \end{pmatrix} y(t) + \begin{pmatrix} \cos t \\ t \end{pmatrix}$ $y(0) = \begin{pmatrix} 1 \\ 1 \end{pmatrix}$

(d) $\begin{cases} y_1'(t) = y_2(t) + y_3(t) \\ y_2'(t) = -4y_1(t) + 4y_2(t) + 2y_3(t) - e^t \\ y_3'(t) = 4y_1(t) - 3y_2(t) - y_3(t) \end{cases}$

with initial conditions $y_1(0) = 1$, $y_2(0) = 0$, $y_3(0) = -1$

(e) $\begin{cases} y_1' = y_1 - y_2 + e^t \cos t \\ y_2' = y_1 + y_2 + e^t \sin t \end{cases}$

4. In (9.E.15) verify that C_2, C_3, and C_4 are eigenvectors belonging to λ_2, λ_3, and λ_4.

5. Suppose that $A = (a_{ij})$ is a triangular $n \times n$ matrix, i.e., $a_{ij} = 0$ if $i < j$.
    (a) Find the eigenvalues of A.
    (b) Show that there is an integer k such that $e^A = I + \sum_{i=1}^{k} A^i/i!$.

6. Show that if A and B are similar $n \times n$ matrices, then λ is an eigenvalue of A if and only if λ is an eigenvalue of B.

7. Show that for any $n \times n$ matrix A, e^A has an inverse.

8. Show that an $n \times n$ matrix A fails to have an inverse if and only if λ is a factor of the characteristic polynomial $\phi(\lambda) = (-1)^n \lambda^n + c_{n-1} \lambda^{n-1} + \cdots + c_1 \lambda + c_0$.

9. Two tanks, T_1 and T_2, are connected as indicated in Fig. 9.4. Tank T_1 contains 20 pounds of salt dissolved in 100 gallons of water and tank T_2 contains 100 gallons of pure water. Pure water enters the system through tank T_1 and the resulting mixtures are circulated as described in Fig. 9.4.

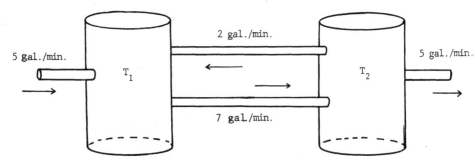

Figure 9.4

(a) Let $y_1(t)$ and $y_2(t)$ denote the amount of salt at time t in tanks T_1

and T_2, respectively, and find y_1 and y_2 by setting up and solving an appropriate system of differential equations.

(b) Find y_1 and y_2 if water containing 1 pound of salt per gallon enters tank T_1, rather than just pure water, and if tank T_2 initially contains 10 pounds of salt.

Section F

1. Use the methods of this section to find e^{tA} if

(a) $A = \begin{pmatrix} 4 & -1 \\ 2 & 1 \end{pmatrix}$

(b) $A = \begin{pmatrix} -1 & 2 \\ -3 & 4 \end{pmatrix}$

(c) $A = \begin{pmatrix} 0 & 1 & 1 \\ 1 & 0 & 1 \\ 1 & 1 & 0 \end{pmatrix}$

(d) $A = \begin{pmatrix} 0 & -4 & 0 \\ 1 & 4 & 0 \\ 0 & 0 & 2 \end{pmatrix}$

2. It can be shown that if A is an $n \times n$ matrix, then there is a nonsingular matrix Q such that

$$Q^{-1}AQ = \begin{pmatrix} J_1 & 0 & 0 \cdots 0 \\ 0 & J_2 & 0 \cdots 0 \\ & \cdots\cdots\cdots \\ 0 & 0 & 0 \cdots J_k \end{pmatrix} = J$$

where J_i is an $n_i \times n_i$ square matrix (called a *Jordan block*) of the form

$$J_i = \begin{pmatrix} \lambda_i & 1 & 0 \cdots 0 \\ 0 & \lambda_i & 1 \cdots 0 \\ & \cdots\cdots\cdots \\ 0 & 0 \cdots 0 & \lambda_i & 1 \\ 0 & 0 \cdots 0 & 0 & \lambda_i \end{pmatrix}$$

and λ_i is an eigenvalue of A; the matrix J is said to be a *Jordan matrix*.

(a) Suppose that

$$J = \begin{pmatrix} \lambda & 1 & 0 \cdots 0 \\ 0 & \lambda & 1 \cdots 0 \\ & \cdots\cdots\cdots \\ 0 & 0 \cdots 0 & \lambda & 1 \\ 0 & 0 \cdots 0 & 0 & \lambda \end{pmatrix}$$

is an $n \times n$ Jordan block and consider the differential equation

$$Y' = JY \tag{*}$$

where

$$Y = \begin{pmatrix} y_1 \\ y_2 \\ \vdots \\ y_n \end{pmatrix}$$

Show that (*) can be written in the form

$$y_1' = \lambda y_1 + y_2$$
$$y_2' = \lambda y_2 + y_3$$
$$\cdots \cdots \cdots \cdots$$

$$y_{n-1}' = \lambda y_{n-1} + y_n$$
$$y_n' = \lambda y_n$$

and explain how you would solve this system.

(b)   Use (a) to find a solution to

$$Y' = \begin{pmatrix} 2 & 1 & 0 \\ 0 & 2 & 1 \\ 0 & 0 & 2 \end{pmatrix} Y.$$

(c)   Show that if

$$J_i = \begin{pmatrix} \lambda_i & 1 & 0 \cdots 0 \\ 0 & \lambda_i & 1 \cdots 0 \\ \cdots \cdots \cdots \cdots \cdots \\ 0 & 0 \cdots \lambda_i & 1 \\ 0 & 0 \cdots 0 & \lambda_i \end{pmatrix}$$

is an $n \times n$ matrix, then

$$e^{J_i t} = e^{\lambda_i t} \begin{pmatrix} 1 & t & \dfrac{t^2}{2!} & \cdots & \dfrac{t^{n-1}}{(n-1)!} \\ 0 & 1 & t & \cdots & \dfrac{t^{n-2}}{(n-2)!} \\ \cdots \cdots \cdots \cdots \cdots \cdots \cdots \cdots \\ 0 & 0 & 0 & \cdots & 1 \end{pmatrix}.$$

(d)   Show that if

$$A = QJQ^{-1}$$

where J is a Jordan matrix of the form

$$\begin{pmatrix} J_1 & 0 \cdots 0 \\ 0 & J_2 \cdots 0 \\ \cdots \cdots \cdots \\ 0 & 0 \cdots J_n \end{pmatrix}$$

then

$$
e^{tA} = Q \begin{pmatrix} e^{J_1 t} & 0 \cdots \cdots 0 \\ 0 & e^{J_2 t} \cdots 0 \\ \cdots \cdots \cdots \cdots \cdots \\ 0 & 0 \cdots e^{J_n t} \end{pmatrix}
$$

(e) Let

$$
A = \begin{pmatrix} 3 & -1 & 0 \\ 1 & -1 & -1 \\ -2 & 4 & 3 \end{pmatrix}
$$

and show that if

$$
Q = \begin{pmatrix} 2 & -1 & 0 \\ 1 & 1 & 1 \\ 0 & 2 & 1 \end{pmatrix}
$$

then $J = Q^{-1}AQ$ is a Jordan matrix; use J to find e^{tA}.

Section G

1. Find

$$
\prod_{n=1}^{\infty} \left(1 + \frac{4}{(n + 2)(n + 6)} \right)
$$

[*Hint*: Show that the nth partial product t_n is equal to $\prod_{k=1}^{n} (k + 4)^2 / (k + 2)(k + 6)$.]

2. Find

(a) $\displaystyle \prod_{n=2}^{\infty} \left(1 - \frac{1}{n^2} \right)$

(b) $\displaystyle \prod_{n=2}^{\infty} \left(1 - \frac{1}{n} \right)$

(c) $\displaystyle \prod_{n=2}^{\infty} \frac{n^3 - 1}{n^3 + 1}$

3. Show that if $a_n > 0$ for each n, then $\prod_{n=1}^{\infty} (a_n + 1)$ converges if and only if $\sum_{n=1}^{\infty} a_n$ converges. [*Hint*: Let $s_n = a_1 + \cdots + a_n$ and $t_n = \prod_{k=1}^{n} (a_k + 1)$. Note that

$$
s_n < t_n \tag{*}
$$

and use the fact that $1 + x \le e^x$ for all x [cf. (5.C.10)] to show that

$$
t_n \le e^{s_n} \tag{**}
$$

Deduce from (*) and (**) that the sequence $\{s_n\}$ is bounded if and only if the sequence $\{t_n\}$ is bounded.]

4. The product $\prod_{n=1}^{\infty} (1 + a_n)$ is said to *converge absolutely* if $\prod_{n=1}^{\infty} (1 + |a_n|)$ converges. Show that $\prod_{n=1}^{\infty} (1 + a_n)$ converges absolutely if and only if $\sum_{n=1}^{\infty} a_n$ converges absolutely.

5. Show that if $\prod_{n=1}^{\infty} (1 + a_n)$ converges, then $\lim_{n \to \infty} a_n = 0$.

6. Let $a_n = (-1)^n (1/\sqrt{n})$. Show that $\prod_{n=1}^{\infty} (1 + a_n)$ diverges even though $\sum_{n=1}^{\infty} a_n$ converges.

7. In the notation of the probability problem given in this section determine whether $Q = 1$ when $p_n = n/(n + 1)$. Do the same when $p_n = n/(n + k)$ for $k > 0$ (write your answer in terms of the Γ function).

REFERENCES

Brauer, F., and J. A. Nohel (1973): *Ordinary Differential Equations: A First Course*, W. A. Benjamin, Menlo Park, Calif.

Coddington, Earl (1961): *An Introduction to Ordinary Differential Equations*, Prentice-Hall, Englewood Cliffs, N.J.

Cullen, Charles (1966): *Matrices and Linear Transformations*, Addison-Wesley, Reading, Mass.

Franklin, J. N. (1968): *Matrix Theory*, Prentice-Hall, Englewood Cliffs, N.J.

Myint-U, Tyn (1978): *Ordinary Differential Equations*, North-Holland, New York.

Putzer, E. J. (1966): *Amer. Math. Monthly, 73*: 2.

Rabenstein, A. L. (1972): *Introduction to Ordinary Differential Equations*, Academic Press, New York.

Thomas, G. B., Jr. (1957): *Amer. Math. Monthly, 64*: 586.

10

AN INTRODUCTION TO FOURIER ANALYSIS

In this chapter we deal with some of the basic elements of Fourier analysis. The first section provides a short historical introduction to the early developments in this field and also includes a discussion of the mathematical analysis of sound. In the next two sections Fourier representations of functions are defined and a number of examples are given. The Gibbs oscillation phenomenon is briefly discussed in Sec. 10.D. Sections 10.E and 10.F are quite theoretical in nature and the less theoretically inclined instructor and/or student may wish to merely note the principal results and continue on. In Sec. 10.G the classical heat equation and its solution are obtained. Also in this section the "formal" solution to the wave equation derived in Chapter 8 is analytically justified. A second formulation of the solution of this equation, d'Alembert's solution, is presented and examined in some detail. Section 10.H deals with a different approach to the problem of summing Fourier series.

A. INTRODUCTION

In Sec. 8.A we arrived at a formal solution to the vibrating string problem or wave equation. During the derivation of this solution the problem arose as to whether or not a suitably large family of functions could be represented on the interval $[0,L]$ by infinite trigonometric sums of the form:

$$\sum_{n=1}^{\infty} c_n \sin \frac{n\pi x}{L}$$

In fact, the entire procedure depended on being able to represent functions in this manner. Daniel Bernoulli produced in 1755 the formal solution given in (8.A.1.b). His work led to considerable controversy at that time, and many of his contemporaries, including Euler and d'Alembert, viewed as untenable his claim to have found a general solution to the vibrating string problem. Their objections stemmed from the obvious fact that if one were to accept Bernoulli's arguments, then it would necessarily follow that the initial position of the string could be represented as an infinite sum of sine functions, which, as Euler observed, would have to be periodic and odd.

In addition, mathematicians of the eighteenth century could not conceive of a sum of differentiable functions (even an infinite sum) that might fail to be differentiable; consequently, since sine functions are clearly differentiable, it would be impossible for a string to be in the initial state illustrated in Fig. 10.1.

Moreover, Euler contended that the initial state of the string could be composed of several connected, but unrelated arcs (Fig. 10.2), and it seemed incomprehensible to him that any one series of the form

$$\sum_{n=1}^{\infty} c_n \sin \frac{n\pi x}{L}$$

could represent such an initial state.

The problem of whether a function could be written as an infinite sum of sine and cosine functions was of major concern in the eighteenth century. Many mathematicians dealt with problems in astronomy, and since astro-

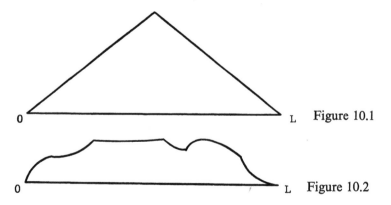

0   L   Figure 10.1

0   L   Figure 10.2

nomical phenomona are frequently periodic, series of sine and cosine functions were of particular significance. Clairaut, studying perturbations due to the sun, went so far as to assert in 1757 that any function could be written in the form

$$f(x) = a_0 + 2 \sum_{n=1}^{\infty} a_n \cos nx$$

Claims and counterclaims abounded, but it was not until the nineteenth century that a sufficient degree of analytic rigor was developed to settle the various points of dispute.

During the eighteenth century, considerable effort was devoted to the study of the vibratory motion of a musical string. It was known that a musical string could be induced to vibrate in such a manner that during the vibrations any two given points of the string would maintain displacements bearing a constant ratio to one another (Fig. 10.3). (We shall refer to such vibrations as *normal vibrations* or *normal modes of vibration*.) For such vibrations, then, there is a constant function c such that

$$\frac{u(x_1, t)}{u(x_2, t)} = c(x_1, x_2) \tag{1}$$

where x_1 and x_2 are arbitrary points of the string and $c(x_1,x_2)$ remains constant for all values of $t > 0$. Equation (1) may be rewritten in the form $u(x_1,t) = c(x_1,x_2)u(x_2,t)$, and, hence, if x_2 is held fixed, we have that the displacements $u(x,t)$ can be expressed in the form $X(x)T(t)$. As we have seen before, solutions to the wave equation will have the form

$$\sum_{n=1}^{\infty} c_n \sin \frac{n\pi x}{L} \cos \frac{n\pi a t}{L} \tag{2}$$

For a given string of length L, the normal mode of vibrations that produces the lowest sound is called the *fundamental mode* and is given by the first term in (2)

$$c_1 \sin \frac{\pi x}{L} \cos \frac{\pi a t}{L} \tag{3}$$

(it can be experimentally verified that the fewer the vibrations the lower the sound, and clearly the number of vibrations per unit time is a minimum

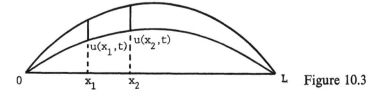

0   x_1 x_2 L   Figure 10.3

when $n = 1$). Note that the initial state of this mode is given by

$$c_1 \sin \frac{\pi x}{L}$$

where c_1 is the amplitude of the vibration (Fig. 10.4). Furthermore, it follows from (3) and (7) of Sec. 7.I that the string vibrates with a period of $2L/a$, where $a = \sqrt{T/\rho}$, T is the tension of the string, and ρ is the density of the string. The pitch we hear is determined by the frequency of the vibration, and this *frequency* is defined as the reciprocal of the period; thus the frequency (or pitch) of the fundamental mode is $(1/2L)\sqrt{T/\rho}$. Observe (as we would expect) that the pitch increases (rises) whenever the string is shortened or tightened or replaced by a string of smaller cross section.

For the musically attuned reader, we mention that the string can be made to vibrate in other normal modes; for instance, the mode corresponding to the first octave is given by the second term in (2)—the inital position of the string is represented by the graph of $c_2 \sin (2\pi x/L)$ and the vibratory motion is governed by

$$c_2 \sin \frac{2\pi x}{L} \cos \frac{2\pi a t}{L}$$

Note that the frequency of the octave is a/L, twice that of the fundamental mode. The mode associated with the musical "fifth" is found by setting $n = 3$.

In general, normal modes of vibration are possible for all positive integers n and have the form

$$c_n \sin \frac{n\pi x}{L} \cos \frac{n\pi a t}{L}$$

where the frequency is $na/2L$ and the amplitude or intensity is c_n. The coefficient c_n essentially determines the loudness of the sound.

Graphs representing the fundamental mode, the first octave, and the corresponding fifth are illustrated in Fig. 10.5.

What causes the difference in sounds emitted at the same frequency by, say, a balalaika and a violin; i.e., what affects the quality of a sound? Bernoulli recognized that, in general, the motion of a string can be described as a composition of various normal vibrations, where these vibrations are superimposed on one another. Hence, it was natural for him to infer that

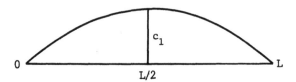

Figure 10.4

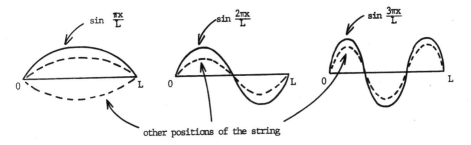

Figure 10.5

all possible motions of the string are merely linear combinations of normal vibrations. For this reason he claimed that every motion of a musical string could be represented in the form

$$\sum_{n=1}^{\infty} c_n \sin \frac{n\pi x}{L} \cos \frac{n\pi at}{L} \tag{4}$$

The terms $c_n \sin (n\pi x/L) \cos (n\pi at/L)$ are called *harmonics* and the series (4) is called a *harmonic series*. The normal vibrations where $n > 1$ are called the "*overtones*" or *harmonics* of the sound—to be distinguished from the fundamental sound that corresponds to $n = 1$. Bernoulli also observed that each normal vibration occurring during the movement of the vibrating string retains its individuality and is not influenced by any of the other normal vibrations that are simultaneously taking place. The timber or quality of a sound depends on the ratio of the intensities that occur in the sound's overtones or harmonics and the fundamental, whereas the pitch depends on the frequency of the fundamental. Perhaps the most remarkable feature of Bernoulli's result is that it gives a complete portrait of the physical properties that make up a sound; in addition, it has aided scientists in discovering how the inner ear decomposes a musical sound into its basic building blocks, the harmonics.

Due, in part, to the many misconceptions concerning functions and infinite series, Bernoulli's ideas lay practically dormant for more than half of a century. It was the work of the French mathematician Joseph Fourier from 1805 to 1822 that served as the catalyst for the revival of interest in the properties of trigonometric series: a *formal trigonometric series* is any series (convergent or divergent) of the form $a_0/2 + \sum_{n=1}^{\infty} (a_n \cos nx + b_n \sin nx)$. Fourier, basing much of his work on that of his predecessors such as Euler and Bernoulli, used these series in the development of a mathematical description of the flow of heat, and from his germinal work have emanated many profound mathematical ideas. In the course of his research he showed that there are trigonometric series that represent broken line graphs such as

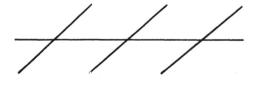

Figure 10.6

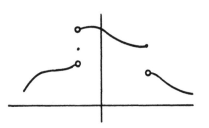

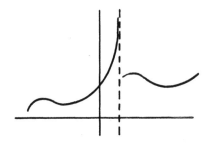

Simple Jump Discontinuities   Not a Simple Jump Discontinuity

Figure 10.7

the one illustrated in Fig. 10.6. Such examples led to a much needed rethinking of the nature of functions. One of Fourier's more notable achievements was to represent the partial sums of trigonometric series by integrals, thus enabling him to establish the convergence of many basic trigonometric series; this idea also proved to be of prime importance in the proof of one of the first significant convergence theorems dealing with trigonometric series, Dirichlet's theorem, one formulation of which is given in (10.A.4).

(10.A.1) *Definition* A function $f: [a,b] \to \mathbf{R}^1$ is said to have a *simple jump discontinuity* at a point $x \in (a,b)$ if both $\lim_{w \to x^-} f(w)$ and $\lim_{w \to x^+} f(w)$ exist but are unequal.

We shall denote $\lim_{w \to x^+} f(w)$ by $f(x + 0)$ and $\lim_{w \to x^-} f(w)$ by $f(x - 0)$.

(10.A.2) *Definition* A function $f: [a,b] \to \mathbf{R}^1$ is said to be *piecewise continuous* if f is continuous on $[a,b]$ except at a finite number of points, at which points the discontinuities are simple jump discontinuities.

(10.A.3) *Examples* Figure 10.7.

(10.A.4) *Theorem* (*Dirichlet*) Suppose that a function f satisfies the *Dirichlet conditions*:

(i) f is piecewise continuous on $[-\pi,\pi]$.

(ii) There is a partition $-\pi = x_0 < x_1 < \cdots < x_n = \pi$ of $[-\pi,\pi]$ such that, on each subinterval $[x_i,x_{i+1}]$, f is either increasing or decreasing.

Then the series

$$\frac{a_0}{2} + \sum_{n=1}^{\infty} (a_n \cos nx + b_n \sin nx) \tag{5}$$

where

$$a_0 = \frac{1}{\pi} \int_{-\pi}^{\pi} f(x)dx \tag{6}$$

$$a_n = \frac{1}{\pi} \int_{-\pi}^{\pi} f(x) \cos nx \, dx \qquad n > 0 \tag{7}$$

$$b_n = \frac{1}{\pi} \int_{-\pi}^{\pi} f(x) \sin nx \, dx \qquad n > 0 \tag{8}$$

converges to

$f(x)$ at each point of continuity $x \in (-\pi, \pi)$

$\dfrac{f(x - 0) + f(x + 0)}{2}$ at each point of discontinuity $x \in (-\pi, \pi)$

$\dfrac{f(\pi - 0) + f(-\pi + 0)}{2}$ at $x = \pm\pi$

(10.A.5) *Observation* Note that at points of discontinuity the series (5) takes on the average value of the left-hand and right-hand limits of f (Fig. 10.8). Furthermore, at each point of continuity we have $[f(x - 0) + f(x + 0)]/2 = f(x)$.

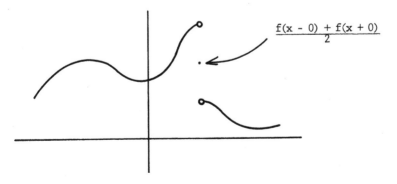

$$\frac{f(x - 0) + f(x + 0)}{2}$$

Figure 10.8

A trigonometric series whose coefficients are determined by (6), (7), and (8) is called a *Fourier series*, and it is these series that provide the focal point of study in this chapter.

The formal definition of a Fourier series is given next.

(10.A.6) Definition Suppose that $f: [-\pi,\pi] \to \mathbf{R}^1$ is integrable. Then the (possibly divergent) series

$$\frac{a_0}{2} + \sum_{n=1}^{\infty} (a_n \cos nx + b_n \sin nx)$$

where

$$a_n = \frac{1}{\pi} \int_{-\pi}^{\pi} f(x) \cos nx \, dx \qquad n \geq 0$$

$$b_n = \frac{1}{\pi} \int_{-\pi}^{\pi} f(x) \sin nx \, dx \qquad n > 0$$

is called the *Fourier series expansion* (or the *Fourier series*) of f.

The Fourier series of f will be denoted by $T(f)$. The constants a_n and b_n are called the *Fourier coefficients*; we shall see in the next section that they arise in a very natural way.

Although thus far we have restricted ourselves to considerations involving the interval $[-\pi,\pi]$, we shall see shortly that other intervals may be used as well.

(10.A.7) Observation At present our notion of the integrability of a function is rather limited, i.e., an integrable function can have no more than a finite number of discontinuities. In the next chapter the concept of integrability will be greatly expanded to include the more general concepts of Riemann and Lebesgue integrability, where an infinite number of discontinuities may be permitted in certain cases. All theorems in the present chapter that mention the integrability of a function f are valid even if f is assumed to be integrable in the more general sense of Lebesgue (and Riemann) integration.

Theorem (10.A.4) and a number of refinements of it helped provide a sound basis for the incipient theory of Fourier series, and the ideas that evolved from this theory quickly captured the imagination of a great many mathematicians. In fact, the problems arising from this theory have played an instrumental role in the development of such areas as set theory, summability theory, and Lebesgue integration. The list of contributors to the study of Fourier analysis is long and impressive.

B. FOURIER SERIES REPRESENTATIONS

The following easily derived result will be of considerable use in the proofs of a number of theorems in this chapter.

(10.B.1) *Theorem* If m and n are nonnegative integers, then

$$\int_{-\pi}^{\pi} \cos mx \cos nx \, dx = \begin{cases} 0 & \text{if } m \neq n \\ \pi & \text{if } m = n \neq 0 \\ 2\pi & \text{if } m = n = 0 \end{cases}$$

$$\int_{-\pi}^{\pi} \sin mx \sin nx \, dx = \begin{cases} 0 & \text{if } m \neq n \\ \pi & \text{if } m = n \neq 0 \\ 0 & \text{if } m = n = 0 \end{cases}$$

$$\int_{-\pi}^{\pi} \sin mx \cos nx \, dx = 0$$

Proof. The proof is immediate from the following trigonometric identities (see problem B.1):

$$\cos mx \cos nx = \tfrac{1}{2}(\cos (m + n)x + \cos (m - n)x)$$
$$\sin mx \sin nx = \tfrac{1}{2}(\cos (m - n)x - \cos (m + n)x)$$
$$\sin mx \cos nx = \tfrac{1}{2}(\sin (m + n)x + \sin (m - n)x)$$

(10.B.2) *Corollary* If for each nonnegative integer n,

$$s_n(x) = \frac{a_0}{2} + \sum_{k=1}^{n} (a_k \cos kx + b_k \sin kx)$$

then

$$\int_{-\pi}^{\pi} s_n(x) \cos kx \, dx = \begin{cases} \pi a_k & \text{if } 0 \leq k \leq n \\ 0 & \text{if } k > n \end{cases} \tag{1}$$

and

$$\int_{-\pi}^{\pi} s_n(x) \sin kx \, dx = \begin{cases} \pi b_k & \text{if } 1 \leq k \leq n \\ 0 & \text{if } k > n \end{cases} \tag{2}$$

In general there is no guarantee that the Fourier series $T(f)$ of a function f will actually converge to f [just as there is no assurance that the Taylor series expansion of a function will converge to the function—recall (8.D.5)]. Nevertheless the partial sums of $T(f)$ do provide in most instances the "best" trigonometric approximations of f. This can be seen as follows.

Suppose that $f: [-\pi,\pi] \to \mathbf{R}^1$ is *square integrable*, i.e., $\int_{-\pi}^{\pi} f^2(x) \, dx$ exists, where $f^2(x) = (f(x))^2$. Let $T(x) = a_0/2 + \sum_{n=1}^{\infty} (a_n \cos nx + b_n \sin nx)$ be a trigonometric series with partial sums $s_n(x)$. We show that

$$\int_{-\pi}^{\pi} (f(x) - s_n(x))^2 \, dx \tag{3}$$

is minimal when $T(x)$ is the Fourier series expansion of f. [Note that the square is necessary here, for without it, $\int_{-\pi}^{\pi} (f(x) - s_n(x)) \, dx = 0$ whenever $f(x) - s_n(x)$ is an odd function (see problem B.2).]

For fixed n, when does $\int_{-\pi}^{\pi} (f(x) - s_n(x))^2 \, dx$ take on a minimal value? For each n and for each $(2n + 1)$-tuple of coefficients of $T(x)$, $(a_0, a_1, \ldots, a_n, b_1, \ldots, b_n)$, the integral $\int_{-\pi}^{\pi} (f(x) - s_n(x))^2 \, dx$ yields a numerical value, and thus, the function

$$\psi(a_0, a_1, \ldots, a_n, b_1, \ldots, b_n) = \int_{-\pi}^{\pi} (f(x) - s_n(x))^2 \, dx$$

is clearly a continuously differentiable function mapping $\mathbf{R}^{2n+1}$ into $\mathbf{R}^1$. Hence by the natural generalization of (7.A.13), a minimal value (if it exists) will be attained at a point in $\mathbf{R}^{2n+1}$ determined by the equations:

$$\frac{\partial \psi}{\partial a_k} = 0 \qquad 0 \le k \le n$$

$$\frac{\partial \psi}{\partial b_k} = 0 \qquad 1 \le k \le n. \tag{4}$$

Equations (4) can be rewritten as

$$2 \int_{-\pi}^{\pi} (f(x) - s_n(x)) \cos kx \, dx = 0 \qquad 0 \le k \le n$$

$$2 \int_{-\pi}^{\pi} (f(x) - s_n(x)) \sin kx \, dx = 0 \qquad 1 \le k \le n. \tag{5}$$

It follows from (1) and (2) that equations (5) have the form

$$\int_{-\pi}^{\pi} f(x) \cos kx \, dx - \pi a_k = 0 \qquad 0 \le k \le n$$

$$\int_{-\pi}^{\pi} f(x) \sin kx \, dx - \pi b_k = 0 \qquad 1 \le k \le n$$

which implies that a minimum of ψ is attained when a_n and b_n are defined by

$$a_n = \frac{1}{\pi} \int_{-\pi}^{\pi} f(x) \cos nx \, dx \qquad n \ge 0$$

$$b_n = \frac{1}{\pi} \int_{-\pi}^{\pi} f(x) \sin nx \, dx \qquad n \ge 1$$

thus, the Fourier series expansion of f does indeed give the best [in the sense that (3) is minimal] trigonometric approximation of f over $[-\pi, \pi]$.

There still remain the basic questions:

What functions $f: [-\pi,\pi] \to \mathbf{R}^1$, can be *represented* by a trigonometric series, i.e., for what functions $f: [-\pi,\pi] \to \mathbf{R}^1$ does there exist a trigonometric series $T(x)$ such that $T(x) = f(x)$ for each $x \in [-\pi,\pi]$? Somewhat more specifically, when can a function $f: [-\pi,\pi] \to \mathbf{R}^1$ be represented by its Fourier series $T(f)$?

Continuity of the function f is not sufficient. In fact, in 1876 P. du Bois-Reymond constructed a continuous function $f: [-\pi,\pi] \to \mathbf{R}^1$ with the property that its Fourier series failed to converge to f at each point in a dense subset of $[-\pi,\pi]$ (a dense subset A of a space X is a subset with the property that $\bar{A} = X$; for example, the rational numbers form a dense subset of $\mathbf{R}^1$). Still, in a large number of instances, convergence is obtained; Secs. 10.E and 10.F are devoted to finding sufficient conditions for this to occur.

We now turn to some specific examples of functions and their Fourier series representations.

C. EXAMPLES

In this section we examine a number of Fourier series representations of functions, and indicate graphically how the lower order partial sums approximate the given functions. In general, we shall forgo attempting to establish the convergence or mode of convergence of the resulting series; considerations of this nature are taken up in subsequent sections.

(10.C.1) *Example* (*The Zigzag Function*) We compute the Fourier series of the zigzag function over the interval $[-\pi,\pi]$. This function (Fig. 10.9) is given by

$$f(x) = \begin{cases} -(x + \pi) & -\pi \leq x \leq -\dfrac{\pi}{2} \\[2mm] x & -\dfrac{\pi}{2} \leq x \leq \dfrac{\pi}{2} \\[2mm] (\pi - x) & \dfrac{\pi}{2} \leq x \leq \pi \end{cases}$$

Since f is an odd function on the interval $[-\pi,\pi]$, so is $f(x) \cos kx$, and, therefore, $a_k = (1/\pi) \int_{-\pi}^{\pi} f(x) \cos kx \, dx = 0$ for each k (problem B.2). Since $f(x) \sin kx$ is an even function, we have

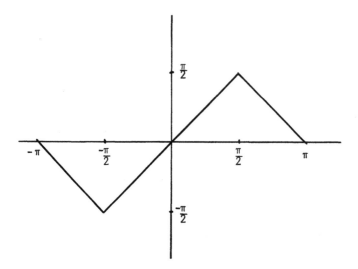

Figure 10.9

$$b_k = \frac{2}{\pi} \int_0^\pi f(x) \sin kx \, dx = \frac{2}{\pi} \left[\int_0^{\pi/2} x \sin kx \, dx + \int_{\pi/2}^\pi (\pi - x) \sin kx \, dx \right]$$

$$= \frac{2}{\pi} \left\{ \frac{-x \cos kx}{k} \bigg|_0^{\pi/2} + \frac{1}{k} \int_\pi^{\pi/2} \cos kx \, dx - (\pi - x) \frac{\cos kx}{k} \bigg|_{\pi/2}^\pi \right.$$

$$\left. - \frac{1}{k} \int_{\pi/2}^\pi \cos kx \, dx \right\}$$

$$= \frac{2}{\pi} \left\{ \frac{-\pi}{2k} \cos \frac{k\pi}{2} + \frac{1}{k^2} \sin kx \bigg|_\pi^{\pi/2} + \frac{\pi}{2k} \cos \frac{k\pi}{2} - \frac{1}{k^2} \sin kx \bigg|_{\pi/2}^\pi \right\}$$

$$= \frac{4}{\pi k^2} \sin \frac{k\pi}{2} = \begin{cases} 0 & \text{if } k = 2n \\ \dfrac{4(-1)^n}{\pi(2n+1)^2} & \text{for } k = 2n + 1, \, n = 0, 1, 2, \ldots \end{cases}$$

Consequently,

$$T(f) = \frac{4}{\pi} \left\{ \sin x - \frac{1}{9} \sin 3x + \frac{1}{25} \sin 5x - \cdots + \frac{(-1)^n}{(2n+1)^2} \sin (2n+1)x + \cdots \right\}$$

[Note that the convergent series $(4/\pi) \sum_{n=0}^\infty 1/(2n+1)^2$ dominates $T(f)$ everywhere, and therefore $T(f)$ converges absolutely (and uniformly) on $[-\pi,\pi]$; moreover, from certain results to come it will follow that $T(f)$ converges to f on $[-\pi,\pi]$.]

In Fig. 10.10 the partial sum s_3 is compared with f.

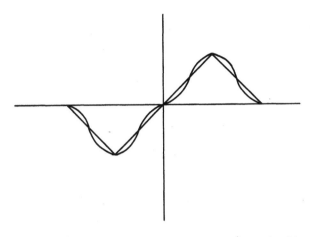

Figure 10.10

(10.C.2) *Example* Let $f: [-\pi,\pi] \to \mathbf{R}^1$ be the identity function $f(x) = x$. We compute the Fourier series over the interval $[-\pi,\pi]$. Since f is odd, we have $a_k = 0$ for each k. For $k \geq 1$ the coefficient b_k is given by

$$b_k = \frac{2}{\pi} \int_0^\pi x \sin kx \, dx = \frac{2}{\pi} \left\{ -x \frac{\cos kx}{k} \Big|_0^\pi + \frac{1}{k} \int_0^\pi \cos kx \, dx \right\} = \frac{2(-1)^{k+1}}{k}$$

Thus,

$$T(f) = 2 \sum_{k=1}^\infty \frac{(-1)^{k+1}}{k} \sin kx.$$

Clearly $T(f)(\pm\pi) = 0$. Frequently, it will be of interest (and in many applications, essential) that f be extended periodically to a larger domain. In order to maintain the convergence property of $T(f)$, this may necessitate redefining f at the endpoints of the original interval. For instance, in the present example, if we should wish to extend f periodically to $\mathbf{R}^1$, we must redefine f at $\pm\pi$ to be 0 (which, for future reference, we observe is equal to $[f(\pi - 0) + f(-\pi + 0)]/2$, where $f(\pi - 0)$ and $f(-\pi + 0)$ are defined as in (10.A.1), [cf. (10.A.4)] (Fig. 10.11).

In the next section we shall see that if f is extended in this manner, then $T(f)$ will converge pointwise to f. It is interesting to note the rather peculiar behavior of the partial sums $s_n(x)$ of $T(f)$ at the points $\pm\pi$. As indicated in Fig. 10.12, the partial sums exhibit a substantial overshoot near these endpoints, and curiously, an increase in n will not diminish the amplitude of the overshoot, although with increasing n the overshoot occurs over smaller and smaller intervals about $\pm\pi$. This phenomenon is called the *Gibbs oscillation phenomenon* and is discussed further in Sec. 10.D.

(10.C.3) *Example* The identity function $f(x) = x$ defined on $[0,2\pi]$. Here, although the domain is no longer $[-\pi, \pi]$, we still define a_k by

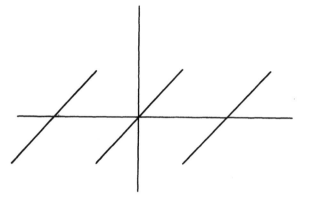

Figure 10.11

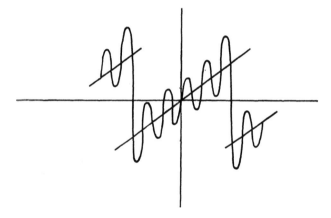

Figure 10.12

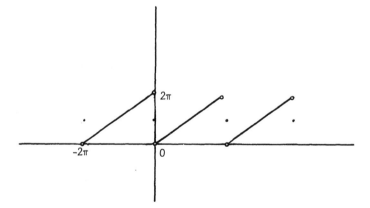

Figure 10.13

$(1/\pi) \int_0^{2\pi} x \cos kx\, dx$ and b_k by $(1/\pi) \int_0^{2\pi} x \sin kx\, dx$, since the arguments found in Sec. 10.B are just as valid for the interval $[0,2\pi]$ as they were for the interval $[-\pi,\pi]$. It is not difficult to show that

$$a_0 = 2\pi$$
$$a_k = 0 \qquad k \ge 1$$
$$b_k = \frac{-2}{k}$$

(see problem C.1). The Fourier series $T(f)$ is given in the usual manner:

$$T(f) = \pi - 2 \sum_{k=1}^{\infty} \frac{\sin kx}{k}$$

Notice that $T(f)(0) = \pi = T(f)(2\pi)$. We show in the next section that $T(f)$ converges to f at each point $x \in (0,2\pi)$. If we wish to extend f periodically from $(0,2\pi)$ to $\mathbf{R}^1$, we must redefine f at 0 and 2π in order that the values of f there equal π. Again, we observe for future reference that $\pi = [f(0 + 0) + f(2\pi - 0)]/2$ (Fig. 10.13). As in the previous example, the partial sums of $T(f)$, where f is the extended function, exhibit a noticeable overshoot near the endpoints 0 and 2π (Fig. 10.14).

It is important to note that the Fourier series representations of the identity function described in examples (10.C.3) differ. This is due to the change in the fundamental interval (from $[-\pi,\pi]$ to $[0,2\pi]$). In fact, as is easily seen from the graphs depicted in Figures 10.11 and 10.13, the two Fourier series expansions represent quite different functions. If f_1 represents the extension of $f(x) = x$ from $(-\pi,\pi)$, as illustrated in Fig. 10.11, and f_2 the extension of $f(x) = x$ from $(0,2\pi)$, as illustrated in Fig. 10.13, then, as expected,

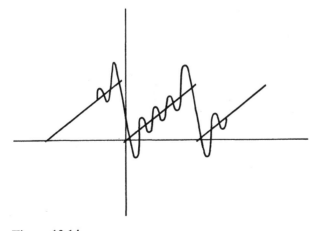

Figure 10.14

$$f_1(x) = f_2(x) \qquad \text{on } (0, \pi)$$

but

$$f_1(x) \neq f_2(x) \qquad \text{on } (\pi, 2\pi)$$

Thus, in dealing with Fourier series expansions, the fundamental interval used as the domain of a given function is of considerable importance, and in working problems, the reader should be quite aware which interval is being employed.

(10.C.4) *Example (The Sawtooth Function over the Interval* $[-\pi,\pi]$*).* (See Fig. 10.15.)

This function can be obtained by shifting the graph of the function described in (10.C.2) π units to the left. If we let $f_1(x)$ be the function whose graph is given in Fig. 10.11, then it follows immediately from (10.C.2) that the Fourier series of the sawtooth function f is given by

$$T(f) = f_1(\pi + x) = 2\left[\sin(\pi + x) - \frac{1}{2}\sin 2(\pi + x) + \frac{1}{3}\sin 3(\pi + x) - \cdots \right]$$

$$= 2\left(-\sin x - \frac{1}{2}\sin 2x - \frac{1}{3}\sin 3x - \cdots \right) = -2 \sum_{k=1}^{\infty} \frac{\sin kx}{k}$$

(10.C.5) *Example*   A quite useful series is obtained from the function

$$S(x) = \begin{cases} \dfrac{-1}{2}(x + \pi) & \text{if } -\pi \leq x < 0 \\ 0 & \text{if } x = 0 \\ \dfrac{1}{2}(\pi - x) & \text{if } 0 < x \leq \pi \end{cases}$$

which is an elementary modification of (10.C.4): $S(x) = -\frac{1}{2}f_1(x + \pi)$ (Fig. 10.16).

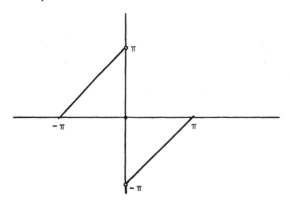

Figure 10.15

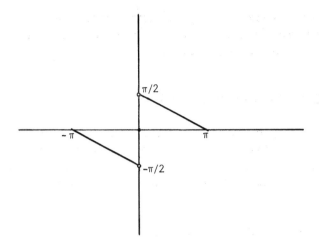

Figure 10.16

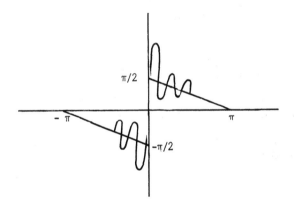

Figure 10.17

It follows from (10.C.4) that the Fourier series of $S(x)$, $T(S)$, is defined by

$$T(S) = \sum_{k=1}^{\infty} \frac{\sin kx}{k}$$

Convergence theorems from Sec. 10.E ensure that $S(x) = T(S)(x)$ on $[-\pi,\pi]$, although the partial sums of $T(S)$ are subject to the Gibbs oscillation phenomenon near $x = 0$ (see Fig. 10.17). An analytical analysis of these oscillations for this function is given in the next section.

It should be apparent from the examples presented in this section that one inherent advantage of a Fourier series expansion over an expansion such as the Taylor expansion is that noncontinuous functions may often be represented by Fourier series. Fourier series are also especially useful in representing periodic functions. A third advantage of Fourier series is their adaptability to problems involving the superposition principle (as illustrated, for instance, in Sec. 10.A when the harmonics were added to obtain an overall mathematical description of the sound produced).

D. THE GIBBS OSCILLATION PHENOMENON

In Sec. 10.C we saw that the partial sums of a Fourier series representation, $T(f)$, deviated markedly from f near points of discontinuity. In this section we examine in some detail the behavior of the partial sums of the Fourier series of the function S described in Example (10.C.5). We begin with the following lemma.

(10.D.1) Lemma   For each $x \in \mathbf{R}^1$ and each integer $n \geq 1$,

$$\frac{1}{2} + \sum_{k=1}^{n} \cos kx = \frac{\sin (n + \frac{1}{2})x}{2 \sin (x/2)} \tag{1}$$

[Here, since

$$\lim_{x \to z} \frac{\sin (n + \frac{1}{2})x}{2 \sin (x/2)} = n + \frac{1}{2}$$

where z is any integral multiple of 2π (see problem D.1), we assume that the right-hand side of (1) takes on this value whenever $x = 0$, 2π, -2π, 4π, -4π, ...]

Proof.   It follows from the trigonometric identity

$$\sin \left(k + \frac{1}{2}\right) x - \sin \left(k - \frac{1}{2}\right) x = 2 \sin \frac{x}{2} \cos kx$$

that the sum $2 \sin (x/2) \sum_{k=1}^{n} \cos kx$ "telescopes" to $\sin (n + \frac{1}{2})x - \sin (x/2)$. Consequently, the result follows if $\sin (x/2) \neq 0$. If $\sin (x/2) = 0$, then x must be some integral multiple of 2π. In this case, however, the left-hand side of (1) is $n + \frac{1}{2}$, and by the above parenthetical remark, the result also holds for these values.

The next step is to replace the partial sums of $T(S)$ with integrals. This procedure is due to Fourier. Note that

$$\frac{x}{2} + s_n(x) = \frac{x}{2} + \sum_{k=1}^{n} \frac{\sin kx}{k} = \int_0^x \left(\frac{1}{2} + \sum_{k=1}^{n} \cos kt\right) dt$$

$$= \int_0^x \frac{\sin ((n + \frac{1}{2})t)}{2 \sin (t/2)} dt$$

$$= \frac{1}{2} \int_0^x \frac{\sin (n + \frac{1}{2})t}{t/2} dt + \frac{1}{2} \int_0^x \left(\frac{1}{\sin (t/2)} - \frac{1}{t/2}\right) \sin \left(n + \frac{1}{2}\right) t \, dt$$

Consequently, we have

$$s_n(x) = \frac{-x}{2} + \frac{1}{2}\int_0^x \frac{\sin(n + \frac{1}{2})t}{t/2}\,dt + \frac{1}{2}\int_0^x \left(\frac{1}{\sin(t/2)} - \frac{1}{t/2}\right)\sin\left(n + \frac{1}{2}\right)t\,dt \tag{2}$$

We use equation (2) to determine the behavior of $s_n(x)$ as $x \to 0$ and $n \to \infty$. Two applications of L'Hospital's rule may be used to show that

$$\lim_{t \to 0}\left(\frac{1}{\sin(t/2)} - \frac{1}{t/2}\right) = 0 \tag{3}$$

Therefore, in small neighborhoods of $x = 0$, we have that $s_n(x)$ closely approximates $\int_0^x [\sin(n + \frac{1}{2})t]/t\,dt$ for all $n \geq 1$. Since

$$\int_0^x \frac{\sin((n + \frac{1}{2})t)}{t}\,dt = \int_0^{(n+\frac{1}{2})x} \frac{\sin t}{t}\,dt \tag{4}$$

it follows that in these neighborhoods

$$s_n(x) \approx \int_0^{(n+\frac{1}{2})x} \frac{\sin t}{t}\,dt \qquad \text{for all } n \geq 1$$

Thus, it is essentially the upper limit $u = (n + \frac{1}{2})x$ which determines the value of $s_n(x)$ in small neighborhoods of $x = 0$. From the graph of $(\sin t)/t$ $(t > 0)$ (Fig. 10.18), it is easy it see that

$$\max_{u>0}\left\{\int_0^u \frac{\sin t}{t}\,dt\right\} = \int_0^\pi \frac{\sin t}{t}\,dt$$

[see (10.D.2)]. The value of this latter integral is known to be $(\pi/2)(1.179\cdots)$. Since for any fixed n, we can find x such that $u = (n + \frac{1}{2})x$ (the larger the

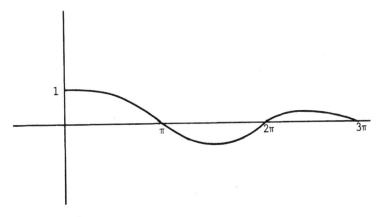

Figure 10.18

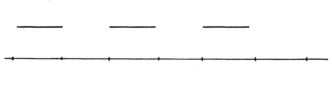

Figure 10.19

value of n, the smaller is x) and since $S(x) \cong \pi/2$ for x near 0, we see that an "overshoot" by $s_n(x)$ of approximately 18% is maintained as $n \to \infty$ (but over smaller and smaller intervals centered at $x = 0$).

This example does not represent a special case: all piecewise smooth functions (functions whose derivatives are piecewise continuous) exhibit this unusual behavior at points of discontinuity, and amazingly the overshoot is always on the order of 18%.

This phenomenon was first observed by Wilbraham in 1848 and then rediscovered in 1898 by Albert Michelson, an American physicist noted for his development of precision instruments. Michelson had invented a device which could physically reproduce the partial sums of a Fourier series up to the 80th order for any graphically given function. It soon came to his attention that although his device yielded generally excellent approximations of the square wave function (Fig. 10.19), there were persistent distortions (a slight high-frequency wiggle) in neighborhoods of the discontinuities. Disturbed that something was wrong with his invention, he wrote Josiah Gibbs, a distinguished mathematical physicist, about this aberration. Gibbs studied the problem and published a reply in *Nature* (1899) that explained the phenomenon in terms of the nonuniform convergence of the Fourier series in neighborhoods of the discontinuities.

(10.D.2)   *Exercise* Show that if $a_n = \int_{n\pi}^{(n+1)\pi} (\sin t)/t \, dt$, then

    (a)   For each n, $|a_n| \geq |a_{n+1}|$
    (b) $\lim_{n \to \infty} a_n = 0$

and, hence, the alternating series $\sum_{n=0}^{\infty} a_n$ converges.

E. POINTWISE CONVERGENCE

From the examples studied thus far it would appear doubtful that we could analyze the convergence properties of Fourier series with the methods used in connection with the convergence of power series. In particular, the comparison test for series is of limited use in dealing with Fourier series. As

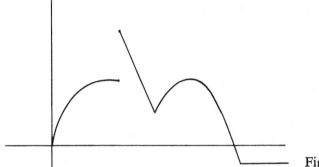

Figure 10.20

mentioned in Sec. 10.A, one quite useful method, due to Fourier, is to express the partial sums of a Fourier series $T(f)$ as integrals. Mathematicians (Riemann and Fejér among others) have since provided other ways of summing Fourier series; Fejér's method will be discussed in Sec. 10.H.

In this section we limit our study of convergence to functions that are piecewise smooth on a given interval. A function f is said to be *piecewise smooth* on an interval I if both f and f' are continuous on I except possibly at a finite number of points in I. At these exceptional points the only discontinuities allowed are simple jump discontinuities.

(10.E.1) *Example* (See Fig. 10.20.)
Our principal result is the following.

(10.E.2) *Theorem* If f is piecewise smooth on $[-\pi,\pi]$, and if $T(f)$ is the Fourier series of f, then

(i) $T(f)$ converges pointwise to f at each point of continuity in $(-\pi,\pi)$.
(ii) $T(f)$ converges to $[f(x - 0) + f(x + 0)]/2$ at each point $x \in (-\pi,\pi)$ of discontinuity of f.
(iii) $T(f)$ converges to $[f(-\pi + 0) + f(\pi - 0)]/2$ at $x = \pm\pi$.

The proof of this theorem is broken up into a number of lemmas. The *Dirichlet kernel* is defined by

$$D_n(x) = \begin{cases} \dfrac{\sin (n + \frac{1}{2})x}{2 \sin (x/2)} & \text{if } x \neq 0, 2\pi, -2\pi, 4\pi, -4\pi, \ldots \\ n + \frac{1}{2} & \text{otherwise} \end{cases}$$

[cf. (10.D.1)]. From (10.D.1) we obtain a useful integral representation of the nth partial sum of $T(f)$.

(10.E.3) Lemma If f is integrable on $[-\pi,\pi]$ and $s_n(x)$ is the nth partial sum of $T(f)$, then

$$s_n(x) = \frac{1}{\pi} \int_{-\pi}^{\pi} f(t)D_n(t - x)\, dt$$

Proof. It follows immediately from the definition of $s_n(x)$ and (10.D.1) that

$$s_n(x) = \frac{1}{\pi} \int_{-\pi}^{\pi} f(t)\left[\frac{1}{2} + \sum_{k=1}^{n}(\cos kt \cos kx + \sin kt \sin kx)\right] dt$$

$$= \frac{1}{\pi} \int_{-\pi}^{\pi} f(t)\left[\frac{1}{2} + \sum_{k=1}^{n}\cos k(t - x)\right] dt$$

$$= \frac{1}{\pi} \int_{-\pi}^{\pi} f(t)D_n(t - x)\, dt$$

(10.E.4) Lemma If $f: [-\pi,\pi] \to \mathbf{R}^1$ is a piecewise smooth function and if $x_0 \in [-\pi,\pi]$, then

$$\lim_{n \to \infty} s_n(x_0) = \begin{cases} \dfrac{f(x_0 - 0) + f(x_0 + 0)}{2} & \text{if } x_0 \in (-\pi, \pi) \\[2mm] \dfrac{f(\pi - 0) + f(-\pi + 0)}{2} & \text{if } x_0 = \pm\pi \end{cases}$$

Proof. For purposes of integration [see (1) below] we assume that f has been extended periodically from $[-\pi,\pi]$ to $\mathbf{R}^1$ (with period 2π). If $x_0 \in [-\pi,\pi]$, then since $f(x_0 - 0)$ and $f'(x_0 - 0)$ exist, we have that on the interval $-\pi \le t < 0$,

$$\frac{f(x_0 + t)}{2 \sin (t/2)} = \frac{f(x_0 + t) - f(x_0 - 0)}{2 \sin (t/2)} + \frac{f(x_0 - 0)}{2 \sin (t/2)}$$

$$= \frac{f(x_0 + t) - f(x_0 - 0)}{t} \cdot \frac{t/2}{\sin (t/2)} + \frac{f(x_0 - 0)}{2 \sin (t/2)}$$

If we let

$$F_1(t) = \frac{f(x_0 + t) - f(x_0 - 0)}{t} \cdot \frac{t/2}{\sin (t/2)}$$

then

$$\frac{f(x_0 + t)}{2 \sin (t/2)} = F_1(t) + \frac{f(x_0 - 0)}{2 \sin (t/2)}$$

Since $F_1(0 - 0) = f'(x_0 - 0)$ (see problem D.2), F_1 can be extended to the closed interval $[-\pi,0]$ as a piecewise continuous function. In a similar manner we find that on the interval $[0,\pi]$ there is a piecewise continuous function F_2 such that

$$\frac{f(x_0 + t)}{2 \sin (t/2)} = F_2(t) + \frac{f(x_0 + 0)}{2 \sin (t/2)}$$

Therefore from (10.E.3) we have (with an appropriate substitution)

$$s_n(x_0) = \frac{1}{\pi} \int_{-\pi}^{\pi} f(x_0 + t)D_n(t) \, dt = \frac{1}{\pi} \int_{-\pi}^{\pi} \frac{f(x_0 + t)}{2 \sin (t/2)} \sin \left(n + \frac{1}{2}\right) t \, dt$$

$$= \frac{1}{\pi} \int_{-\pi}^{0} F_1(t) \sin \left(n + \frac{1}{2}\right) t \, dt + \frac{1}{\pi} \int_{0}^{\pi} F_2(t) \sin\left(n + \frac{1}{2}\right) t \, dt \qquad (1)$$

$$+ f(x_0 - 0)\left(\frac{1}{\pi} \int_{-\pi}^{0} \frac{\sin (n + \frac{1}{2})t}{2 \sin (t/2)} \, dt\right.$$

$$\left. + f(x_0 + 0) \frac{1}{\pi} \int_{0}^{\pi} \frac{\sin (n + \frac{1}{2})t}{\sin (t/2)} \, dt\right)$$

and by the Riemann-Lebesgue Lemma (6.B.6), this sum converges to

$$\frac{f(x_0 - 0) + f(x_0 + 0)}{2}$$

as $n \to \infty$.

This argument shows that if x_0 is a point of continuity of f, then the sequence $\{s_n(x_0)\}$ converges to $f(x_0)$ and, similarly, the sequence $\{s_n(\pm\pi)\}$ converges to $[f(-\pi + 0) + f(\pi - 0)]/2$. This completes the proof of Theorem (10.E.2).

We now extend this result to arbitrary closed intervals centered at the origin. In view of the previous result, it is customary to redefine $f(x)$ at points of discontinuity so that $f(x) = T(f)(x)$ everywhere. This convention is observed in the following corollary.

(10.E.5) Corollary If $f(x)$ is piecewise smooth on $(-l,l)$, then for each $x \in [-l,l]$,

$$f(x) = \frac{a_0}{2} + \sum_{n=1}^{\infty} \left(a_n \cos \frac{n\pi x}{l} + b_n \sin \frac{n\pi x}{l}\right)$$

where

$$a_n = \frac{1}{l} \int_{-l}^{l} f(x) \cos \frac{n\pi x}{l} \, dx \qquad n \geq 0$$

$$b_n = \frac{1}{l} \int_{-l}^{l} f(x) \sin \frac{n\pi x}{l} \, dx \qquad n > 0$$

Proof. Let $\phi: [-\pi,\pi] \to [-l,l]$ be defined by $\phi(t) = (l/\pi)t = x$. Setting $g(t) = f(\phi(t)) = f(x)$, we see immediately that $g(t)$ satisfies the hypothesis of (10.E.2), and therefore, on the interval $-l \le x \le l$,

$$\frac{f(x-0)+f(x+0)}{2} = \frac{f(\phi(t-0))+f(\phi(t+0))}{2} = \frac{g(t-0)+g(t+0)}{2}$$

$$= \frac{a_0}{2} + \sum_{n=1}^{\infty} (a_n \cos nt + b_n \sin nt) \qquad \text{(by 10.E.2)}$$

where

$$a_n = \frac{1}{\pi} \int_{-\pi}^{\pi} g(s) \cos ns \, ds \qquad n \ge 0$$

$$b_n = \frac{1}{\pi} \int_{-\pi}^{\pi} g(s) \sin ns \, ds \qquad n \ge 1$$

Since $t = (\pi/l)x$, we have shown that if $x \in [-l,l]$, then

$$\frac{f(x-0)+f(x+0)}{2} = \frac{a_0}{2} + \sum_{n=1}^{\infty} \left(a_n \cos \frac{n\pi x}{l} + b_n \sin \frac{n\pi x}{l} \right)$$

In addition, if we set $t = ls/\pi$, then we find

$$a_n = \frac{1}{\pi} \int_{-\pi}^{\pi} f\left(\frac{ls}{\pi}\right) \cos ns \, ds = \frac{1}{l} \int_{-l}^{l} f(t) \cos \frac{n\pi t}{l} \, dt$$

Note that in both (10.E.2) and (10.E.5), f is defined on an interval centered at the origin. Quite often however, in applied problems a function f is defined on intervals such as $[0,\pi]$ or more generally $[0,l]$. If, for example, f is defined on $[0,\pi]$, then in order to use our previous results it is necessary to extend f to the interval $[-\pi,\pi]$. Although this can be done in many ways, two methods of extension are of special interest: even extensions and odd extensions. A function f is said to be *extended evenly* from $[0,\pi]$ to $[-\pi,\pi]$ if $f(x) = f(-x)$ for each $x \in [-\pi,0]$. The function f is said to be *extended oddly* from $[0,\pi]$ to $[-\pi,\pi]$ if $f(x) = -f(-x)$ for each $x \in [-\pi,0]$ (Fig. 10.21).

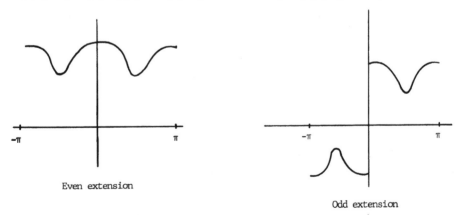

Even extension

Odd extension

Figure 10.21

In the case of an even extension of f, the product function $f(x) \sin nx$ is odd on $[-\pi,\pi]$, and hence, $b_n = 0$ for all $n \geq 1$. Consequently the Fourier series corresponding to an even extension is a pure cosine series. If f is extended as an odd function, then a pure sine series representation of f is obtained. Therefore, on the interval $[0,\pi]$, f can have two quite different representations. Note, moreover, that if f is continuous on $[0,\pi]$, then the even extension of f is continuous, but the odd extension has jump discontinuities at $\pm\pi$ and at 0 [unless $f(0) = 0$ and $f(\pi) = 0$]. It is customary in the case of an odd extension to redefine f by setting $f(\pm\pi) = f(0) = 0$ (although this does not eliminate the problem of discontinuity). Fourier series as a rule converge more quickly if f is continuous everywhere on $[-\pi,\pi]$ and $f(-\pi) = f(\pi)$ (as is indicated when one takes into account the Gibbs phenomenon). Hence, even extensions are often preferred to odd extensions unless $f(0) = f(\pi) = 0$, although frequently the type of extension is dictated by the function under consideration. The proof of the following corollary is left as an exercise to the reader.

(10.E.6) Corollary If f is continuous and piecewise smooth on $[0,\pi]$, then for all x, $f(x) = a_0/2 + \sum_{n=1}^{\infty} a_n \cos nx$, where

$$a_n = \frac{2}{\pi} \int_0^{\pi} f(x) \cos nx \, dx$$

and on $(0,\pi)$, $f(x) = \sum_{k=1}^{\infty} b_n \sin nx$, where

$$b_n = \frac{2}{\pi} \int_0^{\pi} f(x) \sin nx \, dx$$

If f is represented by a cosine series on $[0,\pi]$, then f is considered to be an even extension from $[0,\pi]$ to $[-\pi,\pi]$, and if f is represented by a pure sine series on $[0,\pi]$, then f is considered to be an odd extension from $[0,\pi]$ to $[-\pi,\pi]$ where $f(0) = f(\pm\pi) = 0$.

(10.E.7) Exercise Formulate and prove a corollary similar to (10.E.6) for continuous functions defined on an interval $[0,l]$.

We conclude this section with a standard result concerning the termwise differentiation of $T(f)$.

(10.E.8) Corollary If f and f' are piecewise smooth on $[-\pi,\pi]$ and if f satisfies the boundary conditions $f(-\pi) = f(\pi)$, then

$$T(f')(x) = \frac{d}{dx} T(f)(x)$$

for all x, where it is understood that $dT(f)/dx$ is obtained by termwise differentiation.

Proof. It follows from (10.E.2) that

$$T(f')(x) = \sum_{k=1}^{\infty} (\alpha_k \cos kx + \beta_k \sin kx)$$

where

$$\alpha_k = \frac{1}{\pi} \int_{-\pi}^{\pi} f'(t) \cos kt \, dt$$

$$\beta_k = \frac{1}{\pi} \int_{-\pi}^{\pi} f'(t) \sin kt \, dt$$

Integration by parts together with the boundary conditions imposed on f yield

$$\alpha_k = kb_k \qquad \text{if } k \geq 1$$
$$\beta_k = -ka_k \qquad \text{if } k \geq 1$$

where a_k and b_k are the Fourier coefficients associated with f, i.e., $T(f)(x) = a_0/2 + \sum_{k=1}^{\infty} (a_k \cos kx + b_k \sin kx)$. It now follows that

$$T(f')(x) = \sum_{k=1}^{\infty} k \, (-a_k \sin kx + b_k \cos kx)$$

$$= \sum_{k=1}^{\infty} ((a_k \cos kx)' + (b_k \sin kx)') = \frac{d}{dx} T(f)(x)$$

which concludes the proof.

F.* UNIFORM CONVERGENCE

This section is devoted to proving one major result involving the uniform convergence of Fourier series. Uniform convergence is of concern for many reasons: it permits term-by-term integration and also ensures the continuity of the function being represented. Furthermore, the Gibbs phenomenon is eliminated when the convergence is uniform. Our principal theorem is the following.

(10.F.1) *Theorem* If f is a continuous function on $[-\pi,\pi]$ that satisfies the boundary conditions $f(-\pi) = f(\pi)$ and if f has a piecewise continuous derivative on $[-\pi,\pi]$, then the Fourier series $T(f)$ converges uniformly to f on $[-\pi,\pi]$.

The proof of the theorem is quite intricate. We begin with a lemma involving the convergence properties of the following special function:

$$S(x) = \begin{cases} \frac{-1}{2}(x + \pi) & -\pi \le x < 0 \\ 0 & x = 0 \\ \frac{1}{2}(\pi - x) & 0 < x \le \pi \end{cases} \tag{1}$$

whose Fourier series, $\sum_{k=1}^{\infty} (\sin kx)/k$, was derived in (10.C.5).

(10.F.2) *Definition* A Fourier series $T(f)$ of a function f is said to *converge boundedly* to f on $[-\pi,\pi]$ provided $T(f)$ converges pointwise to f and all partial sums are uniformly bounded on $[-\pi,\pi]$.

(10.F.3) *Lemma* The Fourier series $T(S)$ of the function S defined by (1) converges boundedly to S on $[-\pi,\pi]$ and converges uniformly to S on every closed subinterval of $[-\pi,\pi]$ not containing $x = 0$.

Proof. It follows from (10.E.2) that $T(S)$ converges pointwise to S on $[-\pi,\pi]$. Moreover, as was shown in (2) of Sec. 10.D, the nth partial sum s_n satisfies

$$s_n(x) = \frac{-x}{2} + \frac{1}{2} \int_0^x \frac{\sin (n + \frac{1}{2})t}{(t/2)} \, dt$$
$$+ \frac{1}{2} \int_0^x \left[\frac{1}{\sin (t/2)} - \frac{1}{(t/2)} \right] \sin \left(n + \frac{1}{2} \right) t \, dt$$

and both of the above integrals are bounded on $[-\pi,\pi]$ for all n. Thus, $T(S)$ converges boundedly on $[-\pi,\pi]$. It remains to establish uniform convergence on closed subintervals of $[-\pi,\pi]$ not containing $x = 0$. To show that $\{s_n\}$ converges uniformly, we note that

$$|s_n(x) - s_m(x)| \le \frac{1}{2} \left| \int_0^x \left[\frac{\sin (n + \frac{1}{2})t - \sin (m + \frac{1}{2})t}{t/2} \right] dt \right|$$
$$+ \left| \int_0^x \left[\frac{1}{\sin (t/2)} - \frac{1}{t/2} \right] \left[\frac{\sin (n + \frac{1}{2})t - \sin (m + \frac{1}{2})t}{2} \right] dt \right| \tag{2}$$

By (3) of Sec. 10.D, we have

$$\lim_{t \to 0} \left(\frac{1}{\sin (t/2)} - \frac{1}{t/2} \right) = 0 \tag{3}$$

and therefore, the function g defined by

$$g(t) = \begin{cases} \dfrac{1}{\sin (t/2)} - \dfrac{1}{t/2} & \text{if } t \in [0, \pi], \, t \ne 0 \\ 0 & \text{if } t = 0 \end{cases}$$

is continuous and integrable on $[0,\pi]$. Hence, by the Riemann-Lebesgue Lemma (6.B.6), the second integral in inequality (2) converges to 0 as m and n tend to infinity. Since

$$\int_0^x \frac{\sin (n + \tfrac{1}{2})t}{t}\, dt = \int_0^{(n+\frac{1}{2})x} \frac{\sin t}{t}\, dt$$

it follows that

$$\int_0^x \frac{\sin (n + \tfrac{1}{2})t - \sin (m + \tfrac{1}{2})t}{t}\, dt = \int_0^{(n+\frac{1}{2})x} \frac{\sin t}{t}\, dt - \int_0^{(m+\frac{1}{2})x} \frac{\sin t}{t}\, dt$$

Suppose now that $0 < \eta \le x \le \pi$ and that $m,n \ge N$. Then

$$\left| \int_0^x \frac{\sin (n + \tfrac{1}{2})t - \sin (m + \tfrac{1}{2})t}{t}\, dt \right| = \left| \int_{(m+\frac{1}{2})\eta}^{(n+\frac{1}{2})\eta} \frac{\sin t}{t}\, dt \right|$$

The latter integral tends to zero with increasing N. This follows from the fact that if for each nonnegative integer k, we let $a_k = \int_{k\pi}^{(k+1)\pi} (\sin t)/t\, dt$, then the series $\sum_{k=0}^{\infty} a_k$ is a convergent alternating series ($|a_0| > |a_1| > \cdots$ and $\lim_{k \to \infty} |a_k| = 0$). We can now conclude that the sequence of partial sums $\{s_n\}$ is a Cauchy sequence that converges uniformly on $[-\pi, -\eta] \cup [\eta, \pi]$, where η is any suitably small positive number. This result together with the fact that $\lim_{n \to \infty} s_n(x) = T(f)(x)$ for every x implies that the sequence $\{s_n\}$ converges uniformly to S on $[-\pi, -\eta] \cup [\eta, \pi]$ for small positive values of η.

(10.F.4) *Lemma* Suppose that $\{g_k(x,\theta)\}$ is a sequence of functions defined on a closed rectangle $R: a \le x \le b, c \le \theta \le d$ such that

   (i)   Each g_k is integrable with respect to θ on $[c,d]$ for each fixed $x \in [a,b]$.
   (ii) There is a real number K independent of k such that $|g_k(x,\theta)| \le K$ for each $(x,\theta) \in R$,
   (iii) The sequence $\{g_k\}$ converges pointwise to a function $g: R \to \mathbf{R}^1$ that is integrable with respect to θ on $[c,d]$ for each fixed $x \in [a,b]$.
   (iv) The sequence $\{g_k\}$ is uniformly convergent on every subregion R' of R of the type $a \le x \le b, c < c' \le \theta \le d' < d$.

If for each k, the function $\hat{g}_k$ is defined by

$$\hat{g}_k(x) = \int_c^d g_k(x, \theta)\, d\theta$$

then the sequence $\{\hat{g}_k\}$ converges uniformly to $\hat{g}$ on $[a,b]$, where $\hat{g}$ is defined by

$$\hat{g}(x) = \int_c^d g(x, \theta)\, d\theta$$

Proof. It suffices to show that given $\varepsilon > 0$, there exists a positive integer N such that whenever $n \ge N$ and $x \in [a,b]$,

$$\left| \int_c^d (g_n(x, \theta) - g(x, \theta))\, d\theta \right| < \varepsilon$$

By hypothesis there is a $K > 0$ such that on R,

$$|g_n(x, \theta)| \le K \quad \text{and} \quad |g(x, \theta)| \le K$$

Select $c' < d'$ so that $0 < c' - c < \varepsilon/8K$ and $0 < d - d' < \varepsilon/8K$. Then by (6.B.3) we have

$$\left| \int_c^d (g_n(x, \theta) - g(x, \theta))\, d\theta \right|$$

$$\le \int_c^{c'} |g_n(x, \theta) - g(x, \theta)|\, d\theta + \int_{c'}^{d'} |g_n(x, \theta) - g(x, \theta)|\, d\theta$$

$$+ \int_{d'}^{d} |g_n(x, \theta) - g(x, \theta)|\, d\theta$$

$$\le \int_c^{c'} 2K\, d\theta + \int_{c'}^{d'} |g_n(x, \theta) - g(x, \theta)|\, d\theta + \int_{d'}^{d} 2K\, d\theta$$

$$< \frac{\varepsilon}{4} + \int_{c'}^{d'} |g_n(x, \theta) - g(x, \theta)|\, d\theta + \frac{\varepsilon}{4}$$

Since $g_n \to g$ uniformly on R', there is a positive integer N such that for all $(x, \theta) \in R'$,

$$|g_n(x, \theta) - g(x, \theta)| < \frac{\varepsilon}{2(d - c)}$$

Thus, for each $x \in [a, b]$,

$$\left| \int_c^d (g_n(x, \theta) - g(x, \theta))\, d\theta \right| < \frac{\varepsilon}{4} + \frac{\varepsilon}{2} \frac{(d' - c')}{(d - c)} + \frac{\varepsilon}{4} < \varepsilon$$

which completes the proof.

Proof of (10.F.1).   Using integration by parts and the fact that $f(-\pi) = f(\pi)$, we can rewrite the Fourier coefficients of f in the form

$$a_k = \frac{1}{\pi} \int_{-\pi}^{\pi} f(\xi) \cos k\xi\, d\xi = \frac{1}{k\pi} \left\{ f(\xi) \sin k\xi \Big|_{-\pi}^{\pi} \right\} - \frac{1}{k\pi} \int_{-\pi}^{\pi} f'(\xi) \sin k\xi\, d\xi$$

$$= \frac{-1}{k\pi} \int_{-\pi}^{\pi} f'(\xi) \sin k\xi\, d\xi$$

$$b_k = \frac{1}{\pi} \int_{-\pi}^{\pi} f(\xi) \sin k\xi\, d\xi = \frac{-1}{k\pi} \left\{ f(\xi) \cos k\xi \Big|_{-\pi}^{\pi} \right\}$$

$$+ \frac{1}{k\pi} \int_{-\pi}^{\pi} f'(\xi) \cos k\xi\, d\xi = \frac{1}{k\pi} \int_{-\pi}^{\pi} f'(\xi) \cos k\xi\, d\xi$$

Substitution of these values into the formal Fourier series

$$T(f) - \frac{a_0}{2} = \sum_{k=1}^{\infty} (a_k \cos kx + b_k \sin kx)$$

yields

$$T(f) - \frac{a_0}{2} = \sum_{k=1}^{\infty} \int_{-\pi}^{\pi} f'(\xi) \left[\frac{\cos k\xi \sin kx - \sin k\xi \cos kx}{k\pi} \right] d\xi$$

$$= \sum_{k=1}^{\infty} \int_{-\pi}^{\pi} f'(\xi) \left[\frac{-\sin k(\xi - x)}{k\pi} \right] d\xi$$

Extending f and f' periodically from $[-\pi, \pi)$ to all values of x [note that because of the end conditions, the extension of f with period 2π is continuous for all x, and f' remains bounded for all x (why?)] and setting $\theta = \xi - x$, we obtain from the periodicity of the integrands

$$T(f)(x) - \frac{a_0}{2} = \sum_{k=1}^{\infty} \int_{-\pi}^{\pi} f'(\theta + x) \left(\frac{-\sin k\theta}{k\pi} \right) d\theta \qquad (4)$$

Next we show that the order of appearance of the summation and integration symbols in (4) can be interchanged. Let

$$g_n(x, \theta) = \sum_{k=1}^{n} f'(\theta + x) \left(\frac{-\sin k\theta}{k\pi} \right) \qquad (5)$$

By (10.F.3) and the boundedness of f', the sequence $\{g_n(x,\theta)\}$ converges (and is uniformly bounded) for all values of x and θ; furthermore, this sequence converges uniformly to $f'(\theta + x)[-S(\theta)/\pi]$ on all closed connected regions in the $x\theta$-plane not containing points of the form $(x, 2\pi m)$, where m is an integer; the function S is as defined by (1).

Since the conditions of (10.F.4) are satisfied, it follows that the sequence

$$\left\{ \int_{-\pi}^{\pi} g_n(x, \theta) \, d\theta \right\}$$

converges uniformly to $\int_{-\pi}^{\pi} f'(\theta + x)[-S(\theta)/\pi] \, d\theta$. Moreover, since

$$\int_{-\pi}^{\pi} g_n(x, \theta) \, d\theta = \sum_{k=1}^{n} \int_{-\pi}^{\pi} f'(\theta + x) \left(\frac{-\sin k\theta}{k\pi} \right) d\theta$$

we have

$$\sum_{k=1}^{\infty} \int_{-\pi}^{\pi} f'(\theta + x) \left(\frac{-\sin k\theta}{k\pi} \right) d\theta = \int_{-\pi}^{\pi} f'(\theta + x) \left(\frac{-S(\theta)}{\pi} \right) d\theta$$

(which represents the interchange of $\int$ and $\sum$). We now set

$$G(x) = \frac{-S(x)}{\pi}$$

The graph of G is indicated in Fig. 10.22.

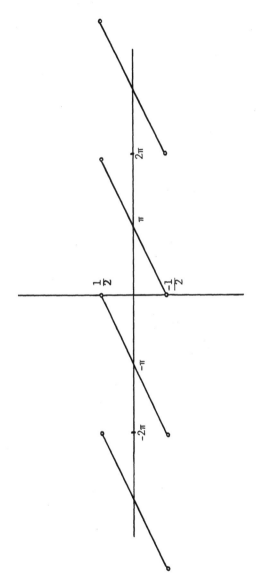

Figure 10.22

We have shown thus far that the Fourier series $T(f)$ of f converges uniformly on $[-\pi,\pi]$ and satisfies the equation

$$T(f)(x) - \frac{a_0}{2} = \int_{-\pi}^{\pi} f'(\theta + x)G(\theta)\, d\theta \tag{6}$$

We conclude the proof by showing that

$$\int_{-\pi}^{\pi} f'(x + \theta)G(\theta)\, d\theta = f(x) - \frac{a_0}{2}$$

We integrate by parts. Because of the discontinuity of $f(x + \theta)G(\theta)$ at $\theta = 0$, we replace $\int_{-\pi}^{\pi}$ by $\int_{-\pi}^{0} + \int_{0}^{\pi}$. Then

$$\int_{-\pi}^{0} f'(x + \theta)G(\theta)\, d\theta = \int_{-\pi}^{0} \frac{d}{d\theta}[f(x + \theta)G(\theta)]\, d\theta - \int_{-\pi}^{0} f(x + \theta)G'(\theta)\, d\theta$$

$$= f(x)G(0 - 0) - f(x - \pi)G(-\pi)$$

$$- \frac{1}{2\pi}\int_{-\pi}^{0} f(x + \theta)\, d\theta$$

$$= \frac{f(x)}{2} - \frac{1}{2\pi}\int_{-\pi}^{0} f(x + \theta)\, d\theta$$

Similarly,

$$\int_{0}^{\pi} f'(x + \theta)G(\theta)\, d\theta = f(x + \pi)G(\pi) - f(x)G(0 + 0) - \frac{1}{2\pi}\int_{0}^{\pi} f(x + \theta)\, d\theta$$

$$= \frac{f(x)}{2} - \frac{1}{2\pi}\int_{0}^{\pi} f(x + \theta)\, d\theta$$

and hence,

$$\int_{-\pi}^{\pi} f'(x + \theta)G(\theta)\, d\theta = f(x) - \frac{1}{2\pi}\int_{-\pi}^{\pi} f(x + \theta)\, d\theta$$

$$= f(x) - \frac{1}{2\pi}\int_{-\pi+x}^{\pi+x} f(w)\, dw \qquad (w = x + \theta)$$

$$= f(x) - \frac{1}{2\pi}\int_{-\pi}^{\pi} f(w)\, dw$$

$$= f(x) - \frac{a_0}{2}$$

as required.

G. PARTIAL DIFFERENTIAL EQUATIONS

In this section we discuss the role played by Fourier series in the determination of solutions to certain partial differential equations. In particular, we shall see how the theory developed thus far lends mathematical validity to the operations that were employed in finding a "formal solution" to the wave equation $u_{tt}(x,t) = a^2 u_{xx}(x,t)$. We begin with a topic that was at the core of Fourier's research: the mathematical formulation of the theory of heat. Fourier was convinced that the development of a mathematical theory of heat was a necessary requisite for the progress of mankind, and certainly, present-day problems involving the re-entry of space vehicles, heat-transfer problems in nuclear reactors, etc., tend to bear out his conviction.

The first problem with which we deal concerns the prediction of the distribution of heat in a homogeneous rod of length l. We shall assume that the rod is insulated laterally so that there is no transfer of heat to or from the rod except at the ends of the rod. For convenience, we suppose that the rod is centered on the x-axis with ends at $x = 0$ and $x = l$ (Fig. 10.23). We let $u(x,t)$ denote the temperature of the cross section of the rod determined by x and at time t. To obtain the heat equation, we shall apply the law of conservation of energy which states that the (signed) rate at which heat enters a region plus the rate at which heat is generated inside the region is equal to the rate at which heat is stored in the region.

We shall assume that the region R under consideration is a cross-sectional slice of the rod with end faces $C(x)$ and $C(x + \Delta x)$; at present we do not need to assume that Δx is small. We let A represent the cross-sectional area of the rod (Fig. 10.24). We shall also assume (since the rod is laterally insulated) that the heat flow is parallel to the axis of the rod and uniformly distributed over each cross section $C(x)$. With these assumptions we can postulate the existence of a function r (of x and t) which has the property that

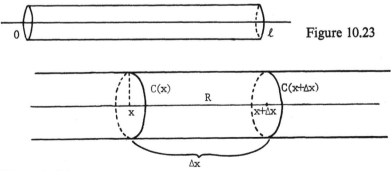

Figure 10.23

Figure 10.24

the rate of flow of heat across a face $C(x)$ at time t in the direction of increasing x is given by $A \cdot r(x,t)$. Thus there is a net flow to the right across $C(x)$ if $r(x,t) > 0$ and a net flow to the left if $r(x,t) < 0$. Note that it follows that the flow of heat into R through $C(x)$ is $A \cdot r(x,t)$ regardless of the sign of r (check this). Similarly the flow of heat into R through $C(x + \Delta x)$ is $-A \cdot r(x + \Delta x, t)$ regardless of the sign of r. Consequently, the rate at which heat enters the slice R is given by $A \cdot r(x,t) - A \cdot r(x + \Delta x, t)$.

Physicists find that the energy stored in a volume of uniform material is proportional to its temperature. If we denote the constant of proportionality by ρc, where ρ is the density of mass and c is heat capacity per unit mass, then the heat energy stored in R at time t is given by

$$\int_x^{x+\Delta x} \rho c u(s, t) A \, ds$$

If we assume that u_t is continuous, then by the Leibniz rule (7.F.1) the rate of increase of the energy stored in R is given by

$$\frac{\partial}{\partial t} \int_x^{x+\Delta x} \rho c u(s, t) A \, ds = \int_x^{x+\Delta x} \rho c u_t(s, t) A \, ds$$

Similarly, if we let g denote the rate at which heat is generated per unit volume, then the heat generated within the slice R is given by

$$\int_x^{x+\Delta x} g(s, t) A \, ds$$

(the heat generated within the rod may be due to electric current, chemical processes, mechanical vibrations, etc.).

From the law of conservation of energy it now follows that

$$-A[r(x + \Delta x, t) - r(x, t)] + A \int_x^{x+\Delta x} g(s, t) \, ds = A \int_x^{x+\Delta x} \rho c u_t(s, t) \, ds$$

$$(1)$$

Fourier's law for heat conduction in one dimension states that

$$r(x, t) = -k u_x(x, t) \qquad (k > 0) \tag{2}$$

for small changes in u (i.e., heat flows in the direction opposite to increasing temperature and is proportional to the rate of change in the temperature). From equations (1) and (2) we have

$$Ak[u_x(x + \Delta x, t) - u_x(x, t)] + A \int_x^{x+\Delta x} g(s, t) \, ds = A \int_x^{x+\Delta x} \rho c u_t(s, t) \, ds$$

If u_{xx} exists and is continuous, then by the fundamental theorem of calculus we have

$$A \int_x^{x+\Delta x} [ku_{xx}(s, t) + g(s, t) - \rho c u_t(s, t)] \, ds = 0 \tag{3}$$

At this point we need the following easily derived result.

(10.G.1) *Exercise* If h is a continuous function, then $\lim_{\Delta x \to 0}$ $(1/\Delta x) \int_x^{x+\Delta x} h(s) \, ds = h(x)$.

Dividing both sides of (3) by Δx and applying (10.G.1), we find that

$$ku_{xx}(x, t) + g(x, t) = \rho c u_t(x, t)$$

In the case when $g \equiv 0$, we arrive at the simplest mathematical formulation of the "heat equation":

$$u_{xx}(x, t) = \frac{1}{a^2} u_t(x, t) \tag{4}$$

the constant $a^2 = k/\rho c$ is called the *diffusivity*.

In order to obtain unique solutions to this equation, we need to impose boundary and initial conditions. First, for simplicity, we assume that the ends of the rod maintain a constant temperature 0, that is, $u(0,t) = u(l,t) = 0$ for all t, and that the initial temperature of the various cross sections is governed by $u(x,0) = f(x)$ (the temperature is constant on each cross section). Thus, the mathematical problem to be solved is that of finding $u(x,t)$ such that

$$a^2 u_{xx} = u_t \quad \text{where } a^2 = k/\rho c \;\; 0 < x < l, t > 0 \text{ (the heat equation). (5a)}$$

$$u(0, t) = 0 = u(l, t) \quad t > 0 \text{ (the boundary conditions).} \tag{5b}$$

$$u(x, 0) = f(x) \quad \text{where } 0 \leq x \leq l \text{ (the initial state condition).} \tag{5c}$$

The boundary conditions described in (5b) are called *homogeneous boundary conditions*.

To find a solution of this problem we employ once again Fourier's separation of variables technique that was introduced in connection with the wave equation; thus we look for a nontrivial solution having the form

$$u(x, t) = X(x)T(t) \tag{6}$$

that satisfies conditions (5a) and (5b). We assume throughout that $t \geq 0$. Substitution of (6) in (5a) yields

$$a^2 X_{xx}(x)T(t) = X(x)T_t(t) \tag{7}$$

for all $(x,t) \in J = \{(x,t) \mid 0 < x < l, t > 0\}$. Equation (7) implies that

$$\frac{X_{xx}(x)}{X(x)} = \frac{T_t(t)}{a^2 T(t)} \tag{8}$$

for all $(x,t) \in J$ such that $X(x)$ and $T(t)$ are not 0. [Justification for momentarily overlooking values of x and t where $X(x)T(t) = 0$ is provided in Exercise (10.G.2).] From (8) we can conclude, as we did in Chapter 8, that $(1/a^2)T_t(t)/T(t)$ is equal to a constant λ; that is,

$$T'(t) = a^2 \lambda T(t) \tag{9}$$

and

$$X_{xx}(x) = \lambda X(x) \tag{10}$$

Since the general solution of (9) is

$$T(t) = Ce^{a^2 \lambda t} \tag{11}$$

it follows that $T(t)$ is never 0 (or always 0, which we reject since we assume that the initial temperature of the rod is not everywhere zero). As a consequence of (5b) and (6)

$$X(0) = 0 = X(l) \tag{12}$$

and the general solution of (10) is now easily seen to be

$$X(x) = C_1 e^{\sqrt{\lambda} x} + C_2 e^{-\sqrt{\lambda} x} \tag{13}$$

(10.G.2)   Exercise   Show by direct substitution that the function $X(x)T(t)$ defined by (11) and (13) is a solution of (5a, b) for $0 < x < l$, $t > 0$.

We now determine those values of λ that yield a *nontrivial* solution of (10) and satisfy the boundary conditions (12). Clearly, $\lambda = 0$ will not do. Therefore, we can suppose that $\sqrt{\lambda} = \alpha + i\beta \neq 0$. Since $0 = X(0) = C_1 + C_2$ we have

$$X(x) = C_1(e^{(\alpha + i\beta)x} - e^{-(\alpha + i\beta)x})$$

where $C_1 \neq 0$. Furthermore, since $X(l) = 0$, it follows that

$$0 = e^{(\alpha + i\beta)l} - e^{-(\alpha + i\beta)l} = e^{\alpha l}(\cos \beta l + i \sin \beta l) - e^{-\alpha l}(\cos \beta l - i \sin \beta l)$$

If we equate real and imaginary parts, we see that

$$(e^{\alpha l} - e^{-\alpha l}) \cos \beta l = 0$$
$$(e^{\alpha l} + e^{-\alpha l}) \sin \beta l = 0 \tag{14}$$

The second equation implies that $\sin \beta l = 0$, which in turn implies that

$$\beta = \frac{n\pi}{l} \qquad \text{for } n = 0, \pm 1, \pm 2, \ldots$$

Since $\cos \beta l = \cos n\pi = \pm 1$, it follows from (14) that $e^{2\alpha l} = 1$; therefore, since $l \neq 0$, it must be the case that $\alpha = 0$. Consequently, we have

$$\sqrt{\lambda} = in \frac{\pi}{l} \qquad n = 0, \pm 1, \pm 2, \ldots$$

and therefore, all nontrivial solutions of (10) that satisfy (13) are of the form

$$X_n(x) = K_n(e^{(in\pi x/l)} - e^{-(in\pi x/l)}) \qquad n = \pm 1, \pm 2, \ldots$$

If we set $C_n = 2iK_n$, then we obtain

$$X_n(x) = C_n \sin \frac{n\pi x}{l} \qquad n = \pm 1, \pm 2, \ldots \qquad (15)$$

Consequently, it follows from (11) and (15) that all nontrivial solutions of (5a) and (5b) have the form

$$u_n(x, t) = C_n e^{-((n\pi a/l)^2 t)} \sin (n\pi x/l) \quad n = \pm 1, \pm 2, \ldots \qquad (16)$$

[Since $\sin (-x) = -\sin x$, we can combine the solutions $u_{-n}(x,t) = C_{-n} e^{-((-n\pi a/l)^2 t)} \sin (-n\pi x/l)$ and $u_{+n}(x,t) = C_{+n} e^{-((n\pi a/l)^2 t)} \sin (n\pi x/l)$ into the single solution $u_n(x,t) = C_n e^{-((n\pi a/l)^2 t)} \sin (n\pi x/l)$, where $C_n = C_{+n} - C_{-n}$.]

In general, no finite combination of the u_n's will satisfy the initial state conditions (5c); hence, (using the superposition principle), we look for possible solutions of the form

$$u(x, t) = \sum_{n=1}^{\infty} u_n(x, t) \qquad (17)$$

There is, of course, no a priori reason that an infinite sum of solutions of equation (5a) should still be a solution of (5a). Before tackling this point, however, we first consider (5c). We obviously need that

$$f(x) = u(x, 0) = \sum_{n=1}^{\infty} u_n(x, 0) = \sum_{n=1}^{\infty} C_n \sin \frac{n\pi x}{l} \qquad \text{for } x \in [0, l] \quad (18)$$

Since $f(0) = f(l)$, it follows from (10.E.2) that if f is continuous and piecewise smooth on the interval $[0,l]$, then f can be represented by the series given in (18), where

$$C_n = \frac{1}{l} \int_{-l}^{l} f(t) \sin \frac{n\pi t}{l} \, dt \qquad (19)$$

Hence, with C_n so defined, the function $u(x,t)$ defined in (17) will satisfy (5b) and (5c) for $x \in [0,l]$ and $t > 0$. It remains to check that $u(x,t)$ is a solution of (5a).

Note that each partial sum $s_n(x,t)$ of $u(x,t)$ is a solution of (5a). We shall approximate u_t by $\partial s_n/\partial t$ and u_{xx} by $\partial^2 s_n/\partial x^2$ uniformly on closed semiinfinite strips

$$R_\delta = \{(x, t)|0 \le x \le l, t \ge \delta\}, \text{ where } \delta > 0.$$

Once this is done, then for each $\varepsilon > 0$, there will exist a positive integer N (dependent on ε and δ) such that whenever $n \ge N$ and $(x,t) \in R_\delta$, then

$$\left| u_t(x, t) - \frac{\partial}{\partial t} s_n(x, t) \right| < \frac{\varepsilon}{2}$$

and

$$\left| u_{xx}(x, t) - \frac{\partial^2}{\partial x^2} s_n(x, t) \right| < \frac{\varepsilon}{2a^2}$$

and consequently, since $a^2 \partial^2 s_n / \partial x^2 - \partial s_n / \partial t = 0$, we shall have

$$|a^2 u_{xx} - u_t| = \left| (a^2 u_{xx} - u_t) - \left(a^2 \frac{\partial^2}{\partial x^2} s_n - \frac{\partial}{\partial t} s_n \right) \right|$$

$$\leq a^2 \left| u_{xx} - \frac{\partial^2}{\partial x^2} s_n \right| + \left| u_t - \frac{\partial}{\partial t} s_n \right| < \frac{\varepsilon}{2} + \frac{\varepsilon}{2} = \varepsilon$$

on R_δ provided that $n \geq N$. Since ε is arbitrary, u must satisfy (5a) on R_δ, and since δ is arbitrary, u must satisfy (5a) in the region $0 \leq x \leq l, t > 0$.

Thus to conclude the proof, it remains to show that u_t and u_{xx} can be approximated by the corresponding partial derivatives of s_n. First we show that the sequence $\{s_n\}$ converges uniformly to u. This convergence is due to the uniform boundedness of the sequence $\{X_k\}$ and the exponential decay of the factor T_k as $k \to \infty$, and may be seen as follows. We have

$$u_k(x, t) = X_k(x) T_k(t)$$

where

$$X_k(x) = C_k \sin \frac{k\pi x}{l}$$

$$T_k(t) = e^{-((k\pi a/l)^2 t)}$$

If we extend f oddly from $[0,l]$ to $[-l,l]$, then f will be continuous on $[-l,l]$ and, hence, bounded there. Then from (19) we have

$$C_k \leq \frac{1}{l} \int_{-l}^{l} |f(t)| \, dt = M \qquad \text{for } k \geq 1$$

and therefore

$$|X_k(x)| \leq M \qquad\qquad k \geq 1$$

Furthermore, for $t \geq \delta$, $T_k(t) \leq T_k(\delta)$ for each $k \geq 1$, and thus on R_δ

$$|u(x, t) - s_n(x, t)| = \sum_{k=n+1}^{\infty} |u_k(x, t)| \leq M \sum_{k=n+1}^{\infty} T_k(\delta) \qquad (20)$$

Since $\lim_{w \to \infty} e^w / w = \infty$, there is an integer N such that $w^{-1} > e^{-w}$ whenever $w \geq N$. If k is chosen sufficiently large in (20) so that $(k^2 \pi^2 a^2 / l^2) > N$, then

$$T_k(\delta) < \frac{l^2}{\pi^2 \delta a^2} \frac{1}{k^2}$$

Therefore, on R_δ we have

$$\sum_{k=N+1}^{\infty} u_k(x, t) \leq \frac{Ml^2}{\pi^2 \delta a^2} \sum_{k=N+1}^{\infty} \frac{1}{k^2}$$

Since $\lim_{N \to \infty} \sum_{k=N+1}^{\infty} 1/k^2 = 0$, it follows that $\lim_{N \to \infty} \sum_{k=N+1}^{\infty} u_k(x,t) = 0$, and this convergence is uniform on R_δ.

The reader may complete the proof by resolving the following exercises.

(10.G.3) Exercises (a) Suppose that $\sum_{n=1}^{\infty} v_n(x)$ converges to $f(x)$ for each $x \in [a,b]$. Suppose further that for each n, v_n' exists and is continuous on $[a,b]$. Show that if $\sum_{n=1}^{\infty} v_n'(x)$ converges uniformly on $[a,b]$, then $\sum_{n=1}^{\infty} v_n'(x) = f'(x)$ for each $x \in [a,b]$. [*Hint*: Let $g(x) = \sum_{n=1}^{\infty} v_n'(x)$ and consider $\int_a^x g(t)\, dt$.] It is not necessary that v_n' be continuous; for a proof of this stronger result see Hardy and Rogosinski (1944).

(b) Mimic in almost every detail that part of the proof of (10.G.2) which showed that the sequence $\{s_n\}$ converges uniformly to u to prove that the sequences $\{\partial s_n/\partial t\}$ and $\{\partial^2 s_n/\partial x^2\}$ converge uniformly on R_δ.

(c) Use (a) and (b) to show that

$$\sum_{n=1}^{\infty} \frac{\partial}{\partial t} u_n(x,\, t) = \frac{\partial}{\partial t} u(x,\, t)$$

$$\sum_{n=1}^{\infty} \frac{\partial}{\partial x} u_n(x,\, t) = \frac{\partial}{\partial x} u\, (x,t)$$

$$\sum_{n=1}^{\infty} \frac{\partial^2}{\partial x^2} u_n(x,\, t) = \frac{\partial^2}{\partial x^2} u(x,\, t)$$

(d) Show that $\lim_{t \to \infty} u(x,t) = 0$.

Note that in the preceding exercise (10.G.3.d) the limit is independent of x. Thus, the temperature of the entire rod tends toward 0. For this reason, the constant function $v(x) = 0$ is often called the *steady state solution* of (5) and $u(x,t)$ the *transient state solution*. Graphically, we have the situation in Fig. 10.25, where the curves represent the temperature across the rod at various times t.

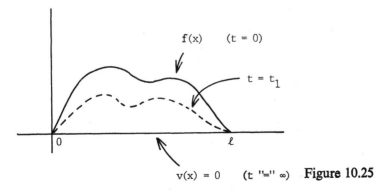

f(x)   (t = 0)

t = t₁

ℓ

v(x) = 0   (t "=" ∞)   **Figure 10.25**

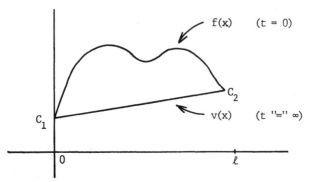

$$\text{Figure } 10.26$$

Now let us see what happens if we have nonhomogeneous boundary conditions. Suppose for instance that the ends of the rod are maintained at constant but different temperatures, i.e.,

$$u(0, t) = C_1 \qquad t > 0$$
$$u(l, t) = C_2 \qquad t > 0$$

The steady state solution $v(x)$ must satisfy (5a) as well as the conditions $v(0) = C_1$ and $v(l) = C_2$. Thus we have $a^2 v_{xx} = v_t$. Since v is independent of time, it follows that $v_t = 0$, which implies that $v_{xx} = 0$. Consequently, v is linear and is given by

$$v(x) = \frac{C_2 - C_1}{l} x + C_1$$

(see Fig. 10.26).

To solve the nonhomogeneous problem $[u(0,t) = C_1, u(l,t) = C_2]$, we reduce it to a homogeneous problem by setting

$$w(x, t) = u(x, t) - v(x)$$

where $w(x,t)$, the solution to be found, is the transient state solution (it is expected to tend to 0 with increasing time) and $v(x)$ is the steady state solution.

It is then clear that since $a^2 u_{xx} = u_t$

$$a^2 w_{xx} = a^2 u_{xx} = u_t = w_t$$

furthermore, we have

$$w(0, t) = 0 = w(l, t) \qquad\qquad t > 0$$
$$w(x, 0) = u(x, 0) - v(x) = f(x) - v(x) \qquad x > 0$$

and this represents a homogeneous boundary problem of the type we have already solved.

In the problem section, the reader is asked to investigate the solution of the heat equation in the case that one of the ends is not maintained at a constant temperature.

We now turn our attention back to the vibrating string problem and the wave equation. We assume as we did in Chapter 8 that the string in equilibrium is stretched along the x axis with ends fastened at $x = 0$ and $x = l$. We consider the case where the string is initially displaced as indicated in Fig. 10.27 and then released.

Under the supposition that each point on the string vibrates along a vertical line segment we seek a function, $u(x,t)$, that locates the vertical displacement at time t of a point x from its equilibrium position. As before, we expect $u(x,t)$ to satisfy

$$a^2 u_{xx} = u_{tt} \qquad\qquad 0 \le x \le l, t > 0 \qquad (21a)$$

$$u(0, t) = u(l, t) = 0 \qquad\qquad t \ge 0 \qquad (21b)$$

$$u(x, 0) = f(x) = \begin{cases} qx & 0 \le x \le \dfrac{l}{2} \\[2ex] q(l - x) & \dfrac{l}{2} \le x \le l \end{cases} \qquad (21c)$$

$$u_t(x, 0) = 0 \qquad\qquad 0 \le x \le l \qquad (21d)$$

where q is the slope of the line indicated in Fig. 10.27. As was found to be the case in Chapter 8, the formal solution of (21) has the form

$$u(x, t) = \sum_{n=1}^{\infty} b_n \cos \frac{n\pi a t}{l} \sin \frac{n\pi x}{l} \qquad (22)$$

where it is presumed that the coefficients are such that

$$f(x) = \sum_{n=1}^{\infty} b_n \sin \frac{n\pi x}{l} \qquad 0 \le x \le l \qquad (23)$$

Since f is continuous and piecewise smooth on $[0,l]$ and, in addition, $f(0) = f(l) = 0$, it follows immediately from (10.E.2) that the representation given in (23) converges everywhere (to f) on $[0,l]$. It is understood of course that

$$b_n = \frac{2}{l} \int_0^l f(x) \sin \frac{n\pi x}{l}\, dx \qquad n \ge 1$$

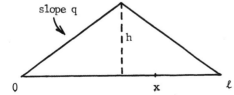

slope q

h

0   x l Figure 10.27

(10.G.4) *Exercise* Use integration by parts to show that

$$b_{2m} = 0 \qquad\qquad m = 0, 1, \ldots$$

$$b_{2m+1} = \frac{4ql}{\pi^2} \frac{(-1)^m}{(2m+1)^2} \qquad m = 0, 1, \ldots \qquad (24)$$

and that

$$u(x,t) = \frac{4ql}{\pi^2} \sum_{m=0}^{\infty} \frac{(-1)^m}{(2m+1)^2} \cos \frac{(2m+1)\pi at}{l} \sin \frac{(2m+1)\pi x}{l} \qquad (25)$$

Observe that series (25) converges absolutely and uniformly for all x and all t since it is dominated by $(4ql/\pi^2) \sum_{m=0}^{\infty} 1/(2m+1)^2$. Furthermore, it is clear from the identity

$$\cos \alpha \sin \beta = \tfrac{1}{2}(\sin (\alpha + \beta) + \sin (\beta - \alpha))$$

that $u(x,t)$ can be written as the sum of two series:

$$u(x, t) = \frac{2ql}{\pi^2} \sum_{m=0}^{\infty} \frac{(-1)^m}{(2m+1)^2} \sin \frac{(2m + 1)\pi}{l} (x + at)$$

$$+ \frac{2ql}{\pi^2} \sum_{m=0}^{\infty} \frac{(-1)^m}{(2m+1)^2} \sin \frac{(2m + 1)\pi}{l} (x - at)$$

If we extend f from $[0,l]$ to all x as an odd function with period $2l$, we find that

$$f(x) = \frac{4ql}{\pi^2} \sum_{m=0}^{\infty} \frac{(-1)^m}{(2m + 1)^2} \sin \frac{(2m + 1)\pi}{l} x$$

for all x. Consequently, we can now write the formal solution in the form

$$u(x, t) = \tfrac{1}{2}(f(x + at) + f(x - at)) \qquad (26)$$

(10.G.5) *Exercise* (a) Take the appropriate partial derivatives of f to show directly that (26) is a solution of (21).

(b) Show that the period of the function $u(x,t)$ given in (26) is $2l/a$.

The formulation given in (26) of the solution to the wave equation is known as d'Alembert's solution and was published by the French mathematician Jean le Rond d'Alembert in 1749. One advantage of this formulation is that it becomes relatively easy to depict graphically the behavior of $u(x,t)$ for small changes in time. Suppose that the initial state of the string is that given in Fig. 10.27. We hold $x = x_0$ fixed and consider a change in time from t to $t + \Delta t$. Under these conditions, the function u changes in value from

$$u(x_0, t) = \tfrac{1}{2}[f(x_0 + at) + f(x_0 - at)]$$

to

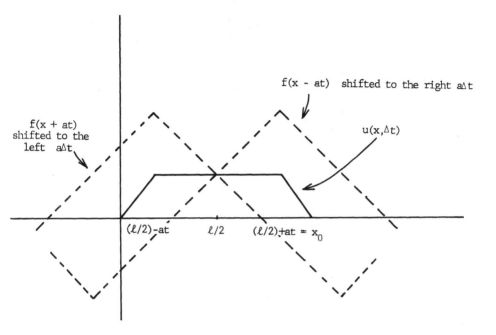

Figure 10.28

$$u(x_0, t + \Delta t) = \tfrac{1}{2}[f(x_0 + at + a\,\Delta t) + f(x_0 - at - a\,\Delta t)]$$

It is evident that $u(x_0, t + \Delta t)$ is produced by shifting $f(x + at)$ to the left (as if it were rigidly constructed) by an amount $a\,\Delta t$, and by shifting $f(x - at)$ to the right by the same amount. This situation is illustrated in Fig. 10.28, where $t = 0$ and Δt is considered to be small. For an "infinite" string the reader might visualize the motion of the string as follows. There are two basic waves, one determined by $f(x_0 - at)$ and the other by $f(x_0 + at)$, that move along the string in opposite directions. The actual position of the string is given by the average of these two waves.

D'Alembert's solution permits an easy calculation of the velocity of the two waves. We consider the movement of a point during the time interval extending from $t = 0$ to $t = t_1$. At time $t = 0$ a typical point on the wave denoted $f(x + at)$ in Fig. 10.28 is located at $(x, f(x + a \cdot 0)) = (x, f(x))$, and at time $t = t_1$ this point is found at $(x - at_1, f((x - at_1) + at_1)) = (x - at_1, f(x))$. For example, the peak $(l/2, h)$ belonging to the wave $f(x + at)$ indicated above is located at time $t = 0$ at $(l/2, f(l/2))$ and at time $t = t_1$ has moved to $((l/2) - at_1, h)$. This implies that in the time period $[0, t_1]$ each point on the wave $f(x + at)$ has moved to the left a distance of $(x - at_1) - x = -at_1$ units. Thus, the velocity of the wave $f(x + at)$ is

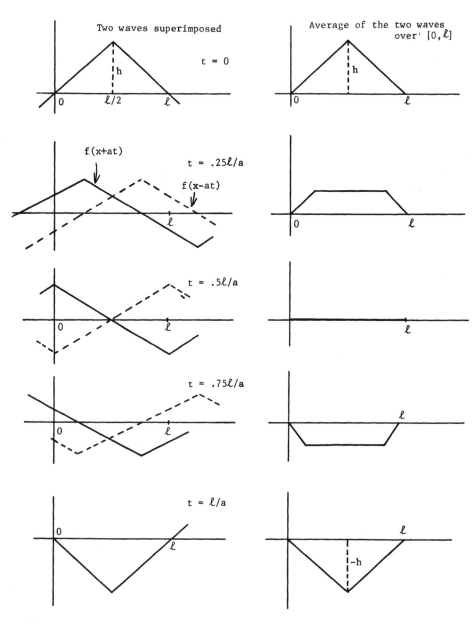

Figure 10.29

$$\frac{-at_1}{t_1} = -a$$

(10.G.6) *Exercise* (a) Show that the velocity of the wave denoted by $f(x - at)$ is $+a$.

(b) Let Δx denote the distance the peak $(l/2,h)$ that belongs to the wave denoted by $f(x - at)$ has moved during the time period $[0, \Delta t]$. Show that

$$\frac{\Delta x}{2l} = \frac{\Delta t}{2l/a}$$

(c) Let Δx denote the distance the peak $(l/2,h)$ belonging to the wave $f(x + at)$ has moved during the time period $[0, \Delta t]$. Show that

$$\frac{\Delta x}{2l} = \frac{-\Delta t}{2l/a}$$

From (10.G.6.a) and (10.G.6.c) we see that if Δt represents α percent of a full oscillation period $2l/a$ for the wave solution $u(x,t)$, then the amount traveled by the two basic waves, denoted by $f(x + at)$ and $f(x - at)$, during this time increment is α percent of $2l$, which, of course, is one full period for f. We use this observation to indicate in Fig. 10.29 various positions (or "snap shots") of the string, given by $u(x,t)$, during one-half period of time (from $t = 0$ to $t = l/a$).

Note that the preceding figures reveal "corners" where u cannot have a derivative; this obviously poses a problem since our solution is clearly differentiable everywhere. Thus, periodically in time at each point $x_0 \in [0,l]$, the solution (22) must fail to satisfy (21a) although at all other instances in time, this equation is satisfied. For a discussion of this problem, see Sagan (1961).

H.* FEJÉR'S METHOD

As we have mentioned before, in 1876 P. du Bois-Reymond constructed a continuous function whose Fourier series failed to converge on a dense subset of $[-\pi,\pi]$. One might expect that this example would eliminate efforts to find a general convergence theorem for Fourier series for the entire class of continuous functions. However, as is often the case, problems of this nature tend to spur mathematicians on to finding ways of overcoming the apparent "pathologies" involved. One solution to this particular problem was devised by Lipót Fejér in 1904. His idea is simple enough. Instead of considering the limit of the partial sums s_n of $T(f)$, Fejér used the limit of the successive arithmetic means $S_n(x)$ of the partial sums:

$$S_1(x) = s_0(x)$$

$$S_2(x) = \frac{s_0(x) + s_1(x)}{2}$$

$$\cdots\cdots\cdots\cdots\cdots\cdots\cdots\cdots$$

$$S_n(x) = \frac{s_0(x) + \cdots + s_{n-1}(x)}{n}$$

He found that $S(x) = \lim_{n \to \infty} S_n(x)$ gave a representation of f for *all* continuous functions f on $[-\pi, \pi]$.

At first it may seem surprising that there should be improvement in convergence using Fejér's method, since on the "average" $s_n(x)$ is a better approximation to $f(x)$ than is $S_n(x)$ (see Sec. 10.B). However, it can be shown that Fejér's method provides a "more uniform" convergence to f.

The key to analyzing the convergence properties of this new method lies in first representing $S_n(x)$ as an integral.

In (10.E.3) we found that if f is integrable on $[-\pi, \pi]$, then $s_n(x)$, the nth partial sum of $T(f)$, can be written (with an appropriate change of variable) in the integral format

$$s_n(x) = \frac{1}{\pi} \int_{-\pi}^{\pi} f(x + t) D_n(t)\, dt$$

where

$$D_n(t) = \frac{\sin(n + \tfrac{1}{2})t}{2 \sin(t/2)}$$

is the Dirichlet kernel. Thus, if we assume that f has been extended periodically (with period 2π), then we have

$$S_n(x) = \frac{1}{\pi} \int_{-\pi}^{\pi} f(x + t) \frac{1}{n}\left[\sum_{k=0}^{n-1} D_k(t)\right] dt$$

$$= \frac{1}{\pi} \int_{-\pi}^{\pi} f(x + t) \hat{M}_n(t)\, dt \tag{1}$$

where

$$\hat{M}_n(t) = \frac{1}{2n \sin(t/2)} \left[\sin \frac{t}{2} + \sin \frac{3t}{2} + \cdots + \sin\left(n - \frac{1}{2}\right)t\right]$$

The finite sums in square brackets can be written in closed form; in fact, this sum equals the imaginary part of

$$e^{it/2} + e^{i3t/2} + \cdots + e^{i(n-\frac{1}{2})t} \tag{2}$$

The complex sum (2) can be rewritten in the form

$$e^{it/2}[1 + e^{it} + e^{2it} + \cdots + e^{i(n-1)t}] = e^{it/2}\left(\frac{1 - e^{int}}{1 - e^{it}}\right)$$

(which holds when $t = 0$ as well, as can be seen by an elementary application of L'Hospital's rule). If we recall that $\sin x = (e^{ix} - e^{-ix})/2i$ (see problem H.1), it becomes an easy exercise to show that (2) equals $e^{int/2} \sin (nt/2)/\sin (t/2)$. Therefore, the imaginary part of (2) is equal to $(\sin (nt/2))^2/\sin (t/2)$ for all real t. Consequently,

$$\hat{M}_n(t) = \frac{1}{2n}\left(\frac{\sin (nt/2)}{\sin (t/2)}\right)^2 \tag{3}$$

for all real t.

Substituting $s = t/2$ into Fejér's integral formula (1) for $S_n(x)$, we obtain

$$S_n(x) = \frac{1}{\pi}\int_{-\pi/2}^{\pi/2} f(x + 2s)\frac{1}{n}\left(\frac{\sin ns}{\sin s}\right)^2 ds$$

Since

$$\frac{1}{\pi}\int_{-\pi/2}^{0} f(x + 2s)\frac{1}{n}\left(\frac{\sin ns}{\sin s}\right)^2 ds = \frac{1}{\pi}\int_{0}^{\pi/2} f(x - 2s)\frac{1}{n}\left(\frac{\sin ns}{\sin s}\right)^2 ds$$

we can write $S_n(x)$ in the form

$$S_n(x) = \frac{1}{\pi}\int_{0}^{\pi/2}\frac{f(x + 2s) + f(x - 2s)}{2} M_n(s)\, ds \tag{4}$$

The term $M_n(s)$, sometimes referred to as the "Fejér kernel," is given by

$$M_n(s) = \frac{2}{n}\left(\frac{\sin ns}{\sin s}\right)^2 \tag{5}$$

(10.H.1) *Theorem*

(i) On each fixed interval $[\delta, \pi/2]$, $\delta > 0$,

$$\frac{1}{\pi}\int_{\delta}^{\pi/2} M_n(s)\, ds \to 0 \qquad \text{as } n \to \infty$$

(ii) $$\frac{1}{\pi}\int_{0}^{\pi/2} M_n(s)\, ds = 1 \text{ for all } n \tag{6}$$

Proof. The proof of (i) is trivial. To establish (ii) we make use of the following trigonometric identities. Since

$$2 \sin \alpha \sin \beta = \cos (\alpha - \beta) - \cos (\alpha + \beta)$$

for all α and β, we have for $\alpha = (k + \frac{1}{2})x$ and $\beta = x/2$,

$$2 \sin \left(k + \frac{1}{2}\right) x \sin \frac{x}{2} = \cos kx - \cos (k + 1)x \qquad (7)$$

for all integers k. From (7) it is then immediate that

$$2 \sin \frac{x}{2} \sum_{k=0}^{n-1} \sin \left(k + \frac{1}{2}\right) x = 1 - \cos nx = 2 \sin^2 \left(\frac{nx}{2}\right) \qquad (8)$$

Furthermore, from (1) in (10.D.1) we have

$$\sin \left(k + \frac{1}{2}\right) x = 2 \sin \left(\frac{x}{2}\right) \left[\frac{1}{2} + \sum_{j=1}^{k} \cos jx\right]$$

and hence,

$$\sum_{k=0}^{n-1} \sin \left(k + \frac{1}{2}\right) x = 2 \sin \left(\frac{x}{2}\right) \left[\frac{n}{2} + (n - 1) \cos x + \cdots + \cos (n - 1)x\right]$$

$$(9)$$

Formulas (8) and (9) imply that

$$2 \left[\frac{n}{2} + (n - 1) \cos x + \cdots + \cos (n - 1)x\right] = \frac{\sin^2 (nx/2)}{\sin^2 (x/2)}.$$

Therefore, it follows that

$$M_n(s) = \frac{2}{n} \left(\frac{\sin ns}{\sin s}\right)^2 = 2 + \frac{4(n - 1)}{n} \cos 2s + \cdots + \frac{4}{n} \cos (n - 1)2s. \qquad (10)$$

Thus, in view of (10),

$$\frac{1}{\pi} \int_0^{\pi/2} M_n(s) \, ds = 1$$

which establishes (6).

We are now ready to prove an important convergence result, due to Fejér, for the class of piecewise continuous functions on $[-\pi,\pi]$. We shall see that at points of discontinuity the sums $S_n(x)$ converge to $[f(x - 0) + f(x + 0)]/2$. We adopt the convention that all piecewise continuous functions on $[-\pi,\pi]$ are to be evaluated according to the rule:

$$f(x) = \begin{cases} \dfrac{f(x - 0) + f(x + 0)}{2} & \text{if } x \in (-\pi,\pi) \\[2mm] \dfrac{f(\pi - 0) + f(-\pi + 0)}{2} & \text{if } x = \pi, -\pi \end{cases} \qquad (11)$$

and that they have been extended periodically (with period 2π) to $\mathbf{R}^1$.

With this convention we can now state the following convergence theorem.

(10.H.2) Theorem (*Fejér's Convergence Theorem*) If f is piecewise continuous on $[-\pi,\pi]$, then

$$\lim_{n \to \infty} S_n(x) = f(x)$$

for each $x \in [-\pi,\pi]$.

Proof. In view of (6) we can write

$$f(x) = \frac{1}{\pi} \int_0^{\pi/2} f(x) M_n(s)\, ds \qquad \text{for all } n \geq 0 \tag{12}$$

Therefore, it follows from (4), (11) and (12) that

$$|S_n(x) - f(x)| = \left| \frac{1}{\pi} \int_0^{\pi/2} \left[\frac{f(x + 2s) + f(x - 2s)}{2} - f(x) \right] M_n(s)\, ds \right|$$

$$\leq \left| \frac{1}{2\pi} \int_0^{\delta} [f(x + 2s) - f(x + 0)] M_n(s)\, ds \right|$$

$$+ \left| \frac{1}{2\pi} \int_0^{\delta} [f(x - 2s) - f(x - 0)] M_n(s)\, ds \right|$$

$$+ \left| \frac{2K}{\pi} \int_{\delta}^{\pi/2} M_n(s)\, ds \right|$$

where K is an upper bound for f on $[-\pi,\pi]$. Applying the mean value theorem for integrals (6.B.7), we obtain

$$|S_n(x) - f(x)| \leq \{|f(x + 2s_1) - f(x + 0)|$$

$$+ |f(x - 2s_2) - f(x - 0)|\} \left(\frac{1}{2\pi} \int_0^{\delta} M_n(s)\, ds \right)$$

$$+ \frac{2K}{\pi} \int_{\delta}^{\pi/2} M_n(s)\, ds$$

where s_1 and s_2 satisfy $0 \leq s_1 \leq \delta$, $0 \leq s_2 \leq \delta$. It follows from (10.H.1.ii) that

$$|S_n(x) - f(x)| \leq \{|f(x + 2s_1) - f(x + 0)| + |f(x - 2s_2) - f(x - 0)|\}$$

$$+ \frac{2K}{\pi} \int_{\delta}^{\pi/2} M_n(s)\, ds \tag{13}$$

for $n \geq 0$ and for *all small* $\delta > 0$. Since f is piecewise continuous, we find that for each $\varepsilon > 0$, there exists $\delta > 0$ such that whenever s_1 and s_2 satisfy $0 \leq s_1 \leq \delta$, $0 \leq s_2 \leq \delta$, then $|f(x + 2s_1) - f(x + 0)| < \varepsilon/3$ and $|f(x - 2s_2) - f(x - 0)| < \varepsilon/3$. Moreover, from (10.H.1.ii) we have

$$\frac{2K}{\pi} \int_{\delta}^{\pi/2} M_n(s) \, ds < \frac{\varepsilon}{3}$$

for such a small fixed $\delta > 0$, provided that n is suitably large. Hence, these observations combined with (13) show that for each $\varepsilon > 0$, there exists an N such that whenever $n \geq N$, then $|S_n(x) - f(x)| < \varepsilon$. This establishes the pointwise convergence of $\{S_n(x)\}$ on $[-\pi,\pi]$.

PROBLEMS

Section A

1. Recall that a function $f: \mathbf{R}^1 \to \mathbf{R}^1$ is periodic if there is a positive number p such that $f(x) = f(x + p)$, for each $x \in \mathbf{R}^1$. Show that if $f: \mathbf{R}^1 \to \mathbf{R}^1$ is periodic with period $p > 0$, then $f(x) = f(x + np)$ for each integer n.
2. Suppose that f and g are periodic function, with common period p. Show that $af + bg$ and $f \cdot g$ are periodic with period p (where a and b are constants).
3. Show that if $f: \mathbf{R}^1 \to \mathbf{R}^1$ is the pointwise limit of a sequence of periodic functions $\{f_n\}$ with common period p, then f is periodic with period p.
4. Suppose that $f: [-\pi,\pi] \to \mathbf{R}^1$ is integrable. Show that each partial sum of the Fourier series $T(f)$ of f is periodic with period 2π.
5. Show that if f satisfies the conditions of Theorem (10.A.4), then $T(f)$ is periodic on $\mathbf{R}^1$ with period 2π.
6. With the aid of (10.A.4) evaluate $T(f)$ and then sketch two or three periods if
    (a) $f(x) = x$ $\qquad -\pi < x < \pi$
    (b) $f(x) = |x|$ $\qquad -\pi < x < \pi$
    (c) $f(x) = x^2$ $\qquad -\pi < x < \pi$
    (d) $f(x) = 2$ $\qquad -\pi < x < \pi$
    (e) $f(x) = \begin{cases} 0 & \text{if } -\pi < x < 0 \\ 1 & \text{if } 0 < x < \pi \end{cases}$
    (f) $f(x) = \begin{cases} \cos x & \text{if } 0 \leq x < \pi \\ 0 & \text{if } -\pi < x < 0. \end{cases}$

Section B

1. Complete the proof of (10.B.1).
2. Show that if f is an integrable odd function on $[-a,a]$, then $\int_{-a}^{a} f(x) \, dx = 0$.

3. Show that if the partial sums s_n of $T(f)$ converge uniformly to f on $[-\pi,\pi]$, then
   (a) f is continuous.
   (b) The sequence $\{\int_{-\pi}^{\pi} s_n(x)\,dx\}$ converges to $\int_{-\pi}^{\pi} f(x)\,dx$.
   (c) The sequence $\{\int_{-\pi}^{\pi} s_n(x) \cos nx\,dx\}$ converges to $\int_{-\pi}^{\pi} f(x) \cos nx\,dx$ and the sequence $\{\int_{-\pi}^{\pi} s_n(x) \sin nx\,dx\}$ converges to $\int_{-\pi}^{\pi} f(x) \sin nx\,dx$.

4. Show that if $f: [-\pi,\pi] \to \mathbf{R}^1$ and $g: [-\pi,\pi] \to \mathbf{R}^1$ are integrable functions over $[-\pi,\pi]$ that differ at most on a finite number of points $x \in [-\pi,\pi]$, then $T(f) = T(g)$.

5. Show that if $f: \mathbf{R}^1 \to \mathbf{R}^1$ is piecewise continuous and periodic with period p, then

$$\int_0^p f(x)\,dx = \int_a^{a+p} f(x)\,dx \qquad \text{for each real number } a$$

6. Show that for any interval $[a, a + 2\pi]$, the orthogonality conditions given in Theorem (10.B.1) continue to hold when $\int_{-\pi}^{\pi}$ is replaced by $\int_a^{a+2\pi}$.

7. Show that if f is continuous on $[a, a + 2\pi]$ and if the sequence $\{s_n\}$, where

$$s_n(x) = \frac{a_0}{2} + \sum_{k=1}^{n} (a_k \cos kx + b_k \sin kx)$$

converges uniformly to f, then

$$a_k = \frac{1}{\pi} \int_a^{a+2\pi} f(x) \cos kx\,dx \qquad n \geq 0$$

$$b_k = \frac{1}{\pi} \int_a^{a+2\pi} f(x) \sin kx\,dx \qquad n \geq 1$$

Section C

1. Complete the details in (10.C.3).
2. Show that $\cos x - \frac{1}{4} \cos 2x + \frac{1}{9} \cos 3x - \frac{1}{16} \cos 4x + \cdots$ is the Fourier series of $\frac{1}{12}\pi^2 - \frac{1}{4}x^2$ on $(-\pi,\pi)$.
3. Show that $\sin x + \frac{1}{3} \sin 3x + \frac{1}{5} \sin 5x + \cdots$ is the Fourier series of

$$f(x) = \begin{cases} \dfrac{\pi}{4} & 0 < x < \pi \\ 0 & x = \pm\pi, 0 \\ \dfrac{-\pi}{4} & -\pi < x < 0 \end{cases}$$

4. Show that

$$\frac{\pi}{4} - \frac{2}{\pi} \sum_{n=1}^{\infty} \frac{\cos{(2n-1)x}}{(2n-1)^2} - \sum_{n=1}^{\infty} \frac{(-1)^n}{n} \sin{nx}$$

is the Fourier series of

$$f(x) = \begin{cases} 0 & -\pi \le x < 0 \\ x & 0 \le x \le \pi. \end{cases}$$

(The Fourier series converges to $\pi/2$ at $\pm\pi$.)

5. Show that $4c \sum_{k=1}^{\infty} [(\sin{k\pi/2})/k] \cos{kx}$ is the Fourier series of

$$f(x) = \begin{cases} -c & -\pi \le x < -\dfrac{\pi}{2} \\ 0 & x = \pm\dfrac{\pi}{2} \\ +c & -\dfrac{\pi}{2} < x < \dfrac{\pi}{2} \\ -c & \dfrac{\pi}{2} < x \le \pi. \end{cases}$$

6. Show (using appropriate trigonometric identities) that the Fourier series representation of $\cos^2 x$ on the interval $(-\pi,\pi)$ is $\frac{1}{2}(1 + \cos{2x})$.

7. Explain analytically why the Fourier series of $f(x) = x$, corresponding to the interval $[-\pi,\pi]$, is not equal to the Fourier series of $f(x) = x$, corresponding to the interval $[0,2\pi]$.

8. Suppose that f is a piecewise continuous function on $[-a,a]$ and let $y = \pi x/a = k(x)$. Define a function g by setting $g(y) = (f \circ k^{-1})(y)$.
   (a) Show that $(g \circ k)(x) = f(x)$.
   (b) Assume that $g(y) = T(g)(y)$, for all but a finite number of points y in $[-\pi,\pi]$, and show that

$$f(x) = \frac{a_0}{2} + \sum_{n=1}^{\infty} \left(a_n \cos{\frac{n\pi x}{a}} + b_n \sin{\frac{n\pi x}{a}} \right) \qquad (*)$$

for all but a finite number of points x in $[-a,a]$, where

$$a_n = \frac{1}{a} \int_{-a}^{+a} f(x) \cos{\frac{n\pi x}{a}} \, dx \qquad n \ge 0 \qquad (**)$$

$$b_n = \frac{1}{a} \int_{-a}^{+a} f(x) \sin{\frac{n\pi x}{a}} \, dx \qquad n \ge 1 \qquad (***)$$

9. Show that for nonnegative integers m and n the following orthogonality properties are satisfied:

$$\int_{-a}^{a} \cos \frac{n\pi x}{a} \cos \frac{m\pi x}{a} \, dx = \begin{cases} 0 & \text{if } m \neq n \\ a & \text{if } m = n \neq 0 \\ 2a & \text{if } m = n = 0 \end{cases}$$

$$\int_{-a}^{a} \sin \frac{n\pi x}{a} \sin \frac{m\pi x}{a} \, dx = \begin{cases} 0 & \text{if } m \neq n \\ a & \text{if } m = n \neq 0 \\ 0 & \text{if } m = n = 0 \end{cases}$$

$$\int_{-a}^{a} \sin \frac{n\pi x}{a} \cos \frac{n\pi x}{a} \, dx = 0$$

10. Using the orthogonality conditions of problem C.9 and the trigonometric expansion (*) for f in problem C.8 deduce formally (via term-by-term integration) the formulas (**) and (***) for a_n and b_n.

11. With the aid of the previous problems find the Fourier series representation of the following functions and compare these series with those obtained over the interval $(-\pi, \pi)$:
    (a) $f(x) = x/2$ $-1 < x < 1$
    (b) $f(x) = 1$ $-1 < x < 1$
    (c) $f(x) = |x|$ $-2 < x < 2$

12. Suppose that $f: \mathbf{R}^1 \to \mathbf{R}^1$ is a piecewise continuous function and that f is periodic with period p. Show that if f satisfies condition (10.A.4.ii) on $[-p/2, p/2]$, then the corresponding Fourier series representation of f on $[a, a + p]$ has the form

$$\frac{1}{2} a_0 + \sum_{n=1}^{\infty} \left(a_n \cos \frac{2\pi n x}{p} + b_n \sin \frac{2\pi n x}{p} \right)$$

where

$$a_n = \frac{2}{p} \int_{a}^{a+p} f(x) \cos \frac{2\pi n x}{p} \, dx = \frac{2}{p} \int_{0}^{p} f(x) \cos \frac{2\pi n x}{p} \, dx$$

and

$$b_n = \frac{2}{p} \int_{a}^{a+p} f(x) \sin \frac{2\pi n x}{p} \, dx = \frac{2}{p} \int_{0}^{p} f(x) \sin \frac{2\pi n x}{p} \, dx$$

13. Suppose that $f: \mathbf{R}^1 \to \mathbf{R}^1$ is a periodic function with period p and that f is integrable over any finite interval. Suppose, further, that

$$f(x) = \frac{a_0}{2} + \sum_{n=1}^{\infty} \left(a_n \cos \frac{2\pi n x}{p} + b_n \sin \frac{2\pi n x}{p} \right)$$

except perhaps on a countable set of points. Assume that if m is a positive integer, then

$$\int_{0}^{mp} f(x) \, dx = \lim_{n \to \infty} \int_{0}^{mp} s_n(x) \, dx$$

where

$$s_n(x) = \frac{a_0}{2} + \sum_{k=1}^{n} \left(a_k \cos \frac{2\pi kx}{p} + b_k \sin \frac{2\pi kx}{p} \right)$$

Show that

$$\frac{\int_0^{mp} f(x)\, dx}{mp} = \frac{\int_0^{p} f(x)}{p} = \frac{a_0}{2}.$$

$[\int_0^{mp} f(x)\, dx/mp$ is the average value of f.]

14. Suppose that $f: (0,\pi) \rightarrow \mathbf{R}^1$ satisfies the Dirichlet conditions of (10.A.4) on $(0,\pi)$. To determine the Fourier series representation over this interval, construct the odd extension of f given by,

$$F(x) = \begin{cases} f(x) & 0 < x < \pi \\ 0 & x = -\pi, 0 \\ -f(-x) & -\pi < x < 0 \end{cases}$$

(a) Show that $T(F)(x) = \sum_{n=1}^{\infty} b_n \sin nx$, where $b_n = (2/\pi) \int_0^{\pi} f(x) \sin nx\, dx$, $n \geq 1$.

(b) Explain why $f(x) = \sum_{n=1}^{\infty} b_n \sin nx$ for every point of continuity of f in $(0,\pi)$. (This expansion is called the *half-range sine series* of f.)

(c) Show that if $0 < x < \pi$, then

$$\cos x = \frac{8}{\pi} \sum_{k=1}^{\infty} \frac{k \sin (2kx)}{4k^2 - 1}$$

is the half-range sine series for f over $(0,\pi)$. Does this series represent $\cos x$ at $x = 0$ and π?

15. Suppose that instead of extending f from $[0,\pi]$ to $[-\pi,\pi]$ as in problem C.14, we extend f *evenly* from $[0,\pi]$ to $[-\pi,\pi]$ by

$$G(x) = \begin{cases} f(x) & 0 \leq x \leq \pi \\ f(-x) & -\pi \leq x < 0 \end{cases}$$

(a) Show that $T(G)(x) = a_0/2 + \sum_{n=1}^{\infty} a_n \cos nx$, where

$$a_n = \frac{2}{\pi} \int_0^{\pi} f(x) \cos nx\, dx \qquad n \geq 0$$

(b) Assume that f satisfies Dirichlet's conditions on $[0,\pi]$ given in (10.A.4) and explain why $f(x) = a_0/2 + \sum_{n=1}^{\infty} a_n \cos nx$ at every point of continuity of f in $[0,\pi]$. (This expansion of f on $[0,\pi]$ is called the *half-range cosine series* of f.)

(c) Show that the half-range cosine series of $f(x) = \cos x$, $0 \leq x \leq \pi$, is exactly equal to $\cos x$.

(d) Find the half-range cosine series for $f(x) = \sin x$, $0 \leq x \leq \pi$.

16. If f is defined on $[0,\pi]$, explain why it is generally better to use the half-range cosine series rather than the half-range sine series for representing f. [*Hint*: Take into account the Gibbs phenomenon and the conclusion of Dirichlet's theorem (10.A.4).]

Section D

1. Show that

$$\lim_{x \to x*} \frac{\sin (n + \frac{1}{2})x}{2 \sin (x/2)} = n + \frac{1}{2}$$

whenever $x^* = 0, 2\pi, -2\pi, 4\pi, -4\pi, \ldots$.
2. Show that if $f: [-\pi,\pi] \to \mathbf{R}^1$ is piecewise smooth, $x_0 \in [-\pi,\pi]$, and

$$F(t) = \frac{f(x_0 + t) - f(x_0 - 0)}{t}$$

then $F(0 - 0) = f'(x_0 - 0)$.
3. Use L'Hospital's rule to show that

$$\lim_{t \to 0} \left(\frac{1}{\sin (t/2)} - \frac{1}{(t/2)} \right) = 0$$

4. Suppose that f satisfies the Dirichlet conditions of (10.A.4) and that f has a jump discontinuity at 0 given by

$$\lim_{x \to 0^-} f(x) = C + M$$
$$\lim_{x \to 0^+} f(x) = C - M$$

where $M \neq 0$. Define $g: [-\pi,\pi] \to \mathbf{R}^1$ by

$$g(x) = \begin{cases} f(x) + \frac{2M}{\pi} [S(x) - C] & \text{if } x \in [-\pi, \pi]\backslash\{0\} \\ 0 & \text{if } x = 0 \end{cases}$$

where $S(x)$ is given in (10.C.5).
(a) Show that g satisfies the Dirichlet conditions given in (10.A.4) on $[-\pi,\pi]$ and that in a small neighborhood about $x = 0$, g is continuous.
(b) Show that $T_n(g) = T_n(f) + (2M/\pi)T_n(S - C)$, where $T_n(\cdot)$ is the nth partial sum of $T(\cdot)$.
(c) Using the fact that the Fourier series representation of a function satisfying Dirichlet's conditions converges uniformly on a closed

subinterval $[-\pi,\dot{\pi}]$ on which the function is continuous, show that $T_n(g) \approx C$ for large n in a small neighborhood about $x = 0$.

(d)   Using results from (b) and (c), show that if x is small, then

$$\frac{T_n(f)(x)}{M} \approx -\frac{T_n(S - C)(x)}{\pi/2}$$

for large values of n.

(e)   Can one conclude that the partial sums $T_n(f)$ have a total overshoot of approximately 18% near $x = 0$?

5.   Suppose that f satisfies the Dirichlet condition stated in (10.A.4) and that $x_1, \ldots, x_n$ are the points of discontinuity of f in $(-\pi,\pi)$. Show that f can be expressed in the form

$$f = g + \sum_{k=1}^{n} f_k + f_\pi$$

where g is continuous on $(-\infty, +\infty)$ and $f_1, \ldots, f_n$, and f_π are sawtooth functions with jumps occurring at $x_1, \ldots, x_n$, and $+\pi$, respectively. Also show that for each k, $f_k = a_k S(x - x_k) + b_k$, where a_k and b_k are appropriate constants, $f_\pi = a_\pi S(x - \pi) + b_\pi$, and S is as defined in (10.C.5).

Section E

1.   (a)   Find a function that is continuous but is not piecewise smooth.

(b)   Show that $f: [-\pi,\pi] \to \mathbf{R}^1$ defined by

$$f(x) = \begin{cases} \sin \dfrac{1}{x} & \text{if } 0 < x \le \pi \\ 0 & \text{if } -\pi \le x \le 0 \end{cases}$$

is not piecewise smooth.

2.   Suppose that $f: (0,a] \to \mathbf{R}^1$ is a continuous and piecewise smooth function. Let $f: \mathbf{R}^1 \to \mathbf{R}^1$ be the even extension of f with period $2a$. Explain why $T(f)(x) = f(x)$ for each $x \in \mathbf{R}^1$.

3.   For which of the following functions does (10.E.2) imply that $T(f)$ represents f?

(a) $f(x) = \begin{cases} \sin \dfrac{1}{x} & \text{if } 0 < x \le \pi \\ 0 & \text{if } -\pi \le x \le 0 \end{cases}$

(b) $f(x) = \begin{cases} 1 & \text{if } -\pi \le x < 0 \\ 0 & \text{if } 0 \le x \le \pi \end{cases}$

$$
\text{(c)} \quad f(x) = \begin{cases} x + 2 & \text{if } -\pi \le x \le 0 \\ x^2 \sin \dfrac{1}{x} & \text{if } 0 < x \le \pi. \end{cases}
$$

4. Show that $x^2 = (\pi^2/3) + 4\sum_{n=1}^{\infty} [(-1)^n/n^2] \cos nx$ if $-\pi \le x < \pi$ and use this result to show that $\sum_{n=1}^{\infty} (1/n^2) = \pi^2/6$. [*Hint*: Consider $x = \pi$.]

5. Differentiate term by term the Fourier series $T(f)$ associated with the function

$$
f(x) = |x| \qquad -\pi < x < \pi
$$

and compare this result with the Fourier series of

$$
f'(x) = \begin{cases} 1 & 0 < x < \pi \\ -1 & -\pi < x < 0. \end{cases}
$$

Are the series identical? Do the same for $f(x) = \pi - x$, $-\pi < x < \pi$.

6. Verify the following assertions, giving special attention to convergence.
   (a) Since $|x| = \frac{1}{2} - (4/\pi^2) \sum_{n=0}^{\infty} [1/(2n + 1)^2] \cos (2n + 1)\pi x$, $-1 < x < 1$, $\pi^2/8 = 1 + \frac{1}{9} + \frac{1}{25} + \cdots$.
   (b) Since $x = 2\sum_{n=1}^{\infty} [(-1)^{n-1}/n] \sin nx$, $-\pi < x < \pi$, $\pi/4 = 1 - \frac{1}{3} + \frac{1}{5} - \frac{1}{7} + \cdots$.
   (c) Since $x^2 = \frac{4}{3}\pi^2 + 4\sum_{n=1}^{\infty} (\cos nx/n^2 - \pi \sin nx/n)$, $0 < x < 2\pi$, $\pi^2/12 = 1 - \frac{1}{4} + \frac{1}{9} - \frac{1}{16} + \cdots$.

7. Show that if $f: (-\pi,\pi) \to \mathbf{R}^1$ is a piecewise continuous function and if $\hat{f}: \mathbf{R}^1 \to \mathbf{R}^1$ is a periodic extension of f with period 2π, then

$$
\int_{-\pi}^{\pi} f(t)D_n(t - x)\, dt = \int_{-\pi}^{\pi} \hat{f}(x + t)D_n(t)\, dt
$$

Section F

1. Show that $\int_n^{\infty} [(\sin t)/t]\, dt \to 0$ as $n \to \infty$. [*Hint*: Integrate by parts.]

2. Show that if f is continuous on $[a,b]$ and f' is piecewise continuous on $[a,b]$, then

$$
\int_a^b f'(t)\, dt = f(b) - f(a)
$$

3. Show that the Fourier series of

$$
f(x) = \frac{\sin x}{x} \qquad -\pi < x < \pi
$$

converges everywhere, and that the convergence is uniform. [*Hint*: Study f' in a neighborhood of $x = 0$.] Sketch $T(f)$.

4. Does the Fourier series of cosh x, $-\pi < x < \pi$, converge uniformly? Sketch $T(f)$.

5. Does the Fourier series of $x^2 - 1$, $-1 < x < 1$, converge uniformly? Sketch $T(f)$.

6. Show that if f is a bounded and continuous function on $-a < x < a$ with a piecewise continuous derivative and if $f(-a + 0) = f(a - 0)$, then $T(f)$ converges uniformly on $[-a,a]$, where $T(f)(\pm a) = f(a - 0)$ $[=f(-a + 0)]$.

7. Suppose that f is a continuous bounded function on $(0,a)$ with a piecewise continuous derivative and such that $\lim_{x \to 0^+} f(x)$ and $\lim_{x \to a^-} f(x)$ exist. Show that the Fourier cosine series of f converges uniformly to f on $0 \leq x \leq a$. (The series converges to $f(0 + 0)$ at $x = 0$, and to $f(a - 0)$ at a.) [*Hint*: Consider the even extension of f with period $2a$.]

8. Suppose that f is a continuous and bounded function on $0 < x < a$ with a piecewise continuous derivative. Show that if $f(0 + 0) = f(a - 0) = 0$, then the Fourier sine series of f converges uniformly to f on the interval $0 \leq x \leq a$. (The series converges to 0 at $x = 0$ and $x = a$.) [*Hint*: Consider the odd extension of f with period $2a$.]

9. Suppose that

$$\frac{a_0}{2} + \sum_{n=1}^{\infty} \left(a_n \cos \frac{n\pi x}{a} + b_n \sin \frac{n\pi x}{a} \right)$$

is a trigonometric series such that $\sum_{n=1}^{\infty} (|a_n| + |b_n|)$ converges. Show that the partial sums $s_n(x)$ converge uniformly on the interval $-a \leq x \leq a$ to a continuous function f.

In the remaining problems the following definition and theorem will be useful.

Definition A function $f: [a,b] \to \mathbf{R}^1$ is said to be of *bounded variation* if $f = u - v$, where u and v are bounded increasing functions on $[a,b]$.

Theorem (*The Second Mean Value Theorem of Integral Calculus*) Suppose that f is a continuous function on the interval $[a,b]$ and suppose that k is a bounded increasing function on $[a,b]$. Let A and B be real numbers such that $A \leq k(a + 0)$, and $B \geq k(b - 0)$. Then there exists a point $x_0 \in [a,b]$ such that $\int_a^b k(x)f(x) \, dx = A \int_a^{x_0} g(x) \, dx + B \int_{x_0}^b g(x) \, dx$. This theorem is due to Ossian Bonnet, and a proof may be found in Hardy and Rogosinski (1944).

10. Show that if $k(x)$ is a nonnegative bounded increasing function then

$$\int_a^b k(x)g(x) \, dx = B \int_{x_0}^b g(x) \, dx$$

for some $x_0 \in [a,b]$.

11. Suppose that $f(x)$ is a bounded function of bounded variation on $[-\pi,\pi]$.
    (a) Show that f is the difference of two bounded increasing functions u and v on $[-\pi,\pi]$, such that $u > 0$ and $v < 0$ on $[-\pi,\pi]$.
    (b) Let B be an upper bound for both $|u|$ and $|v|$ on $[-\pi,\pi]$. Apply Bonnet's theorem to show that

    $$|a_n| = \left|\frac{1}{\pi}\int_{-\pi}^{\pi} f(t)\cos nt\, dt\right| \le \frac{4B}{\pi n} \qquad n \ge 1$$

    $$|b_n| \le \frac{4B}{\pi n} \qquad n \ge 1$$

12. Suppose that f is a continuous function on $[-\pi,\pi]$ that satisfies the boundary condition $f(-\pi) = f(\pi)$. Suppose further that f' is piecewise continuous and of bounded variation on $[-\pi,\pi]$. Show that the Fourier coefficients of f converge to zero; in particular, show that there is a positive number, A, such that

    $$|a_n| \le \frac{A}{n^2} \qquad |b_n| \le \frac{A}{n^2}$$

13. Suppose that the Fourier coefficients of $f\colon [-\pi,\pi] \to \mathbf{R}^1$ satisfy

    $$|a_n| \le \frac{A}{n^2}$$

    $$|b_n| \le \frac{A}{n^2}$$

    where A is a positive number.
    (a) Show that if $T_N(f) = \sum_{k=N+1}^{\infty} (a_k \cos kx + b_k \sin kx)$, then

    $$|T_N(f)| \le \frac{4A}{N^3}$$

    (b) Show that if f satisfies the conditions in problem F.12 [and therefore the hypothesis of Theorem (10.E.2) as well], then

    $$|f(x) - s_N(x)| \le \frac{4A}{N^3}$$

    for each $x \in [-\pi,\pi]$.

Section G

1. We consider a totally insulated bar where not only the lateral portion of the bar is insulated but the ends are insulated as well. We assume that

the bar has a uniform cross section, uniform density, and uniform heat capacity c throughout the bar and that the axis of the bar coincides with the x-axis, with ends located at $x = 0$ and $x = l$. This leads to the following problem:

(i) $a^2 u_{xx} = u_t$ $0 < x < l, t > 0$
(ii) $u_t(0,t) = 0 = u_t(l,t)$ $t > 0$
(iii) $u(x,0) = f(x)$

(a) Let $v(x) = \lim_{t \to \infty} u(x,t)$ be the steady-state solution of this prob-
     lem. Show that $v(x) = \tau$ on $0 < x < l$, where $\tau = (1/l) \int_0^l f(x)\, dx$.
     [*Hint*: The total heat content in the rod is a constant with respect
     to time.] Note that this result shows that the final uniform tem-
     perature is the average of the initial temperature.
(b) Consider a nontrivial solution of (i) and (ii) of the form $X(x)T(t)$.
     (1) Show that X and T satisfy

$$T'(t) + \lambda a^2 T(t) = 0 \qquad t > 0$$

and $$X''(x) + \lambda X(x) = 0 \qquad 0 < x < l$$

$$X'(0) = 0 = X'(l)$$

Moreover, argue why λ must be nonnegative.
     (2) If $\lambda \neq 0$ show that $X(x)$ must have the form $A \cos(\sqrt{\lambda} x)$ where
          $\lambda = (n\pi/l)^2$, $n = 1, 2, \ldots$. If $\lambda = 0$, show that X and T are
          constant functions.
     (3) Explain why any finite linear combination of solutions of (i)
          and (ii) is again a solution of (i) and (ii).
     (4) Indicate formally why the solution of (i), (ii), and (iii) should
          have the form

$$u(x, t) = \tau + \sum_{n=1}^{\infty} a_n \cos\left(\frac{n\pi x}{l}\right) e^{-(n\pi x/l)^2 t}$$

where

$$u(x, 0) = \tau + \sum_{n=1}^{\infty} a_n \cos\frac{n\pi x}{l} = f(x) \qquad 0 < x < l$$

$$\tau = \frac{1}{l} \int_0^l f(x)\, dx \qquad a_n = \frac{2}{l} \int_0^l f(x) \cos\frac{n\pi x}{l}\, dx$$

     (5) Show that the above expression for $u(x,t)$ converges uniformly
          to τ on $0 < x < l$ as $t \to \infty$.
2. Suppose that we have a rod insulated laterally and that this rod has a
    constant temperature at its left end but is subject to convective heat
    transfer at its right end. This problem can be stated in the following way:

(α_1) $a^2 u_{xx}(x, t) = u_t(x, t)$ $\qquad 0 < x < l, t > 0$

(α_2) $u(0, t) = \tau_0$ $\qquad\qquad t > 0$

(α_3) $-ku_x(l, t) = B(u(l, t) - \tau_1)$ $\qquad t > 0$

(where k, B, and τ_1 are constants)

(α_4) $u(x, 0) = f(x)$ $\qquad\qquad 0 < x < l$

[With regard to the boundary condition (α_3) think of τ_1 as representing the temperature of the medium surrounding the right end of the rod.]

(a) Explain why the steady-state solution $v(x)$ satisfies

(s_1) $v''(x) = 0$ $\qquad 0 < x < l$

(s_2) $v(0) = \tau_0$

(s_3) $-kv'(l) = B(v(l) - \tau_1)$

[Hint: Consider $\lim_{t \to \infty} u(x,t) = v(x)$.]
Show that

$$v(x) = \tau_0 + \frac{B[\tau_1 - \tau_0]}{k + Bl} x$$

(b) Show that the transient-state solution $w(x,t) = u(x,t) - v(x)$ satisfies the homogeneous system:

(t_1) $a^2 w_{xx} = w_t$ $\qquad\qquad 0 < x < l, t > 0$

(t_2) $w(0, t) = 0$ $\qquad\qquad\qquad t > 0$

(t_3) $Bw(l, t) + kw_x(l, t) = 0$ $\qquad\quad t > 0$

(t_4) $w(x, 0) = f(x) - v(x) = g(x)$ $\qquad 0 < x < l$

(1) Show that if $\phi_1, \phi_2, \ldots, \phi_n$ are solutions of (t_1), (t_2), and (t_3), then any finite linear combination of these solutions also satisfies the boundary value problem (t_1), (t_2), (t_3).

(2) Let $X(x)T(t)$ be a nontrivial solution of the boundary value problem (t_1), (t_2), (t_3). Show that X and T satisfy

(β_1) $T' + \lambda^2 a^2 T = 0$ $\qquad\qquad t > 0$

(β_2) $X'' + \lambda^2 X = 0$ $\qquad\quad 0 < x < l$

(β_3) $X(0) = 0$

(β_4) $BX(l) + kX'(l) = 0$

(3) Show that the general solution of (β_2), (β_3) is

$$X(x) = b \sin \lambda x.$$

(4) Show that $\lambda = 0$ yields a trivial solution.

(5) Using the solution given in (3) and the right-hand boundary conditions (β_4) show that λ satisfies

$$\tan \lambda l = -\frac{k}{B}\lambda$$

(where it is understood that both λ and b are not zero).

(6) If we sketch the graphs of $\tan \lambda l$ and $-(k/B)\lambda$, we obtain Fig. 10.30. Explain why $\lambda_{-n} = -\lambda_n$, $n = 1, 2, \ldots$.

(7) Give an heuristic argument for concluding that the solution of the transient problem $\beta_1, \beta_2, \beta_3, \beta_4$ is of the form

$$w(x, t) = \sum_{n=1}^{\infty} b_n \sin (\lambda_n x) e^{-\lambda_n^2 a^2 t} \qquad (*)$$

where the coefficients b_n are determined in such a way that

$$w(x, 0) = \sum_{n=1}^{\infty} b_n \sin \lambda_n x = g(x) \qquad 0 < x < l. \qquad (**)$$

Formula $(**)$ is not a Fourier series since the λ_n's are not integral multiples of λ_1. In fact for large n,

$$\lambda_n \cong \frac{2n - 1}{2} \frac{\pi}{l}$$

It can be shown that

$$\int_0^l \sin \lambda_n x \sin \lambda_m x \, dx = 0 \qquad \text{if } m \neq n \qquad (***)$$

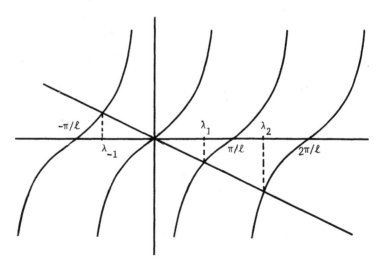

Figure 10.30

(8)   With the aid of formulas (∗∗) and (∗∗∗) use a formal argument
to show that

$$b_k = \frac{\displaystyle\int_0^l g(x) \sin \lambda_k x \, dx}{\displaystyle\int_0^l \sin^2 \lambda_k x \, dx} \qquad k = 1, 2, \ldots \qquad (\ast\ast\ast\ast)$$

Since $u(x,t) = w(x,t) + v(x)$, we expect that the solution will be

$$u(x, t) = \tau_0 + \frac{B[\tau_1 - \tau_0]}{k + Bl} x + \sum_{n=1}^{\infty} b_n \sin (\lambda_n x) e^{-\lambda_n^2 a^2 t}.$$

If formulas (∗∗), (∗∗∗) and (∗∗∗∗) are analytically valid this suggests that
the theory of Fourier series for representing functions yields but one of
many methods for representing functions. The above problem (β_2), (β_3),
(β_4) is an example of a Sturm-Liouville problem, and the theory associated with this problem is of considerable value in applied work. In
the next exercise we shall formally state the regular Sturm-Liouville
problem and ask some basic questions about a simplified problem.

3.   The boundary value problem

(B_1) $(s(x)X'(x))' - q(x)X'(x) + \lambda^2 p(x)X(x) = 0 \qquad a < x < b$

(B_2) $\alpha_1 X(a) - \alpha_2 X'(a) = 0$

(B_3) $\beta_1 X(b) - \beta_2 X'(b) = 0$

is called a Sturm-Liouville problem if the following conditions are satisfied:

(i) λ is a numerical parameter.
(ii) $\alpha_1^2 + \alpha_2^2 > 0$, $\beta_1^2 + \beta_2^2 > 0$
       (this guarantees two boundary conditions).
(iii) $s, s', p,$ and q are continuous on $[a,b]$.
(iv) $s(x) > 0$ and $p(x) > 0$ on $[a,b]$.

If there exists a numerical value λ and a corresponding nontrivial solution
X for (B_1), (B_2) and (B_3), then λ is called an *eigenvalue* and $X(x)$ is
called a corresponding *eigenfunction* of the problem.
(a)   Show that

(A_1) $X''(x) + \lambda^2 X'(x) = 0 \qquad 0 < x < l$
(A_2) $X(0) = 0$
(A_3) $BX(l) + kX'(l) = 0$

is a regular Sturm-Liouville problem.
(b)   Suppose that λ_m and λ_n are distinct eigenvalues with corresponding

eigenfunctions X_m and X_n that are associated with the above problem (A_1), (A_2), (A_3). Show that

$$\int_0^l X_m(x)X_n(x)\, dx = 0$$

[*Hint*: Note that $X_n''X_m - X_m''X_n = (\lambda_m^2 - \lambda_n^2)X_mX_n$ and using integration by parts and the boundary conditions (A_2) and (A_3) show that $\int_0^l (X_n''X_m - X_m''X_n)\, dx = 0$.]

Suppose that f is a piecewise smooth function on $[a,b]$. Then if X_1, $X_2, \ldots$ are eigenfunctions of the regular Sturm-Liouville problem, (B_1), (B_2), (B_3) such that the α's and β's are nonnegative, it can be shown that $f(x)$ can be represented in the form

$$\frac{f(x + 0) + f(x - 0)}{2} = \sum_{n=1}^{\infty} c_n X_n(x) \qquad a < x < b$$

where

$$c_n = \frac{\displaystyle\int_a^b p(x)f(x)X_n(x)\, dx}{\displaystyle\int_a^b p(x)X_n^2(x)\, dx}$$

This result supports the final formulation of $u(x,t)$ given in problem G.2.

Section H

1. Assume that if z is complex number, then $e^z = 1 + z + z^2/2! + z^3/3! + \cdots$, and show that if x is real number, then $\sin x = (e^{ix} - e^{-ix})/2i$.
2. In (10.H.2), show that if f is continuous on $\mathbf{R}^1$, then the convergence is uniform on $\mathbf{R}^1$.
3. (a) Show that the Fourier series expansion

$$1 + \sum_{n=1}^{\infty} 2 \cos nx$$

fails to converge for all x.

(b) Show that

$$\lim_{n \to \infty} S_n(x) = 0$$

for all $x \in (-\pi,\pi)$, where

$$S_n(x) = \frac{s_0(x) + s_1(x) + \cdots + s_{n-1}(x)}{n}.$$

4. Could formula (13) in Sec. 10.H be used to obtain a useful uniform error bound of $|f(x) - S_n(x)|$ in specific problems?

5. Show that

$$S_n(x) = \frac{a_0}{2} + \sum_{k=1}^{n-1} \left(\frac{n-k}{n}\right)(a_k \cos kx + b_k \sin kx)$$

where S_n is as defined in Sec. 10.H.

6. Explain why

$$\int_{-\pi/2}^{0} f(x + 2s)\left(\frac{\sin ns}{\sin s}\right)^2 ds = \int_{0}^{\pi/2} f(x - 2s)\left(\frac{\sin ns}{\sin s}\right)^2 ds$$

7. Suppose that $c = \sum_{n=1}^{\infty} c_n$ is a convergent series of numbers and let s_n be the nth partial sum. Set $\sigma_n = (s_1 + \cdots + s_n)/n$. Show that $\lim_{n \to \infty} \sigma_n = c$.

$$\left[Hint: \sigma_n - c = \frac{(s_1 - c) + \cdots + (s_n - c)}{n} \right.$$

$$= \left. \frac{\sum_{k=1}^{m_0} (s_k - c)}{n} + \frac{\sum_{k=m_0+1}^{n} (s_k - c)}{n} \right]$$

8. Show that the Dirichlet kernel D_n is periodic with period 2π.

REFERENCES

Hardy, G. H., and W. Rogosinski (1944): *Fourier Series*, Cambridge University New York.

Sagan, H. (1961): *Boundary and Eigenvalue Problems in Mathematics*, Wiley, New York.

11

AN INTRODUCTION TO MODERN
INTEGRATION THEORY

The basic intent of this chapter is to provide the reader with an initial look at measure theory and an introduction to an important generalization of the integral defined in Chapter 6.

In the first few sections of this chapter a number of proofs are omitted in order that the reader may quickly gain a general overview of the need for and the uses of a more general theory of integration.

Section 11.A is somewhat historically oriented. The Riemann integral is defined and two criteria for the existence of this integral are given. Certain limitations of the Riemann integral are brought out to call attention to the desirability of an integral that can "integrate" more functions. The definition of this integral, the Lebesgue integral, leads naturally to the problem of measuring the size of arbitrary sets of real numbers. The introduction to measure theory that follows is designed to be more heuristic than detailed; a number of basic results are stated without proof, although some of the more interesting and accessible theorems are proved.

A second approach to extending the notion of the integral begins in Sec. 11.E. This work culminates in the classical and important Lebesgue dominated convergence theorem. In Sec. 11.H we discuss some natural extensions of results associated with the integral of a continuous function.

The chapter concludes with a sampling of what are generally referred to as *Fubini theorems*, theorems that involve the integration of functions of more than one variable.

A. THE RIEMANN INTEGRAL: VIRTUES AND LIMITATIONS

Enormous progress has been made in the field of analysis during this century, progress stemming in large part from the work of an outstanding French mathematician, Henri Lebesgue (1875–1941), at the beginning of the century. Lebesgue developed a new and more general approach to integration and his brilliant contributions brought about a substantial cohesiveness and simplicity to many areas of analysis.

In the years preceding the modern development of the integral a number of unusual examples appeared that seriously undermined the then existing notions of the integral. The advent of the Lebesgue integral helped restore confidence in some of the questioned theories and, in fact, served to both strengthen and expand these ideas. In order to gain a better appreciation of the modern theory of integration we trace briefly the early development of the integral.

Leibniz (1646–1716) considered integration and differentiation as inverse operations and presumed that integration determined the "net area" associated with the graph of the function being integrated. This attitude has prevailed almost intact to the present time and is still seen as the major theme in most elementary treatments of the integral.

One hundred and fifty years after the work of Leibniz, Cauchy, using his precise formulation of the concepts of continuity and the differential, considered the integral of a continuous function f over a finite interval $[a,b]$ as the limit of what are now called *Cauchy sums*, that is, sums of the form

$$\sum_{i=1}^{n} f(x_{i-1})(x_i - x_{i-1}) \tag{1}$$

where $\{x_0, x_1, \ldots, x_n\}$ is a partition of $[a,b]$. Cauchy emphasized that for a continuous function f this process would yield the primitive of f (as we have seen in Chapter 6).

The emphasis on representing arbitrary (not necessarily continuous) functions by Fourier series kindled interest in the possibility of extending the applicability of the integral to a wider class of functions than those previously considered. The first fundamental breakthrough in this direction was achieved by G. F. Riemann (1826–1866). Riemann also considered integrability in terms of Cauchy sums as well as in terms of sums of the form

$$\sum_{i=1}^{n} f(x_i^*)(x_i - x_{i-1})$$

where x_i^* is any point in the interval $[x_{i-1}, x_i]$. The Riemann integral may be defined as follows.

(11.A.1) Definition A function $f: [a,b] \to \mathbf{R}^1$ is *Riemann integrable* if

$$\lim_{|P| \to 0} \sum_{i=1}^{n} f(x_i^*) \, \Delta x_i$$

exists (as usual, $\Delta x_i = x_i - x_{i-1}$). The *Riemann integral* of f over the interval $[a,b]$ is defined to be the value of this limit and is denoted by $\int_a^b f(x) \, dx$. (To say that $L = \lim_{|P| \to 0} \sum_{i=1}^{n} f(x_i^*) \, \Delta x_i$ means that for each $\varepsilon > 0$, there is a $\delta > 0$ such that if $P = \{x_0, x_1, \ldots, x_n\}$ is any partition of $[a,b]$ with mesh less than δ, and if for each i, x_i^* is any point in the interval $[x_{i-1}, x_i]$, then $|L - \sum_{i=1}^{n} f(x_i^*) \, \Delta x_i| < \varepsilon$.)

The reader should compare this definition with the definition of the integral given in Sec. 6.A.

Riemann addressed the question: Under what general conditions (other than continuity) will the integral of a function f exist? Riemann found that one criterion for the existence of this limit could be expressed in terms of the oscillation of a function.

(11.A.2) Definition If $A \subset \mathbf{R}^1$, $f: A \to \mathbf{R}^1$, then the *oscillation* of f on A, $\text{osc}(f;A)$, is defined by

$$\text{osc}\,(f;\,A) = \sup \{f(x) \mid x \in A\} - \inf \{f(x) \mid x \in A\}$$

Riemann's First Criterion for Integrability (I):

A function $f: [a,b] \to \mathbf{R}^1$ is (Riemann) integrable if and only if

$$\lim_{|P| \to 0} (D_1 \, \Delta x_1 + D_2 \, \Delta x_2 + \cdots + D_n \, \Delta x_n) = 0,$$

where $P = \{x_0, x_1, \ldots, x_n\}$ is a partition of $[a,b]$, $\Delta x_i = x_i - x_{i-1}$, and $D_i = \text{osc}(f; [x_{i-1}, x_i])$; this limit is defined as in (11.A.1).

This criterion says essentially that a function f is integrable if and only if large oscillations of f are restricted to "small" subsets of $[a,b]$. This idea is also expressed by a second criterion for integrability, also due to Riemann. In order to formulate this criterion we adopt the following notation.

(11.A.3) *Notation*   Suppose that $f: [a,b] \to \mathbf{R}^1$, $P = \{x_0, \ldots, x_n\}$ is a partition of $[a,b]$, and that σ is a positive number. Then $S_P(\sigma)$ will denote the sum of all the Δx_i such that $D_i \geq \sigma$ (as before, $\Delta x_i = x_i - x_{i-1}$ and $D_i = \mathrm{osc}(f; [x_{i-1}, x_i])$).

Riemann's Second Criterion for Integrability (II):

A function $f: [a,b] \to \mathbf{R}^1$ is integrable on $[a,b]$ if and only if f is bounded and for each pair of positive numbers ε and σ, there exists a $\delta > 0$ such that whenever P is a partition of $[a,b]$ with $|P| < \delta$, then $S_P(\sigma) < \varepsilon$.

Before proving that these criteria do indeed ensure the integrability of a function $f: [a,b] \to \mathbf{R}^1$, we first show that the two criteria are equivalent. In problem A.3 the reader is asked to show that continuous functions satisfy the Riemann criteria. Readers only mildly interested in some of the subtleties of Riemann integration may wish to pass over the proofs of the next few theorems.

(11.A.4) *Theorem*   The Riemann criteria I and II are equivalent.

Proof.   I $\Rightarrow$ II.   First note that it follows immediately from I that f is bounded. (Why?) Let $M_d = \sup \{\sum_{i=1}^n D_i \, \Delta x_i \mid \{x_0, x_1, \ldots, x_n\}$ is a partition of $[a,b]$ with mesh less than $d\}$. Observe that if σ is any positive number, then for every partition P such that $|P| \leq d$, we have from the definition of $S_P(\sigma)$

$$\sigma S_P(\sigma) \leq M_d$$

Therefore, $S_P(\sigma) \leq M_d/\sigma$, and since M_d/σ converges to zero as d approaches 0, the result follows.

II $\Rightarrow$ I.   Suppose that $\varepsilon > 0$ and $\sigma > 0$ are given. Then there is a $\delta > 0$ such that $S_P(\sigma) < \varepsilon$ for every partition P of $[a,b]$ with mesh less than δ. If $P = \{x_0, \ldots, x_n\}$ is such a partition, then it is easy to see that

$$\sum_{i=1}^n D_i \, \Delta x_i < D\varepsilon + \sigma(b - a)$$

where D denotes the oscillation of f on $[a,b]$. Since D is finite and ε is independent of σ it follows that

$$\lim_{|P| \to 0} \sum_{i=1}^n D_i \, \Delta x_i = 0$$

Next we show that if f satisfies the second Riemann criterion (and, hence, also the first Riemann criterion), then f is Riemann integrable. The following two proofs are fairly lengthy and may by omitted without undue loss to the reader.

(11.A.5) *Theorem* If a function $f: [a,b] \to \mathbf{R}^1$ satisfies the second Riemann criterion for integrability, then f is Riemann integrable.

Proof. Since f is bounded, there is an $M > 0$ such that $|f(x)| \le M$ for every $x \in [a,b]$. Let $\varepsilon > 0$ and $\sigma > 0$ be given. By the second Riemann criterion there is a $\delta > 0$ such that if P is a partition with mesh less than δ, then $S_P(\sigma) < \varepsilon$. Let $P_1 = \{x_0, \dots, x_n\}$ and $P_2 = \{y_0, \dots, y_m\}$ be any two partitions of $[a,b]$ with mesh less than δ. Let

$$S_1 = \sum_{k=1}^{n} f(x_k^*)\,\Delta x_k$$

$$S_2 = \sum_{i=1}^{m} f(y_i^*)\,\Delta y_i$$

be any two Riemann sums associated with these partitions. We shall first show that

$$|S_1 - S_2| \le A\sigma + B\varepsilon$$

where A and B are constants independent of ε and σ.

Let $P = \{z_0, \dots, z_r\}$ be the partition obtained from P_1 and P_2 by using all the points in each of these two partitions (P is called the *refinement* of P_1 and P_2) (Fig. 11.1). Let $S_3 = \sum_{q=1}^{r} f(z_q^*)\Delta z_q$ be an arbitrary Riemann sum corresponding to the partition P. Clearly, we have $|S_1 - S_2| \le |S_1 - S_3| +$

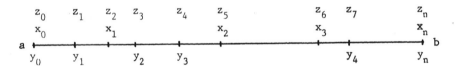

Figure 11.1

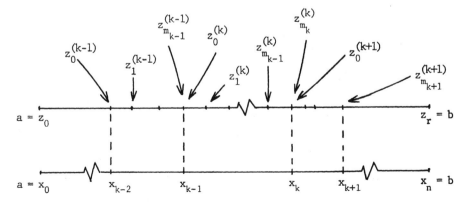

Figure 11.2

$|S_2 - S_3|$. To find $|S_1 - S_3|$, we relabel the points of P according to the scheme in Fig. 11.2. Then S_3 can be rewritten as

$$S_3 = \sum_{k=1}^{n} \left(\sum_{j=1}^{m_k} f(z_j^{*(k)})(z_j^{(k)} - z_{j-1}^{(k)}) \right)$$

and, consequently

$$|S_1 - S_3| = \left| \sum_{k=1}^{n} \left[f(x_k^*)(x_k - x_{k-1}) - \sum_{j=1}^{m_k} f(z_j^{*(k)})(z_k^{(k)} - z_{j-1}^{(k)}) \right] \right| \quad (2)$$

If k denotes a value where $\operatorname{osc}(f; [x_{k-1}, x_k]) \leq \sigma$ on $[x_{k-1}, x_k]$, then the corresponding difference appearing on the right side of (2) is bounded by $\sigma(x_k - x_{k-1})$. If k denotes a value where $\operatorname{osc}(f; [x_{k-1}, x_k]) > \sigma$, then the corresponding difference appearing on the right side of (2) is bounded by $2M(x_k - x_{k-1})$. Since $S_{P_1}(\sigma) < \varepsilon$, it follows immediately from these observations that

$$|S_1 - S_3| \leq \sigma(b - a) + 2M\varepsilon.$$

In a similar fashion we can show that

$$|S_2 - S_3| \leq \sigma(b - a) + 2M\varepsilon$$

and, therefore, we have

$$|S_1 - S_2| \leq 2(b - a)\sigma + 4M\varepsilon.$$

From this result it follows that if $\{P_n\}$ is a sequence of partitions of $[a,b]$ such that $\lim_{n \to \infty} |P_n| = 0$, then any corresponding sequence of Riemann sums $\{S_n\}$ will form a Cauchy sequence of real numbers, which by (4.C.4) converges to a real number L. It is now easy to see that for each $\varepsilon > 0$, there is a $\delta > 0$ such that if $P = \{x_0, \ldots, x_n\}$ is a partition of $[a,b]$ with mesh less than δ and if $\sum_{i=1}^{n} f(x_i^*) \Delta x_i$ is any corresponding Riemann sum, then

$$\left| \sum_{i=1}^{n} f(x_i^*) \Delta x_i - L \right| < \varepsilon,$$

which concludes the proof.

Finally, we show that if $f: [a,b] \to \mathbf{R}^1$ is Riemann integrable, then f satisfies the first (and, hence, also the second) Riemann criterion for integrability.

(11.A.6) *Theorem* If $f: [a,b] \to \mathbf{R}^1$ is Riemann integrable, then f satisfies the first Riemann criterion for integrability.

Proof. Let $\varepsilon > 0$ be given. Since f is Riemann integrable, there is a $\delta > 0$ such that if $P = \{x_0, x_1, \ldots, x_n\}$ is any partition of $[a,b]$ with mesh less than δ, then

$$\left| \sum_{i=1}^{n} f(x_i^*) \, \Delta x_i - \int_a^b f(x) \, dx \right| < \frac{\varepsilon}{6(b-a)} \tag{3}$$

where, for each i, $1 \le i \le n$, x_i^* is an arbitrary point in the interval $[x_{i-1}, x_i]$. Note that from (3) it follows that if x_i^* and $\hat{x}_i^*$ are arbitrary points in $[x_{i-1}, x_i]$, then

$$\sum_{i=1}^{n} |f(x_i^*) - f(\hat{x}_i^*)| \, \Delta x_i < \frac{\varepsilon}{3(b-a)}$$

Let $P = \{x_0, x_1, \ldots, x_n\}$ be a partition of $[a,b]$ such that $|P| < \delta$, and for each i, $1 \le i \le n$, let

$$m_i = \inf \{f(x) \mid x \in [x_{i-1}, x_i]\}$$

and

$$M_i = \sup \{f(x) \mid x \in [x_{i-1}, x_i]\}$$

In each interval $[x_{i-1}, x_i]$ choose points x_i^* and $\hat{x}_i^*$ such that $|f(x_i^*) - m_i| < \varepsilon/3(b-a)$ and $|f(\hat{x}_i^*) - M_i| < \varepsilon/3(b-a)$. Then, if $D_i = \operatorname{osc}(f; [x_{i-1}, x_i])$, we have

$$\sum_{i=1}^{n} D_i \, \Delta x_i = \sum_{i=1}^{n} |M_i - f(\hat{x}_i^*) + f(\hat{x}_i^*) - f(x_i^*) + f(x_i^*) - m_i| \, \Delta x_i$$

$$\le \sum_{i=1}^{n} |M_i - f(\hat{x}_i^*)| \, \Delta x_i + \sum_{i=1}^{n} |f(\hat{x}_i^*) - f(x_i^*)| \, \Delta x_i$$

$$+ \sum_{i=1}^{n} |m_i - f(x_i^*)| \, \Delta x_i$$

$$< \frac{\varepsilon}{3} + \frac{\varepsilon}{3} + \frac{\varepsilon}{3}$$

which proves that

$$\lim_{|P| \to 0} \sum_{i=1}^{n} D_i \, \Delta x_i = 0$$

To illustrate the generality of integrability in terms of I and II, Riemann constructed an ingenious example of an integrable function having a large number of discontinuities. This function is defined as follows. Let $\phi: \mathbf{R}^1 \to (-\frac{1}{2}, \frac{1}{2})$ be defined by

$$\phi(x) = \begin{cases} 0 & \text{if } x = \dfrac{2n+1}{2}, \quad n = 0, \pm 1, \pm 2, \ldots \\[2mm] x - m(x) & \text{if } x \ne \dfrac{2n+1}{2}, \quad n = 0, \pm 1, \pm 2, \ldots \end{cases}$$

where $m(x)$ represents the integer nearest x. A portion of the graph of ϕ is indicated in Fig. 11.3.

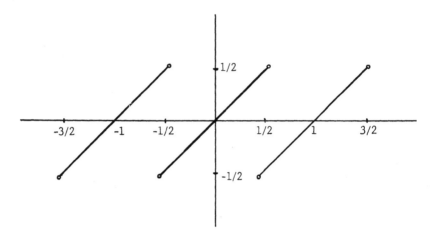

Figure 11.3

Now, let $f: \mathbf{R}^1 \to \mathbf{R}^1$ be defined by

$$f(x) = \phi(x) + \frac{\phi(2x)}{2^2} + \cdots + \frac{\phi(nx)}{n^2} + \cdots$$

Since the series $\phi(x) + \phi(2x)/2^2 + \cdots + \phi(nx)/n^2 + \cdots$ is dominated by the series $\sum_{n=1}^{\infty} 1/n^2$, it follows that f is defined for all x.

Riemann showed (and the reader is asked to show in problem A.10) that

1. f fails to be continuous at each point $\hat{x} \in \mathbf{R}^1$ of the form

$$\hat{x} = \frac{m}{2n}$$

where m and $2n$ are relatively prime integers, i.e., m and $2n$ have no common divisor that is a prime number. In fact, at each such point $\hat{x}$, it can be shown that

$$f(\hat{x} - 0) - f(\hat{x}) = \frac{\pi^2}{16n^2}$$

$$f(\hat{x} + 0) - f(\hat{x}) = -\frac{\pi^2}{16n^2}$$

where as usual $f(\hat{x} - 0) = \lim_{x \to \hat{x}-} f(x)$ and $f(\hat{x} + 0) = \lim_{x \to \hat{x}+} f(x)$.

2. f is continuous at all other points.

It is easy to show that the set of real numbers of the form $m/2n$, where m and $2n$ are relatively prime integers, forms a countably infinite dense subset of $\mathbf{R}^1$ (recall that a set $A \subset X$ is dense in X if $\bar{A} = X$). Thus, we have that for any interval $[a,b]$, f fails to be continuous on a countably infinite dense subset of points. Nevertheless, f is Riemann integrable on $[a,b]$ since given $\sigma > 0$, there are at most a finite number of integers, n, such that $\pi^2/16n^2 > \sigma$, and hence, f clearly satisfies the second Riemann integrability criterion on $[a,b]$. From this example we begin to appreciate the rather substantial way in which the integral introduced in Chapter 6 has been generalized (recall that in Chapter 6 integrability was essentially limited to continuous functions or, at best, to functions with a finite number of discontinuities).

In spite of the far greater latitude achieved by the development of the Riemann integral, this concept of integrability proved to be still too limited to deal adequately with a growing number of problems, especially problems arising in connection with Fourier series. In the derivation of the Fourier coefficients it was necessary to utilize term-by-term integration. Because of the rather delicate convergence properties of trigonometric series, there arises the obvious but important question: Under what conditions is the term-by-term integration of a trigonometric series justified? This question may be formulated in somewhat more abstract terms.

If $\{s_n\}$ is a sequence of integrable functions converging to a function s on an interval $[a,b]$, under what conditions does the sequence $\{\int_a^b s_n(x)\, dx\}$ converge to $\int_a^b s(x)\, dx$?

To see that Riemann integrability is not a sufficient condition for this convergence to occur, we review once again Example (8.C.1.b), which is due to Dirichlet. A sequence $\{u_n\}$ of functions mapping $[0,1]$ into $\mathbf{R}^1$ is defined as follows. Let $r_1, r_2, \ldots$ be the complete list of all rational numbers in $[0,1]$ and for each n, define $u_n: [0,1] \rightarrow \mathbf{R}^1$ by

$$u_n(x) = \begin{cases} 1 & \text{if } x = r_1, r_2, \ldots, r_n \\ 0 & \text{otherwise} \end{cases}$$

Then it is easily seen that $\int_0^1 u_n(x)\, dx = 0$ for each n, and furthermore, the sequence $\{u_n\}$ converges to the function $u: [0,1] \rightarrow \mathbf{R}^1$ defined by

$$u(x) = \begin{cases} 1 & \text{if } x \text{ is rational} \\ 0 & \text{if } x \text{ is irrational} \end{cases}$$

Since the oscillation of u on any subinterval of $[a,b]$ is 1, it follows from the first Riemann integrability criterion that u is not Riemann integrable, and therefore, $\int_0^1 f(x)\, dx$ is not defined, much less equal to 0.

In the next section we explore a method for redefining the integral in such a way that under quite general conditions the integral of the limit of a sequence of integrable functions will be equal to the limit of the integrals of the functions.

B. THE LEBESGUE INTEGRAL

One reason that the limit function u in the Dirichlet example is not Riemann integrable is that on any given subinterval, $[x_{i-1}, x_i]$, of $[a,b]$, no matter how small, there is no one number that serves as a good approximation for u. In some sense, the methods of both Cauchy and Riemann for approximating the values of a function on an interval are too crude to handle examples such as that given by Dirichlet.

An entirely different approach to this problem can be described as follows. Suppose that $f: [a,b] \to \mathbf{R}^1$ is a bounded function whose image is contained in an interval $[c,d]$. Instead of partitioning the interval $[a,b]$, we work with partitions of $[c,d]$. Let $Q = \{y_0, y_1, \ldots, y_n\}$ be such a partition and for $i = 1, 2, \ldots, n - 1$, let $E_i = f^{-1}([y_{i-1}, y_i))$ and for $i = n$, let $E_n = f^{-1}([y_{n-1}, y_n])$. To each set E_i we assign a number, which essentially measures the size of this set. (If E_i is itself an interval $[p,q]$, then the obvious measure of E_i is $q - p$; the measurement of more exotic sets becomes increasingly difficult and will be treated shortly.)

Assume for the moment that we are able to associate an appropriate measure with the sets E_i. For each i, $1 \le i \le n$, let $m(E_i)$ denote this (yet to be defined) measure of E_i. As an approximating sum for a "new" integral, $\int_{[a,b]} f$, we shall use $\sum_{i=1}^n y_i m(E_i)$. In fact, we shall define the integral of f over $[a,b]$ by

$$\int_{[a,b]} f = \lim_{|Q| \to 0} \sum_{i=1}^n y_i m(E_i) \tag{1}$$

provided, of course, that this limit exists and is independent of the partition used. This approach to integration is due to Lebesgue and the integral defined by (1) is customarily referred to as the *Lebesgue integral*. The shaded area in Fig. 11.4 is equal to $y_i m(E_i)$.

Before we tackle the problem of measuring sets, it might be helpful to underscore the difference between the Riemann method and the Lebesgue approach to integration as follows. Suppose that we are given a mound of money to be counted. Riemann would arbitrarily split the mount into a number of smaller mounds and take a representative value $f(x_i^*)$ for each mound so that each mound would be worth $f(x_i^*)(x_i - x_{i-1})$; he would then sum the values of these smaller mounds. Lebesgue, on the other hand, would first divide the original mound of money into smaller mounds according to the value of the currency (over a small range, y_{i-1} to y_i) and then sum the values of these mounds where each mound is worth approximately $y_i m(E_i)$. The reader might note that the Lebesgue method is more systematic, even though in this case both procedures arrive at the same amount. More gen-

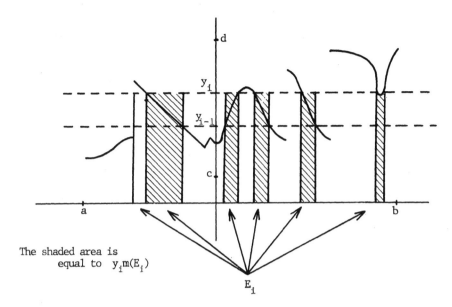

The shaded area is
equal to $y_i m(E_i)$

E_i

Figure 11.4

erally, it can be shown that if a function f is Riemann integrable on an interval $[a,b]$, then f is Lebesgue integrable and the numerical values of the two integrals coincide. Next, we see that there are functions that are Lebesgue integrable but are not Riemann integrable.

We consider once again the Dirichlet example, where $u: [0,1] \to \mathbf{R}^1$ is defined by $u(x) = 1$ if x is rational and $u(x) = 0$ if x is irrational. The range of u, $\{0,1\}$, is obviously contained in $[c,d] = [0,1]$. Each partition $Q = \{y_0, \ldots, y_n\}$ of $[0,1]$ yields precisely two subintervals of $[0,1]$ with nonempty preimages, namely $[y_0,y_1)$ and $[y_{n-1},y_n]$. As before we denote $u^{-1}([y_0,y_1))$ by E_1, $u^{-1}([y_{n-1},y_n])$ by E_n, and, in general, $u^{-1}([y_{i-1},y_i))$ by E_i, for $1 \leq i \leq n - 1$.

Since it is reasonable to define the measure of an empty set to be 0, the sum

$$\sum_{i=1}^{n} y_i m(E_i)$$

reduces to

$$y_1 m(E_1) + 1 \cdot m(E_n)$$

where

$$E_1 = \{x \in [0, 1] \mid x \text{ is irrational}\}$$
$$E_n = \{x \in [0, 1] \mid x \text{ is rational}\}$$

Note that the value y_1 tends to 0 as the mesh of the partition Q approaches 0, and hence, it follows that

$$\int_{[0,1]} u = m(R) \tag{2}$$

where $R = \{x \in [0,1] \mid x \text{ is rational}\}$. Thus the function u will be Lebesgue integrable provided that the measure of R, $m(R)$, can be determined. This we do in the next section (it turns out that the measure of this set is 0).

The Lebesgue integral represents one of the most significant advances in modern analysis. It has an astonishingly wide range of application and is especially important in the theory of probability and in the study of harmonic analysis (a generalization of Fourier analysis).

C. THE MEASUREMENT OF SETS

It should be clear that to apply the Lebesgue approach to integration, a theory of measure must be developed. What criteria then should be a basis for the measurement of sets in such a theory? Four quite natural properties of a measure m might include

1. $m(A) \geq 0$ for each set A.
2. If $A \subset B$, then $m(A) \leq m(B)$.
3. If I is a finite (open, half-open, or closed) interval with left-hand endpoint a and right-hand endpoint b, then $m(I) = b - a$.
4. If $A_1, A_2, \ldots, A_n, \ldots$ are disjoint subsets of $\mathbf{R}^1$, then $m(\bigcup_{i=1}^{\infty} A_i) = \sum_{i=1}^{\infty} m(A_i)$.

With these four properties serving as a temporary guide, let us see how a set such as $R = \{x \in [0,1] \mid x \text{ is rational}\}$ might be measured. First note that the measure of a point is 0; this follows immediately from 1, 2, and 3, since any given point lies in arbitrarily small intervals. Now let $r_1, r_2, \ldots$ be a complete list of the members of R, and for each positive integer i, set $A_i = \{r_i\}$. Then $R = \bigcup_{i=1}^{\infty} A_i$, and we have from 4,

$$m(R) = m\left(\bigcup_{i=1}^{\infty} A_i\right) \leq \sum_{i=1}^{\infty} m(A_i) = 0$$

Observe that it now follows from (2) of Sec. 11.B that if u is the Dirichlet function, then

$$\int_{[0,1]} u = m(R) = 0$$

which is consistent with our wish that

$$\lim_{n \to \infty} \int_{[0,1]} u_n = \int_{[0,1]} u$$

[where we assume that $\int_{[0,1]} u_n = \int_0^1 u_n(x)\, dx$].

Next we show that the measure of

$$B = \{x \in [0, 1] \mid x \text{ is irrational}\}$$

is 1. Since [0,1] is equal to the union of the disjoint sets B and $R = \{x \in [0,1] \mid x \text{ is rational}\}$, it follows from 3 and 4 that

$$1 = m([0, 1]) = m(R) + m(B) = 0 + m(B)$$

and hence, $m(B) = 1$.

Although we have postulated certain properties that a measure should have, we have yet to actually define a particular measure. As a first step in arriving at a satisfactory definition of a measure, we introduce the notion of the outer measure of a set.

(11.C.1)   *Definition*   Let A be a subset of $\mathbf{R}^1$. Then the *outer measure* of A is defined to be $\inf\{\sum_{n=1}^{\infty} (b_n - a_n) \mid (a_1,b_1), (a_2,b_2) \ldots \text{ is a cover of } A\}$. (Recall that a cover of a set A is any family of sets whose union contains A.)

If $\mathscr{A}$ denotes the family of all subsets of $\mathbf{R}^1$, then the function $m^*: \mathscr{A} \to \mathbf{R}^1$ defined by

$$m^*(A) = \text{outer measure of } A$$

is called the (*Lebesgue*) *outer measure*.

(11.C.2)   *Theorem*   The outer measure m^* has the following properties:

(i)   For each $A \subset \mathbf{R}^1$, $0 \le m^*(A) \le \infty$.

(ii)   If $A \subset B$, then $m^*(A) \le m^*(B)$.

(iii)   If I is an interval (closed, open, half-open) with endpoints a and b, then $m^*(I) = b - a$.

(iv)   If $\{A_n \mid n = 1, 2, \ldots\}$ is a countable family of subsets of $\mathbf{R}^1$, then $m^*(\bigcup_{n=1}^{\infty} A_n) \le \sum_{n=1}^{\infty} m^*(A_n)$.

(v)   If A is a countable subset of $\mathbf{R}^1$, then $m^*(A) = 0$.

Proof.   (i)   Obvious.

(ii)   Obvious.

(iii)   We first consider the case where $I = [a,b]$. For any $\varepsilon > 0$, we have $I \subset (a - \varepsilon/2, b + \varepsilon/2)$, and therefore,

$$m^*(I) \le \left(b + \frac{\varepsilon}{2}\right) - \left(a - \frac{\varepsilon}{2}\right) = b - a + \varepsilon$$

Since this is true for each $\varepsilon > 0$, it follows that $m^*(I) \le b - a$.

Showing that $m^*(I) \geq b - a$ is somewhat trickier and involves use of Theorem (4.D.4). Suppose that $\{I_n \mid n \in \mathbf{Z}^+\}$ is any cover of $[a,b]$ by open intervals, where for each n, $I_n = (a_n,b_n)$. We must show that

$$\sum_{n=1}^{\infty} (b_n - a_n) \geq b - a$$

By (4.D.4) there is a finite number of intervals, say $I_1, I_2, \ldots, I_k$, that cover $[a,b]$. Suppose that $a \in I_{n_1}$, where $1 \leq n_1 \leq k$; then $a_{n_1} < a < b_{n_1}$. If $b_{n_1} > b$, then we have $[a,b] \subset (a_{n_1},b_{n_1})$ and, hence,

$$b - a < b_{n_1} - a_{n_1} \leq \sum_{n=1}^{\infty} (b_n - a_n)$$

the desired result. If $b_{n_1} \leq b$, then at least one of the intervals $I_1, I_2, \ldots, I_k$ must contain b_{n_1}, denote it by $I_{n_2} = (a_{n_2},b_{n_2})$, and observe that $a_{n_2} < b_{n_1} < b_{n_2}$. If $b_{n_2} > b$, then we have $a_{n_1} < a < b_{n_2}$ and $[a,b] \subset (a_{n_1},b_{n_1}) \cup (a_{n_2},b_{n_2})$; therefore,

$$\sum_{n=1}^{\infty} (b_n - a_n) \geq (b_{n_1} - a_{n_1}) + (b_{n_2} - a_{n_2}) = b_{n_2} + (b_{n_1} - a_{n_2}) - a_{n_1}$$

$$> b_{n_2} - a_{n_1} > b - a$$

If $b_{n_2} \leq b$, then one of the intervals $I_1, \ldots, I_k$, say $I_{n_3} = (a_{n_3},b_{n_3})$, contains b_{n_2}. We continue this process (a finite number of times) until an interval $I_{n_l} = (a_{n_l},b_{n_l})$ that contains b is reached. Then, since $a_{n_1} < a < b < b_{n_l}$, we have

$$\sum_{n=1}^{\infty} (b_n - a_n) \geq (b_{n_1} - a_{n_1}) + (b_{n_2} - a_{n_2}) + \cdots + (b_{n_l} - a_{n_l})$$

$$= b_{n_l} + (b_{n_{l-1}} - a_{n_l})$$

$$+ (b_{n_{l-2}} - a_{n_{l-1}}) + \cdots + (b_{n_1} - a_{n_2}) - a_{n_1}$$

$$> b_{n_l} - a_{n_1} > b - a$$

which implies that $m^*(I) \geq b - a$; hence, $m^*(I) = b - a$, which establishes (iii) for the case $I = [a,b]$.

Now suppose that $I = (a,b)$ and let $\varepsilon > 0$ be given. Let $\hat{I} = [a + \varepsilon/2, b - \varepsilon/2]$ and $\check{I} = [a - \varepsilon/2, b + \varepsilon/2]$. Then $\hat{I} \subset I \subset \check{I}$, and we have

$$b - a - \varepsilon = \left(b - \frac{\varepsilon}{2}\right) - \left(a + \frac{\varepsilon}{2}\right) = m^*(\hat{I}) \leq m^*(I) \leq m^*(\check{I})$$

$$= \left(b + \frac{\varepsilon}{2}\right) - \left(a + \frac{\varepsilon}{2}\right) = b - a + \varepsilon$$

Thus, for each $\varepsilon > 0$,

$$b - a - \varepsilon \le m^*(I) \le b - a + \varepsilon$$

and hence, it follows that $m^*(I) = b - a$.

A similar argument holds for half-open intervals.

(iv) If $m^*(A_n) = \infty$ for some n, then the result is obvious. Suppose then that $m^*(A_n) = m_n < \infty$ for each n, and let $\varepsilon > 0$ be given. For each n, let $\{I_n^k\}$, $k = 1, 2, \ldots$, be a cover of A_n such that

$$\sum_{k=1}^{\infty} m^*(I_n^k) < m_n + \frac{\varepsilon}{2^n}$$

Then the countable union of countably many sets $\{I_n^k\}$, $n = 1, 2, \ldots$; $k = 1, 2, \ldots$, forms a cover of $\bigcup_{n=1}^{\infty} A_n$ by open intervals; furthermore, by (1.D.7) this cover is still countable. Now note that

$$m^*\left(\bigcup_{n=1}^{\infty} A_n\right) \le \sum_{n=1}^{\infty}\left(\sum_{k=1}^{\infty} m^*(I_n^k)\right) \le \sum_{n=1}^{\infty}\left(m_n + \frac{\varepsilon}{2^n}\right)$$

$$= \left(\sum_{n=1}^{\infty} m_n\right) + \varepsilon$$

$$= \left(\sum_{n=1}^{\infty} m^*(A_n)\right) + \varepsilon$$

Since this is true for each $\varepsilon > 0$, part (iv) is established.

(v) This was essentially proven earlier in our discussion of the measure of subsets of the rationals.

(11.C.3) Exercise (a) Show that if $A \subset \mathbf{R}^1$ and $x \in \mathbf{R}^1$, then $m^*(A) = m^*(A + x)$, where $A + x = \{a + x \mid a \in A\}$.

(b) Show that if $A_1, A_2, \ldots$ is any countable family of mutually disjoint *intervals*, then $m^*(\bigcup_{n=1}^{\infty} A_n) = \sum_{n=1}^{\infty} m^*(A_n)$.

The outer measure has one serious limitation: if A_1 and A_2 are disjoint subsets of $\mathbf{R}^1$, it does not necessarily follow that

$$m^*(A_1 \cup A_2) = m^*(A_1) + m^*(A_2)$$

There are examples where two disjoint sets A_1 and A_2 may be so intertwined that open interval covers of A_1 may overlap A_2 (and vice versa) to such an extent that $m^*(A_1) + m^*(A_2)$ will be strictly greater than $m^*(A_1 \cup A_2)$. A particular example of this is given in the next section. We shall presently declare these sets to be nonmeasurable.

There are many (equivalent) approaches for defining measures and measurable sets. One way to circumvent the problem indicated in the previous paragraph is to define measurable sets as follows.

(11.C.4) *Definition*   A bounded subset A of $\mathbf{R}^1$ is *measurable* if for each bounded open interval I containing A,

$$m^*(A) + m^*(I \backslash A) = m^*(I).$$

The measure of a measurable set A, $m(A)$, is defined to be $m^*(A)$.

What sets are measurable under this definition? First it is clear from the definition and from (11.C.3.b) that if J is a bounded interval, then J is measurable, and $m(J)$ is simply the length of J. In the next theorem we find that all bounded closed subsets of $\mathbf{R}^1$ are measurable. In a subsequent theorem we shall show that all bounded open subsets of $\mathbf{R}^1$ are measurable as well. The proof of the next theorem is fairly complex, and may be omitted; however, the proof does illustrate a number of typical measure theory arguments.

(11.C.5) *Exercise*   (a)   Show that if U is any open subset of $\mathbf{R}^1$, then there is a (possibly finite) sequence of disjoint open intervals, $I_1, I_2, \ldots$ such that $U = \bigcup_{n=1}^{\infty} I_n$.
(b)   Show that if $A \subset \mathbf{R}^1$, then $m^*(A) = \inf \{m^*(U) \mid A \subset U$ and U is open}.

(11.C.6) *Theorem*   If A is a closed and bounded subset of $\mathbf{R}^1$, then A is measurable.

Proof.   Suppose that $I = (a,b)$ is an open interval containing A. Then $I \backslash A$ is an open subset of $\mathbf{R}^1$ and by the previous exercise, there is a sequence of mutually disjoint open intervals $\{I_n\}$ such that $I \backslash A = \bigcup_{n=1}^{\infty} I_n$. By (11.C.3.b) we have

$$m^*(I \backslash A) = \sum_{n=1}^{\infty} m^*(I_n) \qquad (2)$$

The series $\sum_{n=1}^{\infty} m^*(I_n)$ is a convergent series of positive terms; therefore, given $\varepsilon > 0$, there is a positive integer N such that

$$\sum_{n=N+1}^{\infty} m^*(I_n) < \varepsilon. \qquad (3)$$

Note that $A \subset I \backslash \bigcup_{n=1}^{N} I_n$ and, therefore,

$$m^*(A) \leq m^* \left(I \backslash \bigcup_{n=1}^{N} I_n \right) \qquad (4)$$

Furthermore, since $I \backslash \bigcup_{n=1}^{N} I_n$ is composed of a finite number of intervals, it follows that

$$m^*\left(I \setminus \bigcup_{n=1}^{N} I_n\right) + \sum_{n=1}^{N} m^*(I_n) = m^*(I) = b - a \tag{5}$$

From (2) and (3) we have

$$m^*(I \setminus A) = \sum_{n=1}^{N} m^*(I_n) + \sum_{n=N+1}^{\infty} m^*(I_n)$$

$$\leq \sum_{n=1}^{N} m^*(I_n) + \varepsilon \tag{6}$$

Consequently, from (4), (5), and (6) it follows that

$$m^*(A) + m^*(I \setminus A) \leq m^*\left(I \setminus \bigcup_{n=1}^{N} I_n\right) + m^*(I \setminus A) \leq (b - a) + \varepsilon$$

and since ε was arbitrary, we can conclude that

$$m^*(A) + m^*(I \setminus A) \leq b - a \tag{7}$$

Since $I = A \cup (I \setminus A)$, it follows that

$$b - a = m^*(I) \leq m^*(A) + m^*(I \setminus A) \tag{8}$$

and hence, from (7) and (8) we have

$$m^*(A) + m^*(I \setminus A) = m^*(I)$$

which shows that A is measurable.

(11.C.7) Observation In most texts dealing with measure theory the following quite useful result is established: If $A_1, A_2, \ldots$ is any countable family of measurable sets contained in a bounded interval, then the sets $\bigcup_{n=1}^{\infty} A_n$ and $\bigcap_{n=1}^{\infty} A_n$ are measurable, and furthermore, if the sets A_n are mutually disjoint, then

$$m\left(\bigcup_{n=1}^{\infty} A_n\right) = \sum_{n=1}^{\infty} m(A_n)$$

(see Korevaar, 1968). In problem B-C.9, the reader is asked to apply this result to give another proof of Theorem (11.C.6) and of Corollary (11.C.12) below.

(11.C.8) Exercise (a) Show that if $A \subset \mathbf{R}^1$ is a bounded set such that $m^*(A) = 0$, then A is a measurable set, and $m(A) = 0$.

(b) Show that if I is a bounded interval that contains a set A of measure zero, then $I \setminus A$ is measurable and $m(I \setminus A) = m(I)$.

(c) Show that if A and B are two sets of measure 0, then $A \cup B$ is a set of measure 0.

The ability to measure closed sets will also allow us to measure a wide variety of sets, including all bounded open sets, as well as a number of sets that are neither open nor closed. This can be done via the notion of the inner measure of a set.

(11.C.9) *Definition* If A is a bounded subset of $\mathbf{R}^1$, then the *inner measure* of A, $m_*(A)$, is defined by

$$m_*(A) = \sup \{m(F) \mid F \subset A, F \text{ is closed}\}$$
$$= \sup \{m^*(F) \mid F \subset A, F \text{ is closed}\}$$

Note the distinction between the inner and outer measure: the outer measure of a set is obtained by "approximating" the set from the outside with open sets, while the inner measure is derived by "approximating" the set from the inside with closed sets.

(11.C.10) *Observation* The reader may well wonder why closed sets are used in the definition of the inner measure while open sets are employed in the definition of the outer measure. One reason for this change can be seen as follows. Let B be the set of irrational numbers in the interval $[0,1]$. Then by (11.C.8.b) we have that B is a measurable set and $m(B) = 1$. Note, however, that B contains no open sets, and therefore, if we were to use open sets in the preceding definition, we would have $m_*(B) = 0$. Since $m^*(B) = 1$, the use of open sets in the definition of the inner measure would leave us without the following useful and rather remarkable theorem.

(11.C.11) *Theorem* A bounded set A is measurable if and only if $m_*(A) = m^*(A)$.

Proof. We first show that for any subset S of an interval $I = (a,b)$

$$m_*(S) = (b - a) - m^*(I \backslash S); \qquad (9)$$

since S is measurable if and only if

$$m^*(S) = (b - a) - m^*(I \backslash S)$$

the result will follow.

For each closed subset F of S, we have

$$m^*(F) = m(F) \qquad (10)$$

and

$$m^*(I \backslash F) \geq m^*(I \backslash S) \qquad (11)$$

It follows from (10), (11) and (11.C.9) that

$$m_*(S) = \sup \{m(F) \mid F \text{ closed}, F \subset S\} = \sup \{m^*(F) \mid F \text{ closed}, F \subset S\}$$
$$= \sup \{(b - a) - m^*(I \backslash F) \mid F \text{ closed}, F \subset S\}$$
$$= (b - a) - \inf \{m^*(I \backslash F) \mid F \text{ closed}, F \subset S\} \qquad (12)$$
$$= (b - a) - \inf \{m^*(U) \mid U \text{ open}, (I \backslash S) \subset U\}$$
$$= (b - a) - m^*(I \backslash S)$$

The reader is asked to establish equation (12) in problem B-C.4 (see also problem E.4, Chapter 1).

(11.C.12) Corollary If U is a bounded open subset of $\mathbf{R}^1$, then U is measurable.

Proof. Let $\varepsilon > 0$ be given. It suffices to show that $m^*(U) - m_*(U) < \varepsilon$. By (11.C.3.b) there is a sequence of mutually disjoint intervals, (a_1, b_1), $(a_2, b_2), \ldots$, whose union is U. Note that $m^*(U) = \sum_{i=1}^{\infty} (b_i - a_i)$. Choose N so large that

$$\sum_{n=N+1}^{\infty} (b_n - a_n) < \frac{\varepsilon}{2} \qquad (13)$$

and for each n, $1 \leq n \leq N$, let $[c_n, d_n]$ be a closed interval contained in (a_n, b_n) with the property that

$$(b_n - a_n) - (d_n - c_n) < \frac{\varepsilon}{2^{n+1}} \qquad (14)$$

Then if $F = \bigcup_{n=1}^{N} [c_n, d_n]$, we have that F is a closed set, and furthermore, it follows from (13) and (14) that $m^*(U) - m^*(F) < \varepsilon$, and hence, $m^*(U) - m_*(U) < \varepsilon$, which concludes the proof.

In view of the previous theorem and corollary, it is not at all obvious that there might exist nonmeasurable sets. However, such sets do exist. They are in general fairly complicated; one such example is given in the next section.

The idea of obtaining the measure of a set by approximations from the inside and the outside is not a new one. The Greeks, for example, computed the area of a circle by approximating that area with the known areas of polygons that were inscribed in, and circumscribed about, the circle.

The concepts of inner and outer measure can be applied to subsets of $\mathbf{R}^n$ in an entirely analogous fashion. In $\mathbf{R}^2$, for instance, the "measuring sticks" are rectangles, rather than intervals as was the case in $\mathbf{R}^1$. The rectangle R bounded by the lines $x = a$, $x = b$, $y = c$, $y = d$ (Fig. 11.5) has measure $m(R) = (b - a)(d - c)$. (This measure is the same regardless of whether all or some of the boundary points are included.)

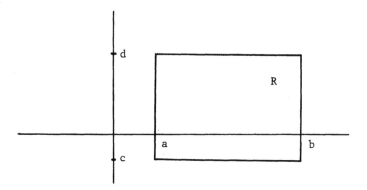

Figure 11.5

If A is a bounded subset of $\mathbf{R}^2$, then the outer measure of A, $m^*(A)$, is defined by

$$m^*(A) = \inf \left\{ \sum_{n=1}^{\infty} m(R_n) \mid R_1, R_2, \ldots \text{ is a cover of } A \text{ by open rectangles} \right\}$$

and the inner measure of A is given by

$$m_*(A) = \sup \left\{ m^*(F) \mid F \text{ closed}, F \subset A \right\}$$

A set A is measurable if $m_*(A) = m^*(A)$. These ideas can obviously be carried over to $\mathbf{R}^3$, $\mathbf{R}^4$, . . . , etc.

In a first calculus course the value of a definite integral is often described as representing the area enclosed by the graph of the function, the x-axis, and the lines $x = a$, $y = b$, where, of course, "area" below the x-axis is considered to be negative (Fig. 11.6).

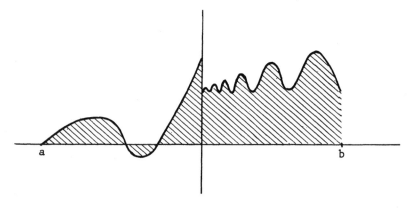

Figure 11.6

Usually in beginning calculus courses the problem of whether these areas actually exist (or how they should be defined) is ignored. For example, what is the area under the graph of the Dirichlet function u described in Sec. 11.A? Or suppose that B is a nonmeasurable subset of the interval $[0,1]$ (such a set is described in the next section) and that $f: [0,1] \to \mathbf{R}^1$ is defined by

$$f(x) = \begin{cases} 1 & \text{if } x \in B \\ 0 & \text{otherwise} \end{cases}$$

What is the area of the region bounded by the graph of this function, the lines $x = 0$ and $x = 1$ and the x-axis? During the years prior to the work of Lebesgue considerable effort was devoted to precisely this type of problem. Two outstanding mathematicians of the period, Giuseppe Peano and Camille Jordan, developed a theory of "content" that dealt satisfactorily with the areas arising from Riemann integrable functions (certain aspects of this theory are discussed in the problem section). In terms of this theory, the Riemann integral of a function $f: [a,b] \to \mathbf{R}^1$ can be defined geometrically as the net area of the region bounded by the graph of the function, the lines $x = a$ and $x = b$ and the x-axis. The theory of content did, however, have one major defect: it did not lend itself well to a theory of convergence. Lebesgue, in his development of a theory of measure, was able to overcome this deficiency. From a geometrical standpoint, Lebesgue's version of the integral paralleled that of Jordan and Peano, but his theory also provided for a number of remarkable convergence properties of sets, functions, and integrals. Lebesgue's geometric solution to the problem of integration can be described as follows. He decreed a nonnegative function f on an interval $[a,b]$ to be integrable if the region R determined by the graph of f, the x-axis, and the lines $x = a$, $x = b$ was measurable. The value of the integral $\int_{[a,b]} f$ was defined to be equal to the measure $m(R)$ of R, and if $m(R)$ was finite, then f was said to be *summable* on $[a,b]$.

If for an arbitrary function f (not necessarily nonnegative) the functions f^+ and f^- are defined by

$$f^+(x) = \max\{f(x), 0\} \qquad f^-(x) = \min\{f(x), 0\}$$

then $f = f^+ + f^-$. Lebesgue defined the integral $\int_{[a,b]} f$ by

$$\int_{[a,b]} f = \int_{[a,b]} f^+ - \int_{[a,b]} (-f^-)$$

provided both f^+ and $-f^-$ are summable (note that f^+ and $-f^-$ are nonnegative functions).

Lebesgue also formulated the analytic definition of the integral that was given in Sec. 11.B, and he showed that the analytic definition and the "geometric" definition just mentioned are equivalent. [Measurability, of course, enters into the analytic definition in order to give meaning to the sums $\sum_{i=1}^{n} y_i m(E_i)$.]

We conclude this section with a very remarkable theorem which asserts that a bounded function $f: [a,b] \to \mathbf{R}^1$ is Riemann integrable if and only if the set of discontinuities of f forms a set of measure 0. To establish this result, we shall need the concept of the oscillation of a function at a point.

(11.C.13) *Definition* Suppose that $f: A \to \mathbf{R}^1$ and that $x \in A$. The *oscillation* of f at x, osc($f;x$), is defined by

$$\text{osc}\,(f; x) = \inf\left\{\text{osc}\,(f; I \cap A) \mid I \text{ is an open interval containing } x\right\}$$

(11.C.14) *Observation* If $f: [a,b] \to \mathbf{R}^1$ and

$$B = \left\{x \in [a, b] \mid \text{osc}\,(f; x) \geq \varepsilon\right\}$$

then B is a closed subset of R^1. To see this, we show that $\mathbf{R}^1 \backslash B$ is open. If $x \in [a,b]$ and $x \notin B$, then there is an open interval I containing x such that osc($f;I \cap A$) < ε. Therefore, we have that $x \in I \subset \mathbf{R}^1 \backslash B$, and, hence, $\mathbf{R}^1 \backslash B$ is open, which, of course, implies that B is closed.

(11.C.15) *Theorem* Suppose that $f: [a,b] \to \mathbf{R}^1$ is a bounded function, and let $B = \{x \in [a,b] \mid f$ is not continuous at $x\}$. Then f is Riemann integrable if and only if B is of measure 0.

Proof. We shall show that if B is of measure 0, then f is integrable. The converse is left as an exercise (see problem B-C.6).

Suppose that B is of measure 0. We shall employ Riemann's first criterion for integrability to establish the integrability of f. Let $\varepsilon > 0$ be given. We must find a $\delta > 0$ such that if $P = \{x_0, x_1, \ldots, x_n\}$ is any partition of $[a,b]$ with mesh less that δ, then

$$D_1 \Delta x_1 + D_2 \Delta x_2 + \cdots + D_n \Delta x_n < \varepsilon \tag{15}$$

where, as before, $\Delta x_i = x_i - x_{i-1}$ and $D_i = \text{osc}(f;[x_{i-1},x_i])$.

Since f is bounded, there is a positive number M such that $|f(x)| < M$ for each $x \in [a,b]$, and consequently, for each such x, we have osc($f;x$) < $2M$. Let $Q = \max \{4M, 1, 2(b - a)\}$, and set $\varepsilon^* = \varepsilon/Q$. The idea behind the remainder of the proof is quite simple. The bad points (points in $[a,b]$ with oscillation greater than or equal to ε^*) can be enclosed in a finite number of closed intervals whose total length is quite small, and the "good" points (the rest of the points) will have such small oscillations under f that even though these points may be more numerous, their influence on (15) will be slight. Now the details.

Let $A = \{x \in [a,b] \mid \text{osc}(f;x) \geq \varepsilon^*\}$. Then since $A \subset B$, it follows that

$m(A) = 0$ and, therefore, there are open intervals $I_1 = (a_1,b_1)$, $I_2 = (a_2,b_2)$, ... such that

$$A \subset \bigcup_{n=1}^{\infty} I_n$$

and

$$\sum_{n=1}^{\infty} m(I_n) < \varepsilon^*$$

By (11.C.14), A is closed in $[a,b]$; therefore, by (4.B.14), A is compact, and hence, there is a finite subfamily of $\{I_n \mid n = 1, 2, \ldots\}$, say $I_1, I_2, \ldots, I_k$, such that $A \subset \bigcup_{n=1}^{k} I_n$. Let $C = [a,b]\backslash\bigcup_{n=1}^{k} I_n$. Then C is compact, and for each $x \in C$, $\text{osc}(f;x) < \varepsilon^*$. It follows from the definition of $\text{osc}(f;x)$ that for each $x \in C$ there is an open interval I_x containing x such that $I_x \cap A = \emptyset$ and $\text{osc}(f;I_x) < \varepsilon^*$. The two families $\{I_x \mid x \in C\}$ and $\{I_n \mid 1 \le n \le k\}$ together form an open cover of $[a,b]$, and since $[a,b]$ is compact, there is by (4.B.8) a Lebesgue number δ corresponding to this cover. We show that if $P = \{x_0, x_1, \ldots, x_n\}$ is any partition of $[a,b]$ with mesh less than δ, then

$$D_1 \Delta x_1 + D_2 \Delta x_2 + \cdots + D_n \Delta x_n < \varepsilon$$

To see this, let $P = \{x_0, x_1, \ldots, x_n\}$ be such a partition and break the sum

$$D_1 \Delta x_1 + D_2 \Delta x_2 + \cdots + D_n \Delta x_n$$

into two parts

$$\sum D_i \Delta x_i \quad \text{where } [x_{i-1}, x_i] \cap C \neq \emptyset \tag{16}$$

$$\sum D_i \Delta x_i \quad \text{where } [x_{i-1}, x_i] \cap C = \emptyset \tag{17}$$

Note that if $[x_{i-1},x_i] \cap C = \emptyset$, then $[x_{i-1},x_i]$ must lie in some interval I_x, and hence, $\text{osc}(f;[x_{i-1},x_i]) < \varepsilon^*$. Therefore, for the sum appearing in (16) we have

$$\sum D_i \Delta x_i \le \varepsilon^*(b - a) < \frac{\varepsilon}{Q}(b - a) \le \frac{\varepsilon}{2}$$

Furthermore, it is clear that for the sum appearing in (17) we have

$$\sum D_i \Delta x_i \le 2M\varepsilon^* < \frac{\varepsilon}{2}$$

and therefore

$$\sum_{i=1}^{n} D_i \Delta x_i < \frac{\varepsilon}{2} + \frac{\varepsilon}{2} = \varepsilon$$

which concludes the proof.

Since any countable set has measure 0, it follows that any function with a countable number of discontinuities is Riemann integrable. There are sets of

measure 0 that are uncountable (the Cantor set being one of the more con-
spicuous examples—see problem B-C.7, and therefore, a function with even
an uncountable number of discontinuities may still be Riemann integrable.
These results show that the Riemann integral substantially generalizes the
integral defined in Chapter 6, just as the Lebesgue integral represents a significant generalization of the Riemann integral.

D.* A NONMEASURABLE SET

The focus of this section is on describing a set that fails to be measurable.
For each $x \in [0,1]$, let $C_x = \{y \in [0,1] \mid |x - y| \text{ is rational}\}$. The reader may
show easily (see problem D.1) that

  1. $x \in C_x$, for each $x \in [0,1]$.
  2. $[0,1] = \cup \{C_x \mid x \in [0,1]\}$.
  3. If $C_{x_1} \cap C_{x_2} \neq \varnothing$, then $C_{x_1} = C_{x_2}$.
  4. For each $x \in [0,1]$, C_x is a countable set.

It follows from the above properties that the sets C_x can be used to form a
partition of $[0,1]$, i.e., there is a subfamily of these sets

$$\mathscr{C} = \{C_{x_\alpha} \mid \alpha \in J\}$$

such that $\bigcup_{\alpha \in J} C_{x_\alpha} = [0,1]$ and if $\alpha \neq \beta$, then $C_{x_\alpha} \cap C_{x_\beta} = \varnothing$. Furthermore,
since $[0,1]$ is uncountable and each C_{x_α} is countable, we have that $\{\alpha \mid \alpha \in J\}$ is
an uncountable set. For each $\alpha \in J$, select a point $w_\alpha \in C_{x_\alpha}$ and let $W = \{w_\alpha \mid$
$\alpha \in J\}$. (From a logical standpoint, this step is a highly nontrivial matter and
requires the use of a powerful and somewhat controversial axiom known as
the *axiom of choice*.) We shall show that W is nonmeasurable. To do this we
first define $w \oplus z$, the "sum" of two points w and z in $[0,1]$, as follows:

$$w \oplus z = \begin{cases} w + z - 1 & \text{if } w + z > 1 \\ w + z & \text{if } w + z \leq 1 \end{cases}$$

If $A \subset [0,1]$ and $w \in [0,1]$, we define $w \oplus A$ by

$$w \oplus A = \{w \oplus a \mid a \in A\}$$

Next we show that if $A \subset [0,1]$ is measurable and if $w \in [0,1]$, then $w \oplus A$ is
also measurable, and moreover, $m(A) = m(w \oplus A)$. To see this, let $A_1 = $
$A \cap [0,1 - w]$ and $A_2 = A \cap (1 - w,1]$. Then by (11.C.7) both A_1 and A_2
are measurable and $m(A) = m(A_1) + m(A_2)$. Since $w \oplus A_1 = w + A_1$, it
follows from (11.C.3) that $w \oplus A_1$ is measurable and that $m(w \oplus A_1) = $
$m(w + A_1) = m(A_1)$. Furthermore, since $w \oplus A_2 = (w - 1) + A_2$, we have
that $w \oplus A_2$ is measurable and that $m(w \oplus A_2) = m(A_2)$. Since $w \oplus A = $

$(w \oplus A_1) \cup (w \oplus A_2)$ and since the sets $w \oplus A_1$ and $w \oplus A_2$ are disjoint, it follows from (11.C.7) that $w \oplus A$ is measurable and

$$m(w \oplus A) = m(w \oplus A_1) + m(w \oplus A_2) = m(A_1) + m(A_2) = m(A)$$

Now let $r_1, r_2, \ldots$ be a complete listing of the rational numbers in $[0,1]$ and consider the sets

$$E_1 = r_1 \oplus W, \qquad E_2 = r_2 \oplus W, \qquad \cdots$$

where W is defined as before. Then, for each i, $m(E_i) = m(W)$. It is easily verified that these sets are mutually disjoint and that $[0,1] = \bigcup_{i=1}^{\infty} E_i$ (see problem D.2). Note that if $m(W) = q > 0$, then $m([0,1]) = \sum_{i=1}^{\infty} m(E_i) = \infty$, which is absurd. On the other hand, if $m(W) = 0$, then we have $m([0,1]) = \sum_{i=1}^{\infty} m(E_i) = 0$, which is equally absurd. Therefore, we conclude that W is nonmeasurable.

The Lebesgue measure is just one of several measures that can be assigned to certain families of subsets of $\mathbf{R}^1$. This leads to the question of whether or not it is possible to find a measure m for *all* subsets of $\mathbf{R}^1$ that satisfies the following conditions:

1. $m(A) \geq 0$ for each $A \subset \mathbf{R}^1$.
2. If $A_1, A_2, \ldots$ is any countable family of mutually disjoint subsets of $\mathbf{R}^1$, then $m(\bigcup_{n=1}^{\infty} A_n) = \sum_{n=1}^{\infty} m(A_n)$.

The answer to this query is not known, although it is known that no such measure exists that also satisfies the additional condition: $m(x + A) = m(A)$ for each $x \in \mathbf{R}^1$ and each $A \subset \mathbf{R}^1$. Thus, the problem of nonmeasurable sets is not restricted to the Lebesgue measure.

E. THE LEBESGUE INTEGRAL: A SECOND APPROACH

In this section, we develop a generalization of the Riemann integral that does not rely to any great extent on the measure theory presented in the previous sections. Although the approach used to define this new integral is quite different from what we have done previously, it can be shown that this integral, which we shall also call the Lebesgue integral, is equivalent to the (Lebesgue) integral as defined in (1), Sec. 11.B. We begin with the following definition.

(11.E.1) *Definition* The *extended real number system* $\mathbf{R}_\infty$ consists of all real numbers together with two symbols, or "ideal numbers," $(+\infty)$ and $(-\infty)$. It is understood that $-\infty < x < +\infty$ for every real number x. The following operations are permitted:

(i) $x + (+\infty) = x + \infty = +\infty, x + (-\infty) = x - \infty = -\infty.$
(ii)   If $x > 0$, then $x\cdot(+\infty) = \infty, x\cdot(-\infty) = -\infty.$
(iii)   If $x < 0$, then $x\cdot(+\infty) = -\infty, x\cdot(-\infty) = +\infty.$

The development of the integral in this section is based almost entirely on the properties of step functions.

(11.E.2) *Definition* Suppose that J is a bounded interval in $\mathbf{R}^1$ [$J = [a,b]$, $(a,b]$, $[a,b)$, or (a,b)]. A function $s: J \to \mathbf{R}^1$ is said to be a *step function* if there is a partition $\{x_0, x_1, \ldots, x_n\}$ of J such that s is equal to a constant, a_i, on each interval $J_i = (x_{i-1}, x_i)$. The *integral* of s over J is defined by

$$\int_J s = \sum_{i=1}^{n} a_i m(J_i)$$

where $m(J_i) = x_i - x_{i-1}$ is the measure (length) of the bounded subinterval J_i (Fig. 11.7).

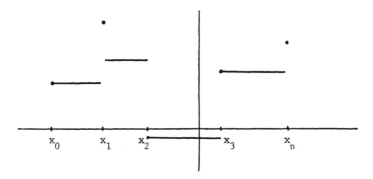

Figure 11.7

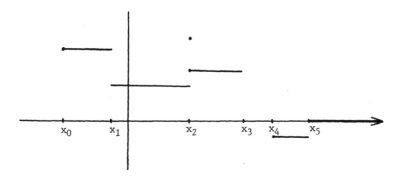

Figure 11.8

(11.E.3) *Observation* The requirement in (11.E.2) that J be bounded can be removed by assuming that the function s is identically 0 outside of a bounded subinterval of J. The integral of such a function s is still defined by (11.E.2) (Fig. 11.8).

The next theorem (whose proof can be safely entrusted to the reader) gives a number of elementary properties of step functions and their integrals. The reader should note that the value of the integral of a step function as defined in (11.E.2) coincides with the value of the corresponding Riemann integral.

(11.E.4) *Theorem* Suppose that J is an interval in $\mathbf{R}^1$.

(i) If $\phi_1: J \to \mathbf{R}^1$ and $\phi_2: J \to \mathbf{R}^1$ are step functions, then $\phi_1 + \phi_2$ is a step function and $\int_J (\phi_1 + \phi_2) = \int_J \phi_1 + \int_J \phi_2$.

(ii) If $\phi: J \to \mathbf{R}^1$ is a step function and if $c \in \mathbf{R}^1$, then $c\phi$ is a step function and $\int_J c\phi = c \int_J \phi$.

(iii) If $\phi_1: J \to \mathbf{R}^1$ and $\phi_2: J \to \mathbf{R}^1$ are step functions such that $\phi_1(x) \leq \phi_2(x)$ for each $x \in J$, then $\int_J \phi_1 \leq \int_J \phi_2$.

(iv) If $\phi: J \to \mathbf{R}^1$ is a step function, then $|\phi|$ is a step function and $|\int_J \phi| \leq \int_J |\phi|$.

The only concept of measure that is needed in the following development of the integral is the idea of a set of measure 0. We shall say that a particular property holds *almost everywhere* (a.e.) on a set X if the property holds everywhere on X except on a subset A of X of measure zero (which may be empty). For example, a sequence $\{s_n\}$ is said to converge a.e. to a function f on a set X if the sequence $\{s_n\}$ converges pointwise to f everywhere on X except on a subset A of X of measure 0.

The definition of the extended integral is based on the following concept.

(11.E.5) *Definition* A sequence of step functions $\{s_n\}$ defined on an interval J is said to be *mean fundamental* if for each $\varepsilon > 0$, there is a positive integer N such that $\int_J |s_n - s_m| < \varepsilon$ whenever $m,n \geq N$.

(11.E.6) *Observation* If $\{s_n\}$ and $\{t_n\}$ are two mean fundamental sequences of step functions on an interval J, and if for each n, $u_n = s_n - t_n$, then the sequence $\{u_n\}$ is also mean fundamental. To see this simply note that

$$|u_n - u_m| = |s_n - t_n - (s_m - t_m)| \leq |s_n - s_m| + |t_n - t_m|$$

and hence, by (11.E.4)

$$\int_J |u_n - u_m| \leq \int_J |s_n - s_m| + \int_J |t_n - t_m|$$

(11.E.7) *Definition*   Suppose that J is an interval in $\mathbf{R}^1$. A function $f: J \to \mathbf{R}_\infty$ is said to be *summable* over J if there is a mean fundamental sequence of step functions $\{s_n\}$ defined on J that converges a.e. to f on J. The *Lebesgue integral* of f over J is defined by

$$\int_J f = \lim_{n \to \infty} \int_J s_n \tag{1}$$

As was mentioned previously, it can be shown that the two integrals that we have denoted as Lebesgue integrals do coincide for all summable functions. In the next result we see that if a function f is summable over an interval J, then $\int_J f$ is finite.

(11.E.8) *Theorem*   If $\{s_n\}$ is a mean fundamental sequence of step functions on an interval J, then

(i)   The sequence $\{\int_J s_n\}$ is a Cauchy sequence.
(ii) $\lim_{n \to \infty} \int_J s_n$ is finite.
(iii) $\lim_{n \to \infty} \int_J |s_n|$ is finite.

Proof.   (i)   Let $\varepsilon > 0$ be given and let N be a positive integer such that for all $m,n \geq N$, $\int_J |s_n - s_m| < \varepsilon$. Then if $m,n \geq N$, we have

$$\left| \int_J s_n - \int_J s_m \right| = \left| \int_J (s_n - s_m) \right| \leq \int_J |s_n - s_m| < \varepsilon$$

and consequently, the sequence $\{\int_J s_n\}$ is a Cauchy sequence.

(ii)   This follows from the fact that all Cauchy sequences in $\mathbf{R}^1$ are convergent.

(iii)   This follows immediately from the inequality

$$\left| \int_J |s_n| - \int_J |s_m| \right| = \left| \int_J (|s_n| - |s_m|) \right| \leq \int_J ||s_n| - |s_m|| \leq \int_J |s_n - s_m|$$

(11.E.9) *Examples*   (a)   If $u: [0,1] \to \mathbf{R}^1$ is the Dirichlet function

$$u(x) = \begin{cases} 1 & \text{if } x \text{ is rational} \\ 0 & \text{if } x \text{ is irrational} \end{cases}$$

then the sequence of step functions $\{s_n\}$, where each s_n is the constant function 0, is a mean fundamental sequence of step functions that converges a.e. to u, and hence, $\int_{[0,1]} u = 0$.

(b)   Suppose that $f: [a,b] \to \mathbf{R}^1$ is any continuous function. Let $\{P_k\}$ be a sequence of partitions of $[a,b]$ such that $\lim_{k \to \infty} |P_k| = 0$. For each k, let $P_k = \{x_0^{(k)}, x_1^{(k)}, \ldots, x_{n_k}^{(k)}\}$ and define a step function $s_k: [a,b] \to \mathbf{R}^1$ by

$$s_k(x) = \begin{cases} f(x^{(k)}_{i-1}) & \text{if } x \in [x^{(k)}_{i-1}, x_i), \, i = 1, \ldots, n_k - 1 \\ f(x^{(k)}_{n_k-1}) & \text{if } x \in [x^{(k)}_{n_k-1}, x^{(k)}_{n_k}] \end{cases}$$

Then it is not difficult to establish that $\{s_k\}$ is a mean fundamental sequence of step functions that converges to f (see problem E.5) and, hence, f is integrable in the sense of (11.E.7). In (11.E.12) we show that all Riemann integrable functions are integrable in the sense of (11.E.7) (see also problem E.9), and furthermore, for these functions, the corresponding values of the Riemann integral and the Lebesgue integral coincide.

(c)   In view of (11.E.8.i) it is natural to inquire if Cauchy sequences of step functions could be used in place of mean fundamental sequences in the definition of the Lebesgue integral (11.E.7), i.e., could we use the Cauchy sequence $\{\int_J s_n\}$ to replace the integral $\int_J |s_n - s_m|$ with the possibly more manageable expression

$$\left| \int_J s_n - \int_J s_m \right|$$

The following example illustrates what would go awry if this were done. Let $J = [0,1]$ and for each positive integer k define $s_k : J \to \mathbf{R}^1$ by

$$s_k(x) = \begin{cases} 0 & \text{if } 0 \le x \le \dfrac{1}{k+1} \\ k(k+1) & \text{if } \dfrac{1}{k+1} < x < \dfrac{1}{k} \\ 0 & \text{if } \dfrac{1}{k} \le x \le 1 \end{cases}$$

Clearly, for each $x \in [0,1]$ the sequence $\{s_k(x)\}$ converges to 0, and furthermore, for each k,

$$\int_J s_k = 1$$

Thus, we have

$$\left| \int_J s_n - \int_J s_m \right| = 0$$

for each m and n; however,

$$\lim_{k \to \infty} \int_J s_k \ne \int_J 0$$

In problem E.6, the reader is asked to show that $\int_J |s_n - s_m| = 2$ whenever $m \ne n$.

A problem arises in connection with Definition (11.E.7) as to whether the summability of a function is a well-defined concept. That is, it must be shown that if $\{s_n\}$ and $\{\hat{s}_n\}$ are two sequences of step functions that converge to f, a.e., then $\lim_{n\to\infty} \int_J s_n = \lim_{n\to\infty} \int_J \hat{s}_n$. The proof of this is unusually complex and the more impatient readers may wish to simply accept the fact that $\int_J f$ is not dependent on the mean fundamental sequence that is used; such readers may omit the following lemma and Theorem (11.E.11).

(11.E.10) *Lemma*   Suppose that J is an interval and that $\{s_n\}$ and $\{\sigma_n\}$ are two mean fundamental sequences of step functions each of which converges a.e. to a function f on J. Then the sequence $\{\tau_n\} = \{s_n - \sigma_n\}$ is a mean fundamental sequence of step functions that converges a.e. to zero on J.

Proof.   Let A and B be subsets of measure 0 of J such that the sequence $\{s_n\}$ converges to f on $J\backslash A$ and the sequence $\{\sigma_n\}$ converges to f on $J\backslash B$. Then the set $C = A \cup B$ is a set of measure 0, and clearly the sequence $\{\tau_n\}$ converges to $f - f = 0$ on $J\backslash C$. Furthermore, by (11.E.6) the sequence $\{\tau_n\}$ is a mean fundamental sequence.

Suppose for the moment that we can show that if $\{\tau_n\}$ is any mean fundamental sequence of step functions on an interval J that converges a.e. to 0, then $\lim_{n\to\infty} \int_J \tau_n = 0$. It will then follow from (11.E.8) and (11.E.10) that if $\{s_n\}$ and $\{\hat{s}_n\}$ are mean fundamental sequences of step functions on J that converge a.e. to f, then

$$\lim_{n\to\infty} \int_J s_n = \lim_{n\to\infty} \int_J \hat{s}_n$$

To see this note that

$$\left| \int_J s_n - \int_J \hat{s}_n \right| = \left| \int_J (s_n - \hat{s}_n) \right| \le \int_J |s_n - \hat{s}_n|$$

and that

$$\lim_{n\to\infty} \int_J |s_n - \hat{s}_n| = 0$$

Thus, to show that the limit given in (11.E.7) is well defined, it suffices to establish the following theorem.

(11.E.11) *Theorem*   If $\{s_n\}$ is a mean fundamental sequence of step functions that converges a.e. to 0 on an interval J, then $\lim_{n\to\infty} \int_J s_n = 0$.

Proof.   We first prove the following special case: If $\{s_n\}$ is a decreasing sequence of nonnegative step functions that converges to 0 a.e. on an interval

J, then $\lim_{n \to \infty} \int_J s_n = 0$. We may assume throughout the remainder of the proof that the bounded interval J is also closed (why?).

First observe that the sequence $\{s_n\}$ is uniformly bounded by a positive number M. This is clear from the fact that s_1 takes on only a finite number of real values $m_1, \ldots, m_q$ on J, and therefore, since the sequence $\{s_n\}$ is a decreasing sequence, it follows that if $M = \max \{m_1, \ldots, m_q\}$, then $0 \le s_i \le M$ for each i.

For each step function s_n, there is a finite set X_n of points of J such that $J \backslash X_n$ consists of a finite number of mutually disjoint open subintervals on each of which s_n is constant. Let $E_1 = \bigcup_{n=1}^{\infty} X_n$, and note that E_1 is a countable set. Let $E_2 = \{x \in J \mid \lim_{n \to \infty} s_n(x) \ne 0\}$. Then E_2 is of measure zero, and hence, $E_0 = E_1 \cup E_2$ is also of measure zero. Therefore, for each $\varepsilon > 0$, there is a countable collection $\{I_n \mid n = 1, 2, \ldots\}$ of open intervals that covers E_0 such that $\sum_{n=1}^{\infty} m(I_n) < \varepsilon$.

Now suppose that $x \in J \backslash E_0$. The sequence $\{s_n(x)\}$ converges to zero, and consequently, there is an integer, N_x, such that $s_{N_x}(x) < \varepsilon$. Since x is not an endpoint of some interval on which s_{N_x} is constant (such endpoints belong to E_0), there is an open set, O_{N_x}, containing x on which s_{N_x} takes on a constant value less than ε. Repeating this procedure for each $x \in J \backslash E_0$, we obtain an open cover, $\{O_{N_x} \mid x \in J \backslash E_0\}$, of $J \backslash E_0$ with the property that $s_{N_x} < \varepsilon$ on each O_{N_x}. The family $\mathcal{U} = \{O_{N_x} \mid x \in J \backslash E_0\} \cup \{I_n \mid n = 1, 2, \ldots\}$ forms an open cover of J. Since J is compact, there is by (4.D.4) a finite subcover of $\mathcal{U}$ consisting of a finite number of sets, $O_{N_{x_1}}, O_{N_{x_2}}, \ldots, O_{N_{x_p}}$, together with a finite number of intervals $I_{n_1}, \ldots, I_{n_q}$. If $N = \max \{N_{x_1}, \ldots, N_{x_p}\}$, then s_N is less than ε on $\bigcup_{k=1}^{p} O_{N_{x_k}}$, and since $\{s_n\}$ is a monotone decreasing sequence, it follows that for all $i \ge N$, s_i is also less than ε on $\bigcup_{k=1}^{p} O_{N_{x_k}}$. Define a step function ϕ by

$$\phi(x) = \begin{cases} M & \text{if } x \in \bigcup_{j=1}^{q} I_{n_j} \\ \varepsilon & \text{otherwise} \end{cases}$$

Then $s_n \le \phi$ on J for all $n \ge N$. Since $\int_J \phi(x) \le \varepsilon[b - a] + M$, where a and b are the endpoints of J, it follows from (11.E.4) that $0 \le \int_J s_n \le \varepsilon[M + b - a]$, for all $n \ge N$. Since $[M + b - a]$ is independent of ε and ε was arbitrary, the proof of the special case is complete.

Now we consider the general case. Suppose then that $\{s_n\}$ is a mean fundamental sequence of step functions that converges a.e. to 0 on J. By (11.E.4) and (11.E.8) we have $\lim_{n \to \infty} \int_J |s_n| = L$, where L is finite and nonnegative. We shall show that $L = 0$. Given $\varepsilon > 0$, it is clear that there is an integer N_1 such that if $n \ge N_1$, then $L - \varepsilon \le \int_J |s_n|$ and $\int_J |s_{N_1} - s_n| < \varepsilon/2$.

Now we construct a decreasing sequence $\{\sigma_n\}$ of step functions mapping J into $[0, \infty]$ such that

(a) $\int_J |s_{N_1} - \sigma_2| < \dfrac{\varepsilon}{2}$

(b) $\int_J |\sigma_{k-1} - \sigma_k| < \dfrac{\varepsilon}{2^{k-1}}$

(c) $\displaystyle\lim_{k\to\infty} \int_J \sigma_k = 0$

Note that once this is accomplished, we shall have

$$L - \varepsilon \le \int_J |s_{N_1}| = \int_J |s_{N_1} - \sigma_2 + \sigma_2 - \cdots - \sigma_k + \sigma_k|$$

$$\le \int_J |s_{N_1} - \sigma_2| + \int_J |\sigma_2 - \sigma_3| + \cdots$$

$$+ \int_J |\sigma_{k-1} - \sigma_k| + \int_J |\sigma_k|$$

$$\le \frac{\varepsilon}{2} + \frac{\varepsilon}{2^2} + \cdots + \frac{\varepsilon}{2^{k-1}} + \int_J |\sigma_k|$$

and, therefore,

$$0 \le L \le 2\varepsilon + \int_J |\sigma_k|$$

From (c) and the fact $\varepsilon > 0$ was arbitrary, we conclude that $L = 0$.

To construct the sequence $\{\sigma_k\}$ with properties (a), (b), and (c), we first determine a suitable subsequence of $\{s_n\}$ as follows. Since $\int_J |s_{N_1} - s_n| < \varepsilon/2$ for all $n \ge N_1$, and since $\{s_n\}$ is a mean fundamental sequence, there is an integer $N_2 > N_1$ such that for each $n \ge N_2$

$$\int_J |s_{N_1} - s_{N_2}| < \frac{\varepsilon}{2}$$

and

$$\int_J |s_{N_2} - s_n| < \frac{\varepsilon}{2^2}$$

Inductively, we construct a subsequence, $\{s_{N_k}\}$, of $\{s_n\}$ such that for $k = 1$, $2, \ldots$,

$$\int_J |s_{N_k} - s_{N_{k+1}}| < \frac{\varepsilon}{2^k}$$

Now let

$$\sigma_1 = |s_{N_1}|$$
$$\sigma_2 = \min\{\sigma_1, |s_{N_2}|\}$$
$$\cdots\cdots\cdots\cdots\cdots\cdots$$
$$\sigma_k = \min\{\sigma_{k-1}, |s_{N_k}|\}$$

Then $\sigma_1, \sigma_2, \ldots$ is a decreasing sequence of step functions that converges to 0 a.e. on J. Moreover, since $\sigma_{k-1}(x) - \sigma_k(x) = 0$ for all x, where $|s_{N_k}(x)| \geq \sigma_{k-1}(x)$ and for all other $x \in J$,

$$\sigma_{k-1}(x) - \sigma_k(x) = \sigma_{k-1}(x) - |s_{N_k}(x)|$$
$$\leq |s_{N_{k-1}}(x)| - |s_{N_k}(x)|$$
$$\leq |s_{N_{k-1}}(x) - s_{N_k}(x)|$$

it follows that

$$0 \leq \int_J (\sigma_{k-1} - \sigma_k) \leq \int_J |s_{N_{k-1}} - s_{N_k}| \leq \frac{\varepsilon}{2^{k-1}}$$

Furthermore, the sequence $\{\sigma_n\}$ is a mean fundamental sequence since whenever $n > m$, we have

$$\int_J (\sigma_m - \sigma_n) = \int_J (\sigma_m - \sigma_{m+1}) + \int_J (\sigma_{m+1} - \sigma_{m+2}) + \cdots + \int_J (\sigma_{n-1} - \sigma_n)$$

$$\leq \frac{\varepsilon}{2^m} + \frac{\varepsilon}{2^{m+1}} + \cdots + \frac{\varepsilon}{2^{n-1}}$$

$$\leq \frac{\varepsilon}{2^m} \left(1 + \frac{1}{2} + \frac{1}{2^2} + \cdots \right)$$

$$= \frac{\varepsilon}{2^m}(1 + 1) = \frac{\varepsilon}{2^{m-1}}$$

Hence, by (11.E.10) we have

$$\lim_{n \to \infty} \int_J \sigma_n = 0,$$

which concludes the proof.

We end this section by establishing some of the basic properties of the Lebesgue integral. Since the integral of a summable function is the limit of a sequence of integrals of step functions, it is reasonable to expect that many of the basic properties of the integrals of step functions will carry over to the integrals of summable functions. In the following theorems we see that, in general, this is the case.

(11.E.12) Theorem If f and g are summable functions on an interval J, and if a and b are real numbers, then the linear conbination $af + bg$ is summable for every a and b; furthermore,

$$\int_J (af + bg) = a \int_J f + b \int_J g$$

Proof. Let $\{s_n\}$ and $\{t_n\}$ be mean fundamental sequences of step functions that converge a.e. to f and g, respectively. For each n, define the step function

$$u_n = as_n + bt_n.$$

Then clearly, the sequence $\{u_n\}$ converges a.e. to $af + bg$, and furthermore, by (11.E.6), this sequence is mean fundamental. Finally, we have

$$\int_J (af + bg) = \lim_{n \to \infty} \int_J u_n = \lim_{n \to \infty} \int_J (as_n + bt_n)$$

$$= a \lim_{n \to \infty} \int_J s_n + b \lim_{n \to \infty} \int_J t_n$$

$$= a \int_J f + b \int_J g$$

(11.E.13) Theorem If f is a summable function on an interval J and if $f \geq 0$ on J, then $\int_J f \geq 0$.

Proof. We shall show that there is a sequence of nonnegative step functions $\{\sigma_n\}$ such that

$$\lim_{n \to \infty} \int_J \sigma_n = \int_J f$$

Then, since $\int_J \sigma_n \geq 0$ for each n, the result will follow.

Since f is summable on J, there is a mean fundamental sequence of step functions $\{s_n\}$ that converges a.e. to f on J. For each n, let $\sigma_n = \max \{s_n, 0\}$. Then each function σ_n is a step function, and clearly the sequence $\{\sigma_n\}$ is a mean fundamental sequence converging to f a.e., and hence,

$$\int_J f = \lim_{n \to \infty} \int_J \sigma_n \geq 0$$

(11.E.14) Corollary If f and g are summable functions on an interval J and if $f \geq g$ on J, then $\int_J f \geq \int_J g$.

Proof. Let $h = f - g$ and apply (11.E.12) and (11.E.13).

(11.E.15) Notation If $A \subset \mathbf{R}^1$ and $f: A \to \mathbf{R}^1$, then $f^+: A \to \mathbf{R}^1$ is defined by $f^+(x) = \max \{f(x), 0\}$, and $f^-: A \to \mathbf{R}^1$ is defined by $f^-(x) = \min \{f(x), 0\}$.

(11.E.16) Theorem If f is a summable function on J, then f^+ and f^- are summable on J.

Proof. The proof that f^+ is summable parallels the proof given in (11. E.13). To see that f^- is summable note that $f^-(x) = \min \{f(x), 0\} = -\max \{-f(x), 0\} = -(-f)^+(x)$ and apply (11.E.12).

(11.E.17) *Theorem* Suppose that f is a summable function on an interval J. Then

(i) $|f|$ is summable.

(ii) $|\int_J f| \le \int_J |f|$.

Proof. Since $|f| = f^+ - f^-$, it follows from (11.E.12) and (11.E.16) that $|f|$ is summable.

Since $f \le |f|$ and $-f \le |f|$, we have from (11.E.14)

$$-\int_J f \le \int_J |f| \quad \text{and} \quad \int_J f \le \int_J |f|$$

and, therefore, $|\int_J f| \le \int_J |f|$.

(11.E.18) *Theorem* If f and g are summable functions on an interval J, then max (f,g) and min (f,g) are summable on J.

Proof. The proof follows immediately from the fact that max $(f,g) = \frac{1}{2}(f + g) + \frac{1}{2}|f - g|$ and min $(f,g) = \frac{1}{2}(f + g) - \frac{1}{2}|f - g|$.

In the next theorem we see that from the standpoint of integration, it is of no consequence if two functions differ on a set of measure zero.

(11.E.19) *Theorem* If f and g are summable functions on an interval J, and if $f = g$ a.e. on J, then $\int_J f = \int_J g$.

Proof. For each n, let s_n be identically zero on J. Then $\{s_n\}$ is a mean fundamental sequence that converges a.e. to $f - g$, and therefore, $\int_J (f - g) = 0$. It now follows from (11.E.12) that $\int_J f = \int_J g$.

(11.E.20) *Exercise* Show that if f is a summable function on an interval J and if $J = J_1 \cup J_2$, where J_1 and J_2 are disjoint subintervals of J on which f is summable, then

$$\int_{J_1} f + \int_{J_2} f = \int_J f$$

We conclude this section with an interesting and instructive example.

(11.E.21) *Example* Let $J = [0,\infty)$ and define $f: J \to \mathbf{R}^1$ by

$$f(x) = \frac{(-1)^{k+1}}{k} \quad \text{if } x \in [k - 1, k); \ k = 1, 2, \ldots$$

It is easy to see that $\lim_{k \to \infty} \int_{[0,k)} f$ is finite (use the alternating series test); however, by (11.E.17), $\int_J f$ does not exist since $\int_J |f|$ fails to exist (it would be equal to $\sum_{n=1}^{\infty} 1/n$).

Note that this example illustrates a major difference between the theory of improper Riemann integrals (see Sec. C, Chapter 6) and possible corresponding results for an "improper" Lebesgue integral.

F. SOME APPROXIMATION THEOREMS FOR INTEGRALS

The following approximation theorems will be of considerable value as we continue our investigation of the properties of the Lebesgue integral. We begin with the definition of another form of convergence.

(11.F.1) *Definition* A sequence of summable functions $\{s_n\}$ is said to be *convergent in the mean* to a summable function f on an interval J if

$$\lim_{n \to \infty} \int_J |f - s_n| = 0$$

(11.F.2) *Theorem* If f is summable on an interval J and if $\{s_n\}$ is a mean fundamental sequence of step functions defined on J that converges a.e. to f, then $\{s_n\}$ is convergent in the mean to f.

Proof. Let $\varepsilon > 0$ be given. Since $\{s_n\}$ is a mean fundamental sequence, there is a positive integer N such that

$$\int_J |s_n - s_m| < \varepsilon \qquad \text{for all } m, n \geq N$$

For each $n \geq N$, $|f - s_n|$ is summable. Furthermore, it follows from (11.E.6) that for each fixed $\hat{n} \geq N$, the sequence $\{|s_n - s_{\hat{n}}|\}$ is a mean fundamental sequence of step functions that converges a.e. to $|f - s_{\hat{n}}|$ on J, and therefore, we have

$$\int_J |f - s_{\hat{n}}| = \lim_{n \to \infty} \int_J |s_n - s_{\hat{n}}| < \varepsilon$$

Since this is true for each $\hat{n} \geq N$, the theorem follows.

Although summable functions can be approximated in a variety of ways by step functions, we see in the next example that uniform convergence is not always assured.

(11.F.3) *Example* If $f: [0,1] \to \mathbf{R}^1$ is defined by

$$f(x) = \begin{cases} x^{-1/2} & \text{if } x \neq 0 \\ 0 & \text{if } x = 0 \end{cases}$$

then f is summable on the interval $[0,1]$ (why?); however, it is easy to see that there does not exist a step function, s, such that $|x^{-\frac{1}{2}} - s(x)| < \varepsilon$ for each $x \in [0,1]$.

We shall show presently that if f is summable, then there are step functions that approximate f uniformly except on arbitrarily small intervals. We first need the following definition.

(11.F.4) *Definition* Suppose that $I_1, I_2, \ldots$ is a countable family of bounded intervals. Then the *total length* of this family is $\sum_{k=1}^{\infty} m(I_k)$, where $m(I_k)$ denotes the length of I_k.

(11.F.5) *Theorem* If $f: J \to \mathbf{R}_{\infty}$ is summable on J, then for each $\varepsilon > 0$ and $\delta > 0$, there is a step function $s: J \to \mathbf{R}_{\infty}$ and a corresponding countable collection of subintervals $I_1, I_2, \ldots$ of J of total length less than δ, such that

$$|f(x) - s(x)| < \varepsilon$$

for each $x \in J \backslash \bigcup_{k=1}^{\infty} I_k$.

Proof. Since f is summable on J, there is a mean fundamental sequence $\{s_n\}$ of step functions that converges a.e. to f on J. Let N be the subset of J of measure zero on which the sequence $\{s_n\}$ fails to converge to f. Let $\{s_{n_k}\}$ be a subsequence of $\{s_n\}$ with the following property:

$$\int_J |s_{n_{k+1}} - s_{n_k}| < \frac{1}{4^k} \tag{1}$$

(in problem F.1, the reader is asked to show that such a subsequence exists). Let u_k denote the step function $s_{n_{k+1}} - s_{n_k}$, and let P_k be the partition associated with u_k. Since $\int_J |u_k| < 1/4^k$, it follows that the set of intervals of P_k on which $|u_k(x)| \geq 1/2^k$ has total length less than $1/2^k$. (The shaded area depicted in Fig. 11.9 cannot exceed $1/4^k$.)

Let $\mathscr{C}_k$ denote this finite collection of subintervals of P_k. For each positive integer r, let $\mathscr{K}(r) = \bigcup_{k=r}^{\infty} \mathscr{C}_k$. It is easy to see that $\mathscr{K}(r)$ is a countable collection of subintervals of J with total length less than or equal to

$$\frac{1}{2^r} + \frac{1}{2^{r+1}} + \cdots = \frac{1}{2^{r-1}}$$

Let $E(r)$ be the union of all the intervals in $\mathscr{K}(r)$. It then follows that for each $x \in J \backslash (E(r) \cup N)$,

$$f(x) = \lim_{k \to \infty} s_{n_k}(x) = s_{n_r}(x) + (s_{n_{r+1}}(x) - s_{n_r}(x)) + (s_{n_{r+2}}(x)$$
$$- s_{n_{r+1}}(x)) + \cdots$$
$$= s_{n_r}(x) + \sum_{k=r}^{\infty} u_k(x)$$

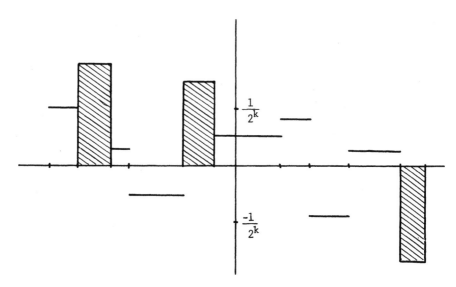

Figure 11.9

and

$$|f(x) - s_{n_r}(x)| \le \sum_{k=r}^{\infty} |u_k(x)| \le \sum_{k=r}^{\infty} \frac{1}{2^k} = \frac{1}{2^{r-1}}$$

Finally, given $\varepsilon > 0$ and $\delta > 0$, there is an integer r_0 such that if $r \ge r_0$, then $1/2^{r-2} < \delta$ and $1/2^{r-1} < \varepsilon$. Hence, for each $r \ge r_0$, we have

$$|f - s_{n_r}| < \varepsilon$$

on $J\setminus(E(r) \cup N)$. Since N is a set of measure zero, we can cover N with a countable number of subintervals of J of total length less than or equal to $1/2^{r-1}$. Therefore, $E(r) \cup N$ is covered by a countable number of subintervals of total length less than or equal to $1/2^{r-1} + 1/2^{r-1} = 1/2^{r-2} < \delta$, which completes the proof.

The next theorem is an easy consequence of (11.F.2) and (11.F.5).

(11.F.6)   Theorem   If f is a summable function on J, then for every $\varepsilon_1 > 0$, $\varepsilon_2 > 0$, and $\delta > 0$, there is a step function $s: J \to \mathbf{R}_\infty$ and a set U of measure less than δ such that:

(i) $\int_J |f - s| < \varepsilon_1$.
(ii) $|f(x) - s(x)| < \varepsilon_2$ for each $x \in J\setminus U$.

In certain situations it will be advantageous to replace mean fundamental sequences of step functions with mean fundamental sequences of summable functions. In the next theorem we see that nothing is lost (or gained) by doing so.

(11.F.7) Theorem If $\{f_k\}$ is a mean fundamental sequence of summable functions that converges a.e. on an interval J to a function $f: J \to \mathbf{R}_\infty$, then:

    (i) f is summable on J.
    (ii) $\int_J f = \lim_{k \to \infty} \int_J f_k$.
    (iii) $\lim_{k \to \infty} \int_J |f - f_k| = 0$ (i.e., the sequence $\{f_k\}$ converges in the mean to f).

Proof. By (11.F.6) each summable function f_k can be approximated on J by a function s_k that has the following properties:

(a) $\displaystyle \int_J |f_k - s_k| \leq \frac{1}{2^k}$

(b) $\displaystyle |f_k(x) - s_k(x)| < \frac{1}{2^k}$ on $J \backslash U_k$

where U_k is a set of measure less than $1/2^k$. We show that the sequence $\{s_k\}$ is a mean fundamental sequence of step functions that converges a.e. to f on J. To see this first note that

$$|s_n - s_m| \leq |s_n - f_n| + |f_n - f_m| + |f_m - s_m|$$

Therefore, we have

$$\int_J |s_n - s_m| \leq \int_J |s_n - f_n| + \int_J |f_n - f_m| + \int_J |f_m - s_m|$$

and hence, by (a) and (b), the sequence $\{s_k\}$ is a mean fundamental sequence. Now, let $D(r) = \bigcup_{k=r}^\infty U_k$, and note that if $x \in J \backslash D(r)$, then it follows from (a) and (b) above that

$$|f_k(x) - s_k(x)| \leq \frac{1}{2^k} \qquad \text{for each } k \geq r \tag{2}$$

Consequently, the sequence $\{s_k\}$ converges uniformly to f on $J \backslash D(r)$. Furthermore, we have

$$m(D(r)) = m\left(\bigcup_{k=r}^\infty U_k\right) \leq \sum_{k=r}^\infty m(U_k) \leq \frac{1}{2^r} + \frac{1}{2^{r+1}} + \cdots = \frac{1}{2^{r-1}}$$

It follows that the subset of J on which the sequence $\{s_k\}$ does not converge to f is of measure zero (since this set is contained in a set of arbitrarily small measure). Therefore, by the definition (11.E.7) of the Lebesgue integral, f is summable on J, and

$$\int_J f = \lim_{k \to \infty} \int_J s_k$$

From (2) it follows that

$$\lim_{k \to \infty} \int_J s_k = \lim_{k \to \infty} \int_J f_k$$

and hence,

$$\int_J f = \lim_{k \to \infty} \int_J f_k$$

Finally, it is easy to see that $\{|f - f_k|\}$ forms a mean fundamental sequence of summable functions that converges a.e. to 0 on J. Combining the above results we have

$$\lim_{k \to \infty} \int_J |f - f_k| = \int_J 0 = 0.$$

G. CONVERGENCE THEOREMS

In the theory of Fourier series we have seen the importance of term-by-term integration of infinite series. We have also seen, however, that in the context of Riemann integration, term-by-term integration is not always permissible. The focus of this section is on finding (in the Lebesgue theory of integration) quite general conditions under which term-by-term integration is valid. More specifically, we shall be interested in the following problem:

If $\{f_n\}$ is a sequence of integrable (summable) functions that converges to a function f, under what conditions is f integrable and when can we assert that

$$\int_J f = \lim_{n \to \infty} \int_J f_n? \tag{1}$$

Theorem (11.F.7) provided one answer to this problem. As we have seen previously, in the context of continuous functions uniform convergence is sufficient to ensure the validity of (1). The most important theorem of this section, the Lebesgue dominated convergence theorem, will provide us with a different (and very easily applied) set of conditions for (1) to hold. We begin with yet another convergence theorem, the monotone convergence theorem.

(11.G.1) Theorem (Monotone Convergence Theorem) Suppose that $f_1 \le f_2 \le f_3 \le \cdots$ is an increasing sequence of summable functions defined on an interval J and that the corresponding integrals, $\int_J f_k$, are uniformly bounded above, i.e., there is a real number M such that $\int_J f_k \le M$ for each k. Then the sequence of functions $\{f_k\}$ converges to a summable function f, and furthermore,

$$\lim_{k \to \infty} \int_J f_k = \int_J f$$

Proof. For each $x \in J$ the increasing sequence of numbers $\{f_k(x)\}$ converges either to a finite value or to $+\infty$. Let $f: J \to \mathbf{R}_\infty$ be defined by $f(x) = \lim_{k \to \infty} f_k(x)$. Note that since the sequence $\{\int_J f_n\}$ is both increasing and bounded, it is a Cauchy sequence. Therefore, given $\varepsilon > 0$, there is a positive integer N such that if $m,n \geq N$, then $|\int_J f_m - \int_J f_n| < \varepsilon$. Thus, if $m > n \geq N$, we have

$$\int_J |f_m - f_n| = \int_J (f_m - f_n) = \int_J f_m - \int_J f_n = \left| \int_J f_m - \int_J f_n \right| < \varepsilon$$

and consequently, the sequence $\{f_k\}$ is a mean fundamental sequence of summable functions that converges pointwise to f on J. Hence, by (11.F.7) the result follows.

(11.G.2) Corollary If $f_1 \geq f_2 \geq f_3 \geq \cdots$ is a decreasing sequence of summable functions on an interval J such that sequence of integrals $\int_J f_k$ is uniformly bounded below, then the sequence $\{f_k\}$ converges to a summable function f, and furthermore, $\lim_{k \to \infty} \int_J f_k = \int_J f$.

Proof. Apply (11.G.1) to the increasing sequence of functions $-f_1 \leq -f_2 \leq -f_3 \leq \cdots$.

(11.G.3) Example Suppose that $-1 < a < 0$ and let $f: [0,1] \to \mathbf{R}^1$ be defined by

$$f(x) = \begin{cases} x & \text{if } 0 < x \leq 1 \\ 0 & \text{if } x = 0. \end{cases}$$

We use the monotone convergence theorem to show that f is summable over $J = [0,1]$. For each $n \in \mathbf{Z}^+$, define $f_n: [0,1] \to \mathbf{R}^1$ by

$$f_n(x) = \begin{cases} x^a & \text{if } \dfrac{1}{n} \leq x \leq 1 \\ 0 & \text{if } 0 \leq x < \dfrac{1}{n} \end{cases}$$

Then we have

(a) $f_1 \leq f_2 \leq f_3 \cdots$

(b) $\displaystyle \int_J f_n = \frac{1}{a+1}\left(1 - \frac{1}{n^{a+1}}\right)$ for each $n \in \mathbf{Z}^+$.

Since for each $n \in \mathbf{Z}^+$, $\int_J f_n \le 1/(a + 1)$, it follows from the monotone convergence theorem that $\int_J f$ exists and is equal to $\lim_{n \to \infty} \int_J f_n = 1/(a + 1)$. Note that since f is unbounded, f is not Riemann integrable over $[0,1]$, although the improper integral $\int_0^1 f(x)\, dx$ does exist (see Sec. 6.C), and is easily seen to be equal to $1/(a + 1)$. What happens in this example if $a \le -1$?

(11.G.4) *Theorem* Suppose that $\{f_k\}$ is a sequence of summable functions on an interval J and that there is a summable function $f: J \to \mathbf{R}_\infty$ such that for each k, $f_k \le f$ a.e. Then

$$h = \sup (f_1, f_2, \ldots)$$

is summable over J.

Proof. By (11.E.18) the supremum of two summable function is summable. Hence, it follows inductively that for each $k = 1, 2, \ldots,$

$$h_k = \sup (f_1, \ldots, f_k)$$

is summable over J. Furthermore, the sequence $\{h_k\}$ is an increasing sequence of summable functions that converges pointwise to h. Since for each k,

$$\int_J h_k \le \int_J f$$

the result follows from the monotone convergence theorem.

The reader should have no trouble establishing the following corollary.

(11.G.5) *Corollary* If $\{f_k\}$ is a sequence of summable functions on an interval J and if there is a summable function $f: J \to \mathbf{R}_\infty$ such that for each k, $f \le f_k$ a.e., then

$$h = \inf (f_1, f_2, \ldots)$$

is summable over J.

We are now in a position to establish the very important classical Lebesgue dominated convergence theorem. This theorem is especially useful in demonstrating the integrability of the limit of a sequence of integrable functions in the case where it is difficult (or impossible) to establish uniform convergence.

(11.G.6) *Theorem* (*Lebesgue Dominated Convergence Theorem*) Suppose that $\{f_k\}$ is a sequence of summable functions on an interval J that converges a.e. to a function $f: J \to \mathbf{R}_\infty$. Suppose further that there is a summable function $g: J \to \mathbf{R}_\infty$ such that $|f_k| \le g$ a.e. for every k; i.e., the sequence $\{f_k\}$ is *dominated* by the function g. Then

(i) f is summable.

(ii) $\int_J f = \lim_{k\to\infty} \int_J f_k.$

Proof. For each positive integer k, let $\beta_k = \sup(f_k, f_{k+1}, \ldots)$. Then by (11.G.4) each function β_k is summable. Note that since $-g \le f_k$ a.e. for each k, we have $\int_J -g \le \int_J \beta_k$, and since the sequence $\{\beta_k\}$ is a decreasing sequence of summable functions converging to f, it follows from (11.G.2) that f is summable and

$$\lim_{k\to\infty} \int_J \beta_k = \int_J f$$

Similarly, if for each positive integer k, we set $\alpha_k = \inf(f_k, f_{k+1}, \ldots)$, then the sequence $\{\alpha_k\}$ is an increasing sequence of summable functions that converges to f, and therefore, it follows from (11.G.1) that

$$\lim_{k\to\infty} \int_J \alpha_k = \int_J f$$

Finally, since for each k,

$$\alpha_k \le f_k \le \beta_k$$

we have

$$\lim_{k\to\infty} \int_J f_k = \int_J f$$

(11.G.7) Example In this example we use the Lebesgue dominated convergence theorem to compute $\int_0^\infty e^{-x^2} dx$. This integral occurs frequently in the theory of probability. First observe that since $\int_0^1 e^{-x^2} dx < \infty$ and $\int_1^\infty e^{-x^2} dx \le \int_1^\infty x e^{-x^2} = \frac{1}{2}$, we have $\int_0^\infty e^{-x^2} dx < \infty$. For each positive integer n, define functions $f_n : [0,\infty) \to \mathbf{R}^1$ and $g_n : [0,\infty) \to \mathbf{R}^1$ by

$$f_n(x) = \begin{cases} \left(1 - \dfrac{x^2}{n}\right)^n & \text{if } 0 \le x \le \sqrt{n} \\ 0 & \text{if } x > \sqrt{n} \end{cases}$$

and

$$g_n(x) = f_n(x)\left(1 - \frac{x^2}{n}\right)^{1/2}$$

In problem G.1 the reader is asked to show that

(a) For each $x \ge 0$ and each positive integer n, $f_n(x) \le e^{-x^2}$.

(b) For each $x \ge 0$, $\lim_{n\to\infty} f_n(x) = e^{-x^2}$.

(c) For each $x \ge 0$ and each positive integer n, $g_n(x) \le e^{-x^2}$.

(d) For each $x \ge 0$, $\lim_{n\to\infty} g_n(x) = e^{-x^2}$.

Moreover, an appropriate substitution and an inductive argument (see problem

G.5) show that

$$\int_{[0,\infty)} f_n = \int_0^\infty f_n(x)\,dx = \sqrt{n}\int_0^{\pi/2} \cos^{2n+1}\theta\,d\theta$$

$$= \sqrt{n}\,\frac{(2n)(2n-2)\cdots 2}{(2n+1)(2n-1)\cdots 3} \tag{2}$$

$$\int_{[0,\infty)} g_n = \int_0^\infty g_n(x)\,dx = \sqrt{n}\int_0^{\pi/2} \cos^{2n+2}\theta\,d\theta$$

$$= \sqrt{n}\,\frac{(2n+1)(2n-1)\cdots 1}{(2n+2)(2n)\cdots 2}\frac{\pi}{2} \tag{3}$$

Now note that the sequences $\{f_n\}$ and $\{g_n\}$ are dominated by the limit function e^{-x^2} and, therefore, by the Lebesgue dominated convergence theorem, we have

$$\lim_{n\to\infty}\int_0^\infty f_n(x)\,dx = \int_0^\infty e^{-x^2}\,dx \tag{4}$$

and

$$\lim_{n\to\infty}\int_0^\infty g_n(x)\,dx = \int_0^\infty e^{-x^2}\,dx \tag{5}$$

Since it follows from (2), (3), (4), and (5) that

$$\left(\int_0^\infty e^{-x^2}\right)^2 = \left[\lim_{n\to\infty}\int_0^\infty f_n(x)\,dx\right]\left[\lim_{n\to\infty}\int_0^\infty g_n(x)\,dx\right]$$

$$= \lim_{n\to\infty}\left[\left(\int_0^\infty f_n(x)\,dx\right)\left(\int_0^\infty g_n(x)\,dx\right)\right]$$

$$= \frac{\pi}{4}$$

we have

$$\int_0^\infty e^{-x^2}\,dx = \frac{\sqrt{\pi}}{2}$$

A number of additional convergence results follow easily from the Lebesgue dominated convergence theorem.

(11.G.8)   Theorem (Bounded Convergence Theorem)   Suppose that $\{f_k\}$ is a sequence of summable functions on a finite interval J that converges a.e. to a function $f: J \to \mathbf{R}_\infty$. Suppose further that there is a positive real number M such that for each k, $|f_k| \le M$ a.e. Then f is summable over J and

$$\lim_{k\to\infty}\int_J f_k = \int_J f$$

Proof. This is an immediate consequence of (11.G.6). Why must J be a finite interval?

(11.G.9) *Theorem* Suppose that $\sum_{k=1}^{\infty} f_k$ is an infinite series of summable functions on an interval J. Suppose further that the partial sums of this series are dominated by a summable function $g: J \to \mathbf{R}_{\infty}$. Then $\sum_{k=1}^{\infty} f_k$ is summable and

$$\int_J \left(\sum_{k=1}^{\infty} f_k \right) = \sum_{k=1}^{\infty} \left(\int_J f_k \right)$$

Proof. This is an immediate consequence of (11.G.6). (Note that if $s_k = f_1 + \cdots + f_k$, then $\int_J s_k = \int_J f_1 + \cdots + \int_J f_k$.)

(11.G.10) *Theorem* Suppose that $\sum_{k=1}^{\infty} f_k$ is a series of summable functions on an interval J. Then either of the two (equivalent) conditions

  (i) $\sum_{k=1}^{\infty} \left(\int_J |f_k| \right)$ is finite.
  (ii) $\sum_{k=1}^{\infty} |f_k|$ converges a.e. to a summable function g on J.

implies that $\sum_{k=1}^{\infty} f_k$ is summable on J and

$$\int_J \left(\sum_{k=1}^{\infty} f_k \right) = \sum_{k=1}^{\infty} \left(\int_J f_k \right)$$

Proof. That (i) implies (ii) is a consequence of the monotone convergence theorem, and that (ii) implies (i) follows from the Lebesgue dominated convergence theorem.

Condition (ii) implies that the partial sums of $\sum_{k=1}^{\infty} f_k$ are dominated by the summable function g, and therefore, the result follows from (11.G.9).

(11.G.11) *Corollary* If f is a nonnegative summable function on an interval J and $\int_J f = 0$, then $f(x) = 0$ a.e. on J.

Proof. Set $f_k = f$ for $k = 1, 2, \ldots$ and observe that the functions f_k satisfy condition (i) of (11.G.10). Consequently, we have that $g = \sum_{k=1}^{\infty} f_k$ is summable over J. It can be shown that if g is any summable function over an interval J, then g is finite a.e. on J. [The proof of this result is surprisingly difficult and is omitted here—see Halmos (1974).] Since $f(x) \neq 0$ implies that $g(x) = \infty$, it follows that f must be equal to zero a.e. on J.

As another application of the Lebesgue dominated convergence theorem, we show that the Lebesgue integral is indeed an extension of the Riemann integral. Specifically, we show that if a function is Riemann integrable, then it is also integrable in the sense of (11.E.7), and moreover, the corresponding values of the two integrals coincide.

(11.G.12) *Theorem* Suppose that $f: [a,b] \to \mathbf{R}^1$ is Riemann integrable. Then f is summable over $[a,b]$, and furthermore,

$$\int_a^b f(x)\, dx = \int_a^b f$$

(In the statement and proof of this theorem we denote the Lebesgue integral $\int_{[a,b]} g$ by $\int_a^b g$.)

Proof. Let $\{P_k = \{x_0^{(k)}, x_1^{(k)}, \ldots, x_{n_k}^{(k)}\}\}$ be any sequence of partitions of $[a,b]$ such that $\lim_{k \to \infty} |P_k| = 0$. For each positive integer k, define a step function $\phi_k: [a,b] \to \mathbf{R}^1$ by

$$\phi_k(x) = \begin{cases} f(x_{i-1}^{(k)}) & \text{if } x \in [x_{i-1}^{(k)}, x_i^{(k)}),\ i = 1, 2, \ldots, n_k - 1 \\ f(x_{n_k-1}^{(k)}) & \text{if } x \in [x_{n_k-1}^{(k)}, x_{n_k}^{(k)}] \end{cases}$$

Then we have

$$\int_a^b \phi_k = \sum_{i=1}^{n_k} f(x_{i-1}^{(k)})\, \Delta x_i = \int_a^b \phi_k(x)\, dx$$

and therefore, since f is Riemann integrable,

$$\lim_{k \to \infty} \int_a^b \phi_k = \lim_{|P| \to 0} \sum_{i=1}^{n_k} f(x_{i-1}^{(k)})\, \Delta x_i = \int_a^b f(x)\, dx \qquad (6)$$

Now note that for each point x where f is continuous, we have

$$\lim_{k \to \infty} \phi_k(x) = f(x)$$

Since f is Riemann integrable, if follows from (11.C.15) that the set of points of discontinuity of f is a set of measure 0, and consequently, the sequence of functions $\{\phi_k\}$ converges to f a.e. Furthermore, the function f is bounded (since f is Riemann integrable), and hence the sequence $\{\phi_k\}$ is dominated by a constant function on $[a,b]$. Thus, the Lebesgue dominated convergence theorem is applicable, and we have that f is summable and

$$\lim_{k \to \infty} \int_a^b \phi_k = \int_a^b f \qquad (7)$$

The result now follows from (6) and (7).

(11.G.13) *Observation* We have just shown that if f is Riemann integrable over a finite interval, then the Riemann integral coincides with the Lebesgue integral. This is not the case with improper integrals; for example, we have seen in (11.E.21) that the function defined by

$$f(x) = \frac{(-1)^{k+1}}{k} \qquad \text{if } x \in [k-1, k),\ k = 1, 2, \ldots$$

is not integrable in the Lebesgue sense although the improper Riemann integral $\int_0^\infty f(x)\,dx$ exists, i.e., it is finite.

A number of additional applications of the convergence theorems are found in Chapter 13.

H.* SOME GENERALIZATIONS

In this section we summarize a number of results that illustrate how certain standard theorems from elementary calculus (fundamental theorem of calculus, integration by parts, etc.) can be phrased in terms of the Lebesgue theory of integration.

The relationship between the (Lebesgue) indefinite integral and its derivative represents one of the most important aspects of Lebesgue's theory of the integral. A function F mapping an interval J into $\mathbf{R}^1$ is said to be an *indefinite integral* if there is a summable function f over J such that

$$F(x) = F(\alpha) + \int_\alpha^x f \qquad (1)$$

for some fixed point $\alpha \in J$. (There may be several functions f that satisfy (1).) Among Lebesgue's most brillant achievements was the establishment of the following result.

(11.H.1) **Theorem** If $F: J \to \mathbf{R}^1$ is an indefinite integral, and if $F(x) = F(\alpha) + \int_\alpha^x f$, then the derivative of F exists and is equal to f a.e. on J.

(Here and in the remainder of this section we use the symbol $\int_\alpha^x f$ to denote the Lebesgue integral $\int_{[\alpha,x]} f$.)

The reader should observe how closely this result parallels the fundamental theorem of elementary calculus: If f is a continuous function on $[a,b]$ then the indefinite integral

$$F(x) = F(a) + \int_a^x f(t)\,dt$$

has a derivative everywhere on (a,b) and this derivative is equal to f.

As the reader is well aware, a primary use of the fundamental theorem of calculus is in the computation of $\int_a^b f(x)\,dx$; a function F is found with the property that $F' = f$ on (a,b), and then the equation

$$\int_a^b f(x)\,dx = F(b) - F(a)$$

is applied. This certainly is a valid procedure if f is continuous on $[a,b]$.

However, some care must be exercised in extending this result to the more general theory. For instance, there are examples of strictly increasing continuous functions F with the property that $F' = 0$ a.e. on an interval $[a,b]$; for such a function F, the derivative F' is summable, yet for *each* $x \in [a,b]$, $F(x) - F(a)$ fails to be equal to $\int_a^x F'$. Nevertheless, with the imposition of additional constraints on F, a "fundamental theorem of calculus" can be formulated in the Lebesgue context.

If F is an indefinite integral on $[a,b]$ and if $F' = f$ a.e. on $[a,b]$, then $F(b) - F(a) = \int_a^x f$.

This result is an immediate consequence of the fact that if F is an indefinite integral on $[a,b]$, then there is a summable function ϕ such that $F(x) - F(a) = \int_a^x \phi$ for all x belonging to $[a,b]$. It follows from (11.H.1) that $F' = \phi$ a.e. on (a,b) and this in turn implies that $f = \phi$ a.e. on (a,b). Thus, f is summable and

$$ F(b) - F(a) = \int_a^b \phi = \int_a^b f. $$

The preceding result indicates the importance of being able to find adequate criteria for determining whether or not a given function can be expressed, at least theoretically, as an indefinite integral over an interval J. One such criterion is based on the notion of absolute continuity.

(11.H.2) *Definition* Suppose that J is an interval and that $F: J \to \mathbf{R}^1$. Then F is said to be *absolutely continuous* on J if corresponding to each $\varepsilon > 0$, there is a $\delta > 0$ such that

$$ \sum_{k=1}^{n} |F(\beta_k) - F(\alpha_k)| < \varepsilon $$

for each finite collection $(\alpha_1, \beta_1), \ldots, (\alpha_n, \beta_n)$ of disjoint subintervals of J with total length less than δ.

It should be clear that absolutely continuous functions are uniformly continuous. The relationship between continuity and absolute continuity is explored in the problem section. A proof of the following theorem may be found in Korevaar (1968).

(11.H.3) *Theorem* Suppose that J is an interval and that $F: J \to \mathbf{R}^1$. Then F is an indefinite integral if and only if F is absolutely continuous on J.

Integration by parts is valid in the Lebesgue theory of integration. If f and g are summable functions on an interval $[a,b]$, and if F and G denote corresponding indefinite integrals, then Fg and fG are summable functions

and

$$\int_a^b Fg = FG \Big|_a^b - \int_a^b fG$$

where $FG |_a^b = F(b)G(b) - F(a)G(a)$.

The change of variables technique can also be described in the Lebesgue context. Suppose that $U(t)$ is an increasing absolutely continuous real-valued function on an interval $\alpha \le t \le \beta$ and that f is any summable function on the interval $a = U(\alpha) < x < U(\beta) = b$. Then the function $f(U(t))U'(t)$ is summable over (α, β) and

$$\int_a^b f = \int_\alpha^\beta f(U)U'$$

The mean value theorems for integrals carry over to the Lebesgue theory as well. For example, if $f: [a,b] \to \mathbf{R}^1$ is continuous on a bounded closed interval $[a,b]$ and if g is a nonnegative summable function over $[a,b]$, then there is a number $c \in (a,b)$ such that

$$\int_a^b fg = f(c) \int_a^b g \qquad (2)$$

(see problem H.7). Classical results involving termwise differentiation also have natural extensions.

Suppose that $\sum_{n=1}^\infty g_n$ is a series of absolutely continuous functions on an interval $J = [a,b]$ and that $\sum_{n=1}^\infty g_n(c)$ converges for some point $c \in [a,b]$. Suppose further that either one of the two following conditions is satisfied:

1. The series $\sum_{n=1}^\infty \int_J |g_n'|$ converges.
2. The series of functions $\sum_{n=1}^\infty |g_n'|$ converges pointwise to a summable function on $[a,b]$.

Then the series $\sum_{n=1}^\infty g_n$ converges pointwise to an absolutely continuous function f on $[a,b]$ and

$$f' = \sum_{n=1}^\infty g_n'$$

a.e. on $[a,b]$. Equality holds at each point of continuity of $\sum_{n=1}^\infty g_n'(x)$.

I. INTEGRATION OF FUNCTIONS OF MORE THAN ONE VARIABLE: SOME FUBINI THEOREMS

Thus far we have dealt primarily with functions of one variable. In this section we extend some of our previous results to deal with the problem of integrating multivariable functions. The approach is somewhat heuristic and we omit a number of proofs.

First we consider the Riemann integrability of a function f mapping a rectangle $[a,b] \times [c,d]$ into $\mathbf{R}^1$. A partition P of such a rectangle is obtained from partitions $\{x_0, x_1, \ldots, x_n\}$ of $[a,b]$ and $\{y_0, \ldots, y_m\}$ of $[c,d]$ and consists of all rectangles of the form $R_{ij} = [x_{i-1}, x_i] \times [y_{j-1}, y_j]$, $1 \le i \le n$, $1 \le j \le m$ (Fig. 11.10). We denote the diameter of each rectangle R_{ij} by $D(R_{ij})$. The *mesh* of the partition is defined to be $\max \{D(R_{ij}) \mid 1 \le i \le n, 1 \le j \le m\}$.

The following definition is an exact analogue of (11.A.1). The area of a rectangle R is denoted by $A(R)$.

(11.I.1) *Definition* A function f mapping a rectangle R into $\mathbf{R}^1$ is *Riemann integrable* if there is a number L such that

$$\lim_{|P| \to 0} \sum_{j=1}^{m} \sum_{i=1}^{n} f(x_i^*, y_j^*) A(R_{ij}) = L \qquad (1)$$

(i.e., given $\varepsilon > 0$, there is a $\delta > 0$ such that if $P = \{R_{ij} \mid 1 \le i \le n, 1 \le j \le m\}$ is a partition of R with mesh less that δ, and if for each pair (i,j), (x_i^*, y_i^*) is an arbitrary point in R_{ij}, then

$$\left| \sum_{j=1}^{m} \sum_{i=1}^{n} f(x_i^*, y_j^*) A(R_{ij}) - L \right| < \varepsilon$$

If the limit (1) exists, we write $\int_R f = L$.

Similar definitions for higher dimensional "rectangles" (cubes, four-dimensional rectangles, etc.) can be given.

Under what conditions does the Riemann integral exist? One criterion is in terms of subsets of $\mathbf{R}^2$ of measure 0.

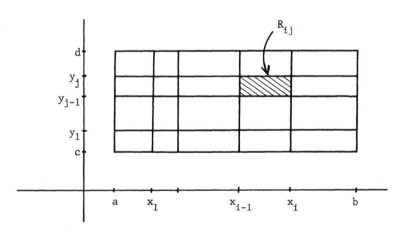

Figure 11.10

(11.I.2) *Definition* A bounded subset B of $\mathbf{R}^2$ is said to be of *measure* 0 if for each $\varepsilon > 0$, there is a countable family of rectangles $S_1, S_2, \ldots$ such that $B \subset \bigcup_{i=1}^{\infty} S_i$ and $\sum_{i=1}^{\infty} A(S_i) < \varepsilon$.

A proof of the next theorem [the higher dimensional analogue of (11.C. 15)] can be found in Royden (1968).

(11.I.3) *Theorem* If R is a rectangle and if $f: R \to \mathbf{R}^1$ is bounded, then $\int_J f$ exists if and only if the set of points of discontinuity of f is of measure 0.

In particular, continuous functions are Riemann integrable.

As is usually the case with definitions the definition of the Riemann integral is of little aid in its computation. We show next that under certain conditions, if f is integrable over a rectangle R, then $\int_R f$ can be computed using successive integrations of functions of one variable. We need the following lemma.

(11.I.4) *Lemma* If f is continuous on the rectangle $R = [a,b] \times [c,d]$, and if $g: [c,d] \to \mathbf{R}^1$ is defined by

$$g(y) = \int_a^b f(x, y)\, dx$$

then g is continuous on $[c,d]$.

Proof. Since f is uniformly continuous on the compact set R, we have that for each $\varepsilon > 0$ there is a $\delta > 0$ such that if $x \in [a,b]$, $y \in [c,d]$, $z \in [c,d]$ and $|y - z| < \delta$, then

$$|f(x, z) - f(x, y)| < \frac{\varepsilon}{b - a}$$

Hence, if $|y - z| < \delta$, we have

$$|g(z) - g(y)| = \left| \int_a^b (f(x, z) - f(x, y))\, dx \right|$$

$$\leq \int_a^b |f(x, z) - f(x, y)|\, dx$$

$$< \left(\frac{\varepsilon}{b - a} \right)(b - a) = \varepsilon$$

which shows that g is uniformly continuous.

A similar result holds for the function $h: [a,b] \to \mathbf{R}^1$ defined by $h(x) = \int_c^d f(x,y)\, dy$.

(11.I.5) *Theorem* If f is continuous on the rectangle $R = [a,b] \times [c,d]$, then

$$\int_a^b \left(\int_c^d f(x, y) \, dy \right) dx = \int_R f = \int_c^d \left(\int_a^b f(x, y) \, dx \right) dy$$

Proof. Let $P = \{R_{ij} \mid 1 \le i \le n, 1 \le j \le m\}$ be a partition of R determined by the partitions $P_x = \{x_0, x_1, \ldots, x_n\}$ and $P_y = \{y_0, y_1, \ldots, y_m\}$ of $[a,b]$ and $[c,d]$, respectively. If $m_{ij} = \inf \{f(x,y) \mid (x,y) \in R_{ij}\}$ and $M_{ij} = \sup \{f(x,y) \mid (x,y) \in R_{ij}\}$, then we have

$$m_{ij}(x_i - x_{i-1}) \le \int_{x_{i-1}}^{x_i} f(x, y) \, dx \le M_{ij}(x_i - x_{i-1}) \qquad (2)$$

for each $y \in (y_{j-1}, y_j)$. By (11.I.4), the integral appearing in (2) is continuous with respect to y, and therefore it follows that

$$m_{ij}(x_i - x_{i-1})(y_j - y_{j-1}) \le \int_{y_{j-1}}^{y_j} \left(\int_{x_{i-1}}^{x_i} f(x, y) \, dx \right) dy$$
$$\le M_{ij}(x_i - x_{i-1})(y_j - y_{j-1})$$

Summing over all the subrectangles, we find that

$$\sum_{j=1}^m \sum_{i=1}^n m_{ij}(x_i - x_{i-1})(y_j - y_{j-1}) \le \int_c^d \left(\int_a^b f(x, y) \, dx \right) dy$$
$$\le \sum_{j=1}^m \sum_{i=1}^n M_{ij}(x_i - x_{i-1})(y_j - y_{j-1})$$

$$(3)$$

As the mesh of the partitions of R tends to 0, the left-hand and right-hand sides of (3) converge to $\int_R f$, and consequently we have

$$\int_R f = \int_c^d \left(\int_a^b f(x, y) \, dx \right) dy$$

A similar argument shows that $\int_a^b (\int_c^d f(x,y) \, dy) \, dx = \int_R f$, and this concludes the proof.

In view of the result just established it is natural (and customary) to write

$$\int_R f = \int_a^b \int_c^d f(x, y) \, dy \, dx = \int_c^d \int_a^b f(x, y) \, dx \, dy$$

Theorem (11.I.5) can be considerably generalized as follows. The proof of the next result is quite sophisticated and is omitted.

(11.I.6)   *Theorem (Fubini's Theorem)* Suppose that $R = [a,b] \times [c,d]$ and that $f: R \to \mathbf{R}^1$ is a summable function. For each $x \in [a,b]$, let $f^x: [c,d] \to \mathbf{R}^1$ be defined by

$$f^x(y) = f(x, y)$$

and for each $y \in [c,d]$, let $f^y \colon [a,b] \to \mathbf{R}^1$ be defined by

$$f^y(x) = f(x, y)$$

Then

(i) f^x is summable for almost every $x \in [a,b]$ (i.e., f^x is summable except on a set of measure 0).

(ii) f^y is summable for almost every $y \in [c,d]$.

(iii) The functions

$$\phi \colon [a, b] \to \mathbf{R}^1$$

$$\psi \colon [c, d] \to \mathbf{R}^1$$

defined by

$$\phi(x) = \begin{cases} \displaystyle\int\!\!\!\int_{[c,d]} f^x & \text{if } f^x \text{ is summable} \\ 0 & \text{otherwise} \end{cases}$$

$$\psi(x) = \begin{cases} \displaystyle\int\!\!\!\int_{[a,b]} f^y & \text{if } f^y \text{ is summable} \\ 0 & \text{otherwise} \end{cases}$$

are summable.

(iv) $\int_{[a,b]} \phi = \int_R f = \int_{[c,d]} \psi$

(11.I.7) *Exercise* Show that (11.I.5) is a special case of Fubini's theorem.

The definition of the Riemann integral can be extended to arbitrary bounded subsets of $\mathbf{R}^2$ as follows.

(11.I.8) *Definition* Suppose that S is a bounded subset of $\mathbf{R}^2$ and that $f \colon S \to \mathbf{R}^1$ is a bounded function. Let R be an arbitrary rectangle that contains S and define $f_S \colon R \to \mathbf{R}^1$ by

$$f_S(x, y) = \begin{cases} f(x, y) & \text{if } (x, y) \in S \\ 0 & \text{if } (x, y) \notin S \end{cases}$$

Then the Riemann integral of f over S, $\int_S f$, is defined by

$$\int_S f = \int_R f$$

provided that $\int_R f$ exists.

Again the question arises of determining conditions to ensure the integrability of f. One important criterion can be given in terms of sets of zero content.

(11.I.9) *Definition* A bounded set $S \subset \mathbf{R}^2$ is said to be of *content* 0 if given $\varepsilon > 0$, there is a *finite* family of rectangles $R_1, R_2, \ldots, R_n$ such that $S \subset \bigcup_{i=1}^{n} R_i$ and $\sum_{i=1}^{n} A(R_i) < \varepsilon$.

The following result is easily established (see problem I.8).

(11.I.10) *Theorem* If S is a bounded subset of $\mathbf{R}^2$ whose boundary has zero content and if $f: S \to \mathbf{R}^1$ is continuous except on a set of zero content, then $\int_S f$ exists.

As the reader is well aware, there are sets S other than rectangles over which the integral $\int_S f$ can be computed as an interated integral. One such possibility is considered in the next theorem.

(11.I.11) *Theorem* Suppose that u_1 and u_2 are continuous functions on an interval $[a,b]$ and that for each $x \in [a,b]$, $u_1(x) \leq u_2(x)$. Let $S = \{(x,y) \mid u_1(x) \leq y \leq u_2(x), a \leq x \leq b\}$. If $f: S \to \mathbf{R}^1$ is summable, then

$$\int_S f = \int_a^b \left(\int_{u_1(x)}^{u_2(x)} f(x, y)\, dy \right) dx$$

Proof. Let $R = [a,b] \times [c,d]$ be a rectangle containing S and as before define $f_S: R \to \mathbf{R}^1$ by

$$f_S(x, y) = \begin{cases} f(x, y) & \text{if } (x, y) \in S \\ 0 & \text{if } (x, y) \notin S \end{cases}$$

Then, by (11.I.5),

$$\int_R f_S = \int_a^b \left(\int_c^d f_S(x, y)\, dy \right) dx$$

But clearly

$$\int_c^d f_S(x, y)\, dy = \int_{u_1(x)}^{u_2(x)} f(x, y)\, dy$$

and this concludes the proof.

(11.I.12) *Exercise* Formulate and prove a similar result for summable functions defined on sets of the form

$$T = \{(x, y) \mid v_1(y) \leq x \leq v_2(y),\, c \leq y \leq d\}$$

(11.I.13) *Example* If S is the region bounded by the curves $y = 3 - 2x^2$ and $y = x^2$ (Fig. 11.11) and if $f: S \to \mathbf{R}^1$ is defined by $f(x,y) = x^3 y$, then

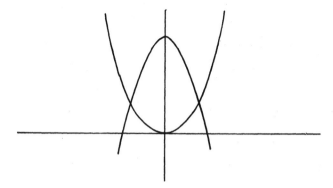

Figure 11.11

$$\int_S f = \int_{-1}^{1} \int_{x^2}^{3-2x^2} x^3 y \, dy \, dx = \int_{-1}^{1} \left(x^3 \left. \frac{y^2}{2} \right|_{x^2}^{3-2x^2} \right) dx$$

$$= \int_{-1}^{1} \left(\frac{9}{2} x^3 - 6x^5 + \frac{3}{2} x^7 \right) dx$$

We mention that a Lebesgue theory of integration for subsets of $\mathbf{R}^2$ (or $\mathbf{R}^n$) can be developed along the lines discussed in the preceding section. A step function s can be defined on $R = [a,b] \times [c,d]$ by forming a partition $\{R_{ij} \mid 1 \le i \le n, 1 \le j \le m\}$ as before and then defining s to be equal to a constant q_{ij} on the interior of each rectangle R_{ij} and to have a finite range on R. If s is such a function, then the integral of s, $\int_R s$, is defined by

$$\int_R s = \sum_{j=1}^{m} \sum_{i=1}^{n} q_{ij} m(R_{ij})$$

where the measure $m(R_{ij})$ is simply the area $A(R_{ij})$ of R_{ij}. If R is an unbounded rectangle (an infinite strip), then a step function s may be defined on R by setting $s = 0$ outside of a bounded subrectangle $\hat{R}$ of R and by defining the step function s on R as we have just indicated. The integral of s over R is given by

$$\int_R s = \int_{\hat{R}} s$$

The concept of a mean fundamental sequence carries over to sequences of step functions of this nature, and an obvious modification of Definition (11.E.7) can be used to define $\int_R f$, where f maps a rectangle R into $\mathbf{R}_\infty$, and is the limit of a mean fundamental sequence of step functions.

Most of the previously derived results (the approximation theorems, the Lebesgue dominated convergence theorem, etc.) carry over directly to the higher dimensional setting; there is essentially no change in the proofs except

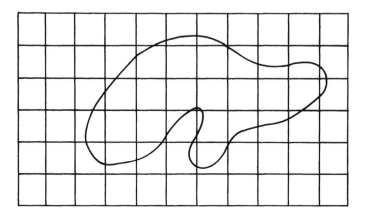

Figure 11.12

to replace words such as "interval" by "rectangle," "length" by "area," etc.

For nonrectangular sets bounded by fairly "nice" curves, grids can also be employed, with the convention that a step function s is defined to be 0 on parts of the grid not lying entirely in the given set (Fig. 11.12). Clearly these ideas can be extended to higher dimensions as well.

There are, however, sets so irregular that the above concepts are not applicable [it may be the case that we wish to integrate over a set D in the plane whose interior is empty, so that for each grid G, the set D would fail to contain any rectangle in G. For more details in this regard the reader is encouraged to consult Royden (1968)].

We conclude this section with the statement of two additional Fubini theorems. These results are used in certain proofs of theorems found in Chapter 13.

(11.I.14) Theorem (Fubini's Reduction Theorem) If $f: \mathbf{R}^2 \rightarrow \mathbf{R}^1$ is a summable function, then $g(x) = \int_{-\infty}^{\infty} f(x,y)\, dy$ exists for almost all x and is a summable function of x over $(-\infty,\infty)$, and $h(y) = \int_{-\infty}^{\infty} f(x,y)\, dx$ exists for almost all y and is a summable function of y over $(-\infty,\infty)$. Furthermore,

$$\int_{-\infty}^{\infty} \left(\int_{-\infty}^{\infty} f(x, y)\, dy \right) dx = \int_{\mathbf{R}^2} f = \int_{-\infty}^{\infty} \left(\int_{-\infty}^{\infty} f(x, y)\, dx \right) dy$$

[Here and below, the integral $\int_{-\infty}^{\infty} f$ denotes the Lebesgue integral of f over $(-\infty,\infty)$.]

(11.I.15) Definition A function $f: J \rightarrow \mathbf{R}_\infty$ is said to be *measurable* over the interval J if there is a sequence of step functions $\{s_n\}$ on J that converges a.e. to f.

(11.I.16) **Theorem (Absolute Convergence Inversion Theorem)** If f is measurable over $\mathbf{R}^2$ and if either interated integral is absolutely convergent, i.e., if either of the interated integrals

$$\int_{-\infty}^{\infty} \left(\int_{-\infty}^{\infty} |f(x, y)| \, dy \right) dx \quad \text{or} \quad \int_{-\infty}^{\infty} \left(\int_{-\infty}^{\infty} |f(x, y)| \, dx \right) dy$$

exists, then f is summable over $\mathbf{R}^2$.

(11.I.17) **Corollary** If a function f satisfies the hypothesis of (11.I.16), then

$$\int_{-\infty}^{\infty} \left(\int_{-\infty}^{\infty} f(x, y) \, dy \right) dx = \int_{-\infty}^{\infty} \left(\int_{-\infty}^{\infty} f(x, y) \, dx \right) dy$$

PROBLEMS

Section A

1. Show that if $f: [a,b] \to \mathbf{R}^1$ is a Riemann integrable function and if c is a constant, then cf is Riemann integrable, and $\int_a^b cf(x) \, dx = c \int_a^b f(x) \, dx$.
2. Show that if $f: [a,b] \to \mathbf{R}^1$ and $g: [a,b] \to \mathbf{R}^1$ are Riemann integrable functions, then $f + g$ is Riemann integrable and $\int_a^b (f + g)(x) \, dx = \int_a^b f(x) \, dx + \int_a^b g(x) \, dx$.
3. Use either of the Riemann criteria to show that if $f: [a,b] \to \mathbf{R}^1$ is a continuous function, then f is Riemann integrable. [*Hint*: Note that f is uniformly continuous.]
4. Use either of the Riemann criteria to show that any bounded function $f: [a,b] \to \mathbf{R}^1$ with at most a finite number of discontinuities is Riemann integrable.
5. Show that if $f: [a,b] \to \mathbf{R}^1$ is Riemann integrable, then so is $|f|$.
6. Show that if $f: [a,b] \to \mathbf{R}^1$ is Riemann integrable and if $g: [a,b] \to \mathbf{R}^1$ is equal to f except at a finite number of points, then g is Riemann integrable and $\int_a^b f(x) \, dx = \int_a^b g(x) \, dx$. Does this hold true if g is equal to f except at a countable number of points?
7. Suppose that $f: [a,b] \to \mathbf{R}^1$ is Riemann integrable. For each partition $P = \{x_0, x_1, \ldots, x_n\}$ of $[a,b]$, let

$$m_i^P = \inf \{f(x) \mid x \in [x_{i-1}, x_i]\} \qquad i = 1, 2, \ldots, n$$
$$M_i^P = \sup \{f(x) \mid x \in [x_{i-1}, x_i]\} \qquad i = 1, 2, \ldots, n$$

Show that

$$\lim_{|P|\to 0} \sum_{i=1}^{n} M_i \, \Delta x_i = \int_a^b f(x) \, dx = \lim_{|P|\to 0} \sum_{i=1}^{n} m_i \, \Delta x_i$$

8. Suppose that $f: [0,1] \to \mathbf{R}^1$ is defined by

$$f(x) = \begin{cases} 0 & \text{if } x \text{ is irrational or } x = 0 \\ \dfrac{1}{n} & \text{if } x = \dfrac{m}{n} \text{ (expressed in lowest terms)} \end{cases}$$

Show that f is Riemann integrable even though f has an infinite number of discontinuities.

9. The following example was given by Darboux in 1875 to illustrate the problems involved in term-by-term integration. For each positive integer k, define $u_k: [0,\infty) \to \mathbf{R}^1$ by

$$u_k(x) = -2k^2 x e^{-k^2 x^2} + 2(k+1)^2 x e^{-(k+1)x^2}$$

and define $u: [0,\infty) \to \mathbf{R}^1$ by

$$u(x) = -2x e^{-x^2}$$

For each n, let $s_n = \sum_{i=1}^{n} u_i$. Show that

(a) $s_{n-1}(x) = -2x e^{-x^2} + 2n^2 x e^{-n^2 x^2}$

(b) $\displaystyle\sum_{n=1}^{\infty} u_n(x) = u(x)$

(c) $\displaystyle\int_0^x u(t) \, dt = e^{-x^2} - 1$

(d) $\displaystyle\sum_{n=1}^{\infty} \int_0^x u_n(t) \, dt = e^{-x^2}$,

and hence, $\displaystyle\sum_{n=1}^{\infty} \int_0^x u_n(t) \, dt \neq \int_0^x \left(\sum_{n=1}^{\infty} u_n(t) \right) dt$

10. This problem is in reference to the Riemann example of an integrable function with a dense set of discontinuities that was introduced in Sec. 11.A.

(a) Let $s_n(x) = \sum_{i=1}^{n} \phi(kx)/k^2$, and show that if $x \in \mathbf{R}^1$ is such that $\phi(kx) \neq 0$ for each k, then $s_n(x - 0) - s_n(x) = 0$, $s_n(x + 0) - s_n(x) = 0$, and therefore, f is continuous at x.

(b) Let $A = \{x \in \mathbf{R}^1 \mid x \text{ is an integer or } \phi(kx) \neq 0 \text{ for each } k\}$. Show that if $\hat{x} \notin A$ and if $\phi(k\hat{x}) = 0$, then $\hat{x} = q/2k$ for some odd integer q.

(c) Express $q/2k$ as $m/2n$, where m and n are relatively prime. Show that the smallest value of k for which $\phi(k\hat{x}) = 0$ is n, the next smallest value is $3n$, the next smallest is $5n$, etc.

(d) From (c) deduce that if $\hat{x} \notin A$, then for each positive integer N,

$$s_N(\hat{x} - 0) - s_N(\hat{x}) = \frac{1}{2}\left(\frac{1}{n^2} + \frac{1}{(3n)^2} + \cdots + \frac{1}{(pn)^2}\right)$$

where pn is the largest odd integer less than or equal to N.

(e) Show that if $\hat{x} \notin A$, then

$$f(\hat{x} - 0) - f(\hat{x}) = \frac{1}{2n^2}\left(1 + \frac{1}{3^2} + \frac{1}{5^2} + \cdots\right)$$

and hence, by problem E.6(a), Chapter 10,

$$f(\hat{x} - 0) - f(\hat{x}) = \frac{\pi^2}{16n^2}.$$

(f) Show in a similar manner that if $\hat{x} \notin A$, then

$$f(\hat{x} + 0) - f(\hat{x}) = \frac{-\pi^2}{16n^2}$$

11. Construct a sequence of Riemann integrable functions $\{f_n\}$ such that
   (1) $f_n: [0, 1] \to \mathbf{R}^1$ for each $n \in \mathbf{Z}^+$
   (2) $\lim_{n \to \infty} f_n(x) = 0$ for each $x \in [0, 1]$

   (3) $\lim_{n \to \infty} \int_0^1 f_n(x)\, dx = \infty$

Sections B-C

1. (a) Show that if m^* is an outer measure, $A \subset \mathbf{R}^1$, and $x \in \mathbf{R}^1$, then $m^*(x + A) = m^*(A)$.
   (b) Show that if $A \subset \mathbf{R}^1$ is a measurable set and if $x \in \mathbf{R}^1$, then $x + A$ is measurable and $m(x + A) = m(A)$.
2. Show that if a set A has outer measure 0 and $B \subset \mathbf{R}^1$ is an arbitrary set, then $m^*(A \cup B) = m^*(B)$.
3. Show that if $A_1, A_2, \ldots$ are sets of measure 0 contained in a bounded interval J, then $A = \bigcup_{n=1}^{\infty} A_n$ is a measurable set and $m(A) = 0$.
4. Establish equation (12) in the proof of (11.C.11).
5. Show that a function $f: [a,b] \to \mathbf{R}^1$ is continuous at a point $x \in [a,b]$ if and only if $\operatorname{osc}(f;x) = 0$.
6. Show that if $f: [a,b] \to \mathbf{R}^1$ is Riemann integrable, then the set of discontinuities of f forms a set of measure 0. [*Hint:* For each $n \in \mathbf{Z}^+$, let $A_n = \{x \in [a,b] \mid \operatorname{osc}(f;x) \geq 1/n\}$ and note that each A_n is compact. Show that for each $n \in \mathbf{Z}^+$, $m(A_n) = 0$, and then use problems B-C.3 and B-C.5.]

7. Show that the Cantor set is an uncountable set of measure 0. Is there a
    closed (in $\mathbf{R}^1$) uncountable set of irrational numbers with measure 0?
8. Show that if $f: [a,b] \to \mathbf{R}^1$ and $g: [a,b] \to \mathbf{R}^1$ are Riemann integrable
    functions, then the product fg is Riemann integrable.
9. (a) Use (11.C.7) to show that bounded open sets are measurable.
    (b) Use part (a) and (11.C.5) to show that bounded closed sets are
         measurable. [*Hint*: Show that if A is a closed subset of a metric
         space, then there is a decreasing sequence of open sets $U_1 \supset U_2 \supset$
         . . . such that $A = \bigcap_{i=1}^{\infty} U_i$.]
The remaining problems are based on the following definitions.

Definition   If $A \subset \mathbf{R}^1$, then $c^*(A)$, *the outer Jordan content of A*, is defined
by

$$c^*(A) = \inf \left\{ \sum_{i=1}^{n} (b_i - a_i) \,|\, A \subset \bigcup_{i=1}^{n} (a_i, b_i) \right\}$$

[i.e., $c^*(A)$ is the greatest lower bound taken over all sums of lengths of *finite*
collections of open intervals covering A].

Definition   If $A \subset \mathbf{R}^1$, then $c_*(A)$, called *the inner Jordan content of A*, is
defined by

$$c_*(A) = \sup \left\{ \sum_{i=1}^{n} (b_i - a_i) \,|\, \bigcup_{i=1}^{n} [a_i, b_i] \subset A \text{ and } [a_i, b_i] \cap [a_j, b_j] = \varnothing \right.$$

$$\left. \text{whenever } i \neq j \right\}$$

Definition   A set $A \subset \mathbf{R}^1$ is said to be *Jordan measurable* if $c_*(A) = c^*(A)$
and $c(A) = c^*(A) = c_*(A)$ is called the *Jordan content of A*.
10. Find a countably infinite set of points with Jordan content equal to 0.
11. Show that is A is a countably dense subset of $[a,b]$, then $c^*(A) = b - a$
     and $c_*(A) = 0$.
12. Show that for any set $A \subset \mathbf{R}^1$, $c_*(A) \leq m_*(A) \leq m^*(A) \leq c^*(A)$.
13. Show that if A is a Jordan measurable set, then A is Lebesgue measurable and $c(A) = m(A)$.

$$c(A) = m(A).$$

Observation   The Jordan theory of content can be extended very easily to
$\mathbf{R}^n$. For instance, in passing from $\mathbf{R}^1$ to $\mathbf{R}^2$ the intervals in $\mathbf{R}^1$ are replaced by
rectangles parallel to the y-axis. The *area of a rectangle*, R, determined by
the vertical lines $x = a$, $x = b$ and the horizontal lines $y = c$, $y = d$, is given
in the obvious manner by $|R| = |b - a||d - c|$. (Note that in this definition

we do not worry about boundary points of the rectangles.) The definitions for c^*, c_*, and c carry over in the obvious fashion from $\mathbf{R}^1$ to $\mathbf{R}^2$ ("interval" is replaced by "rectangle" and "length of interval" by "area of rectangle"). It is easy to see that if $A \subset \mathbf{R}^2$, then

$$c_*(A) \leq m_*(A) \leq m^*(A) \leq c^*(A).$$

14. Suppose that $f: [a,b] \to \mathbf{R}^1$ is a nonnegative Riemann integrable function, and let $P = \{x_0, x_1, \ldots, x_n\}$ be a partition of $[a,b]$.
   (a) Define

$$\bar{s}(P) = \sum_{i=1}^{n} M_i \, \Delta x_i$$

   where $M_i = \sup \{f(x) \mid x \in [x_{i-1},x_i)\}$, $1 \leq i \leq n - 1$, and $M_n = \sup \{f(x) \mid x \in [x_{n-1},x_n]\}$. Show that the region bounded by the graph of the function

$$\bar{a}_P(x) = \begin{cases} M_i & \text{if } x \in [x_{i-1}, x_i), \, i = 1, \ldots, n - 1 \\ M_n & \text{if } x \in [x_{n-1}, x_n] \end{cases}$$

   the vertical lines $x = a$, $x = b$, and the x-axis consists of a finite number of rectangles with total area equal to $\bar{s}(P)$.
   (b) Define

$$\underline{s}(P) = \sum_{i=1}^{n} m_i \, \Delta x$$

   where $m_i = \inf \{f(x) \mid x \in [x_{i-1},x_i)\}$, $1 \leq i \leq n - 1$, and $m_n = \inf \{f(x) \mid x \in [x_{n-1},x_n]\}$. Show that the region bounded by the graph of the function

$$\underline{a}_P(x) = \begin{cases} m_i & \text{if } x \in [x_{i-1}, x_i), \, i = 1, \ldots, n - 1 \\ m_n & \text{if } x \in [x_{n-1}, x_n] \end{cases}$$

   the vertical lines $x = a$, $x = b$, and the x-axis consists of a finite number of rectangles with total area equal to $\underline{s}(P)$.
   (c) If A is the region bounded by the graph of f, the vertical lines $x = a$, $x = b$ and the x-axis, show that for each partition P of $[a,b]$,

$$\underline{s}(P) \leq c_*(A) \leq c^*(A) \leq \bar{s}(P).$$

   (d) Use the fact that f is Riemann integrable to show that as $|P| \to 0$, the difference $\bar{s}(P) - \underline{s}(P)$ also tends to 0. Use this result to show that

$$c(A) = \int_a^b f(x) \, dx$$

Observation From the discussion in Sec. 11.C, we have

$$\int_{[a,b]} f = m(A)$$

Therefore, from the previous problem it follows that if $f: [a,b] \to [0,\infty)$ is Riemann integrable, then

$$\int_a^b f(x)\, dx = \int_{[a,b]} f$$

15. Show that if $f: [a,b] \to \mathbf{R}^1$ is any Riemann integrable function, then

$$\int_a^b f(x)\, dx = \int_{[a,b]} f$$

Section D*

1. Establish properties 1 to 4 given at the beginning of Sec. 11.D.
2. Show that the sets E_i defined in this section are mutually disjoint and that $\bigcup_{i=1}^{\infty} E_i = [0,1]$.
3. Show that if A is a measurable subset of the nonmeasurable set defined in this section, then $m(A) = 0$.

Section E

1. Prove Theorem (11.E.4).
2. Is the product of two step functions a step function?
3. Show that if $s: [a,b] \to \mathbf{R}^1$ is a step function, then $\int_{[a,b]} s = \int_a^b s(x)\, dx$.
4. Find a mean fundamental sequence of step functions that converges everywhere to the Dirichlet function. [*Hint*: Examine carefully the definition of a step function.]
5. Supply the details for Example (11.E.9.b).
6. Show that in (11.E.9.c) $\int_J |s_n - s_m| = 2$ whenever $m \neq n$.
7. Is the function $f: [0,\infty) \to \mathbf{R}^1$ defined by

$$f(x) = \begin{cases} \dfrac{\sin x}{x} & \text{if } x \neq 0 \\ 0 & \text{if } x = 0 \end{cases}$$

summable? Does the improper integral $\int_0^\infty f(x)\, dx$ exist?
7.* Prove or disprove: Suppose that $f: [a,b] \to \mathbf{R}^1$ and that there is a mean fundamental sequence of step functions that converges a.e. to f. Then

there is a mean fundamental sequence of step functions that converges everywhere to f.

8. In this exercise we see that all Riemann integrable functions are summable. Suppose that $f: [a,b] \to \mathbf{R}^1$ is Riemann integrable. Note that by problem B-C.6 of this chapter, the points of discontinuity of f form a set of measure 0. Let $\{P_k = \{x_0^{(k)}, x_1^{(k)}, \ldots, x_{n_k}^{(k)}\}\}$ be a sequence of partitions of $[a,b]$ such that $\lim_{k\to\infty} |P_k| = 0$. For each partition P_k and each i, $1 \le i \le n_k$, select a point $\hat{x}_i^{(k)}$ of continuity of f in the interval $(x_{i-1}^{(k)}, x_i^{(k)})$, and then define a step function $s_k: [a,b] \to \mathbf{R}^1$ by

$$s_k(x) = \begin{cases} f(\hat{x}_i^{(k)}) & \text{if } x \in [x_{i-1}^{(k)}, x_i^{(k)}), \ i = 1, 2, \ldots, n_k - 1 \\ f(\hat{x}_{n_k}^{(k)}) & \text{if } x \in [x_{n_k-1}^{(k)}, x_{n_k}^{(k)}]. \end{cases}$$

(a) Show that $\int_{[a,b]} s_k = \sum_{i=1}^{n_k} f(\hat{x}_i^{(k)}) \, \Delta x_i$.

(b) Show that if f is continuous at x, then $\lim_{k\to\infty} s_k(x) = f(x)$.

(c) Let $E = \{x \in [a,b] \mid \lim_{k\to\infty} s_k(x) \ne f(x)\}$. Show that $m(E) = 0$.

(d) Show that $\{s_k\}$ is a mean fundamental sequence and hence that f is summable. [*Hint:* Show that f is uniformly continuous on an appropriate subset of $[a,b]$.]

Section F

1. Establish the existence of the subsequence described in the proof of (11.F.5).

2. Let $J = [a,\infty)$ and suppose that the function $f: J \to \mathbf{R}_\infty$ is nonnegative on J and summable over every finite subinterval $[a,b]$. Moreover suppose that

$$\lim_{b\to\infty} \int_a^b f = A < \infty$$

Show that f is summable over J and that $\int_J f = A$. [*Hint:* Let $\{b_k\}$ be an unbounded increasing sequence of numbers, and for each k, define

$$f_k(x) = \begin{cases} f(x) & \text{for } a \le x \le b_k \\ 0 & \text{for } x > b_k \end{cases}$$

apply (11.F.7).]

3. Show that if $s: J \to \mathbf{R}^1$ is a step function on an interval J, then $\lim_{\lambda\to 0} \int_J s(x) \cos \lambda x \, dx = 0$ and $\lim_{\lambda\to 0} \int_J s(x) \sin \lambda x \, dx = 0$ [cf. (6.B.6)].

4. Show that if $f: J \to \mathbf{R}_\infty$ is summable on J, then $f(x) \cos \lambda x$ and $f(x) \sin \lambda x$ are also summable on J. [*Hint:* Use (11.F.7).]

5. (Riemann-Lebesgue lemma) Show that if $f: J \to \mathbf{R}_\infty$ is summable on J, then

$$\lim_{\lambda \to 0} \int_J f(x) \cos \lambda x \, dx = 0$$

and

$$\lim_{\lambda \to 0} \int_J f(x) \sin \lambda x \, dx = 0$$

[*Hint*: Let $\varepsilon > 0$ be given and show that $|\int_J f(x) \cos \lambda x \, dx| < \varepsilon$ and $|\int_J f(x) \sin \lambda x \, dx| < \varepsilon$ by choosing suitable step functions as in (11.F.2); make use of the previous two problems.]

6. If $s: [a,b] \to \mathbf{R}^1$ is a step function on $[a,b]$, then $S: [a,b] \to \mathbf{R}^1$ defined by

$$S(x) = C + \int_a^x s(t) \, dt \qquad a \le x \le b$$

where C is a constant, is called an *indefinite integral* on $[a,b]$. Show that S is uniformly continuous on $[a,b]$.

7. Show that if $f: [a,b] \to \mathbf{R}_\infty$ is a summable function on $[a,b]$, then $F(x) = C + \int_a^x f(t) \, dt$ is uniformly continuous on $[a,b]$. [*Hint*: For each $\varepsilon > 0$, use (11.F.2) to find a suitable step function. Then apply problem F.6 to obtain an appropriate $\delta > 0$.]

8.* A function f defined on an interval J is said to be *measurable* if there is a sequence of step functions $\{s_n: J \to \mathbf{R}^1\}$ that converges a.e. to f on J. Show that if $\{f_k: J \to \mathbf{R}_\infty\}$ is a sequence of summable functions that converges a.e. to a function $f: J \to \mathbf{R}_\infty$ on J, then f is measurable. [*Hint*: Use (11.F.5) and an argument similar to that given in the proof of (11.F.7).]

9. Show that if $f: J \to \mathbf{R}_\infty$ is summable over J and if the derivative f' of f exists a.e. on J, then f' is measurable. [*Hint*: Set $f = 0$ outside J and consider a sequence $g_n = [f(x + h_n) - f(x)]/h_n$, where $\lim_{n \to \infty} h_n = 0$. Apply problem F.8.]

10. Suppose that $J = [a,\infty)$ and that $f: J \to \mathbf{R}_\infty$ is summable on every finite subinterval of J. Show that if, in addition, $\lim_{b \to \infty} \int_a^b |f|$ is finite, then f is summable on J. [*Hint*: Note that $0 \le f^+ \le |f|$ and $0 \le -f^- \le |f|$; apply problem F.2.]

Section G

1. Establish properties (a)–(d) of Example (11.G.7). [*Hint*: Use (5.C.10) and L'Hospital's rule.]
2. Suppose that for each positive integer n, $f_n: [0,2] \to \mathbf{R}^1$ is defined by

$$f_n(x) = \begin{cases} n^{-\frac{1}{2}} & \text{if } \dfrac{1}{n} \le x \le \dfrac{2}{n} \\ 0 & \text{otherwise} \end{cases}$$

(a) Show that the sequence of functions $\{f_n\}$ converges pointwise but not uniformly to the constant function $f \equiv 0$.

(b) Show (using the Lebesgue dominated convergence theorem) that

$$\lim_{n \to \infty} \int_0^2 f_n = \int_0^2 \lim_{n \to \infty} f_n = 0$$

3. Use the Lebesgue dominated convergence theorem to establish the following results:

(a) If f is a summable function over $\mathbf{R}^1$, then

$$\int_{\mathbf{R}^1} f = \lim_{b \to \infty} \int_{-b}^b f$$

i.e., the Lebesgue integral $\int_{\mathbf{R}^1} f$ is equal to the Cauchy principal value $\text{CPV}\int_{-\infty}^{\infty} f(x)\,dx$ [see (6.C.12)].

(b) If g is a summable function over an unbounded interval J and if f is a bounded function over J and is summable over every finite subinterval of J, then fg is summable over J.

4. Determine if the following functions are integrable in the sense of (11.E.7) and, if so, evaluate the integral:

(a) $f: [2,4] \to \mathbf{R}^1$, where

$$f(x) = \begin{cases} x^2 & \text{if } x \text{ is irrational} \\ 0 & \text{otherwise} \end{cases}$$

(b) $f: [0,1] \to \mathbf{R}^1$, where

$$f(x) = \begin{cases} x & \text{if } x = \dfrac{m}{n} \text{ (expressed in lowest terms)} \\ 0 & \text{otherwise} \end{cases}$$

(c) $f: [0,1] \to \mathbf{R}^1$, where

$$f(x) = \begin{cases} x^2 & \text{if } x \text{ is a point in the Cantor set} \\ \sin x & \text{otherwise} \end{cases}$$

(d) $f: [0,\infty) \to \mathbf{R}^1$, where

$$f(x) = \begin{cases} x & \text{if } x \text{ is rational} \\ e^{-x^2} & \text{otherwise} \end{cases}$$

5. Establish equations (2) and (3) of Sec. 11.G.

6. Show that the following "converse" of the Lebesgue dominated convergence theorem does not hold: If $\{f_n\}$ is a sequence of summable functions that converges to a summable function f, then there is a summable function g that dominates the sequence $\{f_n\}$. *Hint*: Consider the functions f_n defined by

$$f_n(x) = \begin{cases} \dfrac{1}{x} & \text{if } n - \dfrac{1}{2} < x < n + \dfrac{1}{2} \\ 0 & \text{otherwise} \end{cases}$$

The following concept is used in a number of the problems below. A function f is *measurable* on an interval J if there is a sequence of step functions $\{s_k\}$ that converges a.e. to f on J.

7. Show that if f is a measurable function on an interval J and if $|f| \le g$ a.e. on J, then there is a sequence of step functions $\{\sigma_k\}$ that converges a.e. to f on J with the property $|\sigma_k| \le g$ for each k. [*Hint*: Let $\sigma_k(x) = \sup\{-g(x), \inf\{s_k(x), g(x)\}\}$.]

8. Show that if f is a measurable function on an interval J and if $|f| \le g$ a.e., where g is a summable function on J, then f is summable on J.

9. Show that if f is a measurable function on a finite interval J and if $|f| \le M$ a.e., then f is summable over J.

10. Show that a measurable function f is summable over an interval J if and only if $|f|$ is summable over J.

11. (*Fatou's Lemma*) Suppose that $\{f_k\}$ is a sequence of summable functions mapping an interval J into $\mathbf{R}_\infty$ and that there is a summable function $g: J \to \mathbf{R}_\infty$ such that $g \le f_k$ a.e. for each k. Suppose further that $\int_J f_k \le M$ for each k. Show that if the sequence $\{f_k\}$ converges a.e. to a function f on J, then f is summable and $\int_J f \le M$. [*Hint*: Show that the sequence $\{g_k = \inf\{f_k, f_{k+1}, \ldots\}\}$ is an increasing sequence of summable functions that converges a.e. to f on J.]

12. Show that if f and g are two measurable functions that are finite a.e. on an interval J, then the product fg is measurable on J.

13. Show that if $f: J \to \mathbf{R}_\infty$ is a summable function over an interval J and if $g: J \to \mathbf{R}_\infty$ is a bounded measurable function, then the product function fg is summable on J. [*Hint*: Show that if M is a constant, then Mf is summable, and therefore, $M|f|$ is also summable.]

14. Suppose that f and g are extended real-valued functions on an interval J and that f^2 and g^2 are summable. Show that fg is summable over J.

15. (a) Show that if c is a constant such that $0 < c < 1$, then

$$t^c \le ct + (1 - c) \qquad \text{for all } t > 0$$

[*Hint*: Set $f(t) = -t^c + ct + (1 - c)$ and show that f has exactly one minimum on the open interval $0 < t < \infty$.]

(b) Let p and q be positive real numbers greater than one and such that

$$\frac{1}{p} + \frac{1}{q} = 1$$

Show that if a and b are nonnegative real numbers or equal to $+\infty$, then

$$ab \leq \frac{1}{p} a^p + \frac{1}{q} b^q$$

[*Hint*: Suppose that a and b are positive real numbers and use (a) with $c = 1/p$ and $t = a^p b^{(1-q)/p}$.]

(c) Suppose that f and g are extended real-valued functions on an interval J such that $|f|^p$ and $|g|^q$ are summable over J for a pair of real numbers p and q greater than one and satisfying $1/p + 1/q = 1$. Show that fg is summable over J.

Section H

1. Show that if a function f is absolutely continuous, then f is uniformly continuous.
2. Show that if $f: [a,b] \to \mathbf{R}^1$ is absolutely continuous, then there is a real number M such that if $P = \{x_0, x_1, \ldots, x_n\}$ is any partition of $[a,b]$, then $\sum_{i=1}^{n} |f(x_i) - f(x_{i-1})| \leq M$. (Functions with this property are said to be of *bounded variation*.)
3. Show that if $f: [a,b] \to \mathbf{R}^1$ is summable and if F is an indefinite integral of f, then F is of bounded variation. [*Hint*: First establish this result in the case that f is nonnegative, and then use the fact that $f = f^+ - (-f^-)$, where f^+ and f^- are as defined in Sec. 11.C.]
4. Show that the function $f: [0,1] \to \mathbf{R}^1$ defined by $f(x) = x \sin x$ is uniformly continuous, but is not of bounded variation and, hence, by problem H.2, is not absolutely continuous. [*Hint*: Consider appropriate partitions of $[0,1]$.]
5. A function $f: J \to \mathbf{R}^1$ is said to satisfy a *Lipschitz condition* on J if there is a constant c such that for each $x,y \in J$, $|f(x) - f(y)| \leq c|x - y|$. Show that if a function f satisfies a Lipschitz condition on an interval J, then f is absolutely continuous on J.
6. Show that if a function $f: [a,b] \to \mathbf{R}^1$ is continuous and piecewise smooth, then f satisfies a Lipschitz condition.
7. Establish (2) in Sec. 11.H.

Section I

1. Show that if R is a rectangle, then a bounded function $f: R \to \mathbf{R}^1$ is Riemann integrable on R if and only if for each $\varepsilon > 0$, there are step functions s_1 and s_2 (defined on some partition of R) such that $s_1 \leq f \leq s_2$ almost everywhere and $\int_R s_2 - \int_R s_1 < \varepsilon$.

2. Suppose that $f: \mathbf{R}^2 \to \mathbf{R}^1$ is continuous at a point $z^* \in \mathbf{R}^2$. Show that

$$\lim_{r \to 0} \frac{1}{\pi r^2} \int_{S_r(z*)} f = f(z^*)$$

3. Find $\int_S f$ if:
   (a) $f(x,y) = x^2 y$ and S is the region bounded by $y = |x|$ and $y = 4 - x^2$.
   (b) $f(x,y) = x - y$ and $S = \{(x,y) \mid 0 \le x \le 4 - y^2, -2 \le y \le 2\}$.
4. (a) Explain why it is reasonable to define the area of a set A by $\int_A 1$ (provided this integral exists).
   (b) Find the area of the region lying between the graphs of the functions $y = x^3 - 3x^2 - 2x$ and $y = -3x^2$.
5. Suppose that $f: [a,b] \to \mathbf{R}^1$ and $g: [a,b] \to \mathbf{R}^1$ are integrable. Show that

$$\int_a^b f(x)g(x)\, dx = \frac{1}{b-a}\left(\frac{1}{2}\int_S (f(x) - f(y))(g(x) - g(y))\, dy\, dx \right.$$

$$\left. + (b-a)\left(\int_a^b f(x)\, dx\right)\left(\int_a^b g(x)\, dx\right)\right)$$

6. Results involving a change of variables for integrals of functions of more than one variable are considerably more difficult to establish than the corresponding results for functions of one variable such as (6.B.9). A principal theorem in this direction is the following [a proof of which may be found in Apostol (1974)]. Suppose that S and $\hat{S}$ are open sets in $\mathbf{R}^n$, and that $T: \hat{S} \to S$ is a continuously differentiable bijection. Let J_T be the Jacobian of T (see Sec. 7.G). Then a function $f: \hat{S} \to \mathbf{R}^1$ is integrable on $\hat{S}$ if and only if $f \circ T$ is integrable on S, and furthermore,

$$\int_S f = \int_S |J_T| f \circ T$$

or, as is frequently written,

$$\int_S f(x_1, x_2, \ldots, x_n)\, dx_1\, dx_2 \cdots dx_n$$

$$= \int_S |J_T(u_1, u_2, \ldots, u_n)| f(T(u_1, u_2, \ldots, u_n))\, du_1\, du_2 \cdots du_n$$

The Jacobian essentially assumes the role of the derivative in (6.B.9).
(a) (Polar coordinates) Show that the Jacobian of the transformation $x = r \cos \theta$, $y = r \sin \theta$ ($r \ge 0$, $0 \le \theta \le 2\pi$) is r, and use this change of variables to evaluate

$$\int_S f(x, y)\, dy\, dx = \int_0^{\sqrt{2}/2} \int_0^{\sqrt{4-x^2}} e^{-(x^2+y^2)}\, dy\, dx$$

[Here $r = \sqrt{x^2 + y^2}$ and θ is the angle measured in a counter clock-wise sense from the positive x-axis to the segment joining the origin to (x,y).]

(b) (Spherical coordinates) Show the Jacobian of the transformation

$$x = \rho \sin \phi \cos \theta$$
$$y = \rho \sin \phi \sin \theta$$
$$z = \rho \cos \phi$$

is $\rho^2 \sin \phi$, and use this change of variables to evaluate

$$\int_S \frac{1}{1 + x^2 + y^2 + z^2} \, dx \, dy \, dz$$

where S is the region determined by

$$x^2 + y^2 \le z^2 \qquad x^2 + y^2 + z^2 \le 1 \qquad z \ge 0$$

[Here $\rho = \sqrt{x^2 + y^2 + z^2}$, θ is as defined in (a), and ϕ is the angle between the positive z-axis and the segment joining the origin to (x,y,z).]

(c) Show that if $f: \mathbf{R}^2 \to [0,\infty)$, then

$$\int_{\mathbf{R}^2} f(x, y) \, dx \, dy = \int_0^{2\pi} \int_0^\infty f(r \cos \theta, r \sin \theta) r \, dr \, d\theta$$

(d) Show that the Γ function can be expressed by

$$\Gamma(x) = \int_0^\infty u^{2x-1} e^{-u^2} \, du$$

and with the aid of (c) show that

$$\Gamma(x)\Gamma(y) = 2\Gamma(x + y) \int_0^{\pi/2} \cos^{2x-1}\theta \sin^{2y-1}\theta \, d\theta$$

From this deduce that $\Gamma(1/2) = \sqrt{\pi}$.

(e) Show that $\int_0^\infty e^{-x^2} \, dx = \sqrt{\pi}/2$. [*Hint*: Observe that

$$\left(\int_0^\infty e^{-x^2} \, dx\right)\left(\int_0^\infty e^{-y^2} \, dy\right) = \int_0^\infty \int_0^\infty e^{-x^2} e^{-y^2} \, dx \, dy$$

and then use polar coordinates.]

7. Construct an example in connection with (11.I.1) to show why we used the diameters of rectangles instead of the areas of a rectangles to define the mesh of a partition of a rectangle.

8. Prove Theorem (11.I.10).

9.* Show that if $f: [a,b] \to \mathbf{R}^1$ is continuous, then $\{(x,f(x)) \mid x \in [a,b]\}$ is of content 0. Where was this result tacitly used in the proof of (11.I.11)?

REFERENCES

Apostol, T. M. (1974): *Mathematical Analysis*, Addison-Wesley, Reading, Mass.
Halmos, P. R. (1974): *Measure Theory*, Springer-Verlag, New York.
Korevaar, J. (1968): *Mathematical Methods*, Academic Press, New York.
Royden, H. L. (1968): *Real Analysis*, Macmillan, New York.
Spivak, M. (1965): *Calculus on Manifolds*, W. A. Benjamin, New York.

12

AN INTRODUCTION TO COMPLEX INTEGRATION

A. CONTOUR INTEGRALS

This chapter is intended to provide the reader with a brief introduction to the theory of integration in complex analysis. As we shall see, although there are certain parallels between this theory and that studied in the previous chapter, there are also a number of significant differences. Many of the results that we shall obtain can be cast in terms of the Lebesgue theory; however, for the sake of simplicity, we shall restrict ourselves to integration in the Riemann context (or that discussed in Chapter 6). We begin by recalling the definitions of the derivative and integral of a function that maps an interval $[a,b] \subset \mathbf{R}^1$ into the complex plane $\mathbf{C}$.

(12.A.1) Definition Suppose that $f: [a,b] \to \mathbf{C}$ and that $f = u + iv$ (where $u: [a.b] \to \mathbf{R}^1$ and $v: [a,b] \to \mathbf{R}^1$).

    (a) If $u'(x)$ and $v'(x)$ exist at a point $x \in [a,b]$, then the derivative of f at x is defined by $f'(x) = u'(x) + iv'(x)$.

    (b) If $\int_a^b u(x)\, dx$ and $\int_a^b v(x)\, dx$ exist, then the integral of f over $[a,b]$ is defined by $\int_a^b f(x)\, dx = \int_a^b u(x)\, dx + i \int_a^b v(x)\, dx$.

The notion of a path will be of considerable importance in the ensuing sections.

(12.A.2) *Definition*   A *path* in **C** is any continuous function γ that maps a closed and bounded interval $[a,b]$ into **C**. The *curve* or *contour* in **C** associated with γ is the image of γ, which we denote $I[\gamma]$. The point $\gamma(a)$ is called the *initial point* of γ and $\gamma(b)$ is called the *terminal point* of γ. If $I[\gamma]$ is a subset of a set D, we shall say that γ is in D.

In this chapter we shall deal primarily with *rectifiable* paths, i.e., paths of "finite length." We recall from Sec. 6.D that the length of a path γ is defined by first approximating γ by polygonal paths (whose lengths are readily calculable) and then taking the least upper bound of the lengths of such "approximations" (Fig. 12.1). More formally, we have the following definition.

Figure 12.1

(12.A.3) *Definition*   Suppose that $\gamma\colon [a,b] \to$ **C** is a path. For each partition $P = \{t_0, t_1, \ldots, t_n\}$ of $[a,b]$, let $M_P = \sum_{i=1}^{n} |\gamma(t_i) - \gamma(t_{i-1})|$. Then the *length of* γ, $l(\gamma)$, is defined by

$$l(\gamma) = \sup\{M_P \mid P \text{ is a partition of } [a, b]\}$$

If $l(\gamma)$ is finite, then γ is said to be *rectifiable*. If γ' exists on $[a,b]$, then γ is said to be a *differentiable path*. If γ' is continuous, then the path γ is said to be *continuously differentiable*, and if γ' is piecewise continuous, then γ is a *piecewise continuously differentiable path*.

(12.A.4) *Observation*   In Sec. 6.D, we dealt with functions that mapped a closed and bounded interval $[a,b]$ into $\mathbf{R}^2$, e.g., $f\colon [0,3] \to \mathbf{R}^2$ defined by $f(x) = (x^2, \sin x)$. Such functions may also be viewed as mapping $[a,b]$ into **C**, in which case, for example, we would write $f(x) = x^2 + i \sin x$.

Although, as was indicated in Chapter 6, not all paths are rectifiable, we nevertheless have the following result, which is an immediate consequence of (6.D.1).

(12.A.5) **Theorem** If $\gamma\colon [a,b] \to \mathbf{C}$ is a piecewise continuously differentiable path, then γ is rectifiable and $l(\gamma) = \int_a^b |\gamma'(t)|\, dt$.

In general, integration of complex-valued functions is done over rectifiable paths in $\mathbf{C}$.

(12.A.6) **Definition** If $\gamma\colon [a,b] \to \mathbf{C}$ is a rectifiable path and if $f\colon I[\gamma] \to \mathbf{C}$, then the *contour integral* of f over γ is defined to be

$$\lim_{|P| \to 0} \sum_{i=1}^{n} f(\gamma(w_i))(\gamma(t_i) - \gamma(t_{i-1})) \quad (w_i \in [t_{i-1}, t_i]) \tag{1}$$

provided this limit exists. This limit is denoted by $\int_\gamma f(z)\, dz$ (of course, z may be replaced by any letter).

Note: The limit given in (1) is equal to a complex number K if for each $\varepsilon > 0$, there is a $\delta > 0$ such that if $P = \{t_0, t_1, \ldots, t_n\}$ is any partition of $[a,b]$ with mesh less than δ and if for each i, w_i is an arbitrary point in $[t_{i-1}, t_i]$, then

$$\left| K - \sum_{i=1}^{n} f(\gamma(w_i))(\gamma(t_i) - \gamma(t_{i-1})) \right| < \varepsilon$$

The reader should observe the similarity between definition (12.A.6) and the definition of the Riemann integral (11.A.1). In fact, if $\gamma\colon [a,b] \to \mathbf{R}^1$ is defined by $\gamma(x) = x$ and if f maps $I[\gamma] = [a,b]$ into $\mathbf{R}^1$ (or $\mathbf{C}$), then (12.A.6) coincides with (11.A.1) [or (12.A.1)].

(12.A.7) **Observation** It is occasionally convenient (but incorrect) to speak of an integral over a contour such as a circle, line segment, etc., rather than over a path. Care must be exercised, however, in doing so. For instance, the paths $\gamma_1\colon [0,1] \to \mathbf{C}$ and $\gamma_2\colon [0,1] \to \mathbf{C}$ defined, respectively, by

$$\gamma_1(t) = t + it$$

and

$$\gamma_2(t) = \begin{cases} 2t + i2t & 0 \le t \le \tfrac{1}{2} \\ (1 - 2t) + i(1 - 2t) & \tfrac{1}{2} \le t \le 1 \end{cases}$$

yield precisely the same contour or image curve $I[\gamma_1] = C = I[\gamma_2]$; yet if $f\colon \mathbf{C} \to \mathbf{C}$ is the constant function $f(z) = 3$, then it is easy to see that

$$\int_{\gamma_1} f(z)\, dz = 3$$

while

$$\int_{\gamma_2} f(z)\, dz = 0$$

(see problem A.2). Thus, in problems of this nature we cannot talk about the integral of a function f over a contour without specifying how the contour is obtained, i.e., how it is parameterized. This leads to the following definition.

(12.A.8) *Definition* Two paths γ_1 and γ_2 (perhaps defined over different intervals) are *equivalent* if the following two conditions are satisfied:

(i) $I[\gamma_1] = I[\gamma_2]$.
(ii) For each complex-valued function f for which either $\int_{\gamma_1} f(z)\,dz$ or $\int_{\gamma_1} f(z)\,dz$ exists, both of these integrals exist and are equal in value.

Some conditions under which paths are equivalent are given in (12.A.12). Two basic questions associated with the definition of the contour integral are:

1. Under what conditions does the contour integral exist?
2. If the contour integral does exist, how can it be calculated?

The next theorem provides a partial answer to the first question.

(12.A.9) *Theorem* If $\gamma: [a,b] \rightarrow \mathbf{C}$ is a rectifiable path in $\mathbf{C}$, and if $f: I[\gamma] \rightarrow \mathbf{C}$ is a continuous function, then $\int_\gamma f(z)\,dz$ exists.

Proof. For each partition $P = \{t_0, \ldots, t_n\}$ of $[a,b]$ let

$$S_P = \sum_{i=1}^{n} f(\gamma(w_i))(\gamma(t_i) - \gamma(t_{i-1}))$$

where w_i is an arbitrary point in $[t_{i-1}, t_i]$. We first show that for each $\varepsilon > 0$, there is a $\delta > 0$ such that if P and P' are partitions of $[a,b]$ with mesh less than δ, then $|S_P - S_{P'}| < \varepsilon$. To see this, let $L = l(\gamma)$ be the length of γ, and let $\varepsilon' = \varepsilon/2L$. Note that if $P = \{t_0, \ldots, t_n\}$ and $P' = \{t'_0, \ldots, t'_m\}$ are any two partitions of $[a,b]$ and if $P_R = \{\hat{t}_0, \ldots, \hat{t}_r\}$ is the partition consisting of the ordered union of the points in P and P' (this partition is called the *common refinement* of P and P'), then there is a subsequence $\{k_j\}_{j=0}^{n}$ of the finite sequence $\{0, 1, \ldots, r\}$ such that $\{\hat{t}_{k_j}\}_{j=0}^{n} = \{t_i\}_{i=0}^{n}$; this subsequence identifies the points of P that are found in P_R. Hence, we have

$$S_P = \sum_{i=1}^{n} f \circ \gamma(w_j^*)[\gamma(t_j) - \gamma(t_{j-1})] \qquad (w_j^* \in [t_{j-1}, t_j])$$

$$= \sum_{j=1}^{n} f \circ \gamma(w_j^*)[\gamma(\hat{t}_{k_j}) - (\hat{t}_{k_{j-1}})]$$

$$= \sum_{i=1}^{r} f \circ \gamma(w_i)\, [\gamma(\hat{t}_i) - \gamma(\hat{t}_{i-1})]$$

where w_i is set equal to w_j^* if $[\hat{t}_{i-1}, \hat{t}_i] \subset [\hat{t}_{k_{j-1}}, \hat{t}_{k_j}]$. Since

$$S_{P_R} = \sum_{i=1}^{r} f(\hat{w}_i)[\gamma(\hat{t}_i) - \gamma(\hat{t}_{i-1})] \qquad (\hat{w}_i \in [\hat{t}_{i-1}, \hat{t}_i])$$

it follows that

$$|S_{P_R} - S_P| = \left| \sum_{i=1}^{r} \{f \circ \gamma(\hat{w}_i) - f \circ \gamma(w_i)\}(\gamma(\hat{t}_i) - \gamma(\hat{t}_{i-1})) \right|$$

$$\leq \sum_{j=1}^{n} \text{osc}\,(f \circ \gamma; [\hat{t}_{k_j-1}, \hat{t}_{k_j}])|\gamma(\hat{t}_{k_j}) - \gamma(\hat{t}_{k_j-1})|$$

$$= \sum_{j=1}^{n} \text{osc}\,(f \circ \gamma; [t_{j-1}, t_j])|\gamma(t_j) - \gamma(t_{j-1})|$$

[The oscillation, osc, of a function over an interval is defined in (11.A.2).] If $O_P = \max\{\text{osc}(f \circ \gamma; [t_{i-1}, t_i]) \mid i = 1, \ldots, n\}$, then $|S_{R_P} - S_P| \leq O_P \cdot L$. Since $f \circ \gamma$ is uniformly continuous on $[a,b]$, there is a $\delta > 0$ such that $O_P < \varepsilon'$ whenever the mesh of P is less than δ. Consequently, if P and P' are partitions of $[a,b]$ with mesh less than δ, then

$$|S_P - S_{P'}| \leq |S_{R_P} - S_P| + |S_{R_P} - S_{P'}| < 2\varepsilon'L = \varepsilon$$

Now let $\{P_n\}$ be a sequence of partitions of $[a,b]$ with the property that $\lim_{n \to \infty} |P_n| = 0$. From what we have just shown, $\{S_{P_n}\}$ is a Cauchy sequence in $\mathbf{C}(=\mathbf{R}^2)$, and therefore, by (4.C.4), this sequence converges to a point $T \in \mathbf{C}$. It follows easily that $\lim_{|P| \to 0} S_P = T$, and consequently, $\int_\gamma f(z)\,dz$ exists and is equal to T.

As a general rule, definitions are notably lacking in computational merit, and the definition of the contour integral is no exception. One fairly efficient method for calculating the integral $\int_\gamma f(z)\,dz$ is based on converting it into a Riemann integral.

(12.A.10) Theorem   Suppose that $\gamma: [a,b] \to \mathbf{C}$ is a piecewise continuously differentiable path. If $f: I[\gamma] \to \mathbf{C}$ is a continuous function, then

$$\int_\gamma f(z)\,dz = \int_a^b f(\gamma(t))\gamma'(t)\,dt$$

Proof.   It suffices to consider the case where γ' is continuous. Let $\varepsilon > 0$ be given and let $\varepsilon' = \varepsilon/L$, where $L = l(\gamma)$. Since $f \circ \gamma$ is uniformly continuous on the compact set $[a,b]$, there is a $\delta' > 0$ such that if $\alpha, \beta \in [a,b]$ and $|\beta - \alpha| < \delta'$, then

$$|f(\gamma(\beta)) - f(\gamma(\alpha))| < \varepsilon'$$

Let $P = \{t_0, \ldots, t_n\}$ be a partition of $[a,b]$ with mesh less than δ', and suppose that for each i, w_i is an arbitrary point in $[t_{i-1}, t_i]$. Set $S_P = \sum_{i=1}^{n} (f \circ \gamma)(w_i)(\gamma(t_i) - \gamma(t_{i-1}))$. Then it follows from (6.B.10) and (6.D.2) that

$$\left| \int_b^b f(\gamma(t))\gamma'(t)\,dt - S_P \right| = \left| \sum_{i=1}^n \int_{t_{i-1}}^{t_i} [f(\gamma(t)) - f(\gamma(w_i))]\gamma'(t)\,dt \right|$$

$$\leq \sum_{i=1}^n \int_{t_{i-1}}^{t_i} |[f(\gamma(t)) - f(\gamma(w_i))]||\gamma'(t)|\,dt$$

$$\leq \sum_{i=1}^n \int_{t_{i-1}}^{t_i} \varepsilon'|\gamma'(t)|\,dt$$

$$= \varepsilon' \sum_{i=1}^n \int_{t_{i-1}}^{t_i} |\gamma'(t)|\,dt = \varepsilon'L = \varepsilon$$

Consequently, we have from (12.A.6) that

$$\int_\gamma f(z)\,dz = \int_a^b f(\gamma(t))\gamma'(t)\,dt$$

(12.A.11) *Example* Let $\gamma: [0,2\pi] \to \mathbf{C}$ be the path defined by $\gamma(t) = a + re^{int}$. Geometrically, this path represents n full counterclockwise rotations about the circle with center a and radius r. If $f(z) = 1/(z - a)$, then we have

$$\int_\gamma f(z)\,dz = \int_\gamma \frac{1}{z - a}\,dz = \int_0^{2\pi} \frac{inre^{int}}{re^{int}}\,dt = ni \int_0^{2\pi} dt = 2\pi ni$$

(12.A.12) *Observation* Suppose that $\gamma: [a,b] \to \mathbf{C}$ and $\hat\gamma: [c,d] \to \mathbf{C}$ are piecewise continuously differentiable paths, and suppose further that there is a continuously differentiable bijection $\phi: [a,b] \to [c,d]$ such that $\gamma = \hat\gamma \circ \phi$. Note that $I[\gamma] = I[\hat\gamma]$. If $f: I[\gamma] \to \mathbf{C}$ is any continuous function, then

$$\int_\gamma f(z)\,dz = \int_{\hat\gamma} f(z)\,dz$$

since by (12.A.10) and (6.B.9)

$$\int_\gamma f(z)\,dz = \int_a^b f(\gamma(t))\gamma'(t)\,dt$$

$$= \int_a^b f(\hat\gamma(\phi(t)))\hat\gamma'(\phi(t))\phi'(t)\,dt$$

$$= \int_c^d f(\hat\gamma(s))\hat\gamma'(s))\,ds = \int_{\hat\gamma} f(z)\,dz$$

In particular, since $\phi: [a,b] \to [c,d]$ defined by

$$\phi(t) = \frac{d - c}{b - a}(t - a) + c$$

is clearly continuously differentiable and bijective, it follows that the domain of a path can be readily changed without affecting the value of a contour integral.

(12.A.13) *Exercise* Suppose that $\gamma: [a,b] \to C$ is a path, and that $f: I[\gamma] \to$ C. To find $\int_\gamma f(z)\, dz$, quite often it will be useful to form a partition $\{x_0, x_1, \ldots, x_n\}$ of $[a,b]$ and integrate over each subinterval of this partition. For $i = 1, 2, \ldots, n$ let $\gamma_i = \gamma|_{[x_{i-1},x_i]}$ and $f_i = f|_{I[\gamma_i]}$. Show that if $\int_\gamma f(z)\, dz$ exists, then

$$\int_\gamma f(z)\, dz = \sum_{i=1}^{n} \int_{\gamma_i} f_i(z)\, dz$$

Then use (12.A.12) to show that each interval $[x_{i-1}, x_i]$ may be replaced by the interval $[0,1]$.

Note: In the above notation it will frequently be convenient to write $\gamma = \gamma_1 + \gamma_2 + \cdots + \gamma_n$ or $\gamma = \gamma_1 \cdot \gamma_2 \cdots \gamma_n$.

(12.A.14) *Example* We wish to integrate $f(z) = z - z^2$ on the contour C determined by the contours C_1, C_2, and C_3 indicated in Fig. 12.2. One parameterization of the contour C that could be used is $\gamma: [0,3] \to C$ defined by

$$\gamma(t) = \begin{cases} -t + it & 0 \le t \le 1 \\ -1 + i(2 - t) & 1 \le t \le 2 \\ t - 3 & 2 \le t \le 3 \end{cases}$$

However, by the previous exercise we are also justified in using the following parameterizations of C_1, C_2, and C_3:

$$\gamma_1(t) = -t + it \qquad 0 \le t \le 1$$
$$\gamma_2(t) = -1 + i(1 - t) \qquad 0 \le t \le 1$$
$$\gamma_3(t) = -1 + t \qquad 0 \le t \le 1$$

Since $f(\gamma_1(t)) = (-t + it) - (-t + it)^2$ on C_1, it follows from (12.A.10) that

$$\int_{\gamma_1} f(z)\, dz = \int_0^1 [(-t + it) - (-t + it)^2][-1 + i]\, dt = -\tfrac{2}{3} - \tfrac{5}{3}i.$$

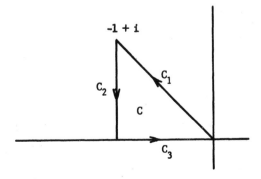

Figure 12.2

Similarly, we find that

$$\int_{\gamma_2} f(z)\, dz = \int_0^1 [(-1 + i(1 - t)) - (-1 + i(1 - t))^2][-i]\, dt = \tfrac{3}{2} + \tfrac{5}{3}i$$

and

$$\int_{\gamma_3} f(z)\, dz = \int_0^1 [(-1 + t) - (-1 + t)^2][1]\, dt = -\tfrac{5}{6}$$

and, therefore, the desired integral is equal to 0.

Next we list some basic properties of contour integrals. Proofs of these properties follow directly from the appropriate definitions and previous results and are left to the reader (see problem A.5).

(12.A.15) Theorem   Suppose that $\gamma: [a,b] \to \mathbf{C}$ is a rectifiable path and that $\int_\gamma f(z)\, dz$ and $\int_\gamma g(z)\, dz$ exist. Then:

(i) $\int_\gamma cf(z)\, dz = c \int_\gamma f(z)\, dz$ for each complex constant c.
(ii) $\int_\gamma (f + g)(z)\, dz = \int_\gamma f(z)\, dz + \int_\gamma g(z)\, dz$.
(iii) $|\int_\gamma f(z)\, dz| \le \int_\gamma |f(\gamma(t))|\, |\gamma'(t)|\, dt$, i.e., $|\int_\gamma f(z)\, dz| \le \int_\gamma |f(z)|\, |dz|$.
(iv) If $M = \sup\{|f(z)|\, |z \in I[\gamma]\}$, then $|\int_\gamma f(z)\, dz| \le Ml(\gamma)$.
(v) If $\gamma: [a,b] \to \mathbf{C}$ is a rectifiable path and if $\hat{\gamma}: [a,b] \to \mathbf{C}$ is defined by $\hat{\gamma}(t) = \gamma(a + b - t)$, then $\int_{\hat{\gamma}} f(z)\, dz = -\int_\gamma f(z)\, dz$. In this case it is customary to write $\hat{\gamma} = -\gamma$.

In the next result we see that the contour integral of the uniform limit of a sequence of integrable functions is equal to the limit of the sequence formed by the contour integrals of these functions.

(12.A.16) Theorem   Suppose that D is an open subset of $\mathbf{C}$ and that $\{f_n\}$ is a sequence of continuous functions (mapping D into $\mathbf{C}$) that converges uniformly on every compact subset of D to a function $F: D \to \mathbf{C}$. Then for each rectifiable path γ in D,

$$\lim_{n \to \infty} \int_\gamma f_n(z)\, dz = \int_\gamma F(z)\, dz$$

Proof.   Let γ be a rectifiable path in D and let $l(\gamma)$ be the length of γ. Since $I[\gamma]$ is a compact subset of D, it follows from the hypothesis that for each $\varepsilon > 0$, there exists an integer N such that whenever $n \ge N$,

$$|F(z) - f_n(z)| < \frac{\varepsilon}{l(\gamma)}$$

for every $z \in I[\gamma]$. Since by (8.C.2) F is continuous on D, it follows that $\int_\gamma F(z)\, dz$ exists, and furthermore we have from the previous theorem (12.A.15.iv) that if $n \ge N$, then

$$\left| \int_{\gamma} F(z)\,dz - \int_{\gamma} f_n(z)\,dz \right| = \left| \int_{\gamma} (F(z) - f_n(z))\,dz \right|$$

$$< \frac{\varepsilon}{l(\gamma)}\, l(\gamma)$$

$$= \varepsilon$$

which concludes the proof.

B. THE CAUCHY-GOURSAT THEOREM

The notion of a primitive plays an important role not only in the integration of functions of a real variable, but in the context of functions of a complex variable as well. A *primitive* of a complex-valued function $f: D \rightarrow \mathbf{C}$ (where $D \subset \mathbf{C}$) is any function $F: D \rightarrow \mathbf{C}$ with the property that $F' = f$. In the next theorem we employ this notion of a primitive to obtain a complex version of the fundamental theorem of calculus. In the remainder of this chapter we shall refer to any open connected subset of $\mathbf{C}$ as a *region* in $\mathbf{C}$.

(12.B.1) *Theorem* Suppose that D is a region and that $f: D \rightarrow \mathbf{C}$ is continuous and has a primitive F on D. If $\gamma: [a,b] \rightarrow D$ is a piecewise continuously differentiable path, then

$$\int_{\gamma} f(z)\,dz = F(\gamma(b)) - F(\gamma(a)).$$

Proof. By (12.A.10), the chain rule (7.E.8), and the fact that $(F \circ \gamma)'$ is piecewise continuous, we have

$$\int_{\gamma} f(z)\,dz = \int_{a}^{b} f(\gamma(t))\gamma'(t)\,dt$$

$$= \int_{a}^{b} (F \circ \gamma)'(t)\,dt$$

$$= F(\gamma(b)) - F(\gamma(a))$$

We shall see presently that, contrary to the case of real-valued functions, there are examples of continuous complex-valued functions that fail to have primitives. In the next theorem we give two necessary and sufficient conditions for a continuous complex function to have a primitive. Before stating this theorem we define the notion of *independent of path*. Suppose that D is a region in $\mathbf{C}$ and that $f: D \rightarrow \mathbf{C}$. Suppose further that for each two points z_1 and z_2 in D, $\int_{\gamma} f(z)\,dz = \int_{\hat{\gamma}} f(z)\,dz$, where γ and $\hat{\gamma}$ are any two rectifiable paths joining z_1 to z_2; i.e., $\int_{\gamma} f(z)\,dz$ depends only on the endpoints of γ.

Then $\int_\gamma f(z)\, dz$ is said to be *independent of path* in D, and we denote this integral by $\int_{z_1}^{z_2} f(z)\, dz$, where z_1 is the initial point of γ and z_2 the terminal point of γ.

(12.B.2) **Theorem** Suppose that D is a region in $\mathbb{C}$ and that $f: D \to \mathbb{C}$ is continuous on D. Then the following conditions are equivalent:

(i) f has a primitive on D.

(ii) $\int_\gamma f(z)\, dz = 0$ for every closed path γ in D (a *closed path* is any path whose initial and terminal points coincide).

(iii) $\int_\gamma f(z)\, dz$ is independent of path in D.

Proof. (i) implies (ii) This is an immediate consequence of (12.B.1).

(ii) implies (iii) If γ_1 and γ_2 have the same initial and terminal points, then the curve $\gamma_1 - \gamma_2$ is a closed path in D, and hence,

$$\int_{\gamma_1} f(z)\, dz - \int_{\gamma_2} f(z)\, dz = \int_{\gamma_1 - \gamma_2} f(z)\, dz = 0$$

which implies that $\int_{\gamma_1} f(z)\, dz = \int_{\gamma_2} f(z)\, dz$.

(iii) implies (i) Let a be an arbitrary fixed point in D and for each $z \in D$, let γ^z be any rectifiable path joining a to z. Note that by hypothesis (iii) the function

$$F(z) = \int_{\gamma^z} f(w)\, dw$$

is well defined; we shall show that F is a primitive of f on D.

Suppose that $z \in D$. Since D is open there is an ε-neighborhood of z, $S_\varepsilon(z)$, that lies entirely in D. For each $h \in \mathbb{C}$, $0 < |h| < \varepsilon$, let $\gamma_h: [0,1] \to S_\varepsilon(z)$ be the path defined by

$$\gamma_h(t) = z + ht$$

Then $I[\gamma_h]$ is the directed straight-line segment joining z and $z + h$ (Fig. 12.3). Since $\gamma_h'(t) = h$ for each t, we have from (12.A.10)

$$\int_{\gamma_h} f(w)\, dw = \int_0^1 f(\gamma_h(t))\gamma_h'(t)\, dt$$

$$= h \int_0^1 f(\gamma_h(t))\, dt \tag{1}$$

Furthermore, it is easy to see from the hypothesis, Fig. 12.3, and (12.A.13) that

$$\frac{F(z+h) - F(z)}{h} = \frac{1}{h} \int_{\gamma_h} f(w)\, dw$$

and therefore it follows from (1) that

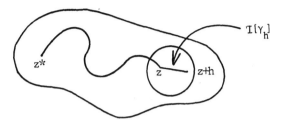

Figure 12.3

$$\frac{F(z + h) - F(z)}{h} = \int_0^1 f(\gamma_h(t))\, dt$$

$$= f(z) + \int_0^1 (f(\gamma_h(t)) - f(z))\, dt \qquad (2)$$

Since $\gamma_h(t) - z = ht$ and since f is continuous, we have that as h tends to 0, $f(\gamma_h(t)) - f(z)$ converges uniformly (for $0 \le t \le 1$) to 0, and therefore,

$$\lim_{h \to 0} \int_0^1 (f(\gamma_h(t)) - f(z))\, dt = 0$$

Consequently, it follows from (2) that

$$F'(z) = \lim_{h \to 0} \frac{F(z + h) - F(z)}{h} = f(z)$$

and hence, F is a primitive of f.

It should be emphasized that the previous theorem shows that for purposes of integration, knowledge of the existence of a primitive is sufficient to allow an arbitrary selection of a continuously differentiable path; for instance, rather than attempting to integrate over the contour γ indicated in Fig. 12.4, it might be more convenient to replace γ with the straight line that connects the points p and q.

At this point the following definition will be helpful.

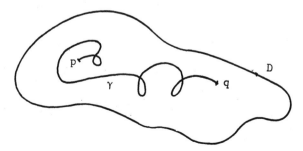

Figure 12.4

A simple closed curve   Figure 12.5

(12.B.3) *Definition*   A *simple closed path* is a closed path $\gamma: [a,b] \to$ C with the property that $\gamma|_{[a,b)}$ is one to one. The image of a simple closed path is called a *simple closed curve* (Fig. 12.5).

One of the most celebrated theorems in analysis and topology is the Jordan curve theorem which states that if S is a simple closed curve in C, then $C \backslash S$ consists of two disjoint connected sets, one bounded and the other unbounded and the curve S is the boundary of each of these sets. The bounded set is called the *interior* of S and the unbounded connected set is called the *exterior* of S. The correct proof of this "obvious" result eluded mathematicians (including Jordan) for many years. The first published error-free proof was apparently given by Veblen in 1905. There are now many proofs: those that start from scratch tend to be long and tedious; shorter, more elegant proofs rely on fairly powerful and sophisticated results. The reader is referred to Christenson and Voxman (1978) for a proof of the former kind.

Simple closed curves may be used to define the notion of simply connected regions.

(12.B.4) *Definition*   A region $D \subset$ C is *simply connected* if whenever S is a simple closed curve in D, then the interior of S is entirely contained in D (Fig. 12.6).

We are now in a position to state what is perhaps the single most important result in complex analysis, the Cauchy-Goursat theorem. Recall from Sec. 7.C that a function f is analytic at a point z if f is differentiable in

Simply connected

Not simply connected

Figure 12.6

a neighborhood of z, and that f is analytic on a set $A \subset C$ if f is differentiable at each point of A.

(12.B.5) *Theorem (Cauchy-Goursat Theorem)* If f is an analytic function on a simply connected region $D \subset C$, then $\int_{\gamma} f(z) \, dz = 0$ for every closed rectifiable (but not necessarily continuously differentiable) path γ in D.

In 1825, Cauchy established this result in the case that f' is continuous (in a paper that was not published until 1874). His work on this problem is generally considered to be one of the finest efforts in the history of mathematics. In 1883, Goursat gave a proof of Cauchy's theorem that avoided the restriction that f' be continuous.

A complete proof of the Cauchy-Goursat theorem is quite difficult and is omitted in our abbreviated introduction to complex integration. The reader may find a proof of this theorem in Ahlfors (1966). As immediate consequences of the Cauchy-Goursat theorem, we obtain the following important corollaries.

(12.B.6) *Corollary* If f is analytic on a simply connected region D, then

$$F(z) = \int_{\gamma^z} f(w) \, dw$$

is a primitive of f [notation as in (12.B.2)].

(12.B.7) *Corollary (Independence of Path)* Suppose that D is a simply connected region, $f: D \to C$ is analytic on D, and $z_1, z_2 \in D$. If γ and $\hat{\gamma}$ are any two rectifiable paths from z_1 to z_2, then

$$\int_{\gamma} f(z) \, dz = \int_{\hat{\gamma}} f(z) \, dz$$

i.e., $\int_{\gamma} f(z) \, dz$ is independent of path in D.

(12.B.8) *Example* If D is a simply connected region containing the points $z_1 = 1 + 2i$ and $z_2 = 3 - i$, and if γ is any rectifiable path from z_1 to z_2 that lies in D, then

$$\int_{\gamma} e^{3z} \, dz = \int_{1+2i}^{3-i} e^{3z} \, dz = \tfrac{1}{3}(e^{9-3i} - e^{3+6i})$$

(12.B.9) *Corollary* Suppose that D is a simply connected region and that $f: D \to C$ is analytic on D. If $\gamma: [a,b] \to D$ is a rectifiable path with initial point z_1 and terminal point z_2, and if F is a primitive of f on D, then

$$\int_\gamma f(z)\, dz = F(z_2) - F(z_1)$$

Proof. If γ is not piecewise continuously differentiable, then it can be replaced by a (piecewise continuously differentiable) polygonal path $\hat{\gamma}$ in D that joins z_1 to z_2. Then by (12.B.7) we have

$$\int_\gamma f(z)\, dz = \int_{\hat{\gamma}} f(z)\, dz]$$

and it follows from (12.B.6) that f has a primitive F on D. The desired result is now a consequence of (12.B.1).

Next we see how the Cauchy-Goursat theorem can be applied to annular regions.

(12.B.10) *Definition* Suppose that C_1 and C_2 are simple closed curves and that C_1 lies in the interior of C_2. Then the region bounded by C_1 and C_2 is called an *annular region*. [It can be shown that there is a continuous bijection ϕ with a continuous inverse that maps an annular region onto the annular region bounded by two concentric circles (Fig. 12.7).]

In the following theorem we need the notion of the *orientation* of a simple closed curve. If $\gamma: [a,b] \to \mathbf{C}$ is a simple closed path, then we shall say that γ (or $I[\gamma]$) is *positively oriented* in the case that as t runs from a to b, the

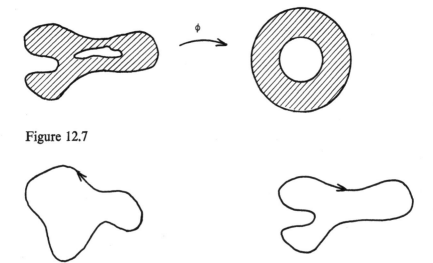

Figure 12.7

Figure 12.8

curve is traversed in a counterclockwise sense, and γ (or $I[\gamma]$) is *negatively oriented* if as t runs from a to b, the curve is traversed in a clockwise sense (Fig. 12.8). Actually, the concept of orientation is more difficult than that indicated by this definition. It is not always clear (from a strictly mathematical point of view) what is meant by "traversing a curve in a counterclockwise or clockwise sense." It suffices to say here that the concept of orientation can be given mathematical rigor, but for our purposes an arrow indicating a given orientation will be adequate.

(12.B.11) *Theorem* Suppose that f is analytic on a region containing an annular region determined by rectifiable paths γ_1 and γ_2, where $I[\gamma_2]$ is contained in the interior of $I[\gamma_1]$. If γ_1 and γ_2 have the same orientation, then

$$\int_{\gamma_1} f(z)\, dz = \int_{\gamma_2} f(z)\, dz$$

Proof. In the proof it will be convenient to "identify" a path with its image, as we indicate in Fig. 12.9. From the comment following (12.B.10) it

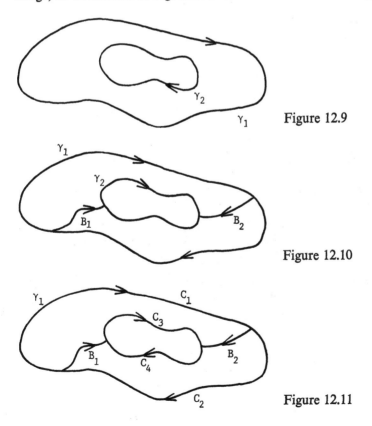

Figure 12.9

Figure 12.10

Figure 12.11

is not difficult to see that rectifiable paths B_1 and B_2 can be found such that $B_1 \cap B_2 = \emptyset$ and B_1 and B_2 join the simple closed curves γ_1 and γ_2 as indicated in Fig. 12.10. Now break the path γ_1 into paths C_1 and C_2 and the path γ_2 into paths C_3 and C_4, as indicated in the Fig. 12.11. Note that $C_1 \cdot B_2 \cdot (-C_3) \cdot (-B_1)$ and $C_2 \cdot B_1 \cdot (-C_4) \cdot (-B_2)$ are simple closed curves and that

$$\int_{C_1 \cdot C_2} f(z)\, dz = \int_{\gamma_1} f(z)\, dz$$

$$\int_{C_3 \cdot C_4} f(z)\, dz = \int_{\gamma_2} f(z)\, dz$$

By the Cauchy-Goursat theorem, (12.A.13), and (12.A.15.v), we have

$$0 = \int_{C_1} f(z)\, dz + \int_{B_2} f(z)\, dz - \int_{C_3} f(z)\, dz - \int_{B_1} f(z)\, dz \qquad (3)$$

and

$$0 = \int_{C_2} f(z)\, dz + \int_{B_1} f(z)\, dz - \int_{C_4} f(z)\, dz - \int_{B_2} f(z)\, dz \qquad (4)$$

Addition of equations (3) and (4) yields

$$\int_{C_1} f(z)\, dz + \int_{C_2} f(z)\, dz = \int_{C_3} f(z)\, dz + \int_{C_4} f(z)\, dz$$

or

$$\int_{\gamma_1} f(z)\, dz = \int_{\gamma_2} f(z)\, dz$$

(12.B.12) *Exercise*   Suppose that f is analytic on a region containing the "multiple annulus" determined by the simple closed rectifiable paths γ, γ_1, $\gamma_2, \ldots, \gamma_n$ indicated in Fig. 12.12. Show that if all these paths have the same orientation, then

$$\int_{\gamma} f(z)\, dz = \int_{\gamma_1} f(z)\, dz + \cdots + \int_{\gamma_n} f(z)\, dz$$

[*Hint*: Connect these paths with appropriate arcs.]

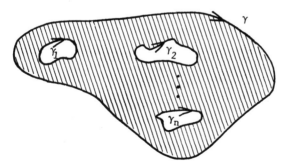

Figure 12.12

Perhaps the most important consequence of the Cauchy-Goursat theorem is the Cauchy integral formula.

(12.B.13) *Theorem (Cauchy Integral Formula)* Suppose that f is analytic on a region D, and let γ be a positively oriented simple closed rectifiable path in D such that the interior of $I[\gamma]$ is contained in D. If z^* is any point in the interior of $I[\gamma]$, then

$$f(z^*) = \frac{1}{2\pi i} \int_\gamma \frac{f(z)}{z - z^*}\, dz \tag{5}$$

To prove this theorem we need the next rather interesting lemma.

(12.B.14) *Lemma* Suppose that D is a simply connected region and that $\hat{z} \in D$. If $f: D\backslash\{\hat{z}\} \to \mathbb{C}$ is analytic and bounded, and if γ is a simple closed rectifiable path in $D\backslash\{\hat{z}\}$, then

$$\int_\gamma f(z)\, dz = 0$$

Proof. We may assume that γ is postively oriented. If $\hat{z}$ is not contained in the interior of $I[\gamma]$, then the result is an immediate consequence of the Cauchy-Goursat theorem. If $\hat{z}$ is in the interior of $I[\gamma]$, then we consider arbitrarily small concentric circles with center $\hat{z}$ (and lying in the interior of $I[\gamma]$) determined by paths of the form

$$\gamma_r(t) = \hat{z} + re^{it} \qquad 0 \le t \le 2\pi,\, r > 0$$

For any such path γ_r it follows from (12.B.11) that

$$\int_\gamma f(z)\, dz = \int_{\gamma_r} f(z)\, dz$$

Since f is bounded, it is clear from (12.A.15.iv) that $\lim_{r\to 0} \int_{\gamma_r} f(z)\, dz = 0$. Thus, $\int_\gamma f(z)\, dz$ must be 0.

Proof of (12.B.13). Note that from the hypothesis it follows that there is a simply connected bounded region E (whose closure is contained in D) that contains the curve $I[\gamma]$ and its interior. Let $g: E \to \mathbb{C}$ be defined by

$$g(z) = \begin{cases} \dfrac{f(z) - f(z^*)}{z - z^*} & \text{if } z \ne z^* \\[2mm] f'(z^*) & \text{if } z = z^* \end{cases}$$

Observe that $\lim_{z\to z^*} g(z) = f'(z^*) = g(z^*)$ (since f is analytic on $\bar{E}$), and therefore it follows that g is continuous on $\bar{E}$. Hence, g is bounded on E, and from (12.B.14) and (12.A.15.ii) we have

$$\int_\gamma \frac{f(z)}{z - z^*}\, dz = \int_\gamma \frac{f(z) - f(z^*)}{z - z^*}\, dz + \int_\gamma \frac{f(z^*)}{z - z^*}\, dz$$

$$= 0 + f(z^*) \int_\gamma \frac{dz}{z - z^*}$$

Since it follows from (12.A.11) and (12.B.11) that

$$f(z^*) \int_\gamma \frac{dz}{z - z^*} = f(z^*) \cdot 2\pi i$$

the proof is complete.

(12.B.15) *Examples* (a) If $f(z) = e^z$ and γ is any simple closed path enclosing 1, then $\int_\gamma [f(z)/(z - 1)]\, dz = 2\pi i f(1) = 2\pi i e$.

(b) To compute $\int_\gamma (\cos z)/(z^2 + 1)\, dz$ over the circle $\gamma(t) = 2e^{it}$, $0 \le t \le 2\pi$, we use partial fractions to obtain

$$\int_\gamma \frac{\cos z}{z^2 + 1}\, dz = \frac{1}{2} i \int_\gamma \frac{\cos z}{z + i}\, dz - \frac{1}{2} i \int_\gamma \frac{\cos z}{z - i}\, dz$$

$$= \frac{1}{2} i (2\pi i \cos (-i)) - \frac{1}{2} i (2\pi i \cos i) = -\pi \cos (-i) + \pi \cos i$$

(12.B.16) *Exercise* Suppose that f is analytic on a region that contains the annulus determined by the positively oriented circular paths

$$\gamma_1(t) = z_0 + r_1 e^{it} \qquad 0 \le t \le 2\pi$$
$$\gamma_2(t) = z_0 + r_2 e^{it} \qquad 0 \le t \le 2\pi$$

$(r_1 < r_2)$. Show that for each point z in this annular region

$$f(z) = \frac{1}{2\pi i} \int_{\gamma_2} \frac{f(w)}{w - z}\, dw = \frac{1}{2\pi i} \int_{\gamma_1} \frac{f(w)}{w - z}\, dw$$

Cauchy's integral formula may be generalized as follows.

(12.B.17) *Theorem* (*Generalized Cauchy Integral Formula*) Suppose that f is analytic on a region D and let γ be a positively oriented simple closed rectifiable path in D such that the interior of $I[\gamma]$ is contained in D. If z^* is any point in the interior of $I[\gamma]$, then $f^{(n)}(z^*)$ exists for $n = 1, \ldots,$ and furthermore,

$$f^{(n)}(z^*) = \frac{n!}{2\pi i} \int_\gamma \frac{f(z)}{(z - z^*)^{n+1}}\, dz$$

Proof. The proof is by induction on n. We have already established this result for $n = 0$ (the Cauchy integral formula). Suppose, then, that the result is true for $k = n - 1$ (and for each $z \in I[\gamma]$); we show that it also holds for $k = n$.

Since $f^{(n)}(z^*) = \lim_{h \to 0} [f^{(n-1)}(z^* + h) - f^{(n-1)}(z^*)]/h$, it suffices to show that

$$\lim_{h \to 0} \left| \frac{f^{(n-1)}(z^* + h) - f^{(n-1)}(z^*)}{h} - \frac{n!}{2\pi i} \int_\gamma \frac{f(z)}{(z - z^*)^{n+1}} \, dz \right| = 0 \qquad (6)$$

By the induction hypothesis, we have

$$\frac{f^{(n-1)}(z^* + h) - f^{(n-1)}(z^*)}{h} = \frac{(n-1)!}{2\pi i} \int_\gamma \frac{1}{h} \left(\frac{f(z)}{(z - z^* - h)^n} - \frac{f(z)}{(z - z^*)^n} \right) dz$$

$$= \frac{(n-1)!}{2\pi i} \int_\gamma \frac{1}{h} f(z) \left(\frac{(z - z^*)^n - (z - z^* - h)^n}{(z - z^* - h)^n (z - z^*)^n} \right) dz. \qquad (7)$$

Since $a^n - b^n = (a - b)(a^{n-1} + a^{n-2}b + \cdots + b^{n-1})$, the right-hand side of equation (7) becomes

$$\frac{(n-1)!}{2\pi i} \int_\gamma \frac{1}{h} f(z) \frac{h((z - z^*)^{n-1} + (z - z^*)^{n-2}(z - z^* - h) + \cdots + (z - z^* - h)^{n-1})}{(z - z^* - h)^n (z - z^*)^n} \, dz$$

Thus, the left-hand side of (6) can be written in the form

$$\lim_{h \to 0} \left| \frac{(n-1)!}{2\pi i} \int_\gamma f(z) \left[\frac{(z - z^*)^{n-1} + (z - z^*)^{n-2}(z - z^* - h) + \cdots + (z - z^* - h)^{n-1}}{(z - z^* - h)^n (z - z^*)^n} \right. \right.$$
$$\left. \left. - \frac{n}{(z - z^*)^{n+1}} \right] dz \right| \leq \frac{(n-1)!}{2\pi} \int_\gamma M \left| \frac{n(z - z^*)^{n-1}}{(z - z^*)^{2n}} - \frac{n}{(z - z^*)^{n+1}} \right| |dz|$$
$$= \frac{(n-1)!}{2\pi} \int_\gamma M \cdot 0 |dz| = 0$$

where M is a real number such that $|f(z)| \leq M$ for each $z \in I[\gamma]$.

(12.B.18) *Example* Suppose that $f(z) = \sqrt{z}$. Then $f^{(3)}(z) = (\frac{3}{8})z^{-\frac{5}{2}}$; hence, $\frac{3}{8} = f^{(3)}(1) = (3!/2\pi i) \int_\gamma \sqrt{z}/(z - 1)^4 \, dz$, and consequently,

$$\int_\gamma \frac{\sqrt{z} \, dz}{(z - 1)^4} = \frac{\pi i}{8}$$

where, say, $\gamma(t) = 1 + \frac{1}{2} e^{it}$, $0 \leq t \leq 2\pi$.

As a consequence of the generalized Cauchy integral theorem, we have the remarkable result that a function analytic on a region D has derivatives of all orders; this result, of course, has no counterpart in the theory of real variables.

(12.B.19) *Theorem* Suppose that f is analytic on a region D. Then f has derivatives of all orders in D.

Proof. For each $z \in D$, there is clearly a positively oriented path γ whose interior lies in D and such that z belongs to the interior of $I[\gamma]$. Thus, by (12.B.17), $f^{(n)}(z)$ exists for each integer $n \geq 1$.

Note that it follows from the preceding theorem that the function f: $D \to C$ defined by $f(z) = |z|$, where D is the unit disk centered at the origin, fails to have a primitive on D even though f is continuous on this set (if F were a primitive of f on D, then $F' = f$, and $F'' = f'$ would exist on D, which is clearly not the case at $z = 0$).

Finally, we obtain a strong converse to the Cauchy-Goursat theorem.

(12.B.20) *Theorem (Morera's Theorem)* Suppose that D is a region and that $f: D \to C$ is a continuous function. If $\int_{\gamma_T} f(z)\, dz = 0$ for every closed triangular path γ_T that together with its interior lie in D, then f is analytic on D (a closed *triangular path* in D is a closed polygonal path in D whose image is a triangle).

Proof. Note that it suffices to assume that D is a disk $\{z \mid |z - p| < r\}$ and to show that f' exists on this disk. For each $z \in D$, let $\gamma_{p,z}: [0,1] \to D$ be the straight line path from p to z defined by $\gamma_{p,z}(t) = p + (z - p)t$. Set

$$F(z) = \int_{\gamma_{p,z}} f(w)\, dw$$

for each $z \in D$, and let z_0 be a fixed but arbitrary point in D. For small values of h, the points $p, z_0, z_0 + h$ determine a triangle (in D), and hence it follows from the hypothesis of this theorem that

$$F(z_0 + h) = \int_{\gamma_{p,z_0+h}} f(w)\, dw = \int_{\gamma_{p,z_0}} f(w)\, dw + \int_{\gamma_{z_0,z_0+h}} f(w)\, dw$$

where $\gamma_{z_0,z_0+h}: [0,1] \to D$ is the straight-line path defined by

$$\gamma_{z_0,z_0+h}(t) = z_0 + ht.$$

Therefore, we have

$$\frac{F(z_0 + h) - F(z_0)}{h} = \frac{1}{h} \int_{\gamma_{z_0,z_0+h}} f(w)\, dw$$

and by the argument given in the proof of (12.B.2), it follows that

$$F'(z_0) = \lim_{h \to 0} \left(\frac{F(z_0 + h) - F(z_0)}{h} \right) = f(z_0)$$

hence, $F'(z) = f(z)$ on D. Thus, F is analytic on D, and consequently by (12.B.19) $f (= F')$ is also analytic on D.

C. POWER SERIES AND ANALYTIC FUNCTIONS

In this section we examine briefly the relationship between analytic functions and power series. The principal result is the following, which states that an analytic function can be represented locally by a power series.

(12.C.1) Theorem If f is analytic at a point z_0, and if D is an open disk centered at z_0 on which f is analytic, then for each $w \in D$

$$f(w) = \sum_{n=0}^{\infty} c_n(w - z_0)^n$$

where

$$c_n = \frac{f^{(n)}(z_0)}{n!}$$

Proof. We prove the theorem for the case $z_0 = 0$. In problem C.7 the reader is asked to make the necessary (and obvious) changes to establish the result for arbitrary z_0.

Suppose that R is the radius of D and $w \in D$. Choose r so that $|w| < r < R$. We shall integrate the left- and right-hand sides of the following easily derived identity (see problem C.6):

$$\frac{1}{2\pi i} \cdot \frac{f(z)}{z - w} = \frac{1}{2\pi i} \cdot \frac{f(z)}{z} \cdot \frac{1}{1 - w/z}$$

$$= \frac{1}{2\pi i} \frac{f(z)}{z} + \frac{1}{2\pi i} \frac{f(z)}{z^2} w + \cdots + \frac{1}{2\pi i} \frac{f(z)}{z^n} w^{n-1} + \frac{1}{2\pi i} \frac{f(z)}{(z - w)z^n} w^n$$

$$\tag{1}$$

over the simple closed path $\gamma(t) = re^{it}$ to obtain from (12.B.17)

$$f(w) = \frac{1}{2\pi i} \int_\gamma \frac{f(z)}{z - w} \, dz$$

$$= \frac{1}{2\pi i} \int_\gamma \frac{f(z)}{z} \, dz + \frac{w}{2\pi i} \int_\gamma \frac{f(z)}{z^2} \, dz + \cdots + \frac{w^{n-1}}{2\pi i} \int_\gamma \frac{f(z)}{z^n} \, dz$$

$$+ \frac{w^n}{2\pi i} \int_\gamma \frac{f(z)}{(z - w)z^n} \, dz$$

$$= f(0) + f'(0)w + \cdots + \frac{f^{(n-1)}(0)}{(n - 1)!} w^{n-1} + \frac{w^n}{2\pi i} \int_\gamma \frac{f(z)}{(z - w)z^n} \, dz$$

To complete the proof, it suffices to show that

$$\lim_{n \to \infty} \frac{w^n}{2\pi i} \int_\gamma \frac{f(z)}{(z - w)z^n} \, dz = 0$$

Since $\gamma(\theta) = re^{i\theta}$, $0 \le \theta \le 2\pi$, it follows from (12.A.10) that

$$\frac{w^n}{2\pi i} \int_\gamma \frac{f(z)}{(z-w)z^n} \, dz = \frac{w^n}{2\pi i} \int_0^{2\pi} \frac{f(re^{i\theta})(ire^{i\theta})}{(re^{i\theta}-w)r^n e^{in\theta}} \, d\theta$$

Furthermore, since $|re^{i\theta} - w| \ge |re^{i\theta}| - |w| = r - |w| > 0$, and since $|f|$ is continuous on the compact set $I[\gamma]$, and, hence, bounded by a real number M, we have from (12.A.15.iv)

$$\left| \frac{w^n}{2\pi i} \int_\gamma \frac{f(z)}{(z-w)z^n} \, dz \right| \le \left(\frac{Mr}{r-|w|} \right) \left(\frac{|w|}{r} \right)^n \tag{2}$$

Since $|w|/r < 1$, the right-hand side of (2) tends to 0 with increasing n, and this establishes the result.

The series

$$f(w) = \sum_{n=0}^{\infty} \frac{f^{(n)}(z_0)}{n!} (w-z_0)^n$$

is called the *Taylor series* for f centered at z_0.

(12.C.2) Examples (a) Suppose that $f(z) = 1/z^2$ and $z_0 = -1$. Then $c_0 = 1$ and for each $n \ge 1$, $c_n = f^{(n)}(-1)/n! = n + 1$; hence, $f(z) = 1 + \sum_{n=1}^{\infty} (n+1)(z+1)^n$, whenever $|z+1| < 1$.

(b) Suppose that $f(z) = (\cos z)^2 = (1 + \cos 2z)/2$ and $z_0 = 0$. Then $c_0 = 1$, $c_{2n} = (-1)^n 2^{2n-1}/(2n)!$, and $c_{2n+1} = 0$; hence, $f(z) = 1 + \sum_{n=1}^{\infty} (-1)^n [2^{2n-1}/(2n)!] z^{2n}$.

Power series representations of functions are especially useful since they are easily subject to many analytic operations. This is essentially due to the fact that a power series converges uniformly on every compact subset of its disk of convergence (8.E.3). In view of (12.C.1) it is natural to ask when does a convergent power series represent an analytic function. An answer to this query is found in the next theorem.

(12.C.3) Theorem If $f(z) = \sum_{n=0}^{\infty} c_n(z-z_0)^n$ has radius of convergence R, then f is analytic on the open disk $\{z \mid |z-z_0| < R\}$; furthermore, the functions f', f'', ... can be found on this disk by differentiating the series term by term. The coefficients c_n are defined by

$$c_n = \frac{f^{(n)}(z_0)}{n!}$$

Proof. Let A be an arbitrary compact subset of the disk of convergence D, and for each k, let $f_k(z) = \sum_{n=0}^{k} c_n(z-z_0)^n$. Then, since the sequence $\{f_k\}$ converges uniformly to f on A, it follows from (12.A.16) that f is continuous on D; furthermore, for each triangular path γ in D

$$0 = \lim_{k \to \infty} \int_\gamma f_k(z)\, dz = \int_\gamma f(z)\, dz$$

and hence by Morera's theorem (12.B.20) f is analytic on D.

We now differentiate the power series $\sum_{n=0}^\infty c_n(z - z_0)^n$ term by term to obtain the derived series $g(z) = \sum_{n=1}^\infty nc_n(z - z_0)^{n-1}$. It follows from (8.E.12) that the original series and the derived series have the same disk of convergence D, and, moreover, on each compact subset A of D, the sequence $\{f_k'\}$ converges uniformly to g. For each k, set $F_k(z) = \int_{\gamma^z} f_k'(w)\, dw$, where γ^z is a rectifiable path in D with initial point z_0 and terminal point z. Then by (12.A.16) we have

$$\lim_{k \to \infty} F_k(z) = \lim_{k \to \infty} \int_{\gamma^z} f_k'(w)\, dw = \int_{\gamma^z} g(w)\, dw$$

From the proof of (12.B.2) we find that F_k is a primitive of f_k' and that $\int_{\gamma^z} g(w)\, dw$ is a primitive of g. Since $F_k(z_0) = 0$, it is clear that $f_k(z) = F_k(z_0) + c_0$, and hence $\int_{\gamma^z} g(z)\, dz = f(z) - c_0$. Thus, $f' - g$ on D.

It we successively differentiate the series for $f(z)$ and set $z = z_0$, then we find (inductively) that for each n, $c_n = [f^{(n)}(z_0)]/n!$, and this concludes the proof.

We finish this section with a variety of interesting and useful results, including Cauchy's estimate formula, Liouville's theorem, and theorems involving the extensions and Taylor series representations of analytic functions.

(12.C.4) Theorem (Cauchy's Estimate Theorem) Suppose that a function f is analytic on a disk $S_\rho(z_0)$ and that $|f(z)| \le M$ for each $z \in S_\rho(z_0)$. Then

$$|f^{(n)}(z_0)| \le \frac{n!M}{\rho^n} \qquad n = 0, 1, \ldots$$

Proof. Let $\gamma: [0, 2\pi] \to \mathbf{C}$ be the path defined by $\gamma(\theta) = z_0 + re^{i\theta}$, $0 < r < \rho$. By (12.B.17) and (12.A.10) we have

$$f^{(n)}(z_0) = \frac{n!}{2\pi i} \int_\gamma \frac{f(z)}{(z - z_0)^{n+1}}\, dz = \frac{n!}{2\pi i} \int_0^{2\pi} \frac{f(re^{i\theta})}{(re^{i\theta})^{n+1}} ire^{i\theta}\, d\theta$$

and hence it follows that

$$|f^{(n)}(z_0)| \le \frac{n!}{2\pi} \int_0^{2\pi} \frac{|f(re^{i\theta})|}{r^n}\, d\theta \le \frac{n!}{2\pi} \int_0^{2\pi} \frac{M}{r^n}\, d\theta = \frac{n!M}{r^n} \tag{3}$$

Since (3) holds for each r, $0 < r < \rho$, the desired result is obtained.

(12.C.5) Definition A function $f: \mathbf{C} \to \mathbf{C}$ is said to be an *entire* function if f is analytic on $\mathbf{C}$.

(12.C.6) *Theorem (Liouville's Theorem)*   If f is a bounded entire function, then f is a constant-valued function on $\mathbf{C}$.

Proof.   Since f is an entire function, f is analytic on every disk $S_\rho(0)$, centered at $z = 0$. By hypothesis there is a real number M such that $|f(z)| \leq M$ for each $z \in \mathbf{C}$. Therefore, it follows from (12.C.4) that $|f^{(n)}(0)| \leq n!M/\rho^n$ for each nonnegative integer n. Thus, since ρ can be made arbitrarily large, we have that $f^{(n)}(0) = 0$ for each positive integer n. Furthermore, by (12.C.1) f can be represented on $\mathbf{C}$ by the power series

$$f(z) = \sum_{n=0}^{\infty} \frac{f^{(n)}(0)}{n!} z^n$$

and, therefore, it follows that $f(z) = f(0)$ for each $z \in \mathbf{C}$.

In Chapter 8 we saw that the convergence of the Taylor series expansion of a function f did not ensure that this series would converge to f (Example (8.D.5)). As we shall see presently, in the context of analytic functions such aberrant examples cannot occur. This is but one of the many reasons why Taylor series expansions are best studied in the setting of complex analysis. We begin our brief investigation of the convergence properties of Taylor series expansions of complex functions with the following concept.

(12.C.7) *Definition*   A subset S of $\mathbf{C}$ is said to be *isolated* if each point $p \in S$ is contained in a neighborhood that excludes all other points of S.

(12.C.8) *Definition*   A set U is said to be a *deleted neighborhood* of a point z_0 if:

(i) $z_0 \notin U$.
(ii) $(S_\varepsilon(z_0))\backslash\{z_0\} \subset U$ for some $\varepsilon > 0$.

Our first major result is the following.

(12.C.9) *Theorem*   Suppose that f is a nonconstant analytic function defined on a region $D \subset \mathbf{C}$ and let $c \in \mathbf{C}$. Then $\{z \in D \mid f(z) = c\}$ is an isolated set.

We first prove a special case of this theorem.

(12.C.10) *Lemma*   Suppose that f is a nonconstant analytic function on an open disk, $S_R(z_0)$, and let $c \in \mathbf{C}$. Then $\{z \in S_R(z_0) \mid f(z) = c\}$ is an isolated set.

Proof.   Suppose first that $c = 0$ and that $f(\hat{z}) = 0$ for some point $\hat{z} \in S_R(z_0)$. Since f is analytic on an open disk $S_\rho(\hat{z}) \subset S_R(z_0)$, it follows from (12.C.1) that f can be represented on $S_\rho(\hat{z})$ by a convergent power series expansion:

$$f(z) = \sum_{k=0}^{\infty} c_k (z - \hat{z})^k \qquad (4)$$

Since $f(\hat{z}) = 0$, the coefficient c_0 must be 0, and therefore, in order to complete the proof of the case where $c = 0$, we must show that $c_k \neq 0$ for some $k > 0$. Suppose, to the contrary, that $c_k = 0$ for each k; this, of course, implies that $f(z) \equiv 0$ on $S_\rho(\hat{z})$. We now show that under this assumption, $f(z) \equiv 0$ on $S_R(z_0)$ (which contradicts the hypothesis of the lemma). Observe that for some positive integer n, there are open disks $S_{\rho_j}(\hat{z}_j) \subset S_R(z_0)$ $(j = 1, \ldots, n)$ such that:

(i) $\hat{z}_1 = \hat{z}, \rho_1 = \rho$.
(ii) $\hat{z}_n = z_0$.
(iii) $\hat{z}_j \in S_{\rho_{j-1}}(\hat{z}_{j-1})$ for $j = 2, \ldots, n$.

For each integer j, $1 \leq j \leq n$, there is by (12.C.1) a power series representation of f on $S_{\rho_j}(\hat{z}_j)$:

$$f(z) = \sum_{k=0}^{\infty} c_k^{(j)} (z - \hat{z}_j)^k$$

Since by our assumption $f(z) \equiv 0$ on $S_{\rho_1}(\hat{z}_1) = S_\rho(\hat{z})$, and $\hat{z}_2 \in S_{\rho_1}(\hat{z}_1)$, it follows that $f(z) \equiv 0$ on a small neighborhood of $\hat{z}_2$. However, this implies that $c_k^{(2)} = 0$ for all k (otherwise, on some deleted neighborhood of $\hat{z}_2$, f would not be zero, which contradicts our assumption). In a similar fashion we can show that for $j = 1, 2, \ldots, n$, $c_k^{(j)} = 0$ for all k. In particular, since $c_k^{(n)} = 0$ for all k, it follows that $f(z) \equiv 0$ in a full neighborhood of z_0. However, this is impossible since by (12.C.1) and the hypothesis of the lemma, f can be represented on $S_R(z_0)$ by a convergent power series

$$f(z) = \sum_{k=0}^{\infty} a_k (z - z_0)^k$$

where not all the a_k are equal to 0, which implies that f cannot be identically 0 on any deleted neighborhood of z_0. Hence, we conclude that not all c_k appearing in (4) are zero, and consequently, on $S_\rho(\hat{z})$, f is represented by

$$f(z) = (z - \hat{z})^k (c_k + c_{k+1}(z - \hat{z}) + \cdots) \qquad (5)$$

where $k > 0$ and $c_k \neq 0$. Since $g(z) = c_k + c_{k+1}(z - \hat{z}) + \cdots$ is analytic and nonvanishing in some neighborhood, $S_\rho(\hat{z})$, of $\hat{z}$, it follows that f is nonvanishing on the deleted neighborhood $S_\rho(z_0) \backslash \{\hat{z}\}$, which shows that the zeros of f are isolated on $S_R(z_0)$. For the case that $c \neq 0$, let $z^* \in S_R(z_0)$ be such that $f(z^*) = c$, and set $h(z) = f(z) - f(z^*)$. Then h is an analytic function on $S_R(z_0)$ and has a zero at z^*. Since this zero is isolated, we have that $f(z) \neq c$ in a deleted neighborhood of z^*, and this concludes the proof.

We now establish (12.C.9). Suppose first that $c = 0$ and that $z_0 \in D$ is a zero of f which is not isolated. Then it follows from (12.C.10) that there is a $\rho > 0$ such that $S_\rho(z_0) \subset D$, and $f(z) = 0$ on $S_\rho(z_0)$. Let z' be any point in D. Since D is a connected subset of C, there is a polygonal path γ contained in D joining z_0 to z'. It is easy to see that there is a set of points $\{z'_1, z'_2, \ldots, z'_n\}$ in $I[\gamma]$ and a set of disks $S_{\rho_j}(z'_j) \subset D$, $1 \leq j \leq n$, such that:

(i) $z'_1 = z_0$, $\rho = \rho_1$.

(ii) $z'_n = z'$.

(iii) $z'_j \in S_{\rho_{j-1}}(z'_{j-1})$ for $j = 2, \ldots, n$

Since $f(z) \equiv 0$ on $S_{\rho_1}(z'_1)$ and $z'_2 \in S_{\rho_1}(z'_1)$, it follows that z_2 is not an isolated zero of f; hence, by (12.C.10), $f(z) \equiv 0$ on $S_{\rho_2}(z'_2)$. Continuing in this way we find that $f(z) \equiv 0$ on $S_{\rho_j}(z'_j)$ for $j = 1, \ldots, n$, and therefore, in particular, $f(z') = f(z'_n) = 0$. Since z' was an arbitrary point of D, it follows that $f(z) \equiv 0$ on D, which contradicts the hypothesis. Consequently, every zero of f on D is isolated. The reader may repeat the argument found at the end of the previous lemma to establish the case where $c \neq 0$.

As an immediate corollary we obtain the following quite remarkable result.

(12.C.11) Corollary If an analytic function can be extended, it can be extended analytically in only one way. That is, if D_1 and D_2 are regions in C such that $D_1 \subset D_2$ and if $f_1: D_1 \to C$, $f_2: D_2 \to C$ are analytic functions such that $f_1 = f_2$ on D_1, then f_1 can be extended to D_2 as an analytic function only as the function f_2.

Proof. Note that if f_2 and $\hat{f}_2$ were two distinct analytic extensions of f_1, then $g = f_2 - \hat{f}_2$ would be a nonconstant analytic function whose zeros do not form an isolated set, which is impossible.

Next we obtain a major theorem in complex analysis, which has no counterpart in the theory of real variables.

(12.C.12) Theorem If f is an analytic function on a disk $S_\rho(z_0)$, and if the Taylor series expansion $T^f_{z_0}$ of f about z_0 converges on $S_R(z_0)$, where $R > \rho$, then f can be extended to $S_R(z_0)$, and furthermore, on $S_R(z_0)$, $T^f_{z_0} = f$.

Proof. By (12.C.1) $f = T^f_{z_0}$ on $S_\rho(z_0)$ and by (12.C.3) the function $g: S_R(z_0) \to C$ defined by $g(z) = T^f_{z_0}(z)$ is analytic on $S_R(z_0)$. Since $g(z) \equiv f(z)$ on $S_\rho(z_0)$, it follows from (12.C.11) that f can be extended analytically to $S_R(z_0)$ in exactly one way, which means that $f = g$ on $S_R(z_0)$; therefore, the extension of f to $S_R(z_0)$ is represented by its Taylor series expansion $T^f_{z_0}$.

The reader is asked to establish the following corollary in problem C.8.

(12.C.13) *Corollary* Suppose that f is infinitely differentiable and that its Taylor series $T^f_{x_0}$ converges on the interval $(x_0 - r, x_0 + r)$. Suppose further that f can be extended into the complex plane as an analytic function on some region A containing $(x_0 - r, x_0 + r)$. Then $f(x) = T^f_{x_0}(x)$ on the interval $(x_0 - r, x_0 + r)$.

D. SINGULARITIES AND THE LAURENT EXPANSION

(12.D.1) *Definition* A function f is said to have an *isolated singularity* at a point z_0 if f is analytic on some deleted neighborhood of z_0.

(12.D.2) *Example* Suppose that $f(z) = (\sin z)/z$. Then $z_0 = 0$ is an isolated singularity of f. Note, however, that since $f(z) = 1 - z^2/3! + z^4/5! - \cdots$ on $\mathbf{C}\backslash\{0\}$, f is clearly extendible to an analytic function on $\mathbf{C}$ (by setting $f(0) = 1$).

(12.D.3) *Example* If $f(z) = 1/z^2$, no analytic extension to $\mathbf{C}$ is possible since $\lim_{z\to 0} 1/z^2 = \infty$. Furthermore, we shall see later that if $f(z) = e^{1/z}$, then $f(z)$ can be made to approach *any* complex number through the use of a suitable choice of values of z near 0. Thus, it would appear that there exists a variety of types of isolated singularities.

(12.D.4) *Definition* A point z_0 is said to be a *removable singularity* of a function f if in some deleted neighborhood of z_0, f is equal to a function g that is analytic in a *full* neighborhood of z_0 [in which case, f is extendible as an analytic function to a full neighborhood of z_0 by setting $f(z_0)$ equal to $g(z_0)$].

As was seen in the previous example, $z_0 = 0$ is a removable singularity of the function $f(z) = (\sin z)/z$. The next theorem provides two analytical characterizations of a removable singularity.

(12.D.5) *Theorem* The following three conditions on a function f are equivalent:

    (i) f has a removable singularity at a point z_0.
    (ii) f is analytic and bounded in some deleted neighborhood of z_0.
    (iii) f is analytic in some deleted neighborhood of z_0 and

$$\lim_{z\to z_0} (z - z_0)f(z) = 0$$

Proof. (i) implies (ii) This follows easily from the definition of a removable singularity.

(ii) implies (iii)   This is clear from the boundedness of f.

(iii) implies (i) Let

$$U = \{z \mid 0 < |z - z_0| < \rho\}$$

denote a deleted neighborhood on which f is analytic. If we set

$$g(z) = \begin{cases} (z - z_0)f(z) & \text{if } z \in U \\ 0 & \text{if } z = z_0 \end{cases}$$

then g is continuous on the full neighborhood $S_\rho(z_0)$ of z_0 and is analytic and bounded on U. Hence, by (12.B.5) $\int_\gamma g(z)\, dz = 0$ on every simple closed rectifiable path in U. Suppose that γ_T is a triangular path in $S_\rho(z_0)$ and that a, b, c are the vertices of $I[\gamma_T]$, where $z_0 = a$. Let $\varepsilon > 0$ be given, and choose points b' and c' on the line segments connecting a to b and a to c, respectively, which are sufficiently close to a so that by (12.A.15.iv), $\int_{\tilde\gamma_T} g(z)\, dz < \varepsilon$, where $\tilde\gamma_T$ is the triangular path with vertices a, b', c' and with the same orientation as γ_T. Now let $\hat\gamma$ be the complementary polygonal path with vertices b, b', c', c and with the same orientation as γ_T. Clearly, since $\hat\gamma$ is a simple closed path in U, we have

$$\int_{\gamma_T} g(z)\, dz = \int_{\tilde\gamma_T} g(z)\, dz + \int_{\hat\gamma} g(z)\, dz = \int_{\tilde\gamma_T} g(z)\, dz$$

and since $\varepsilon > 0$ was arbitrary, it follows that $\int_{\gamma_T} g(z)\, dz = 0$. A similar argument can be made in the case that z_0 is not a vertex of $I(\gamma_T)$, and, therefore, by Morera's theorem we have that g is analytic in a full neighborhood, $S_\rho(z_0)$, of z_0.

Now it follows from (12.C.1) that g can be written as a convergent Taylor series on the disk $S_\rho(z_0)$:

$$g(z) = \sum_{n=1}^{\infty} c_n(z - z_0)^n \qquad \text{(since } g(z_0) = 0\text{)}$$

Hence, on U we have

$$f(z) = \frac{g(z)}{z - z_0} = \sum_{n=1}^{\infty} c_n(z - z_0)^{n-1}$$

Since this series converges on $S_\rho(z_0)$, it represents an analytic function on $S_\rho(z_0)$ (12.C.3), and, consequently, f has a removable singularity at z_0.

We now consider the case where z_0 is an isolated singularity of a function f and f is unbounded in a deleted neighborhood of z_0.

(12.D.6) Definition If z_0 is an isolated singularity of a function f and if $\lim_{z \to z_0} |f(z)| = \infty$, then z_0 is called a *pole* of f. If f is unbounded in a deleted neighborhood of z_0, but $\lim_{z \to z_0} |f(z)| \neq \infty$, then z_0 is called an *essential singularity* of f.

The next theorem provides an important characterization of a pole.

(12.D.7) Theorem A point z_0 is a pole of a function f if and only if z_0 is an isolated singularity of f and there is some integer $m > 0$ such that $\lim_{z \to z_0} (z - z_0)^m f(z)$ exists, is finite, and not equal to zero.

Proof. If z_0 is an isolated singularity and if $\lim_{z \to z_0} (z - z_0)^m f(z) = a$ ($\neq 0$), then $\lim_{z \to z_0} |f(z)| = \lim_{z \to z_0} |a|/|z - z_0|^m = \infty$, and, hence, z_0 is a pole of f.

Conversely, suppose that z_0 is a pole of f. Then there is a deleted neighborhood, $U = S_\rho(z_0) \backslash \{z_0\}$, of z_0 on which f is analytic and $|f(z)| > 1$. Therefore, we can define a bounded analytic function g on U by setting $g(z) = 1/f(z)$. From (12.D.5), we have that z_0 is a removable singularity of g, and, consequently, we may assume that g is analytic on the full disk $S_\rho(z_0)$; note that $g(z) \neq 0$ on U and that $g(z_0) = 0$. It now follows from (12.C.1) and (12.C.3) that there is an integer $m > 0$ such that $g(z) = (z - z_0)^m h(z)$, where $h(z)$ is analytic and not equal to zero anywhere on $S_\rho(z_0)$. Thus, $k(z) = 1/h(z)$ is analytic and nowhere equal to zero on the disk $S_\rho(z_0)$, and, therefore, if $z \in U$, then $f(z) = [1/(z - z_0)^m] k(z)$ [where $k(z_0) \neq 0$]. Hence, we have

$$\lim_{z \to z_0} (z - z_0)^m f(z) = k(z_0) \neq 0$$

which concludes the proof.

Since in the above proof $k(z)$ can be represented by a convergent power series of the form $k(z) = \sum_{n=0}^{\infty} a_n (z - z_0)^n$ ($a_0 \neq 0$), it follows that on U,

$$f(z) = \frac{a_0}{(z - z_0)^m} + \cdots + \frac{a_{m-1}}{(z - z_0)} + a_m + a_{m+1}(z - z_0) + \cdots$$

It is customary to relabel the coefficients so that

$$f(z) = \frac{c_{-m}}{(z - z_0)^m} + \cdots + \frac{c_{-1}}{z - z_0} + \sum_{n=0}^{\infty} c_n (z - z_0)^n \qquad (c_{-m} \neq 0) \quad (1)$$

where $\sum_{n=0}^{\infty} c_n (z - z_0)^n$ is a convergent power series in some neighborhood of z_0.

The finite sum $c_{-m}/(z - z_0)^m + \cdots + c_{-1}/(z - z_0)$ is called the *singular* (or *principal*) *part* of the expansion of f at z_0, and the series $\sum_{n=0}^{\infty} c_n (z - z_0)^n$ is called the *regular part* of this expansion.

If we define $S(\xi) = \sum_{n=1}^{m} c_{-n} \xi^n$ and $R(\xi) = \sum_{n=0}^{\infty} c_n \xi^n$, then (1) may be conveniently expressed by

$$f(z) = S\left(\frac{1}{z - z_0}\right) + R(z - z_0) \tag{2}$$

(12.D.8) *Observation* The expansion given in (2) is unique. To see this, suppose that $S_1(1/(z - z_0)) + R_1(z - z_0)$ is another such expansion of f on U. Then $S_1(1/(z - z_0)) - S(1/(z - z_0)) = R(z - z_0) - R_1(z - z_0)$, and since $R(z - z_0) - R_1(z - z_0)$ is analytic and bounded on $S_\rho(z_0)$, we find immediately that if $\xi = 1/(z - z_0)$, then $S_1(\xi) - S(\xi)$ is analytic and bounded for $|\xi| > 1/\rho$. Since $S_1(\xi) - S(\xi)$ has a power series representation about 0, it follows that $S_1 - S$ is bounded and analytic on all of $\mathbf{C}$. Hence, by Liouville's theorem, $S_1(\xi) - S(\xi)$ is identically equal to a constant a, and therefore, $S_1(\xi) = c_{-m}\xi^m + \cdots + c_{-1}\xi + a$. Since the singular part contains no constant term, it must be the case that $a = 0$; thus, $S_1 \equiv S$ and $R_1 \equiv R$.

(12.D.9) *Definition* A function f is said to have a *pole of order m* at a point z_0 if f can be represented in the form (1).

(12.D.10) *Definition* A function f that is analytic on a region D except for a set of isolated poles is said to be *meromorphic* on D.

(12.D.11) *Exercise* Suppose that f is meromorphic on $\mathbf{C}$ and that z_0 is a pole of f. Then $f(z) = S(1/(z - z_0)) + R(z - z_0)$. Show that $R(z - z_0)$ is analytic on the disk $S_\rho(z_0)$, where ρ is the distance from z_0 to the nearest pole (if one exists); otherwise, $\rho = \infty$. [*Hint*: Consider $f(z) - S(1/(z - z_0))$.]

Suppose that $f(z) = P(z)/Q(z)$ is a rational function, where P and Q have no common factors. If $Q(z) = \prod_{j=1}^n (z - z_j)^{m_j}$, then f has a finite number of poles (which are precisely at the points z_j).

For $j = 1, 2, \ldots, n$, let

$$S_j\left(\frac{1}{z - z_j}\right) = \frac{c_{-m_j}^{(j)}}{(z - z_j)^{m_j}} + \cdots + \frac{c_{-1}^{(j)}}{z - z_j}$$

be the singular part of the expansion of f about z_j. Since

$$g(z) = f(z) - \sum_{j=1}^n S_j\left(\frac{1}{z - z_j}\right)$$

is analytic at $z_1, z_2, \ldots, z_n$ (why?), g is analytic everywhere. Moreover, we assert that g is a polynomial. Clearly, g is a rational function on $\mathbf{C}$ and, hence, may be written as a quotient of polynomials p and q that have no common factors. Since g has a pole whenever q has a zero, it follows that q has no zero. Hence, q must be a constant (nonzero) function, and consequently, g is a polynomial.

Suppose that f is meromorphic on $\mathbf{C}$ and let γ be a simple closed positively oriented path such that

1. $I[\gamma]$ contains no pole of f.
2. There exists exactly one pole (of order m) in the interior of $I[\gamma]$.

Then by (12.D.11) we have $f(z) = S(1/(z - z_0)) + R(z - z_0)$, where $R(z - z_0)$ is analytic on an open disk $S_\rho(z_0)$. If this disk contains the closure of the interior of $I[\gamma]$, then it follows from the Cauchy-Goursat theorem and (12.A.11) that

$$\int_\gamma f(z)\, dz = \int_\gamma S\left(\frac{1}{z - z_0}\right) dz + \int_\gamma R(z - z_0)\, dz = \int_\gamma S\left(\frac{1}{z - z_0}\right) dz$$

$$= \int_\gamma \frac{c_{-m}\, dz}{(z - z_0)^m} + \cdots + \int_\gamma \frac{c_{-1}\, dz}{z - z_0} = 2\pi i c_{-1} \qquad (3)$$

(Actually it can be shown by analytic continuation that $\int_\gamma R(z - z_0)\, dz = 0$ even in the case that the closure of the interior of $I[\gamma]$ is not contained in $S_\rho(z_0)$, and hence, the above equations are valid in this case as well.)

From (3) we see that the calculation of c_{-1} is tantamount to the calculation of the integral of f over the path γ; c_{-1} is called the *residue* of f at z_0 and is denoted by $\text{Res}(f; z_0)$. The question now arises whether c_{-1} can be calculated without first directly calculating $\int_\gamma f(z)\, dz$. In the event that z_0 is a simple pole (a pole of order 1), this is the case since

$$c_{-1} = \lim_{z \to z_0} (z - z_0) f(z)$$

In the next section we shall examine additional instances where such calculations can be effected.

Next we establish the remarkable result that if z_0 is an essential singularity of an analytic function f, and if U is any small neighborhood of z_0, then the function f can be made to approximate (in U) *any* complex number a.

(12.D.12) *Theorem (Casorati-Weierstrass Theorem)* Suppose that z_0 is an essential singularity of a function f that is analytic in a deleted neighborhood U of z_0. Let $\varepsilon > 0$ be given. If a is any complex number, then there is a point $z \in U$ such that $|f(z) - a| < \varepsilon$.

Proof. If no such point z can be found, then $|f(z) - a| \geq \varepsilon$ for each $z \in U$. Therefore, if $g(z) = [f(z) - a]/(z - z_0)$, then $\lim_{z \to z_0} |g(z)| = \infty$, and consequently, g has a pole at z_0. But this implies that

$$f(z) = g(z)(z - z_0) + a$$

either has a pole or a removable singularity at $z = z_0$, which is impossible.

In problem D.4 the reader is asked to show that $z = 0$ is an essential singularity of the function $f(z) = e^{1/z}$.

From our work thus far it would appear likely that if z_0 is an isolated singularity of a function f, then in a deleted neighborhood of z_0,

$$f(z) = S\left(\frac{1}{z - z_0}\right) + R(z - z_0),$$

where $R(z - z_0)$ is analytic in a full neighborhood of z_0 and $S(\xi)$ is analytic on C. In the next theorem we find that such an expansion is valid if f is analytic on an annulus centered at z_0, i.e., on a set of the form $\{z \mid \rho_1 < |z - z_0| < \rho_2\}$.

(12.D.13) *Theorem (Laurent Expansion Theorem)*   If $f: D \to C$ is analytic in a region D that contains an annulus $A = \{z \mid \rho_1 < |z - z_0| < \rho_2\}$, where $0 \le \rho_1 < \rho_2$, then for each $z \in A$,

$$f(z) = \sum_{n=1}^{\infty} c_{-n}(z - z_0)^{-n} + \sum_{n=0}^{\infty} c_n(z - z_0)^n = \sum_{n=-\infty}^{\infty} c_n(z - z_0)^n \quad (4)$$

Furthermore, for $n = 0, \pm 1, \pm 2, \ldots$,

$$c_n = \frac{1}{2\pi i} \int_{\gamma_\rho} \frac{f(z)}{(z - z_0)^{n+1}} \, dz \quad (5)$$

where $\rho_1 < \rho < \rho_2$ and γ_ρ is the path defined by $\gamma_\rho(t) = z_0 + \rho e^{it}$, $0 \le t \le 2\pi$.

If we set $S(1/(z - z_0)) = \sum_{n=1}^{\infty} c_{-n}(z - z_0)^{-n}$ and $R(z - z_0) = \sum_{n=0}^{\infty} c_n(z - z_0)^n$, then S and R are called the *singular* (or *principal*) and *regular parts* of the expansion (4), respectively. The singular part of this expansion converges for all z in the exterior of the circle of radius ρ_1 centered at z_0, and the regular part will converge for all z in the interior of the circle of radius ρ_2 centered at z_0. The paths γ_ρ indicated in the Laurent expansion theorem may be replaced by any simple closed positively oriented path whose image is contained in A. Finally, it can be shown that the expansion given in (4) is unique. A proof of (12.D.13) may be found in Ahlfors (1966).

(12.D.14) *Observation*   It is understood that the doubly infinite series $\sum_{n=-\infty}^{\infty} c_n(z - z_0)^n$ converges at a point z if both the singular part $\sum_{n=1}^{\infty} c_{-n}(z - z_0)^{-n}$ and the regular part $\sum_{n=0}^{\infty} c_n(z - z_0)^n$ converge at z. Note that if $n = -1$ in (5), then $\int_{\gamma_\rho} f(z) \, dz = 2\pi i c_{-1}$, as we would expect. Observe also that if z_0 is an isolated singularity, then the singular part $S(\xi)$ is an entire function. Furthermore, if z_0 is an isolated singularity, then the condition $S(1/(z - z_0)) \equiv 0$ implies that z_0 is a removable singularity. If $S(\xi)$ is a nonzero polynomial, then z_0 is a pole, and if $S(\xi)$ has infinitely many nonzero coefficients, then z_0 is an isolated singularity.

E.   AN INTRODUCTION TO RESIDUE THEORY

In the previous section we have seen that if a function f is analytic in a deleted neighborhood of a point z_0, then

$$f(z) = \sum_{n=-\infty}^{\infty} c_n (z - z_0)^n$$

where for each integer n,

$$c_n = \frac{1}{2\pi i} \int_\gamma \frac{f(z)}{(z - z_0)^{n+1}} \, dz$$

and where γ is any positively oriented simple closed path lying in the deleted neighborhood of z_0 and having z_0 in its interior. Furthermore, we have defined the residue of f at z_0 by

$$\text{Res}\,(f; z_0) = c_{-1} = \frac{1}{2\pi i} \int_\gamma f(z) \, dz$$

With the aid of (12.B.13) we are now in a position to obtain the following fundamental theorem.

(12.E.1) Theorem (General Residue Theorem) Suppose that γ is a positively oriented simple closed path and that f is a function that is analytic on $I[\gamma]$ and its interior except at a finite number of (isolated) singular points $z_1, \ldots, z_n$ which lie in the interior of $I[\gamma]$. Then

$$\int_\gamma f(z) \, dz = 2\pi i [\text{Res}\,(f; z_1) + \cdots + \text{Res}\,(f; z_n)]$$

Proof. For each integer k, $1 \le k \le n$, let γ_k be a positively oriented simple closed path containing z_k in its interior, and which lies in the interior of $I[\gamma]$ as indicated in Fig. 12.13. Then by (12.B.12),

$$\int_\gamma f(z) \, dz = \int_{\gamma_1} f(z) \, dz + \int_{\gamma_2} f(z) \, dz + \cdots + \int_{\gamma_n} f(z) \, dz$$

and since for $k = 1, 2, \ldots, n$,

$$\frac{1}{2\pi i} \int_{\gamma_k} f(z) \, dz = \text{Res}\,(f; z_k)$$

the result follows.

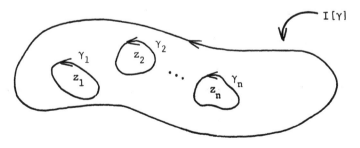

Figure 12.13

The difficulty, of course, in applying the residue theorem lies in the calculation of residues. Although residues are defined as integrals, they frequently can be computed without recourse to integration. For instance, in the next theorem we see that if z_0 is a pole of a function f, then residues are calculated via differentiation.

(12.E.2) *Theorem* Suppose that z_0 is a pole of order m of a function f. Let $\phi(z) = (z - z_0)^m f(z)$. Then

$$\text{Res}\,(f; z_0) = \frac{1}{(m - 1)!} \lim_{z \to z_0} \phi^{(m-1)}(z)$$

Proof. Since z is a pole of order m of f, we have

$$f(z) = \frac{c_{-m}}{(z - z_0)^m} + \cdots + \frac{c_{-1}}{z - z_0} + c_0 + c_1(z - z_0) + \cdots$$

and therefore

$$\begin{aligned}
\phi(z) &= (z - z_0)^m f(z) \\
&= c_{-m} + c_{-m+1}(z - z_0) + c_{-m+2}(z - z_0)^2 + \cdots \\
&\quad + c_{-1}(z - z_0)^{m-1} + \cdots
\end{aligned}$$

is a Taylor series. By (12.C.3) $\phi^{(n)}(z)$ exists for each n, and clearly

$$\phi^{(m-1)}(z_0) = (m - 1)!c_{-1}$$

from which the result follows.

(12.E.3) *Example* The function $f(z) = (\sin z)/(z - \pi/2)^3$ has a pole of order 3 at $\pi/2$. Since $\phi(z) = (z - \pi/2)^3 f(z) = \sin z$, we have $\phi''(z) = -\sin z$ and

$$\text{Res}\left(f; \frac{\pi}{2}\right) = \left(\frac{1}{2}\right) \lim_{z \to \pi/2} (-\sin z) = -\frac{1}{2}$$

For *simple poles* (poles of order 1) we have the following corollary.

(12.E.4) *Corollary* Suppose that $h(z) = f(z)/g(z)$ and that f is analytic at a point z_0 and g is analytic and has a zero of order 1 at z_0. Then

$$\text{Res}\,(h; z_0) = \frac{f(z_0)}{g'(z_0)}$$

Proof. By (12.E.2) we have

$$\text{Res}\,(h; z_0) = \lim_{z \to z_0} (z - z_0)h(z)$$

$$= \lim_{z \to z_0} (z - z_0)\frac{f(z)}{g(z)}$$

$$= \lim_{z \to z_0} \frac{z - z_0}{g(z) - g(z_0)} \cdot f(z)$$

$$= \frac{f(z_0)}{g'(z_0)}$$

What happens if z_0 is not a pole of f? If z_0 is a removable singularity of f, then the Laurent series for f at z_0 reduces to a Taylor series, and consequently, $\text{Res}(f;z_0) = 0$. If z_0 is an essential singularity, then in many instances the residue is found by determining the Laurent series expansion of f.

For example, the function $f(z) = z^{2n}\sin(1/z)$ has an essential singularity at $z = 0$. The residue can be calculated by expanding $\sin(1/z)$ and observing that

$$z^{2n} \sin \frac{1}{z} = z^{2n}\left(\frac{1}{z} - \frac{1}{3!z^3} + \cdots + \frac{(-1)^{n-1}}{(2n-1)!z^{2n-1}} + \frac{(-1)^n}{(2n+1)!z^{2n+1}} + \cdots\right)$$

$$= z^{2n-1} + \cdots + \frac{(-1)^{n-1}}{(2n-1)!} z + \frac{(-1)^n}{(2n+1)!}\frac{1}{z} + \cdots$$

from which it follows that $\text{Res}(f;0) = (-1)^n/(2n+1)!$.

Residue theory can be used in many contexts. In the remainder of this section we examine a few applications of this theory. We first see how residues can be employed in the computation of certain real integrals.

We begin by considering integrals of the form

$$\int_0^{2\pi} f(\cos\theta, \sin\theta)\, d\theta \tag{1}$$

where f is a quotient of polynomials in $\sin\theta$ and $\cos\theta$. We express (1) as a contour integral by setting $\gamma(\theta) = e^{i\theta}$. Since

$$\sin\theta = \frac{e^{i\theta} - e^{-i\theta}}{2i} = \frac{\gamma(\theta) - \dfrac{1}{\gamma(\theta)}}{2i} = \frac{(\gamma(\theta))^2 - 1}{2\gamma(\theta)i}$$

$$\cos\theta = \frac{e^{i\theta} + e^{-i\theta}}{2} = \frac{\gamma(\theta) + \dfrac{1}{\gamma(\theta)}}{2} = \frac{(\gamma(\theta))^2 + 1}{2\gamma(\theta)}$$

and $\gamma'(\theta) = i\gamma(\theta)$, we have

$$\int_0^{2\pi} f(\cos\theta, \sin\theta)\, d\theta = \int_0^{2\pi} f\left(\frac{(\gamma(\theta))^2 + 1}{2\gamma(\theta)}, \frac{(\gamma(\theta))^2 - 1}{2i\gamma(\theta)}\right) d\theta$$

$$= \int_0^{2\pi} \frac{1}{i\gamma(\theta)} f\left(\frac{(\gamma(\theta))^2 + 1}{2\gamma(\theta)}, \frac{(\gamma(\theta))^2 - 1}{2i\gamma(\theta)}\right) \gamma'(\theta)\, d\theta \quad (2)$$

$$= \int_\gamma \frac{1}{iz} f\left(\frac{z^2 + 1}{2z}, \frac{z^2 - 1}{2iz}\right) dz$$

If we let $g(z) = (1/iz)f((z^2 + 1)/2z, (z^2 - 1)/2iz)$, then we obtain

$$\int_0^{2\pi} f(\cos\theta, \sin\theta)\, d\theta = \int_\gamma g(z)\, dz$$

and, hence, evaluation of $\int_0^{2\pi} f(\cos\theta, \sin\theta)\, d\theta$ can be effected in terms on the residues of g.

To illustrate this technique we consider the following example.

(12.E.5) Example We evaluate

$$\int_0^{2\pi} \frac{2d\theta}{4 + 3\cos\theta}.$$

From (2) we find that this integral is equal to

$$\frac{1}{i}\int_\gamma \frac{1}{z}\frac{2dz}{4 + 3\left(\dfrac{z^2 + 1}{2z}\right)} = \frac{1}{i}\int_\gamma \frac{1}{z}\frac{2\cdot 2z\, dz}{8z + 3z^2 + 3}$$

$$= \frac{4}{i}\int_\gamma \frac{dz}{3z^2 + 8z + 3}$$

where $\gamma(t) = e^{it}$, $0 \le t \le 2\pi$. There are two singularities

$$z = \frac{-8 \pm \sqrt{64 - 36}}{6} = \frac{-4 \pm \sqrt{7}}{3}$$

and only the singularity $(-4 + \sqrt{7})/3$ lies in the interior of $I[\gamma]$. Therefore, by (12.E.1) we have

$$\int_0^{2\pi} \frac{2\, d\theta}{4 + 3\cos\theta} = \frac{4}{i} 2\pi i \lim_{z\to(-4+\sqrt{7})/3}\left(z - \left(\frac{-4 + \sqrt{7}}{3}\right)\right)\frac{1}{3z^2 + 8z + 3}$$

$$= \frac{8\pi}{3} \lim_{z\to(-4+\sqrt{7})/3}\left(\frac{1}{z - (-4 - \sqrt{7})/3}\right)$$

$$= \frac{4\pi}{\sqrt{7}}.$$

Integrals of real-valued functions of the form $\text{CPV}\int_{-\infty}^{\infty} f(x)\,dx$ can also be treated in the context of residue theory. The basic idea is as follows. Suppose that f has no singularities on the real axis and (when considered as a function of a complex variable) only isolated singularities in the upper half of the complex plane. We consider oriented paths γ_r whose contours consist of the segment $[-r,r]$, and the semicircle derived from the path Γ_r indicated in Fig. 12.14.

Then $\int_{\gamma_r} f(z)\,dz = \int_{-r}^{r} f(x)\,dx + \int_{\Gamma_r} f(z)\,dz$. Residues are used to calculate $\int_{\gamma_r} f(z)\,dz$, and in many cases $\lim_{r\to\infty} \int_{\Gamma_r} f(z)\,dz$ can be found; thus, from these calculations

$$\text{CP}\int_{-\infty}^{\infty} f(x)\,dx = \lim_{r\to\infty} \int_{-r}^{r} f(x)\,dx$$

can be determined.

(12.E.6) Example We evaluate

$$\text{CP}\int_{-\infty}^{\infty} \frac{x^2}{(x^2 + 1)(x^2 + 9)}\,dx$$

The poles of $f(z) = z^2/(z^2 + 1)(z^2 + 9)$ in the upper-half plane are clearly $z = i$ and $z = 3i$. Thus,

$$\text{Res}\,(f;i) = \lim_{z\to i}\,(z - i)f(z)$$

$$= \lim_{z\to i} \frac{z^2}{(z + i)(z^2 + 9)}$$

$$= \frac{(i)^2}{2i\cdot 8} = \frac{-1}{16i}$$

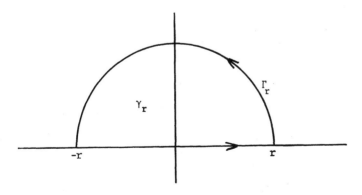

Figure 12.14

and

$$\text{Res} (f; 3i) = \lim_{z \to 3i} (z - 3i)f(z)$$

$$= \lim_{z \to 3i} \frac{z^2}{(z^2 + 1)(z + 3i)}$$

$$= \frac{(3i)^2}{((3i)^2 + 1)(3i + 3i)}$$

$$= \frac{9}{48i}$$

hence,

$$\lim_{r \to \infty} \int_{\gamma_r} f(z) \, dz = 2\pi i(-1/16i + 9/48i) = \pi/4.$$

It remains to calculate

$$\lim_{r \to \infty} \int_{\Gamma_r} f(z) \, dz$$

Note that if $|z| = r$, then for $r > 3$,

$$|z^2| \geq r^2$$
$$|z^2 + 1| \geq r^2 - 1$$
$$|z^2 + 9| \geq r^2 - 9$$

and therefore, since length $\Gamma_r = \pi r$, it follows from (12.A.15.iv) that

$$\int_{\Gamma_r} f(z) \, dz \leq \pi r \left(\frac{r^2}{(r^2 - 1)(r^2 - 4)} \right)$$

Clearly

$$\lim_{r \to \infty} \pi r \left(\frac{r^2}{(r^2 - 1)(r^2 - 4)} \right) = 0$$

and consequently, $\lim_{r \to \infty} \int_{\Gamma_r} f(z) \, dz = 0$, from which it follows that

$$\text{CP} \int_{-\infty}^{\infty} \text{v}\, f(x) \, dx = \frac{\pi}{4}$$

The following result gives a frequently applied criterion for determining if

$$\lim_{r \to \infty} \int_{\Gamma_r} f(z) \, dz = 0$$

(12.E.7) *Theorem* Suppose $f(z) = p(z)/q(z)$, where p and q are polynomials such that $\deg q - \deg p \geq 2$. Then

$$\lim_{r \to \infty} \int_{\Gamma_r} f(z)\, dz = 0$$

where Γ_r is the path described above.

Proof. Note that $\deg z^2 p \leq \deg q$; hence, if $Q(z) = z^2 p(z)/q(z)$, then there are positive real numbers M and N such that if $|z| \geq N$, then

$$|Q(z)| \leq M$$

and therefore,

$$\left|\frac{p(z)}{q(z)}\right| \leq \frac{M}{|z^2|}$$

whenever $|z| \geq N$. Consequently, if $r \geq N$, then for each z on the semicircle determined by Γ_r, we have $|z| = r$ and

$$\left|\frac{p(z)}{q(z)}\right| \leq \frac{M}{r^2}$$

Since by (12.A.15.iv) we have

$$\left|\int_{\gamma_r} f(z)\, dz\right| \leq \pi r \cdot \frac{M}{r^2} = \frac{\pi M}{r}$$

it follows that

$$\lim_{r \to \infty} \int_{\Gamma_r} f(z)\, dz = 0$$

Contours other than those consisting of a segment and a semicircle are often used in the calculation of integrals. Suppose f has a pole on the x-axis, say at $x = 0$. Then contours of the form illustrated in Fig. 12.15 frequently can be of use; we find $\int_{\Gamma_R + \gamma_1 + \Gamma_r + \gamma_2} f(z)\, dz$ and then let R tend to ∞ and r tend to 0.

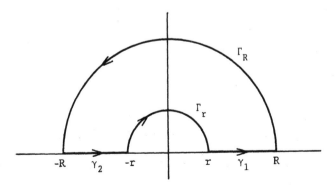

Figure 12.15

Before illustrating this procedure we establish two mildly technical results which are frequently of use in this and similar contexts.

(12.E.8) **Theorem** Suppose that on each semicircular path $\Gamma_R(t) = Re^{it}$, $0 \leq t \leq \pi$, $|f(z)| \leq A/R^k$ for some real constants $k > 0$ and $A > 0$. Then for any $a > 0$

$$\lim_{R \to \infty} \int_{\Gamma_R} e^{iaz} f(z) \, dz = 0$$

Proof. On the path Γ_R we have by (12.A.10)

$$\int_{\Gamma_R} e^{iaz} f(z) \, dz = \int_0^\pi e^{iaRe^{it}} f(Re^{it}) i \, Re^{it} \, dt$$

and therefore, since $e^{it} = \cos t + i \sin t$,

$$\left| \int_{\Gamma_R} e^{iaz} f(z) \, dz \right| \leq \int_0^\pi |e^{iaR \cos t} e^{-aR \sin t} f(Re^{it}) i Re^{it}| dt$$

$$= \int_0^\pi e^{-aR \sin t} |f(Re^{it})| R \, dt$$

$$\leq \frac{A}{R^{k-1}} \int_0^\pi e^{-aR \sin t} \, dt \qquad\qquad (3)$$

To complete the proof, we note that

$$\frac{2}{\pi} \leq (\sin t)/t \leq 1 \text{ whenever } t \in \left(0, \frac{\pi}{2} \right]$$

and

$$\int_0^\pi e^{-aR \sin t} dt = 2 \int_0^{\pi/2} e^{-aR \sin t} \, dt$$

(see Problem E.9), and hence, it follows from (3) that

$$\left| \int_{\Gamma_R} e^{iaz} f(z) \, dz \right| \leq \frac{2A}{R^{k-1}} \int_0^{\pi/2} e^{-2aRt/\pi} \, dt$$

$$= \frac{\pi A}{aR^k} (1 - e^{-aR})$$

which clearly tends to zero with increasing R.

(12.E.9) **Theorem** For each $r > 0$, let Γ_r be the path defined by $\Gamma_r(t) = re^{it}$, $0 \leq t \leq \pi$ and suppose that f has a simple pole at $z = 0$. Then

$$\lim_{r \to 0} \int_{\Gamma_r} f(z) \, dz = \pi i \, \text{Res}(f; 0)$$

Proof. Since the pole at 0 is of order 1, we have from the Laurent expansion of f,

$$f(z) = \frac{\text{Res}(f; 0)}{z} + g(z)$$

where g is analytic (and continuous) at 0. Therefore,

$$\int_{\Gamma_r} f(z)\, dz = \text{Res}(f; 0) \int_{\Gamma_r} \frac{dz}{z} + \int_{\Gamma_r} g(z)\, dz$$

$$= \text{Res}(f; 0) \int_0^\pi \frac{rie^{it}}{re^{it}}\, dt + \int_{\Gamma_r} g(z)\, dz$$

$$= \pi i\, \text{Res}(f; 0) + \int_{\Gamma_r} g(z)\, dz$$

Since g is analytic and continuous at 0, it follows from (12.A.15.iv) that

$$\lim_{r \to 0} \int_{\Gamma_r} g(z)\, dz = 0$$

and this concludes the proof.

(12.E.10) *Example* We use residues to compute $\int_0^\infty [(\sin ax)/x]\, dx$, where $a > 0$. We shall integrate the function $f(z) = e^{iaz}/z$ over the contour γ indicated in Fig. 12.16 and then let $r \to 0$ and $R \to \infty$; observe that the pole $z = 0$ does not lie in the interior of the contour. Clearly from (12.B.5) and (12.A.13),

$$0 = \int_\gamma f(z)\, dz$$

$$= \int_{\Gamma_R} f(z)\, dz + \int_{-R}^{-r} f(x)\, dx + \int_{-\Gamma_r} f(z)\, dz + \int_r^R f(x)\, dx \qquad (4)$$

It is easy to see from (12.E.8) and (12.E.9) that

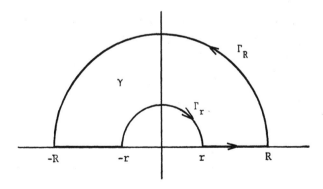

Figure 12.16

$$\lim_{R\to\infty}\int_{\Gamma_R}\frac{e^{iaz}}{z}\,dz=0 \qquad (5)$$

and

$$\lim_{r\to 0}\int_{\Gamma_r}\frac{e^{iaz}}{z}\,dz=-\pi i\,\operatorname{Res}\left(\frac{e^{iaz}}{z};0\right)=-\pi i \qquad (6)$$

Next, one can readily establish that

$$\int_{-R}^{-r}\frac{e^{iax}}{x}\,dx+\int_{r}^{R}\frac{e^{iax}}{x}\,dx=-\int_{r}^{R}\frac{e^{-iax}}{x}\,dx+\int_{r}^{R}\frac{e^{iax}}{x}\,dx$$

$$=2i\int_{r}^{R}\frac{\sin ax}{x}\,dx$$

Therefore,

$$\lim_{\substack{r\to 0\\R\to\infty}}\left(\int_{-R}^{-r}\frac{e^{iax}}{x}\,dx+\int_{r}^{R}\frac{e^{iax}}{x}\,dx\right)=2i\int_{0}^{\infty}\frac{\sin ax}{x}\,dx$$

and, consequently, it follows from (4), (5), and (6) that

$$0=2i\int_{0}^{\infty}\frac{\sin ax}{x}\,dx-\pi i$$

and, hence,

$$\int_{0}^{\infty}\frac{\sin ax}{x}\,dx=\frac{\pi}{2}$$

F. LINE INTEGRALS

Line integrals have many similarities with contour integrals. We shall define the line integral in a three-dimensional setting; completely analogous definitions hold for dimensions other that 3.

A path in $\mathbf{R}^3$ is any continuous function $\gamma\colon [a,b]\to\mathbf{R}^3$; as usual, $I[\gamma]$ will denote the image of γ. Note that for each t we may write $\gamma(t)$ as a vector $(x(t),y(t),z(t))$ in $\mathbf{R}^3$. If $k\colon I[\gamma(t)]\to\mathbf{R}^3$, then k can also be expressed vectorially by $k(\mathbf{w})=(f(\mathbf{w}),g(\mathbf{w}),h(\mathbf{w}))$, where $\mathbf{w}\in\mathbf{R}^3$. The *line integral* $\int_{\gamma}k$ of k over γ is defined as a limit of the sum of inner products as the mesh of partitions of the interval $[a,b]$ tends to 0:

$$\int_{\gamma}k=\lim_{|P|\to 0}\sum_{i=1}^{n}k(\gamma(\tau_i))\cdot(\gamma(t_i)-\gamma(t_{i-1}))\qquad \tau_i\in[t_{i-1},t_i].\qquad (1)$$

As indicated previously, to say that

$$\lim_{|P|\to 0}\sum_{i=1}^{n}k(\gamma(\tau_i))\cdot(\gamma(t_i)-\gamma(t_{i-1}))=L$$

means that given $\varepsilon > 0$, there is a $\delta > 0$ such that if $P = \{t_0, t_1, \ldots, t_n\}$ is any partition of $[a,b]$ with mesh less than δ, and if for each i, τ_i is an arbitrary point in $[t_{i-1}, t_i]$, then

$$\left| \sum_{i=1}^{n} k(\gamma(\tau_i)) \cdot (\gamma(t_i) - \gamma(t_{i-1})) - L \right| < \varepsilon$$

[Here, $|(a,b,c)| = \sqrt{(a,b,c) \cdot (a,b,c)} = \sqrt{a^2 + b^2 + c^2}$.]

Line integrals arise in many contexts including finding the amount of work done by a variable force over a given curve, the calculation of masses of wires and other objects, and the determination of the flux of certain fluid flows.

As was the case with contour integrals, the definition of the line integral is of limited use in obtaining an exact value of the integral. For a piecewise continuously differentiable path [i.e., a path $\gamma(t) = (x(t), y(t), z(t))$ for which the components $x(t)$, $y(t)$, and $z(t)$ are piecewise continuously differentiable functions on the domain of γ] conversion of the line integral to a Riemann integral is possible.

(12.F.1) Theorem If $\gamma \colon [a,b] \to \mathbf{R}^3$ is a piecewise continuously differentiable path and if $k \colon I[\gamma] \to \mathbf{R}^3$ is continuous then

$$\int_{\gamma} k = \int_{a}^{b} k(\gamma(t)) \cdot \gamma'(t)\, dt$$

Proof. Without loss of generality we may assume that γ is continuously differentiable on $[a,b]$. Let $P = \{t_0, t_1, \ldots, t_n\}$ be a partition of $[a,b]$, and for each i, let τ_i be an arbitrary point in $[t_{i-1}, t_i]$. Let $k(\mathbf{w}) = (f(\mathbf{w}), g(\mathbf{w}), h(\mathbf{w}))$ and let $\gamma(t) = (x(t), y(t), z(t))$. Note that by the mean value theorem,

$$k(\gamma(\tau_i)) \cdot (\gamma(t_i) - \gamma(t_{i-1})) = (f(\gamma(\tau_i)), g(\gamma(\tau_i)), h(\gamma(\tau_i))) \cdot (x(t_i) - x(t_{i-1}), y(t_i)$$
$$- y(t_{i-1}), z(t_i) - z(t_{i-1}))$$
$$= (f(\gamma(\tau_i)))(x'(c_i))(t_i - t_{i-1})$$
$$+ (g(\gamma(\tau_i)))(y'(\hat{c}_i))(t_i - t_{i-1})$$
$$+ (h(\gamma(\tau_i)))(z'(\tilde{c}_i))(t_i - t_{i-1})$$

(where τ_i, c_i, $\hat{c}_i$, and $\tilde{c}_i$ all lie in the interval $[t_{i-1}, t_i]$), and therefore,

$$\sum_{i=1}^{n} k(\gamma(\tau_i)) \cdot (\gamma(t_i) - \gamma(t_{i-1}))$$

$$= \sum_{i=1}^{n} [f(\gamma(\tau_i))(x'(c_i)) + g(\gamma(\tau_i))(y'(\hat{c}_i)) + h(\gamma(\tau_i))(z'(\tilde{c}_i))](t_i - t_{i-1})$$

Now if it were the case that

$$\tau_i = c_i = \hat{c}_i = \tilde{c}_i \qquad\qquad (2)$$

for each i (and each partition P), then the limit of the above sums as $|P| \to 0$ would be

$$\int_a^b (f(\gamma(t))x'(t) + g(\gamma(t))y'(t) + h(\gamma(t))z'(t))\, dt = \int_a^b k(\gamma(t)) \cdot \gamma'(t)\, dt \quad (3)$$

In problem F.15 the reader is asked to verify that (3) is also valid in the (likely) event that equality (2) does not hold, and this concludes the proof of the theorem.

Frequently and for fairly obvious reasons, the following notation is used in connection with the line integral

$$\int_\gamma k = \int_a^b [f(\gamma(t))x'(t) + g(\gamma(t))y'(t) + h(\gamma(t))z'(t)]\, dt$$

$$= \int_\gamma f\, dx + g\, dy + h\, dz$$

(12.F.2) **Examples** (a) If $\gamma: [0,2] \to \mathbf{R}^3$ is defined by $\gamma(t) = (2t, t^2, 3t^4)$ and $k: \mathbf{R}^3 \to \mathbf{R}^3$ is defined by $k(x,y,z) = (xy^2, 2yz, 4)$, then since $\gamma'(t) = (2, 2t, 12t^3)$, we have

$$\int_\gamma k = \int_0^2 (2t^5, 6t^6, 4) \cdot (2, 2t, 12t^3)\, dt = \int_0^2 (4t^5 + 12t^7 + 48t^3)\, dt = \frac{1856}{3}$$

Note that we could also write

$$\int_\gamma k = \int_\gamma xy^2\, dx + \int_\gamma 2yz\, dy + \int_\gamma 4\, dz$$

$$= \int_0^2 2t^5(2)\, dt + \int_0^2 (6t^6)(2t)\, dt + \int_0^2 4(12t^3)\, dt$$

$$= \frac{1856}{3}.$$

(b) We find $\int_\gamma k$, where $k(x,y) = (2xy, y^2)$ and γ is the path whose contour is the graph of the parabola $y = x^2$, $-1 \le x \le 2$, i.e., $\gamma: [-1,2] \to \mathbf{R}^2$ is defined by $\gamma(t) = (t, t^2)$. We have

$$\int_\gamma k = \int_\gamma 2xy\, dx + \int_\gamma y^2\, dy = \int_{-1}^2 2t^3(1)\, dt + \int_{-1}^2 t^4(2t)\, dt = \frac{207}{10}$$

(c) We find $\int_\gamma k$, where $k(x,y) = (y^2, xy^2)$ and γ is the path traversing the triangle shown in Fig. 12.17 in a counterclockwise direction. A parameterization of this curve is given by

$$\gamma(t) = \begin{cases} (1 - t, t) & \text{if } 0 \le t \le 1 \text{ (the segment of the line } y = 1 - x) \\ (0, -t + 2) & \text{if } 1 \le t \le 2 \text{ (the segment of the } y\text{-axis)} \\ (t - 2, 0) & \text{if } 2 \le t \le 3 \text{ (the segment of the } x\text{-axis)} \end{cases}$$

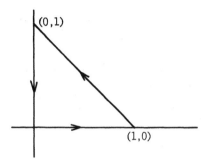

(0,1)

(1,0)

Figure 12.17

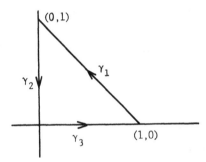

(0,1)

γ_1

γ_2

γ_3 (1,0)

Figure 12.18

and therefore,

$$\int_\gamma k = \int_0^1 (t^2, t^2 - t^3) \cdot (-1, 1)\, dt + \int_1^2 ((-t + 2)^2, 0) \cdot (0, -1)\, dt$$

$$+ \int_2^3 (0^2, 0) \cdot (1, 0)\, dt = -\frac{1}{4}$$

In this case, it is more convenient to consider the path γ as composed of the paths γ_1, γ_2, and γ_3 indicated in Fig. 12.18 where

$$\gamma_1(t) = (1 - t, t) \qquad 0 \le t \le 1$$
$$\gamma_2(t) = (0, -t + 1) \qquad 0 \le t \le 1$$
$$\gamma_3(t) = (t, 0) \qquad 0 \le t \le 1$$

then

$$\int_\gamma k = \int_{\gamma_1} k + \int_{\gamma_2} k + \int_{\gamma_3} k$$

$$= \int_0^1 (t^2, t^2 - t^3) \cdot (-1, 1)\, dt + \int_0^1 ((-t + 1)^2, 0) \cdot (0, -1)\, dt$$

$$+ \int_0^1 (0^2, 0) \cdot (1, 0)\, dt = -\frac{1}{4}$$

That this is possible follows easily from an argument similar to that indicated in (12.A.13).

Now we examine briefly the relationship between line integrals and contour integrals. Here, the line integral is considered in a two dimensional context. Suppose that $\gamma: [a,b] \to \mathbf{C}$ is a continuously differentiable path and that $f: I[\gamma] \to \mathbf{C}$ is a continuous function. Let $f(z) = u(z) + iv(z)$ and let $\gamma(t) = x(t) + iy(t)$. Then by (12.A.10)

$$\int_\gamma f(z)\, dz = \int_a^b f(\gamma(t))\gamma'(t)\, dt$$

$$= \int_a^b (u(\gamma(t)) + iv(\gamma(t)))(x'(t) + iy'(t))\, dt$$

$$= \int_a^b (u(\gamma(t))x'(t) + iu(\gamma(t))y'(t) + iv(\gamma(t))x'(t) - v(\gamma(t))y'(t))\, dt$$

$$= \int_\gamma (u\, dx - v\, dy) + i \int_\gamma (v\, dx + u\, dy)$$

where the last two integrals are line integrals.

The line integral is useful in many physical applications. For example, suppose that an object is to be moved along a path γ in $\mathbf{R}^3$, from the point P to the point Q (Fig. 12.19). We wish to calculate the work done in moving this object. For a constant force F along a straight line segment connecting points P and Q, the work done is given by the inner product

$$F \cdot L$$

where $L = (\Delta x, \Delta y, \Delta z)$ and $F = (f_1, f_2, f_3)$ (Fig. 12.20). In the general case, the force is allowed to vary as the object moves along the given path. If the path γ is defined by $\gamma: [a,b] \to \mathbf{R}^3$, where $\gamma(t) = (x(t), y(t), z(t))$, then each

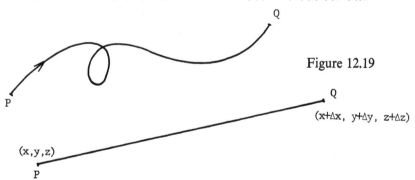

Figure 12.19

Figure 12.20

partition $P = \{t_0, t_1, \ldots, t_n\}$ yields as an approximation of the work done

$$\sum_{i=1}^{n} F(\gamma(\tau_i)) \cdot (\gamma(t_i) - \gamma(t_{i-1})) \qquad \tau_i \in [t_{i-1}, t_i] \qquad (4)$$

since, for each i,

$$F(\gamma(\tau_i)) \cdot (\gamma(t_i) - \gamma(t_{i-1})) = F(\gamma(\tau_i)) \cdot (x(t_i) - x(t_{i-1}), y(t_i) - y(t_{i-1}),$$
$$z(t_i) - z(t_{i-1}))$$

represents work done under the assumption the force is constant on a line segment joining $\gamma(t_{i-1})$ and $\gamma(t_i)$ (Fig. 12.21). The "true" work done is taken to be the limit of the sums (4), where $|P| \to 0$. This, of course, is precisely the definition of the line integral of F over γ, $\int_\gamma F$.

The line integral also appears in the definition of the curl of a vector field, in Ampère's law relating electric currents and the production of magnetic effects, in the calculation of masses, as well as in a host of other contexts.

The line integral and the double integral are tied together by what is known as *Green's theorem in the plane*. George Green (1793–1841) did extensive mathematical work in connection with magnetism and electricity. His contributions had a great deal of influence on a number of brilliant mathematical physicists including Stokes, Maxwell, and others.

Green's theorem is best dealt with in the context of differential forms (which we do not treat in this text). However, we can easily establish certain special cases of this theorem. We shall consider simply connected regions in the plane of the forms shown in Fig. 12.22, where the boundary of each region is a positively oriented path determined by the indicated functions and vertical or horizontal line segments. It is assumed that ϕ_1 and ϕ_2 are piecewise differentiable functions of x and ψ_1 and ψ_2 are piecewise differentiable functions of y.

First, we consider the situation where the region D is bounded as indicated in Fig. 12.22 a, b, c, and g.

(12.F.3) Theorem Suppose that a simply connected region D is bounded by the graphs of two piecewise continuously differentiable functions:

$$\phi_1 \colon [a, b] \to \mathbf{R}^1$$
$$\phi_2 \colon [a, b] \to \mathbf{R}^1$$

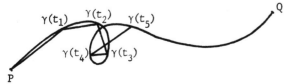

Figure 12.21

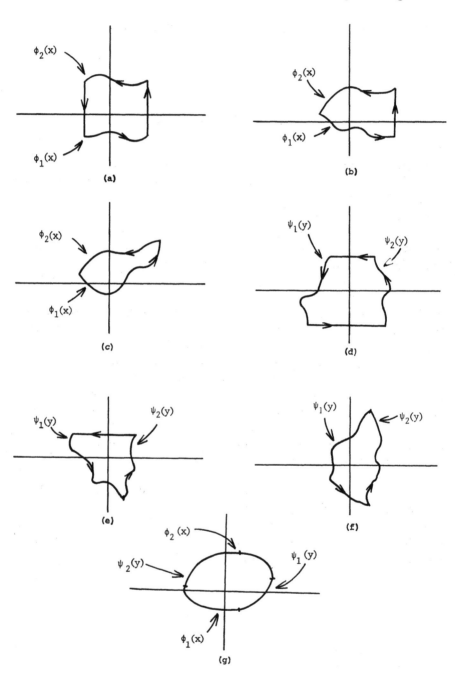

Figure 12.22

[where $\phi_1(x) \leq \phi_2(x)$ for each x] and vertical line segments (or points) $x = a$ and $x = b$. Let $P: \bar{D} \to \mathbf{R}^1$ be continuously differentiable on D and continuous on the closure $\bar{D}$ and let $S: \bar{D} \to \mathbf{R}^2$ be defined by $S(w) = (P(w),0)$. If Γ is the positively oriented path defining the boundary of D, then

$$-\int_\Gamma S = -\int_\Gamma P\,dx - \int_\Gamma 0\,dy$$

$$= \int_D P_y$$

$$= \int_a^b \int_{\phi_1(x)}^{\phi_2(x)} P_y(x,\,y)\,dy\,dx$$

i.e., $-\int_\Gamma P\,dx = \int_a^b (\int_{\phi_1(x)}^{\phi_2(x)} P_y(x,y)\,dy)\,dx$. (Here we assume that Γ has been parameterized so that

$$\int_\Gamma P\,dx = \int_{\Gamma_1} P\,dx + \int_{\delta_1} P\,dx - \int_{\Gamma_2} P\,dx + \int_{\delta_2} P\,dx$$

where if $x \in [a,b]$, then $\Gamma_1(x) = (x,\phi_1(x))$, $\Gamma_2(x) = (x,\phi_2(x))$, and δ_1 and δ_2 represent the right- and left-hand vertical line segments, respectively. Note that since x remains constant on the paths δ_1 and δ_2, we have

$$\int_{\delta_1} P\,dx = 0 = \int_{\delta_2} P\,dx$$

Proof. Clearly, from the Fubini theorem (11.I.6) and the preceding parenthetical remark, we have

$$\int_D P_y = \int_a^b \left(\int_{\phi_1(x)}^{\phi_2(x)} P_y(x,\,y)\,dy \right) dx$$

$$= \int_a^b (P(x,\,\phi_2(x)) - P(x,\,\phi_1(x)))\,dx$$

$$= \int_a^b P(x,\,\phi_2(x))\,dx - \int_a^b P(x,\,\phi_1(x))\,dx = \int_{\Gamma_2} P\,dx - \int_{\Gamma_1} P\,dx$$

$$= -\int_\Gamma P\,dx$$

(12.F.4) *Exercise* Formulate and prove an analogous theorem for the case illustrated by Fig. 12.22 d, e, f, and g to show that if $Q: \bar{D} \to \mathbf{R}^1$ is continuously differentiable on D and continuous on $\bar{D}$, then

$$\int_\Gamma Q\,dy = \int_D Q_x = \int_c^d \left(\int_{\psi_1(y)}^{\psi_2(y)} Q_x\,dx \right) dy$$

Combining these results we obtain a (watered-down) version of Green's theorem.

(12.F.5) *Theorem*   Suppose that D is a simply connected region as indicated in Fig. 12.22 g. Let $S: \bar{D} \to \mathbf{R}^2$ be continuously differentiable on D and continuous on $\bar{D}$, and suppose that $S = (P,Q)$. Then

$$\int_D (Q_x - P_y) = \int_\Gamma P\,dx + Q\,dy$$

where Γ is the positively oriented path around D determined by the boundary of D and parameterized by the four piecewise continuously differentiable functions comprising this boundary.

Proof.   In this case we have from (12.F.3)

$$\int_\Gamma P\,dx = -\int_D P_y \tag{5}$$

and from (12.F.4)

$$\int_\Gamma Q\,dy = \int_D Q_x \tag{6}$$

Addition of the equations (5) and (6) yields the desired result.

(12.F.6) *Example*   We use Green's theorem to evaluate $\int_\gamma (x^2 + y)\,dx + x^2 y\,dy$, where the path γ traverses in a counterclockwise direction the curve formed by $y = x^2$ and $y = 2x$ between the points $(0,0)$ and $(2,4)$ (Fig. 12.23). Here, $P(x,y) = x^2 + y$, $Q(x,y) = x^2 y$, $Q_x(x,y) = 2xy$, $P_y(x,y) = 1$, and hence, by Green's theorem,

$$\int_\gamma (x^2 + y)\,dx + x^2 y\,dy = \int_0^2 \int_{x^2}^{2x} (2xy - 1)\,dy\,dx$$

$$= \int_0^2 (4x^3 - 2x - x^5 + x^2)\,dx = 4$$

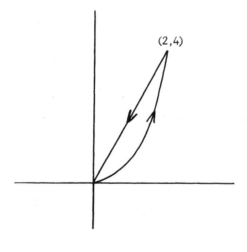

(2,4)

Figure 12.23

It can be shown that Green's theorem is valid for any simple closed rectifiable path [see Korevaar (1968), for example]. In the problem section the reader is asked to give a "proof" of the validity of Green's theorem for regions that can be partitioned into subregions of the types described by Fig. 12.22 (a)–(g) (Fig. 12.24). In problem F.5 the use of a version of Green's theorem for regions of the form shown in Fig. 12.25 is considered.

In our study of contour integrals we found that under certain conditions the value of a contour integral was independent of the path used to join two given points. We conclude this section by finding somewhat analogous conditions to ensure that given two points a and b in a region D and functions P and Q defined on D, then

$$\int_\gamma P\,dx + Q\,dy = \int_{\hat\gamma} P\,dx + Q\,dy$$

for any two paths γ and $\hat\gamma$ in D that join a to b.

Independence of path in a region is equivalent to the line integral along closed paths being equal to 0. The proof of this result (12.F.7) is quite easy and is left to the reader (problem F.11).

(12.F.7) Theorem For each closed path Γ in D

$$\int_\Gamma P\,dx + Q\,dy = 0$$

if and only if for any two paths γ and $\hat\gamma$ in D with identical initial points and

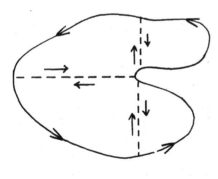

Figure 12.24

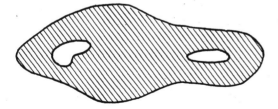

Figure 12.25

identical terminal points,

$$\int_\gamma P\,dx + Q\,dy = \int_{\hat{\gamma}} P\,dx + Q\,dy$$

Another criterion for independence of path can be formulated in terms of the existence of a potential function for P and Q. A differentiable function $F: D \to \mathbf{R}^1$ is said to be a *potential function* for the vector field $G = (P,Q)$ if $-F_x = P$ and $-F_y = Q$, i.e. (in the notation of Chapter 5), $-\nabla F = (P,Q)$.

(12.F.8) Theorem If $P: D \to \mathbf{R}^1$ and $Q: D \to \mathbf{R}^1$ are continuous functions on a region D, then a potential function F for (P,Q) exists in D if and only if for any two piecewise continuously differentiable paths γ and $\hat{\gamma}$ in D with the same initial and the same terminal points

$$\int_\gamma P\,dx + Q\,dy = \int_{\hat{\gamma}} P\,dx + Q\,dy$$

Proof. Suppose that a potential function F for (P,Q) exists and let $\gamma: [a,b] \to D$ be any piecewise continuously differentiable path, where $\gamma(t) = (x(t),y(t))$. Since F_x and F_y are continuous on D, it follows from the chain rule, (12.F.1), and the fundamental theorem of calculus that

$$\int_\gamma P\,dx + Q\,dy = -\int_\gamma F_x\,dx + F_y\,dy$$

$$= -\int_a^b (F_x(\gamma(t))x'(t) + F_y(\gamma(t))y'(t))\,dt$$

$$= -\int_a^b F'(x(t), y(t))\,dt$$

$$= F(x(a), y(a)) - F(x(b), y(b))$$

$$= F(\gamma(a)) - F(\gamma(b)).$$

This shows that the value of $\int_\gamma P\,dx + Q\,dy$ depends only on the value of $\gamma(a)$ and $\gamma(b)$, and hence the value of the line integral over paths in D is independent of (piecewise continuously differentiable) paths joining any two given points in D.

To establish the converse, we must construct the potential function F. This is done in a way somewhat similar to that used in (12.B.2). Fix a point $w = (x^*,y^*) \in D$ and for each point $(x,y) \in D$ let γ be a piecewise continuously differentiable path between (x^*,y^*) and (x,y) and define

$$-F(x, y) = \int_\gamma P\,dx + Q\,dy.$$

We show that $-F_x = P$ (Fig. 12.26).

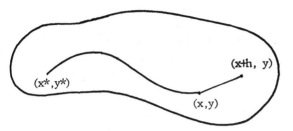

Figure 12.26

Choose $h \neq 0$ small enough so that the line segment connecting (x,y) and $(x + h, y)$ lies in D. Let Γ_h be the path from (x^*,y^*) to $(x + h, y)$ formed by joining the path γ to the path $\gamma_h : [0,1] \to D$ defined by $\gamma_h(t) = (x + th, y)$. Then we have

$$\frac{-F(x + h, y) - (-F(x, y))}{h} = \frac{1}{h}\left(\int_{\Gamma_h} P\,dx + Q\,dy\right) - \frac{1}{h}\left(\int_\gamma P\,dx + Q\,dy\right)$$

$$= \frac{1}{h}\int_{\gamma_h} P\,dx + Q\,dy = \frac{1}{h}\int_{\gamma_h} P\,dx$$

$$= \frac{1}{h}\int_0^1 P(x + th, y)h\,dt$$

$$= \int_0^1 P(x + th, y)\,dt$$

Now we observe that since P is continuous

$$\lim_{h \to 0} \int_0^1 P(x + th, y)\,dt = P(x, y) \tag{7}$$

(see problem F.12) and, therefore,

$$-F_x(x, y) = P(x, y)$$

A similar argument can be supplied by the reader to show that $-F_y = Q$.

From a physical standpoint this theorem asserts that if a force with continuous components P and Q can be derived from a potential function F, then the work done by this force in moving an object from a point A to a point B is simply the difference in potential energy $F(A) - F(B)$.

A continuous vector field $G(P,Q)$ is said to be *conservative* if there is a potential function for G. This terminology is derived from the following consideration. Suppose that $G = (P,Q)$ is a conservative vector field defined on a region D, and let $\gamma : [a,b] \to D$ be a path with piecewise continuously differentiable second derivatives that connects points A and B in D. Then, as we have seen, the work done by the force field in displacing a particle along this path is equal to

$$\int_\gamma G = \int_\gamma P\,dx + Q\,dy = F(A) - F(B) \qquad A = \gamma(a), B = \gamma(b)$$

where F is a potential function for G. In this context Newton's second law of motion takes on the form

$$G(\gamma(t)) = m\gamma''(t)$$

From (12.F.1) we have

$$F(A) - F(B) = \int_\gamma G$$

$$= \int_a^b G(\gamma(t)) \cdot \gamma'(t)\,dt$$

$$= m \int_a^b \gamma''(t) \cdot \gamma'(t)\,dt$$

$$= \frac{m}{2} \int_a^b \frac{d}{dt} (\gamma'(t) \cdot \gamma'(t))\,dt$$

$$= \frac{m}{2} \int_a^b \frac{d}{dt} \|\gamma'(t)\|^2\,dt$$

If we set $v(t) = \|\gamma'(t)\|$, then we have

$$F(A) - F(B) = \int_\gamma G = \frac{m}{2}(v^2(b) - v^2(a))$$

This latter quantity represents the change in kinetic energy of the particle as it moves along the path γ from A to B. Since $F(A) + mv^2(a)/2 = F(B) + mv^2(b)/2$ for all $a, b \in D$, we have shown that the sum of the potential and kinetic energy of a particle remains constant as the particle passes through a conservative force field, i.e., energy is conserved.

Finally, we use Green's theorem to obtain a very easily applied criterion for the existence of a potential function.

(12.F.9) Theorem If D is an open disk and if $P: D \to \mathbf{R}^1$ and $Q: D \to \mathbf{R}^1$ are continuously differentiable functions, then the following are equivalent (we consider only piecewise continuously differentiable paths):

(i) $\int_\gamma P\,dx + Q\,dy = 0$, for each simple closed path γ in D.
(ii) $\int_\gamma P\,dx + Q\,dy = \int_{\hat\gamma} P\,dx + Q\,dy$, whenever the paths γ and $\hat\gamma$ have the same initial and the same terminal points.
(iii) There is a potential function for $G = (P,Q)$.
(iv) $Q_x = P_y$ in D.

Proof. First we show that (i) and (iv) are equivalent. According to

Green's theorem, for each simple closed piecewise continuously differentiable path γ in D, we have $\int_\gamma P\,dx + Q\,dy = \int_{D_\gamma}(Q_x - P_y)$, where D_γ represents the interior of γ. If $Q_x(\hat{x},\hat{y}) - P_y(\hat{x},\hat{y}) \neq 0$ at some point $(\hat{x},\hat{y}) \in D$, then it follows from the continuity of Q_x and P_y that $Q_x - P_y$ is either strictly positive or strictly negative on some suitably small disk, $S_\rho(\hat{x},\hat{y})$, centered at $(\hat{x},\hat{y})$. Hence, if γ is the simple closed path whose image $I[\gamma]$ is the circle of radius $\rho/2$ centered at $(\hat{x},\hat{y})$, then by (6.B.3.iv)

$$0 = \int_\gamma P\,dx + Q\,dy = \int_{S_{\rho/2}(\hat{x},\hat{y})}(Q_x - P_y) \neq 0$$

which is impossible; hence, (i) implies (iv). That (iv) implies (i) is an easy consequence of (12.F.5).

The equivalence of (ii) and (iii) is immediate from (12.F.8). Thus, since it is clear that (ii) implies (i), it only remains to show that (i) implies (iii).

Let $p_0 = (x_0, y_0)$ denote the center of the disk D and let $\gamma_{(x,y)}: [0,1] \to D$ designate the straight-line path defined by $\gamma_{(x,y)}(t) = p_0 + t((x,y) - p_0)$. Set $-F(x,y) = \int_{\gamma_{(x,y)}} P\,dx + Q\,dy$, and note that this function is well defined on D. We claim that $-F$ is the potential function for the field (P,Q) on D. First we show that $-F_x = P$. We have

$$\frac{-F(x + h, y) - (-F(x, y))}{h}$$

$$= \frac{1}{h}\left\{\int_{\gamma_{(x+h,y)}} P\,dx + Q\,dy - \int_{\gamma_{(x,y)}} P\,dx + Q\,dy\right\}$$

and from property (i) and Fig. 12.27, we see that

$$\int_{\gamma_{(x+h,y)}} P\,dx + Q\,dy + \int_{-\gamma_h} P\,dx + Q\,dy + \int_{-\gamma_{(x,y)}} P\,dx + Q\,dy = 0$$

hence,

$$\frac{1}{h}\left(\int_{\gamma_{(x+h,y)}} P\,dx + Q\,dy - \int_{\gamma_{(x,y)}} P\,dx + Q\,dy\right)$$

$$= \frac{1}{h}\int_{\gamma_h} P\,dx + Q\,dy$$

$$= \frac{1}{h}\int_{\gamma_h} P\,dx$$

and since $\lim_{h\to 0}(1/h)\int_{\gamma_h} P\,dx = P(x,y)$, it follows that $-F_x(x,y) = P(x,y)$.

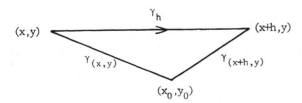

Figure 12.27

Similarly, we can prove that $-F_y = Q$, and this concludes the proof.

With the aid of a classic result (the Riemann mapping theorem, which states that if D is a simply connected region of $\mathbf{C}$, and not equal to $\mathbf{C}$, then there is an analytic bijection from D onto the unit disk), theorem (12.F.9) can be extended to simply connected regions.

(12.F.10) *Example* We consider $\int_\gamma 2x \sin y \, dx + x^2 \cos y \, dy$, where $(2,3)$ is the initial point and $(4,-1)$ is the terminal point of γ. Since $Q_x = 2x \cos y = P_y$ in the entire plane, it follows that if we wish to evaluate this integral, then we may integrate over any path connecting the points $(2,3)$ and $(4,-1)$. Integrating over the curve indicated in Fig. 12.28, we find that

$$\int_\gamma 2x \sin y \, dx + x^2 \cos y \, dy = \int_2^4 2t \sin 3 \, dt + \int_3^{-1} 16 \cos t \, dt$$

$$= 16 \sin(-1) - 4 \sin 3$$

Theorem (12.F.9) yields a weakened, but useful, form of the Cauchy-Goursat theorem.

(12.F.11) *Theorem* If f is analytic and continuously differentiable on an open disk D, and if γ if a simple closed piecewise continuously differentiable path in D, then $\int_\gamma f(z) \, dz = 0$.

Proof. Let $f = u + iv$, where $u : D \to \mathbf{R}^1$ and $v : D \to \mathbf{R}^1$. From the discussion on the top of page 510, we find that

$$\int_\gamma f(z) \, dz = \int_\gamma u(x, y) \, dx + (-v(x, y)) \, dy + i \left\{ \int_\gamma v(x, y) \, dx + u(x, y) \, dy \right\}$$

$$(8)$$

Since u_x, u_y, v_x, and v_y are continuous and satisfy the Cauchy-Riemann conditions $u_y = -v_x$ and $u_x = v_y$, on D, it follows from (12.F.9.iv) that the line integrals on the right side of (8) are both zero, which concludes the proof.

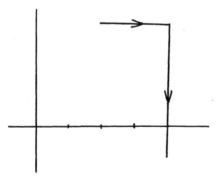

Figure 12.28

The reader may wonder why in (12.F.11) we insisted that f be analytic and continuously differentiable, since we have previously shown that if f is analytic then every derivative is continuously differentiable. However, the proof of this latter result depends upon the Cauchy-Goursat theorem (12.B.5) (which did not require continuity of f'). Therefore, in order that (12.F.11) be independent of (12.B.5), it should not be assumed that f' is continuous on D simply because f is analytic on D. Before the work of Goursat, it was necessary to define an analytic function on an open set U to be a function having a continuous complex derivative on U.

It has only been quite recently that the continuity of f' and the existence of the higher order derivatives have been established without the use of integration methods. (These integration-free proofs are much more difficult than the traditional ones using integration.) For additional details, the reader may consult Connell and Porcelli (1961) and Plunkett (1957).

PROBLEMS

Section A

1. Show by a careful selection of partition points that the path $\gamma: [0,1] \to \mathbf{C}$ defined by

$$\gamma(t) = \begin{cases} t + it \sin \dfrac{1}{t} & \text{if } t \neq 0 \\ 0 & \text{if } t = 0 \end{cases}$$

    has infinite length and that the path $\gamma: [0,1] \to \mathbf{C}$ defined by

$$\gamma(t) = \begin{cases} t + it^2 \sin \dfrac{1}{t} & \text{if } t \neq 0 \\ 0 & \text{if } t = 0 \end{cases}$$

    has length less than 2.

2. Show directly that if $\gamma_1: [0,1] \to \mathbf{C}$ is defined by $\gamma_1(t) = t + it$ and $\gamma_2: [0,1] \to \mathbf{C}$ is defined by

$$\gamma_2(t) = \begin{cases} 2t + i2t & 0 \leq t \leq \frac{1}{2} \\ 1 - 2t + i(1 - 2t) & \frac{1}{2} \leq t \leq 1 \end{cases}$$

    and if f is a function that is identically equal to 3 on $I[\gamma_1] = I[\gamma_2]$, then $\int_{\gamma_1} f(z)\, dz = 3$ but $\int_{\gamma_2} f(z)\, dz = 0$.

3. Calculate the following contour integrals:
    (a) $\int_{\gamma} (z - 2)\, dz$, where $\gamma: [0,3] \to \mathbf{C}$ is defined by $\gamma(t) = t + it^2$.

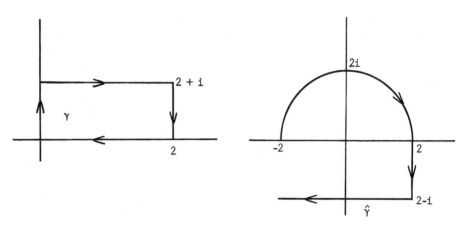

Figure 12.29

(b) $\int_\gamma f(z)\, dz$, where $f(z) = f(x + iy) = x$ and $\gamma: [0,\pi/2] \to \mathbf{C}$ is defined by $\gamma(t) = 1 + \cos t + i \sin t$.

(c) $\int_\gamma [(z - 3)/z]\, dz$, where
    (i) $\gamma: [0,\pi] \to \mathbf{C}$ is defined by $\gamma(t) = 3e^{it}$.
    (ii) $\gamma: [\pi,2\pi] \to \mathbf{C}$ is defined by $\gamma(t) = 3e^{it}$.
    (iii) $\gamma: [-\pi,\pi] \to \mathbf{C}$ is defined by $\gamma(t) = 3e^{it}$.

(d) $\int_\gamma |z|\, dz$, where $\gamma: [0,2\pi] \to \mathbf{C}$ is defined by $\gamma(t) = 2 + 2e^{it}$.

(e) $\int_\gamma (1/z^3)\, dz$, where $\gamma: [0,1] \to \mathbf{C}$ is defined by $\gamma(t) = e^{-2\pi it}$.

4. Choose an appropriate parameterization of the indicated contours in Fig. 12.29 and compute the following integrals:
    (a) $\int_\gamma z^2\, dz$
    (b) $\int_{\hat\gamma} z^2\, dz$
    (c) $\int_\gamma (\sin z + e^z)\, dz$
    (d) $\int_{\hat\gamma} (1/z)\, dz$

5. Prove Theorem (12.A.15).

6. Show (without evaluating the integral) that

$$\left| \int_\gamma e^z\, dz \right| \le 4\pi e$$

where $\gamma: [0,2\pi] \to \mathbf{C}$ is defined by $\gamma(t) = e^{2it}$.

Section B

1. Find
    (a) $\int_\gamma (z^3 + 2z)\, dz$, where γ is any continuously differentiable path joining the points 0 and i.

(b) $\int_\gamma e^{3z} \, dz$, where γ is a parameterization of the circle $|z| = 3$ (oriented counterclockwise).

(c) $\int_\gamma \cos 2z \, dz$, where γ is any continuously differentiable path connecting the points $1 + i$ and $2 - i$.

2. Evaluate

(a) $\int_\gamma \left(\dfrac{e^{3z^2}}{z^3 - z} \right) dz$ where $\gamma(t) = 2e^{it}, \ 0 \le t \le 2\pi$

(b) $\int_\gamma \left(\dfrac{3e^z - \cos z}{z} \right) dz$ where $\gamma(t) = e^{it}, \ 0 \le t \le 2\pi$

(c) $\int_\gamma \dfrac{dz}{z(z - 3)}$ where $\gamma(t) = 2e^{it}, \ 0 \le t \le 2\pi$

(d) $\int_\gamma \left(\dfrac{\sin z}{2z(z^2 - 1)} \right) dz$ where $\gamma(t) = 2e^{it}, \ 0 \le t \le 2\pi$

(e) $\int_\gamma \left(\dfrac{\cos z}{2z(z^2 - 1)} \right) dz$ where $\gamma(t) = \dfrac{1}{2} e^{it}, \ 0 \le t \le 2\pi$

3. Evaluate

(a) $\int_\gamma \left(\dfrac{z^3}{(z + 2)^3} \right) dz$

where $\gamma: [0, 2\pi] \to \mathbf{C}$ is defined by $\gamma(t) = 3 + 4e^{it}$.

(b) $\int_\gamma \left(\dfrac{e^z}{(z - 1)^4} \right) dz$ where $\gamma: [0, 2\pi] \to \mathbf{C}$ is defined by $\gamma(t) = 2e^{it}$.

4. Find

$$\int_\gamma \left(\frac{2}{z - 1} - \frac{1}{z + i} \right) dz$$

where $\gamma(t) = 2e^{it}, \ 0 \le t \le 2\pi$. [*Hint*: Since $f(z) = 2/(z - 1) - 1/(z + i)$ is not analytic at the points $z = 1$ and $z = -i$, enclose these points with appropriate circular paths (see Fig. 12.30) and use (12.B.12) to show that

$$\int_\gamma f(z) \, dz = \int_{\gamma_1} \frac{2}{z - 1} \, dz + \int_{\gamma_1} \frac{-1}{z + i} \, dz + \int_{\gamma_2} \frac{2}{z - 1} \, dz + \int_{\gamma_2} \frac{-1}{z + i} \, dz$$

Evaluate each of the integrals appearing in the right-hand side of this equation.]

5. Use the procedure outlined in the preceding problem to evaluate the following:

(a) $\int_\gamma \left(\dfrac{3}{z^2 - 4} + \dfrac{1}{z^2 + 1} \right) dz$

where $\gamma(t) = 3 + 4e^{it}, \ 0 \le t \le 2\pi$.

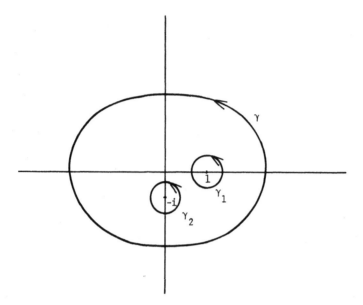

Figure 12.30

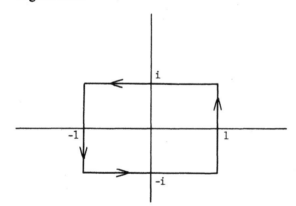

Figure 12.31

(b) $\displaystyle\int_\gamma \left(\cos z - \frac{2}{2z^2 + 1/4}\right) dz$

over the contour shown in Fig. 12.31.

6. Evaluate

(a) $\displaystyle\int_\gamma \frac{dz}{(z - 1)^2 z^4}$

where $\gamma : [0,2\pi] \to \mathbf{C}$ is defined by $\gamma(t) = 4e^{it}$.

(b) $\displaystyle\int_\gamma \left(\frac{e^z}{(z + 3)^3 z^2}\right) dz$

where $\gamma : [0,2\pi] \to \mathbf{C}$ is defined by $\gamma(t) = -2 + 4e^{it}$.

(c) $\displaystyle\int_{\gamma}\left(\frac{1}{z^2 + 1}\right) dz$

where $\gamma: [0,2\pi] \to \mathbf{C}$ is defined by $\gamma(t) = i + e^{it}$.

7.* In this problem the reader is asked to establish the Cauchy-Goursat theorem in the special case that the path of integration is a triangle. Suppose that a function f is analytic on a simply connected region R and that γ is a simple closed path in R such that $I[\gamma]$ is a triangle ABC.

(a) Let D, E, and F be the midpoints of the line segments AB, BC, and AC, respectively. Let $s = \int_{ABC} f(z)\, dz$ and show that

$$s = \int_{ADF} f(z)\, dz + \int_{BED} f(z)\, dz + \int_{CFE} f(z)\, dz + \int_{DEF} f(z)\, dz$$

and, hence, that the value of at least one of the four integrals on the right-hand side of the above equation is greater than or equal to $s/4$. Denote the corresponding triangle of one such integral by $A_1B_1C_1$.

(b) Repeat the procedure in (a) to obtain a sequence of triangles $ABC \supset A_1B_1C_1 \supset A_2B_2C_2 \supset A_3B_3C_3 \supset \cdots$ such that for each n,

$$|s| \leq 4^n \int_{A_nB_nC_n} f(z)\, dz \qquad (*)$$

(Here, of course, we are integrating along the boundary of $A_nB_nC_n$.)

(c) Show that there is a unique point z^* belonging to the intersection of the sequence of triangles defined in (b). Use the fact that f is differentiable at z^* to show that

$$\left| \frac{f(z) - f(z^*)}{z - z^*} - f'(z^*) \right|$$

can be made arbitrarily small for values of z near z^*.

(d) Note that

$$f(z) = f(z^*) - z^*f'(z^*) + f'(z^*)z + (z - z^*)\left(\frac{f(z) - f(z^*)}{z - z^*} - f'(z^*)\right).$$

Now apply (12.B.1) to show that $\int_{A_nB_nC_n} dz = 0$ and $\int_{A_nB_nC_n} z\, dz = 0$, and then use the results of (b) and (c) to show that for large n, $\int_{A_nB_nC_n} f(z)\, dz$ will be sufficiently small so that the desired result will follow from $(*)$.

Section C

1. Find the Taylor series expansion of

$$f(z) = \frac{1}{(1 - z)^2}$$

about $z = 0$ and determine its radius of convergence.

2. Find the Taylor series expansion of

$$f(z) = \frac{\sin z}{1 - z}$$

about $z = 0$. [*Hint*: Find Taylor series expansions of $\sin z$ and $1/(1 - z)$ and multiply.]

3. Find the Taylor series expansion of $g(z) = e^{z^2 + 4z}$ about $z = -2$.

4. (a) Suppose that f is analytic on an open connected set D, $z_0 \in D$, and $\{z_n\}$ is a sequence in D converging to z_0 and with the property that $f(z_n) = 0$ for each n. Show that f is identically 0 on D.

    (b) Show that if f and g are analytic on an open connected set D and if $\{z_n\}$ is a sequence of distinct points in D that converges to $z_0 \in D$ such that for each n, $f(z_n) = g(z_n)$, then $f = g$ on D.

5. Show that if f is analytic on a region D and if f is identically 0 on a straight-line segment or arc lying in D, then f is identically 0 on D.

6. Verify the identity given in (1) of this section.

7. Establish (12.C.1) in the case that $z_0 \neq 0$.

8. Prove Corollary (12.C.13).

9. Suppose that f is analytic on a region D, and let C be a circle with radius r and center z_0 such that C together with its interior lies in D. Let $M = \max\{|f(z)| \mid z \in C\}$.

    (a) Show that

$$|f^{(n)}(z_0)| \leq \frac{Mn!}{r^n}.$$

    (b) Show that $|f|$ cannot have a strict maximum at any point where f is analytic.

10. Prove the *Fundamental Theorem of Algebra*: A nonconstant polynomial $p(z) = c_n z^n + c_{n-1} z^{n-1} + \cdots + c_1 z + c_0$ has at least one root. [*Hint*: Suppose not and show that $1/p(z)$ must be a bounded entire function.]

Section D

1. Find the Laurent series expansion of the following functions:

    (a) $f(z) = \sin (1/z)$ about $z = 0$.

    (b) $g(z) = e^{2z}/z^3$ about $z = 0$.

    (c) $f(z) = (\cos z + z)/z^4$ about $z = 0$.

    (d) $g(z) = 1/[z^4(z + 2)^2]$ about $z = -2$. [*Hint*: Let $w = z + 2$.]

2. Determine and classify the singularities of the following functions:

    (a) $f(z) = \cos z^2 + \sin \dfrac{1}{z^2}$

    (b) $f(z) = \dfrac{\sin z}{z^4}$

    (c) $f(z) = \dfrac{1}{z - i} + \dfrac{1}{z^2 + 1}$

    (d) $f(z) = \dfrac{z^2 - z - 6}{(z - 3)(z - i)^3}$

3. Show that if f has a zero of order $m \geq 1$ at z_0, then there is a $\delta > 0$ such that $f(z) \neq 0$ whenever $0 < |z - z_0| < \delta$.

4. Show that 0 is an essential singularity of $f(z) = e^{1/z}$. [*Hint*: Consider $\lim_{x \to 0^-} e^{1/x}$ and $\lim_{x \to 0^+} e^{1/x}$ for real values of x.]

5. Determine whether the following functions are meromorphic:

    (a) $f(z) = \dfrac{\sin z}{z^2 - 6z - 1}$

    (b) $f(z) = \sin \dfrac{1}{z}$

    (c) $f(z) = e^{-(1/z^2)}$

    (d) $f(z) = \dfrac{1}{\sin z}$

6. Use partial fractions to find the Laurent series expansion of

$$f(z) = \dfrac{1}{(z - 2)(z + 1)}$$

    in the indicated regions:

    (a) $1 < |z| < 2$

    [*Hint*: Write $1/(z - 2)$ as $-\frac{1}{2}(1/(1 - z/2))$ and write the latter fraction as a geometric series; let $\alpha = 1/z$ and expand $1/(z + 1) = \alpha/(1 + \alpha)$ as a geometric series in α.]

    (b) $|z| < 1$

    (c) $|z| > 2$

7. Show that the singularities of $f(z) = 1/(z^2 + 1)$ are not removable.

Section E

1. Find the residues at the indicated poles:

    (a) $f(z) = \dfrac{\sin z}{z^4}$ $z = 0$

    [*Hint*: Use the Laurent expansion of $f(z)$.]

(b) $f(z) = \dfrac{3 + \cos z}{z}$ $z = 0$

(c) $f(z) = \dfrac{1}{z^2 + 1}$ $z = i, z = -i$

(d) $f(z) = \dfrac{e^z}{z^2 + 1}$ $z = i, z = -i$

(e) $f(z) = \dfrac{e^{3z}}{z^2 - 6z + 9}$ $z = 3$

(f) $f(z) = \dfrac{\sin 2z}{(z + 1)^3}$ $z = -1$

(g) $f(z) = \dfrac{4}{z^2(z^2 - i)}$ $z = 0$

2. Use the residue theorem to calculate the following integrals, where $\gamma: [0,2\pi] \to \mathbf{C}$ is defined by $\gamma(t) = e^{it}$.

(a) $\displaystyle\int_\gamma \frac{z}{z + 9z^2}\, dz$

(b) $\displaystyle\int_\gamma \frac{z^3}{2z^4 + 1}\, dz$

(c) $\displaystyle\int_\gamma \frac{e^z}{\sin z}\, dz$

(d) $\displaystyle\int_\gamma \frac{e^z}{z^2}\, dz$

(e) $\displaystyle\int_\gamma \frac{dz}{z^4(z - 3)}$

3. Evaluate:

(a) $\displaystyle\int_0^{2\pi} \frac{d\theta}{3 - \sin\theta}$

(b) $\displaystyle\int_0^{2\pi} \frac{d\theta}{(4 - 2\sin\theta)^2}$

(c) $\displaystyle\int_0^{2\pi} \frac{d\theta}{\cos\theta + \sqrt{5}}$

(d) $\displaystyle\int_0^{2\pi} \frac{d\theta}{1 + 3\cos^2\theta}$

4. Evaluate:

(a) $\text{CP}\displaystyle\int_{-\infty}^{\infty} \frac{dx}{(x^2 + 9)(x^2 + 1)^2}$

(b) $\text{CP}\displaystyle\int_{-\infty}^{\infty} \frac{x^2}{x^6 + 4}\, dx$

(c) $\displaystyle\int_0^\infty \frac{\cos 4x}{x^2 + 4}\, dx$

5. Show that $\int_0^\infty (1 - \cos x)/x^2\, dx = \pi/2$ by using the technique described in (12.E.10).

6. Use the fact that $\int_0^\infty e^{-x^2}\, dx = \sqrt{\pi}/2$ to show that

$$\int_0^\infty \cos x^2\, dx = \frac{1}{2}\sqrt{\frac{\pi}{2}} = \int_0^\infty \sin x^2\, dx$$

[*Hint*: Integrate along the contour indicated in Fig. 12.32. Show that $\lim_{R\to\infty}\int_{C_1} e^{-z^2}\, dz = 0$ and that $\int_{C_2} e^{-z^2}\, dz = e^{i\pi/4}\int_R^0 e^{-it^2}\, dt$. You may use the fact that for $0 \le t \le \pi/4$, $\cos 2t \ge 1 - 4t/\pi$.]

7. Find $\text{CP}\displaystyle\int_{-\infty}^\infty v\, \frac{x^6}{(1 + x^4)^2}\, dx$

8. Use (12.E.8) and contours of the form indicated in Fig. 12.33 to evaluate:

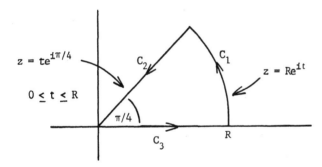

Figure 12.32

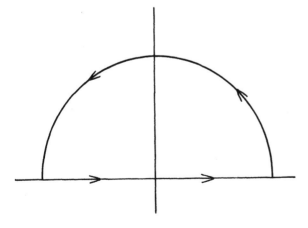

Figure 12.33

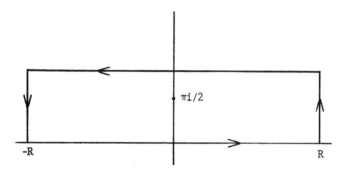

Figure 12.34

(a) $\int_0^\infty [x^2/(x^4 + 1)] \, dx$. [*Hint*: Note that the integrand is an even function and use the fact that $\int_0^R [x^2/(x^4 + 1)] \, dx = \frac{1}{2} \int_{-R}^R [x^2/(x^4 + 1)] \, dx$.]

(b) $\text{cpv}\int_{-\infty}^\infty [e^{ix}/(x^2 + 9)] \, dx$.

(c) $\int_0^\infty [1/(x^2 + 9)^4] \, dx$.

9. (a) Use L'Hospital's rule to show that $2/\pi \le (\sin t)/t \le 1$ if $t \in (0,\pi/2]$.

   (b) Show that $\int_0^\pi e^{-aR\sin t} \, dt = 2 \int_0^{\pi/2} e^{-aR\sin t} \, dt$.

10. Find $\text{cp} \int_{-\infty}^\infty \dfrac{x^2}{(x^2 + 4)(x^2 + 9)} \, dx$

11. Integrate along paths such as the one indicated in Fig. 12.34 to evaluate $\text{cpv}\int_{-\infty}^\infty [\cos 2x/(e^x + e^{-x})] \, dx$.

Section F

1. Evaluate the following line integrals over the indicated paths:

   (a) $\int_\gamma k$, where $\gamma(t) = (t,2t,-t)$ and $k(x,y,z) = (\sin y, xy, x^2 z)$.

   (b) $\int_\gamma k$, where γ traverses the path shown in Fig. 12.35 in a clockwise

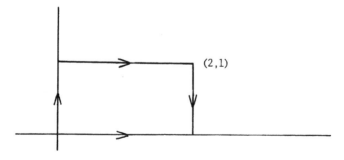

Figure 12.35

direction and $k(x,y) = (e^{2xy}, x^2y)$.

(c) $\int_\gamma (x^2 + y) \, dx + \int_\gamma xy \, dy$, where γ traverses the parabola $y = 3x^2$ from $x = -2$ to $x = 2$.

(d) $\int_\gamma (y^2 \, dx + z \, dy - x \, dz)$, where γ traverses the straight-line segment from $(0,0,0)$ to $(1,3,-2)$.

(e) $\int_\gamma k$, where $k(x,y) = (x^2 + y, x^2 + y^2)$ and γ traverses once the ellipse $x^2/9 + y^2/16 = 1$ in a counterclockwise direction.

2. Use line integrals to evaluate the following contour integrals:

(a) $\int_\gamma f(z) \, dz$, where $\gamma(t) = 2t + it$, $0 \le t \le 1$ and $f(z) = z^2$.

(b) $\int_\gamma f(z) \, dz$, where $\gamma(t) = 2e^{it}$, $0 \le t \le \pi$ and $f(z) = f(x + iy) = \cos x + 2ix^2y$.

3. Calculate the work done by the force field $F(x,y) = (x^2 - y^3, x^2 + y^2)$ in moving a particle once around the circle $x^2 + y^2 = 1$ in a positive direction.

4. Verify Green's theorem in the following cases, i.e., compute the integrals with and also without the aid of Green's theorem:

(a) $\int_\gamma x^3y^4 \, dx + x^4y^3 \, dy$, where γ traverses once the circle $x^2 + y^2 = 9$ in a counterclockwise direction.

(b) $\int_\gamma (x + 2y) \, dx + y \sin x \, dy$, where γ traverses the boundary of the rectangle indicated in Fig. 12.36.

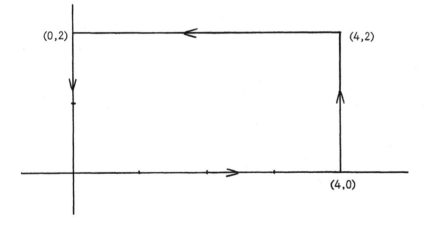

Figure 12.36

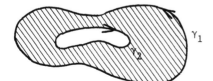

Figure 12.37

(c) $\int_\gamma (x^3 - 2xy)\, dx + (y^2 - x^2)\, dy$, where γ traverses the unit circle once in a counterclockwise direction.

(d) $\int_\gamma (x^2 + 4xy)\, dx + (2x^2 + 3y)\, dy$, where γ traverses once the ellipse $9x^2 + 16y^2 = 144$ in a positive direction.

5. Show that if D is the shaded region indicated in Fig. 12.37, then

$$\int_D Q_x - P_y = \int_{\gamma_1} P\, dx + Q\, dy - \int_{\gamma_2} P\, dx + Q\, dy$$

[*Hint*: Use the indicated auxiliary arcs to form appropriate simple closed curves (Fig. 12.38).] Extend this result to regions D of the form illustrated in Fig. 12.39.

6. Use the result of the preceding problem to evaluate:

(a) $\displaystyle\int_{\gamma_1} x^2 y\, dx + (x + y)\, dy - \int_{\gamma_2} y\, dx - x\, dy,$

where γ_1 traverses once the circle $x^2 + y^2 = 9$ in a positive sense and γ_2 traverses once the circle $x^2 + y^2 = 4$ also in a positive sense.

(b) $\displaystyle\int_{\gamma_1} y\, dx + (2x + y^2)\, dy - \int_{\gamma_2} y\, dx + (2x + y^2)\, dy,$

where γ_1 traverses once the circle $x^2 + y^2 = 4$ in a positive sense and γ_2 traverses once the unit circle also in a positive sense.

7. (a) Show that if S is a simple closed curve that bounds a region for which Green's theorem is applicable, and if γ is a path that traverses S in a counterclockwise direction, then the area of the region bounded by S is equal to $\frac{1}{2}\int_\gamma x\, dy - y\, dx$.

(b) Use part (a) to show that the area enclosed by the ellipse, $x^2/a^2 + y^2/b^2 = 1$ is πab. [*Hint*: Use the parametric representation of the ellipse, $\gamma(t) = (a \cos t, b \sin t)$, $0 \le t \le 2\pi$.]

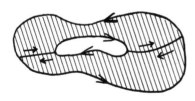

Figure 12.38

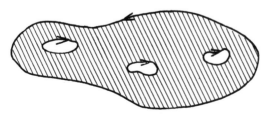

Figure 12.39

(c) Use part (a) to show that the area of the triangle that lies in the plane and with vertices (a_1,b_1), (a_2,b_2), (a_3,b_3) may be given by $\frac{1}{2}[(a_2 - a_1)(b_3 - b_1) - (b_2 - b_1)(a_3 - a_1)]$.

8. (a) Find the work done by the force field $F(x,y) = (x - 2y, x + y)$ in moving a particle once around the ellipse $x^2/9 + y^2/4 = 1$ in a clockwise direction.

   (b) Find the work done by the force field $F(x,y) = (6 - y, x)$ in moving a particle along an arch of the cycloid $k(t) = (3t - 3\sin t, 3 - 3\cos t)$, $0 \le t \le 2\pi$.

9. Give an intuitive argument to show that if P and Q and their first partials are continuous on the region D indicated in Fig. 12.40, and if γ_1 and γ_2 are any two continuously differentiable paths in D that go around the hole once in a counterclockwise direction, then

$$\int_{\gamma_1} P\,dx + Q\,dy = \int_{\gamma_2} P\,dx + Q\,dy$$

[*Hint*: Consider the path Γ formed by removing small pieces of γ_1 and γ_2 and adding the indicated line segments s_1 and s_2. Note that $\int_\Gamma P\,dx + Q\,dy = 0$. Cut out smaller and smaller pieces of γ_1 and γ_2 (Fig. 12.41).]

10. Use problem F.9 to evaluate

$$\int_\gamma \frac{2y\,dy - 2x\,dx}{x^2 + y^2}$$

where γ is the curve

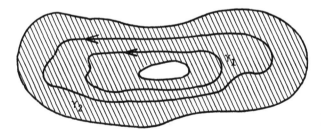

Figure 12.40

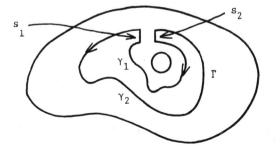

Figure 12.41

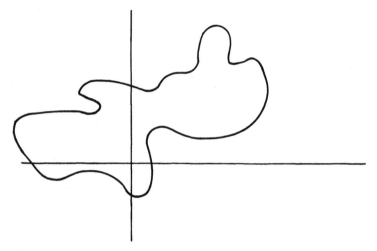

Figure 12.42

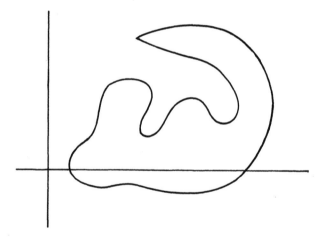

Figure 12.43

   shown is Fig. 12.42 or the curve shown in Fig. 12.43.
11. Prove Theorem (12.F.7).
12. Establish equation (7) in Sec. 12.F.
13. Show that there is (and find) a potential function ϕ for each of the
     following functions:
     (a) $G(x,y) = (e^x \sin y - 2y \sin x, e^x \cos y + 2 \cos x)$
     (b) $G(x,y) = xe^{x^2+y^2}, ye^{x^2+y^2})$
     [*Hint*: First find a function $\hat{\phi}(x,y)$ such that $\hat{\phi}_x = -P$, and then set
     $\phi(x,y) = \hat{\phi}(x,y) + h(y)$; to determine h, recall that $\hat{\phi}_y = -Q$.]

14. Show that the following integrals are independent of path and then evaluate them:
    (a) $\int_\gamma y(e^{xy} + 1)\,dx + x(e^{xy} + 1)\,dy$ from the point $(0,1)$ to the point $(1,0)$.
    (b) $\int_\gamma 2x \sin y\,dx + x^2 \cos y\,dy$ from the point $(0,\pi/2)$ to the point $(\pi/4,\pi)$.
15. Verify (3) assuming that equality (2) does not hold. [*Hint*: Consider

$$\lim_{|P|\to 0} \left| \sum_{i=1}^n \left[f(\gamma(\tau_i^*))(x'(\tau_i^*)) + g(\gamma(\tau_i^*))(y'(\tau_i^*)) + h(\gamma(\tau_i^*))(z'(\tau_i^*)) \right](t_i - t_{i-1}) \right.$$

$$- \sum_{i=1}^n \left[f(\gamma(\tau_i))x'(c_i) + g(\gamma(\tau_i))y'(\hat{c}_i) + h(\gamma(\tau_i))z'(\tilde{c}_i) \right](t_i - t_{i-1}),$$

where for each i, τ_i^* is an arbitrary point in $[t_{i-1}, t_i]$; use the uniform continuity (on $[a,b]$) of all the above functions to show that this limit is equal to zero.]

REFERENCES

Ahlfors, L. (1966): *Complex Analysis*, McGraw-Hill, New York.
Connell, E. H., and P. Porcelli (1961): *Bull. Amer. Math. Soc.*, **67**: 177.
Korevaar, J. (1968): *Mathematical Methods*, Vol. 1, Academic Press, New York.
Plunkett, R. L. (1959): *Bull. Amer. Math. Soc.*, **65**: 1.
Christenson, C., and W. Voxman (1978): *Aspects of Topology*. Marcel Dekker, New York.

13

THE FOURIER AND LAPLACE
TRANSFORMS

A. THE FOURIER INTEGRAL FORMULA

In Chapter 10, we saw that Fourier series expansions were especially useful in finding solutions to certain boundary value problems. In these problems the boundary conditions were specified at two points (for example, the endpoints of a rod), and we found solutions that were valid over a finite interval. In this section, we consider problems that involve infinite intervals. For instance, suppose that we have a very long rod that is insulated laterally and suppose that we have little or no information concerning the boundary (temperature) data at the right end of the rod. In this case, it is convenient to view the rod as extending over the interval $[0,\infty)$. If the available data is limited for each end of the rod, then we shall consider the rod as extending over the entire interval $(-\infty,\infty)$. In both of these cases, it is customary to impose some qualitative behavior constraints on the solution $u(x,t)$. For instance, we may suppose that $\lim_{x \to \pm\infty} u(x,t) = 0$ or that $u(x,t)$ remains bounded as x tends to $\pm\infty$.

To illustrate these ideas, we consider the problem of determining a mathematical formulation that can be used to describe the conduction of heat in the central region of a very long rod. This problem can be resolved by finding a function $u(x,t)$ that satisfies the following conditions:

$$u_{xx} - \frac{1}{a^2} u_t = 0 \qquad -\infty < x < \infty, \, t > 0 \tag{1}$$

$|u(x, t)|$ is bounded as x tends to $\pm\infty$ $\qquad\qquad\qquad\qquad$ (2)

$$u(x, 0) = f(x) \qquad -\infty < x < \infty \tag{3}$$

Note that from a physical standpoint condition (2) is quite reasonable.

To find u, we begin by applying the separation of variables technique used previously; we suppose that there is a solution of the form $X(x)T(t)$ to the boundary value problem determined by equations (1) and (2). Substitution in (1) shows that the functions X and T satisfy

$$X'' - pX = 0 \qquad -\infty < x < \infty \tag{4}$$

$$T' - a^2 pT = 0 \qquad t > 0 \tag{5}$$

for some real constant p.

It is easily checked that solutions to (4) and (5) are given by

$$X(x) = c_1 e^{\sqrt{p}x} + c_2 e^{-\sqrt{p}x}$$

and

$$T(t) = c_3 e^{a^2 pt}$$

respectively (see problem A.7). It is not difficult to see that if $p > 0$, then condition (2) above cannot be satisfied unless c_1 and c_2 are zero or $c_3 = 0$. Since either of these cases yields a trivial result, we make the assumption that $p = -\lambda^2 \le 0$. Equations (4) and (5) then become

$$X'' + \lambda^2 X = 0 \qquad -\infty < x < \infty \tag{6}$$

$$T' + (a\lambda)^2 T = 0 \qquad t > 0 \tag{7}$$

and hence, if for each real number λ, we set

$$X_\lambda(x) = \alpha_\lambda \cos \lambda x + \beta_\lambda \sin \lambda x$$

and

$$T_\lambda(t) = e^{-(a\lambda)^2 t}$$

then we have that

$$X_\lambda(x)T_\lambda(t) = (\alpha_\lambda \cos \lambda x + \beta_\lambda \sin \lambda x)e^{-(a\lambda)^2 t} \tag{8}$$

is a solution of (1) and (2) for each real number λ, and arbitrary constants α_λ and β_λ.

Although $X_\lambda(x)T_\lambda(t)$ satisfies conditions (1) and (2), it is highly unlikely that it would simultaneously satisfy (3), since this would require that $X_\lambda(x) = f(x)$ for all x. In the somewhat analogous situation involving a *finite* rod, we saw (using superposition) that an infinite series of the form

$$\sum_{n=0}^{\infty} X_{\lambda_n}(x)T_{\lambda_n}(t) = \sum_{n=0}^{\infty} (a_n \cos \lambda_n x + b_n \sin \lambda_n x)e^{-(a\lambda_n)^2 t}$$

provided a solution satisfying all of the required conditions (where $\{\lambda_n\}$ was an appropriately chosen sequence of real numbers). In the present case such a "solution" would imply that

$$f(x) = \sum_{n=0}^{\infty} X_{\lambda_n}(x) = \sum_{n=0}^{\infty} (a_n \cos \lambda_n x + b_n \sin \lambda_n x)$$

for all x and suitably chosen coefficients a_n and b_n.

However, if, as before, we were to set $\lambda_n = n\pi/l$ for each n, and then find the corresponding Fourier coefficients a_n and b_n, we would obtain a Fourier series expansion of f that is valid only on the interval $(-l,l)$. (This is characteristic of Fourier series expansions; in general, they are of use only in representing functions over finite intervals.)

Since this approach to finding a solution satisfying (1), (2) and (3) would seem to be of limited interest, we shall pursue another tack, and essentially replace infinite series with integrals. In other words, instead of trying to find a series representation of f, we shall establish that under fairly general conditions f can be represented as an *integral*. To motivate this idea, we first show (heuristically) that if $X_\lambda(t)T_\lambda(t)$ is defined as in (8), then there are solutions of (1) of the form

$$\tilde{u}(x, t) = \int_{-\infty}^{\infty} X_\lambda(x)T_\lambda(t) \, d\lambda \tag{9}$$

that also satisfy both (2) and (3).

Suppose, for the moment, that the partial derivative of an integral is "equal" to the integral of the partial derivative [cf. the Leibniz rule (7.F.1)]. Then since $X_\lambda(x)T_\lambda(t)$ satisfies (1),

$$\frac{\partial^2}{\partial x^2} \tilde{u}(x, t) = \frac{\partial^2}{\partial x^2} \int_{-\infty}^{\infty} X_\lambda(x)T_\lambda(t) \, d\lambda$$

$$= \int_{-\infty}^{\infty} \left(\frac{\partial^2}{\partial x^2} X_\lambda(x)T_\lambda(t) \right) d\lambda$$

$$= \int_{-\infty}^{\infty} \left(\frac{1}{a^2} \frac{\partial}{\partial t} X_\lambda(x)T_\lambda(t) \right) d\lambda$$

$$= \frac{1}{a^2} \frac{\partial}{\partial t} \int_{-\infty}^{\infty} X_\lambda(x)T_\lambda(t) \, d\lambda$$

$$= \frac{1}{a^2} \frac{\partial}{\partial t} \tilde{u}(x, t)$$

and hence, $\tilde{u}$ satisfies (1). We, of course, must also require that $\tilde{u}(x,t)$ remain bounded as $|x| \to \infty$ and that $\tilde{u}(x,0) = \int_{-\infty}^{\infty} X_\lambda(x) \, d\lambda = f(x)$, for all x. In order for this latter condition to hold, it is clear from (8) that we need to find functions C_1 and C_2 of λ so that

$$f(x) = \int_{-\infty}^{\infty} [C_1(\lambda) \cos \lambda x + C_2 (\lambda) \sin \lambda x] \, d\lambda$$

Most of the remainder of this section will be devoted to showing that such functions exist. Again we emphasize that whereas our earlier efforts were directed to finding series representations for certain functions, we now are working toward determining integral representations for functions mapping $\mathbf{R}^1$ into $\mathbf{R}^1$.

First we show that the integral $\int_{-\infty}^{\infty}$ appearing in (9) can be replaced by $\int_0^{\infty}$. If we assume for the moment that functions $C_1(\lambda)$ and $C_2(\lambda)$ exist such that

$$\tilde{u}(x, t) = \int_{-\infty}^{\infty} [C_1(\lambda) \cos \lambda x + C_2(\lambda) \sin \lambda x] e^{-(a\lambda)^2 t} \, d\lambda$$

then since $\sin(-\alpha) = -\sin \alpha$ and $\cos(-\alpha) = \cos \alpha$ for all α, it follows that

$$\begin{aligned}
\tilde{u}(x, t) &= \int_{-\infty}^{\infty} [C_1(\lambda) \cos \lambda x + C_2(\lambda) \sin \lambda x] e^{-(a\lambda)^2 t} \, d\lambda \\
&= \int_0^{\infty} [C_1(\lambda) \cos \lambda x + C_2(\lambda) \sin \lambda x] e^{-(a\lambda)^2 t} \, d\lambda \\
&\quad + \int_{-\infty}^{0} [C_1(\lambda) \cos \lambda x + C_2(\lambda) \sin \lambda x] e^{-(a\lambda)^2 t} \, d\lambda \\
&= \int_0^{\infty} [C_1(\lambda) \cos \lambda x + C_2(\lambda) \sin \lambda x] e^{-(a\lambda)^2 t} \, d\lambda \\
&\quad + \int_0^{\infty} [C_1(-\lambda) \cos(-\lambda x) + C_2(-\lambda) \sin(-\lambda x)] e^{-(a\lambda)^2 t} \, d\lambda \\
&= \int_0^{\infty} [C_1(\lambda) \cos \lambda x + C_2(\lambda) \sin \lambda x] e^{-(a\lambda)^2 t} \, d\lambda \\
&\quad + \int_0^{\infty} [C_1(-\lambda) \cos \lambda x - C_2(-\lambda) \sin \lambda x] e^{-(a\lambda)^2 t} \, d\lambda \\
&= \int_0^{\infty} [A(\lambda) \cos \lambda x + B(\lambda) \sin \lambda x] e^{-(a\lambda)^2 t} \, d\lambda
\end{aligned}$$

where

$$A(\lambda) = C_1(\lambda) + C_1(-\lambda)$$

and

$$B(\lambda) = C_2(\lambda) - C_2(-\lambda)$$

Note that it now follows that $f(x) = \tilde{u}(x,0)$ is defined by

$$f(x) = \int_0^{\infty} [A(\lambda) \cos \lambda x + B(\lambda) \sin \lambda x] \, d\lambda \tag{10}$$

for appropriate functions $A(\lambda)$ and $B(\lambda)$.

How does one compute $A(\lambda)$ and $B(\lambda)$? Fourier answered this question essentially as follows.

Suppose that the function $f: \mathbf{R}^1 \to \mathbf{R}^1$ is piecewise smooth on every bounded interval. Let $\hat{x}$ be an arbitrary point in $\mathbf{R}^1$. Then since $\hat{x} \in (-l,l)$ for all suitably large l, say $l \geq L_{\hat{x}}$, $f(\hat{x})$ can be represented by a Fourier series in the form

$$f(\hat{x}) = \frac{a_0}{2} + \sum_{n=1}^{\infty} \left(a_n \cos \frac{n\pi\hat{x}}{l} + b_n \sin \frac{n\pi\hat{x}}{l} \right) \tag{11}$$

where

$$a_n = \frac{1}{l} \int_{-l}^{l} f(t) \cos \frac{n\pi t}{l} \, dt \quad n \geq 0$$

$$b_n = \frac{1}{l} \int_{-l}^{l} f(t) \sin \frac{n\pi t}{l} \, dt \quad n \geq 1 \tag{12}$$

for each real $l \geq L_{\hat{x}}$. (Here we make the usual assumption that at each point x of discontinuity, $f(x) = [f(x + 0) + f(x - 0)]/2$.) It follows from (11), (12), and the trigonometric identity $\cos(\alpha - \beta) = \cos\alpha \cos\beta + \sin\alpha \sin\beta$ that

$$f(\hat{x}) = \frac{1}{2l} \int_{-l}^{l} f(t) \, dt + \sum_{n=1}^{\infty} \left[\left(\frac{1}{l} \int_{-l}^{l} f(t) \cos \frac{n\pi t}{l} \cos \frac{n\pi\hat{x}}{l} \, dt \right) \right.$$
$$\left. + \left(\frac{1}{l} \int_{-l}^{l} f(t) \sin \frac{n\pi t}{l} \sin \frac{n\pi\hat{x}}{l} \, dt \right) \right]$$
$$= \frac{1}{2l} \int_{-l}^{l} f(t) \, dt + \frac{\pi}{l} \sum_{n=1}^{\infty} \left\{ \frac{1}{\pi} \int_{-l}^{l} f(t) \cos \left[\frac{n\pi}{l} (\hat{x} - t) \right] dt \right\}$$

If we set $\lambda_n = n\pi/l$, $\Delta\lambda_n = \lambda_n - \lambda_{n-1} = \pi/l$, and let

$$g_l(\lambda) = \frac{1}{\pi} \int_{-l}^{l} f(t) \cos [\lambda(\hat{x} - t)] \, dt$$

then we can express $f(\hat{x})$ in the form

$$f(\hat{x}) = \frac{1}{2l} \int_{-l}^{l} f(t) \, dt + \sum_{n=1}^{\infty} g_l(\lambda_n) \Delta\lambda_n \tag{13}$$

Note that

$$P = \{\lambda_0, \lambda_1, \lambda_2, \ldots, \lambda_n, \ldots\} \tag{14}$$

forms an infinite partition of $[0,\infty)$ with mesh π/l. Furthermore, the mesh of such partitions P tends to zero with increasing l. Thus, if we recall the definition of the Riemann integral over a finite interval, it becomes conceivable that the series

$$\frac{1}{2l} \int_{-l}^{l} f(t) \, dt + \sum_{n=1}^{\infty} g_l(\lambda_n) \Delta\lambda_n$$

might be considered as an approximating sum for an integral over $[0,\infty)$. However, since $g_l(\lambda)$ depends not only on λ but on l as well, it is not clear what will occur as l increases. Note, however, that by the definition of g_l we have

$$\lim_{l \to \infty} g_l(\lambda) = \frac{1}{\pi} \, \mathrm{CP} \int_{-\infty}^{\infty} \mathrm{v} \, f(t) \cos [\lambda(\hat{x} - t)] \, dt$$

If we set

$$G(\lambda) = \frac{1}{\pi} \, \mathrm{CP} \int_{-\infty}^{\infty} \mathrm{v} \, f(t) \cos [\lambda(\hat{x} - t)] \, dt = \lim_{l \to \infty} g_l(\lambda)$$

then, in view of the statements just made, we might suspect that

$$\lim_{l \to \infty} \sum_{n=1}^{\infty} g_l(\lambda) \, \Delta\lambda_n = \int_0^{\infty} G(\lambda) \, d\lambda \qquad (15)$$

Moreover, if we assume that

$$\lim_{l \to \infty} \frac{1}{2l} \int_{-l}^{l} f(t) \, dt = 0$$

(a valid assumption if $\mathrm{CP} \mathrm{v} \int_{-\infty}^{\infty} |f(t)| \, dt < \infty$), then by (13) and (15) it is reasonable to conjecture that

$$f(\hat{x}) = \int_0^{\infty} G(\lambda) \, d\lambda + \int_0^{\infty} \left\{ \frac{1}{\pi} \, \mathrm{CP} \int_{-\infty}^{\infty} \mathrm{v} \, f(t) \cos [\lambda(\hat{x} - t)] \, dt \right\} d\lambda. \qquad (16)$$

This heuristic reasoning, although technically questionable, does lead us to a correct result, which we shall prove shortly. The integral representation given by formula (16) is called the *Fourier integral representation of f* and in many instances this representation is valid for all real values x. From the Fourier integral representation of f, we can compute the functions $A(\lambda)$ and $B(\lambda)$ that appear in (10); indeed, we have [by the identity $\cos(\alpha - \beta) = \cos \alpha \cos \beta + \sin \alpha \sin \beta$] that

$$f(x) = \int_0^{\infty} \left\{ \frac{1}{\pi} \, \mathrm{CP} \int_{-\infty}^{\infty} \mathrm{v} \, f(t) \cos \lambda(x - t) \, dt \right\} d\lambda$$

$$= \int_0^{\infty} \left[\left\{ \frac{1}{\pi} \, \mathrm{CP} \int_{-\infty}^{\infty} \mathrm{v} \, f(t) \cos (\lambda t) \, dt \right\} \cos \lambda x \right.$$

$$+ \left. \left\{ \frac{1}{\pi} \, \mathrm{CP} \int_{-\infty}^{\infty} \mathrm{v} \, f(t) \sin \lambda t \, dt \right\} \sin \lambda x \right] d\lambda$$

$$= \int_0^{\infty} [A(\lambda) \cos \lambda x + B(\lambda) \sin \lambda x] \, d\lambda$$

and, therefore,

$$A(\lambda) = \frac{1}{\pi} \, \text{CP} \int_{-\infty}^{\infty} \text{V} \;\; f(t) \cos \lambda t \, dt$$

and

$$B(\lambda) = \frac{1}{\pi} \, \text{CP} \int_{-\infty}^{\infty} \text{V} \;\; f(t) \sin \lambda t \, dt$$

[The reader should compare these coefficients with the Fourier coefficients defined by (12).] Thus, we have shown (intuitively) that there is a solution of the problem determined by (1), (2), and (3) of the form

$$u(x, t) = \int_{0}^{\infty} [A(\lambda) \cos \lambda x + B(\lambda) \sin \lambda x] e^{-(a\lambda)^2 t} \, d\lambda$$

(provided $\text{CPV} \int_{-\infty}^{\infty} |f(t)| \, dt < \infty$).

We now turn to a rigorous proof of Fourier's integral theorem (which the less theoretically inclined reader may wish to omit). Before stating the principal result of this section, we make the following observation.

(13.A.1) *Observation* As we have seen above, the Cauchy principal value interpretation of the integral arises naturally in the context of representing a function f as a Fourier integral. In problem A.9 the reader is asked to use the Lebesgue dominated convergence theorem to show that if $g \in \mathscr{L}_1(-\infty,\infty)$, then $\text{CPV} \int_{-\infty}^{\infty} g(x) \, dx$ is equal to the extended or Lebesgue integral $\int_{(-\infty,\infty)} g$. Furthermore, in problem A.10 it is seen that if a function g is summable over an infinite interval J, and if f is bounded on J and summable over every finite subinterval of J, then fg is summable over J. Thus, in particular, it follows that if $g \in \mathscr{L}_1(-\infty,\infty)$, then $\mathscr{F}^{-1}(g)(x)$, which by definition is the Cauchy principal value $\text{CPV} \int_{-\infty}^{\infty} g(\lambda) e^{-i\lambda x} \, d\lambda$, is also equal to the extended or Lebesgue integral of the function $g(\lambda) e^{-i\lambda x}$ (with respect to λ) over the interval $(-\infty, \infty)$.

In the remainder of this chapter the symbol $\int_{-\infty}^{\infty} g(x) \, dx$ will denote the proper Lebesgue or extended integral of g over $\mathbf{R}^1$; the symbol $\int_{0}^{\infty} g(x) \, dx$ will denote the improper integral, i.e.,

$$\int_{0}^{\infty} g(x) \, dx = \lim_{b \to \infty} \int_{0}^{b} g(x) \, dx$$

(13.A.2) *Theorem* If $f: \mathbf{R}^1 \to \mathbf{R}^1$ is piecewise smooth on every finite interval and if $\int_{-\infty}^{\infty} |f(t)| \, dt < \infty$, then for each $x \in \mathbf{R}^1$

$$\frac{f(x + 0) + f(x - 0)}{2} = \frac{1}{\pi} \lim_{b \to \infty} \int_{0}^{b} \left\{ \int_{-\infty}^{\infty} f(t) \cos \lambda(x - t) \, dt \right\} \, d\lambda$$

$$= \int_{0}^{\infty} \left\{ \frac{1}{\pi} \int_{-\infty}^{\infty} f(t) \cos \lambda(x - t) \, dt \right\} \, d\lambda \qquad (17)$$

The proof is broken into a number of lemmas.

(13.A.3) Lemma

$$\int_0^\pi \frac{\sin (n + \frac{1}{2})x}{\sin (x/2)}\, dx = \pi$$

Proof. By (10.D.1) we have that $1 + 2\sum_{k=1}^n \cos kx = [\sin (n + \frac{1}{2})x]/[\sin (x/2)]$. Integration from 0 to π of both sides of this equation yields the result.

(13.A.4) Lemma

$$\int_0^\infty \frac{\sin x}{x}\, dx = \frac{\pi}{2}$$

Proof. This result can be readily established with the aid of residues [see (12.E.10)]; however, here we present an entirely different proof based on the Riemann–Lebesgue lemma. Since the function $f(x) = 1/(x/2) - 1/[\sin (x/2)]$ is continuous and bounded on $(0,\pi)$ (why?), we can apply the Riemann–Lebesgue lemma (6.B.6) to obtain

$$\lim_{n\to\infty} \int_0^\pi \left[\sin\left(n + \frac{1}{2}\right)x\right]\left[\frac{2}{x} - \frac{1}{\sin (x/2)}\right] dx = 0$$

Hence, it follows from (13.A.3) and by the addition formula for limits, that

$$\lim_{n\to\infty} \int_0^\pi \frac{2\sin (n + \frac{1}{2})x}{x}\, dx = \pi$$

If we set $u = (n + \frac{1}{2})x$, then we have

$$\frac{\pi}{2} = \lim_{n\to\infty} \int_0^\pi \frac{\sin [(n + \frac{1}{2})x](n + \frac{1}{2})}{(n + \frac{1}{2})x}\, dx$$

$$= \lim_{n\to\infty} \int_0^{(n+1/2)\pi} \frac{\sin u}{u}\, du$$

$$= \int_0^\infty \frac{\sin u}{u}\, du$$

which concludes the proof.

The proof of the next lemma is left as an exercise (see problem A.8).

(13.A.5) Lemma If $f: (a,b) \to \mathbf{R}^1$ is a piecewise smooth function and if $c \in (a,b)$, then

$$f'(c + 0) = \lim_{x\to 0^+} \frac{f(c + x) - f(c + 0)}{x}$$

and

$$f'(c - 0) = \lim_{x \to 0^-} \frac{f(c + x) - f(c - 0)}{x}$$

(13.A.6) Lemma If f is a piecewise smooth function on every finite subinterval of $(-\infty,\infty)$, then for each $a > 0$,

$$\lim_{b \to \infty} \int_0^a f(x) \frac{\sin bx}{x} \, dx = \frac{\pi}{2} f(0 + 0)$$

and

$$\lim_{b \to \infty} \int_0^a f(-x) \frac{\sin bx}{x} \, dx = \frac{\pi}{2} f(0 - 0)$$

Proof. Since f is piecewise smooth at $x = 0$, both $f(0 + 0)$ and $f'(0 + 0)$ exist (and are finite). In fact, by the previous lemma,

$$f'(0 + 0) = \lim_{x \to 0^+} \frac{f(x) - f(0 + 0)}{x}$$

which implies that $[f(x) - f(0 + 0)]/x$ is bounded on the finite interval $(0,a)$. Hence, in view of (13.A.4) and the Riemann-Lebesgue lemma, we have

$$\lim_{b \to \infty} \int_0^a f(x) \frac{\sin bx}{x} \, dx = \lim_{b \to \infty} \int_0^a f(0 + 0) \frac{\sin bx}{x} \, dx$$

$$+ \lim_{b \to \infty} \int_0^a \frac{f(x) - f(0 + 0)}{x} \sin bx \, dx$$

$$= f(0 + 0) \lim_{b \to \infty} \int_0^{ab} \frac{\sin u}{u} \, du + 0$$

$$= \frac{\pi}{2} f(0 + 0)$$

A similar proof holds for the second part of the theorem.

We are now ready to prove (13.A.2). In this proof we shall at one point need to change the order of integration, justification for which is found in (11.I.14).

Proof of (13.A.2). For each $b > 0$, we let

$$I_b = \frac{1}{\pi} \int_0^b \left\{ \int_{-\infty}^\infty f(t) \cos \lambda(x - t) \, dt \right\} d\lambda$$

Since, by hypothesis, f is summable over $(-\infty,\infty)$ and since $\cos\lambda(x - t)$ is continuous and uniformly bounded in $\mathbf{R}^2$, we can, according to (11.I.14), interchange the order of integration in I_b to obtain

$$I_b = \frac{1}{\pi} \int_{-\infty}^\infty f(t) \left\{ \int_0^b \cos \lambda(x - t) \, d\lambda \right\} dt$$

Integrating the inner integral (while treating x and t as fixed quantities), we have

$$I_b = \frac{1}{\pi} \int_{-\infty}^{\infty} f(t) \frac{\sin b(x-t)}{x-t} \, dt$$

With the substitution $u = x - t$ (x is held fixed), I_b takes on the form

$$I_b = \frac{1}{\pi} \int_{-\infty}^{\infty} f(x - u) \frac{\sin bu}{u} \, du$$

which, in turn, can be written as

$$I_b = \frac{1}{\pi} \int_{-\infty}^{\infty} g(u) \frac{\sin bu}{u} \, du$$

where $g(u) = f(x - u)$. To find $\lim_{b \to \infty} I_b$, we first split the integral I_b into four parts as indicated below, where δ is a fixed positive number:

$$\int_{-\infty}^{\infty} = \int_{-\infty}^{-\delta} + \int_{-\delta}^{0} + \int_{0}^{\delta} + \int_{\delta}^{\infty}$$

Then we obtain

$$\lim_{b \to \infty} I_b = \frac{1}{\pi} \left\{ \lim_{b \to \infty} \int_{-\infty}^{-\delta} g(u) \frac{\sin bu}{u} \, du + \lim_{b \to \infty} \int_{-\delta}^{0} g(u) \frac{\sin bu}{u} \, du \right.$$

$$\left. + \lim_{b \to \infty} \int_{0}^{\delta} g(u) \frac{\sin bu}{u} \, du + \lim_{b \to \infty} \int_{\delta}^{\infty} g(u) \frac{\sin bu}{u} \, du \right\}$$

Since $g(u)/u$ is piecewise smooth and summable on both $(-\infty, -\delta)$ and (δ, ∞), it follows from the Riemann-Lebesgue lemma (6.B.6) that

$$\lim_{b \to \infty} I_b = \frac{1}{\pi} \left\{ \lim_{b \to \infty} \int_{-\delta}^{0} g(u) \frac{\sin bu}{u} \, du + \lim_{b \to \infty} \int_{0}^{\delta} g(u) \frac{\sin bu}{u} \, du \right\} \quad (18)$$

and since $g(u)$ is piecewise smooth, we have by (13.A.6) that the limit of the second integral on the right side of (18) is equal to $(\pi/2)(g(0 + 0))$. If we set $w = -u$, then the first limit on the right-hand side of (18) becomes

$$\lim_{b \to \infty} \int_{0}^{\delta} g(-w) \frac{\sin bw}{w} \, dw = \frac{\pi}{2} g(0 - 0)$$

and therefore,

$$\lim_{b \to \infty} I_b = \frac{1}{2} [g(0 - 0) + g(0 + 0)]$$

Since $g(u) = f(x - u)$, it follows that

$$\lim_{b \to \infty} I_b = \frac{1}{2} (f(x + 0) + f(x - 0))$$

the desired result.

B. THE FOURIER TRANSFORM

A very important technique used to solve many differential and integral equations is (1) to transform a given equation into a more manageable one, (2) to find the solution to the transformed equation, and (3) by an inverse procedure, use the solution of the transformed equation to determine a solution to the original equation. In this and the next section we shall investigate in some detail the Fourier transform, one of the most commonly used transforms that can be employed in this context. The Fourier transform may be derived easily from the Fourier integral formula by expressing this formula in complex form. To obtain the complex version of (13.A.2), we first note that equation (17) of Sec. 13.A can be written in the form

$$f(x) = \lim_{b \to \infty} \frac{1}{2\pi} \int_{-b}^{b} \left(\int_{-\infty}^{\infty} f(t) \cos \lambda(t - x) \, dt \right) d\lambda \tag{1}$$

since the integral with respect to t is an even function in λ. We also note

$$0 = \lim_{b \to \infty} \frac{1}{2\pi} \int_{-b}^{b} \left(\int_{-\infty}^{\infty} f(t) \sin \lambda(t - x) \, dt \right) d\lambda$$

$$= \frac{1}{2\pi} \text{CP} \int_{-\infty}^{\infty} \left(\int_{-\infty}^{\infty} f(t) \sin \lambda(t - x) \, dt \right) d\lambda \tag{2}$$

since in this case the integral with respect to t is an odd function of λ.

Since for any value y,

$$e^{iy} = \cos y + i \sin y$$

we have

$$f(t) \cos \lambda(t - x) + f(t)i \sin \lambda(t - x) = f(t)e^{i\lambda(t-x)} \tag{3}$$

and, consequently, it follows from (1), (2), and (3) that

$$f(x) = f(x) + 0i = \frac{1}{2\pi} \text{CP} \int_{-\infty}^{\infty} \left(\int_{-\infty}^{\infty} f(t)e^{i\lambda(t-x)} \, dt \right) d\lambda$$

$$= \frac{1}{2\pi} \text{CP} \int_{-\infty}^{\infty} e^{-i\lambda x} \left(\int_{-\infty}^{\infty} f(t)e^{i\lambda t} \, dt \right) d\lambda \tag{4}$$

(13.B.1) *Definition* The function $\hat{f}$ defined by

$$\hat{f}(\lambda) = \int_{-\infty}^{\infty} f(t)e^{i\lambda t} \, dt$$

is called the *Fourier transform* of f; it will frequently be designated by $\mathscr{F}(f)$. With this notation equation (4) can be written

$$f(x) = \frac{1}{2\pi} \, \mathrm{CP} \int_{-\infty}^{\infty} e^{-i\lambda x} \hat{f}(\lambda) \, d\lambda$$

The reader should observe that $\mathscr{F}$ is an operator or function since it maps a class of functions, namely the class of summable functions over $\mathbf{R}^1$ (which we denote $\mathscr{L}_1(-\infty,\infty)$), into another class of functions.

The *inverse Fourier transform* of a function g, $\mathscr{F}^{-1}(g)$, is the function defined by

$$\mathscr{F}^{-1}(g)(x) = \frac{1}{2\pi} \, \mathrm{CP} \int_{-\infty}^{\infty} g(\lambda) e^{-i\lambda x} \, d\lambda$$

In the next theorem we see that $\mathscr{F}$ and $\mathscr{F}^{-1}$ are indeed inverse operations of each other; i.e., if $\hat{f}$ is the Fourier transform of f, then $\mathscr{F}^{-1}(\hat{f}) = f$, or, expressed diagramatically,

$$f \xrightarrow{\mathscr{F}} \hat{f} \xrightarrow{\mathscr{F}^{-1}} f$$

(13.B.2) *Theorem (Fourier Transform Theorem)* Suppose that $f: \mathbf{R}^1 \to \mathbf{R}^1$ is piecewise smooth on every finite interval and that $\int_{-\infty}^{\infty} |f(t)| \, dt < \infty$. Let $\hat{f}$ be the Fourier transform of f. Then for each $x \in \mathbf{R}^1$,

$$\frac{f(x+0) - f(x-0)}{2} = \mathscr{F}^{-1}(\hat{f})(x) = \frac{1}{2\pi} \, \mathrm{CP} \int_{-\infty}^{\infty} \hat{f}(\lambda) e^{-i\lambda x} \, d\lambda$$

where $\hat{f}(\lambda) = \int_{-\infty}^{\infty} f(t) e^{i\lambda t} \, dt$.

Proof. The proof is immediate from (13.A.2) and the above discussion.

Before applying the Fourier transform method to solving certain boundary value problems, we develop a few basic properties of the transform.

(13.B.3) *Theorem* If c_1 and c_2 are constants and if $f, g \in \mathscr{L}_1(-\infty,\infty)$, then

$$\widehat{c_1 f + c_2 g} = c_1 \hat{f} + c_2 \hat{g}$$

(This shows that the Fourier transform operator is a linear operator.)

Proof. Simply observe that

$$\int_{-\infty}^{\infty} (c_1 f + c_2 g)(t) e^{i\lambda t} \, dt = c_1 \int_{-\infty}^{\infty} f(t) e^{i\lambda t} \, dt + c_2 \int_{-\infty}^{\infty} g(t) e^{i\lambda t} \, dt$$

(13.B.4) *Theorem* For each $f \in \mathscr{L}_1(-\infty,\infty)$, $\hat{f}$ is *uniformly* continuous on $\mathbf{R}^1$.

Proof. Since $\hat{f}(\lambda + h) - \hat{f}(\lambda) = \int_{-\infty}^{\infty} e^{i\lambda t}(e^{iht} - 1)f(t) \, dt$, it follows that

$$|\hat{f}(\lambda + h) - \hat{f}(\lambda)| \le \int_{-\infty}^{\infty} |e^{iht} - 1||f(t)| \, dt \qquad (5)$$

Since
$$|e^{iht} - 1||f(t)| \le 2|f(t)|$$
and since $\lim_{h\to 0} |e^{iht} - 1| = 0$, it follows from the Lebesgue dominated convergence theorem that the right-hand side of (5) converges to zero as h tends to zero; furthermore, this convergence is independent of λ.

In applications it is not uncommon to arrive at solutions of boundary value problems of the form
$$\phi(x) = \frac{1}{2\pi} \text{ CP} \int_{-\infty}^{\infty} \hat{f}(\lambda)\hat{g}(\lambda)e^{-i\lambda x} \, d\lambda$$

Thus, it is of interest to investigate the structure of a function h with the property that
$$\hat{h}(\lambda) = \hat{f}(\lambda)\hat{g}(\lambda)$$
We shall proceed somewhat heuristically (by assuming that all of the following steps are valid). Under this assumption we have
$$h(x) = \frac{1}{2\pi} \text{ CP} \int_{-\infty}^{\infty} \hat{h}(\lambda)e^{-i\lambda x} \, d\lambda$$
$$= \frac{1}{2\pi} \text{ CP} \int_{-\infty}^{\infty} \hat{f}(\lambda)\hat{g}(\lambda)e^{-i\lambda x} \, d\lambda$$
$$= \frac{1}{2\pi} \text{ CP} \int_{-\infty}^{\infty} \hat{f}(\lambda)e^{-i\lambda x} \left\{ \int_{-\infty}^{\infty} g(t)e^{i\lambda t} \, dt \right\} \, d\lambda$$
Interchanging the order of integration, we find
$$h(x) = \int_{-\infty}^{\infty} g(t) \left[\frac{1}{2\pi} \text{ CP} \int_{-\infty}^{\infty} \hat{f}(\lambda)e^{-i\lambda(x-t)} \, d\lambda \right] dt$$
$$= \int_{-\infty}^{\infty} g(t)f(x - t) \, dt$$
If we define the *convolution* of two functions f and g by the formula
$$(f*g)(x) = \int_{-\infty}^{\infty} f(x - t)g(t) \, dt$$
then we have that $h(x) = (f*g)(x)$ and $\widehat{f*g} = \hat{f}\hat{g}$. We now establish the existence of the convolution as well as the validity of the steps leading to its definition. The proof is based on two of the Fubini theorems stated in Sec. 11.I.

(13.B.5) Theorem If $f,g \in \mathscr{L}_1(-\infty,\infty)$, then
$$(f*g)(x) = \int_{-\infty}^{\infty} f(x - t)g(t) \, dt$$

exists for almost all x and is a summable function over $(-\infty,\infty)$, i.e., $f*g \in \mathcal{L}_1(-\infty,\infty)$.

Proof. For each fixed t, $\int_{-\infty}^{\infty} |f(x-t)| \, dx = \int_{-\infty}^{\infty} |f(x)| \, dx$. Hence,

$$\int_{-\infty}^{\infty} \left[\int_{-\infty}^{\infty} |f(x-t)g(t)| \, dx \right] dt = \int_{-\infty}^{\infty} |g(t)| \left[\int_{-\infty}^{\infty} |f(x-t)| \, dx \right] dt$$

$$= \int_{-\infty}^{\infty} |g(t)| \left[\int_{-\infty}^{\infty} |f(x)| \, dx \right] dt$$

$$= \left[\int_{-\infty}^{\infty} |g(t)| \, dt \right] \left[\int_{-\infty}^{\infty} |f(x)| \, dx \right] < \infty$$

Therefore, by Fubini's absolute convergence theorem (11.I.16) the function $h(x,t) = f(x-t)g(t)$ is summable over $\mathbf{R}^2$ and, consequently, by Fubini's reduction theorem (11.I.14), we have that $\int_{-\infty}^{\infty} f(x-t)g(t) \, dt$ exists for almost all x and is summable over $-\infty < x < \infty$.

(13.B.6) *Exercise* Show that if f and g belong to $\mathcal{L}_1(-\infty,\infty)$, and if either f or g is bounded, then $(f * g)(x)$ exists for each $x \in \mathbf{R}^1$.

The convolution operation is both commutative and associative.

(13.B.7) *Theorem* If f, g, and k belong to $\mathcal{L}_1(-\infty,\infty)$, then $f * g = g * f$ and $(f * g) * k = f * (g * k)$, a.e.

Proof. Letting $u = x - t$, we have

$$\int_{-\infty}^{\infty} f(x-t)g(t) \, dt = \int_{-\infty}^{\infty} g(x-u)f(u) \, du$$

and hence, $(f * g)(x) = (g * f)(x)$.

To prove associativity, we first observe that by (13.B.5) both $f * g$ and $g * k$ belong to $\mathcal{L}_1(-\infty,\infty)$. Consequently, the convolutions $(f * g) * k$ and $f * (h * k)$ exist a.e. on $(-\infty,\infty)$. At points x where the convolution $(f * g) * k$ exists, we have

$$((f*g)*k)(x) = \int_{-\infty}^{\infty} (f*g)(x-t)k(t) \, dt$$

$$= \int_{-\infty}^{\infty} k(t) \left[\int_{-\infty}^{\infty} f(x-t-s)g(s) \, ds \right] dt$$

and if $u = x - t - s$, then

$$((f*g)*k)(x) = \int_{-\infty}^{\infty} k(t) \left[\int_{-\infty}^{\infty} f(u)g(x-u-t) \, du \right] dt \qquad (6)$$

Now we need to change the order of integration. To do so we must show that the iterated integral on the right-hand side of (6) converges absolutely. To this end, note that $|f|$, $|g|$, $|k| \in \mathscr{L}_1(-\infty,\infty)$, and, therefore, by (13.B.5) $(|k| * (|g| * |f|))(x)$ exists for almost all x. Thus

$$(|k|*(|g|*|f|))(x) = \int_{-\infty}^{\infty} |k|(t)(|g|*|f|)(x-t)\, dt$$

$$= \int_{-\infty}^{\infty} |k(t)| \left[\int_{-\infty}^{\infty} |g(x-t-u)||f(u)|\, du \right] dt$$

which shows that the right-hand side of (6) converges absolutely for almost all x. Thus, for those values of x that yield absolute convergence, we can invert the order of integration [by (11.I.16)] to obtain

$$((f*g)*k)(x) = \int_{-\infty}^{\infty} f(u) \left[\int_{-\infty}^{\infty} g(x-u-t)k(t)\, dt \right] du$$

$$= \int_{-\infty}^{\infty} f(u)\cdot(g*k)(x-u)\, du$$

$$= (f*(g*k))(x) \tag{7}$$

Consequently, we have shown that for almost all x, i.e., except on a set of measure 0,

$$(f*g)*k = f*(g*k),$$

which concludes the proof.

(13.B.8) Theorem If $f, g \in \mathscr{L}_1(-\infty,\infty)$, then

$$\widehat{f*g} = \hat{f}\hat{g}$$

Proof. Let $h = f * g$. By (13.B.5), $h \in \mathscr{L}_1(-\infty,\infty)$, and therefore,

$$\hat{h}(\lambda) = \int_{-\infty}^{\infty} e^{i\lambda t}h(t)\, dt$$

exists for every $\lambda \in \mathbf{R}^1$. Thus, we have

$$\hat{h}(\lambda) = \int_{-\infty}^{\infty} e^{i\lambda t}\left\{ \int_{-\infty}^{\infty} f(t-s)g(s)\, ds \right\} dt \tag{8}$$

In order to change the order of integration, we need to show that the iterated integral appearing in (8) is absolutely convergent. This, however, follows from the proof of (13.B.7) where we noted that $|f| * |g| \in \mathscr{L}(-\infty,\infty)$ which implies that this integral is absolutely convergent. Therefore, we have

$$\hat{h}(\lambda) = \int_{-\infty}^{\infty} g(s)\left[\int_{-\infty}^{\infty} e^{i\lambda t}f(t-s)\, dt \right] ds \tag{9}$$

Letting $t = u + s$ (and holding s temporarily fixed), we obtain

$$\hat{h}(\lambda) = \int_{-\infty}^{\infty} g(s)\left[\int_{-\infty}^{\infty} e^{i\lambda(u+s)}f(u)\, du\right] ds$$

$$= \int_{-\infty}^{\infty} g(s)e^{i\lambda s}\left[\int_{-\infty}^{\infty} e^{i\lambda u}f(u)\, du\right] ds$$

$$= \int_{-\infty}^{\infty} g(s)e^{i\lambda s}\hat{f}(\lambda)\, ds$$

$$= \hat{f}(\lambda) \int_{-\infty}^{\infty} g(s)e^{i\lambda s}\, ds$$

$$= \hat{f}(\lambda)\hat{g}(\lambda)$$

which completes the proof.

In order to apply the Fourier transform directly to the solution of boundary value problems, we need to first see how the transforms of the derivatives of a function are related to the transform of this function.

(13.B.9) *Theorem* If f and its first k derivatives belong to $\mathscr{L}_1(-\infty,\infty)$ and if

$$\lim_{|t| \to \infty} f^{(j)}(t) = 0 \qquad \text{for } j = 0, 1, \ldots, k-1$$

then

$$\widehat{f^{(k)}}(\lambda) = (-i\lambda)^k \hat{f}(\lambda)$$

for every $\lambda \in \mathbf{R}^1$.

Proof. Integrate by parts and apply the fact that $\lim_{|t| \to \infty} f^{(j)}(t) = 0$.

C. APPLICATIONS TO BOUNDARY VALUE PROBLEMS

We begin by showing how the Fourier transform can be applied to the boundary value problem posed in Sec. 13.A. We wish to determine a function that will give the temperature at each point of an infinitely long laterally insulated rod at an arbitrary time $t > 0$; we assume that the temperature distribution is known at time $t = 0$. Thus, we are looking for a solution to

$$u_{xx}(x, t) = \frac{1}{a^2} u_t(x, t) \qquad -\infty < x < \infty, t > 0 \qquad (1)$$

subject to the condition

$$u(x, 0) = f(x) \qquad -\infty < x < \infty \qquad (2)$$

We shall also assume (for reasons that will become clear as we proceed) that the solution $u(x,t)$ satisfies the condition:

$$\lim_{|x| \to \infty} u(x, t) = 0 = \lim_{|x| \to \infty} u_x(x, t) \tag{3}$$

From a physical standpoint this assumption is reasonable since over a finite time interval a stimulus occurring in a neighborhood of $x = 0$ can only be perceived over a finite distance.

We proceed heuristically to obtain a tentative form of the solution. Upon obtaining this "solution," we may then proceed in one of two ways: we can retrace and validate all of the steps taken in finding the heuristic or "formal" solution, or we can check directly to see if the heuristic solution satisfies all of the conditions of the given problem. We shall follow the latter approach.

The basic strategy is to use the Fourier transform to transform the partial differential equation (1) into a *readily solvable ordinary* differential equation, and then, with the aid of the inverse Fourier transform, use this solution to obtain a solution to the original problem. We now proceed with a "formal" argument (one whose steps we do not analytically justify).

Let

$$\hat{u}(\lambda, t) = \int_{-\infty}^{\infty} u(x, t)e^{i\lambda x}\, dx \tag{4}$$

be the Fourier transform of the (yet to be determined) solution of (1). If we multiply both sides of (1) by $e^{i\lambda x}$ and integrate from $-\infty$ to ∞, we obtain

$$\int_{-\infty}^{\infty} a^2 u_{xx}(x, t)e^{i\lambda x}\, dx = \int_{-\infty}^{\infty} u_t(x, t)e^{i\lambda x}\, dx \tag{5}$$

Now note that if we assume that

$$\frac{d}{dt}\int_{-\infty}^{\infty} = \int_{-\infty}^{\infty} \frac{d}{dt}$$

then we have from (4) that

$$\hat{u}_t(\lambda, t) = \int_{-\infty}^{\infty} u_t(x, t)e^{i\lambda x}\, dx \tag{6}$$

Furthermore, by assumption (3) it follows from (13.B.9) that

$$\int_{-\infty}^{\infty} a^2 u_{xx}(x, t)e^{i\lambda x}\, dx = a^2 \hat{u}_{xx}(\lambda, t) = -(a\lambda)^2 \hat{u}(\lambda, t) \tag{7}$$

Thus, from (5), (6), and (7) we obtain the ordinary differential equation

$$\hat{u}_t(\lambda, t) = -(\lambda a)^2 \hat{u}(\lambda, t) \tag{8}$$

The appropriate initial condition corresponding to (8) is derived from (2) and (4) and is easily seen to be

$$\hat{u}(\lambda, 0) = \hat{f}(\lambda) \tag{9}$$

Clearly, the unique solution to (8) that satisfies (9) is given by

$$\hat{u}(\lambda, t) = \hat{f}(\lambda)e^{-(\lambda a)^2 t} \tag{10}$$

If for each fixed t, the right-hand side of (10) is a product of two Fourier transforms, and if we can find a function $g(x,t)$ that yields the transform $e^{-\lambda^2 a^2 t}$, then we can use (13.B.8) to express the solution $u(x,t)$ of system (1) and (2) as a convolution

$$u(x, t) = \int_{-\infty}^{\infty} f(s)g(x - s, t)\, ds \tag{11}$$

At this point we could either consult a table of Fourier transforms in the hope of finding g or apply the Fourier inversion formula to the function $e^{-\lambda^2 a^2 t}$. We take the latter alternative, treating both a and t as constants. We have from the inversion formula that

$$g(x, t) = \frac{1}{2\pi} \operatorname{CP} \int_{-\infty}^{\infty} e^{-\lambda^2 a^2 t} e^{-i\lambda x}\, d\lambda$$

Adding exponents and completing the square, we easily obtain

$$g(x, t) = \frac{e^{-x^2/4a^2 t}}{2\pi} \operatorname{CP} \int_{-\infty}^{+\infty} e^{-a^2 t[\lambda + (ix/(2a^2 t))]^2}\, d\lambda$$

Since a, t, and x are constants in this integration, we can set

$$u = a\sqrt{t}\left(\lambda + \frac{ix}{2a^2 t}\right)$$

and obtain upon substitution

$$g(x, t) = \frac{e^{-x^2/4a^2 t}}{2\pi a\sqrt{t}} \operatorname{CP} \int_{-\infty}^{\infty} e^{-u^2}\, du$$

where it is assumed without loss of generality that $a > 0$. Since $\int_{-\infty}^{\infty} e^{-u^2}\, du = \sqrt{\pi}$ [see (11.G.7)], we have

$$g(x, t) = \frac{e^{-x^2/4a^2 t}}{2a\sqrt{\pi t}}$$

Now that we have obtained the analytical expression for $g(x,t)$, we can apply formula (11) to obtain our tentative solution to the problem defined by (1) and (2). We find that

$$u(x, t) = \frac{1}{2a\sqrt{\pi t}} \int_{-\infty}^{\infty} f(s)e^{-(x-s)^2/4a^2 t}\, ds \tag{12}$$

In order to express (12) in a somewhat more manageable form, we make the substitution $v = (x - s)/2a\sqrt{t}$. Then we have

$$s = x - 2a\sqrt{t}v$$

Substitution in (12) yields

$$u(x, t) = \frac{1}{\sqrt{\pi}} \int_{-\infty}^{\infty} f(x - 2a\sqrt{t}v)e^{-v^2}\, dv \tag{13}$$

Note that the integrand has the factor e^{-v^2} and this will ensure the existence of the integral for a very large family of functions, even for some functions which are unbounded as $|x| \to \infty$! Furthermore, if we set $t = 0$, then we obtain

$$u(x, 0) = \frac{1}{\sqrt{\pi}} \int_{-\infty}^{\infty} f(x)e^{-v^2} \, dv = \frac{f(x)}{\sqrt{\pi}} \int_{-\infty}^{\infty} e^{-v^2} \, dv = f(x).$$

This shows that condition (2) is satisfied by all functions $f: \mathbf{R}^1 \to \mathbf{R}^1$ for which the integral in (13) exists.

(13.C.1) *Exercise* Show by direct substitution that $u(x,t)$ as defined in (12) is a solution to (1) that satisfies (2).

We now apply the Fourier transform to the problem of determining the transverse motion of a string of infinite length where the initial displacement and velocity of the string are known. Specifically, we shall attempt to solve the system

$$\begin{aligned}
u_{tt}(x, t) &= u_{xx}(x, t) & -\infty < x < \infty, t > 0 \\
u(x, 0) &= f(x) & -\infty < x < \infty \\
u_t(x, 0) &= g(x) & -\infty < x < \infty \quad (14)
\end{aligned}$$

with the aid of the Fourier transform. As in the example just discussed, we shall proceed heuristically to obtain a tentative solution, whose validity, of course, must be checked.

If we follow the same sort of reasoning used in the previous example, it is not difficult to see that the transformed system is given by

$$\begin{aligned}
\hat{u}_{tt}(\lambda, t) &= -\lambda^2 \hat{u}(\lambda, t) & -\infty < \lambda < \infty, t > 0 & \quad (15) \\
\hat{u}(\lambda, 0) &= \hat{f}(\lambda) & -\infty < \lambda < \infty & \quad (16) \\
\hat{u}_t(\lambda, 0) &= \hat{g}(\lambda) & -\infty < \lambda < \infty & \quad (17)
\end{aligned}$$

where

$$\hat{u}(\lambda, t) = \int_{-\infty}^{\infty} u(x, t)e^{i\lambda x} \, dx$$

The transformed equation (15) is an ordinary differential equation (with parameter λ) and has the general solution

$$\hat{u}(\lambda, t) = c_1(\lambda) \cos \lambda t + c_2(\lambda) \sin \lambda t$$

It follows from the initial data (16) and (17) that

$$\hat{u}(\lambda, t) = \hat{f}(\lambda) \cos \lambda t + \frac{\hat{g}(\lambda)}{\lambda} \sin \lambda t$$

Applying the inversion formula, we have

$$u(x, t) = \frac{1}{2\pi} \, \text{CP} \int_{-\infty}^{\infty} \hat{u}(\lambda, t) e^{-i\lambda x} \, d\lambda$$

$$= \frac{1}{2\pi} \, \text{CP} \int_{-\infty}^{\infty} \hat{f}(\lambda)(\cos \lambda t) e^{-i\lambda x} \, d\lambda + \frac{1}{2\pi} \, \text{CP} \int_{-\infty}^{\infty} \frac{\hat{g}(\lambda)}{\lambda} (\sin \lambda t) e^{-i\lambda x} \, d\lambda$$

Let $I_1 = (1/2\pi) \, \text{CP}\int_{-\infty}^{\infty} \hat{f}(\lambda) (\cos \lambda t) e^{-i\lambda x} \, d\lambda$ and $I_2 = (1/2\pi) \, \text{CP}\int_{-\infty}^{\infty} (\hat{g}(\lambda)/\lambda)$ $(\sin \lambda t) e^{-i\lambda x} \, d\lambda$. Since $\cos \lambda t = (e^{i\lambda t} + e^{-i\lambda t})/2$, it follows that

$$I_1 = \frac{1}{2}\left[\frac{1}{2\pi} \, \text{CP} \int_{-\infty}^{\infty} (e^{-i\lambda(x-t)} + e^{-i\lambda(x+t)})\hat{f}(\lambda)\right] d\lambda$$

$$= \frac{1}{2} [f(x - t) + f(x + t)]$$

The integral I_2 can be written in the form

$$I_2 = \frac{1}{2\pi} \, \text{CP} \int_{-\infty}^{\infty} \frac{\sin \lambda t}{\lambda} e^{-i\lambda x} \left[\int_{-\infty}^{\infty} g(s) e^{i\lambda s} \, ds\right] d\lambda$$

Changing the order of integration, we find that

$$I_2 = \frac{1}{2\pi} \int_{-\infty}^{\infty} g(s)\left(\text{CP} \int_{-\infty}^{\infty} \frac{\sin \lambda t}{\lambda} e^{-i\lambda(x-s)} \, d\lambda\right) ds$$

$$= \frac{1}{2\pi} \int_{-\infty}^{\infty} g(s)\left[\text{CP} \int_{-\infty}^{\infty} \frac{\sin \lambda t}{\lambda} \cos \lambda(x - s) \, d\lambda\right.$$

$$\left. - \, i \, \text{CP} \int_{-\infty}^{\infty} \frac{\sin \lambda t}{\lambda} \sin \lambda(x - s) \, d\lambda\right] ds$$

$$= \frac{1}{\pi} \int_{-\infty}^{\infty} g(s)\left[\int_{0}^{\infty} \frac{\sin \lambda t}{\lambda} \cos \lambda(x - s) \, d\lambda\right] ds \qquad (18)$$

(the latter equality follows from the fact that for fixed s, t, and x, $[(\sin \lambda t)/\lambda] \sin \lambda(x - s)$ is an odd function in λ and $[(\sin \lambda t)/\lambda] \cos \lambda(x - s)$ is an even function in λ). We now claim that for $t > 0$,

$$\int_{0}^{\infty} \frac{\sin \lambda t}{\lambda} \cos \lambda(x - s) \, d\lambda = \begin{cases} 0 & \text{if } s > x + t \\ \dfrac{\pi}{2} & \text{if } x - t < s < x + t \\ 0 & \text{if } s < x - t \end{cases} \qquad (19)$$

For each nonzero constant c, we define $\text{sgn}(c)$ to be 1 if $c > 0$ and -1 if $c < 0$. From the trigonometric identity

$$\sin \alpha \cos \beta = \frac{\sin (\alpha + \beta) + \sin (\alpha - \beta)}{2}$$

we obtain

$$\int_0^\infty \frac{\sin \lambda t}{\lambda} \cos \lambda(x - s) \, d\lambda = \frac{1}{2}\left(\int_0^\infty \frac{\sin \lambda(t + x - s)}{\lambda} \, d\lambda \right.$$
$$\left. + \int_0^\infty \frac{\sin \lambda(t - x + s)}{\lambda} \, d\lambda \right) \qquad (20)$$

For each constant $c \neq 0$, we have from (13.A.4) or (12.E.10)

$$\int_0^\infty \frac{\sin c\lambda}{\lambda} \, d\lambda = \text{sgn}\,(c) \cdot \frac{\pi}{2} \qquad (21)$$

Now if $s > x + t$ so that $s > x - t$ for $t > 0$, then $t + x - s < 0$ and $t - x + s > 0$, and it follows at once from (20) and (21) that (19) is true whenever $s > x + t$. If s is such that $x - t < s < x + t$, then $t + x - s > 0$ and $t - x + s > 0$, and this clearly establishes formula (19) for this case. Finally, if $s < x - t$, then $t - x + s < 0$; furthermore, since $t > 0$, we have that $s < x - t < x + t$, and, hence, $t + x - s > 0$. Consequently, (19) holds when $s < x - t$, and this proves the claim.

Combining formulas (18) and (19), we obtain

$$I_2 = \frac{1}{2} \int_{x-t}^{x+t} g(s) \, ds$$

Thus, we find that a tentative solution of the problem defined by (14) is of the form

$$u(x, t) = \frac{f(x + t) + f(x - t)}{2} + \frac{1}{2} \int_{x-t}^{x+t} g(s) \, ds \qquad (22)$$

It remains to check the validity of the tentative solution (22). If it is assumed that f'' and g' are continuous on $\mathbf{R}^1$, then direct substitution will establish that the function u defined by (22) is indeed the desired solution (see problem C.3).

D. THE FOURIER SINE TRANSFORM

We consider the heat problem associated with a semiinfinite rod. Mathematically, we want to find a solution, $u(x,t)$, that satisfies the partial differential equation

$$\frac{1}{a^2} u_t(x, t) = u_{xx}(x, t) \qquad (0 < x < \infty, t > 0) \qquad (1)$$

subject to the conditions:

$$u(0, t) = 0 \qquad\qquad (t > 0) \qquad\qquad (2)$$

$$\lim_{x \to \infty} u(x, t) = 0 = \lim_{x \to \infty} u_x(x, t) \qquad (t > 0) \qquad\qquad (3)$$

$$u(x, 0) = f(x) \qquad\qquad (0 < x < \infty) \qquad\qquad (4)$$

In order to apply the (complex) Fourier transform method used in the previous section, we must first extend the definition of f to all real x. For this purpose we use the odd extension f_o of f given by

$$f_o(x) = \begin{cases} f(x) & \text{if } x > 0 \\ 0 & \text{if } x = 0 \\ -f(-x) & \text{if } x < 0 \end{cases}$$

If f is piecewise smooth on $(0,\infty)$ and if $\int_0^\infty |f(t)|\, dt < \infty$, then the extension f_o satisfies the conditions of (13.B.1). Hence, in this case we have

$$f_o(x) = \frac{1}{2\pi}\, \text{CP}\!\int_{-\infty}^{\infty} \hat{f}_o(\lambda)e^{-i\lambda x}\, d\lambda \qquad\qquad (5)$$

where

$$\hat{f}_o(\lambda) = \int_{-\infty}^{\infty} f_o(t)e^{i\lambda t}\, dt \qquad\qquad (6)$$

(As usual we assume that at each point of discontinuity $f_o(x) = [f_o(x + 0) + f_o(x - 0)]/2$.) Since f_o is odd over $\mathbf{R}^1$, it follows that

$$\hat{f}_o(\lambda) = i\int_{-\infty}^{\infty} f_o(t) \sin \lambda t\, dt = 2i \int_{0}^{\infty} f(t) \sin \lambda t\, dt \qquad (7)$$

From (7) it is clear that $\hat{f}_o$ is an odd function in λ, and, therefore, it follows from (5) that

$$f_o(x) = \frac{1}{2\pi}\, \text{CP}\!\int_{-\infty}^{\infty} \hat{f}_o(\lambda)e^{-i\lambda x}\, d\lambda$$

$$= -\frac{i}{2\pi}\, \text{CP}\!\int_{-\infty}^{\infty} \hat{f}_o(\lambda) \sin \lambda x\, d\lambda$$

$$= \frac{-i}{\pi} \int_{0}^{\infty} \hat{f}_o(\lambda) \sin \lambda x\, d\lambda \qquad\qquad (8)$$

From (7) and (8) we have

$$f_o(x) = \frac{2}{\pi} \int_{0}^{\infty} \left[\int_{0}^{\infty} f(t) \sin \lambda t\, dt \right] \sin \lambda x\, d\lambda$$

We designate the Fourier sine transform by

$$\hat{f}_S(\lambda) = \int_{0}^{\infty} f(t) \sin \lambda t\, dt \qquad (\lambda \geq 0) \qquad\qquad (9)$$

The following theorem summarizes the preceding discussion.

(13.D.1) *Theorem (Fourier Sine Formula)* If f is a piecewise smooth function on $(0,\infty)$ and if $\int_0^\infty |f(t)|\, dt < \infty$, then for each $x \in (0,\infty)$,

$$\frac{f(x+0) + f(x-0)}{2} = \frac{2}{\pi} \int_0^\infty \hat{f}_S(\lambda) \sin \lambda x \, d\lambda$$

where

$$\hat{f}_S(\lambda) = \int_0^\infty f(t) \sin \lambda t \, dt$$

The expression $(2/\pi) \int_0^\infty \hat{f}_S(\lambda) \sin \lambda x \, d\lambda$ is called the *Fourier sine representation* of f.

(13.D.2) *Observation* The function f_S is uniformly continuous on $(0,\infty)$ and $\lim_{\lambda \to \infty} \hat{f}_S(\lambda) = 0$. This follows from the fact that by (7) and (9) $\hat{f}_S(\lambda) = (1/2i)\hat{f}_o(\lambda)$, $\hat{f}_o$ is uniformly continuous, and $\lim_{\lambda \to \infty} \hat{f}_o(\lambda) = 0$.

We are now ready to apply the Fourier sine transform to obtain a solution of the problem stated at the beginning of this section. We again follow a heuristic (formal) procedure. In the sequel we shall assume as before that

$$\lim_{x \to \infty} u(x, t) = 0 = \lim_{x \to \infty} u_x(x, t) \tag{10}$$

To transform equation (1), we multiply both sides of (1) by $\sin \lambda x$ and then integrate from 0 to ∞ to obtain

$$\frac{1}{a^2} \int_0^\infty u_t(x, t) \sin \lambda x \, dx = \int_0^\infty u_{xx}(x, t) \sin \lambda x \, dx$$

Setting

$$\hat{u}_S(\lambda, t) = \int_0^\infty u(x, t) \sin \lambda x \, dx \tag{11}$$

and integrating by parts twice, we find [taking (2), (3), and (10) into account] that

$$\frac{1}{a^2} \frac{d}{dt} \hat{u}_S(\lambda, t) = \frac{1}{a^2} \int_0^\infty u_t(x, t) \sin \lambda x \, dx = \int_0^\infty u_{xx}(x, t) \sin \lambda x \, dx$$

$$= u_x(x, t) \sin \lambda x \Big|_0^\infty - \lambda \int_0^\infty u_x(x, t) \cos \lambda x \, dx$$

$$= -\lambda \int_0^\infty u_x(x, t) \cos \lambda x \, dx$$

$$= -\lambda u(x, t) \cos \lambda x \Big|_0^\infty - \lambda^2 \int_0^\infty \int_0^\infty u(x, t) \sin \lambda x \, dx$$

$$= -\lambda^2 \hat{u}_S(\lambda, t)$$

Thus, (1) is transformed into the ordinary differential equation

$$\frac{d}{dt}\hat{u}_S(\lambda, t) = -a^2\lambda^2\hat{u}_S(\lambda, t) \tag{12}$$

The initial condition

$$\hat{u}_S(\lambda, 0) = \hat{f}_S(\lambda) \tag{13}$$

is derived from (11). Since the solution to (12) and (13) is given by

$$\hat{u}_S(\lambda, t) = \hat{f}_S(\lambda)e^{-a^2\lambda^2 t}$$

we can anticipate a solution of (1)–(4) of the form

$$u(x, t) = \frac{2}{\pi}\int_0^\infty \hat{u}_S(\lambda, t) \sin \lambda x \, d\lambda$$

$$= \frac{2}{\pi}\int_0^\infty \hat{f}_S(\lambda)e^{-a^2\lambda^2 t} \sin \lambda x \, d\lambda$$

$$= \frac{2}{\pi}\int_0^\infty \left[\int_0^\infty f(w) \sin \lambda w \, dw\right]e^{-a^2\lambda^2 t} \sin \lambda x \, d\lambda \tag{14}$$

(13.D.3) *Exercise* Verify directly that

$$u(x, t) = \frac{2}{\pi}\int_0^\infty \left[\int_0^\infty f(w) \sin \lambda w \, dw\right]e^{-a^2\lambda^2 t} \sin \lambda x \, d\lambda$$

is a solution of (1)–(4) and that

$$\lim_{x\to\infty} u_t(x, t) = 0$$

for $t > 0$. Supply sufficient conditions on f to ensure the validity of the analytical steps used in deriving this solution.

As a second illustration of the application of the Fourier sine transform, we consider the problem of determining the conduction of heat in a semi-infinite rod insulated laterally where the temperature is varied at the finite end of the rod. We assume that the initial temperature of the rod is 0. Thus, we wish to find a solution $u(x,t)$ of the system:

$$\frac{1}{a^2}u_t(x, t) = u_{xx}(x, t) \qquad (x > 0, t > 0) \tag{15}$$

$$u(0, t) = f(t) \qquad (t > 0) \tag{16}$$

$$u(x, 0) = 0 \qquad (x > 0) \tag{17}$$

Once again we proceed formally. As before, we shall find it convenient to assume that $\lim_{x\to\infty} u(x,t) = 0 = \lim_{x\to\infty} u_x(x,t)$. Setting

$$\hat{u}_S(\lambda, t) = \int_0^\infty u(x, t) \sin \lambda x \, dx$$

we obtain from equation (15)

$$\frac{1}{a^2} \frac{d}{dt} \hat{u}_S(\lambda, t) = \frac{1}{a^2} \int_0^\infty u_t(\lambda, t) \sin \lambda x \, dx$$

$$= \int_0^\infty u_{xx}(x, t) \sin \lambda x \, dx$$

$$= u_x(x, t) \sin \lambda x \Big|_{x=0}^{x=\infty} - \lambda \int_0^\infty u_x(x, t) \cos \lambda x \, dx$$

$$= -\lambda \int_0^\infty u_x(x, t) \cos \lambda x \, dx$$

$$= -\lambda u(x, t) \cos \lambda x \Big|_{x=0}^{x=\infty} - \lambda^2 \int_0^\infty u(x, t) \sin \lambda x \, dx$$

$$= \lambda f(t) - \lambda^2 \hat{u}_S(\lambda, t)$$

Hence, the transformed equation is

$$\frac{d}{dt} \hat{u}_S(\lambda, t) = a^2 \lambda f(t) - a^2 \lambda^2 \hat{u}_S(\lambda, t) \tag{18}$$

Transformed initial data from (17) can be used to find that

$$\hat{u}_S(\lambda, 0) = 0 \tag{19}$$

It is easily seen that the general solution of (18) has the form

$$\hat{u}_S(\lambda, t) = c(\lambda) e^{-a^2 \lambda^2 t} + a^2 \lambda e^{-a^2 \lambda^2 t} \int_0^t e^{a^2 \lambda^2 u} f(u) \, du \tag{20}$$

From (19) we have that $c(\lambda) = 0$, and therefore, it follows that the anticipated solution has the form

$$u(x, t) = \frac{2}{\pi} \int_0^\infty \hat{u}_S(\lambda, t) \sin \lambda x \, d\lambda$$

$$= \frac{2}{\pi} \int_0^\infty a^2 \lambda e^{-a^2 \lambda^2 t} \left[\int_0^t e^{a^2 \lambda^2 w} f(w) \, dw \right] \sin \lambda x \, d\lambda$$

By interchanging the order of integration we find that

$$u(x, t) = \frac{2a^2}{\pi} \int_0^t f(w) \left[\int_0^\infty \lambda e^{-a^2 \lambda^2 (t-w)} \sin \lambda x \, d\lambda \right] dw$$

The integral appearing in brackets is a Fourier sine integral and is listed in Doetsch (1961). From this reference we find that

$$\int_0^\infty \lambda^2 e^{-a^2\lambda^2(t-w)} \sin \lambda x \, d\lambda = \frac{1}{4}\sqrt{\pi}[a^2(t-w)]^{3/2}xe^{(-1/4)[1/a^2(t-w)]x^2}$$

and consequently, the solution $u(x,t)$ is given by

$$u(x,t) = \frac{a^5x}{2\sqrt{\pi}}\int_0^t f(w)(t-w)^{3/2}e^{-[(x^2/4a^2)(1/(t-w))]}\,dw$$

E. THE LAPLACE TRANSFORM

In this section we shall to a limited extent make use of contour integrals which were discussed in the preceding chapter. However, readers unfamiliar with this concept should have little difficulty in assimilating the essential ideas of this section.

In many applications involving time-related problems one deals with functions mapping the interval $(0,\infty)$ into $\mathbf{R}^1$. It is often convenient to extend such a function f by setting $f(t) = 0$ for all $t < 0$. If we apply the complex Fourier integral representation to this extended function, we find that

$$\frac{1}{2\pi} \, \mathrm{CP}\int_{-\infty}^\infty \left[\int_0^\infty f(u)e^{i\lambda u}\,du\right]e^{-i\lambda t}\,d\lambda = \begin{cases} f(t) & \text{if } t > 0 \\ 0 & \text{if } t < 0 \end{cases} \qquad (1)$$

Naturally, for equation (1) to hold, certain conditions must be imposed on f. By (13.A.2), we have that (1) is valid if f is piecewise smooth on every finite subinterval of $(0,\infty)$, $\int_0^\infty |f(t)|\,dt < \infty$, and if $f(t) = [f(t+0) + f(t-0)]/2$ at each jump discontinuity. It turns out, however, that the restriction $\int_0^\infty |f(t)|\,dt < \infty$ is unacceptable in many applications. Nevertheless, it is the case that most functions arising in applied work for which $\int_0^\infty |f(t)|\,dt = \infty$, do satisfy the condition

$$\int_0^\infty |f(t)e^{-ct}|\,dt < \infty$$

for some positive constant c. For such a function f, we have

$$\frac{1}{2\pi} \, \mathrm{CP}\int_{-\infty}^\infty \left[\int_0^\infty f(u)e^{-cu}e^{i\lambda u}\,du\right]e^{-i\lambda t}\,d\lambda = \begin{cases} f(t)e^{-ct} & \text{if } t > 0 \\ 0 & \text{if } t < 0 \end{cases}$$

and, consequently,

$$\frac{e^{ct}}{2\pi} \, \mathrm{CP}\int_{-\infty}^\infty \left[\int_0^\infty f(u)e^{-cu}e^{i\lambda u}\,du\right]e^{-i\lambda t}\,d\lambda = \begin{cases} f(t) & \text{if } t > 0 \\ 0 & \text{if } t < 0 \end{cases} \qquad (2)$$

If we set $s = c - i\lambda$, then we obtain from (2)

$$\frac{e^{ct}}{2\pi} \operatorname{CP} \int_{-\infty}^{\infty} V \left[\int_{0}^{\infty} f(u)e^{-cu}e^{i\lambda u} \, du \right] e^{-i\lambda t} \, d\lambda$$

$$= \frac{e^{ct}}{2\pi} \lim_{b \to \infty} \int_{-b}^{b} \left[\int_{0}^{\infty} f(u)e^{-cu}e^{i\lambda u} \, du \right] e^{-i\lambda t} \, d\lambda$$

$$= \frac{-e^{ct}}{2\pi i} \lim_{b \to \infty} \int_{c+ib}^{c-ib} \left[\int_{0}^{\infty} f(u)e^{-su} \, du \right] e^{(s-c)t} \, ds$$

$$= \frac{1}{2\pi i} \operatorname{CP} \int_{c-i\infty}^{c+i\infty} V \left[\int_{0}^{\infty} f(u)e^{-su} \, du \right] e^{st} \, ds \qquad (3)$$

(The integral $\int_{c-ib}^{c+ib}$ is the contour integral, where integration is in the complex plane along the path indicated in Fig. 13.1.)

The function $F: [c,\infty) \to \mathbf{R}^1$ defined by

$$F(s) = \int_{0}^{\infty} e^{-su} f(u) \, du$$

is called the *Laplace transform* of f, where it is understood that $\int_{0}^{\infty} |f(u)| e^{-su} \, du < \infty$.

(13.E.1) Observation If there is a positive constant $c > 0$ such that

$$\int_{0}^{\infty} |f(u)e^{-cu}| \, du < \infty \qquad (4)$$

then it is easy to see that $\int_{0}^{\infty} |f(u)e^{-su}| \, du < \infty$ for all complex numbers s, where $\operatorname{Re}(s) \geq c$. Hence, one can also consider $F(s) = \int_{0}^{\infty} f(u)e^{-su} \, du$ as a function defined on the half plane $\operatorname{Re}(s) \geq c$ (Fig. 13.2).

Therefore, the Laplace transform yields a mapping between a class of functions satisfying inequality (4) and a family of functions defined on half

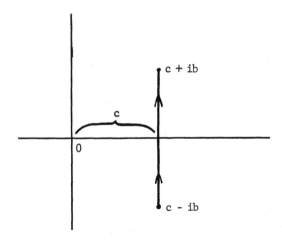

Figure 13.1

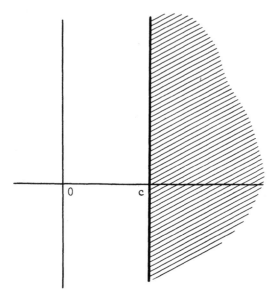

Figure 13.2

planes $\text{Re}(s) \geq c$. It is often convenient to use the symbol $\mathscr{L}$ to denote the Laplace transform; hence, with this notation we have that $\mathscr{L}(f)(s) = F(s)$ [or $\mathscr{L}(f) = F$].

The *inverse Laplace transform* $\mathscr{L}^{-1}$ of a function g is defined by

$$\mathscr{L}^{-1}(g)(t) = \frac{1}{2\pi i} \int_{c-i\infty}^{c+i\infty} g(s)e^{st} \, ds$$

With this notation, what we have shown in (2) and (3) is that $\mathscr{L}^{-1}(\mathscr{L}(f)) = f$, i.e., the inverse Laplace transform of the Laplace transform of a function f is just f. We express this somewhat more formally in the following theorem.

(13.E.2)   Theorem   If $f: (0,\infty) \to \mathbf{R}^1$ is a piecewise smooth function on every finite subinterval of $(0,\infty)$ and if there exists a positive constant $c > 0$ such that

$$\int_0^\infty |f(t)e^{-ct}| \, dt < \infty \tag{5}$$

then

$$\frac{1}{2\pi i} \int_{c-i\infty}^{c+i\infty} F(s)e^{st} \, ds = \begin{cases} \dfrac{f(t+0) + f(t-0)}{2} & \text{if } t > 0 \\ 0 & \text{if } t < 0 \end{cases} \tag{6}$$

where

$$F(s) = \int_0^\infty e^{-su} f(u)\, du \qquad \text{for Re } (s) \geq c \tag{7}$$

is the Laplace transform of f.

One of the most important aspects of the Laplace transform and the inverse Laplace transform is that they are linear operators, as the reader is asked to establish in the next exercise.

(13.E.3) *Exercise* Show that if $\mathscr{L}(f)$ and $\mathscr{L}(g)$ exist, then

$$\mathscr{L}(c_1 f + c_2 g) = c_1 \mathscr{L}(f) + c_2 \mathscr{L}(g)$$

for arbitrary constants c_1 and c_2. Prove an analogous result for $\mathscr{L}^{-1}$.

The Laplace transform is especially useful in solving many initial value problems. As an easy illustration of this we consider the following problem (which can also be readily solved using techniques described in Chapter 2):

$$y'(t) - y(t) = 1 \qquad y(0) = 2 \tag{8}$$

If we take the Laplace transform of both sides of (8), i.e., if we multiply both sides of (8) by e^{-st} and integrate from $t = 0$ to $t = \infty$, then we obtain

$$\int_0^\infty y'(t) e^{-st}\, dt - \int_0^\infty y(t) e^{-st}\, dt = \int_0^\infty e^{-st}\, dt \tag{9}$$

Setting $Y(s) = \int_0^\infty y(t) e^{-st}\, dt$ and integrating the first integral by parts, we see that (9) becomes

$$y(t) e^{-st}\Big|_0^\infty + s Y(s) - Y(s) = \frac{-e^{-st}}{s}\Big|_0^\infty$$

Now if we assume that $\lim_{t\to\infty} y(t) e^{-st} = 0$ [which is the case here provided Re(s) is suitably large], we have $-2 - (1 - s) Y(s) = 1/s$.

Thus, the original problem has now been reduced to an algebraic one. We find that

$$Y(s) = \frac{2s + 1}{-s(1 - s)} = \frac{-3}{1 - s} - \frac{1}{s} \tag{10}$$

and an application of the inverse Laplace transform to both sides of (10) shows

$$y(t) = \mathscr{L}^{-1}(Y(s)) = \mathscr{L}^{-1}\left(\frac{3}{s - 1}\right) - \mathscr{L}^{-1}\left(\frac{1}{s}\right)$$

From the table of Laplace transforms in the appendix we find that

$$\mathscr{L}^{-1}\left(\frac{3}{s-1}\right) = 3\mathscr{L}^{-1}\left(\frac{1}{s-1}\right) = 3e^t \qquad \mathscr{L}^{-1}\left(\frac{1}{s}\right) = 1$$

and therefore, $y(t) = 3e^t - 1$.

Note that in the procedure used in this example, we have essentially assumed the uniqueness of the Laplace transform (if two *different* functions were to have the same Laplace transform, then use of the inverse Laplace transform would not necessarily yield a unique solution to a given problem). In Theorem (13.E.7) we see that, under fairly mild conditions, uniqueness of the inverse transform is assured.

(13.E.4) Definition A function $f: (0,\infty) \to \mathbf{R}^1$ is said to be of *exponential type* if there are positive constants M and α such that $|f(t)| \leq Me^{\alpha t}$, for all $t > 0$.

(13.E.5) Definition A function defined on an unbounded interval I is said to be *piecewise smooth* if it is piecewise smooth on every finite subinterval of I.

(13.E.6) Exercise Show that if $f: (0,\infty) \to \mathbf{R}^1$ is piecewise smooth and of exponential type, then there is an $a > 0$ such that $\int_0^\infty |f(t)|e^{-ct}\, dt < \infty$, for all $c \geq a$.

(13.E.7) Theorem (Uniqueness Theorem) Suppose that $f: (0,\infty) \to \mathbf{R}^1$ and $g: (0,\infty) \to \mathbf{R}^1$ are two piecewise smooth functions of exponential type and that $\mathscr{L}(f) = \mathscr{L}(g)$. Then $f(t) = g(t)$ for every $t \in (0,\infty)$. (As usual, we assume that $f(t) = [f(t + 0) + f(t - 0)]/2$ and $g(t) = [g(t + 0) + g(t - 0)]/2$ at each jump discontinuity.)

Proof. By (13.E.2) and (13.E.6) there is a positive constant $c > 0$ such that

$$f(t) - g(t) = \frac{1}{2\pi i}\int_{c-i\infty}^{c+i\infty} [F(s) - G(s)]e^{st}\, ds$$

for all $t \in (0,\infty)$. Since $F(s) = G(s)$ for all s of the form $c + i\lambda$, $(-\infty < \lambda < \infty)$, it follows that $f(t) = g(t)$ on $(0,\infty)$.

We review briefly the procedure used to find a solution to

$$y'(t) - y(t) = 1 \qquad y(0) = 2 \tag{11}$$

In this procedure we first tentatively assumed that the problem had a solution for all $t > 0$. We then took the Laplace transform of both sides of (11), and since Laplace transforms (if they exist) of identically equal functions are equal, we found that $Y(s) = 3/(s - 1) - 1/s$. Theorem (13.E.7) now justifies our use of the table to determine a (unique) piecewise smooth function of exponential type associated with this transform.

The next theorem will enable us to deal with higher order differential equations.

(13.E.8) *Theorem* Suppose that $f, f', \ldots, f^{(n-1)}$ are continuous real-valued functions on $[0,\infty)$. Suppose further that each of these functions is of exponential type and that $f^{(n)}$ is piecewise continuous on $(0,\infty)$. Then

$$\mathscr{L}(f^n)(s) = s^n \mathscr{L}(f)(s) - s^{n-1}f(0) - \cdots - sf^{(n-2)}(0) - f^{(n-1)}(0)$$

for all complex numbers in some half plane $\mathrm{Re}(s) > \alpha$.

Proof. The proof is by induction. First, suppose that $f: [0,\infty) \to \mathbf{R}^1$ is continuous, piecewise smooth, and of exponential type. We show that $\mathscr{L}(f')$ exists and, furthermore, that

$$\mathscr{L}(f')(s) = s\mathscr{L}(f)(s) - f(0)$$

for all complex s belonging to a half plane $\mathrm{Re}(s) > \alpha$.

On each finite interval $[0,a]$ there exists only a finite number of points $t_1, \ldots, t_n$, where f' is discontinuous. Since the discontinuities are simple jump discontinuities, we can use integration by parts over each subinterval $[t_{i-1},t_i]$ to obtain

$$\int_0^a f'(t)e^{-st}\,dt$$

$$= e^{-st}f(t)\Big|_0^{t_1} + e^{-st}f(t)\Big|_{t_1}^{t_2} + \cdots + e^{-st}f(t)\Big|_{t_n}^a$$

$$+ s\left[\int_0^{t_1} f(t)e^{-st}\,dt + \int_{t_1}^{t_2} f(t)e^{-st}\,dt + \cdots + \int_{t_n}^a f(t)e^{-st}\,dt\right] \quad (12)$$

The function f is continuous, and therefore

$$\int_0^a f'(t)e^{-st}\,dt = e^{-sa}f(a) - f(0) + s\int_0^a f(t)e^{-st}\,dt$$

Since f is of exponential type, there exist positive constants M and α such that $|f(t)| \le Me^{\alpha t}$ for all t satisfying $0 \le t \le \infty$; consequently, we have

$$|e^{-sa}f(a)| \le Me^{\mathrm{Re}(\alpha-s)a}$$

Therefore, if $\mathrm{Re}(s) > \alpha$, then

$$\lim_{a\to\infty} Me^{\mathrm{Re}(\alpha-s)a} = 0$$

from which it follows that

$$\mathscr{L}(f')(s) = \int_0^\infty f'(t)e^{-st}\,dt = \lim_{a\to\infty}\int_0^a f'(t)e^{-st}\,dt$$

$$= -f(0) + \lim_{a\to\infty} s\int_0^a f(t)e^{-st}\,dt$$

$$= s\mathscr{L}(f)(s) - f(0)$$

This establishes the theorem for the case $n = 1$. To complete the induction we assume that the result is valid for $k = n - 1$, and show that it is true for $k = n$. We have

$$\mathscr{L}(f^n)(s)$$

$$= \mathscr{L}((f^{n-1})')(s)$$

$$= s\mathscr{L}(f^{n-1})(s) - f^{(n-1)}(0)$$

$$= s[s^{n-1}\mathscr{L}(f)(s) - s^{n-2}f(0) - \cdots - sf^{n-3}(0) - f^{n-2}(0)] - f^{(n-1)}(0)$$

$$= s^n\mathscr{L}(f)(s) - s^{n-1}f(0) - \cdots - sf^{n-2}(0) - f^{(n-1)}(0)$$

It is now clear in view of (13.E.3) and (13.E.8) that we can transform an initial value problem of the type

$$a_0 y^{(n)}(t) + \cdots + a_{n-1}y'(t) + a_n y(t) = f(t) \qquad 0 \le t < \infty$$

where $a_0, \ldots, a_n$ are constants, and with initial condition $y^{(j)}(0) = c_j$, $j = 0, \ldots, n - 1$, into an algebraic equation of the form

$$(a_0 s^n + \cdots + a_{n-1}s + a_n)Y(s)$$

$$= F(s) + b_0 s^{n-1} + \cdots + b_{n-(k+1)}s^k + \cdots + b_{n-2}s + b_{n-1}$$

where $Y(s)$ and $F(s)$ are the Laplace transforms of $y(t)$ and $f(t)$, and

$$b_{n-(k+1)} = \sum_{j=k+1}^n a_{n-j}c_{j-(k+1)}$$

[provided that $\mathscr{L}(f)$ exists]. Thus, by setting $H(s) = (a_0 s^n + \cdots + a_{n-1}s + a_n)^{-1}$ we find

$$Y(s) = F(s)H(s) + \left(\frac{b_0 s^{n-1} + \cdots + b_{n-(k+1)}s^k + \cdots + b_{n-2}s + b_{n-1}}{a_0 s^n + \cdots + a_{n-1}s + a_n}\right)$$

Since the inverse Laplace transform is a linear operator, we obtain

$$y(t) = \mathscr{L}^{-1}(Y(s))(t) = \mathscr{L}^{-1}(F(s)H(s))(t) + \mathscr{L}^{-1}\left(\frac{b_0 s^{n-1} + \cdots + b_{n-1}}{a_0 s^n + \cdots + a_n}\right)(t)$$

Once $\mathscr{L}^{-1}(H)$ is known, we can apply the convolution theorem (13.E.10) below to obtain $\mathscr{L}^{-1}(FH)$. Then by a partial fraction decomposition of $(b_0 s^{n-1} + \cdots + b_{n-1})/(a_0 s^n + \cdots + a_n)$, we can compute its inverse and finally obtain our solution.

(13.E.9) *Example* We apply the foregoing method to solve

$$y''' - 4y'' + y' - 6y = 0 \qquad y(0) = 1 \qquad y'(0) = 4 \qquad y''(0) = -3$$

By (13.E.8) we have

$$\mathscr{L}(y''')(s) = s^3 \mathscr{L}(y)(s) - s^2 y(0) - s y'(0) - y''(0)$$
$$\mathscr{L}(y'')(s) = s^2 \mathscr{L}(y)(s) - s y(0) - y'(0)$$
$$\mathscr{L}(y')(s) = s \mathscr{L} y(s) - y(0)$$

and therefore,

$$s^3 \mathscr{L}(y)(s) - s^2 - 4s + 3 - 4(s^2 \mathscr{L}(y)(s) - s - 4)$$
$$+ s\mathscr{L}(y)(s) - 1 - 6\mathscr{L}(y)(s) = 0$$

from which it follows that

$$\mathscr{L}(y)(s) = \frac{s^2 + 8s + 14}{s^3 - 4s^2 + s - 6} = \left(\frac{7}{12}\right)\frac{1}{s+1} + \left(\frac{47}{20}\right)\frac{1}{s-3} + \left(\frac{2}{5}\right)\frac{1}{s+2}. \quad (13)$$

Applying the inverse Laplace transform to (13) we find from the table of Laplace transforms that

$$y(t) = \frac{7}{12} e^{-t} + \frac{47}{20} e^{3t} + \frac{2}{5} e^{-2t}$$

(13.E.10) *Theorem* If $f: (0,\infty) \to \mathbf{R}^1$ and $g: (0,\infty) \to \mathbf{R}^1$ are piecewise smooth functions of exponential type, then

$$\mathscr{L}^{-1}(F(s)G(s))(t) = \int_0^t f(t - u)g(u)\, du$$

for $t > 0$. Moreover, the function $K(t) = \int_0^t f(t-u)g(u)\, du$ is piecewise smooth and of exponential type.

Proof. Let

$$H(t) = \begin{cases} 1 & \text{if } t > 0 \\ 0 & \text{if } t < 0 \end{cases}$$

and define functions ϕ and ψ by

$$\phi(t) = f(t)H(t)e^{-ct}$$
$$\psi(t) = g(t)H(t)e^{-ct}$$

Let c be a sufficiently large constant so that ϕ and ψ belong to $\mathscr{L}_1(-\infty,\infty)$. [We assume that f and g have been extended to $(-\infty,\infty)$ in an arbitrary way.] By setting $s = c + i\lambda$, we find that

$$F(s) = \int_0^\infty f(t)e^{-st}\, dt = \text{CP}\!\int_{-\infty}^{+\infty} f(t)H(t)e^{-ct}e^{i(-\lambda)t}\, dt = \hat{\phi}(-\lambda)$$

where $\hat{\phi}$ is the (complex) Fourier transform of ϕ. In a similar manner, we obtain

$$G(s) = \hat{\psi}(-\lambda)$$

By (13.B.8), it follows that

$$F(s)G(s) = \hat{\phi}(-\lambda)\hat{\psi}(-\lambda) = \widehat{\phi*\psi}(-\lambda) \tag{14}$$

Furthermore, from (13.B.5) we have that $\phi * \psi$ exists, and by the definition of the convolution, we see that

$$(\phi*\psi)(t) = {}_{CP}\int_{-\infty}^{\infty} {}_{V} \phi(t-u)\psi(u)\, du$$

$$= {}_{CP}\int_{-\infty}^{\infty} {}_{V} f(t-u)H(t-u)e^{-c(t-u)}g(u)H(u)e^{-cu}\, du \tag{15}$$

$$= \begin{cases} e^{-ct}\int_{0}^{t} f(t-u)g(u)\, du & \text{if } t > 0 \\ 0 & \text{if } t < 0 \end{cases}$$

This last equality follows from the fact that if $t > 0$, then

$$H(t-u) = \begin{cases} 1 & \text{if } 0 < u < t \\ 0 & \text{if } u > t \end{cases}$$

and, if $t < 0$, then $H(t-u) = 0$ for all $u > 0$.

Since f and g are piecewise smooth functions on $0 < t < \infty$, it follows easily from (15) that $\phi * \psi$ is a piecewise smooth function on $(-\infty,\infty)$, and by (13.B.5) $\phi * \psi$ belongs to $\mathscr{L}_1(-\infty,\infty)$. Consequently, from (13.B.1), we have

$$(\phi*\psi)(t) = \frac{1}{2\pi}\, {}_{CP}\int_{-\infty}^{\infty} {}_{V} \widehat{\phi*\psi}(\lambda)e^{-i\lambda t}\, d\lambda$$

$$= \frac{1}{2\pi}\, {}_{CP}\int_{-\infty}^{\infty} {}_{V} \widehat{\phi*\psi}(-\lambda)e^{i\lambda t}\, d\lambda$$

$$= \frac{1}{e^{ct}2\pi i}\, {}_{CP}\int_{-\infty}^{\infty} {}_{V} \widehat{\phi*\psi}(-\lambda)e^{(c+i\lambda)t}i\, d\lambda$$

$$= \frac{1}{e^{ct}2\pi i}\, {}_{CP}\int_{-\infty}^{\infty} {}_{V} F(c+i\lambda)G(c+i\lambda)e^{(c+i\lambda)t}i\, d\lambda$$

$$= \frac{1}{e^{ct}2\pi i}\, {}_{CP}\int_{c-i\infty}^{c+i\infty} {}_{V} F(s)G(s)e^{st}\, ds \tag{16}$$

Finally, from (15) and (16) it follows that

$$K(t) = \int_{0}^{t} f(t-u)g(u)\, du = \frac{1}{2\pi i}\, {}_{CP}\int_{c-i\infty}^{c+i\infty} {}_{V} F(s)G(s)e^{st}\, ds = \mathscr{L}^{-1}(F(s)G(s))(t)$$

for all $t > 0$.

(13.E.11) *Theorem* If f and g satisfy the conditions of (13.E.10), then $\mathscr{L}(\int_0^t f(t - u)g(u) \, du)(s) = \mathscr{L}(f)(s) \cdot \mathscr{L}(g)(s)$ for all s belonging to some half plane $\text{Re}(s) > \alpha$.

Proof. From (14) and (15) we have

$$F(s)G(s) = \int_{-\infty}^{\infty} \left[\int_0^t f(t - u)g(u) \, du \right] H(t) e^{-ct} e^{-i\lambda t} \, dt$$

where $s = c + i\lambda$. Hence,

$$F(s)G(s) = \int_0^{\infty} \left[\int_0^t f(t - u)g(u) \, du \right] e^{-st} \, dt$$

$$= \mathscr{L}\left(\int_0^t f(t - u)g(u) \, du \right)$$

(13.E.12) *Example* We consider the problem

$$y'(t) - ay(t) = f(t) \qquad 0 < t; \, y(0) = c \qquad (17)$$

where f is a piecewise smooth function of exponential type. Taking the Laplace transform of both sides of (17), we find that

$$sY(s) - c - aY(s) = F(s)$$

therefore,

$$Y(s) = \frac{F(s)}{s - a} + \frac{c}{s - a}$$

and hence,

$$y(t) = \mathscr{L}^{-1}(Y) = \mathscr{L}^{-1}\left(F(s) \frac{1}{s - a} \right) + c\mathscr{L}^{-1}\left(\frac{1}{s - a} \right)$$

From the table we find that $\mathscr{L}^{-1}(1/(s - a)) = e^{at}$, and consequently by (13.E.10), it follows that

$$\mathscr{L}^{-1}\left(F(s) \frac{1}{s - a} \right) = \int_0^t e^{a(t-u)} f(u) \, du$$

from which we conclude that

$$y(t) = \int_0^t e^{a(t-u)} f(u) \, du + ce^{at} \qquad (18)$$

(13.E.13) *Exercise* Show directly that the function defined by (18) is a solution of (17).

(13.E.14) *Example* We consider Abel's integral equation

$$f(t) = \int_0^t y(u)(t - u)^\alpha \, du \qquad (0 < \alpha < 1) \tag{19}$$

where f is assumed to be differentiable and of exponential type over $[0,\infty)$. Here y is the unknown function. Applying (13.E.11) and noting from the table that

$$\mathscr{L}\left(\frac{1}{t^\alpha}\right) = \frac{\Gamma(1 - \alpha)}{s^{1-\alpha}} \qquad 0 < \alpha < 1$$

we find that the Laplace transform of equation (19) is given by

$$F(s) = Y(s) \frac{\Gamma(1 - \alpha)}{s^{1-\alpha}}$$

and hence,

$$Y(s) = \frac{sF(s)}{\Gamma(1 - \alpha)s^\alpha}$$

Since $f(0) = 0$, it follows from (13.E.8) that $sF(s) = \mathscr{L}(f'(t))(s)$. From the table of Laplace transforms, we find that $\mathscr{L}(t^{\alpha-1})(s) = \Gamma(\alpha)/s^\alpha$, and therefore

$$Y(s) = \frac{sF(s)}{\Gamma(\alpha)\Gamma(1 - \alpha)} \cdot \frac{\Gamma(\alpha)}{s^\alpha}$$

$$= \mathscr{L}\left(\frac{f'(t)}{\Gamma(\alpha)\Gamma(1 - \alpha)}\right)(s) \cdot \mathscr{L}(t^{\alpha-1})(s)$$

Consequently, by (13.E.11)

$$y(t) = \frac{1}{\Gamma(\alpha)\Gamma(1 - \alpha)} \int_0^t f'(t - u)u^{\alpha-1} \, du$$

F.* THE δ-FUNCTION: AN INTRODUCTION TO GENERALIZED FUNCTIONS

We begin our brief study of the δ-function by considering the nth-order differential equation with homogeneous initial conditions

$$a_0 y^{(n)}(t) + a_1 y^{(n-1)}(t) + \cdots + a_{n-1} y^{(1)}(t) + a_n y(t) = f(t) \quad t > 0$$
$$y^{(j)}(0) = 0 \qquad 0 \leq j \leq n - 1 \tag{1}$$

where the coefficients $a_0, a_1, \ldots, a_n$ are constants. The Laplace transform equation associated with the problem is given by

$$(a_0 s^n + \cdots + a_{n-1}s + a_n)Y(s) = F(s)$$

where $Y(s)$ and $F(s)$ are the Laplace transforms of the functions $y(t)$ and $f(t)$, respectively. If we set $G(s) = [a_0 s^n + \cdots + a_{n-1}s + a_n]^{-1}$, then we have

$$Y(s) = G(s)F(s) \qquad (2)$$

The inverse Laplace transform of $G(s)$ generates a unique function $g(t)$, and it follows from the convolution theorem (13.E.10) that the solution, $y(t)$, to (1) is defined by

$$y(t) = \int_0^t g(t - \tau)f(\tau)\,d\tau \qquad (3)$$

Since the convolution commutes (13.B.7) we also have

$$y(t) = \int_0^t f(t - \tau)g(\tau)\,d\tau \qquad (4)$$

Equations of the form given by (1) arise in many applied problems. For instance, in electrical engineering (1) may be used in describing an electrical network or system where $f(t)$ denotes the voltage at the input terminals and $y(t)$ denotes the resulting current in the network. Hence, in this context $f(t)$ can be considered as the input (or *excitation*) and $y(t)$ the corresponding output (or *response*) to the stimulus f within the network, where it is understood that both f and y are "inactive" before time $t = 0$.

It is not difficult to see that $G(s)$, the generating function for $g(t)$, depends only upon the system (or network) and not upon the excitation f; therefore, G is often referred to as the *system's generating function* (or simply the *system's function*). The corresponding function $g(t)$ is frequently called *Green's function*. This function is independent of the excitation f and is determined by the system.

In the system just described a very interesting response is obtained when the excitation consists of a strong burst of energy over a short duration of time. For instance, a *unit impulse* of energy due to an extremely high voltage acting over an extremely short period of time can be described by

$$\phi_\varepsilon(t) = \begin{cases} 0 & \text{if } t < 0 \\ \dfrac{1}{\varepsilon} & \text{if } 0 < t < \varepsilon \\ 0 & \text{if } t > \varepsilon \end{cases}$$

where ε is presumed to be a very small positive number. The response to the excitation $f(t) = \phi_\varepsilon(t)$ is given by (4)

$$y(t) = \int_0^t \phi_\varepsilon(t - \tau)g(\tau)\,d\tau = \frac{1}{\varepsilon}\int_{t-\varepsilon}^t g(\tau)\,d\tau$$

It can be shown that Green's function g is continuous for all $t > 0$ (see, e.g.,

Doetsch (1961)), and hence, it follows from the mean value theorem for integrals that

$$y(t) = g(t_\varepsilon)$$

for some point $t_\varepsilon \in (t - \varepsilon, t)$, $t > 0$. For small values of ε this solution serves as an excellent approximation of $g(t)$ over any closed and bounded interval contained in $(0,\infty)$. In fact, as ε tends to 0 the responses to the system produced by the corresponding excitations ϕ_ε approach g for $t > 0$.

Now we consider the problem of trying to define or describe an "instantaneous" impulse. We do this indirectly by introducing an "ideal" or "generalized" function as a certain kind of limit that depends on the family of excitation functions ϕ_ε. This limit "function" (which is not a function in the usual sense of the word) is to describe a unit impulse of energy acting *instantaneously* at time $t = 0$. The total energy associated with this "function" is given by the "integral" of the "function" and is to be equal to 1.

This (yet to be defined) entity is called the *Dirac delta function* (or δ-function) and is symbolized by $\delta(t)$. As we have mentioned, the δ-function is not a function and is certainly not an ordinary limit of the family $\phi_\varepsilon(t)$ as $\varepsilon \to 0$. For if it were, then $\delta(t)$ would be defined by

$$\delta(t) = \begin{cases} 0 & \text{if } t \neq 0 \\ \infty & \text{if } t = 0 \end{cases} \tag{5}$$

However, this would mean that δ is Lebesgue summable and that $\int_{\mathbf{R}^1} \delta = 0$. Physically this interpretation of δ is not the desired one since it would imply that the total energy associated with δ is zero rather than one. Nevertheless, the definition of δ given by (5) would be the only logical interpretation of δ if we were to insist that δ be a mapping from $\mathbf{R}^1$ into the extended reals ($\mathbf{R}_\infty$). Since this interpretation of δ (though convenient!) is of little physical use, we shall pursue the possibility of a new construct, a "function" that is not a function in the usual sense of the word.

We have observed that the unit impulse function $\phi_\varepsilon(t)$, acting for a short period of time from $t = 0$ to $t = \varepsilon$, produces the response

$$y_\varepsilon(t) = \int_0^t \phi_\varepsilon(\tau) g(t - \tau) \, d\tau$$

We also noted that

$$y_\varepsilon(t) = g(t_\varepsilon)$$

where $t_\varepsilon \in (t - \varepsilon, t)$; furthermore, this effect is maintained as $\varepsilon \to 0^+$. That is to say, the "limiting action" given by $\lim_{\varepsilon \to 0^+} y_\varepsilon(t) = g(t)$ preserves the "continuing presence" of the unit impulse of energy associated with ϕ_ε. Thus, we might interpret the delta function through the "symbolic integration" defined by

$$\int_0^t \delta(\tau)g(t - \tau)\, d\tau = \lim_{\varepsilon \to 0^+} \int_0^t \phi_\varepsilon(\tau)g(t - \tau)\, d\tau = g(t) \qquad (6)$$

On the other hand, since

$$\int_0^t \phi_\varepsilon(t - \tau)g(\tau)\, d\tau = \int_0^t \phi_\varepsilon(\tau)g(t - \tau)\, d\tau$$

it would appear that δ has the property

$$\int_0^t \delta(t - \tau)g(\tau)\, d\tau = \lim_{\varepsilon \to 0^+} \int_0^t \phi_\varepsilon(t - \tau)g(\tau)\, d\tau = g(t)$$

Since Green's function g is defined for all t and is equal to 0 for $t < 0$ [see (13.E.2)], we have

$$\int_{-\infty}^\infty \phi_\varepsilon(t - \tau)g(\tau)\, d\tau = \int_0^t \phi_\varepsilon(t - \tau)g(\tau)\, d\tau$$

this suggests the more symmetrical from

$$\int_{-\infty}^\infty \delta(t - \tau)g(\tau)\, d\tau = \lim_{\varepsilon \to 0^+} \int_{-\infty}^\infty \phi_\varepsilon(t - \tau)g(\tau)\, d\tau = g(t) \qquad (7)$$

Even though we have dealt with Green's function associated with (1), it is clear that the right-hand side of (7) is satisfied for any function g defined on $(-\infty, \infty)$ that is continuous in a neighborhood of $\tau = t$, and, hence, the definition (7) might be extended to all such functions.

The *action* of an impulse force ϕ_ε on a continuous function g (whether or not it is a Green's function) can be interpreted as the *value of the integral* $\int_0^\infty \phi_\varepsilon(\tau)g(\tau)\, d\tau$. Since

$$\int_0^\infty \phi_\varepsilon(\tau)g(\tau)\, d\tau = \frac{1}{\varepsilon} \int_0^\varepsilon g(\tau)\, d\tau$$

this integral can be interpreted as determining the *average output* of g over the interval $[0, \varepsilon]$. Hence, we can interpret the *action* of the idealized unit impulse force δ on a function g as the "average output of g over an infinitesimal period of time at $t = 0$". If one is interested in a particular impulse at time $t = t_0$, then the symbol $\delta(t_0 - t)$, is used to indicate that the "instantaneous action" is at time $t = t_0$. In this case the action on $g(t)$ by $\delta(t_0 - t)$ is given by $g(t_0)$ and can be interpreted as "the average output of g over an infinitesimal period of time $t = t_0$."

The idea of computing the average output of a function over a short period of time is of considerable importance. In the laboratory, physical measurements are made of voltages, temperatures, velocities, etc., and these measurements are frequently expressed mathematically in the form

$$\int w(t_0, \tau) g(\tau) \, d\tau$$

where $w(t_0, \tau)$ denotes a function that is nearly constant in a small neighborhood of $t = t_0$ and vanishes elsewhere.

The δ-function is an example of what is called a *generalized function*. There are a number of ways of developing the idea of a generalized function; we introduce briefly (and modify slightly) an approach due to J. Korevaar (1968). In this technique the "new functions" are essentially identified with Cauchy sequences of ordinary functions. (This idea is similar to one very standard method of developing the irrational numbers from the rational numbers.) We begin with an interval $J = [a,b]$, and then augment it to a slightly larger interval $J_\eta = [a - \eta, b + \eta]$, where η is an arbitrarily small positive number. We then introduce a vector space $V(J_\eta)$ consisting of all (ordinary) functions defined on the interval J_η that are summable over every compact subinterval $[\alpha, \beta]$ of J_η. We let $C_0^\infty (J_\eta)$ denote the family of all infinitely differentiable functions ϕ defined on J_η with the property that ϕ is zero outside of a compact subinterval of J_η. Such a function is said to have *compact support* in J_η. The members of $C_0^\infty (J_\eta)$ are called *test functions* and they are used to define convergence in $V(J_\eta)$ as follows. A sequence of functions $\{f_k\}$ in $V(J_\eta)$ is said to *converge* to a function $f \in V(J_\eta)$ if $\lim_{k \to \infty} \int_{J_\eta} f_k \phi = \int_{J_\eta} f \phi$ for every test function $\phi \in C_0^\infty (J_\eta)$.

A sequence $\{f_k\}$ in J_η is said to be *fundamental* (or Cauchy) with respect to this mode of convergence, if the sequence $\{\int_{J_\eta} f_k \phi\}$ is a Cauchy sequence of real numbers for each $\phi \in C_0^\infty(J_\eta)$. (This, of course, implies that $\lim_{k \to \infty} \int_{J_\eta} f_k \phi$ exists and is finite for each $\phi \in C_0^\infty(J_\eta)$.) Two fundamental sequences $\{f_k\}$ and $\{g_k\}$ are said to be *equivalent over J* (not J_η) if for all suitably small $\tilde{\eta}$, say $0 < \tilde{\eta} < \eta$, and for all $\tilde{\phi} \in C_0^\infty(J_{\tilde{\eta}})$

$$\lim_{\tilde{\eta} \to 0} \left(\lim_{k \to \infty} \int_{J_{\tilde{\eta}}} f_k \tilde{\phi} \right) = \lim_{\tilde{\eta} \to 0} \left(\lim_{k \to \infty} \int_{J_{\tilde{\eta}}} g_k \tilde{\phi} \right). \tag{8}$$

It can be shown that (8) defines an equivalence relation; the corresponding equivalence classes are called *distributions over J*.

If $\{f_k\}$ represents a distribution over J, then its *action* on a continuous function g with compact support in J_η is given by $\lim_{\eta \to 0} (\lim_{k \to \infty} \int_{J_\eta} f_k g)$. The existence of this limit is established in Korevaar (1968). The δ-function (distribution) can be represented on $J = [0, \infty)$ by a Cauchy sequence $\{\phi_{\varepsilon_k}\}$, where for $k = 1, 2, \ldots$

$$\phi_{\varepsilon_k}(t) = \begin{cases} 0 & \text{if } t < 0 \\ \dfrac{1}{\varepsilon_k} & \text{if } 0 < t < \varepsilon_k \\ 0 & \text{if } t > \varepsilon_k \end{cases}$$

and $\lim_{k \to \infty} \varepsilon_k = 0$. Note that for every function g that is continuous in a neighborhood of $t = 0$,

$$\lim_{\eta \to 0}\left[\lim_{k \to 0}\int_{-\eta}^{\infty}\phi_{\varepsilon_k}(\tau)g(\tau)\right] = \lim_{k \to 0}\int_{0}^{\infty}\phi_{\varepsilon_k}(\tau)g(\tau)\,d\tau = g(0)$$

Hence, for such functions g it is customary to write this action (over $[0,\infty)$) symbolically by the expression

$$\int_{0}^{\infty}\delta(\tau)g(\tau)\,d\tau = g(0)$$

In particular, $\int_{0}^{\infty}\delta(t)e^{-st}\,dt = e^{-s0} = 1$, and consequently, the Laplace transform of the δ-function is usually defined by $\mathscr{L}(\delta(t)) = 1$.

If we replace $f(t)$ in (1) by $\delta(t)$, then we can take Laplace transforms of both sides of the resulting equation to obtain

$$(a_0 s^n + \cdots + a_n)Y(s) = 1 \tag{9}$$

and hence

$$Y(s) = [a_0 s^n + \cdots + a_n]^{-1} = G(s)$$

This procedure, while appearing likely to work, is in need of some formal justification [see Korevaar (1968), for example].

In the slightly more general case ($t = t_0$), we have

$$\mathscr{L}(\delta(t - t_0))(s) = e^{-st_0} \tag{10}$$

(see problem F.5).

(13.F.1) *Example* We consider

$$y'' + 4y' + 8y = \delta\left(t - \frac{\pi}{2}\right) \qquad y(0) = 0 \qquad y'(0) = 0$$

(An equation such as this might arise in connection with the investigation of current flow, where a voltage impulse occurs at $t = \pi/2$.) By (9) and (10) we have

$$(s^2 + 4s + 8)Y(s) = e^{-(\pi/2)s}$$

and hence

$$Y(s) = \frac{e^{-(\pi/2)s}}{s^2 + 4s + 8} = \frac{1}{2}\left(\frac{2}{(s + 2)^2 + 4}\right)e^{-(\pi/2)s}$$

From the table of Laplace transforms, we have

$$\mathscr{L}^{-1}\left(\frac{2}{(s + 2)^2 + 4}\right) = e^{-2t}\sin 2t$$

and, therefore, it follows that

$$y(t) = \begin{cases} \dfrac{1}{2} e^{-2(t-\pi/2)} \sin 2\left(t - \dfrac{\pi}{2}\right) & \text{if } t \geq \dfrac{\pi}{2} \\ \\ 0 & \text{if } t < \dfrac{\pi}{2} \end{cases}$$ (11)

(see problem F.6).

PROBLEMS

Section A

1. Show that if $f: \mathbf{R}^1 \to \mathbf{R}^1$ is identically zero outside of a finite interval $[a,b]$ and is piecewise smooth on (a,b), then

$$\frac{1}{\pi} \int_0^\infty \left(\int_a^b f(t) \cos \lambda(x - t) \, dt \right) d\lambda = \begin{cases} 0 & \text{if } x < a \\ \dfrac{f(a+0)}{2} & \text{if } x = a \\ \dfrac{f(x-0) + f(x+0)}{2} & \text{if } a < x < b \\ \dfrac{f(b-0)}{2} & \text{if } x = b \\ 0 & \text{if } x > b \end{cases}$$

2. Show that the regular pulse function

$$f(x) = \begin{cases} 1 & \text{if } |x| < 1 \\ 0 & \text{if } |x| > 1 \end{cases}$$

is represented by the integral

$$\int_0^\infty \frac{2 \sin \lambda}{\pi \lambda} \cos \lambda x \, d\lambda = \begin{cases} 0 & \text{if } |x| > 1 \\ 1 & \text{if } |x| < 1 \\ \dfrac{1}{2} & \text{if } |x| = 1 \end{cases}$$

3. Let

$$\frac{f(x+0) + f(x-0)}{2} = \int_0^\infty (A(\lambda) \cos \lambda x + B(\lambda) \sin \lambda x) \, d\lambda$$

be the Fourier integral representation of f. Show that if $f \in \mathcal{L}_1(-\infty,\infty)$ and if f is an even function, then $B(\lambda) \equiv 0$; show also that if $f \in \mathcal{L}_1(-\infty, \infty)$ and is an odd function, then $A(\lambda) \equiv 0$. Conclude that if f is a piecewise smooth even function that is summable on $(-\infty,\infty)$, then

$$\frac{f(x+0)+f(x-0)}{2} = \frac{2}{\pi}\int_0^\infty\left(\int_0^\infty f(t)\cos\lambda t\,dt\right)\cos\lambda x\,d\lambda$$

and if f is a piecewise smooth odd function that is summable on $(-\infty, \infty)$, then

$$\frac{f(x+0)+f(x-0)}{2} = \frac{2}{\pi}\int_0^\infty\left(\int_0^\infty f(t)\sin\lambda t\,dt\right)\sin\lambda x\,d\lambda$$

4. Show that if f is a piecewise smooth function that is summable on $(-\infty,\infty)$, then f can be written as the sum of an even function f_e and an odd function f_o, and that

$$\frac{f(x+0)+f(x-0)}{2} = \frac{2}{\pi}\int_0^\infty\left(\int_0^\infty f_e(t)\cos\lambda t\,dt\right)\cos\lambda x\,d\lambda$$

$$+ \frac{2}{\pi}\int_0^\infty\left(\int_0^\infty f_o(t)\sin\lambda t\,dt\right)\sin\lambda x\,d\lambda$$

5. Show that for all x

$$e^{-|x|} = \frac{2}{\pi}\int_0^\infty\frac{\cos\lambda x}{1+\lambda^2}\,d\lambda$$

6. Show that if $f(x) = 1/(1+x^2)$, then f is summable on $(-\infty,\infty)$ and $1/(1+x^2) = \int_0^\infty e^{-\lambda}\cos\lambda x\,d\lambda$.
7. Show that $X(x) = c_1 e^{\sqrt{p}x} + c_2 e^{-\sqrt{p}x}$ and $T(t) = c_3 e^{a^2 pt}$ are solutions of (4) and (5) in Sec. 13.A, respectively.
8. Prove Lemma (13.A.5). [*Hint*: Apply the mean value theorem to the numerator of the quotients.]
9. Use the Lebesgue dominated convergence theorem to show that if $g \in \mathscr{L}_1(-\infty,\infty)$, then $\text{cpv}\int_{-\infty}^\infty g(x)\,dx = \int_{(-\infty,\infty)} g$.
10. Suppose that g is summable over an infinite interval J and that f is bounded on J and summable over every finite subinterval of J. Show that the product fg is summable over J.

Section B

1. Find the Fourier transforms (if they exist) of the following functions:

(a) $f(x) = \begin{cases} 1 & \text{if } 0 < x < 1 \\ 0 & \text{otherwise} \end{cases}$

(b) $f(x) = \begin{cases} 1 & \text{if } x \in [2n, 2n+1] \quad n = 0, \neq 1, \neq 2, \ldots \\ 0 & \text{otherwise} \end{cases}$

(c) $f(x) = \begin{cases} \cos x & \text{if } x \in (0,\pi) \\ 0 & \text{otherwise} \end{cases}$

2. If the Fourier transform of f is defined by $\hat{f}(x) = (\sin 2x)/x$, find f.
3. Find the Fourier transform of $f(x) = e^{-x^2}$.
4. Show that if $f, g \in \mathcal{L}_1(-\infty,\infty)$, and if $f(x) = 0 = g(x)$ for each $x < 0$, then $(f * g)(t) = \int_0^t f(x)g(t - x)\,dx$.
5. If

$$\hat{f}(\lambda) = \begin{cases} 0 & \text{if } |\lambda| \geq a \\ k & \text{if } |\lambda| < a \end{cases}$$

find f.

6. Show that if $f \in \mathcal{L}_1(-\infty,\infty)$, then $\lim_{|\lambda| \to \infty} \hat{f}(\lambda) = 0$. [*Hint*: Use the Riemann-Lebesgue lemma.]
7. Show that the convolution theorem holds if $f(x) = e^{-x^2} = g(x)$.
8. Show that $f * (g + h) = (f * g) + (f * h)$.

Section C

1. Show that the function u defined by (12) is a solution of (1).
2. Establish equations (15), (16), and (17).
3. Show that if f'' and g' are continuous, then the function u defined by (22) is a solution of (14).
4. Use the separation of variables technique and the Fourier integral representation to show that

$$u(x, t) = \frac{1}{\pi} \int_0^\infty \int_{-\infty}^\infty f(w) \cos \lambda(w - x)e^{-\lambda t}\,dw\,d\lambda$$

is a solution to the boundary value problem:

$$u_{xx} + u_{tt} = 0 \qquad -\infty < x < \infty, t > 0$$
$$u(x, 0) = f(x)$$
$$u(x, t) \leq M \qquad -\infty < x < \infty, t > 0$$

5. Solve the boundary problem given in problem C.4 in the case that u is bounded and

$$f(x) = e^{-|x|}.$$

Section D

1. Let $f: (0,\infty) \to \mathbf{R}^1$ be defined by

$$f(x) = \begin{cases} 1 & \text{if } 0 < x < b \\ 0 & \text{if } x > b \end{cases}$$

for some positive number b. Show that the Fourier sine integral representation of f is given by

$$f(x) = \frac{2}{\pi} \int_0^\infty \frac{1 - \cos b\lambda}{\lambda} \sin \lambda x \, d\lambda \qquad (x > 0)$$

2. Use the Fourier sine integral formula to show that

$$e^{-x} \cos x = \frac{2}{\pi} \int_0^\infty \frac{\lambda^3 \sin \lambda x}{\lambda^4 + 4} \, d\lambda \qquad (x > 0)$$

3. Suppose that $f \in \mathcal{L}_1(0,\infty)$, f, f', and f'' are continuous on $[0,\infty)$ and that $\lim_{x \to \infty} f(x) = \lim_{x \to \infty} f'(x) = 0$. Show that

$$\hat{f}_s'' = -\lambda^2 \hat{f}_s + \lambda f(0)$$

where $\hat{f}_s$ is the Fourier sine transform of f.

4. Use problem D.3 to show that the Fourier sine integral representation of e^{-cx}, where $c > 0$ and $x > 0$ is given by

$$e^{-cx} = \frac{2}{\pi} \int_0^\infty \frac{\lambda \sin \lambda x}{\lambda^2 + c^2} \, d\lambda$$

5. Show that if $f: (0,\infty) \to \mathbf{R}^1$ is defined by $f(x) = x/(x^2 + c^2)$, then $f \notin \mathcal{L}_1(0,\infty)$, but that, nevertheless,

$$f(x) = \frac{2}{\pi} \int_0^\infty \left(\int_0^\infty \frac{t}{t^2 + c^2} \sin \lambda t \, dt \right) \sin \lambda x \, dx$$

if $x > 0$ and $c > 0$. [*Hint:* Use problem D.4.]

6. Suppose that f is a piecewise smooth function on $(0,\infty)$ and that $\int_0^\infty |f| < \infty$. Use the even extension of f over $(-\infty,\infty)$ to derive the Fourier cosine integral representation of f

$$\frac{f(x + 0) + f(x - 0)}{2} = \frac{2}{\pi} \int_0^\infty \hat{f}_C(\lambda) \cos(\lambda x) \, d\lambda$$

where $\hat{f}_C$ is the Fourier cosine transform of f defined by

$$\hat{f}_C(\lambda) = \int_0^\infty f(t) \cos \lambda t \, dt$$

Find the Fourier cosine integral representation of $f(x) = e^{-x}$ $(x > 0)$.

7. Find the Fourier sine and cosine integral representations for each of the following functions:

(a) $f(x) = \begin{cases} \dfrac{1}{a} & \text{if } 0 < x < a \\ 0 & \text{if } a > x \end{cases}$

(b) $f(x) = \begin{cases} a - x & \text{if } 0 < x < a \\ 0 & \text{if } a > x \end{cases}$

8. Use the Fourier cosine integral formula to solve the problem

$$u_{xx}(x, t) = \frac{1}{a^2} u_t(x, t)$$

$$u_x(0, t) = 0 \qquad 0 < t$$

$$u(x, 0) = f(x) \qquad 0 < x$$

9. In this problem use the Fourier cosine integral formula to compute $\int_0^\infty e^{-a^2 x^2} \, dx$. First show that

$$\int_0^\infty e^{-a^2 x^2} \cos \lambda x \, dx = c(a) e^{-\lambda^2/4a^2}$$

by showing that if $g(\lambda) = \int_0^\infty e^{-a^2 x^2} \cos \lambda x \, dx$, then $g'(\lambda) = -\lambda g(\lambda)/2a^2$. Note that $c = g(0) = \int_0^\infty e^{-a^2 x^2} \, dx$. Find c by using the Fourier cosine integral representation of $e^{-a^2 x^2}$.

Section E

1. Use the Laplace transform to solve the following differential equations:
   (a) $y'' + y = 1$, where $y(0) = 0$ and $y'(0) = 1$.
   (b) $y'' + y = 1$, where $y(0) = 1$ and $y'(0) = 0$.
   (c) $y'' + y = t$, where $y(0) = 1$ and $y'(0) = 0$.
   (d) $y'' - 4y = \sin t$, where $y(0) = 0$ and $y'(0) = 0$.
2. (a) Given that $\mathcal{L}(f(t))(s) = F(s)$, show that

   $$\mathcal{L}(e^{at} f(t))(s) = F(s - a)$$

   (This is called the *shifting theorem*.)
   (b) Find the inverse Laplace transforms of the following functions by completing the squares and using the shifting theorem:

   (i) $\dfrac{1}{s^2 + 4s}$

   (ii) $\dfrac{s}{s^2 + 2as + b^2}$

   (iii) $\dfrac{1}{s^2 + 2s + 5}$

3. Solve the following differential equations:
   (a) $y'' + 4y' + 5y = 0 \qquad y(0) = 0, y'(0) = 1$
   (b) $y'' + 2ay' + b^2 y = 0 \qquad y(0) = 1, y'(0) = 0$
   (c) $y'' + 4y' = \sin t \qquad y(0) = 0, y'(0) = 0$

4. (a) There is a useful partial fraction decomposition formula for rational functions $P(s)/Q(s)$, where the degree of P is less than the degree of Q, whenever Q has no multiple roots. In such a case Q factors into

$$Q(s) = c \prod_{k=1}^{n} (s - r_k)$$

$r_i \neq r_j$ if $i \neq j$, and

$$\frac{P(s)}{Q(s)} = \sum_{k=1}^{n} \frac{A_k}{s - r_k} \qquad (*)$$

Show under these conditions that $A_k = P(r_k)/Q'(r_k)$. [*Hint*: Multiply both sides of $(*)$ by $s - r_k$ and take the limit as s tends to r_k. Use L'Hospital's rule.]

(b) Use part (a) to find the above partial fraction decomposition of

(i) $\dfrac{1}{(x^2 - x - 2)(x^2 + x - 2)}$

(ii) $\dfrac{s^2 + 1}{s(s^2 - 4)}$

5. (a) If $f: [0,\infty) \to \mathbf{R}^1$ is periodic with period T and is piecewise smooth show that

$$\mathscr{L}(f(t))(s) = \frac{\displaystyle\int_0^T e^{-st} f(t)\, dt}{1 - e^{-st}}$$

$$\left[\textit{Hint}: \text{Show that } \int_{nT}^{(n+1)T} e^{-st} f(t)\, dt = e^{-snt} e^{-su} f(u)\, du. \right]$$

(b) Find the Laplace transform of the sawtooth function,

$$f(t) = t \qquad \text{for } 0 \leq t \leq a, f(t + a) = f(t)$$

(c) Find the Laplace transform of the rectified sine wave

$$f(t) = \sin t \qquad 0 \leq t \leq \pi, f(t + \pi) = f(t)$$

6. For each $c \geq 0$, let

$$u_c(t) = \begin{cases} 0 & \text{if } t < c \\ 1 & \text{if } t \geq c \end{cases}$$

(a) Show that $\mathscr{L}(u_c(t))(s) = e^{-ct}/s$

(b) Show that if $f: [0,\infty) \to \mathbf{R}^1$ is piecewise smooth and of exponential type, then

$$\mathcal{L}(u_c(t)f(t - c))(s) = e^{-cs}F(s)$$

where

$$\mathcal{L}(f(t))(s) = F(s)$$

(notation as in problem E.6).

(c) Find $\mathcal{L}(u_c(t) \sin (t - c))$.

7. Suppose that a constant force is applied to a spring-mass system which is in motion for a short period of time from t_0 to t_1, $(0 < t_0)$. Assume that the equation of motion for the system is given by

$$y''(t) + a^2 y(t) = f(t) \qquad y(0) = y_0 \qquad y'(0) = y_1$$

$$f(t) = \begin{cases} 0 & \text{for } t < t_0 \\ k & \text{for } t_0 \le t \le t_1 \\ 0 & \text{for } t > t_1. \end{cases}$$

(a) Show that the transformed equation is

$$Y(s) = \frac{sy_0 + y_1}{s^2 + a^2} + \frac{1}{s^2 + a^2} F(s)$$

(b) Show that

$$y(t) = y_0 \cos at + y_1 \frac{\sin at}{a} + \int_0^t \frac{\sin a(t - u)}{a} f(u) \, du$$

8. (a) Show that if f is piecewise smooth and of exponential type ($|f(t)| \le ke^{at}$ for $t > M$), then

$$\frac{d}{ds} \int_0^\infty e^{-st}f(t) \, dt = \int_0^\infty \frac{d}{ds} (e^{-st}f(t)) \, dt$$

provided $\text{Re}(s) > a$. [Hint: Consider

$$\int_0^\infty \frac{e^{-(s+h)t} - e^{-st}}{h} f(t) \, dt$$

and use the mean value theorem and the Lebesgue dominated convergence theorem.]

(b) Show that $|t^n f(t)| \le ke^{bt}$ for $t > M$, where $b = \max\{a, \varepsilon\}$ and $\varepsilon > 0$ is any arbitrarily small number.

(c) Let $\mathcal{L}(f(t))(s) = F(s)$ and show that

$$F'(s) = \mathcal{L}(-tf(t))(s)$$

and

$$F^{(n)}(s) = \mathcal{L}((-t)^n f(t))(s)$$

if $a > 0$ and $\text{Re}(s) > a$.

(d) Use (c) to show that for each positive integer n, $\mathcal{L}(t^n)(s) = n!/s^{n+1}$.

9. (a) Since $\sin t = \sum_{n=0}^\infty (-1)^n t^{2n+1}/(2n + 1)!$ and $\mathcal{L}(\sin t) = 1/(s^2 + 1)$,

it is natural to ask if

$$\mathcal{L}(\sin t) = \sum_{n=0}^{\infty} \mathcal{L}\left(\frac{(-1)^n t^{2n+1}}{(2n+1)!}\right)$$

Show that the answer is yes if $s > 1$.

(b) Part (a) is a particular illustration of the following: Suppose that $\{f_n\}$ is a sequence of functions such that $F_n(s) = \mathcal{L}(f_n)(s)$ exists for all $n \geq 0$ in a common half plane $\mathrm{Re}(s) \geq c_0$. Suppose, in addition, that the functions

$$A_n(s) = \int_0^{\infty} e^{-st}|f_n(t)|\, dt \qquad n \geq 0$$

also exist in this half plane $[\mathrm{Re}(s) \geq c_0]$ and that $\sum_{n=0}^{\infty} A_n(c_0)$ converges. Show that $\sum_{n=0}^{\infty} f_n(t)$ converges absolutely a.e. in $(0,\infty)$ to a function f with the property that $F(s) = \mathcal{L}(f(t))(s) = \sum_{n=0}^{\infty} F_n(s)$ for all s in the half plane $\mathrm{Re}(s) \geq c_0$. [*Hint*: Use the monotone convergence theorem (11.G.1).]

10. (a) Show that the (Laplace) transform of Bessel's differential equation of order zero, $ty''(t) + y'(t) + ty(t) = 0$, $t > 0$, is

$$(s^2 + 1)Y'(s) + sY(s) = 0 \qquad (**)$$

(b) Show that $Y(s) = c(s^2 + 1)^{-\frac{1}{2}}$ is a solution of $(**)$, where c is an arbitrary constant.

(c) Show that $Y(s) = (c/s)(1 + 1/s^2)^{-\frac{1}{2}}$.

(d) The binomial expansion theorem states that for each complex number, the series $\sum_{n=0}^{\infty} \binom{\alpha}{n} z^n$, where $\binom{\alpha}{n} = \alpha(\alpha - 1)\cdots(\alpha - n + 1)/n!$, converges absolutely to $(1 + z)^{\alpha}$ for all $|z| < 1$ (and diverges for all $|z| > 1$). Use this theorem to show that for all $|s| > 1$,

$$Y(s) = c \sum_{n=0}^{\infty} \frac{(-1)^n}{2^{2n}(n!)^2} \frac{(2n)!}{s^{2n+1}} \qquad (***)$$

Hint: Show that

$$\frac{(2n)!}{2^{2n}(n!)^2} = \frac{2n - 1}{2n} \cdots \frac{7}{8}\frac{5}{6}\frac{3}{4}\frac{1}{2}$$

$$= (-1)^n \binom{-\frac{1}{2}}{n}$$

(e) Show that $(2n)!/2^{2n}(n!)^2 < 1$, for all $n \geq 1$, and then show that the series $(***)$ converges absolutely for all $|s| > 1$.

(f) Use the results from (b) and (d), to show that the inverse Laplace transform of $Y(s)$ is given by

$$c \sum_{n=0}^{\infty} \frac{(-1)^n}{2^{2n}(n!)^2} t^{2n} = c \sum_{n=0}^{\infty} \frac{(-1)^n}{(n!)^2} \left(\frac{t}{2}\right)^{2n}$$

(g) Show that the series in (f) converges absolutely for all t.

(h) Reverse the steps you used in (a) to show that the series in (f) must represent a solution of Bessel's equation of order zero for all $t > 0$ and that this solution is bounded as $t \to 0^+$.

11. (a) Show that if f is piecewise smooth and of exponential type on $[0,\infty)$, and if, in addition, $|f(t)/t|$ remains bounded as $t \to 0^+$, then

$$\int_0^\infty \left[\int_0^\infty |f(t)| e^{-ut} \, du \right] e^{-st} \, dt$$

converges for all large $s > 0$.

(b) Use (a) and (11.I.16) (Fubini's absolute convergence inversion theorem) to prove that if f satisfies the conditions in (a), then for all large positive s

$$\mathscr{L}\left(\frac{f(t)}{t}\right)(s) = \int_0^\infty \frac{f(t)}{t} e^{-st} \, dt = \int_0^\infty F(u) \, du$$

where $F(s) = \mathscr{L}(f(t))(s)$.

(c) Show that $\mathscr{L}((\sin t)/t)(s) = \pi/2 - \tan^{-1}(s) = \tan^{-1}(1/s)$. [*Hint*: $e^{-st}/t = \int_s^\infty e^{-ut} \, du$.]

Section F

1. Solve the following differential equations:
    (a) $y' + 3y = \delta(t - 1)$ $y(0) = 0$
    (b) $y'' + 5y' + 6y = 3\delta(t - x)$ $y(0) = 1$
    (c) $y'' + 2y' + y = e^{-t} + 3\delta(t - 1)$ $y(0) = 0$ $y'(0) = 0$.

2. (a) In some treatments the Laplace transform of the δ-function is defined by

$$\mathscr{L}(\delta(t)) = \lim_{\varepsilon \to 0} \mathscr{L}(\phi_\varepsilon(t))$$

Show that under this definition

$$\mathscr{L}(\delta(t)) = 1$$

(b) Use this procedure to establish

$$\mathscr{L}(\delta(t - t_0))(s) = e^{-t_0 s}$$

3. Establish the following equation

$$\int_a^b f(t) \, \delta(t - t_0) \, dt = \begin{cases} f(t_0) & \text{if } a \le t_0 \le b \\ 0 & \text{otherwise} \end{cases}$$

and find $\int_1^3 e^{\sqrt{3}t} \delta(t - 2) \, dt$.

4. Hooke's law states that if a spring is stretched (or compressed) beyond

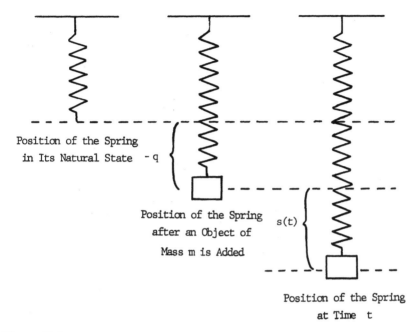

Position of the Spring
in Its Natural State -q

Position of the Spring
after an Object of
Mass m is Added

s(t)

Position of the Spring
at Time t

Figure 13.3

its natural length, then the spring exerts a restoring force proportional
to the displacement of the spring: $F(x) = -kx$, where k is the spring
constant. Suppose that a spring is displaced q units by the addition of
an object of mass m, and then an additional external force $f(t)$ (which
varies with time) is applied to the spring. Let $s(t)$ denote the displacement
of the spring at time t (Fig. 13.3).

Newton's second law of motion asserts that the rate of change of the
momentum of an object is proportional to the resultant force F acting
on the object and is in the direction of this force (momentum is the product of the mass of an object and its velocity). With suitable units the
mathematical expression of this law is given by $mv'(t) = F(t)$. Since the
derivative of the velocity function $v(t)$ is the acceleration $a(t)$, we have
that $ma(t) = F(t)$; if $s(t)$ represents the displacement of the object at
time t, then $ms''(t) = F(t)$.

(a) Show, using Hooke's law and Newton's second law of motion, that
the spring system just described is governed by

$$ms''(t) = -mg + kq - ks(t) + f(t) = -ks(t) + f(t) \qquad s(0) = 0 = s'(t)$$

where g is the gravitational constant.

(b) Suppose that an object of mass 4 is added to a spring and then after
2 seconds the object is struck from above by a force yielding an

instantaneous unit impulse. Use the Laplace transform of the δ-function to describe the displacement of the spring; i.e., find $s(t)$.

(c) Find $s(t)$ if the object is struck from below by a force yielding an instantaneous impulse of 3 units.

5. Establish (10) of Sec. 13.F.

6. Verify (11) of Sec. 13.F.

REFERENCES

Bateman, H. (1954): *Bateman Manuscript Project*, McGraw-Hill, New York.

Doetsch, G. (1961): *Guide to the Applications of Laplace Transforms*, Van Nostrand, London.

Korevaar, J. (1968): *Mathematical Methods*, Academic Press, New York.

14

A SAMPLING OF NUMERICAL ANALYSIS

Perhaps no branch of analysis has received more attention or has assumed such importance in the past two decades as has the field of numerical analysis. Although this in itself constitutes sufficient grounds for including an introductory treatment of numerical methods in a text of this nature, further pedagogical justification is derived from the interesting and often intricate ways that many of the elementary theoretical results (Rolle's theorem, the mean value theorems, Taylor series expansion, etc.) are used in developing and evaluating standard numerical techniques.

In spite of the recent upsurge of interest in numerical analysis, this branch of analysis scarcely constitutes a new domain of mathematics. The Greeks were specialists in approximating "noncomputable" areas of various geometric figures with "computable" areas of polygons. We have already noted the utility of Newton's method for finding roots of equations, and, in fact, many results that serve as the basis of present-day work in numerical analysis bear the names of such eighteenth- and nineteenth-century mathematicians as Lagrange, Euler, and Gauss.

What then is numerical analysis? The term is a relatively new one, and as might be expected in such a highly active field, a clear definition has yet to emerge. In general terms, numerical analysis involves the finding of constructive methods or algorithms that may be used to obtain approximate solutions to problems whose exact solutions are known to exist but are extremely difficult or impossible to obtain. Many of the methods used in arriving at these approximations involve recursive or iterative procedures, where a basic step or operation is used repeatedly until an acceptable approximation is found. Such procedures are customarily programmed for computer use.

The "rate of convergence" of an iterative method is of considerable importance, and the study of such rates forms an integral part of numerical analysis. Error analysis also plays a major role in this context; sources of errors are manifold and include computer round-off error, errors due to truncations, and errors inherent in mathematical models of physical systems. We shall, from time to time, focus on the problem of determining error bounds; however, the reader should be aware that our treatment gives only scant indication of the complexities involved.

It should be clear that there is an extremely close interplay between numerical analysis and computer science. The use of machines to effect numerical calculations dates back to the work of an Englishman, Charles Babbage (1792–1871), who devoted an extraordinary amount of energy to the construction of a "difference engine." Although this machine was never completed, it would have been capable of performing various mathematical operations and storing information. Babbage's brilliance was duly recognized, but his eccentricities were the subject of even more scrutiny (among his more notable achievements was the invention of the locomotive cowcatcher). Lively accounts of this relatively unknown (and unknowing) predecessor of the computer age can be found in Eves (1969).

Needless to say, the progress made during the last 25 years in the development of computers and hand-held calculators is truly astonishing, if not alarming. This progress has been the basis for much of the renewed interest and consequent advances that have occurred in numerical analysis.

A. NEWTON'S METHOD AND ITS VARIATIONS

In this section we first reexamine Newton's method for locating the zeros of a function. Then we shall introduce a variant of this method called the *Regula Falsi* (false position) method, and finally we shall compare the two methods with respect to their efficiency and accuracy.

Recall that Newton's method was based on the following procedure. Given a function $f: \mathbf{R}^1 \to \mathbf{R}^1$, a point x_0 was chosen (as close to an unknown zero of f as possible) and a sequence $\{x_n\}$ in $\mathbf{R}^1$ was defined inductively by

$$x_{n+1} = x_n - \frac{f(x_n)}{f'(x_n)} \tag{1}$$

In Chapter 5 it was indicated that under reasonable conditions this sequence would converge to a zero of f, although an example was presented there that showed that an unfortunate choice of x_0 could lead to a highly divergent sequence. We shall now present three results that ensure the convergence of this sequence.

(14.A.1) *Theorem* Suppose that $f\colon \mathbf{R}^1 \to \mathbf{R}^1$ and let $\phi(x) = x - f(x)/f'(x)$. Let $x_0 \in \mathbf{R}^1$ and suppose that there is a positive number r and an $\alpha \in [0,1)$ such that

(i) f' and f'' exist on $I = [x_0 - r, x_0 + r]$ (and, hence, f' is continuous on I).

(ii) $f'(x) \neq 0$ for each $x \in I$ (and, hence, $f' > 0$ on I or $f' < 0$ on I).

(iii) $|\phi'(x)| = |f(x)f''(x))/(f'(x))^2| \leq \alpha$ for each $x \in I$.

(iv) $|f(x_0)/f'(x_0)| \leq (1 - \alpha)r$.

Then the sequence $\{x_n\}$ defined by (1) converges to a zero x^* of f in I and x^* is the unique zero of f in this interval.

Proof. We first show that the function ϕ, when restricted to I, is a contraction map (4.E.8). First note that if $x \in I$, then by the mean value theorem and (iii)

$$|\phi(x) - \phi(x_0)| \leq \alpha|x - x_0| \leq \alpha r$$

and, hence, by property (iv),

$$|\phi(x) - x_0| \leq |\phi(x) - \phi(x_0) + \phi(x_0) - x_0|$$
$$\leq \alpha r + (1-\alpha)r = r$$

therefore, ϕ maps the complete metric space I into itself. To see that ϕ is a contraction map simply observe that if $p_1, p_2 \in I$, then by the mean value theorem and (iii), we have

$$|\phi(p_1) - \phi(p_2)| \leq \alpha|p_1 - p_2|$$

Therefore, by (4.E.10), there is a point $x^* \in I$ such that $\phi(x^*) = x^*$. Thus, we have

$$x^* = x^* - \frac{f(x^*)}{f'(x^*)} \tag{2}$$

and hence, $f(x^*) = 0$. Since for each n, $\phi(x_n) = x_{n+1}$, where x_{n+1} is defined by (1), and since ϕ is a contraction map, it follows as in the proof of (4.E.10) that the sequence $\{x_n\}$ is a Cauchy sequence converging to the point x^*. That x^* is the unique zero of f is a consequence of (ii) (f is either an increasing or a decreasing function).

We note that for each n the error $|x^* - x_n|$ associated with x_n is bounded by $\alpha^n r$. This may be seen as follows. First we use induction to show that for each k, $|x_{k+1} - x_k| \leq \alpha^k(1 - \alpha)r$. By (iv) $|x_1 - x_0| = |f(x_0)/f'(x_0)| \leq (1 - \alpha)r$; to complete the induction note that if $|x_k - x_{k-1}| \leq \alpha^{k-1}(1 - \alpha)r$, then

$$|x_{k+1} - x_k| = |\phi(x_k) - \phi(x_{k-1})| \leq \alpha|x_k - x_{k-1}| \leq \alpha^k(1 - \alpha)r$$

Now we have

$$|x^* - x_n| = \lim_{m \to \infty} |x_m - x_n| \leq \lim_{m \to \infty} \sum_{j=0}^{m-n-1} |x_{n+j+1} - x_{n+j}|$$

$$\leq \lim_{m \to \infty} \sum_{j=0}^{m-n-1} \alpha^{n+j}(1 - \alpha)r$$

$$= \alpha^n(1 + \alpha + \alpha^2 + \cdots)(1 - \alpha)r = \alpha^n r$$

The reader may easily establish the following corollary of (14.A.1), which essentially shows that if the initial estimate of a zero of f is sufficiently accurate, then under suitable conditions, the convergence of the sequence defined by (1) is guaranteed.

(14.A.2) Corollary Suppose that z is a zero of f and that there is an open interval I containing z such that f'' is continuous on I and $f'(x) \neq 0$ for each $x \in I$. Show that there exists an $\varepsilon > 0$ such that if $|x_0 - z| < \varepsilon$, then the sequence determined by (1) converges to z.

Theorem (14.A.1) is perhaps of more theoretical than practical consequence since in general it may be difficult to determine the numbers α and r for which conditions (iii) and (iv) of this theorem will be satisfied. The result obtained in (14.A.2) is also of limited use in applications, since it is often not clear how close x_0 must be to z. The next result, however, can be readily applied in a number of circumstances.

(14.A.3) Theorem Suppose that $f: [a,b] \to \mathbf{R}^1$ is a continuous function that satisfies the following conditions (these conditions are frequently referred to as the *Fourier conditions*):

(i) $f(a)f(b) < 0$, i.e., $f(a)$ and $f(b)$ have different signs, and hence, f has at least one zero in the interval (a,b).

(ii) f'' is continuous on (a,b) and does not change sign in this interval, i.e., f is concave up or concave down on the interval (a,b).

(iii) $f'(x) \neq 0$ for each $x \in [a,b]$; hence, by (ii) f is either strictly increasing or strictly decreasing on $[a,b]$ and this implies that f has a unique zero in (a,b).

(iv)   If c is the endpoint of $[a,b]$ at which $|f'(x)|$ is smaller, then

$$\left|\frac{f(c)}{f'(c)}\right| \leq b - a.$$

Then if $x_0 \in (a,b)$, the sequence defined by (1) converges to the unique zero of f in the interval $[a,b]$.

Proof.   Because of the hypothesis, there are a number of cases that must be considered. We shall establish one such case, and the reader is asked in problem A.7 to make the necessary modifications to verify the remaining cases.

Suppose that $f(a) > 0$ and $f(b) < 0$ and $f''(x) \leq 0$ on the interval $[a,b]$. Observe that in this case it follows from the hypothesis that $f'(x) < 0$ for each $x \in [a,b]$, and therefore, f is a strictly decreasing function on this interval (Fig. 14.1).

Let x^* be the unique zero of f and let x_0 be an arbitrary point in (a,b). We first suppose that $x_0 > x^*$ and show using an inductive argument that the sequence $\{x_n\}$ defined by (1) is a decreasing sequence that is bounded below by x^*, i.e., we show that for each n,

(a) $x_n \geq x^*$
(b) $x_n \geq x_{n+1}$

To start the induction, we note that $f(x_0) < 0$, since x_0 lies to the right of x^*; furthermore, as we have already observed, $f'(x_0) < 0$, and, therefore, since

$$x_1 = x_0 - \frac{f(x_0)}{f'(x_0)}$$

it follows that $x_1 < x_0$. Thus we have verified (a) and (b) in the case $n = 0$.

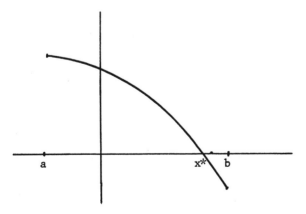

Figure 14.1

To complete the induction, suppose that (a) and (b) hold for $n = k$, i.e., $x_k \geq x^*$ and $x_k \geq x_{k+1}$; we must show that $x_{k+1} \geq x^*$ and $x_{k+1} \geq x_{k+2}$. By the mean value theorem (5.B.7) we have

$$f(x_k) - f(x^*) = f'(\hat{x})(x_k - x^*) \tag{3}$$

for some $\hat{x}$, where $x^* \leq \hat{x} \leq x_k$. Since $f(x^*) = 0$, it follows from (3) that

$$\frac{f(x_k)}{f'(\hat{x})} = x_k - x^*$$

Since $f(x_k) \leq 0$ and since $f'(x_k) \leq f'(\hat{x}) < 0$, we find that

$$x_{k+1} = x_k - \frac{f(x_k)}{f'(x_k)} \geq x_k - \frac{f(x_k)}{f'(\hat{x})} = x_k - (x_k - x^*) = x^*$$

Furthermore, since $f(x_{k+1}) \leq 0$ and $f'(x_{k+1}) < 0$, it follows that

$$x_{k+2} = x_{k+1} - \frac{f(x_{k+1})}{f'(x_{k+1})} \leq x_{k+1} \tag{4}$$

which concludes the induction argument.

Since the sequence $\{x_n\}$ defined by (1) is a decreasing sequence that is bounded below by x^*, it follows from (3.B.10.e) that this sequence converges to a point $\hat{x} \geq x^*$. An argument similar to that employed in the proof of (4.E.10) shows that $\phi(\hat{x}) = \hat{x}$, where the function ϕ is defined by $\phi(x) = x - (f(x)/f'(x))$. Consequently, $\hat{x}$ is a zero of f in I, and since x^* is the unique zero of f in I, we have $\hat{x} = x^*$; hence, the sequence $\{x_n\}$ converges to x^*.

Suppose next that $x_0 < x^*$. We shall show that $b \geq x_1 \geq x^*$, where $x_1 = x_0 - f(x_0)/f'(x_0)$; the previous argument can then be applied (with x replaced by x_1) to show that the sequence $\{x_n\}$ defined by (1) converges to x^*. By the mean value theorem we have

$$\frac{f(x^*) - f(x_0)}{x^* - x_0} = f'(\tilde{x})$$

(where $x_0 < \tilde{x} < x^*$), which implies that $f(x_0)/f'(\tilde{x}) = x_0 - x^*$. Since $f(x_0) \geq 0$ and since f' is decreasing and negative on (a,b), it follows that

$$x_1 = x_0 - \frac{f(x_0)}{f'(x_0)} > x_0 - \frac{f(x_0)}{f'(\tilde{x})} = x_0 - (x_0 - x^*) = x^*$$

It remains to check that $x_1 \leq b$. By yet another application of the mean value theorem, we have

$$f(x_0) - f(a) = f'(\bar{x})(x_0 - a)$$

where $a \leq \bar{x} \leq x_0$. Note that since $|f'(a)| \leq |f'(x)|$ on $[a,b]$, property (iv) implies that $-f(a)/f'(a) \leq b - a$. Therefore, since $f'(\bar{x}) \leq f'(a) < 0$ and

$f(x_0) - f(a) < 0$, it follows that

$$x_0 - a = \frac{f(x_0) - f(a)}{f'(\bar{x})} \le \frac{f(x_0) - f(a)}{f'(a)} \le \frac{f(x_0)}{f'(a)} + (b - a)$$

Consequently,

$$x_0 \le \frac{f(x_0)}{f'(a)} + b$$

and thus

$$x_1 = x_0 - \frac{f(x_0)}{f'(x_0)} \le \frac{f(x_0)}{f'(a)} + b - \frac{f(x_0)}{f'(x_0)} \le b$$

(14.A.4) *Example* We consider the problem of finding a real zero of $f(x) = x^7 - 1421$. It is easy to see that $2^7 < 1421 < 5^7$ and, hence, $f(2)f(5) < 0$. Furthermore, it is clear that $f'(x) > 0$ and $f''(x) > 0$ for each $x \in [2,5]$ and that

$$\left| \frac{f(2)}{f'(2)} \right| = \frac{|2^7 - 1421|}{7.2^6} = 2.8861 \cdots < 5 - 2$$

$$\left| \frac{f(5)}{f'(5)} \right| = \frac{|5^7 - 1421|}{7.5^6} = 0.1527 \cdots < 5 - 2$$

Therefore by (14.A.3) if we choose any value $x \in (2,5)$ the sequence defined by (1) will converge to the seventh root of 1421.

One variation of Newton's method, the secant method, is based on an algorithm that is obtained from equation (1) by replacing the derivative $f'(x_n)$ with the quotient $[f(x_n) - f(x_{n-1})]/(x_n - x_{n-1})$. (Some justification for this substitution stems from the definition of the derivative.) The algorithm on which the secant method is based is defined by

$$x_{n+1} = x_n - f(x_n) \frac{x_n - x_{n-1}}{f(x_n) - f(x_{n-1})} \tag{5}$$

[compare with (1)]. Two points x_0 and x_1 [with $f(x_0)f(x_1) < 0$] are used to initiate this iterative process. This algorithm has the advantage that no knowledge of the derivative of f is necessary, and thus at each step of the iterative procedure, evaluation of only one function is required rather than the two calculations needed in Newton's method.

The secant method yields fairly rapid convergence to a zero (provided that convergence occurs), although as we shall see, the convergence is not as rapid as that obtained with Newton's method. One of the serious shortcomings of the secant method is that if x_n and x_{n-1} are close together, then there are (computer) problems in obtaining significant estimates of the quotient

$[f(x_n) - f(x_{n-1})]/(x_n - x_{n-1})$. [In fact, it is at this stage that Newton's method would be more useful since the evaluation of $f'(x)$ is usually more reliable than the calculation of the corresponding quotient.] Newton's method and the secant method share one fundamental flaw: in practice there is generally no way of determining how close x_n is to x^*, since although $|x_{n+1} - x_n|$ may be quite small [or $f(x_n)$ may be quite close to 0], the distance from x_n to x^* may still be relatively large. [Nevertheless, in applications, a program based on Newton's method will usually stop when $f(x_n)$ is sufficiently close to zero or when the difference $|x_n - x_{n-1}|$ is less than some predetermined $\varepsilon > 0$.]

There is then a need for methods that yield reliable estimates of the distance between x_n and x^*. For continuous functions, the *bisection method* and the *Regula Falsi method* are particularly useful in this regard.

The idea behind the bisection method is quite simple. To locate a zero of a function f, we first find an interval $I_0 = [a_0, b_0]$ such that $f(a_0)f(b_0) \leq 0$ (this ensures that the continuous function f has a zero in I_0). Let M_0 denote the midpoint of I_0, i.e., $M_0 = (a_0 + b_0)/2$. If $f(a_0)f(M_0) \leq 0$, then we set $a_1 = a_0$, $b_1 = M_0$; otherwise, we let $a_1 = M_0$ and $b_1 = b_0$. A zero of f must now belong to $I_1 = [a_1, b_1]$. Using I_1 in place of I_0, we repeat this procedure to obtain an interval I_2 that contains a zero of f. Note that the size of the interval obtained in this fashion decreases by a factor of one half at each stage of the process. The entire procedure is brought to a halt when the interval I_n is sufficiently small. Theoretically, of course, each interval I_n must contain a zero; nevertheless, as $f(a_n)$ and $f(b_n)$ approach zero, round-off errors can become a serious problem in deciding whether $f(a_n)f(M_n)$ is less than zero (M_n is the midpoint of I_n). There are, however, computer programs for this method which take the round-off problem into account and shut down before an infinite "do loop" is initiated.

The *Regula Falsi* method is somewhat of a hybrid of the bisection method and the secant method. It is like the bisection method in that it traps a zero in each member of a decreasing sequence of intervals ($I_n = [a_n, b_n]$) [where for each n, $f(a_n)f(b_n) \leq 0$], and it is like the secant method in that in each interval I_n, it locates a point w_n defined by

$$w_n = a_n - \frac{f(a_n)(b_n - a_n)}{f(b_n) - f(a_n)} = \frac{f(b_n)a_n - f(a_n)b_n}{f(b_n) - f(a_n)} \tag{6}$$

This method works as follows. Let $I_0 = [a_0, b_0]$ be an interval such that $f(a_0)f(b_0) < 0$. Define $w_0 \in I$ by

$$w_0 = a_0 - \frac{f(a_0)(b_0 - a_0)}{f(b_0) - f(a_0)}$$

Note that w_0 is the x intercept of the line joining the points $(a_0, f(a_0))$ and $(b_0, f(b_0))$. If $f(a_0)f(w_0) \leq 0$, then we set $a_1 = a_0$ and $b_1 = w_0$. Otherwise, we set $a_1 = w_0$ and $b_1 = b_0$. (This procedure ensures that a zero of f lies in the interval $I_1 = [a_1,b_1]$.) We let w_1 be the point lying between a_1 and b_1 that is defined by

$$w_1 = a_1 - \frac{f(a_1)(b_1 - a_1)}{f(b_1) - f(a_1)}$$

this process is then repeated to obtain sequences $\{a_n\}$, $\{b_n\}$, $\{w_n\}$, where for each n, $a_n \le w_n \le b_n$ and w_n is defined by (6). In practice this program customarily stops when $|f(w_n)|$ is sufficiently small. We shall show below that if f is continuous on I_0, then the sequence $\{w_n\}$ must converge to a zero even though the length of the interval $I_n = [a_n,b_n]$ may not tend to 0.

Although round-off errors are still a problem, the *Regula Falsi* method is superior to the bisection method in that a more rapid rate of convergence is obtained; furthermore, in computer runs the *Regula Falsi* method provides a greater likelihood that the intervals I_n will contain a zero of f than does the secant method.

We show now that if f is continuous on the starting interval I_0 and if the *Regula Falsi* method is employed, then either the sequence of left-hand endpoints $\{a_n\}$ or the sequence of right-hand endpoints $\{b_n\}$ of the intervals I_n will converge to a zero. Clearly for each n, $I_{n+1} \subset I_n$, and hence, the sequences $\{a_n\}$ and $\{b_n\}$ converge to points a and b, respectively, where $a_0 \le a_1 \le \cdots \le a \le b \le \cdots \le b_1 \le b_0$. If $a = b$, there is nothing more to prove since a must then be a zero of f [note that since $f(a_n)f(b_n) \le 0$, it follows from the intermediate value theorem that each I_n contains a zero of f]. If $a < b$, then either a or b is a zero of f—if not, $f(a)$ and $f(b)$ would be nonzero and have opposite signs, and it would follow that if $(a_n,f(a_n))$ and $(b_n,f(b_n))$ are quite close to $(a,f(a))$ and $(b,f(a))$, respectively, then the straight-line segment joining $(a_n,f(a_n))$ and $(b_n,f(b_n))$ would intersect the x-axis somewhere between a and b. However, this point of intersection is w_n, which is impossible, since w_n cannot lie between a and b (why?). Hence, either $f(a)$ or $f(b)$ must be zero as claimed.

(14.A.5) Example We use the *Regula Falsi* method to approximate a zero of the function f defined by $f(x) = x^5 + 0.75x^4 + 1.75x^3 + 2.5x^2 + 3.25x - 1$. If we set $x_0 = 0$ and $x_1 = 1$, then the following values are obtained:

$x_2 = 0.1081081$	$x_5 = 0.2251051$	$x_8 = 0.2457230$
$x_3 = 0.1701789$	$x_6 = 0.2361474$	$x_9 = 0.2476250$
$x_4 = 0.2053478$	$x_7 = 0.2423003$	$x_{10} = 0.2486815$

It is readily verified that $x^* = 0.2500000$ is a zero of the given function.

We now briefly consider the speed of convergence of Newton's method and of the *Regula Falsi* method. Suppose that z is a zero of f and that x_{n+1} is defined by (1). By (8.D.7), the Taylor expansion about x_n (evaluated at z) yields

$$0 = f(z) = f(x_n) + f'(x_n)(z - x_n) + \tfrac{1}{2}f''(\alpha_n)(z - x_n)^2 \qquad (7)$$

where α_n lies between z and x_n. From (7) we obtain

$$z = x_n - \frac{f(x_n)}{f'(x_n)} - (z - x_n)^2 \frac{f''(\alpha_n)}{2f'(x_n)}$$

$$= x_{n+1} - (z - x_n)^2 \frac{f''(\alpha_n)}{2f'(x_n)}$$

and, therefore,

$$|z - x_{n+1}| = (z - x_n)^2 \left|\frac{f''(\alpha_n)}{2f'(x_n)}\right|$$

Thus, with suitable conditions placed on f'' and f' we see that the $(n + 1)$st term $|z - x_{n+1}|$ is roughly proportional to the square of the nth error term, and hence, if $|z - x_n|$ is small, the next error has undergone a substantial decrease. Because of the square factor, this type of convergence is often called *quadratic convergence*. The rapidity of convergence found with Newton's method is precisely due to its quadratic nature.

(14.A.6) *Example* We consider again the function $f(x) = x^5 + 0.75x^4 + 1.75x^3 + 2.5x^2 + 3.25x - 1$, given in Example (14.A.5). If $x_0 = 0$, then Newton's method yields the following values:

$x_1 = 0.3076923$	$x_4 = 0.2500000$	$x_7 = 0.2500000$
$x_2 = 0.2528453$	$x_5 = 0.2500000$	$x_8 = 0.2500000$
$x_3 = 0.2500070$	$x_6 = 0.2500000$	$x_9 = 0.2500000$
$x_{10} = 0.2500000$		

The reader should compare these values with those obtained in (14.A.4); note the extraordinary rapidity of convergence of the sequence determined by Newton's method to the true zero $x^* = 0.2500000$.

We limit our analysis of convergence obtained from the *Regula Falsi* method to the case where f'' is continuous and not equal to zero on some interval containing a zero z of f. We need the following result.

(14.A.7) *Exercise* Show that if $f: [a,b] \to \mathbf{R}^1$, $f(a) > 0$, $f(b) < 0$, and $f''(x) > 0$ [or $f''(x) < 0$] for each $x \in (a,b)$, then

    (a)   There is a unique point $c \in (a,b)$ such that $f(c) = 0$.
    (b)   The line L through $(a,f(a))$ and $(b,f(b))$ crosses the x-axis at a point between c and b [or at a point between a and c if $f''(x) < 0$].

[*Hint*: Let the line L be defined by $h(x)$ and let $g(x) = f(x) - h(x)$. Note that if $f''(x) > 0$ on $[c,d]$, then g' is strictly increasing on (a,b) and $g'(d) = 0$ for some $d \in (a,b)$. Use this information to show that $g(x) < 0$ for each $x \in (a,b)$.]

Formulate similar results for the case where $f(a) < 0$ and $f(b) > 0$.

Now observe that it follows from (14.A.7) that if f'' is continuous and nonzero on an interval I containing a zero z of f, then eventually all the terms w_n in the sequence determined by the *Regula Falsi* method will lie on just one side of the zero z; i.e., one of the endpoints of the intervals used in this procedure will eventually be fixed. If we let c denote this endpoint, then from (7) we have

$$w_{n+1} = \frac{cf(w_n) - w_n f(c)}{f(w_n) - f(c)}$$

and, hence,

$$w_{n+1} - z = \frac{(c - z)f(w_n) - f(c)(w_n - z)}{f(w_n) - f(c)} \tag{8}$$

If we expand f about z, we obtain

$$f(w_n) = f(z) + f'(\alpha_n)(w_n - z) = f'(\alpha_n)(w_n - z) \tag{9}$$

$$f(c) = f(z) + f'(\alpha_c)(c - z) = f'(\alpha_c)(c - z) \tag{10}$$

Substitution of (9) and (10) in (8) yields

$$w_{n+1} - z = \frac{(c - z)[f'(\alpha_n)(w_n - z)] - [f'(\alpha_c)(c - z)][w_n - z]}{f'(\alpha_n)(w_n - z) - f'(\alpha_c)(c - z)}$$

where α_n and α_c are between z and w_n or c, respectively.

If we assume that $f(c) \neq 0$ (and, hence, $z \neq c$) and that $f'(\alpha_c) \neq 0$, then it follows that

$$w_{n+1} - z = \frac{(w_n - z)(f'(\alpha_n) - f'(\alpha_c))}{f'(\alpha_n)[(w_n - z)/(c - z)] - f'(\alpha_c)}$$

$$= (w_n - z)\frac{f'(\alpha_n) - f'(\alpha_c)}{f'(\alpha_c)} \cdot \frac{f'(\alpha_c)}{f'(\alpha_n)[(w_n - z)/(c - z)] - f'(\alpha_c)}$$

If f'' is continuous at z and if n is suitably large, then the term just obtained is approximately equal to

$$- (w_n - z)\frac{f'(z) - f'(\alpha_c)}{f'(\alpha_c)}$$

Thus, for large values of n, the $(n + 1)$st error term is essentially equal to the nth error term multiplied by a constant, which implies that the errors tend to decrease linearly. This is a substantially slower rate of convergence than that found in Newton's method.

B. NEWTON'S METHOD FOR TWO VARIABLES

We now consider the problem of finding values of x and y that satisfy simultaneously the equations

$$f(x, y) = 0 \qquad g(x, y) = 0 \tag{1}$$

where f and g are suitably differentiable functions of two variables. To do this we utilize a method patterned somewhat after Newton's method (and called *Newton's method for two variables*). Since the derivation of Newton's method was based in part on the Taylor series expansion of one variable, it is not unreasonable to work with a useful analogue, the Taylor series expansion for functions of more than one variable. The two dimensional version of such an expansion is given in the next theorem. To simplify matters we adopt the following notation: if A is a subset of $\mathbf{R}^1$ and if $f: A \to \mathbf{R}^1$, then $D_{ij}f$ will denote the function obtained by taking the partial derivative of f i times with respect to x and j times with respect to y. In the theorem that follows the order of taking partials is immaterial, and hence, the expression $D_{ij}f$ is unambiguous [see (7.A.11)].

(14.B.1) *Theorem* Suppose that $U \subset \mathbf{R}^2$ is an open disk, $f: U \to \mathbf{R}^1$, and that $D_{ij}f$ exists and is continuous on U for all $i, j, 1 \leq i, j \leq n + 1$. Let (x^*, y^*) be an arbitrary point in U. Then for each $(x,y) \in U$

$$\begin{aligned}
f(x, y) = {} & f(x^*, y^*) + D_{10}f(x^*, y^*)(x - x^*) + D_{01}f(x^*, y^*)(y - y^*) \\
& + D_{20}f(x^*, y^*)\frac{(x - x^*)^2}{2!} + D_{11}f(x^*, y^*)(x - x^*)(y - y^*) \\
& + D_{02}f(x^*, y^*)\frac{(y - y^*)^2}{2!} + \cdots + D_{i0}f(x^*, y^*)\frac{(x - x^*)^i}{i!} \\
& + \cdots + D_{i-j\,j}f(x^*, y^*)\frac{(x - x^*)^{i-j}(y - y^*)^j}{(i - j)!\,j!} + \cdots \\
& + D_{0i}f(x^*, y^*)\frac{(y - y^*)^i}{i!} + \cdots + D_{n0}f(x^*, y^*)\frac{(x - x^*)^n}{n!} \\
& + \cdots + D_{0n}f(x^*, y^*)\frac{(y - y^*)^n}{n!} + R_n(x, y)
\end{aligned} \tag{2}$$

where

$$R_n(x, y) = D_{n+1\ 0}f(\hat{x},\hat{y}) \frac{(x - x^*)^{n+1}}{(n + 1)!} + \cdots$$

$$+ D_{n+1-j\ j}f(\hat{x}, \hat{y}) \frac{(x - x^*)^{n+i-j}(y - y^*)^j}{(n + 1 - j)!\,j!} + \cdots$$

$$+ D_{0\ n+1}f(\hat{x}, \hat{y}) \frac{(y - y^*)^{n+1}}{(n + 1)!}$$

and the point $(\hat{x},\hat{y})$ lies on the line segment connecting (x,y) and (x^*,y^*).

Proof. To prove the theorem, we define an appropriate function of one variable and then apply to it an appropriate combination of the one-dimensional version of Taylor's Theorem (8.D.7) and the chain rule (7.E.4). Note that if we write (x,y) as $(x^* + h, y^* + k)$, then the line passing through (x,y) and (x^*,y^*) can be described by $\{(x^* + th, y^* + tk) \mid t \in \mathbf{R}^1\}$. Furthermore, since U is open, it follows that if h and k are suitably small, then there is a number $\varepsilon > 0$ such that for each $t \in [-\varepsilon, 1 + \varepsilon]$, the point $(x^* + th, y^* + tk)$ lies in U. Let $\phi : [-\varepsilon, 1 + \varepsilon] \to \mathbf{R}^1$ be defined by

$$\phi(t) = f(x^* + th, y^* + tk) \tag{3}$$

Then by (8.D.7), if we expand ϕ about 0, we have

$$\phi(t) = \phi(0) + \phi'(0)\,t + \phi''(0)\frac{t^2}{2!} + \cdots + \phi^{(n)}(0)\frac{t^n}{n!}$$

$$+ \phi^{(n+1)}(\hat{t})\frac{t^{n+1}}{(n + 1)!} \tag{4}$$

where $\hat{t}$ lies between 0 and t. To evaluate $\phi^{(i)}(0)$ for $1 \le i \le n$ and $\phi^{(n+1)}(\hat{t})$, we repeatedly apply the chain rule (7.E.4) to (3) to obtain

$$\phi'(t) = D_{10}f(x^* + th, y^* + tk)h + D_{01}f(x^* + th, y^* + tk)k$$
$$\phi''(t) = D_{20}f(x^* + th, y^* + tk)h^2 + 2D_{11}f(x^* + th, y^* + tk)hk$$
$$\qquad + D_{02}f(x^* + th, y^* + tk)k^2$$
$$\phi'''(t) = D_{30}f(x^* + th, y^* + tk)h^3 + 3D_{21}f(x^* + th, y^* + tk)h^2k$$
$$\qquad + 3D_{12}f(x^* + th, y^* + tk)hk^2 + D_{03}f(x^* + th, y^* + tk)k^3$$

and, in general, for $1 \le i \le n + 1$,

$$\phi^{(i)}(t) = D_{i0}f(x^* + th, y^* + tk)h^i$$
$$\qquad + iD_{i-1\ 1}f(x^* + th, y^* + tk)h^{i-1}k + \cdots$$
$$\qquad + \frac{i}{(i - j)!\,j!} D_{i-j\ j}f(x^* + th, y^* + tk)h^{i-j}k^j + \cdots$$
$$\qquad + D_{0i}f(x^* + th, y^* + tk)k^i.$$

Since $h = (x - x^*)$, $k = (y - y^*)$, $\phi(0) = f(x^*,y^*)$, and for $1 \leq i \leq n$

$$\phi^{(i)}(0) = D_{i0}f(x^*, y^*)\,\frac{(x - x^*)^i}{i!} + \cdots$$

$$+ D_{i-j\,j}f(x^*, y^*)\,\frac{(x - x^*)^{i-j}(y - y^*)^j}{(i - j)!\,j!} + \cdots$$

$$+ D_{0i}f(x^*, y^*)\,\frac{(y - y^*)^i}{i!}$$

if we let $(\hat{x},\hat{y}) = (x^* + \hat{i}h,\, y^* + \hat{i}k)$, then the result follows from (3), (4), and the observation that $f(\hat{x},\hat{y}) = \phi(1)$.

Before turning to Newton's method for two variables, we consider in some detail a fixed-point problem for functions of two variables. Given functions $F(x,y)$ and $G(x,y)$, we investigate under what conditions there is a point $\mathbf{w}^* = (x^*,y^*)$ such that $x^* = F(x^*,y^*)$ and $y^* = G(x^*,y^*)$. If we use the notation

$$H(\mathbf{w}) = \begin{pmatrix} F(\mathbf{w}) \\ G(\mathbf{w}) \end{pmatrix} = \begin{pmatrix} F(x, y) \\ G(x, y) \end{pmatrix}$$

[here and elsewhere we shall abuse the notation slightly, but harmlessly, by letting $\mathbf{w}$ represent both $\binom{x}{y}$ and (x,y)], then the problem is to find a point $\mathbf{w}^*$ such that $H(\mathbf{w}^*) = \mathbf{w}^*$, i.e., a fixed point of H.

One method for determining a fixed point of H is based on constructing a sequence $\{\mathbf{w}_n\}$ of points defined by the algorithm

$$\mathbf{w}_{n+1} = H(\mathbf{w}_n) \qquad n = 0, 1, 2, \ldots \tag{5}$$

where $\mathbf{w}_0$ is an arbitrary point and then showing that under suitable conditions, this sequence will converge to the desired fixed point. Sufficient conditions for this to occur are given in the next theorem.

(14.B.2) Theorem Suppose that $\Omega = \{(x,y) \mid a \leq x \leq b, c \leq y \leq d\}$ is a closed rectangular region in $\mathbf{R}^2$ and that $F: \Omega \to \mathbf{R}^1$ and $G: \Omega \to \mathbf{R}^1$ are a pair of continuous functions defined on Ω such that the function H defined by

$$H(\mathbf{w}) = \begin{pmatrix} F(x, y) \\ G(x, y) \end{pmatrix} \qquad \mathbf{w} = (x, y)$$

maps Ω into Ω. Let a norm, $\|\cdot\|$, on Ω be defined by $\|\mathbf{w}\| = \sqrt{x^2 + y^2}$ if $\mathbf{w} = (x,y)$, and suppose that there is a constant L, $0 < L < 1$, such that if $(\mathbf{w}_1,\mathbf{w}_2) \in \Omega \times \Omega$, then

$$\|H(\mathbf{w}_2) - H(\mathbf{w}_1)\| \leq L\|\mathbf{w}_2 - \mathbf{w}_1\|. \tag{6}$$

Then there is unique point $\mathbf{w}^* = (x^*,y^*) \in \Omega$ such that $\mathbf{w}^* = H(\mathbf{w}^*)$. Moreover, if $\mathbf{w}_0$ is an arbitrary point in Ω, then $\mathbf{w}^*$ is the limit of the sequence $\{\mathbf{w}_n\}$ defined by (5), and for each n, $\|\mathbf{w}_n - \mathbf{w}^*\| \leq (L^n/(1 - L))\|\mathbf{w}_1 - \mathbf{w}_0\|$.

Proof. It is an easy consequence of (6) that if d is the metric induced by the norm $\|\cdot\|$ [that is, $d(\mathbf{w}_1,\mathbf{w}_2) = \|\mathbf{w}_1 - \mathbf{w}_2\|$], then the metric space (Ω,d) is complete and H is a contraction map on this space. The reader may finish the proof by mimicking step by step the proof of Theorem (4.E.10).

Inequality (6) is referred to as a *Lipschitz condition* on the function H over Ω, and the constant L is called a *Lipschitz constant* (such a constant need not be less than 1). In the next theorem, conditions are given for the existence of such a constant. The following two inequalities will be used in the proof:

$$(x_1 x_2 + y_1 y_2)^2 \leq (x_1^2 + y_1^2)(x_2^2 + y_2^2) \tag{7}$$

$$\left(\int_a^b f(x)g(x)\, dx \right)^2 \leq \int_a^b f^2(x)\, dx \int_a^b g^2(x)\, dx \tag{8}$$

where f and g are integrable functions mapping $[a,b]$ into $\mathbf{R}^1$. Inequality (7) is the Cauchy-Schwarz inequality (2.A.21); the reader is asked to establish (8) in problem B.4.

(14.B.3) *Theorem* Let Ω be defined as in the preceding theorem and suppose that $F: \Omega \to \mathbf{R}^1$ and $G: \Omega \to \mathbf{R}^1$ have continuous first partial derivatives on Ω. Then

$$L = \max \left\{ \sqrt{F_x^2(x,y) + F_y^2(x,y) + G_x^2(x,y) + G_y^2(x, y)} \,\middle|\, (x, y) \in \Omega \right\} \tag{9}$$

is a Lipschitz constant for $H = \binom{F}{G}$ on Ω.

Proof. Let $\mathbf{w}_1 = \binom{x_1}{y_1}$ and $\mathbf{w}_2 = \binom{x_2}{y_2}$ be any two points of Ω and let $\mathbf{w}_2 - \mathbf{w}_1 = \binom{h}{k}$. Since Ω is a convex region, the line segment joining $\mathbf{w}_1$ to $\mathbf{w}_2$, given parametrically by $\mathbf{l}(t) = \mathbf{w}_1 + t(\mathbf{w}_2 - \mathbf{w}_1)$, $0 \leq t \leq 1$, is contained in Ω. Hence, the functions f and g defined by $f(t) = F(\mathbf{l}(t))$ and $g(t) = G(\mathbf{l}(t))$ are continuously differentiable functions defined on $(0,1)$, and furthermore, $F(\mathbf{w}_2) - F(\mathbf{w}_1) = \int_0^1 f'(t)\, dt$ and $G(\mathbf{w}_2) - G(\mathbf{w}_1) = \int_0^1 g'(t)\, dt$. It follows from inequalities (7) and (8) that

$$(F(\mathbf{w}_2) - F(\mathbf{w}_1))^2 = \left(\int_0^1 f'(t)\, dt \right)^2 = \left(\int_0^1 [F_x(\mathbf{l}(t))h + F_y(\mathbf{l}(t))k]\, dt \right)^2$$

$$\leq \left(\int_0^1 \sqrt{(h^2 + k^2)} \sqrt{(F_x^2(\mathbf{l}(t)) + F_y^2(\mathbf{l}(t))}\, dt \right)^2$$

$$\leq (h^2 + k^2) \int_0^1 [F_x^2(\mathbf{l}(t)) + F_y^2(\mathbf{l}(t))]\, dt \int_0^1 1^2\, dt$$

Similarly, we find that $(G(\mathbf{w}_2) - G(\mathbf{w}_1))^2 \leq (h^2 + k^2) \int_0^1 (G_x^2(\mathbf{l}(t)) + G_y^2(\mathbf{l}(t)))\, dt$. We now have that if L is defined by (9), then since $\|\mathbf{w}_2 - \mathbf{w}_1\| = \sqrt{h^2 + k^2}$,

$$\|H(\mathbf{w}_2) - H(\mathbf{w}_1)\| = \sqrt{(F(\mathbf{w}_2) - F(\mathbf{w}_1))^2 + (G(\mathbf{w}_2) - G(\mathbf{w}_1))^2}$$

$$\leq \|\mathbf{w}_2 - \mathbf{w}_1\| \sqrt{\int_0^1 [F_x^2(\mathbf{l}(t)) + F_y^2(\mathbf{l}(t)) + G_x^2(\mathbf{l}(t)) + G_y^2(\mathbf{l}(t))]\,dt}$$

$$\leq \|\mathbf{w}_2 - \mathbf{w}_1\|\,L$$

which concludes the proof.

The reader may establish the following corollary.

(14.B.4) *Corollary* If the functions F and G in Theorem (14.B.3) have continuous first-order partial derivatives that vanish simultaneously at a fixed point $\mathbf{w}^* = (x^*, y^*)$ of H, then H is a contraction map on some closed neighborhood Ω^* of $\mathbf{w}^*$.

We shall now show that if F and G have all the properties stated in (14.B.4) and if, in addition, these functions have continuous second-order partial derivatives in Ω^*, then the sequence defined by (5) converges quadratically (provided that the starting value $\mathbf{w}_c$ is sufficiently close to the point $\mathbf{w}^*$). Since $\binom{x^*}{y^*} = \binom{F(\mathbf{w}^*)}{G(\mathbf{w}^*)}$ and since

$$\begin{pmatrix} x_{n+1} \\ y_{n+1} \end{pmatrix} = \begin{pmatrix} F(x_n, y_n) \\ G(x_n, y_n) \end{pmatrix}$$

it follows from (14.B.1) that the error function

$$E(n) = \begin{pmatrix} x_n - x^* \\ y_n - y^* \end{pmatrix}$$

satisfies

$$E(n + 1) = \begin{pmatrix} F_x(\mathbf{w}^*) & F_y(\mathbf{w}^*) \\ G_x(\mathbf{w}^*) & G_y(\mathbf{w}^*) \end{pmatrix} E(n) + B_n$$

where B_n is a matrix whose norm $\|\cdot\|$, defined by

$$\|(a_{ij})\| = \max\{|a_{ij}| \mid i = 1, 2; j = 1, 2\}$$

is less than $C\|E(n)\|^2$ for some nonnegative constant C. Note that the matrix

$$\begin{pmatrix} F_x(\mathbf{w}^*) & F_y(\mathbf{w}^*) \\ G_x(\mathbf{w}^*) & G_y(\mathbf{w}^*) \end{pmatrix}$$

is the Jacobi matrix $J(\mathbf{w}^*)$ of F and G (evaluated at $\mathbf{w}^*$). Since by the hypothesis of (14.B.4), $J(\mathbf{w}^*) = (0)$, it follows that under the above conditions, quadratic convergence is obtained.

To motivate the idea underlying Newton's method for two variables, we briefly reconsider the basic idea behind Newton's method for a single variable. Suppose that f is a differentiable function and x^* is a zero of f that we wish to determine. If $f(x_0) \neq 0$, then an increment $\Delta x_0 = x^* - x_0$ is required so that $f(x_0 + \Delta x_0) = 0$. By Taylor's Theorem, $0 = f(x_0 + \Delta x_0) =$

$f(x_0) + f'(x_0)\,\Delta x_0 + R(\hat{x}_0)$. If we ignore the remainder term $R(\hat{x}_0)$, then we can compute Δx_0 by setting $f(x_0) + f'(x_0)\,\Delta x_0 = 0$. Generally, if $R(\hat{x}_0)$ is small, then $x_0 + \Delta x_0$ will better approximate the zero x^* of f than does x_0. This idea can be carried over directly to a system of two functions

$$f(x, y) = 0 \qquad g(x, y) = 0$$

as follows. Let $\mathbf{w}_0 = \begin{pmatrix} x_0 \\ y_0 \end{pmatrix}$; if $\begin{pmatrix} f(\mathbf{w}_0) \\ g(\mathbf{w}_0) \end{pmatrix} \neq \begin{pmatrix} 0 \\ 0 \end{pmatrix}$, we use the linear part of the Taylor series expansion of f and g (14.B.1) to compute increments Δx_0 and Δy_0 such that

$$0 = f(\mathbf{w}_0) + f_x(\mathbf{w}_0)\,\Delta x_0 + f_y(\mathbf{w}_0)\,\Delta y_0$$
$$0 = g(\mathbf{w}_0) + g_x(\mathbf{w}_0)\,\Delta x_0 + g_y(\mathbf{w}_0)\,\Delta y_0 \tag{10}$$

Solving for Δx_0 and Δy_0 under the assumption that $\det J(\mathbf{w}_0) = \det \begin{pmatrix} f_x(\mathbf{w}_0) & f_y(\mathbf{w}_0) \\ g_x(\mathbf{w}_0) & g_y(\mathbf{w}_0) \end{pmatrix} \neq 0$, we find

$$\Delta x_0 = \frac{\det \begin{pmatrix} -f(\mathbf{w}_0) & f_y(\mathbf{w}_0) \\ -g(\mathbf{w}_0) & g_y(\mathbf{w}_0) \end{pmatrix}}{\det J(\mathbf{w}_0)}$$

$$\Delta y_0 = \frac{\det \begin{pmatrix} f_x(\mathbf{w}_0) & -f(\mathbf{w}_0) \\ g_x(\mathbf{w}_0) & -g(\mathbf{w}_0) \end{pmatrix}}{\det J(\mathbf{w}_0)}$$

As was the case in Newton's method for a single variable, we expect that

$$\mathbf{w}_1 = \begin{pmatrix} x_0 \\ y_0 \end{pmatrix} + \begin{pmatrix} \Delta x_0 \\ \Delta y_0 \end{pmatrix}$$

will better approximate a zero $\mathbf{z}$ of f and g than does $\mathbf{w}_0$ (provided that $\mathbf{w}_0$ is suitably close to $\mathbf{z}$).

To initiate Newton's method for two variables, we choose a point $\mathbf{w}_0 = \begin{pmatrix} x_0 \\ y_0 \end{pmatrix}$ (as near as possible to a zero of f and g) and then taking (10) into account we define inductively a sequence of vectors

$$\mathbf{w}_n = \begin{pmatrix} x_n \\ y_n \end{pmatrix}$$

$$\mathbf{w}_{n+1} = \begin{pmatrix} x_n + \zeta(\mathbf{w}_n) \\ y_n + \eta(\mathbf{w}_n) \end{pmatrix}$$

where the functions $\zeta: \Omega \to \mathbf{R}^1$, $\eta: \Omega \to \mathbf{R}^1$ satisfy the matrix equation

$$J(\mathbf{w}) \begin{pmatrix} \zeta(\mathbf{w}) \\ \eta(\mathbf{w}) \end{pmatrix} = - \begin{pmatrix} f(\mathbf{w}) \\ g(\mathbf{w}) \end{pmatrix} \tag{11}$$

Define a function ϕ by

$$\phi(\mathbf{w}) = \begin{pmatrix} x + \zeta(\mathbf{w}) \\ y + \eta(\mathbf{w}) \end{pmatrix} = \begin{pmatrix} \alpha(\mathbf{w}) \\ \beta(\mathbf{w}) \end{pmatrix} \tag{12}$$

If w^* is a fixed point of ϕ, then clearly

$$\begin{pmatrix} \xi(w^*) \\ \eta(w^*) \end{pmatrix} = \begin{pmatrix} 0 \\ 0 \end{pmatrix}$$

and consequently, by (11)

$$\begin{pmatrix} f(w^*) \\ g(w^*) \end{pmatrix} = \begin{pmatrix} 0 \\ 0 \end{pmatrix}$$

which shows that w^* is a solution of (1). Our principal theorem is the following.

(14.B.5) Theorem If w^* is a zero of the functions f and g, if the second-order partial derivatives of f and g exist in a neighborhood of w^*, and if $\det J(w^*) \neq 0$, then there is a neighborhood Ω^* of w^* such that the sequence $\{w_n\}$ determined by (5) (where $w_0 \in \Omega^*$) converges to w^*.

   If, in addition, f and g have continuous second-order partial derivatives on Ω^*, then the sequence $\{w_n\}$ converges quadratically to w^*.

   For the proof of this theorem we need the following lemma.

(14.B.6) Lemma Suppose that w^* is a fixed point of the function ϕ defined by (12). Suppose further that the second-order partial derivatives of f and g exist in a neighborhood of w^* and that $\det J(w^*) \neq 0$. Then

$$\begin{pmatrix} \alpha_x(w^*) & \alpha_y(w^*) \\ \beta_x(w^*) & \beta_y(w^*) \end{pmatrix} = \begin{pmatrix} 0 & 0 \\ 0 & 0 \end{pmatrix}$$

where α and β are defined by (12).

   Proof. Since the first-order partial derivatives of f and g are continuous on Ω^* and since $\det J(w^*) \neq 0$, it follows that in some neighborhood Ω^* of w^*, $\det J(w) \neq 0$. By (11) we have

$$f_x(w)\xi(w) + f_y(w)\eta(w) = -f(w)$$
$$g_x(w)\xi(w) + g_y(w)\eta(w) = -g(w) \tag{13}$$

and, therefore, by Cramer's rule,

$$\xi(w) = \frac{\det \begin{pmatrix} -f(w) & f_y(w) \\ -g(w) & g_y(w) \end{pmatrix}}{\det J(w)} \tag{14}$$

$$\eta(w) = \frac{\det \begin{pmatrix} f_x(w) & -f(w) \\ g_x(w) & -g(w) \end{pmatrix}}{\det J(w)} \tag{15}$$

for all $w \in \Omega^*$. Since we can assume that the second-order partial derivatives of f and g exist on Ω^*, the above formulas for ξ and η show that these last two functions have first-order partial derivatives on Ω^*. Consequently, we can differentiate the formulas in (13) with respect to x to obtain

$$f_{xx}(w)\xi(w) + f_x(w)\xi_x(w) + f_{yx}(w)\eta(w) + f_y(w)\eta_x(w) = -f_x(w)$$
$$g_{xx}(w)\xi(w) + g_x(w)\xi_x(w) + g_{yx}(w)\eta(w) + g_y(w)\eta_x(w) = -g_x(w) \tag{16}$$

for each $w \in \Omega^*$. In particular, if $w = w^*$, then since $\det J(w^*) \neq 0$ and $f(w^*) = g(w^*) = 0$, it follows immediately from (14) and (15) that

$$\xi(w^*) = \eta(w^*) = 0$$

and hence, from (16) we have

$$f_x(w^*)\xi_x(w^*) + f_y(w^*)\eta_x(w^*) = -f_x(w^*)$$
$$g_x(w^*)\xi_x(w^*) + g_y(w^*)\eta_x(w^*) = -g_x(w^*)$$

Since $\det J(w^*) \neq 0$, this system of equations has a unique solution, which by inspection is seen to be $\xi_x(w^*) = -1$, $\eta_x(w^*) = 0$. In a similar way (differentiating with respect to y) we find $\xi_y(w^*) = 0$, $\eta_y(w^*) = -1$, and, consequently, in view of (12),

$$\begin{pmatrix} \alpha_x(w^*) & \alpha_y(w^*) \\ \beta_x(w^*) & \beta_y(w^*) \end{pmatrix} = \begin{pmatrix} 1 + \xi_x(w^*) & \xi_y(w^*) \\ \eta_x(w^*) & 1 + \eta_y(w^*) \end{pmatrix} = \begin{pmatrix} 0 & 0 \\ 0 & 0 \end{pmatrix}$$

Proof (*of* (14.B.5)). It follows from the preceding lemma and (14.B.4) that under the conditions of the lemma the function ϕ defined by (12) maps Ω^* into Ω^* and is a contraction map on this set. Consequently, if w_0 is an arbitrary point in Ω^*, then by (14.B.2) the sequence $\{w_n\}$ defined by $w_{n+1} = \phi(w_n)$ converges to the unique fixed point w^* of ϕ in Ω^*, and this, as indicated previously, implies that $f(w^*) = 0 = g(w^*)$.

If f and g have continuous second-order partial derivatives in Ω^*, then it follows from the discussion after (14.B.4) that the sequence $\{w_n\}$ converges quadratically to w^*.

(14.B.7) *Example* We consider the system of equations:

$$\sin x - \cos y = 0$$
$$-e^x + 2y = 0$$

The Jacobi matrix is given by

$$J(x, y) = \begin{pmatrix} \cos x & \sin y \\ -e^x & 2 \end{pmatrix}$$

If we use as an initial approximation $w_0 = (x_0, y_0) = (.5, .5)$, then we have

$$\begin{pmatrix} \cos 0.5 & \sin^{0.5} 2 \\ -e^{0.5} & 2 \end{pmatrix}\begin{pmatrix} \xi(\mathbf{w}_0) \\ \eta(\mathbf{w}_0) \end{pmatrix} = -\begin{pmatrix} \sin 0.5 - \cos^{0.5} \\ 1 - e^{0.5} \end{pmatrix}$$

$$\begin{pmatrix} 0.87758 & 0.47943 \\ -1.64872 & 2 \end{pmatrix}\begin{pmatrix} \xi(\mathbf{w}_0) \\ \eta(\mathbf{w}_0) \end{pmatrix} = -\begin{pmatrix} 0.39815 \\ 0.64872 \end{pmatrix}$$

Routine calculations show that

$$x_1 = \xi(\mathbf{w}_0) + x_0 = 0.06501$$

$$y_1 = \eta(\mathbf{w}_0) + y_0 = -0.18295$$

The values $\xi(\mathbf{w}_1)$ and $\eta(\mathbf{w}_1)$ are found by solving

$$\begin{pmatrix} \cos 0.06501 & \sin 0.06501 \\ -e^{0.06501} & 2 \end{pmatrix}\begin{pmatrix} \xi(\mathbf{w}_1) \\ \eta(\mathbf{w}_1) \end{pmatrix} = -\begin{pmatrix} \sin 0.06501 - \cos 0.06501 \\ 1 - e^{0.06501} \end{pmatrix}$$

and then x_2 and y_2 are determined by setting $x_2 = \xi(\mathbf{w}_1) + x_1$ and $y_2 = \eta(\mathbf{w}_1) + y_1$, etc.

The reader might note that the equations [compare with (10)]

$$z = (x - x^*)f_x(x_0, y_0) + (y - y^*)f_y(x_0, y_0) + f(x_0, y_0)$$

$$z = (x - x^*)g_x(x_0, y_0) + (y - y^*)g_y(x_0, y_0) + g(x_0, y_0)$$

represent planes tangent to the graphs of $f(x,y)$ and $g(x,y)$ at the points $(x_0, y_0, f(x_0, y_0))$ and $(x_0, y_0, g(x_0, y_0))$, respectively. The point (x_1, y_1) is simply the intersection of these two planes with the plane $z = 0$. Thus, as mentioned in Chapter 7, tangent planes have replaced the tangent lines which were used in determining the points x_n in Newton's method for one variable.

C. NUMERICAL INTEGRATION AND DIFFERENTIATION

Suppose we wish to manufacture a wire whose shape conforms to the graph of the function $y = \sin 2x$, $0 \le x \le 10$ (Fig. 14.2). For this purpose we shall use a straight piece of wire and bend it into the desired shape. To find the length of this straight piece, it suffices by (12) Sec. 14.D, Chapter 6, to evaluate

$$\int_0^{10} \sqrt{1 + 4\cos^2 2x}\, dx$$

however, the value of this integral cannot be calculated using standard integration techniques. This situation is typical of many integrals that arise in applications, and hence, much reliance must be placed on numerical methods for approximating the value of such integrals.

Many techniques for the approximation of integrals have been developed,

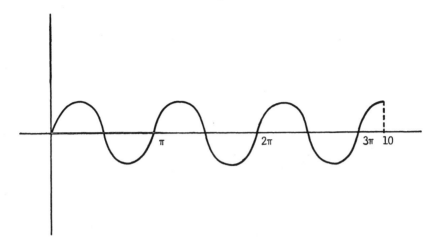

Figure 14.2

a substantial number of which are based on first approximating a function with a polynomial and then integrating the polynomial. Interestingly, this can be accomplished without actually computing the coefficients of the particular polynomial used. We have already examined two classes of polynomials that are of some use in numerical integration: the Taylor polynomials and the Bernstein polynomials. It turns out, however, that both of these polynomials have rather serious drawbacks; the Bernstein polynomials are computationally cumbersome, and the Taylor polynomials yield good approximations generally only over small intervals (and their calculation may also be excessively difficult). We begin this section with a brief study of another class of polynomials, the Lagrange polynomials, which are much better adapted to computer use.

For a function f that does not oscillate violently, it is geometrically plausible that polynomials passing through a prescribed number of points of the graph of f should yield reasonably good approximations of f. In fact, with some luck, integration errors (indicated in Fig. 14.3 by the shaded area) might even tend to cancel one another out.

Suppose then that $f: [a,b] \rightarrow \mathbf{R}^1$ and let $\{x_0, x_1, \ldots, x_n\}$ be a partition of $[a,b]$. We shall define a polynomial p of degree $\leq n$ that maps $[a,b]$ into $\mathbf{R}^1$ and has the property that $p(x_i) = f(x_i)$ for each i, $0 \leq i \leq n$. Furthermore, we shall show that p is the only polynomial of degree less than or equal to n with this property. The reader should note that the next theorem is actually independent of any function f—only $n + 1$ points in $\mathbf{R}^2$ with distinct x coordinates are needed.

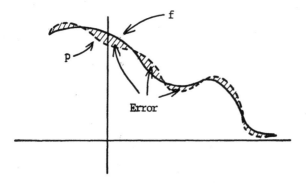

Figure 14.3

(14.C.1)   *Theorem*   If (x_0,y_0), (x_1,y_1), . . . , (x_n,y_n) are $n + 1$ points in $\mathbf{R}^2$ such that $x_i \neq x_j$ whenever $i \neq j$, then there is a unique polynomial p of degree less than or equal to n such that $p(x_i) = y_i$ for each i, $0 \le i \le n$.

Proof.   First we show that such a polynomial exists. For each i, $0 \le i \le n$, let

$$\pi_i(x) = \frac{x - x_0}{x_i - x_0} \frac{x - x_1}{x_i - x_1} \cdots \frac{x - x_{i-1}}{x_i - x_{i-1}} \frac{x - x_{i+1}}{x_i - x_{i+1}} \cdots \frac{x - x_n}{x_i - x_n}$$

$$= \prod_{\substack{k=0 \\ k \neq i}}^{n} \frac{x - x_k}{x_i - x_k}$$

Note that for each i, $0 \le i \le n$

(A) π_i is a polynomial of degree n.
(B) $\pi_i(x_i) = 1$.
(C) $\pi_i(x_j) = 0$ if $i \neq j$.

The polynomial p defined by

$$p(x) = y_0\pi_0(x) + y_1\pi_1(x) + \cdots + y_n\pi_n(x)$$

is clearly the desired polynomial. To see that p is unique, suppose that $\hat{p}$ is another polynomial of degree less than or equal to n such that $\hat{p}(x_i) = y_i$ for $0 \le i \le n$. Then $q(x) = p(x) - \hat{p}(x)$ is a polynomial of degree less than or equal to n and $q(x_i) = 0$ for each i, $0 \le i \le n$. Thus $q(x)$ is a polynomial of degree less than or equal to n with $n + 1$ distinct zeros, an impossibility (unless q is identically zero).

Polynomials of degree less than or equal to n that pass through $n + 1$ given points are called *Lagrange interpolating polynomials of order n*. We shall denote this class of polynomials by $\mathscr{L}_n$.

What error is involved when the Lagrange polynomials are used to approximate functions? It should be apparent that no general error bound exists since arbitrary (continuous) functions may have extremely high (and low) points that are not taken into account by the interpolating polynomials. However, some degree of control is gained if knowledge of the higher derivatives of a function are available.

(14.C.2) *Theorem* Suppose that $f: [a,b] \to \mathbf{R}^1$ and that $\{x_0, x_1, \ldots, x_n\}$ is a set of distinct points in $[a,b]$ such that $x_i = a$ and $x_j = b$ for some i and j. Let $p \in \mathscr{L}_n$ be the Lagrange interpolating polynomial of order n such that $p(x_i) = f(x_i)$ for $0 \le 1 \le n$. Let $q: [a,b] \to \mathbf{R}^1$ be the polynomial defined by $q(x) = (x - x_0)(x - x_1)\cdots(x - x_n)$. If $f^{(n+1)}$ exists on $[a,b]$, then for each point $x \in [a,b]$, there is a point $\alpha_x \in (a,b)$ such that

$$f(x) - p(x) = [1/(n + 1)!]q(x)f^{(n+1)}(\alpha_x) \tag{1}$$

Proof. We can assume that $x_0 < x_1 < \cdots < x_n$. Suppose that $\hat{x} \in (x_{i-1}, x_i)$ and define $\phi: [a,b] \to \mathbf{R}^1$ by

$$\phi(x) = f(x) - p(x) - \frac{q(x)}{q(\hat{x})}(f(\hat{x}) - p(\hat{x}))$$

Note that $\phi(\hat{x}) = 0$ and that $\phi(x_j) = 0$ for each j, $0 \le j \le n$. Thus, ϕ has at least $n + 2$ zeros in the interval $[a,b]$. Application of Rolle's theorem to each of the intervals $[x_0, x_1], [x_1, x_2], \ldots, [x_{i-1}, \hat{x}], [\hat{x}, x_i], [x_i, x_{i+1}], \ldots, [x_{n-1}, x_n]$ yields points $x_1^* \in (x_0, x_1)$, $x_2^* \in (x_1, x_2)$, $\ldots$, $x_{i_1}^* \in (x_{i-1}, \hat{x})$, $x_{i_2}^* \in (\hat{x}, x_i)$, $x_{i+1}^* \in (x_i, x_{i+1})$, $\ldots$, $x_n^* \in (x_{n-1}, x_n)$, each of which is a zero of ϕ'. Therefore, ϕ' has at least $n + 1$ zeros in (a,b). Rolle's theorem may now be applied to ϕ' on the intervals $[x_1^*, x_2^*], [x_2^*, x_3^*], \ldots, [x_{i_1}^*, x_{i_2}^*], \ldots, [x_{n-1}^*, x_n^*]$ to yield n zeros of ϕ'' in (a,b). Continuing in this fashion we find that $\phi^{(n+1)}$ has at least one zero $\alpha_{\hat{x}}$ in the interval (a,b).

Now observe that since p is a polynomial of degree less than or equal to n, the $(n + 1)$st derivative of p is identically 0; furthermore, from the definition of q it is clear that the $(n + 1)$st derivative of q is $(n + 1)!$. Consequently, we have

$$0 = \phi^{(n+1)}(\alpha_{\hat{x}}) = f^{(n+1)}(\alpha_{\hat{x}}) - \frac{(n + 1)!}{q(\hat{x})}(f(\hat{x}) - p(\hat{x})) \tag{2}$$

and, hence,

$$f(\hat{x}) - p(\hat{x}) = \frac{1}{(n + 1)!}q(\hat{x})f^{(n+1)}(\alpha_{\hat{x}})$$

(14.C.3) *Exercise* Show that if $q(x) = (x - x_0)(x - x_1)\cdots(x - x_n)$, then

$$q'(x) = \sum_{i=0}^{n} (x - x_0)(x - x_1)\cdots(x - x_{i-1})(x - x_{i+1})\cdots(x - x_n)$$

Suppose now that $f: [a,b] \to \mathbf{R}^1$ is an integrable function and that $\int_a^b f(x)\, dx$ is to be estimated. The simplest (although not necessarily the best) partition of $[a,b]$ is obtained by choosing subintervals each of length $(b - a)/n$. We shall henceforth denote the Lagrange interpolation polynomial corresponding to f and to this partition by L_n^f. In order to approximate $\int_a^b f(x)\, dx$ by $\int_a^b L_n^f(x)\, dx$, it is necessary to find a fairly efficient way of evaluating the latter integral. We show that $\int_a^b L_n^f(x)\, dx$ is equal to the sum

$$\sum_{i=0}^{n} \frac{b - a}{n}\, w_i f(x_i) \tag{3}$$

where the constants w_i (called *weighting factors*) are independent of f. Calculation of the weighting factors allows us to avoid the tedious computations involved in explicitly finding the coefficients of the Lagrange polynomials for large values of n. Formulas of the form (3), where the points x_i are equally spaced, are called the *Newton-Cotes formulas*. As we shall see presently, special cases of these formulas include the familiar trapezoid rule and Simpson's rule.

The weights found in (3) can be calculated as follows. By (14.C.1)

$$L_n^f(x) = \sum_{i=0}^{n} f(x_i)\pi_i(x)$$

where

$$\pi_i(x) = \prod_{\substack{k=0 \\ k \neq i}}^{n} \frac{x - x_k}{x_i - x_k}$$

and hence

$$\int_a^b L_n^f(x)\, dx = \sum_{i=0}^{n} f(x_i) \int_a^b \pi_i(x)\, dx$$

Next, we simplify the expression $\int_a^b \pi_i(x)\, dx$. First note that if $h = (b - a)/n$ and t is arbitrary, then

$$\pi_i(a + th) = \prod_{\substack{k=0 \\ k \neq i}}^{n} \frac{a + th - (a + kh)}{a + ih - (a + kh)} = \prod_{\substack{k=0 \\ k \neq i}}^{n} \left(\frac{t - k}{i - k}\right)$$

For each i, $0 \leq i \leq n$ define the polynomial $\tilde{\pi}_i$ by

$$\tilde{\pi}_i(t) = \prod_{\substack{k=0 \\ k \neq i}}^{n} \left(\frac{t - k}{i - k}\right)$$

Observe that if $t = 0$, then $a + th = a$, and if $t = n$, then $a + th = a + n(b - a)/n = b$, and therefore by (6.B.9) [let $x = a + th$ and note that $\pi_i(a + th) = \tilde{\pi}_i(t)$],

$$\int_a^b \pi_i(x)\,dx = h \int_0^n \tilde{\pi}_i(t)\,dt$$

Consequently, we have

$$\int_a^b L_n(x)\,dx = h \sum_{i=0}^n w_i f(x_i)$$

where

$$w_i = \int_0^n \tilde{\pi}_i(t)\,dt$$

is independent of f.

In particular, let us see what form (3) takes when $n = 1$. In this case,

$$w_0 = \int_0^1 \frac{t-1}{0-1}\,dt = \int_0^1 (1-t)\,dt = \frac{1}{2}$$

$$w_1 = \int_0^1 \frac{t-0}{1-0}\,dt = \frac{1}{2}$$

and therefore, we have

$$\int_a^b L_1^f(x)\,dx = \frac{h}{2}(f(x_1) + f(x_0)) \qquad \text{(trapezoid rule)}$$

The graph of L_1^f is the straight line connecting the two points $(x_0, f(x_0))$ and $(x_1, f(x_1))$, and $(h/2)(f(x_0) + f(x_1))$ is the area of the trapezoid T indicated in Fig. 14.4.

In the case that $n = 2$ routine calculations show that

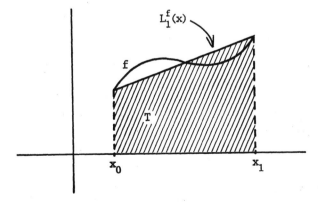

Figure 14.4

$$w_0 = \int_0^2 \frac{t-1}{0-1} \frac{t-2}{0-2} \, dt = \frac{1}{3}$$

$$w_1 = \int_0^2 \frac{t-0}{1-0} \frac{t-2}{1-2} \, dt = \frac{4}{3}$$

$$w_2 = \int_0^2 \frac{t-0}{2-0} \frac{t-1}{2-1} \, dt = \frac{1}{3}$$

Thus, we have

$$\int_a^b L_2^f(x \, dx) = \frac{h}{3}(f(x_0) + 4f(x_1) + f(x_2)) \qquad \text{(Simpson's rule).}$$

where $x_1 = (x_0 + x_2)/2$.

Error estimates in connection with the trapezoid rule and Simpson's rule may be derived as follows. If $n = 1$ (and if f'' is continuous), then it follows from (14.C.2) that

$$\int_{x_0}^{x_1} f(x) \, dx - \int_{x_0}^{x_1} L_1^f(x) \, dx = \tfrac{1}{2} \int_0^1 f''(\alpha_x)(x - x_0)(x - x_1) \, dx$$

Let $g(x) = f''(\alpha_x)$; it follows from (1) and the continuity of f'' that g is continuous except possibly at the points x_0 and x_1, and an easy argument using L'Hospital's rule [see the proof of (14.C.6)] shows that g can be extended as a continuous function on $[a,b]$. Hence, by (6.B.8) there is a number β, $x_0 \le \beta \le x_1$ such that

$$\frac{1}{2} \int_{x_0}^{x_1} g(x)(x - x_0)(x - x_1) \, dx = \frac{g(\beta)}{2} \int_{x_0}^{x_1} (x - x_0)(x - x_1) \, dx$$

$$= \frac{g(\beta)}{12}(x_0 - x_1)^3 = \frac{f''(\alpha_\beta)}{12}(x_0 - x_1)^3$$

Therefore, if f'' is continuous on $[x_0,x_1]$ and $M = \sup\{|f''(x)| \mid x \in [x_0,x_1]\}$, then the error arising from the use of the trapezoid rule is bounded by $(M/12)(x_1 - x_0)^3$.

Finding the error involved when Simpson's rule is applied is somewhat more subtle. We follow the proof given in Brand (1955).

(14.C.4) Theorem Suppose that $f: [x_0,x_2] \to \mathbf{R}^1$ is continuous and that $f^{(4)}$ exists on $[x_0,x_2]$. Let $h = (x_2 - x_0)/2$ and let $x_1 = x_0 + h$ be the midpoint of the interval $[x_0,x_2]$. Then there is a point $\alpha^* \in (x_0,x_2)$ such that

$$\int_{x_0}^{x_2} f(x) \, dx - \frac{h}{3}[f(x_0) + 4f(x_1) + f(x_2)] = \frac{-h^5}{90} f^{(4)}(\alpha^*)$$

Proof. Let F be an antiderivative of f and define a function $E: [-h, h] \to \mathbf{R}^1$ by

$$E(x) = F(x_1 + x) - F(x_1 - x) - \frac{x}{3}[f(x_1 - x) + 4f(x_1) + f(x_1 + x)]$$

Note that $E(h)$ is precisely the error resulting from the approximation of $\int_{x_0}^{x_2} f(x)\,dx$ by Simpson's rule. The reader may show directly that

$$E'(x) = \frac{2}{3}[f(x_1 + x) + f(x_1 - x) - 2f(x_1)]$$

$$- \frac{x}{3}[f'(x_1 + x) - f'(x_1 - x)] \qquad (4)$$

$$E''(x) = \frac{1}{3}[f'(x_1 + x) - f'(x_1 - x)] - \frac{x}{3}[f''(x_1 + x) + f''(x_1 - x)] \quad (5)$$

$$E'''(x) = -\frac{x}{3}[f'''(x_1 + x) - f'''(x_1 - x)] \qquad (6)$$

(see problem C.2). For no immediately apparent reason we now consider the function ϕ defined by

$$\phi(x) = E(x) - \frac{E(h)x^5}{h^5} \qquad (7)$$

Since $\phi(0) = 0 = \phi(h)$, it follows from Rolle's theorem there is a point $\alpha_1 \in (0, h)$ such that $\phi'(\alpha_1) = 0$. Since $\phi'(0) = 0 = \phi'(\alpha_1)$, it follows again by Rolle's theorem that there is a point $\alpha_2 \in (0, \alpha_1)$ such that $\phi''(\alpha_2) = 0$, and finally, a third application of Rolle's theorem yields a point $\alpha_3 \in (0, \alpha_2)$ such that $\phi'''(\alpha_3) = 0$. From (7) we have

$$E(x) = \phi(x) + \frac{E(h)x^5}{h^5}$$

and hence,

$$E'''(x) = \phi'''(x) + \frac{5 \cdot 4 \cdot 3 E(h)x^2}{h^5}$$

Thus,

$$E'''(\alpha_3) = \frac{60(\alpha_3)^2 E(h)}{h^5} \quad \text{and} \quad E(h) = \frac{E'''(\alpha_3)h^5}{60(\alpha_3)^2}$$

On the other hand, application of the mean value theorem to (7) yields

$$E'''(\alpha_3) = \frac{-\alpha_3}{3}[f'''(x_1 + \alpha_3) - f'''(x_1 - \alpha_3)] = \frac{-2\alpha_3^2}{3} f^{(4)}(x_1 + \alpha_4)$$

where $\alpha_4 \in (-\alpha_3, \alpha_3)$. Therefore, if we set $\alpha^* = x_1 + \alpha_4$, we have

$$E(h) = \frac{-2}{3}(\alpha_3)^2 f^{(4)}(\alpha^*) \frac{h^5}{60(\alpha_3)^2}$$

$$= \frac{-h^5 f^{(4)}(\alpha^*)}{90}$$

Extensive tables that list error bounds and weights for various values of n may be found in Abramowitz and Stegun (1964). We present here an abbreviated version of such a table.

n	w_0	w_1	w_2	w_3	w_4	w_5	E_n
1	$\frac{1}{2}$	$\frac{1}{2}$					$-\left(\frac{1}{12}\right)h^3 f''(x)$
2	$\frac{1}{3}$	$\frac{4}{3}$	$\frac{1}{3}$				$-\left(\frac{1}{90}\right)h^5 f^{(4)}(x)$
3	$\frac{3}{8}$	$\frac{9}{8}$	$\frac{9}{8}$	$\frac{3}{8}$			$-\left(\frac{3}{80}\right)h^5 f^{(4)}(x)$
4	$\frac{14}{45}$	$\frac{64}{45}$	$\frac{24}{45}$	$\frac{64}{45}$	$\frac{14}{45}$		$-\left(\frac{8}{845}\right)h^7 f^{(6)}(x)$
5	$\frac{95}{288}$	$\frac{375}{288}$	$\frac{250}{288}$	$\frac{250}{288}$	$\frac{250}{288}$	$\frac{375}{288}$	$\frac{-(275)}{12096}h^7 f^{(6)}(x)$
6	$\frac{5257}{17280}$	$\frac{25039}{17280}$	$\frac{9261}{17280}$	$\frac{20823}{17280}$	$\frac{20823}{17280}$	$\frac{9261}{17280}$	$\frac{-9}{1400}h^9 f^{(8)}(x)$

A few observations concerning this table are in order. First we note that for each n, the weights are independent of f, and hence, direct calculation of the coefficients of the interpolating Lagrange polynomials is unnecessary. Secondly, it should be observed that the error bounds do not form a strictly decreasing sequence as n increases (compare $n = 2$ and $n = 3$, for example), and furthermore, it is less than clear (nor is it even the case) that these bounds will necessarily converge to 0. In fact, there are examples of continuous functions for which the polynomials L_n fail to converge to f as n increases. For these and other reasons, the Newton-Cotes formulas are rarely used for values of n greater that 7 or 8.

In practice, the basic strategy often used in connection with the Newton-Cotes formulas is to first divide the given interval $[a,b]$ into a prescribed number of subintervals and then apply a lower order Newton-Cotes formula (such as the trapezoid rule or Simpson's rule) to each subinterval. The resulting formulas are called *composite formulas*. For instance, in the case of the trapezoid rule, if $\{x_0, x_1, \ldots, x_k\}$ is an equally spaced partition of $[a,b]$ and $h = (b - a)/k$, then

$$\int_a^b f(x)\, dx = \int_{x_0}^{x_1} f(x)\, dx + \int_{x_1}^{x_2} f(x)\, dx + \cdots + \int_{x_{k-1}}^{x_k} f(x)\, dx$$

An application of the trapezoid rule to each subinterval and use of the result established on p. 614 yields the composite trapezoid rule

$$\int_a^b f(x)\,dx = \frac{h}{2}(f(x_0) + f(x_1)) - \frac{h^3}{12}f''(\alpha_1) + \frac{h}{2}(f(x_1) + f(x_2))$$

$$- \frac{h^3}{12}f''(\alpha_2) + \cdots + \frac{h}{2}(f(x_{k-1}) + f(x_k)) - \frac{h^3}{12}f''(\alpha_k)$$

$$= \frac{h}{2}(f(x_0) + 2f(x_1) + \cdots + 2f(x_{k-1}) + f(x_k))$$

$$- \frac{h^3}{12}\sum_{i=1}^k f''(\alpha_i)$$

where $\alpha_i \in (x_{i-1}, x_i)$. Note that since

$$\inf\{f''(\alpha) \mid \alpha \in [a,b]\} \le \sum_{i=1}^k \frac{f''(\alpha_i)}{k} \le \sup\{f''(\alpha) \mid \alpha \in [a,b]\},$$

it follows that there is an $\alpha^* \in [a,b]$ such that $\sum_{i=1}^k f''(\alpha_i) = kf''(\alpha^*)$ (provided that f'' is continuous on $[a,b]$). Since $h^3/12 = (b-a)^3/12k^3$, we find that

$$\int_a^b f(x)\,dx = \frac{h}{2}(f(x_0) + 2f(x_1) + \cdots + 2f(x_{k-1}) + f(x_k)) - \frac{(b-a)^3}{12k^2}f''(\alpha^*)$$

where $\alpha^* \in [a,b]$. It follows that the error term

$$E(k) = -\frac{(b-a)^3}{12k^2}f''(\alpha^*) \tag{8}$$

is essentially proportional to $1/k^2$.

(14.C.5) Exercise Do a similar analysis for Simpson's rule to show that if n is even then

$$\int_v^b f(x)\,dx = \frac{h}{3}(f(x_0) + 4f(x_1) + 2f(x_2) + 4f(x_3) + \cdots$$

$$+ 4f(x_{k-1}) + f(x_k)) - \frac{(b-a)^5}{90k^4}f^{(4)}(\alpha^*)$$

where $a \le \alpha^* \le b$.

For many purposes the accuracy obtained by using partitions of relatively small mesh will suffice. However, for problems involving such phenomena as long term satellite trajectories, an extraordinarily high degree of precision is required. Rather than utilizing indefinite increases of k to obtain such precision, it is frequently more efficient to calculate certain values of k and then use these values to find a still better estimate. Suppose for instance that the trapezoid rule is used and that calculations have been made for $k = k_1$ and $k = k_2$. Denote the corresponding trapezoid sums by S_1 and S_2 and the error terms by E_1 and E_2. Thus, we have

$$\int_a^b f(x)\,dx = S_1 + E_1 = S_2 + E_2$$

We wish to approximate E_2 in terms of E_1. We have from (8)

$$\frac{E_2}{E_1} = \frac{[(b-a)^3/12k_2^2]f''(\alpha_2)}{[(b-a)^3/12k_1^2]f''(\alpha_1)}$$

and, therefore, if (and this is a big if), $f''(\alpha_1)$ and $f''(\alpha_2)$ are approximately equal, it will follow that $E_2/E_1 \cong (k_1/k_2)^2$ and $E_2 \cong (k_1/k_2)^2 E_1$. Therefore, for $k_2 = 2k_1$, we have $E_2 \cong \tfrac{1}{4}E_1$, which implies that

$$\int_a^b f(x)\,dx = \tfrac{4}{3}\int_a^b f(x)\,dx - \tfrac{1}{3}\int_a^b f(x)\,dx = (\tfrac{4}{3}S_2 + \tfrac{4}{3}E_2) - (\tfrac{1}{3}S_1 + \tfrac{1}{3}E_1)$$

$$\cong \tfrac{4}{3}S_2 - \tfrac{1}{3}S_1$$

since $\tfrac{4}{3}E_2 - \tfrac{1}{3}E_1 = 0$. This procedure is known as the *Richardson extrapolation technique*. To avoid the inherent supposition that f'' is essentially constant over $[a,b]$, a refinement of Richardson's method—the Romberg integration—is often employed. A good discussion of this procedure may be found in Burden, Faires and Reynolds (1978) or in more detail in Ralston (1965).

Numerical differentiation proceeds somewhat along the lines of numerical integration: a Lagrange polynomial is used to approximate a given function f and the derivative of this polynomial is then used as an approximation of the derivative of f. If we are interested in the derivative of f at a particular point x_1, it is frequently convenient to let that point be a member of the partition used to define the interpolating Lagrange polynomial.

As usual, once we have decided which function is to be used, the question of an approximation error bound must then be confronted. In the next theorem we use (14.C.2) to obtain one such error estimate (which, however, has the disadvantage of requiring some knowledge of the higher derivatives of f).

(14.C.6) *Theorem* Suppose that $f: [a,b] \to \mathbf{R}^1$ and that $f^{(n+1)}$ exists and is continuous on $[a,b]$. Let $\{x_0, x_1, \ldots, x_n\}$ be a set of distinct points in $[a,b]$ such that $x_i = a$ and $x_j = b$ for some i and j, and let p be the interpolating polynomial of degree less than or equal to n corresponding to f and this set of points. Then for each i, $0 \le i \le n$,

$$f'(x_i) - p'(x_i) = \frac{1}{(n+1)!}\,f^{(n+1)}(\hat{\alpha}_i)\prod_{\substack{k=0 \\ k \ne i}}^{n}(x_i - x_k)$$

for some point $\hat{\alpha}_i \in [a,b]$.

Proof. First we consider equation (1) in somewhat more detail. Observe that it follows from (1) that if $x \in [a,b]$, and $x \neq x_i$ for $i = 0, 1, \ldots, n$, then

$$f^{(n+1)}(\alpha_x) = \frac{(n + 1)!(f(x) - p(x))}{q(x)} \qquad (9)$$

Furthermore, by L'Hospital's rule we have that for each i,

$$\lim_{x \to x_i} f^{(n+1)}(\alpha_x) = (n + 1)! \frac{f'(x_i) - p'(x_i)}{q'(x_i)}$$

Since $q'(x_i) \neq 0$ (why?), the function $\phi : [a,b] \to \mathbf{R}^1$ defined by

$$\phi(x) = \begin{cases} f^{(n+1)}(\alpha_x) & \text{if } x \neq x_i,\ i = 0, 1, \ldots, n \\ (n + 1)! \dfrac{f'(x_i) - p'(x_i)}{q'(x_i)} & \text{if } x = x_i,\ i = 0, 1, \ldots, n \end{cases} \qquad (10)$$

is continuous on $[a,b]$.

By (1) and the continuity of f, p, and ϕ, we find

$$f(x) - p(x) = \frac{1}{(n + 1)!} q(x) f^{(n+1)}(\alpha_x) = \frac{1}{(n + 1)!} q(x)\phi(x)$$

for *all* $x \in [a,b]$. Since $q(x) \neq 0$ on the set $D = [a,b] \backslash \{x_0, \ldots, x_n\}$ it follows that the right-hand side of (9) is differentiable on D. Therefore, ϕ is differentiable on D, and, hence, if $x \in D$,

$$f'(x) - p'(x) = \frac{1}{(n + 1)!} (q'(x)\phi(x) + q(x)\phi'(x)) \qquad (11)$$

which implies that

$$(n + 1)!(f'(x) - p'(x)) - q'(x)\phi(x) = q(x)\phi'(x)$$

Next we observe that it follows from the previous equation and the continuity of ϕ that $\lim_{x \to x_i} q(x)\phi'(x) = 0$. Thus, by (11) we have

$$\frac{q'(x_i)\phi(x_i)}{(n + 1)!} = \lim_{x \to x_i} \frac{q'(x)\phi(x)}{(n + 1)!} = f'(x_i) - p'(x_i) \qquad (12)$$

To complete the proof we use the continuity of $f^{(n+1)}$ to find a point $\hat{\alpha}_i$ such that $\phi(x_i) = f^{(n+1)}(\hat{\alpha}_i)$. If $m = \inf\{f^{(n+1)}(x) \mid x \in [a,b]\}$ and $M = \sup\{f^{(n+1)}(x) \mid x \in [a,b]\}$, then by the continuity of $f^{(n+1)}$ and the intermediate value theorem, we have

$$f^{(n+1)}([a, b]) = [m, M] \qquad (13)$$

Since $\phi(x_i) = \lim_{x \to x_i} f^{(n+1)}(\alpha_x)$, it follows from (13) that there is a point $\hat{\alpha}_i \in [a,b]$ such that $\phi(x_i) = f^{(n+1)}(\hat{\alpha}_i)$, and, consequently, we have from (12) and (14.C.3) that

$$f'(x_i) - p'(x_i) = \frac{1}{(n+1)!} q'(x_i)\phi(x_i)$$

$$= \frac{1}{(n+1)!} \prod_{\substack{k=0 \\ k \neq i}}^{n} (x_i - x_k) f^{(n+1)}(\hat{a}_i)$$

which completes the proof.

(14.C.7) **Corollary** If f satisfies the properties given in (14.C.6) and if p_n is the interpolation polynomial for f associated with the set of distinct points $\{\alpha, x_1, \ldots, x_n\}$ in $[a,b]$, then for $i = 1, 2, \ldots, n$,

$$f'(\alpha) = p_n'(\alpha) + \frac{1}{(n+1)!} f^{(n+1)}(\hat{a}) \prod_{k=1}^{n} (\alpha - x_k) \tag{14}$$

where $\hat{a} \in [a,b]$.

(14.C.8) **Examples** (a) Suppose that certain experimental results yield:

$$f(0.5) = 2.41 \qquad f(0.6) = 2.46 \qquad f(0.8) = 2.49$$

We wish to approximate $f'(0.7)$. In general, the Lagrange interpolating polynomial of order two that passes through the points $(x_0, f(x_0))$, $(x_1, f(x_1))$, and $(x_2, f(x_2))$ is given by

$$p(x) = f(x_0) \frac{(x - x_1)(x - x_2)}{(x_0 - x_1)(x_0 - x_2)} + f(x_1) \frac{(x - x_0)(x - x_2)}{(x_1 - x_0)(x_1 - x_2)}$$
$$+ f(x_2) \frac{(x - x_0)(x - x_1)}{(x_2 - x_0)(x_2 - x_1)}$$

and the derivative of p is easily seen to be defined by

$$p'(x) = f(x_0) \frac{2x - x_1 - x_2}{(x_0 - x_1)(x_0 - x_2)} + f(x_1) \frac{2x - x_0 - x_2}{(x_1 - x_0)(x_1 - x_2)}$$
$$+ f(x_2) \frac{2x - x_0 - x_1}{(x_2 - x_0)(x_2 - x_1)}$$

Direct substitution of the values $x_0 = 0.5$, $x_1 = 0.6$, $x_2 = 0.8$, $f(x_0) = 2.41$ $f(x_1) = 2.46$, and $f(x_2) = 2.49$ yields $p'(0.7) = 0.15$.

(b) We approximate $f'(1.5)$, where $f(x) = (e^x - 2x)^4$. If we set $x_0 = 1.4$, $x_1 = 1.5$, and $x_2 = 1.6$, then a straightforward calculation shows that $f(x_0) = 0.0547096$, $f(x_1) = 0.0852808$, and $f(x_2) = 0.11978593$, from which it follows that $p'(1.5) = 0.32538$. A direct calculation of the derivative of f at the point 1.5 yields $f'(1.5) = 0.33092$. There are, of course, obvious round-off errors.

Although the interpolation polynomial p defined by (14.C.1) is unique, there are a number of ways that it can be expressed. One particularly useful formulation, Newton's form, has the advantage that it is defined progressively so that as interpolating points are added, it is not necessary to rework previous calculations (which is not true for Largrange's form of the interpolating polynomial).

To derive Newton's form, suppose that corresponding to distinct points $\{x_0, x_1, \ldots, x_{j-1}\}$ the interpolating polynomial p_{j-1} has been constructed for a given function f. We want to use p_{j-1} to construct the interpolating polynomial p_j corresponding to the set of distinct points $\{x_0, x_1, \ldots, x_{j-1}, x_j\}$.

If

$$Q_j(x) = p_j(x) - p_{j-1}(x) \tag{15}$$

then Q_j is a polynomial of degree less than or equal to j such that

$$Q_j(x_i) = 0 \quad \text{for } i = 0, 1, \ldots, j-1$$

Since the points $\{x_0, \ldots, x_{j-1}\}$ are distinct, Q_j must have the form

$$Q_j(x) = a_j(x - x_0)\cdots(x - x_{j-1}) \tag{16}$$

Differentiating (15) j times [and using (16)], we find that

$$a_j = \frac{p_j^{(j)}(x)}{j!} \tag{17}$$

We can calculate a_j from (17) by first writing p_j in the Lagrange form:

$$p_j(x) = \sum_{i=0}^{j} f(x_i) \prod_{\substack{k=0 \\ k \neq i}}^{j} \frac{x - x_k}{x_i - x_k}$$

$$= \left[\sum_{i=0}^{j} f(x_i) \prod_{\substack{k=0 \\ k \neq i}}^{j} \frac{1}{x_i - x_k} \right] x^j + \text{polynomial of degree} < j \tag{18}$$

The symbol $f[x_0, \ldots, x_j]$ is frequently used to denote a_j; it is called the jth *divided difference* of f at the points $x_0, \ldots, x_j$ ($f[x_0]$ is defined to be $f(x_0)$). With this notation we have from (17) and by taking the jth derivative of (18),

$$f[x_0, \ldots, x_j] = \sum_{i=0}^{j} f(x_i) \prod_{\substack{k=0 \\ k \neq i}}^{j} \frac{1}{x_i - x_k}$$

$$p_j(x) = p_{j-1}(x) + f[x_0, \ldots, x_j] \prod_{k=0}^{j-1} (x - x_k)$$

Newton's form of the interpolation polynomial p_n (corresponding to a function f and the set of distinct points $\{x_0, x_1, \ldots, x_n\}$) is given by

$$p_n(x) = f[x_0] + f[x_0, x_1](x - x_0) + f[x_0, x_1, x_2](x - x_0)(x - x_1) + \cdots$$
$$+ f[x_0, x_1, \ldots, x_n] \prod_{k=0}^{n-1} (x - x_k)$$

(14.C.9) *Exercise* Show that $f[x_0, x_1] = (f(x_1) - f(x_0))/(x_1 - x_0)$ and

$$f[x_0, x_1, x_2] = \frac{f[x_1, x_2] - f[x_0, x_1]}{x_2 - x_0}$$

(14.C.10) *Example* For $n = 1$, $x_0 = \alpha$, and $x_1 = \alpha + h$, where $h > 0$, we have

$$p_1(x) = f[\alpha] + f[\alpha, \alpha + h](x - \alpha)$$

and

$$p_1'(x) = f[\alpha, \alpha + h] = \frac{f(\alpha + h) - f(\alpha)}{h}$$

If we assume that f'' is continuous, it follows from (14.C.7) that

$$f'(\alpha) = \frac{f(\alpha + h) - f(\alpha)}{h} - \frac{f''(\hat{\alpha})h}{2}$$

where $\hat{\alpha} \in [\alpha, \alpha + h)$ and $f''(\hat{\alpha})h/2$ is the "error" term.

If $n = 2$, $x_0 = \alpha$, $x_1 = \alpha - h$, and $x_2 = \alpha + h$ ($h > 0$), then
$$p_2(x) = f[\alpha] + f[\alpha, \alpha - h](x - \alpha) + f[\alpha, \alpha - h, \alpha + h](x - \alpha)(x - \alpha + h)$$
$$p_2'(x) = f[\alpha, \alpha - h] + f[\alpha, \alpha - h, \alpha + h](2x - 2\alpha + h)$$

and

$$p_2'(\alpha) = f[\alpha, \alpha - h] + f[\alpha, \alpha - h, \alpha + h]h$$
$$= \frac{1}{2h}[f(\alpha + h) - f(\alpha - h)]$$

and therefore, by (14.C.7)

$$f'(\alpha) = \frac{f(\alpha + h) - f(\alpha - h)}{2h} - \frac{f'''(\hat{\alpha})h^2}{6}$$

where $\hat{\alpha} \in (\alpha - h, \alpha + h)$ and $f'''(\hat{\alpha})h^2/6$ is the "error" term.

One might suppose that as h tends toward 0, the corresponding approximations $p_2'(\alpha)$ of $f'(\alpha)$ would improve. However, computer calculations do not normally show this. This phenomenon may be explained as follows.

Suppose that we wish to approximate $f'(x)$ by $p_2'(\alpha) = (f(\alpha + h) - f(\alpha - h))/2h$. The computer will most likely calculate $f(\alpha + h) + e^+$ and $f(\alpha - h) + e^-$ instead of $f(\alpha + h)$ and $f(\alpha - h)$, where e^+ and e^- denote small error terms. Thus, instead of $p_2'(\alpha)$, the value $P_2'(\alpha) = p_2'(\alpha) + (e^+ - e^-)/2h$ is obtained, which means that the computed value $P_2'(\alpha)$ differs from $f'(\alpha)$ by the amount

$$-\left(\frac{e^+ - e^-}{2h} + \frac{f'''(\hat{\alpha})h^2}{6}\right) \tag{19}$$

where $\hat{\alpha} \in (\alpha - h, \alpha + h)$.

As h tends to 0, we cannot expect that $e^+ - e^-$ will approach zero, and therefore, we see that $|(e^+ - e^-)/2h|$ may very well become infinite as $h \to 0$! What values of h should be used then? Obviously the optimal value of h is obtained when (19) takes on a minimal value. If we suppose that near α, $|f'''(x)| \cong M$ and that $|e^+ - e^-|$ is bounded by B, and if in view of (19) we set

$$R(h) = \frac{B}{2h} + \frac{M}{6}h^2 \qquad (h > 0)$$

then we find that R assumes a minimum value at $h = \sqrt[3]{3B/2M}$.

(14.C.11) *Example* If $B = 2 \times 10^{-8}$ and $M = 1$, then

$$h = \sqrt[3]{3 \times 10^{-8}} = \sqrt[3]{30} \times 10^{-3} \cong 3 \times 10^{-3}$$

which indicates that a good value for h is 0.003.

Additional and more refined results concerning numerical differentiation can be found in Burden, Faires and Reynolds (1978) and Hildebrand (1974).

D. NUMERICAL SOLUTIONS OF DIFFERENTIAL EQUATIONS

Finding approximations to solutions of differential equations is of major importance in applied mathematics. In many instances the determination of an explicit solution to a given differential equation can verge on the impossible; for example, to explicitly solve even such an innocuous appearing equation as $x^2y'' + (3x - 1)y' + y = 0$ would require considerable effort. Satisfactory resolution of such problems is frequently limited to finding adequate estimates of the true solutions.

In this section we shall consider a few of the basic strategies that are currently used to approximate solutions to first-order differential equations. It should be noted that, as in the case of most other areas of numerical analysis, it is impossible to single out the "best" line of attack; although the methods to be outlined can be readily applied to a wide class of equations, it is not difficult to find equations for which these procedures must be modified or abandoned.

We commence with a rather naive approach, due to Euler, for approximating solutions of first-order differential equations. Suppose that we are to find a solution of

$$y'(x) = f(x, y(x)) \tag{1}$$

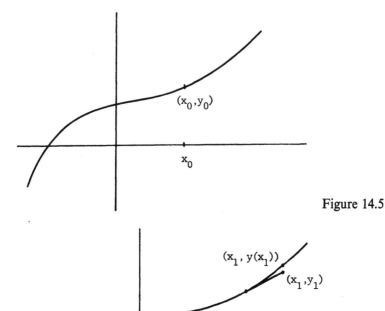

(x_0, y_0)

x_0

Figure 14.5

$(x_1, y(x_1))$

(x_1, y_1)

h

x_0 x_1

Figure 14.6

satisfying the initial condition $y(x_0) = y_0$. In the graph in Fig. 14.5 the true (but unknown) solution $y(x)$ is represented.

Euler noted that although the solution $y(x)$ of (1) is unknown, the derivative of y at the point x_0 is given: $y'(x_0) = f(x_0, y_0)$. From this observation we might expect that if $x_1 = x_0 + h$ and $y_1 = y_0 + hf(x_0, y_0)$, then the line segment L_1 [with slope $y'(x_0) = f(x_0, y_0)$] that connects the points (x_0, y_0) and (x_1, y_1) will serve as a good approximation to the graph of the solution $y(x)$ on the interval $[x_0, x_1]$ (at least for small values of h). In particular, the point (x_1, y_1) should provide a fairly good approximation of the point $(x_1, y(x_1))$ that lies on the graph of the true solution (Fig. 14.6).

To extend this approximate "solution," we next construct a line segment L_2 that emanates from the point (x_1, y_1). Ideally, the slope of this segment should be the value $y'(x_1) = f(x_1, y(x_1))$. However, since this value is most likely not known, we set the slope of L_2 equal to the *calculable* value $f(x_1, y_1)$; if (x_1, y_1) is quite close to $(x_1, y(x_1))$, then relatively little error is introduced

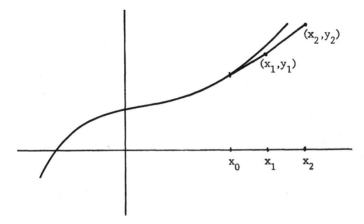

Figure 14.7

here. The terminal point of the segment L_2 is the point (x_2,y_2), where $x_2 = x_1 + h$ and $y_2 = y_1 + hf(x_1,y_1)$.

Although, in general, it cannot be expected that L_2 will yield as good an approximation to the graph of the true solution $y(x)$ as did L_1; still, for small values of h, such an approximation may be adequate. Repeating this process a finite number of times, we obtain points $(x_1,y_1), (x_2,y_2), \dots, (x_n,y_n)$, where for each k, $1 \le k \le n$, $x_k = x_0 + kh$ and $y_k = y_{k-1} + hf(x_{k-1},y_{k-1})$. The line segments that connect these points give rise to a polygonal approximation of the graph of the true solution.

(14.D.1) *Exercise* Sketch solution curves for which Euler's method would fail badly even if small values of h are used.

The Euler procedure (which for fairly obvious reasons is rarely used in practice) can be considered as an especially simple case of a method based on Taylor series expansions. If the solution y of (1) is expanded about x_0 and evaluated at $x_1 = x_0 + h$, then we have

$$y(x_1) = y(x_0) + y'(x_0)h + y''(x_0)\frac{h^2}{2!} + \cdots \tag{2}$$

Note that truncation of the right-hand side of (2) after two terms yields precisely the Euler estimate. In general, one would expect to obtain increasingly better approximations of $y(x_1)$ by retaining more terms of the expansion. Since $y'(x) = f(x,y(x))$, it follows from (7.E.4) and (7.A.11) that

$$y''(x) = f_x(x, y(x)) + f_y(x, y(x))y'(x) = f_x(x, y(x))$$
$$+ f_y(x, y(x))f(x, y(x)) \tag{3}$$

and therefore, in view of (2), it is likely that

$$y_1 = y_0 + hf(x_0, y_0) + \frac{h^2}{2!}\left(f_x(x_0, y_0) + f_y(x_0, y_0)f(x_0, y_0)\right) \qquad (4)$$

will result in a better approximation of $y(x_1)$ than does the corresponding value determined by Euler's method. In spite of this, however, expression (4) has certain disadvantages: it requires the calculation of the first partial derivatives of f and also necessitates the computation of three different functions at each step of the recursive process that would be used to find the succeeding values $y_2, y_3, \ldots$.

The method that we shall now consider avoids these problems since it enables us to replace the partial derivatives found in (4) with appropriate machine calculable values of $f(x,y)$. First we observe that (4) can be rewritten in the form

$$y_1 = y_0 + h(f(x_0, y_0) + \frac{h^2}{2!}[f_x(x_0, y_0) + f_y(x_0, y_0)f(x_0, y_0)]$$

$$= y_0 + h[f(x_0,y_0) + \frac{h}{2}\frac{df}{dx}(x_0, y_0)] \qquad (5)$$

where df/dx represents the total derivative of $f(x,y(x))$. The next result will provide us with a way of approximating the sum found in the square brackets of the latter term.

(14.D.2)   *Theorem*   Suppose that $f(x,y)$ has continuous second-order partial derivatives and that $y(x)$ is a solution of the initial value problem

$$\phi'(x) = f(x, \phi(x)); \; \phi(x_0) = y_0 \qquad a \le x \le b$$

Then $f(x,y(x)) + (h/2) \, df(x,y(x))/dx = \frac{1}{2}(f(x,y(x)) + f(x + h, y(x) + hf(x$ $y(x))) + O(h^2)$ for all x and h such that $x \in [a,b]$ and $x + h \in [a,b]$. [Here $O(h^2)$ represents a function ψ of h with the property that $|\psi(h)| \le Ch^2$ for some constant C. The constant C is independent of both x and h.]

Proof.   Let $g(x) = f(x,y(x))$. It follows readily from the hypothesis that y'' exists and is continuous on $[a,b]$, and hence, both g' and g'' are continuous on $[a,b]$ as well. Therefore, if x and $x + h$ lie in the interval $[a,b]$, we have from the Taylor expansion of g (14.B.1) that

$$g(x + h) = g(x) + g'(x)h + \frac{g''(\hat{x})}{2} h^2 \qquad (6)$$

where $\hat{x}$ lies between x and $x + h$. This implies that

$$g'(x) = \frac{g(x + h) - g(x)}{h} - \frac{g''(\hat{x})h}{2} \qquad (7)$$

and, consequently,

$$f(x, y(x)) + \frac{h}{2}\frac{df}{dx}(x, y(x)) = g(x) + \frac{h}{2}g'(x) = g(x) + \frac{1}{2}(g(x + h)$$

$$- g(x)) - \frac{g''(\hat{x})}{4}h^2 = \frac{1}{2}(f(x, y(x))$$

$$+ f(x + h, y(x + h))) + O(h^2) \qquad (8)$$

where $O(h^2) = (-g''(\hat{x})/4)h^2$. Note that the continuity of g'' on $[a,b]$ implies that there is a constant C such that $|g''(x)/4| \leq C$ for all $x \in [a,b]$.

We now consider $f(x + h, y(x + h))$. Applying the Taylor expansion to the true solution $y(x)$, we find

$$y(x + h) = y(x) + y'(x)h + \frac{y''(\tilde{x})}{2}h^2$$

where $\tilde{x}$ lies between x and $x + h$. From (6) and an application of the Taylor expansion theorem to f, we obtain

$$f(x + h, y(x + h)) = f(x + h, y(x) + hy'(x)) + f_y(x + h, y^*)\frac{y''(\tilde{x})}{2}h^2 \qquad (9)$$

where y^* lies between $y(x + h)$ and $y(x) + hy'(x)$. Observe that since $y(x)$ and $y(x) + hy'(x)$ are continuous, and hence, bounded on $[a,b]$, it follows that y^* also remains bounded for varying values of x and $x + h$; consequently, points of the form $(x + h, y^*)$ can be assumed to lie in a closed and bounded region on which f_y is continuous. Thus, we can write

$$f_y(x + h, y^*)\frac{y''(\tilde{x})}{2}h^2 = O(h^2). \qquad (10)$$

From (6), (1), and (8) we have

$$f(x, y(x)) + \frac{h}{2}\frac{df}{dx}(x, y(x)) = \frac{1}{2}(f(x, y(x)) + f(x + h, y(x) + hy'(x))) + O(h^2)$$

the desired result.

In view of (5) and (14.D.2) it is now reasonable to replace (4) by

$$y_1 = y_0 + \frac{h}{2}(f(x_0, y_0) + f(x_0 + h, y_0 + hf(x_0, y_0)))$$

Recursively, we can then obtain a finite set of points (x_0, y_0), (x_1, y_1), ..., (x_n, y_n), where for each k,

$$x_k = x_0 + kh$$

and

$$y_k = y_{k-1} + \frac{h}{2}\left(f(x_{k-1}, y_{k-1}) + f(x_k, y_{k-1} + hf(x_{k-1}, y_{k-1}))\right) \quad (11)$$

These points often yield very good approximations to the points $(x_0, y(x_0))$, $(x_1, y(x_1))$, ..., $(x_n, y(x_n))$ that lie on the graph of the true solution $y(x)$. The method just described is called a *simplified* (or *second-order*) *Runge-Kutta procedure*.

As we have remarked earlier, the primary advantage of the simplified Runge-Kutta algorithm (11) over the algorithm arising from (4) is that no partial derivatives appear in (11). Furthermore, it can be shown that the magnitude of error is of the same order for both methods; i.e., in either method there is a constant M such that (with the above notation) $|y_k - y(x_k)| \leq Mh^2$, where M is independent of k and the step size h. Thus, this method yields increasingly better approximations as h tends to 0. A method having this property is said to be *stable*.

(14.D.3) *Example* We consider the first-order differential equation

$$y' + 3xy = 2x \qquad y(0) = 1 \tag{12}$$

If we set $h = 0.1$, then we obtain the following data (note that $y = \frac{1}{3}e^{(-\frac{3}{2})x^2} + \frac{2}{3}$ is the exact solution of (12)),

x_k	Approximate solution: y_k	Exact solution: $y(x_k)$
0.1	1.000000000	1.000000000
0.2	0.995000000	0.995037313
0.3	0.980520500	0.980588178
0.4	0.957828868	0.957905304
0.5	0.928829113	0.928875954
0.6	0.895796644	0.895763093
0.7	0.861083453	0.860916084
0 8	0 826846657	0.826501820
09	0.794842695	0.794297629
1.0	0.766310711	0.765570005
1.1	0.741941142	0.741043387
1.2	0.721960617	0.720945969
1.3	0.706168664	0.705108374
1.4	0.694128456	0.684288576
1.5	0.685255552	0.684288576

There are higher order Runge-Kutta methods, the derivations of which become increasingly tedious, and are omitted here. The standard (or fourth-order) Runge-Kutta method is perhaps the most commonly used procedure of this type. This method is based on the following algorithm:

$$y_{i+1} = y_i + \frac{h}{6}(a_1 + 2a_2 + 2a_3 + a_4)$$

where

$$a_1 = f(x_i, y_i)$$

$$a_2 = f(x_i + \frac{h}{2}, y_i + \frac{1}{2}ha_1)$$

$$a_3 = f(x_i + \frac{h}{2}, y_i + \frac{1}{2}ha_2)$$

$$a_4 = f(x_i + h, y_i + ha_3)$$

It can be shown that the fourth-order Runge-Kutta method is stable and, furthermore, the error factor that results from the use of this method is bounded in absolute value by Mh^4 for a suitable constant M.

(14.D.4) Example We again consider the first-order differential equation given in (14.D.3)

$$y' + 3xy = 2x \qquad y(0) = 1$$

The fourth-order Runge-Kutta formula yields the following results:

x	Approximate Solution	Exact Solution
0.0	1.000000000	1.000000000
0.1	0.995037313	0.995037313
0.2	0.980588171	0.980588178
0.3	0.957905288	0.957905304
0.4	0.928875939	0.928875954
0.5	0.895763131	0.895763093
0.6	0.860916285	0.860916084
0.7	0.826502358	0.826501820
0.8	0.794298723	0.794297629
0.9	0.765571875	0.765570005
1.0	0.741046197	0.741043387
1.1	0.720949777	0.720945969
1.2	0.705113099	0.705108374
1.3	0.693092672	0.684288576
1.4	0.684294405	0.684288576
1.5	0.678078580	0.678072706

In applied situations the Runge-Kutta methods are often used in combination with a general class of procedures referred to as predictor-corrector methods. Basically, this works as follows. Runge-Kutta formulas are employed to calculate the values y_{i+1} which were discussed above. These values are called *predictors* and are used to find presumably still better approximations (called *correctors*), which in turn can be used as predictors to derive increasingly refined estimates. To get an idea of how this is done, we return briefly to the Euler method and the second-order Runge-Kutta algorithm. One obvious source of error in the Euler method arose from the fact that the derivative was considered constant over each interval; for instance, the slope calculated at (x_0, y_0) was assumed to hold over the entire interval $[x_0, x_1]$. However, by taking some sort of average of the derivatives at x_0 and x_1, we might improve on Euler's estimate. Thus, we set

$$y_1 = y_0 + \frac{h[f(x_0, y_0) + f(x_1, y_1)]}{2}$$

where y_1 is to approximate $y(x_1)$. The problem here, of course, is that we are using y_1 to define y_1. To avoid this difficulty, we could use the Euler method to predict y_1 (the predicted value we denote by $y_1^{[p]}$) and then define the corrected value $y_1^{[c]}$ by

$$y_1^{[c]} = y_0 + \frac{h[f(x_0, y_0) + f(x_1, y_1^{[p]})]}{2}$$

In general, if y_i were an acceptable estimate, then we can obtain a predicted value $y_{i+1}^{[p]}$ and define the "corrected" value $y_{i+1}^{[c]}$ by

$$y_{i+1}^{[c]} = y_i + \frac{h[f(x_i, y_i) + f(x_{i+1}, y_{i+1}^{[p]})]}{2} \tag{13}$$

Note that the expression $h[f(x_i, y_i) + f(x_{i+1}, y_{i+1}^{[p]})]/2$ is close to the trapezoid rule used for approximating the integral

$$\int_{x_i}^{x_{i+1}} f(x, y(x)) \, dx$$

This suggests a somewhat more general technique. First observe that if

$$y(x) = y_0 + \int_{x_0}^{x} f(t, y(t)) \, dt$$

then $y'(x) = f(x, y(x))$ and $y(x_0) = y_0$; thus, $y(x)$ is a solution of (1). However this solution, as it stands, is of no particular use since there is no direct way of calculating $\int_{x_0}^{x} f(t, y(t)) \, dt$. Nevertheless, we can employ the approximation techniques discussed in Sec. 14.C for this purpose. Suppose that corresponding to points x_{i-2}, x_{i-1}, and x_i we have found acceptable estimates y_{i-2}, y_{i-1}, and y_i, of $y(x_{i-2})$, $y(x_{i-1})$, and $y(x_i)$, respectively; i.e., we have (presumably) successfully approximated $y'(x_{i-2})$, $y'(x_{i-1})$, and $y'(x_i)$ by $z_{i-2} = f(x_{i-2}, y_{i-2})$, $z_{i-1} = f(x_{i-1}, y_{i-1})$, and $z_i = f(x_i, y_i)$, respectively. We wish to estimate $y'(x_{i+1})$ (Fig. 14.8). Note that if $y(x) = y_{i-2} + \int_{x_{i-2}}^{x} f(t, y(t)) \, dt$, then $y'(x_{i+1}) = f(x_{i+1}, y(x_{i+1}))$. To approximate the integral $\int_{x_{i-2}}^{x} f(t, y(t)) \, dt$,

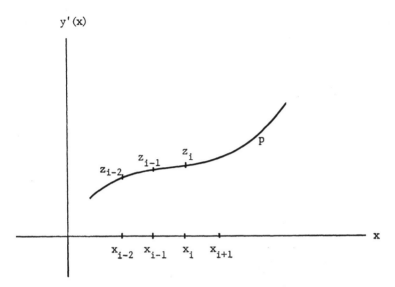

Figure 14.8

we first fit an interpolating polynomial p through the points (x_{i-2}, z_{i-2}), (x_{i-1}, z_{i-1}), and (x_i, z_i). Thus, p is used to approximate $f(t, y(t))$ on the interval $[x_{i-2}, x_{i+1}]$ and the predicted value, $y_{i+1}^{[p]}$, is defined to be

$$y_{i-2} + \int_{x_{i-2}}^{x_{i+1}} p(t)\, dt$$

The value $y_{i+1}^{[p]}$ can now be substituted in (13) to obtain a corrector value, and this latter value can itself be used as a new predictor and then be recycled through this procedure. One check on the procedure is to compute $y_{i+1}^{[p]} - y_{i+1}^{[c]}$. If small, we move on; if large, it may be necessary to use x values that are closer together and/or recycle several times. It should be noted, however, that continued recycling will not guarantee convergence to the true value $y(x_{i+1})$.

The predictor-corrector method that we have discussed is a particularly simple one and is not exceptionally accurate. There are many interesting refinements and variations of this procedure that are commonly used, including the Adams-Moulton and Adams-Bashford methods [see Ralston (1965)]. A widely used procedure due to Milne, which is quite similar to the one just studied, is described in problem D.8.

(14.D.5) Example We consider once again the first-order differential equation

$$y' = 2x - 3xy \qquad y(0) = 1$$

To initiate the predictor-corrector method we use the values

$$y_0 = 1.000000000$$
$$y_1 = 0.995037313$$
$$y_2 = 0.980588171$$

obtained in (14.D.4) from the fourth-order Runge-Kutta precedure. We find that

$$z_0 = f(x_0, y_0) = f(0, 1) = 0$$
$$z_1 = f(x_1, y_1) = f(0.1, 0.995037313) = -0.098511104$$
$$z_2 = f(x_2, y_2) = f(0.2, 0.980588171) = -0.188352903$$

A direct calculation shows that if p is the interpolating polynomial of order 2 that passes through the points (x_0,z_0), (x_1,z_1), (x_2,z_2), where $x_0 = 0$. $x_1 = h$, and $x_2 = 2h$, then

$$\int_0^{3h} p(x)\, dx = \frac{3h}{4}(z_0 + 3z_2) \tag{14}$$

and hence in the present example we have

$$y_3^{[p]} = y_2 + \int_0^{3h} p(x)\, dx = 0.980588171 + \int_0^3 p(x)\, dx = 0.938208768$$

From (13) we find the corrector value to be

$$y_3^{[c]} = y_2 + \frac{h[f(x_2,y_2) + f(x_3,y_3^{[p]})]}{2} = 0.973157721$$

This procedure can be repeated to find $y_4^{[p]}$ and $y_4^{[c]}$. In problem D.8 the reader is asked to find a general formula similar to (13) that will facilitate the calculation of y_{i+1} given the values y_{i-2}, y_{i-1}, and y_i.

One final comment is perhaps in order: it is always possible to find examples that neither a Runge-Kutta nor a Runge-Kutta-Predictor-Corrector method can adequately handle. As a natural consequence, this area of numerical analysis still serves as a focus for a good deal of ongoing active research.

E.* Fourier Series

There are many numerical problems that arise in connection with the Fourier series representations of a function: approximations of the Fourier coefficients, rates of convergence, etc. In this section we shall concentrate on just one theorem—a fairly deep result involving error bounds associated with truncated Fourier series. The proof of the principal result is long and intricate, but the interested reader is encouraged to proceed at his own pace through the details. The main theorem is the following.

(14.E.1) *Theorem* Suppose that $f: [-\pi, \pi] \to \mathbf{R}^1$ is a continuous function with a bounded piecewise continuous derivative on $(-\pi, \pi)$ and such that $f(-\pi) = f(\pi)$. Let $S_n(x)$ denote the nth partial sum of the Fourier series of f and let $E_n(x) = f(x) - S_n(x)$. If $M = \max\{|f'(x)| \mid x \in [-\pi, \pi]\}$, then for $n \geq 2$,

$$|E_n(x)| \leq MK_n$$

where

$$K_n = \frac{14 + \ln[(n + \frac{1}{2})\pi]}{n + \frac{1}{2}}$$

Proof. The notation of Sec. F, Chapter 10, will be employed throughout the proof. By (10.F.4), we have $T(f) = f$ on $[-\pi, \pi]$. Furthermore, it follows from the proof of (10.F.4) that

$$E_n(x) = f(x) - S_n(x) = T(f)(x) - S_n(x)$$

$$= \sum_{k=n+1}^{\infty} (a_k \cos kx + b_k \sin kx) \qquad (1)$$

$$= \sum_{k=n+1}^{\infty} \int_{-\pi}^{\pi} f'(\theta + x) \left(\frac{-\sin k\theta}{k\pi} \right) d\theta$$

$$= \int_{-\pi}^{\pi} f'(\theta + x) \left(\sum_{k=n+1}^{\infty} \frac{-\sin k\theta}{k\pi} \right) d\theta$$

$$= \int_{-\pi}^{\pi} f'(\theta + x) G_n(\theta) \, d\theta$$

where

$$G_n(\theta) = \frac{-1}{\pi} \sum_{k=n+1}^{\infty} \frac{\sin k\theta}{k}$$

(Recall that in Sec. F of Chapter 10 it was shown that the operations of integration and summation can be taken in either order: note where this property was used in establishing the previous equalities.)

Since f' has period 2π and is bounded, we have

$$|E_n(x)| \leq \left(\max_{|x| \leq \pi} |f'(x)| \right) \left(\int_{-\pi}^{\pi} |G_n(\theta)| \, d\theta \right)$$

and, hence, it remains to find a constant K_n such that $\int_{-\pi}^{\pi} |G_n(\theta)| \, d\theta \leq K_n$.

The estimate of K_n will proceed in two stages. We shall first develop a rather general formula useful in error estimates and then apply it to a particular function in order to find a bound for

$$\int_{-\pi}^{\pi} |G_n(\theta)| \, d\theta \tag{2}$$

Suppose that $\phi(x)$ is a real-valued continuous function defined on $[\frac{1}{2},\infty)$. We are especially interested in those functions ϕ that vary slowly as $x \to \infty$. Our first goal is to obtain a representation of

$$\sum_{k=n+1}^{\infty} e^{i\theta k}\phi(k) \qquad -\pi \le \theta \le \pi$$

in terms of the sum of an improper integral and a rapidly converging series. Thus, we essentially wish to replace a slowly converging infinite series with an improper integral whose value can be easily estimated. We begin by considering

$$\int_{-\frac{1}{2}}^{\frac{1}{2}} \phi(k + t)e^{i\theta(k+t)} \, dt$$

where k is a positive integer. Clearly, since $e^{ix} = \cos x + i \sin x$, we have

$$\int_{-\frac{1}{2}}^{\frac{1}{2}} \phi(k + t)e^{i\theta(k+t)} \, dt = e^{ik\theta} \int_{-\frac{1}{2}}^{\frac{1}{2}} \phi(k + t) \cos \theta t \, dt$$

$$+ ie^{ik\theta} \int_{-\frac{1}{2}}^{\frac{1}{2}} \phi(k + t) \sin \theta t \, dt. \tag{3}$$

We apply the mean value theorem for integrals to the integrals appearing on the right side of (3). Since $\theta \in [-\pi,\pi]$, we have $\cos \theta \ge 0$ on $[-\frac{1}{2},\frac{1}{2}]$, and, hence, (by (6.B.8)) there exists $\hat{k}$, $k - \frac{1}{2} \le \hat{k} \le k + \frac{1}{2}$, such that

$$\int_{-\frac{1}{2}}^{\frac{1}{2}} \phi(k + t) \cos \theta t \, dt = \phi(\hat{k}) \int_{-\frac{1}{2}}^{\frac{1}{2}} \cos \theta t \, dt$$

$$= \phi(\hat{k}) \frac{2}{\theta} \sin \frac{\theta}{2} \tag{4}$$

If $0 \le \theta \le \pi$, then by (6.D.8), there are constants k_1 and k_2, $k < k_1 < k + \frac{1}{2}$, $k - \frac{1}{2} < k_2 < k$ such that

$$\int_{-\frac{1}{2}}^{\frac{1}{2}} \phi(k + t) \sin \theta t \, dt = \int_{0}^{\frac{1}{2}} \phi(k + t) \sin \theta t \, dt$$

$$- \int_{-\frac{1}{2}}^{0} \phi(k + t) (-\sin \theta t) \, dt$$

$$= \phi(k_1) \int_{0}^{\frac{1}{2}} \sin \theta t \, dt - \phi(k_2) \int_{-\frac{1}{2}}^{0} (-\sin \theta t) \, dt$$

$$= \phi(k_1) \left\{ \frac{-\cos \theta t}{\theta} \bigg|_0^{\frac{1}{2}} \right\} - \phi(k_2) \left\{ \frac{\cos \theta t}{\theta} \bigg|_{-\frac{1}{2}}^0 \right\}$$

$$= \left\{ \frac{1 - \cos (\theta/2)}{\theta} \right\} (\phi(k_1) - \phi(k_2))$$

Since $\sin x$ is an odd function, it follows immediately that if $\theta \in [-\pi, 0]$, then

$$\int_{-\frac{1}{2}}^{\frac{1}{2}} \phi(k + t) \sin \theta t \, dt = -\int_{-\frac{1}{2}}^{\frac{1}{2}} \phi(k + t) \sin |\theta| t \, dt$$

$$= \frac{-(1 - \cos(|\theta|/2))}{|\theta|} (\phi(k_1) - \phi(k_2))$$

$$= \frac{1 - \cos (\theta/2)}{\theta} (\phi(k_1) - \phi(k_2))$$

Hence, for $\theta \in [-\pi, \pi]$, we have

$$\int_{-\frac{1}{2}}^{\frac{1}{2}} \phi(k + t) \sin \theta t \, dt = \frac{1 - \cos (\theta/2)}{\theta} (\phi(k_1) - \phi(k_2)) \tag{5}$$

From (3), (4), and (5) we obtain

$$\int_{-\frac{1}{2}}^{\frac{1}{2}} \phi(k + t) e^{i\theta(k+t)} \, dt = e^{i\theta k} \phi(\hat{k}) \frac{2}{\theta} \sin \frac{\theta}{2}$$

$$+ i e^{i\theta k} \frac{1 - \cos (\theta/2)}{\theta} (\phi(k_1) - \phi(k_2)) \tag{6}$$

where

$$-\tfrac{1}{2} + k < \hat{k} < k + \tfrac{1}{2}$$
$$-\tfrac{1}{2} + k < k_2 < k$$
$$k < k_1 < k + \tfrac{1}{2} \tag{7}$$

Addition and subtraction of $e^{i\theta k} \phi(k)$ to and from equation (6) yields

$$e^{i\theta k} \phi(k) = \frac{\theta/2}{\sin (\theta/2)} \int_{-\frac{1}{2}}^{\frac{1}{2}} \phi(k + t) e^{i\theta(k+t)} \, dt + e^{i\theta k} [\phi(k) - \phi(\hat{k})]$$

$$- i e^{i\theta k} \frac{1 - \cos (\theta/2)}{\theta} \frac{\theta/2}{\sin (\theta/2)} (\phi(k_1) - \phi(k_2)) \tag{8}$$

where $\theta \in [-\pi, \pi]$, k is a positive integer and $\hat{k}$, k_1, and k_2 satisfy the inequalities given in (7).

Finally, substitution for $k + t$ in (8) results in the basic formula

$$\sum_{k=n+1}^{\infty} e^{i\theta k}\, \phi(k) = \frac{\theta/2}{\sin(\theta/2)} \int_{n+\frac{1}{2}}^{\infty} \phi(t) e^{i\theta t}\, dt + \sum_{k=n+1}^{\infty} e^{i\theta k}\, (\phi(k) - \phi(\hat{k}))$$

$$- i\, \frac{1 - \cos(\theta/2)}{\theta}\, \frac{\theta/2}{\sin(\theta/2)} \sum_{k=n+1}^{\infty} e^{i\theta k}\, (\phi(k_1) - \phi(k_2))$$

$$= \frac{\theta/2}{\sin(\theta/2)} \int_{n+\frac{1}{2}}^{\infty} \phi(t) \cos \theta t\, dt$$

$$+ \sum_{k=n+1}^{\infty} (\cos \theta k)\, (\phi(k) - \phi(\hat{k}))$$

$$+ \frac{1 - \cos(\theta/2)}{\theta}\, \frac{\theta/2}{\sin(\theta/2)} \sum_{k=n+1}^{\infty} (\sin \theta k)(\phi(k) - \phi(\hat{k}))$$

$$+ i\left[\frac{\theta/2}{\sin(\theta/2)} \int_{n+\frac{1}{2}}^{\infty} \phi(t) \sin \theta t\, dt \right. \tag{9}$$

$$+ \sum_{k=n+1}^{\infty} (\sin \theta k)(\phi(k) - \phi(\hat{k}))$$

$$\left. - \frac{1 - \cos(\theta/2)}{\theta}\, \frac{\theta/2}{\sin(\theta/2)} \sum_{k=n+1}^{\infty} (\cos \theta k)(\phi(k_1) - \phi(k_2)) \right].$$

Before applying (9) we pause briefly to let the reader establish the following results.

(14.E.2) *Exercises* (a) Show that $\max_{0 \le x \le \pi/2} \{x/\sin x\} = \pi/2$. [*Hint*: Show that $x/(\sin x)$ is increasing on $(0, \pi/2)$.]

(b) Show that $|(1 - \cos(x/2))/x| < \pi/8$, if $x \in [-\pi, \pi]$. [*Hint*: Consider the Taylor expansion of $(1 - \cos(x/2))/x$ for $0 \le x \le \pi$.]

(c) Show that

$$\left| \frac{1 - \cos(x/2)}{x} \right| \left| \frac{x/2}{\sin(x/2)} \right| < 1 \qquad \text{if } x \in [-\pi, \pi]$$

(d) If $\phi(x) = -1/\pi k$, show that

$$\sum_{k=n+1}^{\infty} |\phi(k) - \phi(\hat{k})| + \sum_{k=n+1}^{\infty} |\phi(k_1) - \phi(k_2)| \le \frac{2}{\pi(n-1)}$$

where $\hat{k}$, k_1, and k_2 satisfy (7). [*Hint*: Show that each term is dominated by $1/\pi(k-1)^2$ and use an integral estimate.]

We now begin our error analysis by applying (9) to the function $\phi(x) = -1/\pi x$, $x \ne 0$. For this function we have

$$\sum_{k=n+1}^{\infty} e^{i\theta k}\, \phi(k) = \frac{-1}{\pi} \sum_{k=n+1}^{\infty} \frac{\cos \theta k}{k} - \frac{i}{\pi} \sum_{k=n+1}^{\infty} \frac{\sin \theta k}{k} \tag{10}$$

Equating imaginary parts in (10) with those in (9) [with $\phi(x) = -1/\pi x)$], we find that if $\theta \in [-\pi,\pi]$, then

$$G_n(\theta) = \frac{-1}{\pi} \sum_{k=n+1}^{\infty} \frac{\sin \theta k}{k} = \frac{-1}{\pi} \frac{\theta/2}{\sin (\theta/2)} \int_{n+\frac{1}{2}}^{\infty} \frac{\sin \theta t}{t}\, dt$$

$$+ \sum_{k=n+1}^{\infty} (\sin \theta k)\, (\phi(k) - \phi(\hat{k}))$$

$$- \frac{1 - \cos (\theta/2)}{\theta} \frac{\theta/2}{\sin (\theta/2)} \sum_{k=n+1}^{\infty} (\cos \theta k)(\phi(k_1) - \phi(k_2))$$

It now follows from (a), (c), and (d) of (14.F.2) that if $\theta \in [-\pi,\pi]$, and $n \geq 2$, then

$$|G_n(\theta)| \leq \frac{1}{2}\left| \int_{n+\frac{1}{2}}^{\infty} \frac{\sin \theta t}{t}\, dt \right| + \frac{2}{\pi(n-1)} \tag{11}$$

To calculate $\int_{n+1/2}^{\infty} (\sin \theta t)/t\, dt$ we first substitute $\xi = \theta t$ to obtain

$$\int_{n+\frac{1}{2}}^{\infty} \frac{\sin \theta t}{t}\, dt = \int_{(n+\frac{1}{2})\theta}^{\infty} \frac{\sin \xi}{\xi}\, d\xi \tag{12}$$

(here it is assumed $0 < \theta < \pi$). An integration by parts yields

$$\int_{(n+\frac{1}{2})\theta}^{\infty} \frac{\sin \xi}{\xi}\, d\xi = \left. \frac{-\cos \xi}{\xi} \right|_{(n+\frac{1}{2})\theta}^{\infty} - \int_{(n+\frac{1}{2})\theta}^{\infty} \frac{\cos \xi}{\xi^2}\, d\xi \tag{13}$$

and integrating by parts once more, we find that

$$- \int_{(n+\frac{1}{2})\theta}^{\infty} \frac{\cos \xi}{\xi^2}\, d\xi = \left. \frac{-\sin \xi}{\xi^2} \right|_{(n+\frac{1}{2})\theta}^{\infty} - 2 \int_{(n+\frac{1}{2})\theta}^{\infty} \frac{\sin \xi}{\xi^3}\, d\xi \tag{14}$$

Thus, from (12), (13), and (14) (where $0 < \theta < \pi$), we have

$$\int_{n+\frac{1}{2}}^{\infty} \frac{\sin \theta t}{t}\, dt = \frac{\cos (n + \frac{1}{2})\theta}{(n + \frac{1}{2})\theta} + Z(n,\theta) \tag{15}$$

where

$$Z(n,\theta) \leq \frac{2}{[(n + \frac{1}{2})\theta]^2} \tag{16}$$

Hence in view of (11), (15), and (16) we have that if $0 < \theta < \pi$, and if $n \geq 2$, then

$$|G_n(\theta)| \leq \frac{1}{2}\left[\frac{1}{(n + \frac{1}{2})\theta} + \frac{2}{[(n - \frac{1}{2})\theta]^2} \right] + \frac{2}{\pi(n-1)} \tag{17}$$

We now integrate $|G_n(\theta)|$. Since G_n is an odd function, it follows immediately that $|G_n|$ is even. Consequently, we have

$$\int_{-\pi}^{\pi} |G_n(\theta)|\, d\theta = 2 \int_0^{\pi} |G_n(\theta)|\, d\theta$$

$$= 2 \int_0^{1/(n+\frac{1}{2})} |G_n(\theta)|\, d\theta + 2 \int_{1/(n+\frac{1}{2})}^{\pi} |G_n(\theta)|\, d\theta. \quad (18)$$

We shall show presently that if $x \in [0, 1/(n + \frac{1}{2})]$, then $|G_n(x)| < 1$, from which it follows that

$$\int_0^{1/(n+\frac{1}{2})} |G_n(\theta)|\, d\theta < \frac{1}{n + \frac{1}{2}}. \quad (19)$$

Before establishing (19) we consider $\int_{1/(n+\frac{1}{2})}^{\pi} G_n(\theta)\, d\theta$. In view of (17), we have

$$\int_{1/(n+\frac{1}{2})}^{\pi} |G_n(\theta)|\, d\theta \le \frac{1}{2(n + \frac{1}{2})} \int_{1/(n+\frac{1}{2})}^{\pi} \frac{d\theta}{\theta}$$

$$+ \frac{1}{(n + \frac{1}{2})^2} \int_{1/(n+\frac{1}{2})}^{\pi} \frac{d\theta}{\theta^2} + \frac{2}{\pi(n - 1)} \int_{1/(n+\frac{1}{2})}^{\pi} d\theta \quad (20)$$

Furthermore, since

$$\int_{1/(n+\frac{1}{2})}^{\pi} \frac{d\theta}{\theta} = \ln (n + \tfrac{1}{2})\pi \quad (21)$$

it follows immediately from (19) and (21) that

$$\int_{1/(n+\frac{1}{2})}^{\pi} |G_n(\theta)|\, d\theta \le \frac{1}{2} \frac{\ln (n + \frac{1}{2})\pi}{n + \frac{1}{2}} + \frac{1}{(n + \frac{1}{2})^2} \left((n + \tfrac{1}{2}) - \frac{1}{\pi} \right) + \frac{2}{n - 1} \quad (22)$$

Combining (18), (19), and (22), we obtain the desired result. For $n \ge 2$,

$$\int_{-\pi}^{\pi} |G_n(\theta)|\, d\theta \le 2 \left[\frac{1}{n + \frac{1}{2}} + \frac{1}{2} \frac{\ln (n + \frac{1}{2})\pi}{(n + \frac{1}{2})} \right.$$

$$\left. + \frac{1}{n + \frac{1}{2}} \left(1 - \frac{1}{(n + \frac{1}{2})\pi} \right) + \frac{2}{n - 1} \right]$$

$$\le \frac{2}{n + \frac{1}{2}} \left(1 + \frac{1}{2} \ln \left(n + \frac{1}{2} \right)\pi + \frac{2(n + \frac{1}{2})}{n - 1} \right)$$

$$\le \frac{2}{n + \frac{1}{2}} \left(7 + \frac{1}{2} \ln \left(n + \frac{1}{2} \right)\pi \right)$$

$$= \left(\frac{1}{n + \frac{1}{2}} \right) \left(14 + \ln \left[\left(n + \frac{1}{2} \right)\pi \right] \right).$$

Thus to conclude the proof it remains to show that $G_n(x) < 1$ for each $x \in [0, 1/(n + \frac{1}{2})]$.

By (10.C.5) it follows that

$$\left|\sum_{k=n+1}^{\infty} \frac{\sin kx}{x}\right| \le \frac{\pi}{2} + \left|\sum_{k=1}^{n} \frac{\sin kx}{x}\right|$$

and from the discussion on pages 346 and 347, and, in particular, from formula (10.D.2), we find

$$\left|\sum_{k=n+1}^{\infty} \frac{\sin kx}{x}\right| \le \frac{\pi}{2} + \left|\frac{-x}{2} + \frac{1}{2}\int_0^x \frac{\sin(n+\frac{1}{2})t}{t/2} dt\right|$$

$$+ \left|\frac{1}{2}\int_0^x \left(\frac{1}{\sin(t/2)} - \frac{1}{t/2}\right)\sin\left(\left(n+\frac{1}{2}\right)t\right) dt\right|$$

If $0 < t \le 1/(n+\frac{1}{2})$, then $\sin(n+\frac{1}{2})t/(t/2) > 0$, and, therefore, on the interval $(0, 1/(n+\frac{1}{2})]$ we have

$$0 < \frac{1}{2}\int_0^x \frac{\sin(n+\frac{1}{2})t}{t/2} dt \le \frac{1}{2}\int_0^{1/(n+\frac{1}{2})} \frac{\sin(n+\frac{1}{2})t}{t/2} dt = \int_0^1 \frac{\sin u}{u} du < 1$$

hence, if $x \in [0, 1/(n+\frac{1}{2})]$, then

$$\left|\frac{-x}{2} + \frac{1}{2}\int_0^x \frac{\sin(n+\frac{1}{2})t}{t/2} dt\right| < 1$$

Since $\cos t > \frac{1}{2}$ whenever $0 \le t \le 1/(n+\frac{1}{2})$, it follows that on the interval $[0, 1/(n+\frac{1}{2})]$, $(2 \sin t)' = 2 \cos t > 1$, which implies that $2 \sin t \ge t$. Thus, if $t \in [0, 1/(n+\frac{1}{2})]$, then $1 \le (t/2)/\sin(t/2) \le 2$, and therefore, on this interval, $0 \le ((t/2)/\sin(t/2)) - 1 \le 1$. Consequently, we have

$$0 \le \frac{1}{2}\int_0^{1/(n+\frac{1}{2})} \left(\frac{1}{\sin(t/2)} - \frac{1}{t/2}\right)\sin\left(\left(n+\frac{1}{2}\right)t\right) dt$$

$$< \frac{1}{2}\int_0^{1/(n+\frac{1}{2})} \frac{\sin(n+\frac{1}{2})t}{t/2} dt = \frac{1}{2}\int_0^1 \sin u \, du < \frac{1}{2}$$

and, hence, if $x \in [0, 1/(n+\frac{1}{2})]$, then

$$|G_n(x)| = \frac{1}{\pi}\left|\sum_{k=n+1}^{\infty} \frac{\sin kx}{k}\right| \le \frac{1}{\pi}\left(\frac{\pi}{2} + 1 + \frac{1}{2}\right) = \frac{1}{2} + \frac{3}{\pi}\cdot\frac{1}{2} < 1$$

For additional results involving numerical methods and Fourier series the reader should consult Acton (1970) and Hamming (1962).

PROBLEMS

Section A

1. Establish the following result: Suppose that r is a zero of a function f and let x_n be the nth approximation of r that is determined by Newton's

method. Suppose further that there is a closed interval I containing r such that

(i) f'' is continuous on I.

(ii) $f'(x)$ and $f''(x)$ are never zero on I.

(iii) If c is the endpoint of I at which $|f'|$ is the smaller, then $|f(c)/f'(c)| \leq$ length I.

(iv) There are constants m and M such that $0 < m \leq |f'(x)|$ and $|f''(x)| \leq M$ for each $x \in I$.

Then, $|r - x_{n+1}| \leq (M/2m)(r - x_n)^2$ (where $x_0 \in I$). [*Hint*: Note that $0 = f(x_n) + (x_{n+1} - x_n)f'(x_n)$ and $0 = f(r) = f(x_n) + f'(x_n)(r - x_n) + (f''(z)/2)(r - x_n)^2$ (where z lies between x_n and r); subtract.]

2. Find the first few terms of the iterations involved in Newton's method and in the *Regula Falsi* method to find the real zeros of the following functions:

(a) $f(x) = x^4 - 2x + 2 \qquad x_0 = 1.0$

(b) $f(x) = x^3 - 2x^2 - 1 \qquad x_0 = 2.2$

Make an error analysis of your results (use problem A.1).

3. Suppose that $f: [a,b] \rightarrow [a,b]$ is a continuous function and that f is differentiable on (a,b) and has the property that $|f'(x)| \leq q < 1$ for each $x \in (a,b)$. Show that f has a *unique* fixed point in $[a,b]$. [Recall from (4.E.7) that f has at least one fixed point.]

4. Assume that $f: [a,b] \rightarrow [a,b]$ satisfies the hypotheses of problem A.3. Let $x_0 \in (a,b)$ and show that the sequence $\{x_n\}$ defined recursively (inductively) by $x_{n+1} = f(x_n)$, $n = 0, 1, \ldots$, converges to the unique fixed point x^* of f.

5. Show that if $f: [a,b] \rightarrow [a,b]$ satisfies the hypotheses of problem A.3, and if the sequence $\{x_n\}$ is defined as in problem A.4, then for each n

$$|x_n - x^*| < \frac{q^n}{1 - q} |x_0 - x_1|$$

(notation as in problems A.3 and A.4).

6. (a) Use the previous problems to find an approximation of a root of

$$5^{-x} - x = 0$$

that is accurate to within 10^{-4}. [*Hint*: Let $f(x) = 5^{-x}$ and choose an appropriate interval.]

(b) Do the same as in part (a) for

$$3 + \frac{2x - 1}{x + 4} - x = 0$$

7. Resolve the remaining cases in (14.A.3):

(i) $f(a) > 0, f(b) < 0, f''(x) \geq 0$
(ii) $f(a) < 0, f(b) > 0, f''(x) \leq 0$
(iii) $f(a) < 0, f(b) > 0, f''(x) \geq 0$

by either replacing f by $-f$ or x by $-x$.

8. Show that by (14.A.3), to ensure the convergence of the sequence used in Newton's method in finding the nth root of a positive number q, it suffices to use an interval $[a,b]$

$$0 < a < q^{1/n} < b$$

where b is appropriately large.

9. Find an interval $[a,b]$ such that the following functions satisfy the conditions of (14.A.3) on this interval:
   (a) $f(x) = x^3 - x + 3$
   (b) $f(x) = e^{-2x} - x$
   (c) $f(x) = 1/x - k$ $(k > 0)$

10. Show that under the hypotheses of problem A.4, if $x_0 \leq x^*$, then $x_n \leq x^*$ for each n and if $x_0 \geq x^*$, then $x_n \geq x^*$ for each n.

11. Indicate how you would approximate

$$4 + \{4 + [4 + (4 + 4^{1/4})^{1/4}]^{1/4}\}^{1/4} \cdots$$

where this process is carried out indefinitely. [*Hint*: Consider problem A.4.]

Section B

1. Find the first five terms of the Taylor expansion of the following functions:
   (a) $f(x,y) = x \sin y$ about the point $(1,0)$.
   (b) $f(x,y) = x^2 e^{xy}$ about the point $(1,0)$.

2. Use Newton's method for two variables to derive at least the second approximation to a solution of

$$x^2 + y^2 - 25 = 0$$
$$x^2 y - 2xy + 4 = 0$$

starting with $x_0 = 0.6$, $y_0 = 4.9$.

3. Use Newton's method for two variables to derive at least the second approximation to a solution of

$$3x^2 + xy - 15 = 0 \qquad \sqrt{xy} - 3 = 0$$

starting with $x_0 = 2$, $y_0 = 3$.

4. Establish (8) in Sec. 14.B. [*Hint*: Show that for any $\alpha > 0$, $2|\int_a^b f(x)g(x)\,dx| \leq \alpha \int_a^b f^2(x)\,dx + (1/\alpha) \int_a^b g^2(x)\,dx$ (consider $(\alpha f + g)^2$), and then choose a value of α so that the terms on the right-hand side of the inequality become equal.]

5. The method of *steepest descent* can be described as follows. To approximate a solution of

$$f(x,y) = 0 \qquad g(x,y) = 0 \qquad\qquad (*)$$

we define a function ψ by

$$\psi(x,y) = \tfrac{1}{2}(f^2(x,y) + g^2(x,y))$$

A solution to $(*)$ will clearly occur when ψ is minimal. To approximate the value (x^*,y^*) that will yield a minimum of ψ, we begin with an initial ordered pair (x_0,y_0) and then consider the gradient of ψ at (x_0,y_0): $\nabla\psi(x_0,y_0) = (\psi_x(x_0,y_0),\psi_y(x_0,y_0))$. Recall that the gradient gives the direction of greatest change in ψ (which we assume indicates a decrease in ψ). Now we set

$$x_1 = x_0 + t\psi_x(x_0,y_0)$$
$$y_1 = y_0 + t\psi_y(x_0,y_0)$$

To find an appropriate value t, we choose that value of t that will minimize $\psi(x_0 + t\psi_x(x_0 y_0), y_0 + t\psi_y(x_0,y_0))$; the value is found by setting

$$\frac{d}{dt}\left(\psi(x_0 + t\psi_x(x_0,y_0), y_0 + t\psi_y(x_0,y_0))\right)$$

equal to 0. This procedure is then repeated to find values (x_2,y_2), (x_3, y_3), ...

(a) Use the method of steepest descent to find an approximation to a solution of

$$x^3 - y^2 + 2 = 0 \qquad y^2x + 2x^2y - 5 = 0$$

Start with $x_0 = 1$, $y_0 = 2$.

(b) Use the method of steepest descent to find an approximation to a solution of

$$\ln(x - y) = 0 \qquad xy - 1 = 0$$

Start with $x_0 = 1$, $y_0 = 1$

Section C

1. Find the Lagrange interpolation polynomial of order 3 that passes through the indicated points:

(a) $(1,2)$, $(-1,3)$, $(2,4)$, $(0,0)$.

(b) $(x_i, f(x_i))$ where $f(x) = x^3$ and $x_i = i$, $i = 0, 1, 2, 3$.

2. Establish equations (4), (5), and (6) of Sec. 14.C.

3.* Establish the following two results:

(a) Suppose that $f: [a,b] \to \mathbf{R}^1$ and that f'' is continuous on $[a,b]$. Form a partition of $[a,b]$ consisting of n equal subintervals, and let T_n be the approximation of $\int_a^b f(x)\, dx$ using the trapezoid rule (with respect to this partition). Show that

$$\left| \int_a^b f(x)\, dx - T_n \right| \le \frac{M}{12} \frac{(b-a)^3}{n^2}$$

where M is any upper bound for $|f''|$ on $[a,b]$.

(b) Suppose that $f: [a,b] \to \mathbf{R}^1$ and that $f^{(4)}$ is continuous on $[a,b]$. Form a partition of $[a,b]$ consisting of n equal subintervals, and let S_n be the approximation of $\int_a^b f(x)\, dx$ using Simpson's rule (with respect to this partition). Show that if $h = (b-a)/n$, then

$$\int_a^b f(x)\, dx - S_n = - \frac{h^4(b-a)}{180} f^{(4)}(z)$$

for some value $z \in (a,b)$.

4. Use problem C.3 to find the smallest value of n that can be used to approximate $\int_1^2 e^{3x}\, dx$ to within 10^{-4} using the trapezoid rule, and using Simpson's rule. Do the same for $\int_1^2 e^x/x\, dx$.

5. Use Simpson's rule to approximate to within 10^{-2} the length of the graph of $y = e^x$ from $x = 0$ to $x = \frac{1}{2}$.

6. If we recall the definition of the integral $\int_a^b f(x)\, dx$, then it is apparent that the sums $t_n = \sum_{i=0}^{n-1} f(x_i) h$, where $\{x_0, x_1, \ldots, x_n\}$ is a partition of $[a,b]$ and $x_i - x_{i-1} = h$ (for each i) can be considered as approximations of this integral. Show that if f is differentiable, then $|\int_a^b f(x)\, dx - t_n| \le M(b-a)^2/n$, where $M \ge |f'(x)|$ for each $x \in [a,b]$.

7. Suppose that $x_0 < x_1 < x_2 < x_3$ and $x_i - x_{i-1} = h$ for $i = 1, 2, 3$. Then the $\frac{3}{8}$-rule is

$$\int_{x_0}^{x} f(x)\, dx = \frac{3}{8} h(f(x_0) - 3f(x_1) + 3f(x_2) + f(x_3)) + R$$

Show that $|R| \le (3/80)h^5 M$, where $M \ge |f^{(4)}(x)|$ for each $x \in [x_0, x_3]$.

8. Approximate $\int_0^\infty e^{-x^2}\, dx$ to within 10^{-5} by showing that $\int_q^w e^{-x^2}\, dx < e^{-q^2}/2q$ $(q > 0)$ and then choosing an appropriately large value of q. [Hint: Note that $qe^{-x^2} \le xe^{-x^2}$ if $x \ge q$.]

9. Use a Taylor expansion to establish the following error estimates (in both cases $M \ge |f'''(x)|$ for each $x \in [x_0 - h, x_0 + h]$).

(a) If $[f(x_0 + h) - f(x_0 + h)]/2h$ is used to approximate $f'(x_0)$, then the error is less than or equal to $Mh^2/6$.

(b) If $[f(x_0 + h) + f(x_0 - h) - 2f(x_0)]/h^2$ is used to approximate $f''(x_0)$, then the error is less than $Mh/3$.

10. Construct an example to illustrate the inadequacies of the procedures introduced in this section for approximating the derivative of a function.

Section D

1. Use the Euler method with $h = 0.1$ to find an approximation to the solution of

$$y'(x) = \frac{\cos y}{1 + x} - \frac{1}{2} y^2 \qquad y(0) = 0$$

at $x = 0.4$.

2. Use the Euler method with $h = 0.1$ to find an approximate solution to

$$y'(x) = x + 2y \qquad y(0) = 0$$

on the interval $[0,1]$. Compare your answer to the exact solution of this differential equation.

3.* Give a geometrical interpretation of the values a_1, a_2, a_3, and a_4 used in the fourth-order Runge-Kutta formula.

4. Suppose that $y' = f(x)$, i.e., f does not depend on y. Observe that in this case the values a_1, a_2, a_3, a_4 used in the fourth-order Runge-Kutta formula reduce to $a_1 = f(x_i)$, $a_2 = f(x_i + \frac{1}{2}h) = a_3$, and $a_4 = f(x_i + h)$, and show that $y_{i+1} - y_i = (h/6)(f(x_i) + 4f(x_i + h/2) + f(x_i + h))$. Note that this coincides with Simpson's rule, and deduce an error estimate.

5. Apply the predictor- corrector method described in Sec. 14.D to obtain approximate solutions to the following differential equations:
(a) $y' = -xy^2 + 2x \qquad y(0) = 1$
(b) $y' = x^2 y^{1/3} \qquad y(0) = 0$
(c) $y' = (\cos y)/(1 + x) - y^2/2 \qquad y(0) = 0$
Use step size $h = 0.1$

6. In problem D.5 compare your result with the result obtained if only the predictor is used, i.e., ignore the corrector value.

7. At each step in the predictor-corrector method successive corrector values are possible. For example in problem D.5, we might have used

$$y_{n+1}^{[0]} = y_n + f(x_n, y_n)$$

$$y_{n+1}^{[1]} = y_n + \frac{h}{2} [f(x_n, y_n) + f(x_{n+1}, y_{n+1}^{[0]})]$$

$$y_{n+1}^{[2]} = y_n + \frac{h}{2}[f(x_n,y_n) + f(x_{n+1},y_{n+1}^{[1]})]$$

. . . .

The sequence of these corrector values at any one step can be programmed to stop whenever two successive corrector values differ by
less than some predetermined value $\varepsilon > 0$ (or when some fixed number
of corrector values has been determined). Rework problem D.5 using
two corrector values at each step.

8. Use the Lagrange interpolating polynomial of order 2 passing through
the points (x_{i-2}, z_{i-2}), (x_{i-1}, z_{i-1}), (x_i, z_i), where $x_k = x_0 + kh$ for
$k = 1, 2, \ldots$, to find an expression similar to that obtained in (13) of
Sec. 14.D that may be used to express $\int_{x_{i-2}}^{x_{i+1}} p(x)\, dx$ in terms of the
values z_{i-2}, z_{i-1}, and z_i.

9. The Milne predictor-corrector method is similar to that developed in
Sec. 14.D except that four initial points are used to initiate the procedure
instead of three.

(a) Suppose that for a given differential equation $y' = f(x,y)$ acceptable approximate values y_i, y_{i-1}, y_{i-2}, and y_{i-3} have been found
that correspond to points x_i, x_{i-1}, x_{i-2}, and x_{i-3}, respectively,
where, as usual, $x_k = x_{k-1} + h$. Let $z_k = f(x_k, z_k)$, $k = i - 3$,
$i - 2, i - 1, i$, and let p be the interpolating polynomial of order 3
that passes through the points (x_{i-3}, z_{i-3}), (x_{i-2}, z_{i-2}), (x_{i-1}, z_{i-1}),
and (x_i, z_i). Show that

$$\int_{x_{i-3}}^{x_{i+1}} p(x)\, dx = \frac{4h}{3}(2z_i - z_{i-1} + 2z_{i-2})$$

The predictor value $y_{i+1}^{[p]}$ is defined in this procedure by

$$y_{i+1}^{[p]} = y_{i-3} + \int_{x_{i-3}}^{x_{i+2}} p(x)\, dx$$

(b) In Milne's method, the corrector value $y_{i+1}^{[c]}$ is defined by

$$y_{i+1}^{[c]} = y_{i-1} + \frac{h}{3}(z_{i-1} + 4z_i + f(x_{i+1}, y_{i+1}^{[p]}))$$

Interpret and derive this formula with the aid of Simpson's rule.

10. Apply Milne's method to the differential equation given in (14.D.3).

REFERENCES

Abramowitz, M., and I. Stegun (eds.) (1964): *Handbook of Mathematical Functions*,
Applied Math. Series 55, National Bureau of Standards.

Acton, F. S. (1970): *Numerical Methods that Usually Work*, Harper and Row, New York.

Brand, L. (1955): *Advanced Calculus*, Wiley, New York.

Burden, R., J. Faires, and A. Reynolds (1978): *Numerical Analysis*, Prindle, Weber, and Schmidt, Boston.

Eves, H. (1969): *In Mathematical Circles*, Prindle, Weber, and Schmidt, Boston.

Hamming, R. W. (1962): *Numerical Methods for Scientists and Engineers*, McGraw-Hill, New York.

Hildebrand, F. B. (1974): *Introduction to Numerical Analysis*, McGraw-Hill, New York.

Isaacson, E., and H. B. Keller (1966): *Analysis of Numerical Methods*, Wiley, New York.

Myint-U, T. (1978): *Ordinary Differential Equations*, North-Holland, New York.

Ralston, A. (1965): *A First Course in Numerical Analysis*, McGraw-Hill, New York.

ANSWERS TO SELECTED PROBLEMS

CHAPTER 1

Section B

1. $A \times B = \{(2,3),(2,\frac{1}{4}),(3,3),(3,\frac{1}{4}),(-\frac{1}{2},3),(-\frac{1}{2},\frac{1}{4})\}$
3. (a) Dom $f = [2,5]$ (d) Dom $f = [4,9]$
    Range $f = [2,11]$ Range $f = [5,6]$
    $f^{-1}(y) = (y + 4)/3$ $f^{-1}(y) = (y - 3)^2$
4. (a) $\{-5,7\}$; (e) $(\sqrt{7},\sqrt{15}]$
8. $(f \circ g)(2) = 9\frac{1}{3}$
11. (a) $\{(k + 1)^2 + 1\} = \{5,10,17,26, \ldots\}$
12. (a) $\pi/2$; (b) 2π
14. (a) Neither; (c) even; (e) neither

Section C

1. $A \cap B = \{3,5,7,11,13,17,19\}$
    $B \backslash A = \{1,9,15\}$
2. $\bigcap\limits_{n=1}^{\infty} A_n = [0,2]$

Section D

5. No

Section E

1. (a) $\sup = \sqrt{3}$; (d) $\sup = \sqrt{6}$, $\inf = -\sqrt{6}$

Section F

1. $2 + 6i + 4 - 2i = 6 + 4i$

   $(2 + 6i)(4 - 2i) = 20 + 20i$

   $\dfrac{2 + 6i}{4 - 2i} = -\dfrac{1}{5} + \dfrac{7}{5}i$

2. $|3 - 6i| = 3\sqrt{5}$

3. (a) $2(\cos \pi/3 + i \sin \pi/3)$

13. $\dfrac{\pi}{2} + 2\pi k$, where $k = 0, \pm 1, \pm 2, \cdots$

14. $2xy - \dfrac{y}{x^2 + y^2} - y - e^x \sin y$

CHAPTER 2

Section A

3. Yes
13. (a) Yes
14. (a) No
15. Independent

Section B

1. (a) $x = \frac{11}{6}, y = \frac{25}{6}, z = \frac{23}{6}$

2. (a) $\begin{pmatrix} \frac{2}{6} & \frac{1}{6} \\ 0 & \frac{3}{6} \end{pmatrix}$

Section C

1. (a) Stiffness matrix $\begin{pmatrix} -1.5 & 1 & 0 & 0 \\ 1 & -5.5 & 2.5 & 2 \\ 0 & 2.5 & -5.5 & 3 \\ 0 & 2 & 3 & -9 \end{pmatrix}$

   (b) Stiffness matrix $\begin{pmatrix} -6 & 3 & 1 \\ 3 & -4 & 1 \\ 1 & 1 & -4 \end{pmatrix}$

2. (a) 6

Section D

1. (b) $y(t) = \frac{1}{2} + \frac{1}{2}e^{-t^2}$
2. $t = 2\dfrac{\ln 35/110}{\ln 85/110}$
4. (d) $y(t) = c(1 + t) - 1$
7. (b) $y(t) = c_1 e^{t/2} + c_2 t e^{t/2}$
10. (a) Independent; (e) dependent
15. $y_1(t) = e^{-2t}$, $y_2(t) = e^t$, $y_3(t) = e^{-t}$
17. (d) $y(t) = (c_1 + c_2 t)e^{-2t} + [(2t - 1)/32]e^{2t}$

Section E

1. (a) Is a linear transformation. (b) Is not a linear transformation.
2. (b) Is a linear transformation. (c) Is not a linear transformation.
12. $-2t^2 + t + 14$
13. (a) $\begin{pmatrix} 1 & 1 & 0 \\ -\frac{1}{2} & -\frac{1}{2} & \frac{3}{2} \end{pmatrix}$

CHAPTER 3

Section A-B

1. (a) (X,d) is a metric space. (b) (X,d) is a metric space. (d) (X,d) is not a metric space.
5. (a) $\frac{3}{2}$; (b) limit does not exist.

9. All metrics except the discrete metric
17. Each point in $\mathbf{R}^2$ is a cluster point of the sequence $\{u_n\}$.
21. No

CHAPTER 4

Section A

2. $\delta = 1/20$ (if $f(x) = 4x - 7$, $a = 2$, $\varepsilon = \frac{1}{5}$).
5. False; yes.
9. (a) The function f is continuous at $(0,2)$ but not at $(0,0)$.
10. (a) No; (b) Yes

Section B

1. (a) $\{x \mid x \in [1,8]\}$
   (c) $\{x \mid x \in A\}$
4. (a) Compact; (c) compact
11. (a) No; (c) no

Section C

3. No

Section E

12. (b) No
14. (a) Does not have the fixed point property. (c) Has the fixed point property. (d) Has the fixed point property. (e) Does not have the fixed point property. (h) Has the fixed point property.

CHAPTER 5

Section A

2. (a) 9; (b) limit does not exist.
5. (b) No

Section B

5. $c = 5$

Section C

1. (a) 0; (b) $\frac{1}{6}$
2. (b) e^{ac}
   (d) 0

Section D

1. (a) $5\frac{2}{75}$
2. (a) $x_3 = -\frac{3}{4}$

CHAPTER 6

Section A

2. (a) Uniformly continuous; (b) not uniformly continuous
6. (a) $63\frac{3}{4} - 2\ln 4$

Section B

2. (a) -2
6. $101\frac{1}{4}$ degrees

Section C

1. (b) Diverges; (e) $-6 \cdot 2^{\frac{1}{3}}$
4. (a) 0
7. $\dfrac{-8\sqrt{\pi}}{15}$
8. $\displaystyle\int_0^1 x^2 \ln^3 (1/x)\, dx = \int_0^\infty e^{-3u} u^3\, du = (1/3^4)\Gamma(4) = \dfrac{2}{27}$

Section D

3. (b) $227/24$
6. $\sqrt{10}$

CHAPTER 7

Section A

1. (a) $-\sqrt{5}/5$
2. $f_x(0,0) = 0 = f_y(0,0)$
4. $z - 7 + 25(x + 1) - 6(y - 3)$
5. $f_{xy}(0,1) = 2 = f_{yx}(0,1), f_{xxy}(0,1) = 6, f_{yxy}(0,1) = 0$
7. (b) Absolute maximum: $6\frac{1}{8}$, at $(0,\frac{1}{4})$, $(1,\frac{1}{4})$
   Local minimum: $5\frac{3}{4}$, at $(\frac{1}{2},0)$
   Absolute minimum: $-\frac{1}{8}$, at $(\frac{1}{2},2)$
   Saddle point: $(\frac{1}{2},\frac{1}{4})$
12. (c) Absolute maximum at $(6,2)$

Section B

1. (b) $(0 \quad -1/\pi)$
2. (a) $(2,\pi/4)$
5. $z = -2x + 7$
6. $\dfrac{1}{\sqrt{2}}\left[\left(1 - \dfrac{3\pi^2}{16}\right) + \left(\dfrac{\pi}{4} - 2\left(\dfrac{\pi}{4}\right)^3\right)\right]$

Section C*

1. $f'(z) = 2x + 2iy$
2. (a) $-e^{-x}(\cos y - i \sin y)$
   (d) $2x + 3 + i2y$

Section D

1. $D_2 f(1,-1,2,1) = (-4 \quad 3)$
2. $D_2 f(-\pi,2,0,0) = 12\pi(e^{-8\pi} - 1)$

3. $D_2 f(1,2,1) = \frac{1}{2}$
4. $d_w f(3,0,-2,1) = (4 \quad 19)$
6. (a) $\begin{pmatrix} 3 & 0 & 0 & 0 \\ 12 & 6 & 0 & 3 \end{pmatrix}$

Section E

1. $40 + 14e^2$
3. (a) $(-8,0,36,0)$
4. $\dfrac{4096}{625}$

Section F

1. (a) $(\sin x)/x$
3. (a) $3 \sin x^3 + (2/x^3) \cos x^3 - (2/x) \sin x^2 - (2/x^3) \cos x^2$
4. (b,iii) $y(t) = c/t$
   (b,iv) $y(t) = 1 + ce^{-t^2/2-t}$

Section G

3. Yes, since $\det \begin{pmatrix} 2x + 2y & 2x \\ 1 & -1 \end{pmatrix} \neq 0$ if $x = 2, y = 1$.
5. No
7. $(-1,1)$: yes, $(0,0)$: no
9. (c) $y + e^{ty} = c$

Section H

1. $x = 2/\sqrt{3}, y = 3/\sqrt{3}, z = 4/\sqrt{3}$
4. 3
7. $15\sqrt{2}/8$
8. $72/7$

Section I

2. (b) Parabolic
   (d) Elliptic

CHAPTER 8

Section A

1. (a) $A''(x) - \lambda x A'(x) - \lambda A(x) = 0$
      $B'(t) + \lambda B(t) = 0$

2. (a) $u(x, t) = \sin \dfrac{n\pi x}{L} \cos \dfrac{n\pi t}{L}$

Section B

2. (a) Converges; (c) converges; (e) diverges
6. 50 feet
9. (b) Converges
14. (b) 1
15. (c) $-\infty$
17. Converges if $x < \frac{2}{3}$; diverges if $x > \frac{3}{2}$.

Section C

3. (a) Converges pointwise, but not uniformly.
4. (a) Converges uniformly on $[-1,0]$.

Section D

1. (b) $\displaystyle\sum_{n=0}^{\infty} (-1)^n \frac{x^{6n+3}}{(2n + 1)!}$

Section E

1. (b) Radius of convergence: 3; interval of convergence: $(-1,5)$; the series does not converge at either endpoint.
2. (c) Radius of convergence: 1; interval of convergence $(0,2)$; the series converges at $x = 2$, diverges at $x = 0$; the derivative of the series diverges at both endpoints.
5. (a) The series converges uniformly on this interval. (b) The series does not converge uniformly on this interval.

CHAPTER 9

Section B

1. (b) 0
3. (b) ∞

Section D

1. (a) $y(t) = a_0 \left(1 + \dfrac{t^2}{2} + \dfrac{t^4}{2 \cdot 4} + \dfrac{t^3}{2 \cdot 4 \cdot 6} + \cdots \right)$

   $+ a_1 \left(t + \dfrac{t^3}{3} + \dfrac{t^5}{3 \cdot 5} + \dfrac{t^7}{3 \cdot 5 \cdot 7} + \cdots \right)$

   (d) $y = a_0 \left(1 - \dfrac{1}{2} t^2 - \dfrac{1}{4!} t^4 - \sum_{n=3}^{\infty} \dfrac{3 \cdot 5 \cdots (2n - 3)}{(2n)!} t^{2n} \right)$

   $+ a_1 t + \sum_{n=1}^{\infty} \dfrac{2 \cdot 4 \cdots (2n - 2)}{(2n + 1)!} t^{2n+1}$

2. $y_1(t) = 1 - \dfrac{\alpha}{2!} t^2 + \dfrac{(4 - \alpha)\alpha}{4!} t^4$

   $- \dfrac{(8 - \alpha)(4 - \alpha)\alpha}{6!} t^6 - \cdots$

   $\dfrac{-[4(n - 1) - \alpha] \cdots (4 - \alpha)\alpha] t^{2n}}{(2n)!} - \cdots$

   $y_2(t) = t + \dfrac{2 - \alpha}{3!} t^3 + \dfrac{(6 - \alpha)(2 - \alpha)}{5!} t^5$

   $+ \dfrac{(10 - \alpha)(6 - \alpha)(2 - \alpha)}{7!} t^7 + \cdots$

   $+ \dfrac{(4n - 2 - \alpha) \cdots (2 - \alpha)}{(2n + 1)!} t^{2n+1} + \cdots$

4. (b) $y_1(t) = 1 + 2t + \dfrac{2}{3} t^2 + \dfrac{4}{45} t^3 + \cdots$ where the recursion

   formula is $a_{n+1} = \dfrac{2a_n}{(2n + 1)(n + 1)}$

   and

   $y_2(t) = \sqrt{t} \left(1 + \dfrac{2}{3} t + \dfrac{2}{15} t^2 + \dfrac{4}{315} t^3 + \cdots \right)$ where the recursion formula is $a_{n+1} = \dfrac{2a_n}{(2n+3)(n+1)}$

(c) $y_1(t) = \sqrt{t}e^t$

$y_2(t) = \sqrt{t}e^{-t}$

(d) $y_1 = t \displaystyle\sum_{n=0}^{\infty} (-1)^n \left(\dfrac{2^n}{1 \cdot 3 \cdot 5 \cdots (2n+1)} \right) t^n$

$y_2 = t^{1/2} \displaystyle\sum_{n=0}^{\infty} (-1)^n \dfrac{t^n}{n!} = t^{1/2}e^{-t}$

8. $y_1(t) = t$; $y_2(t) = t(-t + 1/4t^2 - 1/18t^3 + (3/16 \cdot 18)t^4 + \cdots)$
   $+ t \ln |t|$, where $b_{n+1} = -nb_n/[n(n+2)+1]$

9. (c) $y_1(t) = t^2 + \dfrac{t^3}{4} + \dfrac{t^4}{40} + \cdots$ (if $a_0 = 1$)

$y_2(t) = \dfrac{1}{12} y_1(t) \ln t + \left(\dfrac{1}{t} - \dfrac{1}{2} + \dfrac{1}{4}t + \cdots \right)$ (if $b_0 = 1$)

Section E

3. (b) $Y(t) = \begin{pmatrix} -3te^{2t} - \frac{1}{3}t - \frac{5}{18} \\ -6te^{2t} - \frac{4}{3}t + 6e^{2t} - \frac{7}{9} \end{pmatrix}$

(e) $y_1 = c_1 e^t \cos t + c_2 e^t \sin t + te^t \cos t$

$y_2 = c_1 e^t \sin t - c_2 e^t \cos t + te^t \sin t$

9. (a) $y_1' = \dfrac{2y_2}{100} - \dfrac{7y_1}{100}$ $y_1(0) = 20$

with initial conditions

$y_2' = \dfrac{7y_1}{100} - \dfrac{7y_2}{100}$ $y_2(0) = 0$

Section F

1. (b) $\begin{pmatrix} 3e^t - 2e^{2t} & -2e^t + 2e^{2t} \\ 3e^t - 3e^{2t} & -2e^t + 3e^{2t} \end{pmatrix}$

Section G

2. (a) $\frac{1}{2}$; (c) $\frac{2}{3}$

CHAPTER 10

Section A

6. (c) $\dfrac{\pi^2}{3} + 4 \sum\limits_{n=1}^{\infty} \dfrac{(-1)^n}{n^2} \cos nx$

   (e) $\dfrac{1}{2} + \dfrac{2}{\pi} \sum\limits_{n=1}^{\infty} \dfrac{\sin (2n-1)x}{2n-1}$

   (f) $\dfrac{\cos x}{2} + \sum\limits_{n=1}^{\infty} \dfrac{4n}{\pi(4n^2-1)} \sin 2nx$

Section C

11. (a) $\dfrac{1}{\pi} \sum\limits_{n=1}^{\infty} \dfrac{(-1)^{n+1}}{n} \sin n\pi x$

   (c) $1 + \dfrac{4}{\pi^2} \sum\limits_{n=1}^{\infty} \dfrac{(-1)^n}{n} \cos \dfrac{n\pi x}{2}$

15. (d) $\dfrac{2}{\pi} - \dfrac{4}{\pi} \sum\limits_{n=1}^{\infty} \dfrac{\cos 2nx}{(2n+1)(2n-1)}$

Section E

3. (a) No
   (b) Yes

Section F

5. Yes

CHAPTER 11

Section E

2. Yes

Section G

4. (a) $\frac{56}{3}$; (b) 0

Section I

3. (b) -8
4. (b) 2
6. (b) $\pi\left(1 - \dfrac{\pi}{4}\right)\left(2 - \sqrt{2}\right)$

CHAPTER 12

Section A

3. (a) $-42 + 9i$; (b) $-\frac{3}{2} + ((4 + \pi)/4)i$; (e) 0
4. (a) 0; (c) 0

Section B

1. (a) $-\frac{3}{4}$; (c) $\frac{1}{2}(\sin 4 - \sin 2) - i \sinh 2$
2. (a) $2\pi i(e^3 - 1)$; (c) $-2\pi i/3$; (e) $-\pi i$
3. $e\pi i/3$
4. $2\pi i$
5. (b) 0
6. (b) $\pi e^{-3}i/3$

Section C

1. $T_0^f(z) = \displaystyle\sum_{n=0}^{\infty} (n + 1)z^n \qquad R = 1$

3. $T_2^f(z) = e^{-4}\left(\displaystyle\sum_{n=0}^{\infty} \frac{(z + 2)^{2n}}{n!}\right)$

Section D

1. (b) $\dfrac{1}{z^3} + \dfrac{2}{z^2} + \dfrac{2}{z} + \displaystyle\sum_{n=0}^{\infty} \frac{2^{n+3}}{(n + 3)!}\, z^n$

(d) $\dfrac{1}{16}\left(\dfrac{1}{z + 2}\right)^2 + \dfrac{1}{8}\dfrac{1}{z + 2}$

$+ \dfrac{1}{6 \cdot 2^6} \displaystyle\sum_{n=0}^{\infty} \frac{(n + 6)(n + 5)\cdots(n + 1)}{2^n}(z + 2)^n$

2. (b) $z = 0$ is a pole of order 3. (d) $z = 1$ is a pole of order 3; $z = 3$ is a removable singularity.
5. (a) Meromorphic; (c) not meromorphic
6. (b) $\dfrac{-1}{3} \sum\limits_{n=0}^{\infty} \left((-1)^n + \dfrac{1}{2^{n+1}}\right) z^n$

Section E

1. (a) $-\frac{1}{3}$; (c) $-i/2$, $i/2$; (d) $e^i/2i$, $e^{-i}/2i$; (f) $2 \sin 2$
2. (b) $2\pi i$; (d) $2\pi i$
3. (a) $-\sqrt{2}\pi/2$; (d) π
4. (a) $\pi/192$
8. (a) $\pi/\sqrt{2}$; (c) $\pi/108$
10. $\pi/5$

Section F

1. (b) $(e^4 - 9)/2$; (c) $64 \cdot 59/15$
2. (a) $\frac{2}{3} + \frac{11}{3} i$
3. $3\pi/4$
4. (a) 0; (d) 0
6. (b) 15π
13. (b) $\phi(x,y) = -\frac{1}{2}(e^{x^2+y^2})$
14. (a) 0

CHAPTER 13

Section B

1. (c) $\hat{f}(\lambda) = \dfrac{i\lambda}{\lambda^2 - 1} (e^{i\lambda\pi} + 1)$

2. $f(t) = \begin{cases} \frac{1}{2} & \text{if } |x| < 2 \\ 0 & \text{if } |x| > 2 \end{cases}$

5. $f(x) = \dfrac{k \sin ax}{\pi x}$

Section C

5. $u(x,t) = \dfrac{2}{\pi} \displaystyle\int_0^{\infty} \dfrac{\sin \lambda x}{1 + \lambda^2} e^{-\lambda t}\, d\lambda$

Section D

7.   (a) $f(x) = \dfrac{2}{\pi} \displaystyle\int_0^\infty \dfrac{\sin \lambda a}{\lambda a} \cos \lambda x \, d\lambda$

8. $u(x,t) = \dfrac{2}{\pi} \displaystyle\int_0^\infty \hat{f}_C(\lambda) e^{-\lambda^2 a^2 t} \cos \lambda x \, d\lambda$

9. $c = \dfrac{\sqrt{\pi}}{2|a|}$

Section E

1.   (d) $y(t) = \tfrac{1}{2} \int_0^t \sinh 2(t - x) \sin x \, dx$
2.   (a) $e^{-2t}(\cosh 2t - \sinh 2t)$
3.   (a) $y(t) = e^{-2t} \sin t$
5.   (c) $\dfrac{1}{s^2 + 1} \left(\dfrac{1 + e^{-\pi s}}{1 - e^{-\pi s}} \right) = \dfrac{1}{s^2 + 1} \coth \left(\dfrac{\pi s}{2} \right)$

Section F

1.   (c) $y(t) = e^{-t}(t^2/2 + 3e(t - 1)u_c(t))$, where

$$u_c(t) = \begin{cases} 0 & \text{if } t < c \\ 1 & \text{if } t \geq c \end{cases}$$

4.   (b) $s(t) = 1/(2\sqrt{k})u_2(t) \sin [(\sqrt{k}/2)(t - 2)]$

CHAPTER 14

Section A

2.   (b) Note that $f(2.2) = -0.0320$ and $f(2.3) = 0.5870$, and consequently a zero of f lies between 2.2 and 2.3. It is easy to see that the conditions of problem A.1 are satisfied (with $M/2m \leq 0.0514$). Therefore, we have

      $x_0 = 2.2$ error ≤ 0.1

      $x_1 = 2.2056$ error ≤ 0.0006

      $x_2 = 2.20558$ error ≤ 0.00000036

      Regula Falsi method

      $a_0 = 2.2$ $b_0 = 2.3$ $w_0 = 2.20052$

$a_1 = 2.20052$ $\quad b_1 = 2.3$ $\qquad w_1 = 2.20521$
$a_2 = 2.20521$ $\quad b_2 = 2.3$ $\qquad w_2 = 2.20554$
$a_3 = 2.20554$ $\quad b_3 = 2.3$ $\qquad w_3 = 2.20557$
$a_4 = 2.20554$ $\quad b_4 = 2.20557$

6. (a) 0.4696
9. (b) [0.4,0.5]

Section B

2. $f(x_0,y_0) = -0.63000$, $\quad g(x_0,y_0) = -0.13400$, $\quad x_1 = 0.55074$, $\quad y_1 = 4.97032$; $f(x_1,y_1) = 0.00737$, $g(x_1,y_1) = 0.03283$
5. (b) $x_1 = 1.38378$, $\quad y_1 = 0.61622$, $\quad f(x_1,y_1) = -0.14729$, $\quad g(x_1,y_1) = -0.29140$

Section C

1. (a) $P_3 = -\frac{5}{6}x^3 + \frac{5}{2}x^2 + \frac{1}{3}x$
4. For $\int_1^2 (e^x/x)\,dx$, $n = 48$ (trapezoid rule), $n = 4$ (Simpson's rule).
5. For $n = 8$, length $\cong 0.934$.

Section D

1. $x_0 = 0$ $\qquad y_0 = 0$
   $x_1 = 0.1$ $\qquad y_1 = 0.1$
   $x_2 = 0.2$ $\qquad y_2 = 0.189955$
   $x_3 = 0.3$ $\qquad y_3 = 0.269985$
   $x_4 = 0.4$ $\qquad y_4 = 0.340477$

n	x_n	y_n	$f(x_n,y_n)$	$y_{n+1}^{[p]}$	$f(x_{n+1},y_{n+1}^{[p]})$	$y_{n+1}^{[c]}$
0	0	0	1	0.1	0.899549	0.094977
1	0.1	0.094977	0.900483	0.185026	0.801992	0.180101
2	0.2	0.180101	0.803636	0.260465	0.709364	0.255751
3	0.3	0.255751	0.711506	0.326902	0.623026	0.322478
4	0.4	0.322478				

Table of Laplace Transforms

Function $f(t)$	Transform $F(s)$	
1	$\dfrac{1}{s}$	$s > 0$
t^n, n a nonnegative integer	$\dfrac{n!}{s^{n+1}}$	$s > 0$
t^α, $\alpha > -1$	$\dfrac{\Gamma(\alpha + 1)}{s^{\alpha+1}}$	$s > 0$
$\sin at$	$\dfrac{a}{s^2 + a^2}$	$s > 0$
$\cos at$	$\dfrac{s}{s^2 + a^2}$	$s > 0$
$t \sin at$	$\dfrac{2as}{(s^2 + a^2)^2}$	$s > 0$
$t \cos at$	$\dfrac{s^2 - a^2}{(s^2 + a^2)^2}$	$s > 0$
$\sinh at$	$\dfrac{a}{s^2 - a^2}$	$s > |a|$
$\cosh at$	$\dfrac{s}{s^2 - a^2}$	$s > |a|$
e^{at}	$\dfrac{1}{s - a}$	$s > a$
$t^n e^{at}$, n a nonnegative integer	$\dfrac{n!}{(s-a)^{n+1}}$	$s > a$
$e^{at} \sin bt$	$\dfrac{b}{(s - a)^2 + b^2}$	$s > a$
$e^{at} \cos bt$	$\dfrac{s - a}{(s - a)^2 + b^2}$	$s > a$
$u_c(t)$	$\dfrac{e^{-cs}}{s}$	$s > 0$
$\displaystyle\int_0^t f(t - u)g(u)\,du$	$F(s)G(s)$	
$\delta(t - c)$	e^{-cs}	

SYMBOLS USED IN THE TEXT

W	Wronskian	51
A_ϕ	matrix belonging to the linear transformation ϕ	56
(X, d)	metric space	73
$S_\varepsilon(x)$	ϵ-neighborhood about x	77
$S_\varepsilon^d(x)$	ϵ-neighborhood about x in (X, d)	77
$\bar{A}$	closure of A	101
$\lim_{x \to x^*} f(x)$	limit of f as x approaches x^*	126
$\lim_{x \to c^+} f(x)$	right-hand limit of f at c	126
$\lim_{x \to c^-} f(x)$	left-hand limit of f at c	126
$d_{\hat{x}} f$	differential of f at $\hat{x}$	137
$P = \left\{ x_0, \ldots, x_n \right\}$	partition	148
$\lvert P \rvert$	mesh of the partition P	148
$\lim_{\lvert P \rvert \to 0} \left\{ \cdots \right\}$	limit of $\left\{ \ldots \right\}$ as mesh to P tends to 0	148
$\text{CP} \int_{-\infty}^{\infty} \text{v} \, f(x) \, dx$	Cauchy principal value	164
Γ	gamma function	164
B	beta function	167
$D_{\mathbf{u}} f(\mathbf{w})$	directional derivative of f	189
$\nabla f(z)$	gradient of f	200
$J_p(t)$	Bessel function of order p	301
$\prod_{i=1}^{\infty} c_i$	infinite product	319
$f(x + 0)$	right-hand limit	334
$f(x - 0)$	left-hand limit	334
$T(f)$	Fourier series of f	336
$D_n(t)$	Dirichlet kernel	349
$M_n(s)$	Fejér kernel	375
$\text{osc}(f; A)$	oscillation of f on A	397
$m^*(A)$	outer measure of A	407
$m_*(A)$	inner measure of A	412
$\text{osc}(f; x)$	oscillation of f at x	416

a.e.	almost everywhere	421
$\int_J f$	Lebesgue integral of f over J	422
$\mathrm{Res}(f; z_0)$	residue of f at z_0	497
$\hat{f}$	Fourier transform of f	547
$\mathscr{F}(f)$	Fourier transform of f	547
$\mathscr{F}^{-1}(f)$	inverse Fourier transform	548
$\mathscr{L}$	Laplace transform	564
$\mathscr{L}^{-1}$	inverse Laplace transform	564

INDEX

9 780367 452018